Varactor Applications

Varactor Applications

Paul Penfield, Jr.

Assistant Professor of Electrical Engineering
Massachusetts Institute of Technology

Robert P. Rafuse

Assistant Professor of Electrical Engineering
Massachusetts Institute of Technology

The M.I.T. Press

Massachusetts Institute of Technology
Cambridge, Massachusetts
1962

Library of Congress Catalog Card Number: 62-19758
Printed in the United States of America

ACKNOWLEDGMENT

This is Special Technical Report Number 6 of the Research Laboratory of Electronics of the Massachusetts Institute of Technology.

The Research Laboratory of Electronics is an interdepartmental laboratory in which faculty members and graduate students from numerous academic departments conduct research.

The research reported in this book was made possible in part by support extended the Massachusetts Institute of Technology, Research Laboratory of Electronics, jointly by the U. S. Army (Signal Corps), the U. S. Navy (Office of Naval Research), and the U. S. Air Force (Office of Scientific Research) under Signal Corps Contract DA 36-039-sc-78108, Department of the Army Task 3-99-20-001 and Project 3-99-00-000; and in part by Signal Corps Contract DA-SIG-36-039-61-G14; additional support was received from the National Science Foundation (Grant G-9330) and from Lincoln Laboratory, a center for research operated by the Massachusetts Institute of Technology with the joint support of the U. S. Army, Navy, and Air Force under Air Force Contract AF19(604)-7400.

This research was also supported in part by the U. S. Army (Signal Corps) under Contract DA36-039-sc-78236 and in part by the U. S. Navy (Bureau of Ships) under Contract NObsr-77621, both with Microwave Associates, Inc.

Reproduction in whole or in part is permitted for any purpose of the United States Government.

PREFACE

Solid-state rectifiers have always exhibited a nonlinear capacitance under reverse bias. Shortly after the invention of the transistor, the theory of this capacitance was worked out to essentially the same form it has today. In the early 1950's diode capacitors using this effect were made for voltage tuning and similar operations. These had relatively high losses, however, and could not work at microwave frequencies.

High-quality diode capacitors were developed in the middle 1950's under the direction of A. Uhlir, Jr., then of Bell Telephone Laboratories, Inc. Partly to distinguish these devices from the more lossy diodes then available, the name varactor‡ was given to them. Varactors are now made commercially by over a dozen companies, and in smaller quantities for in-house or developmental purposes by probably a hundred or so more. The varactor business is rapidly growing; present sales are at a rate of about one or two million dollars per year, and within a few years will probably be ten or twenty times this amount (nevertheless, varactors account for a very small fraction of total semiconductor sales).

Varactors are used in low-noise microwave parametric amplifiers, in low-noise small-signal frequency converters, and as frequency multipliers, dividers, mixers, and modulators for frequency control and comparison systems. In addition, they are useful in power-conversion applications, as frequency multipliers (for example, in solid-state microwave sources), as modulators (for example, in small transmitters), as power limiters, and as switches.

Ever since diode capacitors were first made, it has been known that the nonlinear capacitance is accompanied by a parasitic loss. At high frequencies the only significant loss is in the form of a series resistance, which for most varactors has been found to be constant over a wide range of bias and frequency. The early theories of varactor applications usually neglected this parasitic loss, and somewhat later theories accounted for it only in an approximate way. In a nutshell, the purpose of the present book is to give a unified theory of the effect of the series resistance on the various varactor applications.

This work began in the summer of 1959, when one of the authors worked at Microwave Associates, Inc., and developed the theory of parametric amplifiers using varactors with series resistance. Because the loss element is in series, he chose to use a series, or impedance, type of analysis, instead of the more common paral-

‡Named originally by M. E. Hines, then of Bell Telephone Laboratories, Inc.

lel, or admittance, type. This approach yielded somewhat simpler formulas, accurate to higher pumping levels and higher frequencies. Meanwhile, the other author, who had been interested in microwave amplifiers, recognized the series approach as a powerful one. He did a similar analysis of frequency converters, and began the analyses of many other applications, including harmonic multiplication. This work became so absorbing that he decided to write his doctoral thesis in 1960 on varactor applications. Since then the authors have refined, extended, and consolidated all these results for publication here.

This book is in every sense of the word a research report. Since the research is still in progress, the book cannot be expected to be in finished, polished form. In some places our analyses are not complete, and in others they are too complete. If this book were to be written a few years from now, it might be quite different. We have tried to point out profitable extensions of this research, and in particular we are now working on these specific topics: diode model, symmetrical circuits, principles of frequency separation, charge-storage effects, high-dynamic-range applications, noise in multipliers and dividers, bandwidth in amplifiers, converters, and multipliers, noise in degenerate parametric amplifiers, and the use of varactors in other applications, such as limiting, switching, voltage tuning, and so forth.

All our analyses are based on a very simple model for the varactor. This model, of course, has limitations; it does not accurately describe real varactor diodes under all conditions. Our critics will say that we have developed too far the theory based on this model — that is, that we have found too many optimizations, carried out too many calculations to meaningless third significant figures, and in general predicted effects that may not occur for real varactors because the model fails. Perhaps this is true. These varactor applications are among the very few nonlinear circuit problems of any consequence that can be solved exactly, and there is always a temptation to get carried away and solve such problems more thoroughly than necessary. Nevertheless, we believe that it is important, for at least three reasons, to extend the theory relatively far. First, if our theory fails to account for observed phenomena, it is probably because the model is inaccurate. If the range of validity of our theory covered only the range of validity of the model, then one would not know where to look for improvements — the model or the theory. Second, in the future improved varactors may become available to make the more detailed calculations meaningful. And third, we decided that it was better to include too much than too little, and that this extra detail can be tolerated in a research report.

Obviously, we could not have done this work without help from many people. For technical discussions and/or guidance, we are particularly indebted to Professors H. A. Haus, R. B. Adler, W. D. Jackson, and E. J. Baghdady, all of M. I. T., and to Dr. A.

Uhlir, Jr., of Microwave Associates, Inc. We are grateful to the M. I. T. Research Laboratory of Electronics, and to its directors, past and present, Professors J. B. Wiesner and H. J. Zimmermann, for making research time, equipment, and facilities available to us. We have had stimulating discussions of various phases of this work with M. E. Hines, of Microwave Associates, Inc., with N. Houlding, of National Co., and with F. S. Harris, of Microwave Associates, Inc., who did things experimentally that "couldn't be done" and thus piqued our curiousity. Portions of this work were presented to the Harmonic Generation Study Group, an informal evening study group of engineers in the Boston area, and the comments of these people were very helpful. In particular, one of them, S. L. Johnson of the Raytheon Company, devised, with W. Blaisdell, also of Raytheon, some of the nomographs in Appendix D.

We have used in the book many results derived by students who wrote theses under our supervision. We are especially grateful to them for having allowed us to publish their work: B. L. Diamond, Soheir T. Meleika, M. Greenspan, S. J. Sohn, and R. K. Breon. Other students who, by their discussions, enhanced our understanding of varactors included R. W. Markley, E. A. Patrick, G. Lewis, J. F. Banzhaf, III, M. W. A. Wanas, and M. E. Goff.

We are indebted for some of the associated experimental work to R. Catalan, J. Edwards, and E. Hug.

Many other people have helped us in the preparation of the book, and we are grateful to all of them. Miss Elizabeth J. Campbell and Miss Sylvia K. Healy made many of the calculations and plotted many of the graphs. Other calculations were obtained from the following computers: M. I. T. Department of Civil Engineering IBM 650; M. I. T. Computation Center IBM 709 and 7090; M. I. T. Lincoln Laboratory IBM 7090. Portions of the manuscript were typed by Miss Marguerite A. Daly and Miss Audrey M. Katz. The drafting was done by Mr. C. Navedonsky, Mrs. Jean M. Porter, and by members of M. I. T. Illustration Service. We are indebted to Mrs. Miriam C. Smythe of the R. L. E. Publications Office for kindly reading some of the manuscript. Finally, we are grateful to the staff of The M. I. T. Press, especially to our editor, Miss Constance D. Boyd, and to our typist, Mrs. Alison J. Wager.

This work was supported in part by Ford Foundation post doctoral fellowships held by Penfield from 1960 to 1962. Portions of this book are based on the following theses submitted in partial fulfillment of the requirements of the indicated degrees in the Department of Electrical Engineering at the Massachusetts Institute of Technology: Robert P. Rafuse, Sc. D. (September, 1960); Bliss L. Diamond, S. M. (February, 1961); Roy K. Breon, S. B. (May, 1961); Soheir T. Meleika, S.M. (June, 1962); Marshall Greenspan, S. M. (June, 1962); Sung J. Sohn, S. M. (June, 1962).

Cambridge, Massachusetts
June, 1962

Paul Penfield, Jr.
Robert P. Rafuse

This book incorporates, either completely or by omitting very unimportant details, the results of a number of previous reports, which are now superseded:

Haus, H. A., and P. Penfield, Jr., On the Noise Performance of Parametric Amplifiers, Energy Conversion Group Unpublished Internal Memorandum No. 19, Department of Electrical Engineering, M. I. T. (August 11, 1959).

Penfield, P., Jr., The High-Frequency Limit of Varactor Amplifiers, Unpublished private memorandum (August 27, 1959).

Penfield, P., Jr., Interpretation of Some Varactor Amplifier Noise Formulas, Unpublished private memorandum (September 1, 1959).

Rafuse, R. P., "Measurement of Absolute Noise Performance of Parametric Amplifiers, Digest of Technical Papers, 1960 International Solid-State Circuits Conference, Philadelphia, Pa. (February 10-12, 1960).

Rafuse, R. P., Some Observations on Noise Figure, Exchangeable Power Gain, and Synthesis Procedures for Devices Utilizing Time-Varying, Lossy, and Nonlinear Capacitors, Unpublished private memorandum (April, 1960).

Penfield, P., Jr., Fundamental Limits of Varactor Parametric Amplifiers, Microwave Associates, Inc., Burlington, Mass. (August 15, 1960).

Rafuse, R. P., "Parametric Applications of Semiconductor Capacitor Diodes," Sc.D. Thesis, Department of Electrical Engineering, Massachusetts Institute of Technology, Cambridge (September, 1960).

Rafuse, R. P., "An Exact Solution of the Abrupt-Junction Divide-by-Two Circuit," Unpublished private memorandum (October 20, 1960).

Penfield, P., Jr., Small-Signal Theory of Varactor Frequency Dividers, Energy Conversion Group Unpublished Internal Memorandum No. 40, Department of Electrical Engineering, M. I. T. (November 17, 1960).

Rafuse, R. P., The Large-Signal Exact Solution for the Abrupt-Junction Divide-by-Two Circuit, Energy Conversion Group Unpublished Internal Memorandum No. 42, Department of Electrical Engineering, M. I. T. (December 16, 1960).

Rafuse, R. P., "High-Power Operation of Varactor Devices", Digest of Technical Papers, 1962 International Solid-State Circuits Conference, Philadelphia, Pa. (February, 1962).

CONTENTS

Varactor Applications

Chapter 1

INTRODUCTION

A varactor‡ is a semiconductor junction diode with a useful nonlinear reverse-bias capacitance. Varactors can be used at frequencies up to, and including, the microwave region; their useful frequency range is limited primarily by a parasitic series resistance (and, in some instances, parasitic lead inductance). This series resistance degrades the performance of the varactor in its various applications and provides fundamental limits on the quality of performance achievable at high frequencies. This book is devoted primarily to a theoretical study of the role of the series resistance in dictating these limits.

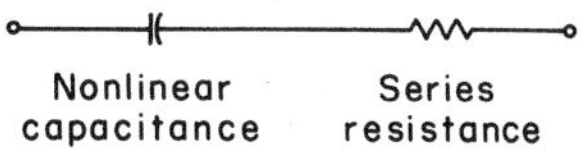

Figure 1.1. Simple ac equivalent circuit of a varactor, first given by Uhlir.[12]‡‡

The most common ac equivalent circuit of a varactor, Figure 1.1, shows only a nonlinear capacitance and the series resistance. Use of this equivalent circuit is justified in Chapter 4 for the analyses of varactor applications in Chapters 5, 6, 8, 9, and 10.

1.1. Varactor Applications

The nonlinear capacitance of a varactor is useful in many ways. Generally speaking, these involve conversion of power from one rf frequency to another.[10] Two, three, four, or more frequencies may interact in the varactor, and of those, some may be useful inputs or outputs, while the others are idlers that, although they may be necessary to the operation of the device, are not part of any input or output.

One of the simplest applications is harmonic generation, or frequency multiplication. The varactor is excited at a frequency‡‡‡ ω_0

‡Also called a capacitive diode, or reactance diode, or parametric diode, or known by trade names such as Varicap.

‡‡Superscript numerals denote references listed at the end of each chapter.

‡‡‡In this book we use radian frequencies like ω_0, rather than ordinary frequencies such as f_0. They are related by $\omega_0 = 2\pi f_o$.

and power is delivered to a load at frequency $2\omega_o$, or $3\omega_o$, or in general $\ell\omega_o$ for some integer ℓ. Harmonic generators are useful in frequency control and comparison systems and for generation of moderate amounts of microwave power. A chain of varactor multipliers, driven by a VHF transistor oscillator or amplifier, makes a very practical solid-state source of microwave power. Typical solid-state sources deliver from a few milliwatts to a few watts of power at 1 Gc to 10 Gc or higher, the varactor stages themselves often having as high as 90 per cent efficiency. The series resistance limits the efficiency at high frequencies. The analysis of harmonic multipliers is very difficult, and in Chapter 8 we limit our discussion to those made from a particularly simple type of varactor, the abrupt-junction type, although we discuss the graded-junction doubler in Appendix F.

Another simple application of varactors is in frequency division, or subharmonic generation. The varactor is excited at ω_o, and power is delivered to a load at $\omega_o/2$ or ω_o/ℓ, for some integer ℓ (the cases with ℓ equal to 2, 4, 8, 16, and so on, are of special importance). Frequency dividers are useful in frequency control and comparison systems, and for possible use as fast bistable elements, parametrons, in computers. The series resistance limits the frequencies at which division can take place and the efficiency of power conversion, and it limits the speed with which subharmonic oscillations can grow. These limits are discussed in Chapter 9.

Varactors can be used to generate other frequencies. If power is supplied at ω_o, power can be delivered to a load at frequencies of the form $n\omega_o/\ell$ for integers n and ℓ (and especially for ℓ = 2, 4, 8, 16, and so on). This rational fraction generation is not discussed in detail in this book. In addition, oscillations can occur at frequencies unrelated to the input frequency, although usually such oscillations are undesirable.

Some other varactor applications require three or more useful frequencies. If a large current at frequency ω_p and a small current at frequency ω_s are put through a varactor, sidebands with frequencies of the form $n\omega_p + \omega_s$, for n a positive or a negative integer, are generated. If power flows at one of these frequencies (usually n = 1 or n = -1) to a load, the varactor is said to be a frequency converter. If the output frequency is larger than ω_s, it is an upconverter; otherwise, it is a downconverter. If the excitation at ω_s and at the generated sidebands is small, the conversion is linear, and can be used for communications purposes. The series resistance limits the frequencies at which this conversion can take place profitably; at high frequencies it limits the gain and introduces noise. These limitations are discussed in detail for several frequency converters in Chapter 5.

The process of passing the large current at frequency ω_p through the varactor is known as pumping, and the pumped varactor behaves like a time-varying capacitance, rather than a nonlinear one. The series resistance dissipates power because of the pumping current, and the allowable swing in capacitance (or in elastance, the reciprocal of the capacitance) is limited by the maximum elastance of the varactor. These limits are given in detail in Chapter 7. It is convenient to describe frequency converters in two steps: (1) a pumping analysis, and (2) the utilization of the pumped varactor.

Another important application of pumped varactors is in parametric amplifiers. The varactor is pumped at frequency ω_p, and a signal is introduced at frequency ω_s. If the varactor is terminated properly at the sidebands (especially the frequency $\omega_p - \omega_s$), the varactor behaves at ω_s as if it were an impedance with a negative real part. This negative resistance can be used as an amplifying mechanism. The varactor series resistance limits the frequencies ω_p and ω_s for which amplification can be achieved, and it introduces noise. These limits are discussed in Chapter 6.

We have described several applications of varactors that involve conversion of power from one frequency to another. It is convenient, in comparing these various applications, to use a mnemonic to remember the frequency of the input, of the out-

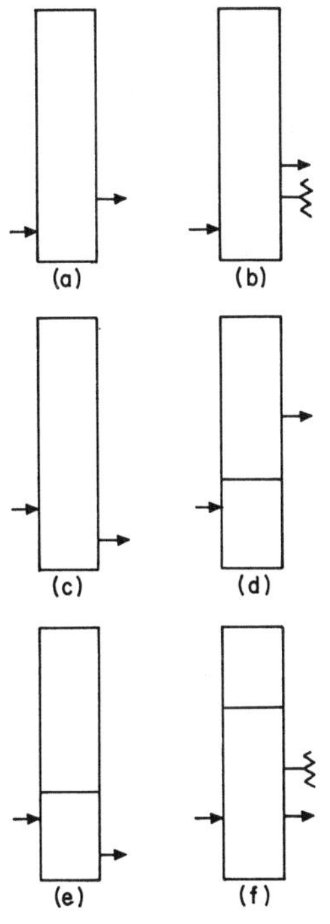

Figure 1.2. Schematic representation of the relations between input, output, pump, and idler frequencies. In each case, the input is at the left and the output at the right. The pump, if any, is indicated by a line inside the box, which represents the varactor or the varactor as pumped. Idlers, if any, are denoted by resistor symbols. The heights of the lines indicate the magnitudes of the frequencies. In future plots of this type, the frequencies may be indicated by abbreviations, such as s = signal, p = pump, and the like.

The following applications are shown here: (a) multiply-by-two circuit; (b) multiply-by-three circuit with an idler at the second harmonic; (c) divide-by-two circuit; (d) upconverter; (e) downconverter; (f) parametric amplifier.

put, of the pump, and of the idlers, if any. For each of the different applications, a convenient way of representing the relations between the various frequencies is shown in Figure 1.2. Roughly speaking, the boxes in each case represent the varactor and, perhaps, some associated circuitry. On the left are the input frequencies. If the varactor is pumped, a line within the box represents the pump frequency. Idlers are denoted by resistor-like symbols on the right, and the output frequency or frequencies are shown with arrows on the right. The heights of the lines indicate their relative frequencies; thus, in Figure 1.2f, the idler frequency is larger than the signal.

Most of the applications we describe are represented in Figure 1.2, although in some cases different relationships between the frequencies are also possible, and other idlers may be present. Figures like these are used throughout the book to indicate succinctly the device under discussion. Figures 5.1, 6.1, 8.1, and 8.2 are more detailed collections of similar diagrams.

1.2. Analyses of Varactor Applications

The analyses of varactor applications in this book have some important features. First, they are all based on the series equivalent circuit of Figure 1.1, and use impedance rather than admittance concepts. Second, we use frequency-domain concepts exclusively. Third, specific circuits are not solved; instead, the circuits are kept as general as possible, so that the performance limits obtained are fundamental in the sense that they cannot be violated by some ingenious choice of circuit. Where there is a choice, we try to use mks units rather than the so-called "engineering" units.

1.2.1. Impedance Approach

The varactor model, Figure 1.1, has two elements in series; thus it is natural to express its voltage v(t) in terms of its current i(t), rather than vice versa. Therefore,

$$v(t) = R_s i(t) + \int S(t) i(t)\, dt, \tag{1.1}$$

where S(t) is the incremental elastance as a function of time,

$$S(t) = \frac{1}{C(t)}$$

$$= \frac{dv'}{dq} \tag{1.2}$$

Two things are important about Equation 1.1. First, the nonlinear capacitance is represented as an elastance, not a capacitance, by means of Equation 1.2, which gives S as the reciprocal of C. The elastance used is the incremental elastance, the slope of the voltage-charge relation, rather than the ratio of voltage to charge.[1] The incremental quantity is the one that is measured physically. And second, in Equation 1.2, v' is the voltage across the nonlinear capacitive part of the varactor, "inside the series resistance."

Many people have made analyses of varactor applications in which admittance rather than impedance concepts are used. The nonlinear capacitive part of the varactor model of Figure 1.1 is represented by the formula,

$$i(t) = C(t) \frac{dv'(t)}{dt}, \tag{1.3}$$

which is, so far, perfectly correct. Then, one or both of these assumptions are made: (1) that the series resistance is represented by a frequency-dependent equivalent parallel resistance; (2) that at certain frequencies the varactor is short-circuited inside the series resistance. Neither assumption is valid. The capacitance is nonlinear, and a series resistance cannot be represented as an equivalent parallel conductance, no matter what frequency dependence is assumed. The varactor cannot be short-circuited inside the series resistance because this point is not accessible (if it were, we should introduce all signals there, and bypass the loss altogether). Perhaps in some applications, especially for small pumping and small loss, the errors introduced are not serious. However, the admittance analysis has no benefit to compensate for its inaccuracy; on the contrary, many of the fundamental limits cannot be as easily related to measurable varactor parameters.‡

It should be noted that Equation 1.3 can be used for some applications if a certain infinite matrix is inverted before eliminating the undesired sidebands. Leenov[6] has used this approach for the upconverter, but his analysis is unnecessarily restricted because the capacitance, as pumped, must be sinusoidal. In general, a sinusoidal capacitance, as pumped, is not desirable.

Kurokawa and Uenohara[5] have compared the series and parallel approaches as applied to parametric amplifiers and upconverters.

---

‡This is because the capacitance can become quite large. This fact is difficult to account for with the admittance analysis. The impedance analysis is not affected, because the elastance merely approaches zero.

1.2.2. Use of Frequency Domain

The equation that describes the varactor, Equation 1.1, is written in the time domain. It is a nonlinear differential equation, which must be solved together with the equation(s) describing the varactor termination. Expressing any but the most trivial terminating equations in the time domain is impossible. Thus, for practical purposes, solving Equation 1.1 in the time domain is out of the question. We turn to the frequency domain.

In linear circuit analysis, the frequency domain is useful because linear algebraic equations can be solved instead of linear differential equations. This same simplification works for the nonlinear varactor equation: If we express Equation 1.1 in the frequency domain, we have, instead of the one nonlinear differential equation, a set of coupled nonlinear algebraic equations. In most cases these are much easier to solve than the nonlinear differential equation, especially because the terminating network is invariably described already in the frequency domain. Throughout this book we solve Equation 1.1 using frequency-domain techniques.

1.2.3. General Circuits; Fundamental Limits

Most analyses of varactor (and other) circuits make use of a specific circuit, perhaps with some adjustable parameters. A seemingly "logical" circuit is analyzed, with the hope that the results also apply to other circuits. In contrast to this approach, we do not solve specific circuits, but instead keep the circuit almost completely unspecified. Thus, for example, in their analysis of parametric amplifiers, Knechtli and Weglein[3] state that their purpose is to relate the noise figure and pump power to measurable parameters of the varactor and the circuit. On the contrary, our purpose in this book is to relate noise figure and other performance criteria to measurable varactor parameters and not to circuit parameters.

In our analysis of the varactor multipliers in Chapter 8, we do not inquire as to what type of circuit is used to couple the varactor to the source and to the load. Instead, we calculate the powers and efficiency at the varactor, with the knowledge that no possible choice of circuitry can lead to a higher efficiency or more power capabilities. Similarly, in Chapter 6 the parametric amplifier is analyzed with no mention of tuned circuits and filters. The only aspect of the varactor terminating network that is essential is the impedance at an idler frequency. Thus the resulting limits are fundamental in the sense that no change in circuitry can be used to bypass them.

What we call in this book analysis of varactor applications is really closer to synthesis. In network theory, analysis and synthesis differ in part because analysis usually has to do with specific circuits, and synthesis with conditions of realizability and methods for constructing what is realizable. In the same sense, we are concerned with performance limits, and our derivations show conditions that must be met to achieve these limits. Perhaps we should call this work "synthesis" rather than "analysis."

1.3. Outline of the Book

For the general orientation of the reader, we give a brief outline of this book and state for each chapter the one or two most important results.

In Chapter 2 we discuss very briefly some basic concepts that are used many times in the book. Various noise terms are explained, and various types of gain are contrasted. For the remainder of the book, the gain of most importance is the exchangeable gain, and to describe amplifier noise we use the effective input noise temperature.

In Chapter 3 we treat the varactor without the series resistance. The Manley-Rowe formulas[7,9] are used to determine properties of such ideal-varactor frequency converters, parametric amplifiers, multipliers, and dividers. It is tempting to think of these as approximate properties for the varactor with finite series resistance, but synthesis procedures based on Chapter 3 are apt to be very misleading.

In Chapter 4 we review the physics of varactors and the approximations that lead to the model of Figure 1.1. The most important source of noise is the series resistance, which exhibits thermal noise. We give the frequency-domain equations for a variety of applications, based on Equation 1.1. The most natural characterizations for varactors are the cutoff frequency ω_c and the normalization power P_{norm}. These convenient combinations of measurable varactor parameters enter naturally into the fundamental limits we derive.

In Chapter 5 we investigate frequency converters, especially those limits that are imposed by the varactor series resistance. The two most important concepts introduced are those of an optimum pump frequency and a minimum noise temperature. In various circumstances, the reasons for choosing the pump frequency are different, and many configurations have (approximately) the same fundamental limit on noise.

In Chapter 6 we take up parametric amplifiers. The varactor series resistance limits the pump and signal frequencies for which gain can be achieved and provides a minimum noise temperature (the same as the minimum for the frequency converters). This minimum noise temperature is obtained only if the pump frequency is chosen equal to an optimum pump frequency. The degenerate amplifier is discussed somewhat less completely.

In Chapter 7 we treat the pumping of varactors. Bounds for a modulation ratio (defined in Chapters 5 and 6) are derived, and formulas given for pump power. It is shown that lower power is often required if the pumping current, rather than the pumping voltage, is sinusoidal. Optimization of parametric amplifiers with limited pump power is possible through the use of the design charts developed in Chapter 7 and given in Appendix D.

In Chapter 8 we discuss multipliers and derive limits on efficiency and power capabilities of abrupt-junction multiply-by-two, -three, -four, -five, -six, and -eight circuits.‡

In Chapter 9 we investigate dividers, both the small-signal theory and the large-signal theory. The rise time and possible frequency range of parametrons are limited by the series resistance, and for large signals the efficiency and frequency range are restricted. We treat explicitly the abrupt-junction divide-by-two and divide-by-four circuits.

In Chapter 10 we take up various aspects of the large-signal theory of frequency converters and amplifiers, that is, when the signal becomes comparable in amplitude with the pump. Limits on efficiency of power conversion, dynamic range, and some effects of pump-power variation are discussed.

Many of our results have direct application in design problems, and many types of design charts and nomographs are given in Appendix D. The other appendices cover specific subjects, including calculation of varactor elastance, effect of nonconstant series resistance, negative-resistance amplifier circuitry, and voltage tuning.

Many important topics are not discussed in this book. Varactors can be used for many purposes other than conversion of frequency. Varactor limiters[11, 8, 13, 2] are potentially useful, and we have not derived any limits on their performance. Voltage tuning is treated only very briefly (Appendix E). Although switching with varactors is promising, we have not investigated it. Even many frequency-conversion applications have not been covered, for lack of space and time. Varactors will undoubtedly serve as modulators, and although this application is basically one of frequency conversion, we have not looked at it explicitly.

We do not explore the problem of bandwidth in any varactor devices. Surely there exist fundamental limits to the bandwidth of varactor converters,[4] amplifiers, and multipliers, but we have not calculated them.

There are many practical problems associated with varactor circuits that we do not attempt to solve. Although we do give a few circuits for realizing the performance limits, these are by

‡We also cover the graded-junction multiply-by-two circuit in Appendix F.

and large not well-designed circuits. For example, one of the most effective techniques for separation of frequencies is by means of symmetrical circuits, such as bridges. We do not discuss this technique in any detail. This does not mean, however, that the theory is not "practical." We believe that it is, in the best sense of the word "practical." It indicates fundamental limits on performance and leads to (somewhat unusual) circuit conditions that must be met to achieve this performance.

The material in this book is almost wholly theoretical. We have decided not to present experimental evidence of its validity, because certain areas are better documented than others. The parametric amplifier formulas have been available for some time, and have been used for amplifier design by many groups. As far as we are aware, no serious discrepancies have been found, although probably the more detailed formulas have not been employed. The formulas for multipliers and dividers have been used for the design of many solid-state sources, and the limits have proved to be accurate, except for an enhancement of power capabilities because of high-frequency failure of forward conduction (see Section 4.1.4). Again, the more complicated formulas have probably not been used.

REFERENCES

1. Heffner, H., "Capacitance Definitions for Parametric Operation," IRE Trans. on Microwave Theory and Techniques, MTT-9, 98-99 (January, 1961).
2. Ho, I. T., and A. E. Siegman, "Passive Phase-Distortionless Parametric Limiting with Varactor Diodes," IRE Trans. on Microwave Theory and Techniques, MTT-9, 459-472 (November, 1961).
3. Knechtli, R. C., and R. D. Weglein, "Low-Noise Parametric Amplifier," Proc. IRE, 48, 1218-1226 (July, 1960).
4. Kuh, E. S., "Theory and Design of Wide-Band Parametric Converters," Proc. IRE, 50, 31-38 (January, 1962).
5. Kurokawa, K., and M. Uenohara, "Minimum Noise Figure of the Variable-Capacitance Amplifier," Bell System Tech. J., 40, 695-722 (May, 1961).
6. Leenov, D., "Gain and Noise Figure of a Variable-Capacitance Up-Converter," Bell System Tech. J., 37, 989-1008 (July, 1958).
7. Manley, J. M., and H. E. Rowe, "Some General Properties of Nonlinear Elements. — Part I. General Energy Relations," Proc. IRE, 44, 904-913 (July, 1956).
8. Olson, F. A., C. P. Wang, and G. Wade, "Parametric Devices Tested for Phase-Distortionless Limiting," Proc. IRE, 47, 587-588 (April, 1959).

9. Penfield, P., Jr., Frequency-Power Formulas, The Technology Press and John Wiley & Sons, Inc., New York (1960).
10. Rowe, H. E., "Some General Properties of Nonlinear Elements II. Small Signal Theory," Proc. IRE, 46, 850-860 (May, 1958).
11. Siegman, A. E., "Phase-Distortionless Limiting by a Parametric Method," Proc. IRE, 47, 447-448 (March, 1959).
12. Uhlir, A., Jr., "The Potential of Semiconductor Diodes in High-Frequency Communications," Proc. IRE, 46, 1099-1115 (June, 1958).
13. Wolf, A. A., and J. E. Pippin, "A Passive Parametric Limiter," Digest of Technical Papers, 1960 International Solid-State Circuits Conference, Philadelphia, Pa. (February 10-12, 1960), pp. 90-91.

Chapter 2

BASIC CONCEPTS

In Chapter 2 we review some of the basic ideas that will be used many times in this book. The topics covered are relatively unconnected. They are treated rather briefly, without formal derivations. More details are given in the references that are cited.

The topics are the following: the Manley-Rowe formulas for nonlinear reactances (Section 2.1), the concepts of imbedding and separation of frequencies (Section 2.2), the various descriptions of gain in amplifiers (Section 2.3), gain criteria for linear amplifiers (Section 2.4), and noise (Section 2.5).

2.1. Manley-Rowe Formulas

The frequency-power formulas of Manley and Rowe[17, 18]‡ apply to nonlinear reactances, including nonlinear capacitors. These formulas are used in Chapter 3 to derive many properties of circuits using such nonlinear reactances. The reader is specifically cautioned, however, that the varactor with series resistance does not obey the Manley-Rowe formulas, and neglect of this fact can sometimes lead to serious miscalculations. Nevertheless, the Manley-Rowe formulas are important for intuitive understanding of frequency converters, parametric amplifiers, and multipliers and dividers.

Consider a nonlinear capacitor excited with two frequencies f_1 and f_0 or radian frequencies $\omega_1 = 2\pi f_1$ and $\omega_0 = 2\pi f_0$. Because of the nonlinearities, sidebands with frequencies of the form $m\omega_1 + n\omega_0$, for positive and negative integral values of m and n, are generated. In general, the capacitor exchanges power with its terminations at all such sidebands.

Since the voltage and current of the capacitor have components at all these sidebands, each can be written in a Fourier series as

$$v(t) = \sum_{m=-\infty}^{+\infty} \sum_{n=-\infty}^{+\infty} V_{mn} e^{j(m\omega_1 + n\omega_0)t} \qquad (2.1)$$

‡Superscript numerals denote references listed at the end of each chapter.

and

$$i(t) = \sum_{m=-\infty}^{+\infty} \sum_{n=-\infty}^{+\infty} I_{mn} e^{j(m\omega_1 + n\omega_0)t}. \quad (2.2)$$

The time-averaged power delivered to the nonlinear capacitor is zero, because the capacitor, although nonlinear, is lossless. However, the nonlinearities convert power from one frequency to another. Let P_{mn} stand for the power into the capacitor at frequency $m\omega_1 + n\omega_0$,

$$P_{mn} = 2 \text{ Re } V_{mn} I^*_{mn}. \quad (2.3)$$

Then, the fact that the capacitor is lossless is expressed by

$$\sum_{m,n} P_{mn} = 0, \quad (2.4)$$

where this summation extends only over combinations of m and n such that $m\omega_1 + n\omega_0$ is positive.‡

The Manley-Rowe formulas provide information in addition to Equation 2.4. Not all combinations of P_{mn} that obey Equation 2.4 are possible with nonlinear reactances, but only those for which certain weighted sums of the P_{mn} vanish. The Manley-Rowe formulas are

$$\sum_{m,n} \frac{mP_{mn}}{m\omega_1 + n\omega_0} = 0 \quad (2.5)$$

and

$$\sum_{m,n} \frac{nP_{mn}}{m\omega_1 + n\omega_0} = 0, \quad (2.6)$$

where these sums are over the same combinations of m and n as in Equation 2.4. If we multiply Equation 2.5 by ω_1 and Equation 2.6 by ω_0 and add, we obtain Equation 2.4. Thus the Manley-Rowe formulas, Equations 2.5 and 2.6, imply that the capacitor is lossless and, therefore, together constitute one relation in addition to the condition of losslessness.

‡The sum can be over, instead, any combinations of m and n such that only one of each pair of frequencies $m\omega_1 + n\omega_0$ and $-(m\omega_1 + n\omega_0)$ is included for each m and n. For example, we may sum m from $-\infty$ to $+\infty$, and n from 0 to ∞.

These formulas hold if two incommensurate frequencies are applied. If more than two frequencies are applied,[18] there are correspondingly more relations similar to Equations 2.5 and 2.6. For our purposes, however, Equations 2.5 and 2.6 are adequate.

Consider a large-signal parametric amplifier or frequency converter that is pumped at frequency ω_p and has a signal introduced at frequency ω_s. The pertinent Manley-Rowe formulas are

$$\sum_{m,n} \frac{mP_{mn}}{m\omega_s + n\omega_p} = 0 \tag{2.7}$$

and

$$\sum_{m,n} \frac{nP_{mn}}{m\omega_s + n\omega_p} = 0. \tag{2.8}$$

On the other hand, suppose that the signal introduced is small in amplitude compared with the pump. Then, the powers exchanged at sidebands of frequency $m\omega_s + n\omega_p$ for m different from 1 and 0 are negligible. One of the Manley-Rowe equations involves only the small-signal powers P_n at the sidebands of the form $\omega_s + n\omega_p$, for n a positive or negative integer. It is, from Equation 2.7,

$$\sum_{n=-\infty}^{\infty} \frac{P_n}{\omega_s + n\omega_p} = 0. \tag{2.9}$$

This Manley-Rowe formula is pertinent to small-signal parametric amplifiers and frequency converters. Note that the sum is over positive and negative (and zero) values of n, so that some of the denominators in the sum are negative.

If only one driving frequency is used, we can obtain the pertinent Manley-Rowe formula by setting m equal to zero in Equations 2.5 and 2.6. Thus, if P_n is now defined as the power input at frequency $n\omega_0$,

$$\sum_{n=1}^{\infty} P_n = 0. \tag{2.10}$$

This formula describes frequency multipliers, dividers, and rational-fraction generators made from lossless nonlinear reactances. It looks as if we have repeated Equation 2.4, the condition of losslessness, but Equation 2.10 is slightly different, in that dc power is excluded.

The Manley-Rowe formulas are obeyed by many physical systems aside from the simple nonlinear capacitor.‡ Among these are nonlinear (and linear) inductors; ideal transformers, lossless cables, waveguides, waveguide junctions, cavities, circulators, and other lossless linear devices; certain nonlinear electromagnetic media[6] (including ferrites without loss); irrotational, longitudinal,[7] and certain other types of electron clouds and beams;[3] certain conservative electromechanical transducers, including rotating machines, microphones, and so forth; ideal acoustic media;[24, 18] networks of elements, each of which obeys the formulas.[18] Any derivation based exclusively on the Manley-Rowe formulas holds for all these systems.

2.2. Specification of Circuitry

A varactor is always used in a circuit of some kind. A description of this circuit is, of course, important for analysis of the varactor application. If we were to draw the circuit completely, we could, with some trouble, solve this circuit, either in the time domain or the frequency domain. If, however, we choose not to draw it, the analysis will demand from us knowledge of those aspects of the circuit that are important and will not require those aspects that are not. In this way, we can concentrate on the important properties of the circuit and perhaps adjust these to optimize the performance.

To follow this second, more desirable course, we do not draw specific circuits to analyze, but instead draw the circuit in a very general way. We use a box to represent unspecified circuitry or else to represent the varactor plus unspecified circuitry. This technique is, of course, not new, but it is very useful, and we wish to discuss two aspects of it now. The first is the concept of imbedding, and the second is that of frequency separation.

2.2.1. Imbedding

We say a device is imbedded in a network when the only way to excite the device is through this network. Thus, for example, the general technique of making a two-port amplifier from a one-port negative resistance is to imbed this device in a three-port network; the other two ports are used as the amplifier input and output. This particular imbedding‡‡ is shown in Figure 2.1. To know the properties of the resulting amplifier, it is not necessary to know in detail the construction of the imbedding network. Only the terminal behavior at the frequency of interest is important.

This imbedding increases the number of terminal pairs, from

‡A more complete discussion is given elsewhere.[18]

‡‡This imbedding is discussed at length in Appendix C.

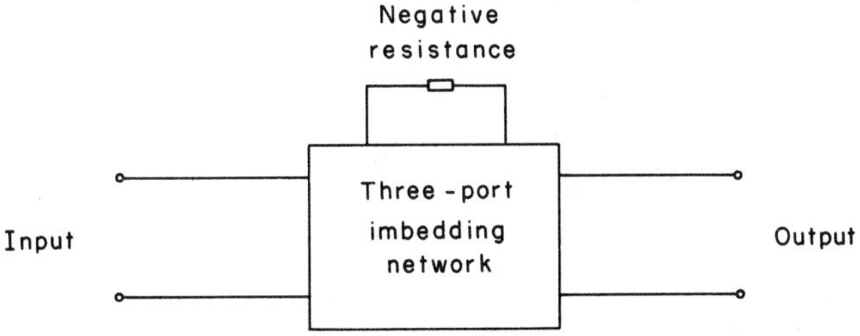

Figure 2. 1. Imbedding of a negative resistance to form a two-port amplifier. This circuit is discussed in detail in Appendix C.

one to two. Other imbeddings may reduce (or not change) the number of terminal pairs. For example, in Figure 2. 2 we show an imbedding of three negative resistances to form an amplifier; this imbedding reduces the number of terminal pairs. Haus and Adler[10] have based much of their general noise theory on the concept of imbedding.

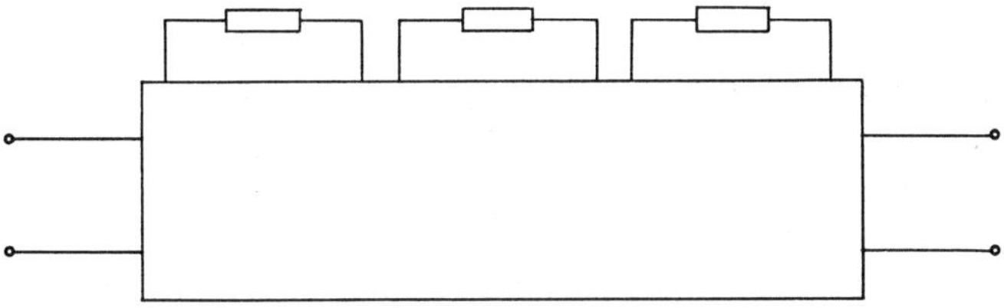

Figure 2. 2. Imbedding reducing the number of terminal pairs.

Imbedding of a varactor is more difficult than this, because several frequencies must be considered simultaneously.

2.2.2. Separation of Frequencies

Varactor applications considered in this book involve converting power from one frequency to another. The varactor is imbedded in a network that has other ports. This network has many purposes. It must terminate the varactor properly at some frequencies. It must couple the varactor at other frequencies to sources and loads.

An imbedding network (or in fact any frequency-converting network or device) has a number of ports, as indicated in Figure 2. 3. These are physical ports, in the sense that each is at a separate point in space. For example, if the varactor is included within the box of Figure 2. 3, these terminals might be the input, output, pump input, and dc bias input of a varactor parametric amplifier.

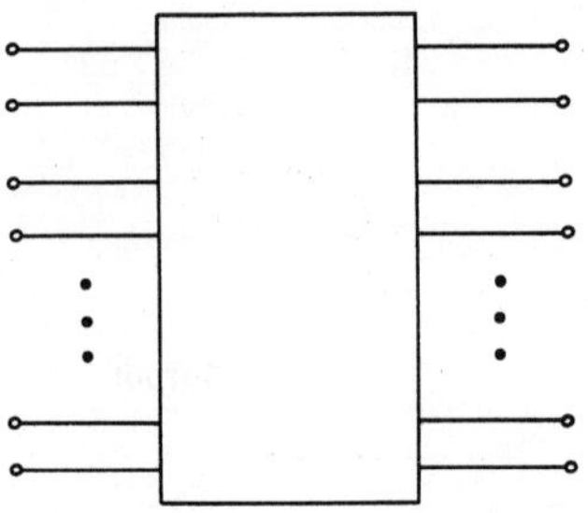

Figure 2. 3. A frequency-converting network with several physical ports.

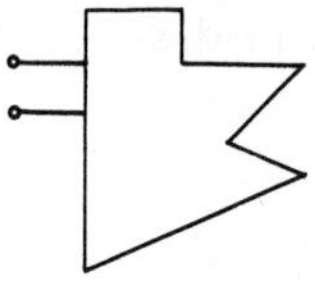

Figure 2.4. Close-up view of one of the physical ports of the network of Figure 2. 3.

In general, power at a number of frequencies flows through each port, either by design or accidentally. Consider, for example, one of the ports, shown separately in Figure 2.4. This port has a voltage and a current that have components at several frequencies. Let us now consider the voltage and current to be divided into separate frequency components and draw an "equivalent terminal pair" for each, as shown in Figure 2. 5. If only one frequency passes through the original port, only one of the new type is necessary, but if more than one did, several ports of the new type are required, one for each frequency. The number of ports has increased, and each of the new ports passes power at only one frequency. The equivalent voltage and current at each of the new ports is the proper Fourier coefficient of the voltage and current at the old port. The new ports are nonphysical in the sense that one cannot point to the actual circuit and find that number of ports. This concept of equivalent ports is very valuable for visualizing the action of several frequencies simultaneously. In the terminations of these "ports" in Figure 2. 5, some physical elements may appear more than once if current with more than one frequency component passes through them.

A varactor is a one-port device. If it exchanges power at more than one frequency, however, it is convenient to represent it as a multiport device, in accordance with these ideas. Thus, if the model of Figure 1. 1 is assumed, the varactor can be rep-

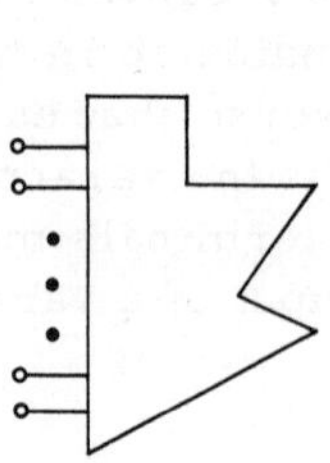

Figure 2. 5. Several fictitious ports have been added to the portion of the system shown in Figure 2. 4; each exchanges power at only one frequency.

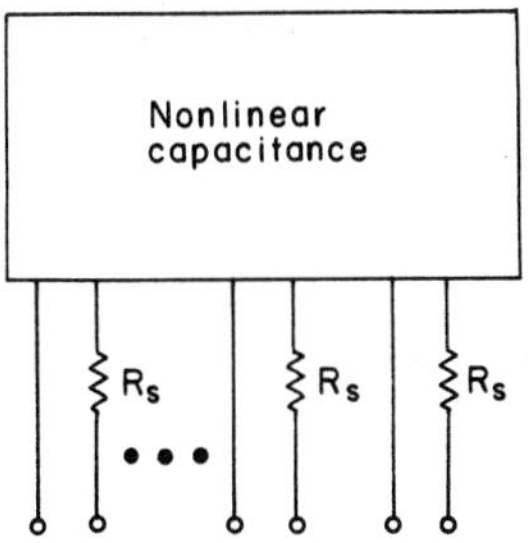

Figure 2.6. Representation of a (one-port) varactor exchanging power at several frequencies, based on the model of Figure 1.1. Note that the linear series resistance R_s appears in series with each port, since current at each frequency passes through it.

resented in the form of Figure 2.6, with several separate ports, one for each frequency of interest.

In general, the varactor is terminated by a circuit. If this circuit is linear, it can be represented separately at each frequency of interest. Thus, at some frequencies, the termination of the varactor may be an open circuit, or a short circuit, or some terminating impedance. At other frequencies, there may be sources. Each of these equivalent terminations can be attached separately to the corresponding fictitious "port" of Figure 2.6. In many cases the same physical components may appear in several of these terminations. In each case, the terminating impedance is determined by looking outward from the varactor, that is, by "sitting on the varactor" and looking out at the outside world. For the analysis of many applications of varactors, this sort of information (equivalent termination at each frequency) is the only description of the varactor imbedding that is necessary.

If many varactors are used (see, for example, Section 6.3), then the preceding ideas are extended in an obvious way. The termination of the array of varactors at each frequency of interest can be described by an impedance matrix with equivalent sources.

In practice, the actual separation of frequencies must ultimately be done physically. This separation, however, requires symmetries or filters that add to the complexity of the analysis, so that the fictitious separation described here is a useful analytical tool.

2.3. Power Gains

There are many ways of describing the amplification of an amplifier. Besides the current gain and the voltage gain, which are ratios of, respectively, currents and voltages, there are many ratios of powers that one can use. In different circumstances different types of gains are more useful. Some are easily measured, whereas others may be of more benefit theoretically.

Consider an amplifier, Figure 2.7, with a source at the input and a load at the output. In general, the source and load

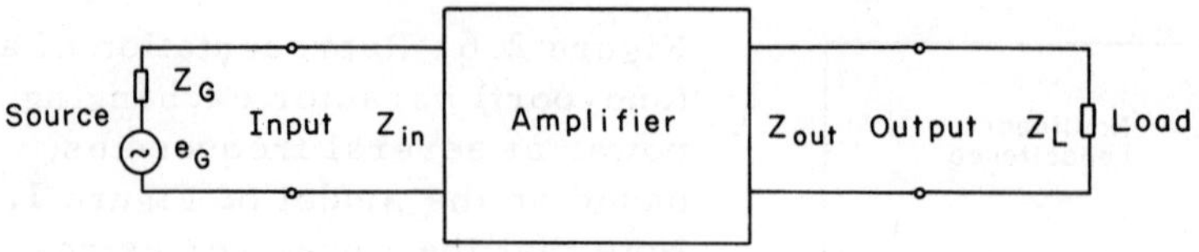

Figure 2.7. Amplifier (or filter or attenuator) with source and load impedances Z_G and Z_L and input and output impedances Z_{in} and Z_{out}. The input frequency may or may not be equal to the output frequency.

are not matched to the amplifier; the amplifier output impedance depends upon the source impedance, and its input impedance depends on the load impedance. There are at least five different ratios of powers that have been called "gains."

First, the ratio of power dissipated in the load to power delivered to the input of the amplifier is called simply the power gain. It is the product of the voltage gain and the current gain. The power gain depends upon both Z_G and Z_L. This power gain is of limited use.

Second, the source has an available power

$$\frac{\overline{e_G^2}}{4 \text{ Re } Z_G}, \tag{2.11}$$

which is, in general, different from the amplifier input power. This available power is the maximum power that can be delivered by the source into any impedance that terminates it. This power is actually delivered if the impedance chosen is Z_G^*. In Expression 2.11, the bar denotes an average, which may be necessary if the source has statistical properties. The ratio of amplifier output power to source available power is called the transducer gain. This gain depends upon both Z_G and Z_L.

Third, for a given source impedance Z_G, the amplifier output has a certain available power, which is in general different from the actual output power. The ratio of output available power to source available power is called the available gain. The available gain depends upon Z_G but not upon Z_L.

Fourth, if the amplifier output impedance Z_{out} has a negative real part, the available power is not customarily defined, because any amount of power can come out of the amplifier. The simplest way around this difficulty is to define the exchangeable power of a source, such as the generator of Figure 2.7, as

$$P_{ex} = \frac{\overline{e_G^2}}{4 \text{ Re } Z_G} \tag{2.12}$$

whether or not Z_G has a positive real part.[8,9,10] Thus, when Re $Z_G > 0$, the exchangeable power is the same as the available power and is positive. However, when Re $Z_G < 0$, the exchangeable power is negative, and the available power is not defined. The exchangeable gain of an amplifier is the ratio of output exchangeable power to input exchangeable power, and is defined whether the real parts of the source and output impedances are positive or negative. The exchangeable gain is the logical generalization of the available gain when some of the resistances may be negative. The exchangeable gain, like the available gain, depends upon Z_G but not upon Z_L.

Fifth, one sometimes hears the term <u>insertion gain</u>. This is the ratio of output power to the power that would be dissipated in the load if the amplifier were, in some sense, not present. Thus, when the amplifier is "inserted," the output power increases by a factor of the insertion gain. The difficulty with this definition is that the circuit without the amplifier is often not stated. Sometimes, when the circuit is removed, a pair of wires from input to output is assumed to take its place. On other occasions, the amplifier is "removed" by shutting off its power source or by physically removing one or more critical components. The value of the insertion gain clearly depends upon the state of the system just before the "insertion." The concept of insertion gain is too vague for our use.

For almost all our purposes, the exchangeable gain is the most convenient. We shall calculate the exchangeable gain of the frequency converters discussed in Chapter 5, and we shall base all our noise parameters on exchangeable gain, so that frequency converters and amplifiers with negative input and output impedances are included more or less automatically.

2.4. Conditions for Gain in Amplifiers

Suppose we have a linear network, such as that in Figure 2.8, with an impedance matrix $\underset{\sim}{Z}$. Can this network perform as an amplifier? We shall derive a simple criterion for telling, from the matrix $\underset{\sim}{Z}$, whether or not the device can amplify.

The essential aspect of amplification is the generation of real power. Increase of a voltage or of a current is not sufficient,

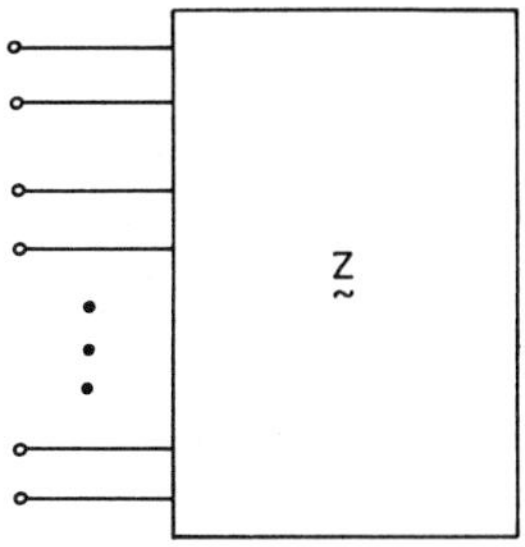

Figure 2.8. Multiport network considered for possible use as an amplifier.

since a passive transformer or a quarter-wave transmission line can do that. Only devices that can have a net output of power at the desired frequency can serve as amplifiers. Also, by a variety of schemes all such devices can be made into amplifiers. Therefore, in considering a given network, it is pertinent to examine its ability to generate power at the frequency of interest.

The real power into the device in Figure 2.8 is, in terms of the voltage and current column matrices $\underset{\sim}{V}$ and $\underset{\sim}{I}$:

$$\text{Power input} = (\underset{\sim}{V}^\dagger \underset{\sim}{I} + \underset{\sim}{I}^\dagger \underset{\sim}{V}), \tag{2.13}$$

where the dagger indicates the complex conjugate transpose of a matrix. Since the impedance matrix is $\underset{\sim}{Z}$,

$$\underset{\sim}{V} = \underset{\sim}{Z}\underset{\sim}{I}, \tag{2.14}$$

so that Equation 2.13 becomes

$$\text{Power input} = \underset{\sim}{I}^\dagger(\underset{\sim}{Z} + \underset{\sim}{Z}^\dagger)\underset{\sim}{I}. \tag{2.15}$$

This quadratic form is nonnegative regardless of $\underset{\sim}{I}$ if the matrix $\underset{\sim}{Z} + \underset{\sim}{Z}^\dagger$ is positive semidefinite.[12] On the other hand, if $\underset{\sim}{Z} + \underset{\sim}{Z}^\dagger$ is not positive semidefinite,‡ then some combination of currents at the various ports can be selected to get power out. The ports can be rigged together in such a way that this combination of currents is excited, and the resulting circuit can be used as an amplifier.

In summary, the network of Figure 2.8 can be made into an amplifier if and only if $\underset{\sim}{Z} + \underset{\sim}{Z}^\dagger$ is not positive semidefinite, that is, if $\underset{\sim}{Z} + \underset{\sim}{Z}^\dagger$ has at least one negative eigenvalue. This gain criterion is used directly in Section 6.3.1.

If the network in Figure 2.8 has only one port, then this matrix analysis is unnecessary; the power input is

$$\text{Power input} = 2|I|^2 \operatorname{Re} Z, \tag{2.16}$$

where Z is the impedance of the network at the frequency of interest. Thus power can come out, so that the device can amplify, if and only if Re Z is negative. This gain criterion applies to

---

‡That is, if $\underset{\sim}{Z} + \underset{\sim}{Z}^\dagger$ has at least one negative eigenvalue.

parametric amplifiers using only one varactor; it is used in Section 6.2.1.

2.5. Noise

One of the major advantages of varactors is their relatively low noise. To describe their noise, we use the concepts of noise figure, noise measure, and noise temperature. In Section 2.5 we give the basic properties of these terms, discuss two noise equivalent circuits for amplifiers, and discuss noise in negative-resistance amplifiers.

2.5.1. Noise Equivalent Circuits

We now give some noise equivalent circuits: first, thermal noise in a simple resistor and in a more general passive network; second, any type of noise in a negative resistance, and then in a two-port amplifier, frequency converter, or other two-port network.

2.5.1(a). Resistor. A resistor R in thermal equilibrium at temperature T_r generates noise. A convenient equivalent circuit of the resistor, Figure 2.9, shows the resistor and its Thévenin equivalent open-circuit noise voltage e_n. The voltage

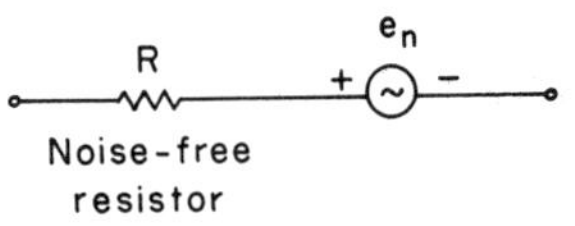

Figure 2.9. Thévenin equivalent circuit of a resistor at temperature T_r.

e_n is a random variable, and cannot be characterized as a mathematical time function. Expressed in the frequency domain, its statistics are such that the spectrum at any frequency f is uncorrelated with the spectrum at any other frequency f', or

$$\overline{e_n(f)e_n^*(f')} = 0, \tag{2.17}$$

and that the mean-squared voltage component located in a range Δf about the frequency f is[23]

$$\overline{|e_n(f)|^2} = 4kT_rR\,\Delta f, \tag{2.18}$$

where k is Boltzmann's constant.

We have followed the customary procedure of calling e_n the effective rms noise voltage. The signal voltages in this book are written as half-amplitudes rather than as rms values. Strictly speaking, a factor of $1/\sqrt{2}$ should accompany all noise voltages when they appear in the same equations with signal voltages. Except in Section 6.6, however, we have omitted these factors because the noise and signal are never directly compared. Reference is made to this apparent discrepancy in many footnotes in Chapters 4, 5, and 6.

Actually, the spectrum given in Equation 2.18 is only approximately true. It is valid from zero up through microwave frequencies, but for still higher frequencies and extremely low temperatures, Equation 2.18 should be modified by quantum effects.[23] For most practical purposes, these effects can be neglected, since Equation 2.18 is valid if

$$kT_r \gg hf, \tag{2.19}$$

where h is Planck's constant.

The available noise power from the resistor is $kT_r\Delta f$.

2.5.1(b). Passive Network. A more general passive network, containing capacitors and inductors, and nonreciprocal elements such as circulators, gyrators, and isolators, has a similar Thévenin equivalent circuit for thermal noise, Figure 2.10. In

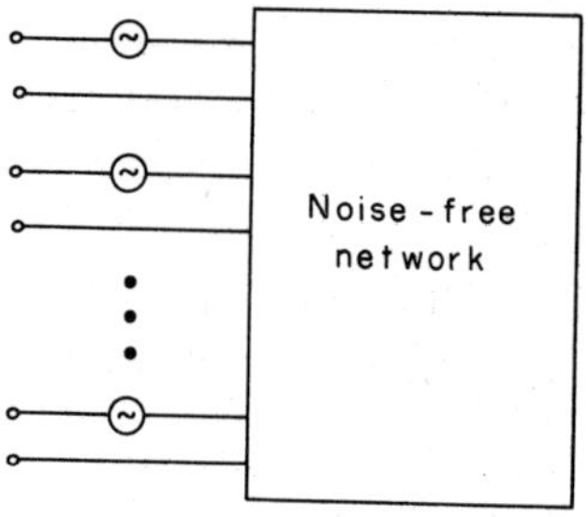

Figure 2.10. Thévenin equivalent circuit for thermal noise in a more general passive network at temperature T_r.

matrix notation, the noise column matrix $\underset{\sim}{E}_n(f)$ is uncorrelated with noise at another frequency f',

$$\overline{\underset{\sim}{E}_n(f)\underset{\sim}{E}_n{}^\dagger(f')} = 0, \tag{2.20}$$

and the various entries of the noise column matrix are correlated with each other at the same frequency,

$$\overline{\underset{\sim}{E}_n(f)\underset{\sim}{E}_n{}^\dagger(f)} = 2kT_r(\underset{\sim}{Z} + \underset{\sim}{Z}^\dagger)\,\Delta f, \tag{2.21}$$

where $\underset{\sim}{Z}$ is the impedance matrix of the network. Equations 2.20 and 2.21 are logical extensions of Equations 2.17 and 2.18.

2.5.1(c). Negative Resistance. A negative resistance invariably has some noise associated with it. A Thévenin equivalent circuit for the negative resistance is shown in Figure 2.11, with

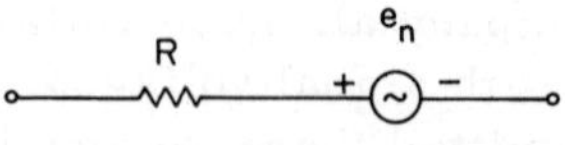

Figure 2.11. Thevenin equivalent circuit of a noisy negative resistance.

an open-circuit voltage and a negative resistance. We may formally define the "temperature" T at frequency f for this device by the formula

$$\overline{|e_n|^2} = -4kTR\,\Delta f, \tag{2.22}$$

where T is positive and R is negative. An interpretation of this "temperature" is given in Section 2.5.5. Note that we have not specified the physical source of this noise.

<u>2.5.1(d). Rothe-Dahlke Model for Amplifiers.</u> Consider a two-port amplifier with noise of unspecified physical origin. This two-port network‡ can be represented by a number of different noise equivalent circuits.[16] For example, we can use a Thevenin equivalent circuit or a Norton equivalent circuit.

An equivalent circuit of particular value is that of Rothe and Dahlke.[22] This circuit, shown in Figure 2.12, has both of the required noise generators at the input. For many purposes,

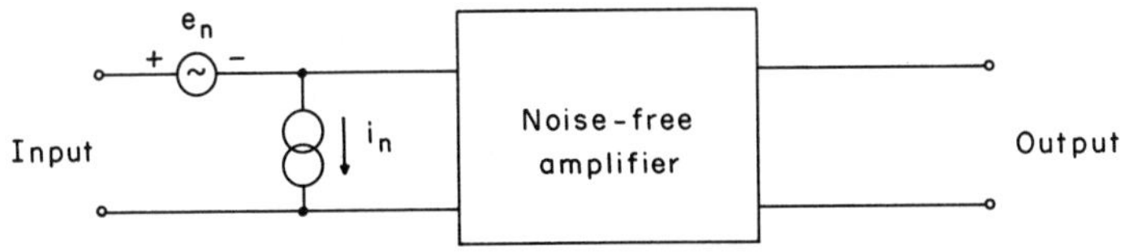

Figure 2.12. Rothe-Dahlke noise model for a noisy two-port network, such as an amplifier or a frequency converter.

especially calculating noise figure, this is convenient, since the noiseless amplifier that follows does not affect the noise figure, except insofar as it affects the generators e_n and i_n.

It is convenient to define the quantities R_n, G_n, and ρ:

$$R_n = \frac{\overline{|e_n|^2}}{4kT_0\,\Delta f}, \tag{2.23}$$

$$G_n = \frac{\overline{|i_n|^2}}{4kT_0\,\Delta f}, \tag{2.24}$$

and

---

‡The network need not be an amplifier. In fact, the output may be at a frequency different from that of the input.

$$\rho = \frac{\overline{i_n e_n^*}}{\sqrt{\overline{|i_n|^2}\,\overline{|e_n|^2}}}, \tag{2.25}$$

where T_0 is standard temperature, 290° K. The correlation coefficient ρ is a complex number less than or equal to 1 in magnitude.

The noise figure of an amplifier or of a twoport is defined in Section 2.5.2. If the amplifier of Figure 2.12 is run from a source with impedance $Z_o = R_o + jX_o$, the noise figure F of this amplifier is, in terms of this model,

$$F = 1 + \frac{\overline{|e_n + Z_o i_n|^2}}{4kT_0 R_o \,\Delta f}$$

$$= 1 + \frac{R_n + |Z_o|^2 G_n + 2\sqrt{R_n G_n}\,\mathrm{Re}\,(\rho Z_o)}{R_o}. \tag{2.26}$$

As Z_o is varied over all values with R_o positive, the noise figure has a minimum value

$$F_{min} = 1 + 2\sqrt{R_n G_n}\,[\sqrt{1 - (\mathrm{Im}\,\rho)^2} + \mathrm{Re}\,\rho], \tag{2.27}$$

which occurs for the <u>optimum source impedance</u>

$$Z_{o,opt} = R_{o,opt} + jX_{o,opt} \tag{2.28}$$

given by

$$|Z_{o,opt}|^2 = \frac{R_n}{G_n} \tag{2.29}$$

$$X_{o,opt} = \sqrt{\frac{R_n}{G_n}}\,\mathrm{Im}\,\rho. \tag{2.30}$$

The noise figure for nonoptimum Z_o can be expressed in terms of F_{min}:

$$F = F_{min} + \frac{G_n}{R_o}|Z_o - Z_{o,opt}|^2. \tag{2.31}$$

In case the source resistance R_o is negative, then F is less than 1 and may be negative. A value of Z_o can be chosen to make F a maximum (that is, as close as possible to 1). The behavior of F as a function of R_o is shown in Figure 2.13 for the simple case with ρ real.

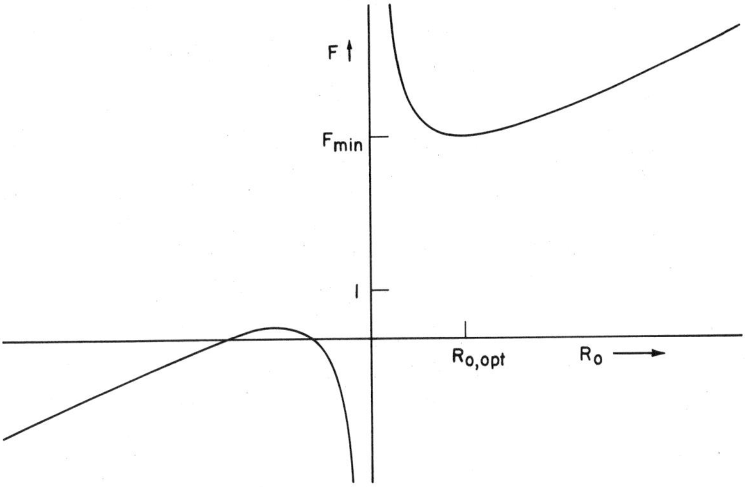

Figure 2.13. Plot of noise figure F as a function of source resistance R_o, both positive and negative, for the particular case that ρ is real.

2.5.1(e). Bauer-Rothe Model for Amplifiers. The Rothe-Dahlke noise model has two correlated noise generators at the input of the amplifier or two-port device. An alternative equivalent circuit, proposed by Bauer and Rothe,[2,20] also has two noise generators at the input, but they are made to be uncorrelated. Because of this, the expression for noise figure has a particularly simple form.

This new model uses wave, or scattering,[4,25] variables. The incoming and outgoing waves at the input of the amplifier are

$$a_1 = \frac{V_1 + Z_\nu I_1}{\sqrt{Z_\nu + Z_\nu^*}} \tag{2.32}$$

and

$$b_1 = \frac{V_1 - Z_\nu^* I_1}{\sqrt{Z_\nu + Z_\nu^*}}, \tag{2.33}$$

where V_1 and I_1 are the input voltage and current half-amplitudes, and Z_ν is a complex number with positive real part, the normalization impedance.[25] In a similar way, the incoming and outgoing wave amplitudes at the output are a_2 and b_2, defined with a normalization impedance that is, in general, different from Z_ν, which was used at the input.

Wave noise generators are defined at the input, in terms of the generators in the Rothe-Dahlke model, Figure 2.12, as

$$a_n = -\frac{e_n + Z_\nu i_n}{2\sqrt{\operatorname{Re} Z_\nu}} \tag{2.34}$$

and

$$b_n = \frac{e_n - Z_\nu^* i_n}{2\sqrt{\operatorname{Re} Z_\nu}}. \tag{2.35}$$

Thus, the amplifier equations, including noise, are, in terms of the two-port scattering matrix $\underset{\sim}{S}$,

$$\begin{bmatrix} b_1 - b_n \\ b_2 \end{bmatrix} = \begin{bmatrix} S_{11} & S_{12} \\ S_{21} & S_{22} \end{bmatrix} \begin{bmatrix} a_1 + a_n \\ a_2 \end{bmatrix}. \tag{2.36}$$

An equivalent circuit of this representation is shown in Figure 2.14. The wave generators are represented by directional couplers and ordinary sources; this schematic is accurate in the limit of small coupling to the line, and large strengths of the sources. Note that this "transmission line" in Figure 2.14 has a complex "characteristic impedance" Z_ν, but it is a lossless line.

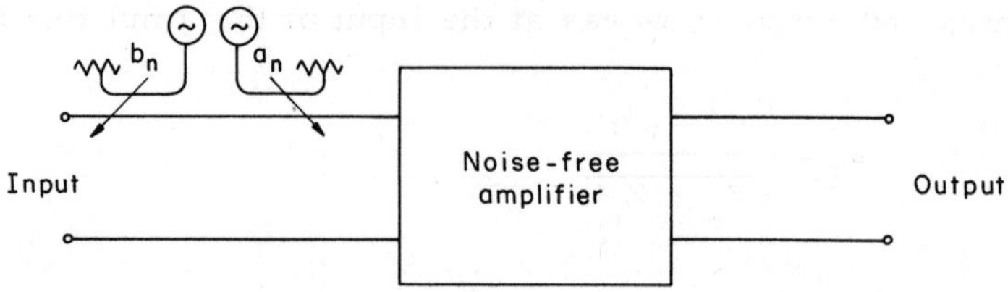

Figure 2.14. Bauer-Rothe wave model of a noisy two-port network, such as an amplifier or a frequency converter. The two generators are made uncorrelated by properly choosing the normalization impedance Z_ν.

In general, the sources a_n and b_n are correlated. However, it is possible to choose a value for Z_ν to make them uncorrelated, so that

$$\overline{a_n b_n^*} = 0. \tag{2.37}$$

Applying Equation 2.37 to Equations 2.34 and 2.35, we find that the required normalization impedance is simply $Z_{o,\,opt}$ from Equation 2.28. This point is important: If the optimum source impedance $Z_{o,\,opt}$ is used as the normalization impedance for defining a_n and b_n, then these two noise wave generators are uncorrelated.

The strengths of a_n and b_n define temperatures T_a and T_b:

$$T_a = \frac{\overline{|a_n|^2}}{k\,\Delta f} \tag{2.38}$$

$$T_b = \frac{\overline{|b_n|^2}}{k\,\Delta f}. \tag{2.39}$$

In terms of the Rothe-Dahlke model, these are

$$T_a = T_0(F_{min} - 1)$$

$$= 2T_0\sqrt{R_n G_n}\,\left[\sqrt{1 - (\mathrm{Im}\,\rho)^2} + \mathrm{Re}\,\rho\right] \tag{2.40}$$

$$T_b = 2T_0\sqrt{R_n G_n}\,\left[\sqrt{1 - (\mathrm{Im}\,\rho)^2} - \mathrm{Re}\,\rho\right]. \tag{2.41}$$

Written in terms of the uncorrelated wave generators, the expression for noise figure, Equation 2.26, is very simple. The reflection coefficient Γ_o of the source with respect to the normalization impedance $Z_\nu = Z_{o,\,opt}$ is

$$\Gamma_o = \frac{Z_o - Z_\nu}{Z_o + Z_\nu^*}. \tag{2.42}$$

The noise figure is

$$F = 1 + \frac{T_a + T_b\,|\Gamma_o|^2}{T_0(1 - |\Gamma_o|^2)}, \tag{2.43}$$

which is obviously made a minimum by choosing the source impedance so that $\Gamma_o = 0$. Equation 2.43 holds whether the real part of the source impedance Z_o is positive or negative, that is, whether Γ_o is less than or greater than 1 in magnitude.

This wave representation fails[20] only when $R_{o,opt} = 0$; that is, only when $R_n = 0$, $G_n = 0$, or $\rho = \pm j$. In each of these cases, $F_{min} = 1$, and only one equivalent noise generator is necessary if it is properly placed.

This wave model is convenient to describe second-stage noise. We use it for that purpose in Appendix C.

2.5.2. Noise Figure

The noise figure F of an amplifier (or any noisy twoport, such as a frequency converter) is intended as an indication of its noisiness. The lower the noise figure, the less noise is contributed by the amplifier, and the less the amplifier degrades the quality of the signal passing through it.

To tell how much the quality of a signal is degraded by an amplifier, we must have some measure of the quality of a signal with background noise. The most common measure is the ratio of signal power to noise power in a band of frequencies Δf, or, for short, the signal-to-noise ratio. Thus, for example, if the signal comes from a noisy source, with equivalent noise voltage e_n, as indicated in Figure 2.15, then the signal-to-noise ratio is

$$\frac{S}{N} = \frac{\overline{e_s^{\,2}}}{\overline{e_n^{\,2}}} . \tag{2.44}$$

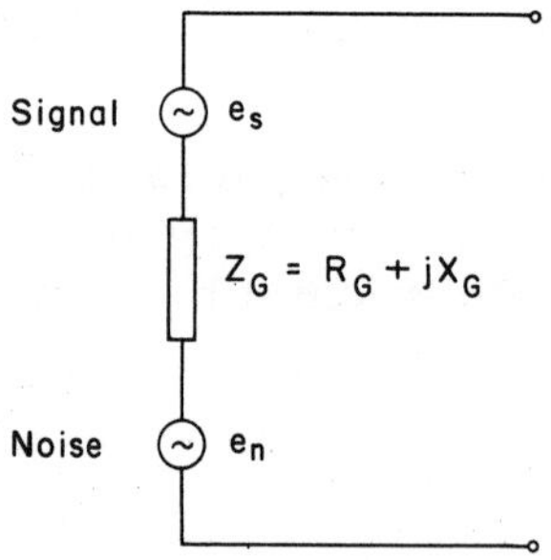

Figure 2.15. A noisy source. The signal is e_s, and the noise is represented by the upper generator e_n. If e_n is caused entirely by thermal noise, the signal-to-noise ratio is Equation 2.45; in any case it is given by Equation 2.44.

This is the ratio of exchangeable signal power to exchangeable noise power.‡ Alternatively, if this source is terminated by a

‡We implicitly assume that the noise and the signal are not correlated. If they are, then the term "noise" is not appropriate.

load, it is the ratio of: (1) power dissipated in the load because of the signal to (2) power dissipated in the load because of the noise.‡

If the noise is taken to be thermal noise of the source at temperature T_s, then Equation 2.44 becomes

$$\frac{S}{N} = \frac{\overline{e_s^2}}{4kT_s R_G \, \Delta f}. \tag{2.45}$$

If a signal plus some noise, with power ratio $(S/N)_{in}$, is passed through an amplifier, the output signal-to-noise ratio $(S/N)_{out}$ is less than $(S/N)_{in}$, because of noise generated within the amplifier. The <u>noise figure</u>, or <u>noise factor</u>, is defined[5, 13, 14, 15] as

$$F = \frac{(S/N)_{in}}{(S/N)_{out}} \tag{2.46}$$

when the input is a source with added thermal noise at temperature T_0.

An alternative formula for F is convenient. The signals and noises in Equation 2.46 can be either actual powers, or else exchangeable powers. If we interpret them as exchangeable powers, and let the exchangeable gain of the amplifier be G_e, then

$$S_{out} = G_e S_{in} \tag{2.47}$$

and

$$N_{out} = G_e N_{in} + N_{internal} \tag{2.48}$$

where $N_{internal}$ is the noise output caused by noise generated internally by the amplifier.‡‡ Equations 2.47 and 2.48 transform Equation 2.46 into

$$F = 1 + \frac{N_{internal}}{N_{out} - N_{internal}}, \tag{2.49}$$

where these noises are exchangeable powers. Note that there is

‡In making this statement, we neglect any noise generated by the load.

‡‡We assume this noise is not correlated with either S_{in} or N_{in}.

no contribution to F from noise entering the amplifier at the output. The noise (N_{out} - $N_{internal}$) is merely the contribution to output noise caused by input noise.

Bear in mind that the noise figure is defined with the source temperature equal to standard temperature, $T_0 = 290°$ K. It therefore does not depend upon the actual noise characteristics of the source. It depends upon the source impedance but not on the load impedance.

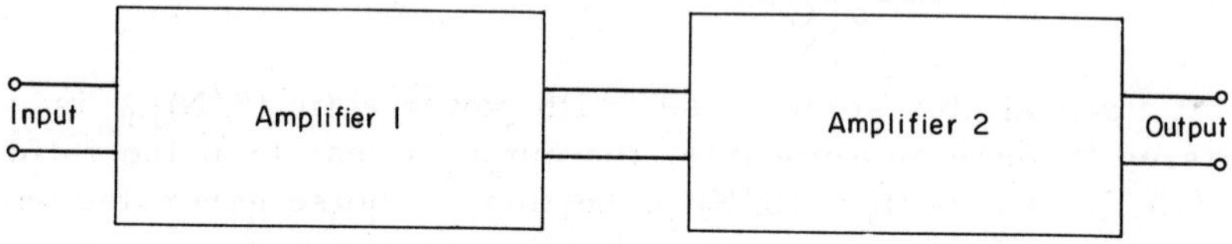

Figure 2.16. Cascade of two noisy amplifiers (or two-port circuits or frequency converters).

If two amplifiers are cascaded, as indicated in Figure 2.16, the over-all noise figure F_{12} can be calculated from the cascade formula.[5] The noise figure F_1 of the first stage depends on the source impedance, and the noise figure F_2 of the second stage depends upon its source impedance, which is simply the output impedance of the first stage. The cascade formula is

$$F_{12} = F_1 + \frac{F_2 - 1}{G_{e1}}, \tag{2.50}$$

where G_{e1} is the exchangeable gain of the first stage.‡

2.5.3. Noise Measure

Ultimately, noise figures for individual amplifier stages are important because the over-all system noise figure can be calculated from them by use of the cascade formula, Equation 2.50. If the first-stage exchangeable gain is high, then the system noise figure reduces to the noise figure of the first stage, and contributions from succeeding stages can be neglected. Thus, the single-stage noise figure is a valuable criterion for amplifiers of high gain.

For amplifiers of low gain, however, the noise figure F is not adequate by itself. An amplifier with excellent noise figure but very little (or no) gain may be less useful than an amplifier

‡We assume that the noise generated internally by the two stages is not correlated.

with higher F but higher gain, because the latter amplifier may lead to a lower cascade noise figure.

A mathematical combination of the noise figure and the gain that is useful for amplifiers of low gain is Haus and Adler's noise measure[9,10] M:

$$M = \frac{F - 1}{1 - \frac{1}{G_e}}. \tag{2.51}$$

For amplifiers with high gain G_e, this reduces to simply F - 1.

The fact that M is a better criterion than F is indicated by the cascade formula for noise measure

$$M_{12} = M_1 + (M_2 - M_1)\frac{G_{e2} - 1}{G_{e1}G_{e2} - 1}, \tag{2.52}$$

which shows that the cascade noise measure can be as low as, but no lower than, the smaller of M_1 and M_2. If two amplifiers with equal values of M are cascaded, it does not matter which is placed first. To choose between two or more amplifiers to be used in cascade in an amplifying system of high gain, the one with the smaller M should be chosen, not necessarily the one with the smaller F.

The noise measure depends upon the source impedance (but not the load impedance), and upon the presence of any feedback. For all source impedances and any feedback arrangement, the lowest positive value of M is called M_{opt}. This number is the lowest possible noise figure at high gain of an amplifier made from individual stages, each with this value of M_{opt}.

For most amplifiers, F can be used alone to judge the quality of noise performance, since the gain will ordinarily be high. For many amplifiers, however, including negative-resistance amplifiers, the noise measure M and especially its lowest positive value M_{opt} are better.

2.5.4. Noise Temperature

If an amplifier is run from a source with positive internal resistance, then the noise figure is greater than 1. Amplifiers of low noise have a noise figure that is close to 1, and it is convenient to sometimes use the excess noise figure F - 1 for more accuracy. This excess noise figure is the ratio of internally generated noise to output noise caused by a source of temperature T_0 at the input. It is convenient to think of the internally generated noise as coming from an equivalent noise generator at the input, with effective input noise temperature[15] T. Then the excess noise figure would be simply the ratio of output noise caused by this equivalent source to that caused by the actual source at temperature T_0, or simply the ratio of T to T_0. Thus, the effective input noise temperature T of the amplifier is

$$T = T_0(F - 1). \tag{2.53}$$

Like the noise figure, T depends upon the source impedance.

If an amplifier is run from a source with noise temperature T_s different from T_0, it is easy to compare the effects of the source noise and those of the amplifier noise. Simply compare the source temperature T_s with the amplifier's noise temperature T. In practice, T_s is usually different from T_0; for example, an antenna pointing toward the sky at microwave frequencies sees an effective noise temperature[21] that can be considerably below T_0.

The characterization of amplifier noise by this effective input noise temperature has, like that using the noise figure, the fault that it is misleading for amplifiers of low gain. A more logical definition of noise temperature might be T_0M, where M is the noise measure, or T_0M_{opt}. Unfortunately, however, these definitions have not been used. In Chapter 5 we describe the noisiness of frequency converters by using T of Equation 2.53. In Chapter 6 we evaluate T for parametric amplifiers at high gain, which is equivalent to using T_0M as the definition of T.

The noise temperature T_{12} of a cascade is, in terms of the noise temperatures T_1 and T_2 of the two stages,

$$T_{12} = T_1 + \frac{T_2}{G_{e1}}. \tag{2.54}$$

2.5.5. Noise Temperature of Negative-Resistance Amplifiers

When an amplifier is made from a negative resistance by imbedding this resistance in a three-port network, as shown in Figure 2.1, the noise measure (if positive) is always greater than the positive quantity

$$\frac{T}{T_0} = \frac{-P_e}{kT_0\ \Delta f} \tag{2.55}$$

(which is also given in Equation 2.22), where P_e is the exchangeable noise power of the negative resistance and so is inherently negative. If the imbedding network is lossless, then M is simply

T/T_0, independent of the characteristics of the lossless network.[19] The optimum noise measure M_{opt} is in this case equal to the value of M achieved. Proofs of these facts are given in the derivations of Appendix C (see Equation C-35).

Thus T, given in Equation 2.55, is the lowest possible value for the noise temperature at high gain, and if the imbedding network is lossless, then T is exactly the noise temperature at high gain. In Chapter 6, when we examine noise in parametric amplifiers, we use Equation 2.55.

2.5.6. Noise Temperature of Other Amplifiers

We now wish to give a lower bound for the noise temperature at high gain of an amplifier made by imbedding a multiport network, as indicated in Figure 2.17. For example, this multiport

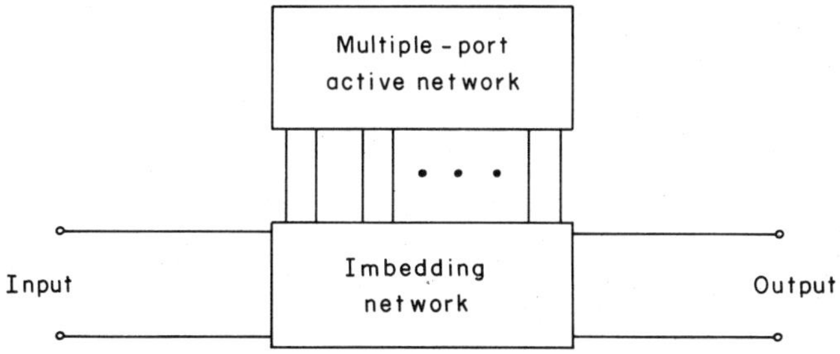

Figure 2.17. Amplifier made out of a k-port active network by imbedding it in a (k + 2)-port imbedding network.

network might be an array of varactors, with communicating idlers, as discussed in Section 6.3.

The noise measure of the resulting amplifier is, if positive, always greater than a number that is characteristic of the multiport network. From the impedance matrix $\underset{\sim}{Z}$ of this network and the Thévenin equivalent noise voltage column matrix $\underset{\sim}{E}$ form the characteristic noise matrix $\underset{\sim}{N}$:

$$\underset{\sim}{N} = -\frac{1}{2}(\underset{\sim}{Z} + \underset{\sim}{Z}^\dagger)^{-1}\underset{\sim}{E}\underset{\sim}{E}^\dagger . \qquad (2.56)$$

Then, Haus and Adler have shown[9, 10] that the noise measure of the amplifier of Figure 2.17 is, if positive, no less than the smallest positive eigenvalue of $\underset{\sim}{N}$, divided by $kT_0\,\Delta f$. For some imbedding networks, this lower bound can be achieved; for others, it cannot.

The lowest possible noise temperature T at high gain is therefore the lowest positive eigenvalue of $\underset{\sim}{N}$, divided by $k\,\Delta f$. This fact is useful in Section 6.3.2.

2.5.7. Extensions of the Noise Definitions

Many of the terms discussed here, including Noise Figure, Noise Factor, Standard Noise Temperature (290°), Noise Temperature (at a Port), and Effective Input Noise Temperature, have been standardized.[13,14,15] Any attempts to extend these concepts beyond their standard definitions are sure to be controversial.

If negative resistances are present, these terms, which are based on available powers and gains, can be defined using the exchangeable powers and gains. This extension, which we use in this book, was introduced by Haus and Adler.[8,9,10]

Many people have attempted to extend these definitions by explicitly including, among the noise sources that contribute to F, noise generated by the amplifier load. The definitions do not include such noise, and there is no clear reason why it would be desirable to do so. There is an intuitive feeling that if the load is not matched to the output impedance of the amplifier, then noise generated in the load gets reflected back toward the load. Actually, however, it is not load noise that is important. Rather, it is noise generated by the second stage and coming out the input of the second stage. Certainly this physical effect of reflection by an imperfectly matched first stage does occur, and using the standard definitions, we account for this noise as part of the second-stage noise figure. It contributes to system noise figure by way of the second term in the cascade formula, Equation 2.50. If this noise is included as part of the first-stage noise figure, then the cascade formula does not work. We do not use this "extension" in this book.

If an amplifier has a single output but many inputs, then the noise description is more complicated.[1] The concept of effective input noise temperature can be extended to this case, as simply the source temperature (the same at all the inputs) that yields the same noise as that generated internally. It is assumed in this extension, of course, that the noises at the various inputs are not correlated. The various inputs may be at different frequencies, or some may be at the same frequency but at different points in space, as, for example, from several antennas.

If an amplifier has more than one output,‡ either at different frequencies (with spectra either overlapping or not) or at different physical terminals, then the concepts of noise figure, noise measure, and noise temperature do not apply. In fact, because the output noises are in general correlated, even the concept of signal-to-noise ratio is vague. Serious mistakes

---

‡In most applications, the degenerate parametric amplifier has more than one output. See Section 6.6.

can be made by using these terms without further inquiry into their meaning. For this reason, in Section 6.6 we do not define a noise temperature or noise figure (aside from a figure of merit T_{eff}) for the double-sideband degenerate parametric amplifier.

REFERENCES

1. Adler, R., R. S. Engelbrecht, H. A. Haus, S. W. Harrison, M. T. Lebenbaum, and W. W. Mumford, "Description of the Noise Performance of Amplifiers and Receiving Systems," Proc. IRE. (forthcoming, 1962).
2. Bauer, H., and H. Rothe, "Der äquivalente Rauschvierpol als Wellenvierpol," Arch. Elektrischen Übertragung, 10, 241-252 (1956).
3. Bers, A., and P. Penfield, Jr., "Conservation Principles for Plasmas and Relativistic Electron Beams," IRE Trans. on Electron Devices, ED-9, 12-26 (January, 1962).
4. Carlin, H. J., "The Scattering Matrix in Network Theory," IRE Trans. on Circuit Theory, CT-3, 88-97 (June, 1956).
5. Friis, H. T., "Noise Figures of Radio Receivers," Proc. IRE, 32, 419-422 (July, 1944).
6. Haus, H. A.,"Power-Flow Relations in Lossless Nonlinear Media," IRE Trans. on Microwave Theory and Techniques, MTT-6, 317-324 (July, 1958).
7. Haus, H. A., "The Kinetic Power Theorem for Parametric, Longitudinal, Electron-Beam Amplifiers," IRE Trans. on Electron Devices, ED-5, 225-232 (October, 1958).
8. Haus, H. A., and R. B. Adler, "An Extension of the Noise Figure Definition," Proc. IRE, 45, 690-691 (May, 1957).
9. Haus, H. A., and R. B. Adler, "Optimum Noise Performance of Linear Amplifiers," Proc. IRE, 46, 1517-1533 (August, 1958).
10. Haus, H. A., and R. B. Adler, Circuit Theory of Linear Noisy Networks, The Technology Press and John Wiley & Sons, Inc., New York (1959).
11. Hildebrand, F. B., Methods of Applied Mathematics, Prentice-Hall, Inc., Englewood Cliffs, N. J. (1952).
12. Hildebrand, F. B., ibid., Chapter 1.
13. "IRE Standards on Receivers: Definitions of Terms, 1952," 52 IRE 17.S1, Proc. IRE, 40, 1681-1685 (December, 1952).
14. "IRE Standards on Electron Tubes: Definitions of Terms,1957," 57 IRE 7.S2, Proc. IRE, 45, 983-1010 (July, 1957).
15. "IRE Standards on Methods of Measuring Noise in Linear Twoports, 1959," 59 IRE 20.S1, Proc. IRE, 48, 60-68 (January, 1960).
16. IRE Subcommittee 7.9 on Noise, H. A. Haus, Chairman, "Representation of Noise in Linear Twoports," Proc. IRE, 48, 69-74 (January, 1960).
17. Manley, J. M., and H. E. Rowe, "Some General Properties of Nonlinear Elements. – Part I. General Energy Relations," Proc. IRE, 44, 904-913 (July, 1956).

18. Penfield, P., Jr., Frequency-Power Formulas, The Technology Press and John Wiley & Sons, Inc., New York (1960).
19. Penfield, P., Jr., "Noise in Negative-Resistance Amplifiers," IRE Trans. on Circuit Theory, CT-7, 166-170 (June, 1960).
20. Penfield, P., Jr., "Wave Representation of Amplifier Noise," IRE Trans. on Circuit Theory, CT-9, 84-86 (March, 1962).
21. Rafuse, R. P., "Characterization of Noise in Receiving Systems," in Lectures on Communication System Theory, E. J. Baghdady, Editor, McGraw-Hill Book Co., Inc., New York (1961), pp. 369-399.
22. Rothe, H., and W. Dahlke, "Theory of Noisy Fourpoles," Proc. IRE, 44, 811-818 (June, 1956).
23. van der Ziel, A., Noise, Prentice-Hall, Inc., New York (1954).
24. Wagner, R. R., "A Lossless Acoustic Power Theorem," S.B. Thesis, Department of Electrical Engineering, M.I.T. Cambridge (May, 1959).
25. Youla, D. C., "On Scattering Matrices Normalized to Complex Port Numbers," Proc. IRE, 49, 1221 (July, 1961).

Chapter 3

LOSSLESS NONLINEAR REACTANCE DEVICES

In Chapters 5 to 10 we shall examine in detail the use of lossy varactors in frequency-converting systems, including upconverters, downconverters, parametric amplifiers, harmonic multipliers, and harmonic dividers, using the model discussed in Chapter 4. Here, we consider the use of a lossless varactor in such systems. Although all of these results can be obtained from the later results by simply letting the series resistance vanish, we prefer to derive them independently, so that they can be used for comparison throughout the rest of the book.

The lossless varactor obeys the Manley-Rowe formulas,[19, 21] ‡ discussed in Section 2.1. This property is all we use in discussing the lossless case. The results of the present chapter then hold for frequency-converting systems using other nonlinear reactances, or using any physical systems that obey the Manley-Rowe formulas, including some nonlinear electromagnetic media,[8, 21, 2] ferrites,[8, 21] and electron-beam devices.[9, 21, 2]

The results obtained in this chapter, from the Manley-Rowe formulas, are of basic importance in understanding the frequency-converting systems. On the other hand, the varactor with loss does not obey the formulas, and consequently, the results of Chapters 5 to 10 are significantly different. The role of the series resistance is important even at frequencies one-thousandth of the cutoff frequency; the reader is specifically cautioned against basing any practical designs on the information of this chapter. Synthesis procedures based on the lossless case are often very misleading.

For harmonic generation, only frequencies of the form $n\omega$ are present, and the pertinent Manley-Rowe formula is Equation 2.10, which is used in Sections 3.4 and 3.5. The results derived there for harmonic multipliers and dividers are used for comparison of the results of Chapters 8 and 9. For large-signal amplifiers and converters, the device is excited with frequencies of the form $n\omega_p + m\omega_s$, where ω_p is the pump frequency and ω_s the signal frequency. The pertinent Manley-Rowe formulas are Equations 2.7 and 2.8. If the pumping amplitude is much larger than the amplitudes of the signal and

---

‡Superscript numerals denote references listed at the end of each chapter.

generated sidebands, then no power is exchanged for values of m different from 1 or 0; the Manley-Rowe formula relating the small-signal variables is Equation 2.9. This formula is used in Sections 3.1 and 3.2 in discussing small-signal frequency converters and parametric amplifiers. The results obtained there will be compared with the results of Chapters 5 and 6.

The actual manipulations in Sections 3.1 and 3.2 are in terms of scattering variables; it is convenient to define and discuss these variables here, and to express the Manley-Rowe formulas in terms of the scattering matrix. Each of the small-signal systems in Sections 3.1 and 3.2 can be represented in a form similar to Figure 3.1. The box encloses the nonlinear reactance, as pumped, as well as miscellaneous linear reactances; therefore, it obeys the Manley-Rowe formulas. In accordance with the ideas of Section 2.2, we show a number of ports, each of which exchanges energy at only one frequency. Simple upconverters would require only two such ports (one at the input frequency, and one at the output frequency), but parametric amplifiers and upconverters with idlers would require more.

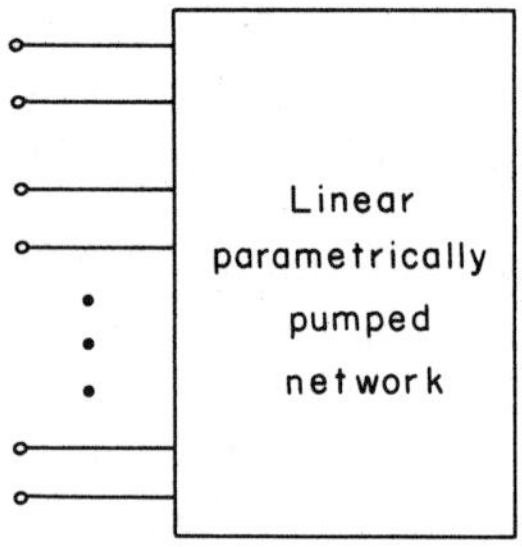

Figure 3.1. Schematic representation of a nonlinear network, as pumped. In accordance with the ideas of Section 2.2, a number of ports are shown, each of which exchanges energy at only one frequency. Physically, the system may have a smaller number of ports, each exchanging energy at several frequencies.

Each of the ports in Figure 3.1 is assumed to be terminated in an impedance Z_k (with a positive real part) and an equivalent voltage E_k, which may represent either a signal or a noise. Since the frequency converter is assumed to be linear, there are linear equations of motion that relate the half-amplitude voltages V_k and currents I_k at the various ports. However, instead of using the impedance or admittance representations, we define for each port[3, 28]

$$a_k = \frac{V_k + Z_k I_k}{\sqrt{Z_k + Z_k^*}} \tag{3.1}$$

and

$$b_k = \frac{V_k - Z_k^* I_k}{\sqrt{Z_k + Z_k^*}}. \tag{3.2}$$

The equations relating the a and b variables are linear, of the form

$$\underset{\sim}{b} = \underset{\sim}{S}\underset{\sim}{a}, \tag{3.3}$$

where $\underset{\sim}{b}$ is a column matrix of the b_k, and $\underset{\sim}{a}$ is a column matrix of the a_k. The square matrix $\underset{\sim}{S}$ is the <u>scattering matrix</u>, and the a_k and b_k are known, respectively, as incoming and outgoing wave amplitudes.

There are only a few facts (which the reader can easily verify for himself) that we need to know about these scattering variables. The power input into the kth port is

$$2 \text{ Re } V_k I_k^* = |a_k|^2 - |b_k|^2, \tag{3.4}$$

in accordance with the interpretation of the a_k and b_k as wave amplitudes. The exchangeable power from any of the terminations is

$$|a_k|^2, \tag{3.5}$$

and the exchangeable power from one of the ports is

$$\frac{|b_j|^2}{1 - |S_{jj}|^2}, \tag{3.6}$$

so that the exchangeable gain from the kth port to the jth port is

$$\frac{|S_{jk}|^2}{1 - |S_{jj}|^2}. \tag{3.7}$$

In view of Equation 3.4, the Manley-Rowe formula involving the small-signal variables, Equation 2.9, becomes

$$\Sigma_k \frac{|b_k|^2 - |a_k|^2}{\omega_k} = 0, \tag{3.8}$$

where ω_k is the angular frequency of the power that is exchanged at port k. In terms of the diagonal matrix $\underset{\sim}{K}$ with entries $1/\omega_k$ along the main diagonal, Equation 3.8 is written in matrix form:

$$\underset{\sim}{a}^\dagger \underset{\sim}{K}\underset{\sim}{a} - \underset{\sim}{b}^\dagger \underset{\sim}{K}\underset{\sim}{b} = 0 \tag{3.9}$$

where the dagger indicates the complex conjugate transpose (Hermitian conjugate) of a matrix. Thus, from Equations 3.3,

$$0 = \underset{\sim}{a}^{\dagger}(\underset{\sim}{K} - \underset{\sim}{S}^{\dagger}\underset{\sim}{K}\underset{\sim}{S})\underset{\sim}{a}, \tag{3.10}$$

and since this holds for any combination of incoming waves, we conclude that

$$\underset{\sim}{K} = \underset{\sim}{S}^{\dagger}\underset{\sim}{K}\underset{\sim}{S}. \tag{3.11}$$

This matrix equation, a constraint on the scattering matrix, is an expression of the Manley-Rowe formulas. We obtain a somewhat more useful expression by premultiplying by $\underset{\sim}{S}\underset{\sim}{K}^{-1}$ and postmultiplying by $\underset{\sim}{S}^{-1}\underset{\sim}{K}^{-1}$ (noting in the process that both inverses exist) to get

$$\underset{\sim}{S}\underset{\sim}{K}^{-1}\underset{\sim}{S}^{\dagger} = \underset{\sim}{K}^{-1}. \tag{3.12}$$

This form of the Manley-Rowe formulas is used in Sections 3.1 and 3.2. A variety of similar matrix constraints has been derived elsewhere;[22] Equations 3.12 have also been used by Kurokawa and Hamasaki.[17, 18]

3.1. Frequency Converters

Several writers[19,23,25,5] have discussed simple frequency converters in terms of the Manley-Rowe formulas. The nonlinear reactance is pumped at frequency ω_p, and small-signal information is exchanged at frequencies of the form $n\omega_p + \omega_s$ (we allow n to be negative). The input is at frequency ω_s, and the output at some other frequency. If the output frequency is higher than the input frequency, the device is an <u>upconverter</u>; otherwise it is a <u>downconverter</u>. If the output frequency is the <u>upper sideband</u> $\omega_u = \omega_p + \omega_s$, we have an upper-sideband upconverter (or downconverter if $0 > \omega_p > -\omega_s$). If the output frequency is the <u>idler</u> frequency, or <u>lower sideband</u> $\omega_i = \omega_p - \omega_s$, we have a lower-sideband upconverter or downconverter. More complicated frequency converters may have extra idlers, and may take output power from other sidebands as well.

In Section 3.1 we limit our discussion to small-signal frequency converters; therefore, the equations are linear, and we can use the scattering-matrix approach. For a discussion of large-signal converters, see Section 3.6.

3.1.1. The Upper-Sideband Upconverter

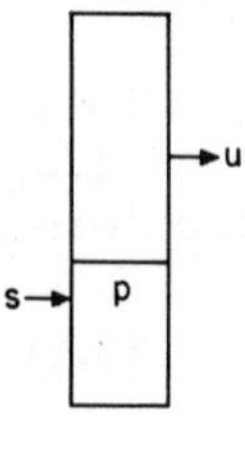

In the simple upper-sideband upconverter, the signal is put in at frequency ω_s and taken out to a load or a following amplifier at frequency $\omega_u = \omega_p + \omega_s$. We assume that no power is exchanged at any other frequencies (except, of course, the pump). The Manley-Rowe formulas predict (from Equation 2.9)

$$\frac{P_s}{\omega_s} + \frac{P_u}{\omega_u} = 0, \tag{3.13}$$

where P_s and P_u are the power inputs at the signal and upper-sideband frequencies. Ignoring circuit losses, we see that the actual power gain is therefore

$$\frac{-P_u}{P_s} = \frac{\omega_u}{\omega_s} \tag{3.14}$$

and is greater than 1. The nonlinear reactance upper-sideband upconverter then can act as an amplifying upconverter (the nonlinear resistance upconverter cannot[20]).

The exchangeable power gain G_e is also equal to the ratio of output to input frequencies. To demonstrate this, we use the simplified diagram of the upconverter shown in Figure 3.2.

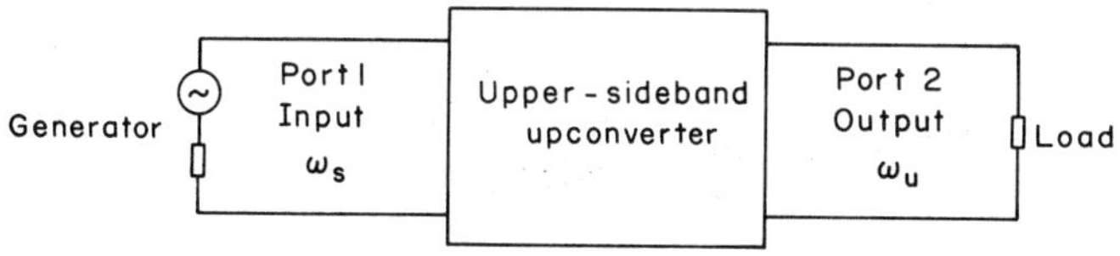

Figure 3.2. Simple upper-sideband upconverter with input at frequency ω_s and output to a load at frequency ω_u. The load and generator impedances need not be simply resistors.

Since small-signal power flows only at two frequencies, we show only two ports. The nonlinear reactance and various other reactances are inside the box, which obeys the Manley-Rowe formulas and, in particular, obeys Equations 3.12, of which the 2,2 element is

$$\omega_s |S_{21}|^2 + \omega_u |S_{22}|^2 = \omega_u. \tag{3.15}$$

Thus the exchangeable gain is, from Expression 3.7,

$$G_e = \frac{|S_{21}|^2}{1 - |S_{22}|^2} = \frac{\omega_u}{\omega_s}. \tag{3.16}$$

We also notice from Equation 3.13 that power must flow in one of the ports if it is to flow out the other; hence, if the terminations are passive, the input and output impedances have a positive real part. Therefore, the upper-sideband upconverter is stable. Stability is also predicted from Equation 3.15, which implies that the magnitude of S_{22} must be less than 1.

The noise performance of the upper-sideband upconverter is also of interest. This cannot be calculated from the Manley-Rowe formulas, because they do not suggest any noise mechanism. The noise temperature of a nonlinear reactance upconverter that obeys the Manley-Rowe formulas is zero. In practice, whatever noise is present arises from physical effects that violate the Manley-Rowe formulas (for example, thermal noise from the series resistance of a varactor). In the absence of such effects, there is no noise generated internally by the upconverter; the noise temperature of actual high-quality upconverters may be very small.

To summarize the results for the upper-sideband upconverter: The power gain is always equal to the exchangeable gain ω_u/ω_s; the output and input impedances have positive real parts; and there is no noise generated internally by the frequency converter.

3.1.2. The Lower-Sideband Upconverter

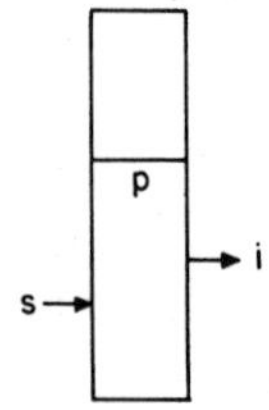

The lower-sideband upconverter is more difficult to understand because the input and output impedances must be negative. The signal is applied at frequency ω_s, and appears at the output at frequency $\omega_i = \omega_p - \omega_s$ or at the negative frequency $-\omega_i = \omega_s - \omega_p$. In applying Equations 3.12, the negative frequency must be used, rather than ω_i; this is because the frequencies are assumed to be of the form $n\omega_p + \omega_s$, with n possibly negative (in this case -1).

The Manley-Rowe formulas predict (from Equation 2.9)

$$\frac{P_s}{\omega_s} - \frac{P_i}{\omega_i} = 0, \tag{3.17}$$

where P_s and P_i are the power inputs to the upconverter at the corresponding frequencies. Notice the surprising implication of Equation 3.17: If the output is terminated in a load that absorbs power, power must flow out the input terminals; thus the input impedance must have a negative real part. Similarly, the output impedance must have a negative real part. This fact leads to possible instabilities in lower-sideband upconverters; it also leads to the use of the negative resistance at signal frequency as a mechanism for gain in its own right (this is parametric amplification).

A simple lower-sideband upconverter is shown in Figure 3.3; the Manley-Rowe formulas, Equations 3.12, predict

$$\omega_s |S_{21}|^2 - \omega_i |S_{22}|^2 = -\omega_i, \tag{3.18}$$

so that the exchangeable gain becomes, from Expression 3.7,

$$G_e = \frac{|S_{21}|^2}{1 - |S_{22}|^2} = \frac{-\omega_i}{\omega_s}. \tag{3.19}$$

The fact that G_e is negative indicates that the output resistance is negative.

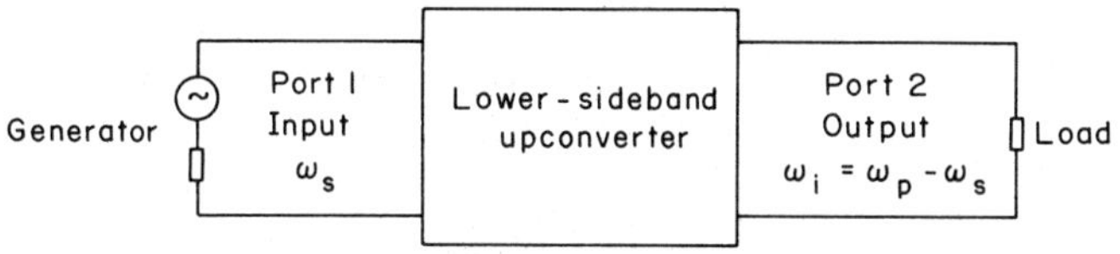

Figure 3.3. Lower-sideband upconverter with input at frequency ω_s and output to a load or following stage at frequency $\omega_i = \omega_p - \omega_s$. The generator and load impedances need not be purely resistive.

Since there is no apparent internal noise source in our simplified upconverter, the noise temperature is zero, as in the upper-sideband upconverter.

In summary, the exchangeable gain of the nonlinear-reactance lower-sideband upconverter is negative; the input and output resistances have a negative real part; and there is no noise generated internally.

3.1.3. Downconverters

Here we discuss two simple downconverters that are similar to the upconverters just described: the upper-sideband downconverter, and the lower-sideband downconverter.

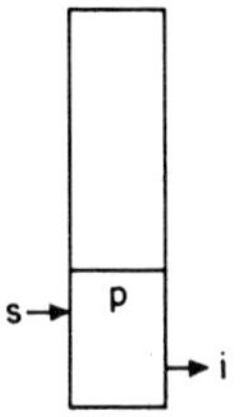

The lower-sideband frequency converter is an upconverter when $\omega_i > \omega_s$; it becomes a downconverter when $\omega_i < \omega_s$ or when the pump frequency is less than $2\omega_s$ but greater than ω_s. The Manley-Rowe formulas still predict Equations 3.17 and 3.19, so that the input and output resistances are negative. There is a possibility of obtaining a large transducer gain (actual output power divided by source available power) because the exchangeable gain is negative, but only at the price of possible instabilities. As in the upconverters, there is no internal noise source suggested for the downconverter.

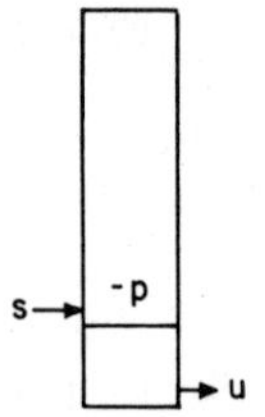

The upper-sideband frequency converter might appear to be an upconverter only, because the output frequency is larger than the input frequency. However, if the pump frequency is allowed to be negative, the output frequency can be positive and less than ω_s; alternatively, if the pump frequency is less than ω_s, the upper-sideband output is at frequency $\omega_u = \omega_s - \omega_p$. Looked at either way, the upper-sideband downconverter output is at a frequency smaller than the input. The Manley-Rowe formulas predict both Equations 3.13 and 3.16. The exchangeable gain is less than 1, so there is no possibility for power gain. But since the input and output impedances are positive, as for the upper-sideband upconverter, the device is stable. The upper-sideband downconverter may be useful in spite of its loss for conversion to a frequency at which low-noise amplification, relatively lossless narrow-band filtering, or other specialized operations are possible.

Like the upconverters, it has no internal noise.

3.1.4. Frequency Converters with Idlers

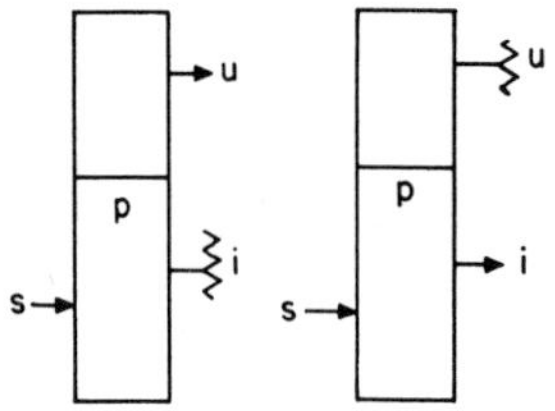

We have just seen, in Sections 3.1.1 to 3.1.3 that upper-sideband upconverters and downconverters are stable but have a limited gain; and that lower-sideband upconverters and downconverters can have arbitrarily high gain but are apt to be unstable because of the negative input and output impedances. The use of an extra idler (that is, idling power flow at some extra frequency) can increase the gain of upper-sideband devices and can help make the input impedance of the lower-sideband devices positive.

We discuss the simplest case here: the device that is pumped at frequency ω_p and exchanges small-signal power at frequencies ω_s, $\omega_i = \omega_p - \omega_s$, and $\omega_u = \omega_p + \omega_s$. This double-sideband frequency converter has its input at ω_s but can have an output at either ω_i or ω_u (or both).

The Manley-Rowe formulas for this device predict

$$\frac{P_s}{\omega_s} + \frac{P_u}{\omega_u} - \frac{P_i}{\omega_i} = 0. \tag{3.20}$$

If power is taken out (either into an idling circuit or to a load) at frequencies ω_u and ω_i, then power enters at the signal frequency (so the input impedance is positive) only if the upper-sideband output power is large enough:

$$\frac{-P_u}{-P_i} > \frac{\omega_u}{\omega_i} . \tag{3.21}$$

Thus, having an idler at the upper sideband, and dissipating enough power there, can make the input impedance of a lower-sideband upconverter positive. As a function of the ratio of P_u to P_i, the various power gains have been given by Adams.[1]

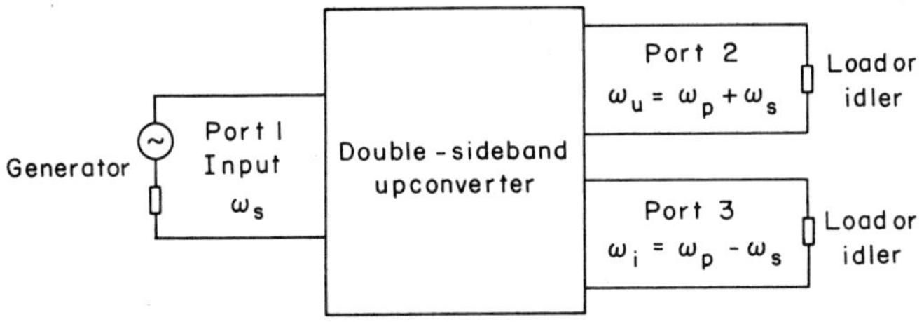

Figure 3.4. The double-sideband upconverter with output at either ω_u or ω_i (or both). The terminating impedances need not be purely resistive.

The double-sideband upconverter (or downconverter) is shown in Figure 3.4. Three ports are indicated. The Manley-Rowe formulas, Equations 3.12, predict, among other relations,

$$\omega_s|S_{21}|^2 + \omega_u|S_{22}|^2 - \omega_i|S_{23}|^2 = \omega_u \tag{3.22}$$

and

$$\omega_s|S_{31}|^2 + \omega_u|S_{32}|^2 - \omega_i|S_{33}|^2 = -\omega_i . \tag{3.23}$$

The exchangeable gain from signal to upper sideband ($G_{e,u}$) and the exchangeable gain from signal to lower sideband ($G_{e,i}$) can be calculated from Equations 3.22 and 3.23:

$$G_{e,u} = \frac{|S_{21}|^2}{1 - |S_{22}|^2} = \frac{\omega_u}{\omega_s} + \frac{\omega_i}{\omega_s} G_e(i,u), \tag{3.24}$$

and

$$G_{e,i} = \frac{|S_{31}|^2}{1 - |S_{33}|^2} = -\frac{\omega_i}{\omega_s} - \frac{\omega_u}{\omega_s} G_e(u,i), \tag{3.25}$$

where $G_e(i,u)$ is the exchangeable gain from frequency ω_i to frequency ω_u with ω_s considered to be an idler:

$$G_e(i,u) = \frac{|S_{23}|^2}{1 - |S_{22}|^2}; \tag{3.26}$$

and where $G_e(u,i)$ is the exchangeable gain from frequency ω_u to frequency ω_i, with ω_s considered to be an idler:

$$G_e(u,i) = \frac{|S_{32}|^2}{1 - |S_{33}|^2}. \tag{3.27}$$

It is apparent from Equation 3.24 that an idler at frequency ω_i increases the exchangeable gain of an upper-sideband upconverter. Also, an idler at the upper-sideband frequency makes the lower-sideband exchangeable gain (which is negative) smaller in magnitude.

Frequency converters with idlers have a finite noise temperature that can be predicted from the circuit of Figure 3.4. For example, consider the upper-sideband. The idling power at frequency ω_i is dissipated in a load. Let us say that this load is at a temperature T_i, so that it has an available noise power of $kT_i\ \Delta f$. Then $\overline{|a_3|^2} = kT_i\ \Delta f$, and the exchangeable power to the load caused by the idler (this is considered as internal to the upconverter) is $G_e(i,u)kT_i\ \Delta f$. Similarly, the exchangeable noise power to the load caused by a noise source of temperature T_0 at the input is $G_{e,u}kT_0\ \Delta f$. The excess noise figure $F - 1$ is the ratio of these two quantities; thus the noise temperature $T = T_0(F - 1)$ is

$$T = T_i \frac{G_e(i,u)}{G_{e,u}} = T_i \frac{|S_{23}|^2}{|S_{21}|^2}. \tag{3.28}$$

It is interesting to relate the noise temperature to the exchangeable gain of the upconverter by eliminating $G_e(i,u)$ from Equations 3.24 and 3.28. Thus the formula

$$T = T_i \frac{G_{e,u} - \frac{\omega_u}{\omega_s}}{\frac{\omega_i}{\omega_s} G_{e,u}} \tag{3.29}$$

shows how high a noise temperature must be expected if the gain is to be improved a specified amount. Alternatively, if we solve for $G_{e,u}$, we can determine how much gain is possible for a given cost in noise:

$$G_{e,u} = \frac{\omega_u}{\omega_s - \omega_i \dfrac{T}{T_i}}. \tag{3.30}$$

If the gain is to be made very high, we must expect a noise temperature of at least $T_i \omega_s / \omega_i$.

In a similar way, the noise for the lower-sideband upconverter can be predicted. If the idling power at frequency ω_u is dissipated in a load at temperature T_u, then the lower-sideband upconverter noise temperature T is

$$T = T_u \frac{G_e(u,i)}{G_{e,i}} = T_u \frac{|S_{32}|^2}{|S_{31}|^2}. \tag{3.31}$$

Eliminating $G_e(u,i)$ from Equations 3.25 and 3.31, we obtain

$$T = T_u \frac{G_{e,i} + \dfrac{\omega_i}{\omega_s}}{-\dfrac{\omega_u}{\omega_s} G_{e,i}} \tag{3.32}$$

and

$$G_{e,i} = \frac{-\omega_i}{\omega_s + \omega_u \dfrac{T}{T_u}}. \tag{3.33}$$

The double-sideband upconverter can also be operated by taking power out both the sidebands into a synchronously pumped mixer. This configuration, although we do not discuss it further, offers advantages of high gain and stability.

3.2. Parametric Amplifiers

The derivations that follow for the minimum noise temperature of a parametric amplifier are the work of H. A. Haus[10,11] and are included here with his kind permission. These original derivations, made with great elegance and generality, originally inspired the present writers to start the investigations reported in this book. This work has not been published before but is very similar to Herrmann's analysis.[15]

In a parametric amplifier, the nonlinear reactance is pumped at frequency ω_p, and a signal is applied at frequency ω_s. Among the sidebands the idler $\omega_i = \omega_p - \omega_s$ is of special importance, although power may flow at other frequencies (say, $\omega_n = n\omega_p + \omega_s$). The action of the dissipated power at ω_i is to produce a negative resistance at the input port, as in the lower-sideband upconverter. This negative resistance is used as a gain mechanism in the parametric amplifier.

In the ordinary parametric amplifier, ω_s and ω_i are not the same. This nondegenerate amplifier is discussed first in Section 3.2.1. If the idler is equal to, or close to, the signal frequency, special care is required. This degenerate amplifier is treated in Section 3.2.2.

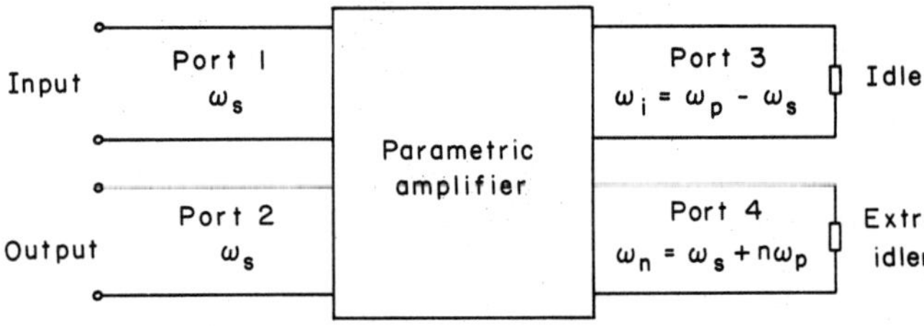

Figure 3.5. Parametric amplifier model with idling power at two frequencies. The input and output are at the same frequency. The idler terminations need not be purely resistive.

In both cases, the amplifier is assumed to be of the form shown in Figure 3.5, with two ports at signal frequency (an input and an output) and ports at the idler frequency and one other frequency. More complicated amplifiers, such as those with more idlers, can be treated similarly.

We assume that the system inside the box obeys the Manley-Rowe formulas. If there is known to be dissipation at signal frequency, we represent it by another port with an external resistance.

3.2.1. The Nondegenerate Amplifier

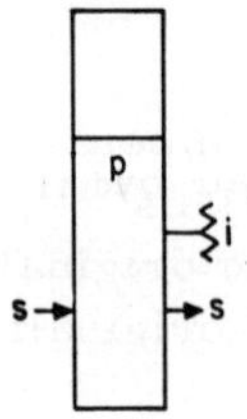

There is no limit to the exchangeable gain that can be obtained from a negative resistance, but the Manley-Rowe formulas predict a nontrivial limit on the obtainable noise temperature. The Manley-Rowe formula we need is the 2,2 element of Equations 3.12:

$$\omega_s\left(|S_{21}|^2 + |S_{22}|^2\right) - \omega_i|S_{23}|^2 + \omega_n|S_{24}|^2 = \omega_s. \quad (3.34)$$

To calculate the noise temperature, we assume that the only noise present is thermal noise generated in the idler terminations. The termination at frequency ω_i is assumed to be at a temperature T_i, and the termination at frequency ω_n at a temperature T_n. The excess noise figure F-1 is the noise power from port 2 caused by the terminations of ports 3 and 4, divided by the noise power from port 2 caused by a noise source of temperature T_0 at the input. Since the noise sources are uncorrelated, and since each available noise power is proportional to the corresponding temperature, the excess noise figure becomes

$$F - 1 = \frac{T_i|S_{23}|^2 + T_n|S_{24}|^2}{T_0|S_{21}|^2}. \quad (3.35)$$

If we eliminate S_{23} from Equation 3.35 by using the Manley-Rowe formula of Equation 3.34, the result is

$$F - 1 = \frac{T_i}{T_0}\frac{\omega_s}{\omega_i}\frac{|S_{21}|^2 + |S_{22}|^2 - 1}{|S_{21}|^2} + \left(\frac{T_i}{T_0}\frac{\omega_n}{\omega_i} + \frac{T_n}{T_0}\right)\frac{|S_{24}|^2}{|S_{21}|^2}. \quad (3.36)$$

But the exchangeable gain of the amplifier is

$$G_e = \frac{|S_{21}|^2}{1 - |S_{22}|^2}, \quad (3.37)$$

so that the noise temperature T is (see Section 2.5.4)

$$T = \frac{T_0(F - 1)}{1 - \dfrac{1}{G_e}} = \frac{\omega_s}{\omega_i}T_i + \left(T_i\frac{\omega_n}{\omega_i} + T_n\right)\frac{|S_{24}|^2}{|S_{21}|^2 + |S_{22}|^2 - 1}. \quad (3.38)$$

We note that for gain, G_e must be either negative or larger than 1, so that $|S_{21}|^2 + |S_{22}|^2 - 1$ must be positive. Thus if ω_n is positive, the last term in Equation 3.38 is positive; for lowest noise temperature, we want to avoid coupling to any resistance at ω_n, that is, make S_{24} vanish. If such coupling is avoided, the noise temperature becomes

$$T = T_i \frac{\omega_s}{\omega_i}, \tag{3.39}$$

a fundamental limit that has been reported before.[14,26] The way to achieve a low noise temperature seems to be to refrigerate the idler and operate with a high idler frequency. The varactor series resistance modifies this result, however (see Chapter 6).

If the extra idler frequency ω_n is negative, and if

$$\omega_n < -\omega_i \frac{T_n}{T_i}, \tag{3.40}$$

then the last term in Equation 3.38 is negative. A lower limit on the noise if found by decoupling from any loss at the normal idler frequency (that is, by forcing S_{23} to vanish), whereupon we have

$$T = T_n \frac{\omega_s}{-\omega_n}. \tag{3.41}$$

Physically, this corresponds to using frequency $(-\omega_n)$ as the actual idler, instead of ω_i.

3.2.2. The Degenerate Amplifier

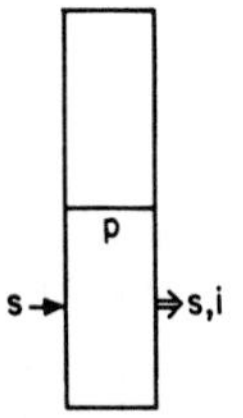

In Section 3.2.1, it was tacitly assumed that the signal and idler frequencies were different. If the pump frequency is close to $2\omega_s$, however, the idler and signal frequencies are close together, and their spectra may, in fact, even overlap. This degenerate amplifier is easier to build (the idler and signal can use the same circuit), but there is some confusion about the noise figure.

This confusion arises from two facts. First, there are many different applications for the degenerate amplifier, and these require many different noise characterizations. Ordinary concepts of noise figure, temperature, and measure do not apply. Second, there are essentially many different schemes for measuring the noise in such an amplifier; a person arrives at different numbers depending on whether he does or does not feed in noise at ω_i as well as ω_s during the test. One of the clearest discussions of this problem is that of Haun,[7] who points out these two reasons for wanting different characterizations of noise. However, Haun neglects the subsequent detection process, and as a result he is not able to arrive at a satisfactory characterization of noise. The degradation in signal-to-noise ratio depends upon the noise generated internally by the degenerate amplifier, upon the type of detector used, and upon the statistics of the signal. Unlike ordinary amplifiers, the degenerate amplifier cannot be characterized by a simple noise temperature.

This problem is discussed in detail in Section 6.6 for a degenerate varactor parametric amplifier of high gain, using a circulator. Usual concepts of noise figure, noise temperature, and noise measure are not used. Instead, the degradation in signal-to-noise ratio is calculated directly, and from these calculations rules are derived for comparing various amplifiers.

The results are, basically, that two degenerate amplifiers can be compared with each other by comparing their values of "effective noise temperature" T_{eff}, also called "double-sideband noise temperature" T'_{DBS}. In general, however, a degenerate amplifier cannot be compared with a nondegenerate amplifier in such a simple way.

When the input signal is separate from, although close to, the idler frequency and when the succeeding stages are able to distinguish between them, then the degradation in signal-to-noise ratio is characterized by a different "noise temperature" T'_{SSB} (denoting single-sideband). In this case a degenerate amplifier can be compared with an ordinary amplifier by comparing T'_{SSB} with the ordinary noise temperature. These two noise "temperatures," or the corresponding noise "figures," are often used to describe degenerate amplifiers (the former, being always lower than the latter, and being easier to measure, is used more often). The relation between them is now derived.

First, we derive the single-sideband noise temperature. If the signal and idler spectra do not overlap, then a noise "figure" can be used that is the same as Equation 3.35 (we assume the amplifier is again represented by Figure 3.5):

$$F'_{SSB} - 1 = \frac{T_i}{T_0}\frac{|S_{23}|^2}{|S_{21}|^2} + \frac{T_n}{T_0}\frac{|S_{24}|^2}{|S_{21}|^2}\,. \tag{3.42}$$

Another application is amplification of two incoherent signals at frequencies ω_s and ω_i, with output at frequency ω_s only. The previous definition of noise figure is not pertinent now, because the idler port 3 of Figure 3.5 is considered as part of the input. This application occurs in radio astronomy, in which the noise in a small band centered about half of the pump frequency is amplified. A logical excess noise figure for characterizing a parametric amplifier for such applications would be the ratio of noise power output caused by internal sources, divided by the noise output caused by incoherent noise, at both inputs, of temperature T_0. This is, in fact, what is actually measured by conventional means using a broadband noise source. It is called the double-sideband noise "figure." In terms of Figure 3.5, this double-sideband noise figure is

$$F'_{DSB} - 1 = \frac{T_n}{T_0} \frac{|S_{24}|^2}{|S_{21}|^2 + |S_{23}|^2} \tag{3.43}$$

because the only noise that is considered as internal to the amplifier is that at port 4. Note that this is always less than the single-sideband noise figure, and in fact, if S_{24} is small, the double-sideband noise can be extremely good. (This result is modified by the varactor series resistance.)

To relate the two noise "temperatures,"

$$T'_{SSB} = T_0(F'_{SSB} - 1) \tag{3.44}$$

and

$$T_{eff} = T'_{DSB} = T_0(F'_{DSB} - 1), \tag{3.45}$$

we merely eliminate S_{24} from Equations 3.42 and 3.43. Thus,

$$T'_{SSB} = T_i \frac{|S_{23}|^2}{|S_{21}|^2} + T'_{DSB} \frac{|S_{21}|^2 + |S_{23}|^2}{|S_{21}|^2}. \tag{3.46}$$

If the ratio of $|S_{23}|^2$ to $|S_{21}|^2$ is known (and this can be easily measured because it is the ratio of the transducer gain from port 3 to port 2, to that from port 1 to port 2), then the relation between the two noise temperatures and the idler temperature is given by Equation 3.46.

If ω_s and ω_i are very close together, we might expect that $|S_{21}|$ would be equal to $|S_{23}|$. In this case,

$$T'_{SSB} = T_i + 2T'_{DSB}. \tag{3.47}$$

Note that if T_i is very small (this can occur, for example, if ports 1 and 3 are an antenna viewing a low-noise sky), then we may double T'_{DSB} to obtain T'_{SSB}. On the other hand, if the idler termination is at room temperature, so that $T_i = T_0$, then we should double the noise "figure" F'_{DSB} to obtain F'_{SSB}. This is the so-called "3-db rule" often given for converting from measured noise figure to single-sideband noise figure.[4,12]

We should not assume without good cause that $|S_{21}|$ and $|S_{23}|$ are equal. As Haun[7] has pointed out, even for quasi-degenerate operation (that is, where ω_s and ω_i are almost equal), $|S_{21}|^2$ and $|S_{23}|^2$ can differ by factors of 10; therefore, caution should be exercised in using Equation 3.47.

Although Equation 3.46 was derived from Figure 3.5, the result also holds for cases with more noise sources, such as more idlers, loss at signal and idler frequencies, noise from the pump, and the like. The additional noise sources simply add to the noise term $|S_{24}|^2 T_n$ in Equations 3.42 and 3.43. Also, the reader will note that we have not used the Manley-Rowe formulas in the derivation of Equation 3.46. The result holds for other frequency converters with nearly equal signal and idler frequencies, whether or not gain is achieved.

3.3. Pumping

The results about pumping derived from the Manley-Rowe formulas are quite misleading. Since the over-all power input to the nonlinear reactance vanishes, the pump supplies exactly the net power that leaves. In the upper-sideband upconverter, the pump supplies the difference between $-P_u$ and P_s; in the lower-sideband upconverter and downconverter, the pump supplies the power output at frequencies ω_s and ω_i; and in the upper-sideband downconverter, the pump actually receives power from the signal source! In the parametric amplifier, the pump supplies both the net signal power and the idler power.

It would appear that the only power needed from the pump is of the same order of magnitude as the signal power. Actually, of course, in practical devices inevitable losses, either parasitic to the nonlinear reactance or in the coupling circuits, dissipate most of the power supplied by the pump. Realistically, pump power requirements can be discussed only in terms of these losses. The amount of power exchanged with the small signals is insignificant.

3.4. Harmonic Multipliers

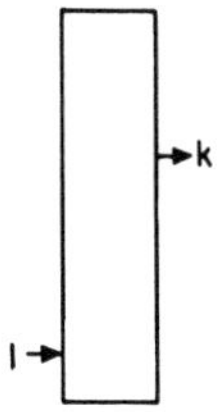

In a harmonic multiplier, power is put in the nonlinear reactance at frequency ω and is taken out at $k\omega$, where k is an integer. The pertinent Manley-Rowe formula is

$$\Sigma_n P_n = 0, \tag{3.48}$$

which appears to be the law of conservation of energy. Actually, the sum in Equation 3.48 does not include dc power, so the Manley-Rowe formula predicts that the output at any harmonic cannot be increased by supplying dc power.

The efficiency of multiplication, which is the ratio of power out at $k\omega$ to the power in at ω, is less than 100 per cent, as is clear from Equation 3.48.

An exception to these rules must be made for some electromechanical devices (particularly electron beams) that obey the Manley-Rowe formulas. Although the total power output at $k\omega$ must be less than the power input at ω, the electric power output may be greater because mechanical power flowing out is negative. This happens, for example, in electron beams that give a harmonic conversion gain.

Needless to say, losses (either parasitic to the nonlinear reactance or in the circuits) reduce the efficiency. The effect of the varactor series resistance is indicated in Chapter 8.

3.5. Harmonic Dividers

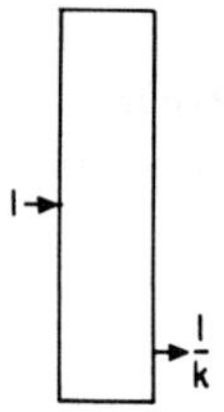

In harmonic multipliers, we put power in at ω and get it out at $k\omega$. In harmonic dividers, we put power in at ω and get it out at ω/k. (This process is also known as subharmonic generation, frequency division, and parametric excitation.) With nonlinear reactances this division is possible (in nonlinear resistances,[20] it is not), and in fact, rational-fraction multiplication, with output frequency m/n times the input frequency, is also possible. Harmonic dividers and rational fraction multipliers are limited by the Manley-Rowe formulas to 100 per cent efficiency, and no increase by dc power is possible.

Harmonic dividers are important in building frequency standards and are also of potential importance in computer technology as bistable oscillators, known as parametrons;[27,16,6,24] the role of the varactor series resistance in limiting the power, efficiency, and build-up time of dividers is discussed in Chapter 9.

3.6. Large-Signal Amplifiers and Converters

The results for frequency converters and parametric amplifiers that were discussed in Sections 3.1 and 3.2 were derived from the Manley-Rowe formulas, under the assumption that the small-signal variables were linearly related. A large-signal theory of such devices is primarily important in indicating the range of validity of the small-signal theory, that is, in determining the devices' dynamic range.

The noise theory results and the scattering matrix arguments are pertinent only to the small-signal devices. However, the Manley-Rowe formulas, in the form of Equations 3.13, 3.14, 3.17, and 3.20, still apply to the large-signal devices, provided that power flow is blocked at all but the indicated frequencies.

The actual power gains (not the exchangeable gains) are unchanged in the large-signal case. This does not imply, of course, that the devices are linear for large signals.

Because of the complicated nonlinearities, the large-signal frequency converters and parametric amplifiers cannot be discussed easily. A few simple results for the varactor with series resistance are given in Chapter 10.

REFERENCES

1. Adams, D. K., "An Analysis of Four-Frequency Nonlinear Reactance Circuits," IRE Trans. on Microwave Theory and Techniques, MTT-8, 274-283 (May, 1960).
2. Bers, A., and P. Penfield, Jr., "Conservation Principles for Plasmas and Relativistic Electron Beams," IRE Trans. on Electron Devices, ED-9, 12-26 (January, 1962).
3. Carlin, H. J., "The Scattering Matrix in Network Theory," IRE Trans. on Circuit Theory, CT-3, 88-97 (June, 1956).
4. Cohn, S. B., "The Noise Figure Muddle," Microwave J., 2, 7-11 (March, 1959).
5. Duinker, S., "General Energy Relations for Parametric Amplifying Devices," Tijdschrift van het Nederlands Radiogenootschap, 24, 287-310 (1959).
6. Goto, E., "The Parametron, a Digital Computing Element which Utilizes Parametric Oscillation," Proc. IRE, 47, 1304-1316 (August, 1959).
7. Haun, R. D., Jr., "Summary of Measurement Techniques of Parametric Amplifier and Mixer Noise Figure," IRE Trans. on Microwave Theory and Techniques, MTT-8, 410-415 (July, 1960).
8. Haus, H. A., "Power-Flow Relations in Lossless Nonlinear Media," IRE Trans. on Microwave Theory and Techniques, MTT-6, 317-324 (July, 1958).
9. Haus, H. A., "The Kinetic Power Theorem for Parametric, Longitudinal, Electron-Beam Amplifiers," IRE Trans. on Electron Devices, ED-5, 225-232 (October, 1958).
10. Haus, H. A., On the Noise Performance of Parametric Amplifiers, Unpublished private memorandum, Department of Electrical Engineering, M.I.T. (March 13, 1959).
11. Haus, H. A., and P. Penfield, Jr., On the Noise Performance of Parametric Amplifiers, Energy Conversion Group Unpublished Internal Memorandum No. 19, Department of Electrical Engineering, M.I.T. (August 11, 1959).
12. Heffner, H., "Masers and Parametric Amplifiers," Microwave J., 2, 33-40 (March, 1959).
13. Heffner, H., and G. Wade, "Minimum Noise Figure of a Parametric Amplifier," J. Appl. Phys., 29, 1262 (August, 1958).
14. Heffner, H., and G. Wade, "Gain, Band Width, and Noise Characteristics of the Variable-Parameter Amplifier," J. Appl. Phys., 29, 1321-1331 (September, 1958).
15. Herrmann, G., "Idler Noise in Parametric Amplifiers," Proc. IRE, 48, 2021-2022 (December, 1960).
16. Hilibrand, J., and W. R. Beam, "Semiconductor Diodes in Parametric Subharmonic Oscillators," RCA Rev., 20, 229-253 (June, 1959).
17. Kurokawa, K., and J. Hamasaki, "Mode Theory of Lossless Periodically Distributed Parametric Amplifiers," IRE Trans. on Microwave Theory and Techniques, MTT-7, 360-365 (July, 1959).

18. Kurokawa, K., and J. Hamasaki, "An Extension of the Mode Theory to Periodically Distributed Parametric Amplifiers with Losses," IRE Trans. on Microwave Theory and Techniques, MTT-8, 10-18 (January, 1960).
19. Manley, J. M., and H. E. Rowe, "Some General Properties of Nonlinear Elements. – Part I. General Energy Relations," Proc. IRE, 44, 904-913 (July, 1956).
20. Page, C. H., "Frequency Conversion With Positive Nonlinear Resistors," J. Res. Natl. Bur. Standards, 56, 179-182 (April, 1956).
21. Penfield, P., Jr., Frequency-Power Formulas, The Technology Press and John Wiley & Sons, Inc., New York (1960).
22. Penfield, P., Jr., ibid., Section 8.3.
23. Rowe, H. E., "Some General Properties of Nonlinear Elements. II. Small Signal Theory," Proc. IRE, 46, 850-860 (May, 1958).
24. Sterzer, F., "Microwave Parametric Subharmonic Oscillators for Digital Computing," Proc. IRE, 47, 1317-1324 (August, 1959).
25. Uhlir, A., Jr., "The Potential of Semiconductor Diodes in High-Frequency Communications," Proc. IRE, 46, 1099-1115 (June, 1958).
26. van der Ziel, A., "Noise Figure of Reactance Converters and Parametric Amplifiers," J. Appl. Phys., 30, 1449 (September, 1959).
27. Wigington, R. L., "A New Concept in Computing," Proc IRE, 47, 516-523 (April, 1959).
28. Youla, D. C., "On Scattering Matrices Normalized to Complex Port Numbers," Proc. IRE, 49, 1221 (July, 1961).

Chapter 4

THE VARACTOR MODEL

The varactor model that we shall use is shown in Figure 4.1. In Figure 4.2a is the schematic symbol for a varactor; this symbol has been adopted as a standard.[13, 14]‡ The alternate symbols shown in Figure 4.2b have also been used.

The model of Figure 4.1 was originally proposed by Uhlir,[57] and has only two elements: a nonlinear capacitance and a constant series resistance. Although we intend to discuss only varactors that use the back-bias capacitance of a diode as the nonlinear element, the analysis in Chapters 5 to 10 holds equally well for other physical devices described by this model.

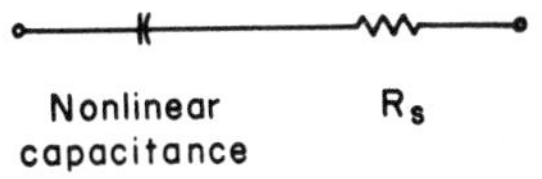

Figure 4.1. The Uhlir[57] model of a varactor has only two elements: a nonlinear capacitance and a constant parasitic series resistance.

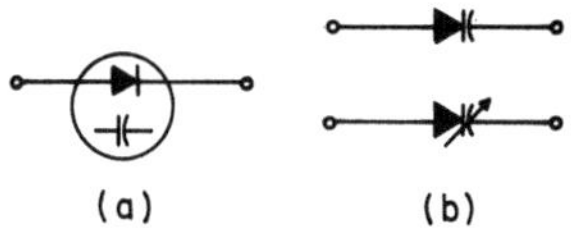

Figure 4.2. Schematic symbols for a varactor. (a) Standardized version.[13, 14] The circle may be omitted. (b) Other common symbols.

In order to justify the use of this model, we point out its limitations; this discussion is given in Sections 4.1 and 4.2. Varactors contribute noise to the circuits in which they are used; the physical sources of noise are discussed in Section 4.3. The large-signal, nonlinear, frequency-domain equations of motion are given in matrix form in Section 4.4. In many applications the varactor is heavily pumped at one frequency, and attention is focused on small-signal currents at other frequencies. Pumping, and the resulting small-signal equations of motion, are discussed in Section 4.5.

The analysis of small-signal frequency converters and parametric amplifiers (Chapters 5 and 6) indicates a figure of merit for the pumped varactor. This is discussed in Section 4.6. In order

‡Superscript numerals denote references listed at the end of each chapter.

to characterize varactors without the pumping, we shall determine (in Section 4.7) which varactors are inherently capable of being pumped so as to have a higher pumped figure of merit. This inquiry leads to a cutoff frequency as a logical characterization of varactor quality; this same quantity also enters into the analysis of large-signal devices (Chapters 8, 9, and 10).

This cutoff frequency and other quantities that characterize varactors are discussed in Section 4.7. We do not cover design or manufacture of varactors, although our conclusions about which varactor parameters are important can (and should) be used as a basis for varactor design.

4.1. p-n Junctions

Modern practical varactors are made from semiconductor junction diodes. All p-n junctions exhibit nonlinear barrier capacitance; if this effect "predominates" in some sense, then the junction can be used to advantage as a varactor diode. Many diodes manufactured for other purposes (for example, power rectification or switching) work well as varactors, although many others do not.

Excellent discussions of p-n junctions and junction capacitance are available.[47, 50, 26, 35, 9, 55, 60] Instead of giving a rigorous treatment here, we introduce only enough background for the discussion of junction capacitance in Section 4.2.

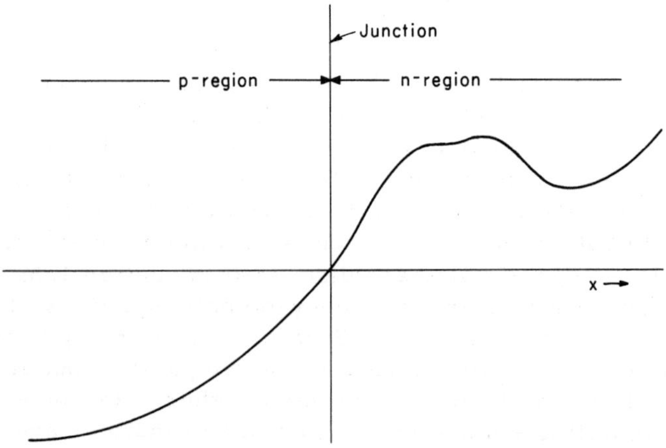

Figure 4.3. Plot of net doping density N(x) as a function of distance, in the region near a p-n junction. For $x > 0$, the density of donors is greater than the density of acceptors, so the net fixed charge density is positive, as shown; for $x < 0$, the opposite effect occurs.

A p-n semiconductor junction is a surface that separates a p-type semiconductor (that is, one with an excess of fixed negatively charged acceptor impurities) from an n-region (that is, one with an excess of fixed positively charged donors). For simplicity, consider a one-dimensional case, indicated in Figure 4. 3, which shows the fixed charge density caused by the acceptors and donors, as a function of distance.

If the charge density pictured in Figure 4. 3 were the only charge, it would set up a large electric field. The effect of this field would be to attract mobile charge carriers (positively charged holes and negatively charged excess electrons) to compensate the fixed charge density. In practice, the semiconductor has mobile charges that almost exactly balance the fixed charges; the resulting crystal is therefore almost uncharged.

This device can act as a rectifier, passing current in only one direction. For a grossly simplified picture of this rectification, suppose a battery were connected to the ends of this piece of semiconductor, with its positive terminal at the left. The effect of the battery voltage would be to push the mobile charge carriers on the left (which are positively charged holes) toward the right; similarly, the electrons on the right would be pushed toward the left. Both types of carriers are pushed toward the junction, near which they recombine.‡ (The carriers that actually cross the junction exist as minority carriers for a short time before they recombine; the average time before recombination is known as the lifetime, and may be in the approximate range from 10^{-10} to 10^{-2} second, depending on the purity of the crystal). Thus considerable current flows. On the other hand, if the battery polarity were reversed, the carriers of each type would be drawn away from the junction, and no net current would flow. The p-n junction is therefore a rectifier.

In practice, of course, the junction does inhibit forward current, and a little reverse current does flow. An analysis valid for small voltages predicts a current i (measured positive in the reverse direction)

$$i = I_{sat}\left(1 - e^{-ev/kT_d}\right), \qquad (4.1)$$

---

‡A hole is actually the result of an electron missing from the crystal structure; a curious result of quantum mechanics is that a hole can be treated as a particle in its own right. When an electron and a hole meet, the electron often fills up the missing place, so that both the electron and the hole disappear. This process is known as recombination; the inverse process, spontaneous creation of an electron and a hole, is known as generation.

where e is the electronic charge, v is the junction voltage‡ (assumed positive in the reverse direction), k is Boltzmann's constant, T_d is the diode temperature,‡‡ and I_{sat} is the <u>saturation current</u> (a diode parameter). This function is plotted in Figure 4.4 for a typical value of I_{sat} for a room-temperature diode. Note that I_{sat} is the current at high reverse voltages.

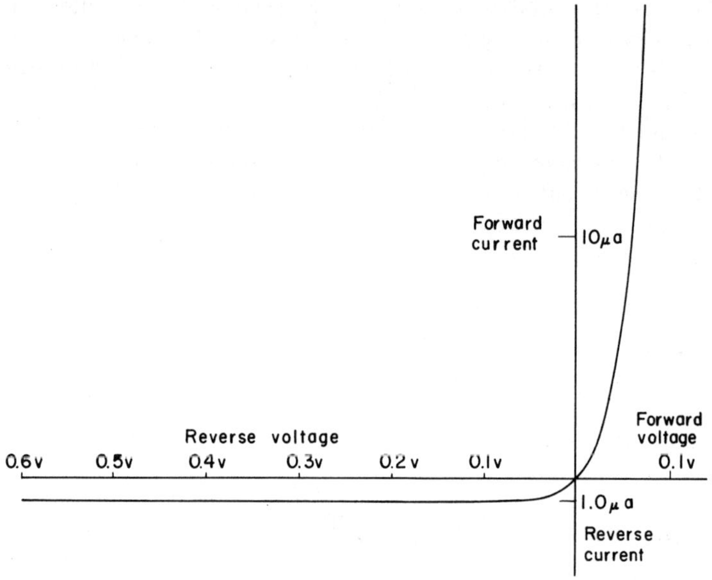

Figure 4.4. Volt-ampere curve for a typical p-n junction, taken from Equation 4.1. We have assumed $T_d = 290°$ K and $I_{sat} = 1.0$ μa. Simple changes of scale make this graph applicable to other temperatures and saturation currents.

This volt-ampere curve makes the p-n junction useful when a nonlinear resistor is called for, as in a rectifier or resistive mixer.

In practice, the dc diode characteristics are modified in three important ways: by excess reverse current, by breakdown at high reverse voltage, and by effects at high forward current. In addition, rectification fails at high frequencies because of the finite lifetime for minority carriers. We now discuss these four effects.

---

‡This voltage is the voltage that appears across the junction. The instantaneous voltage across the varactor is this voltage plus a drop across the series resistance of the diode.

‡‡At room temperature $kT_d/e = 0.025$ volt.

4.1.1. Excess Reverse Current

Practical diodes have a nonsaturating reverse current, caused by any of a number of physical effects. Of these, probably the most important is leakage of current between the two terminals, either on the package or, more likely, over the surface of the semiconductor.[44] Surface leakage increases the reverse-bias dc current and also contributes loss that degrades the ac performance of the varactor.

Another cause of high reverse current is generation of holes and electrons near the junction. Diodes with such generation have a reverse current that does not saturate.[43] Armstrong[1] has discussed other reasons for nonsaturating reverse current in diodes, including some for two- and three-dimensional junctions.

These effects are minimized in high-quality varactor diodes but can be important if other diodes, especially power rectifiers, are used. Occasionally the excess current is so large that maintaining a reverse bias is difficult.

4.1.2. Reverse Breakdown

The simple diode characteristic of Figure 4.4 shows low current even at very high reverse voltages. In practice, however, at some high reverse voltage the varactor "breaks down" and conducts, as illustrated in Figure 4.5. If the breakdown is roughly as pictured in the left-hand curve of Figure 4.5, it is said to be hard and the voltage at which it occurs is said to be

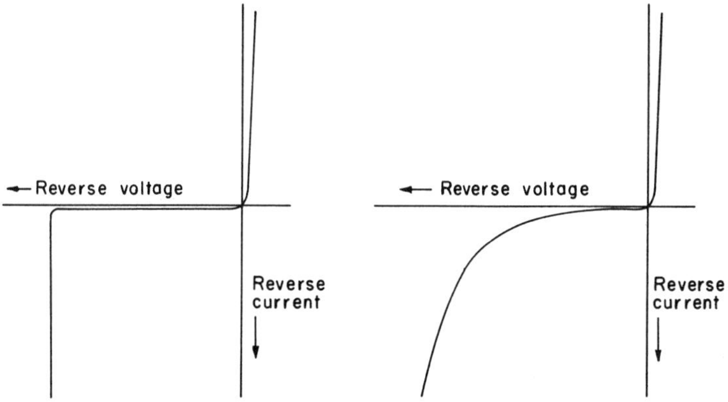

Figure 4.5. Two diode volt-ampere curves showing reverse breakdown. Left, "hard" breakdown, characteristic of the avalanche process; right, "soft" breakdown, probably consisting of avalanche breakdown preceded by surface leakage, generation current, or Zener field-emission current.

the breakdown voltage V_B. On the other hand, for diodes with a soft breakdown, as indicated in the right half of Figure 4.5, the definition of a breakdown voltage is somewhat vague.

It is believed that the major cause of breakdown is an avalanche multiplication process[34] in which holes and electrons are generated (by high-speed existing carriers) so rapidly that the resulting current is large. The soft reverse breakdown, however, is not explained this way; it is usually thought that some other physical effect accounts for the excess reverse current at lower voltages. A probable explanation for diodes with very large doping densities is that this current is due to field emission (the Zener effect).[6]

In the remainder of this book, we shall assume that the varactors have a breakdown voltage V_B which cannot be exceeded.

4.1.3. Contact Potential

For most varactors, I_{sat} is so small that it is possible to apply some forward voltage and still not have too high a dc current. However, there is a limit.

In the brief discussion of the p-n junction, we said that the fixed charge density is almost compensated by the mobile charges. This is not quite true; the compensation is not complete near the junction, and even with no applied voltage there is a slight difference in electric potential across the junction; this is known as the contact potential ϕ. Equation 4.1 holds only for forward voltages much less than ϕ; more sophisticated derivations show[36, 10] that for forward junction voltages approaching ϕ the current becomes very large, and in fact is limited only by the dc resistance of the semiconductor material. The effect of the junction almost disappears, and very large currents can flow. This voltage is therefore a fundamental limitation on the forward applied voltage.

If the saturation current is quite small, the magnitude of the forward current may not be significant until the voltage gets close to $-\phi$. For many practical varactors, then, the voltage can be driven over the full range from $-\phi$ to V_B. For most varactors, ϕ is in the neighborhood of half a volt, rarely much more than 1 volt.

4.1.4. High-Frequency Failure of Rectification

In our explanation of rectification it was observed that mobile charge carriers (holes or electrons) that crossed the junction waited, on the average, some time (known as the lifetime) before recombining. This recombination, of course, is a random process; some carriers recombine immediately, and some remain for much longer. In general, the lifetimes on the two sides of the junction are different.

Suppose that a forward voltage is applied and is then reversed before the carriers have a chance to recombine. Most of the carriers will be drawn back across the junction, and very few will have recombined. The dc current is made up only from those that do recombine; thus the rectification mechanism in a diode is seen to fail at frequencies whose period is considerably below the lifetime. This effect has been used to make a type of diode pulse amplifier.[29]

The implications for high-frequency operation of varactors are quite interesting. In practice the forward voltage is not limited by the contact potential ϕ, but instead can be somewhat higher.[17] Combined with a multiplication of the returning carriers, the effect produces a bias anomaly that we discuss in Section 4.5.2. Finally, the power capabilities of varactor multipliers can be exceeded; this is discussed in more detail in Chapter 8.

4.2. Varactor Equivalent Circuit

The dc characteristics of a p-n junction were discussed in Section 4.1. Now we turn to the ac properties: the junction capacitance, the parasitic series resistance, and the varactor ac equivalent circuit.

4.2.1. Junction Capacitance

Varactors are of interest only because of their nonlinear capacitance. An examination in Appendix A, for arbitrary doping distributions, indicates that depletion-layer capacitance is the most important for practical varactors, except for low or forward bias. For these biases, two correction capacitances are usually important: The first accounts for the fact that the depletion layer is not well defined because of imperfect shielding of the majority carriers; the other[48, 54] accounts for charge stored in injected minority carriers, and is called diffusion capacitance. (Among additional evidence, the careful measurements of Muss[40] and theoretical computations by Morgan and Smits[37] support the hypothesis that depletion-layer capacitance predominates except for low bias.) For simplicity, we neglect these corrections and describe the physical source of depletion-layer capacitance, especially for abrupt-junction and graded-junction varactors.

Recall that the back-biased diode does not conduct appreciable current, because the mobile charge carriers are drawn away from the junction. As they withdraw, they leave the fixed charge density of the acceptors and donors exposed so that close to the junction there is a layer in which a net charge density exists. This depletion layer has a width (call it D) that is a function of the applied voltage. For large reverse voltages, it is very large; for smaller voltages, it is smaller. At an applied voltage of zero, the junction has an electric potential ϕ across it, and the depletion layer has some finite width. For forward voltages the depletion layer becomes still thinner, and finally when $v = -\phi$, the depletion layer disappears. (This description is not wholly accurate,‡ but it is helpful in obtaining a simple intuitive description of junction capacitance.)

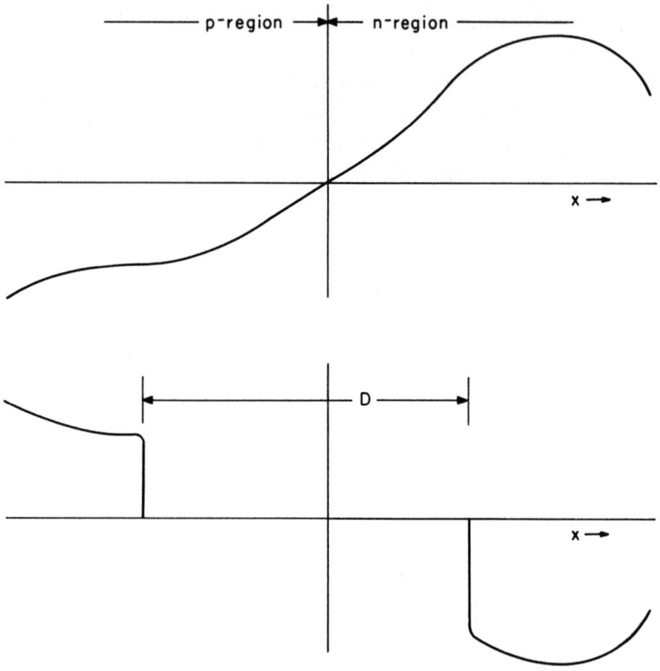

Figure 4.6. A back-biased junction, showing (top) the fixed charge density caused by the impurities (acceptors and donors), and (bottom) the mobile charge density caused by the electrons and holes. Note the depletion layer of thickness D, in which the impurity charges are uncompensated. If the applied voltage is increased, more mobile charges are stripped away, and the depletion layer is made wider.

‡The inaccuracies, as mentioned earlier, arise from the impossibility of precisely defining the edge of the depletion layer, because the mobile charge density does not change abruptly. As a result, at low voltages a more sophisticated theory is necessary.

Figure 4.6 shows a junction with net doping density N(x) back-biased so that it has a depletion layer of thickness D. Note that the total amount of mobile charge that was removed in applying this voltage is found by integrating N(x) from x = 0 to the edge of the depletion layer (in either direction, since the same quantity of charge must have been removed from each side). This charge is a function of the applied voltage;‡ hence, the device behaves like a capacitor with a nonlinear charge-voltage curve. If the voltage is changed a small amount Δv, the charge changes an amount Δq, and their ratio defines, as a function of either voltage or charge, the incremental junction capacitance C or its reciprocal, the incremental junction elastance S:

$$S = \frac{1}{C} = \frac{\Delta v}{\Delta q} . \tag{4.2}$$

In the one-dimensional model pictured here, this elastance is found from the parallel-plate capacitor formula

$$S = \frac{D}{\epsilon A}, \tag{4.3}$$

where D is the depletion-layer thickness, ϵ is the dielectric constant of the semiconductor, and A is the area of the junction. For a proof, see Appendix A. This formula seems reasonable when we remember that the charges removed in making the small change Δv come from the edges of the depletion layer, a distance D apart, just as in a parallel-plate capacitor.

Observe that regardless of the doping distribution, the elastance vanishes for $v = -\phi$; and when $v = V_B$, it is a maximum, called $S_{max} = 1/C_{min}$.

From Equation 4.3, the elastance S is seen to be the same function of voltage or charge as the depletion-layer thickness. This fact enables us to predict the dependence of S on charge quite easily for some simple doping distributions. We discuss them now.

4.2.1(a). The Abrupt-Junction Diode. First, consider the abrupt-junction diode, in which the doping changes abruptly from an excess of donors to an excess of acceptors. This is pictured in Figure 4.7. It is clear from this diagram that the charge q removed is proportional to the depletion-layer thickness, so that the abrupt-junction varactor has an elastance proportional to the charge. In Chapters 8, 9, and 10, this result

‡ This is the voltage across the junction. The instantaneous voltage across the varactor is this voltage plus a drop in the series resistance (see Figure 4.1).

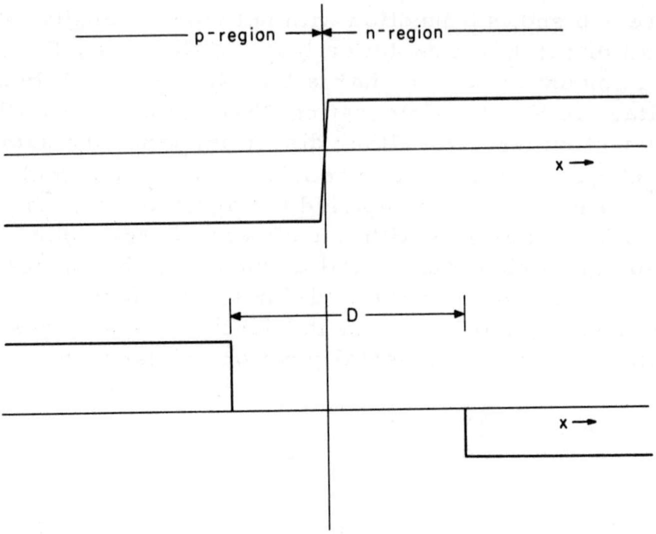

Figure 4.7. Abrupt-junction varactor, back-biased. The doping on each side is independent of x. The top diagram shows the fixed charge density caused by the impurities; the bottom diagram shows the mobile charge density caused by the electrons and holes.

will prove very important in the analysis of large-signal abrupt-junction devices.

To express the elastance of this varactor as a function of the voltage, instead of the charge, we merely integrate the equation

$$dv = S\,dq \tag{4.4}$$

to find that the elastance is proportional to v plus some constant, raised to the one-half power.‡ This constant is approximately (but not exactly[40]) the contact potential ϕ. At breakdown voltage V_B, the elastance is equal to S_{max} and is as large as it can get. Thus, for the abrupt-junction varactor,

$$S = S_{max}\left(\frac{v + \phi}{V_B + \phi}\right)^{1/2}. \tag{4.5}$$

‡Remember that v is the voltage across the junction, which differs from the voltage across the varactor because of the series resistance.

It should be stressed that this formula is not exact for small values of S, that is, for appreciable forward bias. It is, however, a good approximation, especially if the value of ϕ is chosen empirically.

4.2.1(b). The Graded-Junction Diode. As a second important example, let us calculate the elastance of a linearly graded junction, pictured in Figure 4.8. Since the total charge removed q

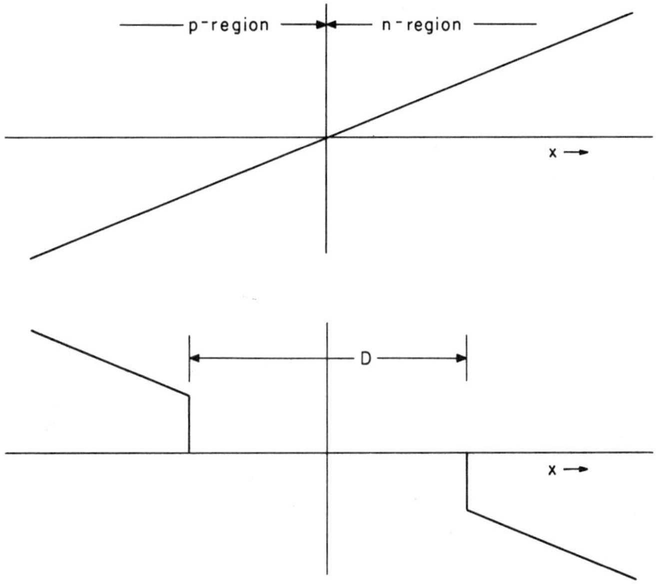

Figure 4.8. Graded-junction varactor, back-biased, with doping density proportional to the coordinate x. The top graph shows the doping density. The bottom graph shows the mobile charge density, which approximately compensates the fixed charge density, except within the depletion layer.

is clearly proportional to the square of the depletion-layer thickness, the elastance varies as the square root of the charge. Integration of Equation 4.4 indicates that the elastance varies as the cube root of the voltage; thus,‡

$$S = S_{max} \left(\frac{v + \phi}{V_B + \phi} \right)^{1/3}, \tag{4.6}$$

‡Remember that v, the voltage across the junction, differs from the voltage across the varactor because of the series resistance.

which is identical to the abrupt-junction formula, but with a different exponent.

Equation 4.6 is not exact for low or forward bias[37] but, in practice, is usually accurate enough, especially if ϕ is chosen empirically.

To give the reader some feeling for these relations, we have plotted in Figure 4.9 the elastance as a function of the applied

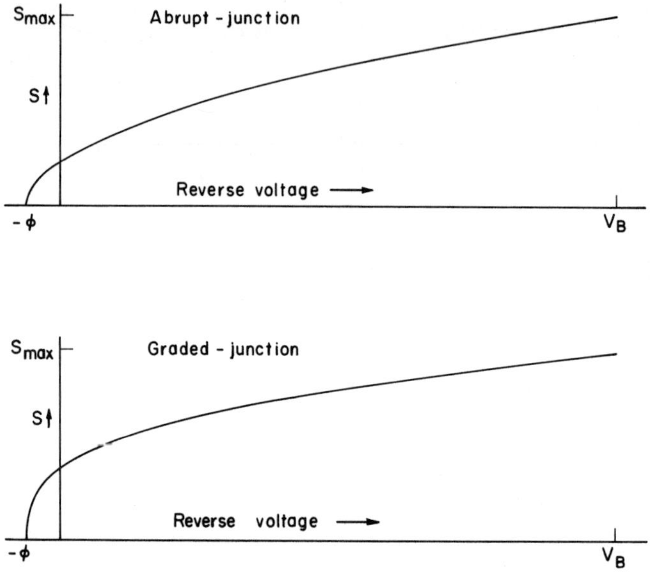

Figure 4.9. Elastance plotted as a function of voltage for (top) abrupt-junction and (bottom) graded-junction varactor. Taken from Equations 4.5 and 4.6. Note that in each case the maximum elastance S_{max} is achieved at the reverse breakdown voltage V_B.

voltage for both an abrupt-junction and a graded-junction varactor.

4.2.1(c). The Diffused Diode. In practice, many varactors are neither exactly abrupt nor exactly graded. Diffused varactors have in theory a doping distribution given by an error function; a typical plot is given in Figure 4.10. Their capacitance-voltage curves are more complicated.[45, 32, 7] At low voltages, the elastance is proportional to the cube root of the voltage, but at higher voltages it follows a square-root law. In practice, high quality diffused varactors seldom operate in the region of the square-root law.

4.2.1(d). Arbitrary Doping Distributions. For still other doping distributions, formulas that relate the capacitance (or elastance) to the voltage[11, 22, 30] are available. Formulas for

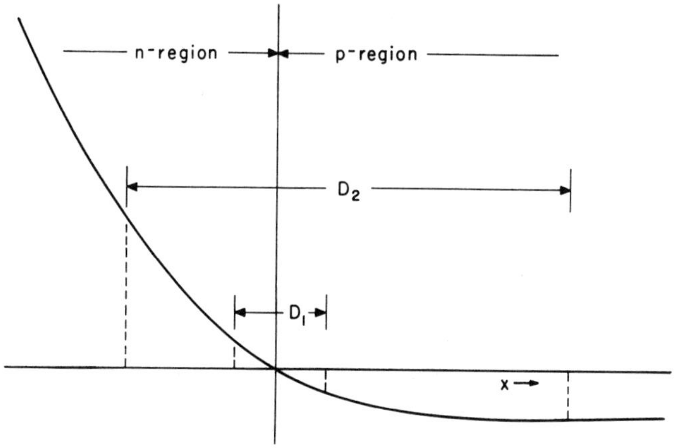

Figure 4.10. Net doping distribution for a diffused junction. For small applied voltages, the depletion-layer is confined to a region in which the doping density is approximately proportional to x; the device behaves like a graded-junction varactor. But for higher applied voltages, the right-hand edge of the depletion layer (now of thickness, say, D_2) is in a region where the doping is uniform, and the left-hand edge does not move appreciably with changes in voltage. Thus the device behaves, for larger voltages, like an abrupt-junction varactor.

the general case are derived in Appendix A. Measurement of back-bias capacitance is widely used as a tool for determining the doping distribution in experimental diodes.

4.2.2. Series Resistance

The most important feature of the varactor diode is, of course, its nonlinear capacitance. The second most important, however, is the parasitic series resistance. This is formed physically of the bulk resistance of the semiconductor and the resistance of the leads.

In calculating the series resistance by integrating the semiconductor bulk resistivity over the path of the current, we should integrate over only the region outside the depletion layer. Thus the series resistance depends on the depletion-layer thickness: As the elastance increases, the series resistance decreases. In particular, for a planar abrupt-junction diode, one side of which is much more heavily doped than the other, the series resistance R_s is

$$R_s = R_{s,min} + \rho\,\epsilon(S_{max} - S), \qquad (4.7)$$

where ρ is the resistivity of the lightly doped side, and $R_{s, min}$ is the series resistance at breakdown voltage. Other formulas that relate R_s to S for other doping distributions can be derived (see Appendix A).

A rigorous examination of the effect of the varactor series resistance should take these variations into account. The frequency conversion occuring in the varying resistance should not be ignored. However, for most present-day varactors the variation in R_s with applied voltage is small; we neglect such variation in Chapters 5 to 10. In the future, as improved varactors with smaller series resistances become available, the variations in series resistance will become more important. At that time it will be necessary to re-examine our findings, taking exact account of the varying series resistance.

It should be pointed out that in practice the dc forward current of a junction diode is limited not only by the junction but also by a series resistance (the latter predominates at very high forward currents). This "dc series resistance" is usually considerably lower than what appears at high frequencies; in this book when we speak of the "series resistance R_s," we have in mind the ac series resistance.

4.2.3. Varactor Ac Equivalent Circuit

So far we have discussed three physical effects that should be included in the varactor ac equivalent circuit: the nonlinear junction elastance, the series resistance, and the conduction across the junction predicted from the dc analysis of Section 4.1. We represent this junction conduction approximately by a nonlinear conductance G(v), where v is the instantaneous voltage across the junction.‡ In the absence of leakage current, breakdown, and carrier generation, G(v) can be found by differentiating Equation 4.1:

$$G(v) = \frac{di}{dv} = \frac{e}{kT_d} I_{sat} e^{-ev/kT_d}. \tag{4.8}$$

In practice, however, a major contribution toward G(v) is leakage of current between the terminals of the varactor, both over the surface of the semiconductor and along the package. At very high frequencies, G(v) is reduced because of rectification failure, as discussed in Section 4.1.4. If any of the other complicating factors mentioned in Section 4.1 is present, G(v) is suitably modified.

---

‡This differs from the instantaneous voltage across the varactor primarily because of the series resistance.

Besides these three elements, we should include in an equivalent circuit any inductance L_{lead} of wires leading to the semiconductor and any stray capacitance C_{case} of the package enclosing the varactor. The resulting equivalent circuit is shown in Figure 4.11.

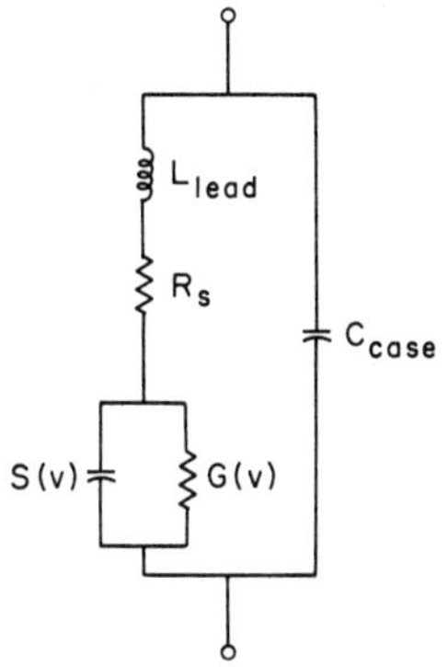

Figure 4.11. Equivalent circuit of a varactor for ac purposes.

In the remainder of this book, we neglect G(v) because in high-quality varactors it is effectively shunted by S(v) at radio frequencies and above.

We also neglect L_{lead} and C_{case}. In many high-quality varactors they are negligible in value, but we do not use this fact as justification. We neglect them because, even if retained, they would not enter into the fundamental limits that we derive. We do not wish to imply that they are unimportant. But because their effect can be removed at a finite number of frequencies by other lossless elements, their presence does not fundamentally degrade the achievable performance of the varactor (although it may make this performance harder to attain).

Neglecting G(v), C_{case}, and L_{lead}, and assuming that R_s is constant, we reduce our equivalent circuit to the Uhlir model of Figure 4.1. The discussion of Sections 4.1 and 4.2 constitutes a justification, on intuitive physical grounds, of this model, together with an indication of when it might fail.

4.2.4. Summary of Varactor Formulas

For future reference we list the formulas that relate elastance, charge, and voltage for both abrupt-junction and graded-junction varactors. They are all obtained from Equations 4.5 and 4.6,

$$S = S_{max}\left(\frac{v + \phi}{V_B + \phi}\right)^{\gamma}, \tag{4.9}$$

where $\gamma = 0.5$ for abrupt-junction varactors and 0.33 for graded-junction varactors. Since S is the incremental elastance dv/dq, we may integrate the equation

$$dq = \frac{dv}{S} = \frac{dv}{S_{max}\left(\frac{v + \phi}{V_B + \phi}\right)^{\gamma}} \tag{4.10}$$

to obtain the relations

$$S = S_{max}\left(\frac{v + \phi}{V_B + \phi}\right)^{\gamma} = S_{max}\left(\frac{q + q_\phi}{Q_B + q_\phi}\right)^{\gamma/(1-\gamma)}, \quad (4.11)$$

$$\left(\frac{v + \phi}{V_B + \phi}\right) = \left(\frac{q + q_\phi}{Q_B + q_\phi}\right)^{1/(1-\gamma)} = \left(\frac{S}{S_{max}}\right)^{1/\gamma}, \quad (4.12)$$

$$\left(\frac{q + q_\phi}{Q_B + q_\phi}\right) = \left(\frac{v + \phi}{V_B + \phi}\right)^{1-\gamma} = \left(\frac{S}{S_{max}}\right)^{(1-\gamma)/\gamma}, \quad (4.13)$$

and the simpler result

$$q + q_\phi = \frac{v + \phi}{(1 - \gamma)S}, \quad (4.14)$$

where the constant of integration $(-q_\phi)$ is the charge at voltage $-\phi$:

$$q_\phi = \frac{\phi}{(1 - \gamma)S_{max}}\left(\frac{V_B + \phi}{\phi}\right)^{\gamma}, \quad (4.15)$$

and where Q_B is the charge at breakdown voltage. The charge q is, of course, zero at zero volt.

For ease in using Equations 4.11 through 4.13, the exponents are given in Table 4.1.

Table 4.1. Exponents Used in Equations 4.11 through 4.13

Abrupt-Junction	$\gamma = 0.5$ $1/(1-\gamma) = 2.0$ $1-\gamma = 0.5$	$\gamma/(1-\gamma) = 1.0$ $1/\gamma = 2.0$ $(1-\gamma)/\gamma = 1.0$
Graded-Junction	$\gamma = 0.33$ $1/(1-\gamma) = 1.5$ $1-\gamma = 0.67$	$\gamma/(1-\gamma) = 0.5$ $1/\gamma = 3.0$ $(1-\gamma)/\gamma = 2.0$

The voltage v that appears in Equations 4.9 through 4.14 is the instantaneous voltage across the nonlinear-capacitance part of the varactor, inside of the series resistance.

4.3. Noise Sources

Varactors contribute noise to the circuits in which they are used. The most important physical source of noise in a varactor is the parasitic series resistance; the simple assumption that this resistance exhibits thermal noise has far-reaching consequences, which are discussed in Chapters 5 and 6. Another source of noise, of far less importance, is shot noise associated either with dc current flowing through the varactor or with the pumping current.

4.3.1. Thermal Noise

The series resistance of the varactor is thought to consist of physical resistances, such as lead resistance and bulk resistance of the semiconductor, and should exhibit thermal noise. If the temperature of the diode is T_d, then the thermal noise can be represented at spot frequencies by equivalent voltage generators in series with the resistance, with mean-squared voltage

$$\overline{|e_n|^2} = 4kT_d R_s \,\Delta f. \tag{4.16}$$

For more details, see Section 2.5.1(a).

The series resistance is assumed to contribute thermal noise at all frequencies, even those that are not necessary in the frequency-conversion process. To obtain the best noise performance from varactor devices, it is usually necessary to block the flow of noise currents at these extra frequencies.

4.3.2. Shot Noise

Shot noise is not of primary importance in varactor devices. Uhlir[56, 58] has investigated the effect of shot noise on diode frequency converters using the nonlinear resistance or the nonlinear capacitance. He concludes that for the nonlinear capacitor case the contribution of shot noise can be negligibly small:[58] "A frequency converter using an ideal nonlinear capacitor has no shot noise output regardless of the local-oscillator waveform, the source and load admittances, and the image terminations."

Of course, practical varactors fall short of being "ideal nonlinear capacitors." They pass a saturation current I_{sat} in the reverse direction and a conduction current $I_{sat} - I_0$ in the forward direction, where I_0 is the average diode current (reckoned as positive if it is reverse current). The effective current, for the purpose of calculating shot noise, is the sum of these two, $2I_{sat} - I_0$. The shot noise can be represented in a circuit by a spot-noise current generator in parallel with the junction, supplying noise at each important frequency, of mean-squared value

$$\overline{|i|^2} = 2e(2I_{sat} - I_0)\,\Delta f. \tag{4.17}$$

Some detailed calculations[41] have shown that shot noise is negligible compared with thermal noise in a single-idler parametric amplifier, provided that

$$\frac{(2I_{sat} - I_0)R_s}{2kT_d/e} \ll \left(\frac{R_s\omega_i}{S_0}\right)^2 \tag{4.18}$$

and

$$\frac{(2I_{sat} - I_0)R_s}{2kT_d/e} \ll \left(\frac{R_s\omega_s}{S_0}\right)^2, \tag{4.19}$$

where ω_s and ω_i are the signal and idler frequencies, and S_0 is the average elastance. If these inequalities are not satisfied, the shot noise may be comparable with the thermal noise; for typical varactors this can happen only when the thermal noise is very low. We feel that we are justified in ignoring shot noise in this book.

4.3.3. Other Noise Sources

There are a few other physical effects that may contribute important noise. The avalanche mechanism that causes breakdown introduces high noise.[34,62] If the varactor is driven into reverse breakdown, this noise is important.

As explained in Section 4.1.4, the forward conduction process fails at high frequencies because of long recombination time. The shot noise from forward conduction current was described in Section 4.3.2. However, if the pump frequency is high enough so that the forward conduction mechanism fails, the carriers that return unrecombined usually trigger additional avalanche multiplication (see Section 4.5.2). Such multiplication produces noise and should be avoided in low-noise operation.

Noise is also caused by fluctuations in the ionization state of donors and acceptors within the depletion layer. Giacoletto[12] has shown that in typical cases this noise, when compared to thermal noise of the series resistance, is significant only at quite low temperatures.

A common type of noise in semiconductor devices is 1/f noise, so named because its intensity increases as the reciprocal of frequency at low frequencies.[61] This type of noise does not appear to be important in varactors; for example, Biard[2] has built low-noise low-frequency varactor upconverters without appreciable 1/f noise.

In some circuits there may be additional sources of noise.[19] For example, the pump may have imperfections that contribute a random noise to the circuit, at any of the sideband frequencies. For simplicity, in our analyses in Chapters 5 and 6 we ignore all noise except thermal noise.

4.4. Large-Signal Equations

The equations that govern the varactor (that is, those that govern the varactor model we use) are obtained from Figure 4.1:

$$v(t) = \int S(t)i(t)\,dt + R_s i(t) + e_n, \tag{4.20}$$

where S(t) is the incremental elastance as a function of time, and e_n is a noise voltage (see Section 4.3). Observe that the elastance appears inside the integral sign because it is the incremental elastance.[18] Equation 4.20 is nonlinear because the actual elastance waveform depends on the charge, as for example by Equation 4.11.

For reasons given in Section 1.2.2, we use a frequency-domain type of analysis in this book. Assuming that a steady state is maintained, we work with relations among the Fourier coefficients of the variables v, i, and S. Therefore, in place of Equation 4.20 we seek relations among these Fourier coefficients.

4.4.1. Harmonic Multipliers and Dividers

In harmonic multipliers and dividers, only frequencies of the form $k\omega_0$ are present; the voltage, current, and incremental elastance can be written in the form

$$v(t) = \Sigma_k V_k e^{jk\omega_0 t}, \tag{4.21}$$

$$i(t) = \Sigma_k I_k e^{jk\omega_0 t}, \tag{4.22}$$

and

$$S(t) = \Sigma_k S_k e^{jk\omega_0 t}, \tag{4.23}$$

where the sums extend from $k=-\infty$ to $+\infty$, and where, because v(t), i(t), and S(t) are real,

$$V_{-k} = V_k^*, \tag{4.24}$$

$$I_{-k} = I_k^*, \tag{4.25}$$

and

$$S_{-k} = S_k^*. \tag{4.26}$$

The asterisk indicates the complex conjugate.

From Equation 4.20, each Fourier coefficient V_k can be calculated in terms of the Fourier coefficients of current and elastance, and the spot-noise voltages‡ E_{nk}. Thus

$$V_k = \left\langle v(t)e^{-jk\omega_o t}\right\rangle$$

$$= R_s \left\langle i(t)e^{-jk\omega_o t}\right\rangle + \frac{1}{jk\omega_o}\left\langle S(t)i(t)e^{-jk\omega_o t}\right\rangle + \left\langle e_n e^{-jk\omega_o t}\right\rangle$$

$$= R_s I_k + \frac{1}{jk\omega_o}\Sigma_\ell \left\langle I_\ell S(t)e^{j\ell\omega_o t}e^{-jk\omega_o t}\right\rangle + E_{nk}$$

$$= R_s I_k + \frac{1}{jk\omega_o}\Sigma_\ell\, I_\ell\, S_{k-\ell} + E_{nk}, \tag{4.27}$$

where the brackets denote a time average. This is easily expressed in matrix form. If we let $\underset{\sim}{V}$ be the column matrix of V_k, in order from $k = -\infty$ to $+\infty$, and let $\underset{\sim}{I}$ and $\underset{\sim}{E}_n$ be the similar matrices of I_k and E_{nk}, then

$$\underset{\sim}{V} = \underset{\sim}{Z}\,\underset{\sim}{I} + \underset{\sim}{E}_n, \tag{4.28}$$

where the square matrix $\underset{\sim}{Z}$ has diagonal components

$$Z_{kk} = R_s + \frac{S_0}{jk\omega_o} \tag{4.29}$$

and off-diagonal terms

$$Z_{k\ell} = \frac{S_{k-\ell}}{jk\omega_o}. \tag{4.30}$$

---

‡The noise voltages are effective rms values instead of half-amplitudes. See Section 2.5.1(a).

Writing a few terms of this matrix equation, we have Equations 4. 31:

$$
\begin{bmatrix} \vdots \\ V_k^* \\ \vdots \\ V_2^* \\ V_1^* \\ V_1 \\ V_2 \\ \vdots \\ V_k \\ \vdots \end{bmatrix}
=
\begin{bmatrix}
\ddots & \vdots & & \vdots & \vdots & \vdots & \vdots & & \vdots & \\
\cdots & R_s - \frac{S_0}{jk\omega_o} & \cdots & -\frac{S_{k-2}^*}{jk\omega_o} & -\frac{S_{k-1}^*}{jk\omega_o} & -\frac{S_{k+1}^*}{jk\omega_o} & -\frac{S_{k+2}^*}{jk\omega_o} & \cdots & -\frac{S_{2k}^*}{jk\omega_o} & \cdots \\
 & \vdots & & \vdots & \vdots & \vdots & \vdots & & \vdots & \\
\cdots & -\frac{S_{k-2}}{j2\omega_o} & \cdots & R_s - \frac{S_0}{j2\omega_o} & -\frac{S_1^*}{j2\omega_o} & -\frac{S_3^*}{j2\omega_o} & -\frac{S_4^*}{j2\omega_o} & \cdots & -\frac{S_{k+2}^*}{j2\omega_o} & \cdots \\
\cdots & -\frac{S_{k-1}}{j\omega_o} & \cdots & -\frac{S_1}{j\omega_o} & R_s - \frac{S_0}{j\omega_o} & -\frac{S_2^*}{j\omega_o} & -\frac{S_3^*}{j\omega_o} & \cdots & -\frac{S_{k+1}^*}{j\omega_o} & \cdots \\
\cdots & \frac{S_{k+1}}{j\omega_o} & \cdots & \frac{S_3}{j\omega_o} & \frac{S_2}{j\omega_o} & R_s + \frac{S_0}{j\omega_o} & \frac{S_1^*}{j\omega_o} & \cdots & \frac{S_{k-1}^*}{j\omega_o} & \cdots \\
\cdots & \frac{S_{k+2}}{j2\omega_o} & \cdots & \frac{S_4}{j2\omega_o} & \frac{S_3}{j2\omega_o} & \frac{S_1}{j2\omega_o} & R_s + \frac{S_0}{j2\omega_o} & \cdots & \frac{S_{k-2}^*}{j2\omega_o} & \cdots \\
 & \vdots & & \vdots & \vdots & \vdots & \vdots & \ddots & \vdots & \\
\cdots & \frac{S_{2k}}{jk\omega_o} & \cdots & \frac{S_{k+2}}{jk\omega_o} & \frac{S_{k+1}}{jk\omega_o} & \frac{S_{k-1}}{jk\omega_o} & \frac{S_{k-2}}{jk\omega_o} & \cdots & R_s + \frac{S_0}{jk\omega_o} & \cdots \\
 & \vdots & & \vdots & \vdots & \vdots & \vdots & & \vdots & \ddots
\end{bmatrix}
\begin{bmatrix} \vdots \\ I_k^* \\ \vdots \\ I_2^* \\ I_1^* \\ I_1 \\ I_2 \\ \vdots \\ I_k \\ \vdots \end{bmatrix}
+
\begin{bmatrix} \vdots \\ E_{nk}^* \\ \vdots \\ E_{n2}^* \\ E_{n1}^* \\ E_{n1} \\ E_{n2} \\ \vdots \\ E_{nk} \\ \vdots \end{bmatrix}
\tag{4.31}
$$

The dots in Equations 4. 31 indicate missing terms. Observe that the bottom half of this matrix equation is unnecessary; each relation is the complex conjugate of one appearing above. This matrix equation, upon which the analyses in Chapters 8 and 9 are based, is nonlinear, because the S_k depend upon the electrical variables. For the abrupt-junction varactor with elastance proportional to the charge, S_k is nonzero when and only when I_k is nonzero (for $k \neq 0$). For other varactors, the dependence is more complicated.

In the application of Equations 4. 31 in Chapters 8, 9, and 10, the noise terms are neglected, because they do not affect the efficiency or power capabilities.

4.4.2. Amplifiers and Frequency Converters

In large-signal amplifiers and frequency converters, two incommensurate frequencies (say, ω_p and ω_s) are used to excite the varactor, and in general, frequencies of the form $\ell\omega_p + m\omega_s$ are generated (where m and ℓ are integers). As a specific example, suppose that current flows only at the four frequencies ω_s, ω_p, $\omega_i = \omega_p - \omega_s$, and $\omega_u = \omega_p + \omega_s$. Then the large-signal equations that relate the voltage coefficients at these frequencies to the currents are given in matrix form in Equations 4.32:‡

$$
\begin{bmatrix} V_u^* \\ V_p^* \\ V_i^* \\ V_s^* \\ V_s \\ V_i \\ V_p \\ V_u \end{bmatrix}
=
\begin{bmatrix}
R_s - \frac{S_0}{j\omega_u} & -\frac{S_s^*}{j\omega_u} & -\frac{S_{2s}^*}{j\omega_u} & -\frac{S_p^*}{j\omega_u} & -\frac{S_{p+2s}^*}{j\omega_u} & -\frac{S_{2p}^*}{j\omega_u} & -\frac{S_{2p+s}^*}{j\omega_u} & -\frac{S_{2u}^*}{j\omega_u} \\
-\frac{S_s}{j\omega_p} & R_s - \frac{S_0}{j\omega_p} & -\frac{S_s^*}{j\omega_p} & -\frac{S_i^*}{j\omega_p} & -\frac{S_u^*}{j\omega_p} & -\frac{S_{2p-s}^*}{j\omega_p} & -\frac{S_{2p}^*}{j\omega_p} & -\frac{S_{2p+s}^*}{j\omega_p} \\
-\frac{S_{2s}}{j\omega_i} & -\frac{S_s}{j\omega_i} & R_s - \frac{S_0}{j\omega_i} & -\frac{S_{p-2s}^*}{j\omega_i} & -\frac{S_p^*}{j\omega_i} & -\frac{S_{2i}^*}{j\omega_i} & -\frac{S_{2p-s}^*}{j\omega_i} & -\frac{S_{2p}^*}{j\omega_i} \\
-\frac{S_p}{j\omega_s} & -\frac{S_i}{j\omega_s} & -\frac{S_{p-2s}}{j\omega_s} & R_s - \frac{S_0}{j\omega_s} & -\frac{S_{2s}^*}{j\omega_s} & -\frac{S_p^*}{j\omega_s} & -\frac{S_u^*}{j\omega_s} & -\frac{S_{p+2s}^*}{j\omega_s} \\
\frac{S_{p+2s}}{j\omega_s} & \frac{S_u}{j\omega_s} & \frac{S_p}{j\omega_s} & \frac{S_{2s}}{j\omega_s} & R_s + \frac{S_0}{j\omega_s} & \frac{S_{p-2s}^*}{j\omega_s} & \frac{S_i^*}{j\omega_s} & \frac{S_p^*}{j\omega_s} \\
\frac{S_{2p}}{j\omega_i} & \frac{S_{2p-s}}{j\omega_i} & \frac{S_{2i}}{j\omega_i} & \frac{S_p}{j\omega_i} & \frac{S_{p-2s}}{j\omega_i} & R_s + \frac{S_0}{j\omega_i} & \frac{S_s^*}{j\omega_i} & \frac{S_{2s}^*}{j\omega_i} \\
\frac{S_{2p+s}}{j\omega_p} & \frac{S_{2p}}{j\omega_p} & \frac{S_{2p-s}}{j\omega_p} & \frac{S_u}{j\omega_p} & \frac{S_i}{j\omega_p} & \frac{S_s}{j\omega_p} & R_s + \frac{S_0}{j\omega_p} & \frac{S_s^*}{j\omega_p} \\
\frac{S_{2u}}{j\omega_u} & \frac{S_{2p+s}}{j\omega_u} & \frac{S_{2p}}{j\omega_u} & \frac{S_{p+2s}}{j\omega_u} & \frac{S_p}{j\omega_u} & \frac{S_{2s}}{j\omega_u} & \frac{S_s}{j\omega_u} & R_s + \frac{S_0}{j\omega_u}
\end{bmatrix}
\begin{bmatrix} I_u^* \\ I_p^* \\ I_i^* \\ I_s^* \\ I_s \\ I_i \\ I_p \\ I_u \end{bmatrix}
+
\begin{bmatrix} E_{nu}^* \\ E_{np}^* \\ E_{ni}^* \\ E_{ns}^* \\ E_{ns} \\ E_{ni} \\ E_{np} \\ E_{nu} \end{bmatrix}.
\tag{4.32}
$$

There we denote the voltage, current, and elastance coefficients in an obvious way (the subscripts s, i, p, u, 2s, 2i, 2p, 2u, p+2s, p-2s, 2p+s, and 2p-s denote half-amplitudes at frequencies ω_s, ω_i, ω_p, ω_u, $2\omega_s$, $2\omega_i$, $2\omega_p$, $2\omega_u$, $\omega_p + 2\omega_s$, $\omega_p - 2\omega_s$, $2\omega_p + \omega_s$, and $2\omega_p - \omega_s$). Both the voltage and the elastance have components at other frequencies. Equations 4.32 are nonlinear, because the elastance coefficients depend on the charge coefficients in a complicated way. For an abrupt-junction varactor, all elastance coefficients vanish except for S_0, S_s, S_i, S_p, and S_u, so that Equations 4.32 are considerably simplified. In Chapter 10, the analysis of large-signal upconverters and amplifiers is based on Equations 4.32.

4.5. Pumping

In many applications the varactor is pumped strongly at some frequency ω_p (and its harmonics) by a pump or local oscillator. This pumping requires voltage and current components at frequency ω_p and its harmonics. If there are, in addition, other small signals, we may be interested in the equations that relate only the small-signal parts of the voltage and current. In Section 4.5.1 we obtain the small-signal equations of motion, and in Section 4.5.2 we discuss an anomaly that arises in pumping practical varactors.

‡The noise voltages are effective rms values instead of half-amplitudes. See Section 2.5.1(a).

4.5.1. Small-Signal Equations

The instantaneous varactor voltage $v(t)$ can be written, from Equation 4.20, as some function f of the charge q, plus the drop across the series resistance, plus the noise term:

$$v(t) = f[q(t)] + R_s i(t) + e_n. \tag{4.33}$$

We let v, i, and q all be composed of a pumping part and a small-signal part:

$$v(t) = v_p(t) + v_{ss}(t), \tag{4.34}$$

$$q(t) = q_p(t) + q_{ss}(t), \tag{4.35}$$

and

$$i(t) = i_p(t) + i_{ss}(t), \tag{4.36}$$

where each of the small-signal variables is small (in a sense to be indicated). What we desire now is an equation relating the small-signal parts.

Since the current is the time derivative of the charge, i_{ss} is the time derivative of q_{ss}. From Equation 4.33,

$$v_p + v_{ss} = f(q_p + q_{ss}) + R_s(i_p + i_{ss}) + e_n. \tag{4.37}$$

Since the small-signal variables are small, we expand $f(q)$ about q_p as an equilibrium point (because q_p is a function of time, this expansion is time-varying):

$$f(q_p + q_{ss}) = f(q_p) + \left(\frac{df}{dq}\right)_p q_{ss} + \frac{1}{2}\left(\frac{d^2f}{dq^2}\right)_p q_{ss}^{\;2} + \cdots, \tag{4.38}$$

where the subscript p indicates evaluation for $q_p(t)$. But as the first derivative of f with respect to q is the varactor incremental elastance S, taking the first-order part of Equation 4.37 yields

$$v_{ss}(t) = (S)_p\, q_{ss} + R_s i_{ss}(t) + e_n, \tag{4.39}$$

or letting $S(t)$ be the varactor incremental elastance, as pumped, we obtain

$$v_{ss}(t) = S(t) \int i_{ss}(t)\, dt + R_s i_{ss}(t) + e_n. \tag{4.40}$$

This equation relates small-signal variables only; the effect of the pump enters only to determine $S(t)$. These small-signal equations are time-varying but linear; they fail when the neglected parts of the Taylor series for $f(q)$ are appreciable, that is, when the change in elastance caused by the small-signal charge is comparable with the elastance as pumped. Observe that the form of Equation 4.40 differs from Equation 4.20, because the elastance appears outside the integral sign.

These small-signal equations of motion can be expressed in frequency-by-frequency form. We suppose the pumping proceeds at frequency ω_p (and its harmonics); therefore, the elastance, as pumped, can be written in a Fourier series

$$S(t) = \sum_{k=-\infty}^{\infty} S_k e^{jk\omega_p t}. \tag{4.41}$$

If the small signals are at frequencies‡ of the form $\ell\omega_p + \omega_s$ (in particular, including $-\omega_i = \omega_s - \omega_p$ and $\omega_u = \omega_p + \omega_s$), we find, in matrix form, Equations 4.42:

$$\begin{bmatrix} \vdots \\ V_i^* \\ V_s \\ V_u \\ \vdots \\ V_\ell \\ \vdots \end{bmatrix} = \begin{bmatrix} \ddots & \vdots & \vdots & \vdots & & \vdots & \\ \cdots & R_s - \frac{S_0}{j\omega_i} & \frac{S_1^*}{j\omega_s} & \frac{S_2^*}{j\omega_u} & \cdots & \frac{S_{\ell+1}^*}{j\omega_\ell} & \cdots \\ \cdots & -\frac{S_1}{j\omega_i} & R_s + \frac{S_0}{j\omega_s} & \frac{S_1^*}{j\omega_u} & \cdots & \frac{S_\ell^*}{j\omega_\ell} & \cdots \\ \cdots & -\frac{S_2}{j\omega_i} & \frac{S_1}{j\omega_s} & R_s + \frac{S_0}{j\omega_u} & \cdots & \frac{S_{\ell-1}^*}{j\omega_\ell} & \cdots \\ & \vdots & \vdots & \vdots & \ddots & \vdots & \\ \cdots & -\frac{S_{\ell+1}}{j\omega_i} & \frac{S_\ell}{j\omega_s} & \frac{S_{\ell-1}}{j\omega_u} & \cdots & R_s + \frac{S_0}{j\omega_\ell} & \cdots \\ & \vdots & \vdots & \vdots & & \vdots & \ddots \end{bmatrix} \begin{bmatrix} \vdots \\ I_i^* \\ I_s \\ I_u \\ \vdots \\ I_\ell \\ \vdots \end{bmatrix} + \begin{bmatrix} \vdots \\ E_{ni}^* \\ E_{ns} \\ E_{nu} \\ \vdots \\ E_{n\ell} \\ \vdots \end{bmatrix}, \tag{4.42}$$

where the subscript ℓ (for voltage and current) refers to the frequency $\ell\omega_p + \omega_s$. Note that Equations 4.42 are linear,

---

‡Equation 4.40 is usually used when the small signals have frequencies that are not commensurate with the pump frequency, in particular, not multiples of it or rational fractions of it. However, no such restriction is inherent in Equation 4.40, which is used in Chapter 9 for small signals at half the pump frequency.

because the S_k are assumed to have been specified in advance by the pumping.

4.5.2. Pumping Anomaly

One characteristic of varactors has not yet been discussed because it appears only under conditions of high-frequency pumping. As mentioned in Section 4.1.4, the dc forward characteristics do not apply instantaneously at high frequencies, because they depend on the relatively slow process of carrier recombination. When pumping at high frequencies, we may apply an instantaneous voltage greater than the contact potential. If the voltage reverses before an appreciable number of carriers recombine, carriers are drawn back across the junction. We might expect that not all of the carriers that cross the junction would come back, but in practice it is found that many more do!

The probable explanation for this high-frequency pumping anomaly has been given by Hefni[20, 21] and Siegel.[51] Apparently the carriers coming back are multiplied by an avalanche process similar to, but not as strong as, the avalanche process that leads to breakdown at high reverse voltages. This pumping anomaly is observed[51, 20, 33, 21] in practically all varactors, although in various pump-frequency ranges. In practice, the multiplication probably introduces noise;[34] this effect has been discussed by Weglein.[62] It can be avoided by biasing the varactor away from the forward conduction region, but this remedy reduces the available range of elastance, especially because it removes a region where the variation in elastance is particularly great (see Figure 4.9).

4.6. Characterization of Pumped Varactors

The series resistance of a varactor clearly degrades its performance. It is important to have a way of knowing which of two (or more) pumped varactors is "better." Our analyses of applications of pumped varactors (Chapters 5 and 6) indicate a suitable figure of merit for pumped varactors. This pumped figure of merit is discussed here, and in Section 4.7 we show a logical way of characterizing the quality of varactors without the pumping.

In Chapters 5 and 6, the quantity (with dimension of frequency‡)

---

‡ This figure of merit depends upon the strength of the pumping, but not directly upon the pump frequency. In particular, it is generally not equal to the pump frequency, and may lie either above or below ω_p. In Chapters 5, 6, and 10 it is shown that pumping at frequency $m_1 \omega_c$ is sometimes desirable, but more generally ω_p has no direct relation to $m_1\omega_c$.

$$\frac{|S_1|}{R_s}, \tag{4.43}$$

which we call $m_1\omega_c$, is important in determining the noise and gain performance of upconverters and parametric amplifiers. In fact, in most of the fundamental limits derived there, it is the only quantity used to characterize the varactor. Thus two pumped varactors may have different S_{max}, R_s, and so on, but if they have the same value of the ratio of Expression 4.43, they are both capable of the same quality of performance. Therefore, we choose this quantity as a pumped figure of merit. A higher value means that the pumped varactor is inherently capable of better noise performance and more gain. Note that this proposed figure of merit is firmly tied to the applications of the pumped varactor. This same figure of merit was used by Boyd,[5] who ascribed it to C. S. Kim.

Our proposed figure of merit is similar to other ones that have been proposed,[27, 38, 39, 28, 62, 15, 4, 3] but is based on the series equivalent circuit, and is therefore a better approximation. It leads more logically and simply to a characterization of quality for a varactor without the pumping.

Kurokawa and Uenohara[31] used both Expression 4.43 and an approximate figure of merit, in an effort to compare them. The fact that the differences are not trivial was shown by Siegel,[52] who calculated both an approximate figure of merit M and another, M_s, which is much closer to $m_1\omega_c$. Siegel found (in one case) that as the pumping level was increased, M went through a maximum and started down again, whereas M_s (and presumably also $m_1\omega_c$) kept increasing as would be expected. If one were to use M as a figure of merit, he would not pump fully, and so would not realize the full potential of his varactor. This comparison illustrates the dangers inherent in using the approximate parallel analyses to describe large pumping.

It should be emphasized that this figure of merit does not tell everything necessary about a pumped varactor. Also important in characterizing pumped varactors are some indication of the impedance level (such as R_s, S_0, or S_{max}) and the ratios

$$\frac{|S_k|}{R_s} \tag{4.44}$$

for values of k different from 1, and the phases of the various S_k. In addition, the ratio $|S_1|/S_0$ is important in bandwidth and rise-time problems.

4.7. Characterization of Varactors

Expression 4.43 is a valid figure of merit for pumped varactors. It does not, however, indicate which of two or more varactors without the pumping is of higher quality. What is required is a characterization of unpumped varactors. This is derived in Section 4.7.1. Characterization of varactors for power-handling capabilities is discussed in Section 4.7.2, and in Section 4.7.3 we discuss minor characteristics.

4.7.1. Cutoff Frequency

In comparing two varactors, we should logically ask which is inherently capable of being pumped so as to have a higher pumped figure of merit, Expression 4.43. We wish to define a characterization of varactor quality, the cutoff frequency, to answer this question. The varactor with higher cutoff frequency can be pumped so as to have a higher pumped figure of merit.

The original characterization of varactor quality[58] was Uhlir's "cutoff frequency"

$$\frac{1}{2\pi R_s C_{min}} = \frac{S_{max}}{2\pi R_s}, \tag{4.45}$$

which is the largest frequency at which the small-signal quality factor of the varactor can exceed 1. This definition, however, has been criticized; it is argued that since the varactor is seldom at breakdown voltage, where C_{min} is achieved, a better characterization would be a cutoff frequency defined with the capacitance at zero bias or at some specified bias. Other "improvements" have been to include some measure of the nonlinearity of the varactor, such as the rate of change of capacitance with voltage.[42, 23, 46]

The characterization of any device should be based on applications. Certainly, the user of the varactor should be the one to say what constitutes varactor quality; the characterization proposed here is based on our analysis of varactor applications, in Chapters 5, 6, 8, 9, and 10.

The trick is to think of the pumped figure of merit, Expression 4.43, which depends on both the pumping and the varactor, as a product of two factors: one depending only on the varactor, and the other a bounded function describing the "efficiency of pumping." We multiply and divide Expression 4.43 by a capacitance, to be specified:

$$\left(\frac{1}{R_s C}\right) \left(|S_1| C\right). \tag{4.46}$$

We want the first factor to be characteristic of the varactor and the second factor to be bounded. To make the problem realistic, we assume that the actual varactor incremental capacitance is restricted to lie between C_{min} and some maximum value C_{max}; alternatively, we say $S_{min} \leq S \leq S_{max}$, where $S_{min} = 1/C_{max}$.

Among the capacitances we might think of using in Expression 4.46 are these:

1. C_{min}.
2. C_{max}.
3. $C_{max} - C_{min}$.
4. $C_{max} + C_{min}$.
5. $\dfrac{(C_{min})^2}{C_{max}}$.
6. Zero-bias capacitance.
7. Zero-bias capacitance $\times \dfrac{C_{max} + C_{min}}{C_{max} - C_{min}}$.
8. Capacitance at a specified bias, say 1 volt.
9. Capacitance at the dc bias point used.
10. Average capacitance, as pumped.

Or we might use the reciprocal of one of the following elastances:

11. $S_{max} - S_{min}$.
12. $S_{max} + S_{min}$.
13. Average elastance, as pumped.
14. $v \dfrac{dS}{dv}$.

In general these are all different.

If the first factor in Expression 4.46 is to be independent of the bias and pumping, we cannot use choices 9, 10, 13, or 14. If the second factor is to be bounded (independent of the values of S_{min} and the zero-bias capacitance), we cannot use choices 2, 3, 4, 6, 7, or 8. This limits the selection to 1, 5, 11, and 12.

If S_{min} is zero or much smaller than S_{max}, then choices 1, 11, and 12 are all the same, and all lead to Uhlir's cutoff frequency, Expression 4.45. If S_{min} is finite, however, the choices are different. The second factor of Expression 4.46 in each case is bounded, but for choices 1 and 12, and also for 5, the achievable bounds depend on S_{min}. On the other hand, for choice 11, many of the achievable bounds (derived in Chapter 7) are independent of S_{min}. Thus if a choice must be made between 1, 11, and 12 (and this is not necessary with varactors for which $S_{min} \ll S_{max}$), choice 11 is preferable.

As an illustration, let us use the various choices to select the best of these three varactors given in Table 4.2. We use arbitrary units for capacitance and set $R_s = 1$, since our only purpose is comparison. Neglecting those choices that depend on the operating point or some arbitrary voltage — 8, 9, 10, 13, and 14 — we show the relative cutoff frequencies in Table 4.3.

Table 4.2. Varactors Used for Illustration

Varactor I	Varactor II	Varactor III
$C_{min} = 2$	$C_{min} = 1$	$C_{min} = 1$
$C_{max} = \infty$	$C_{max} = 2$	$C_{max} = 5$
$C_{zero\text{-}bias} = 3$	$C_{zero\text{-}bias} = 1.5$	$C_{zero\text{-}bias} = 4$

Table 4.3. Relative Figures of Merit for the Three Varactors of Table 4.2, Using Nine Choices for C.

Choice	Varactor I	Varactor II	Varactor III
1	0.5	1.0	1.0
2	0	0.5	0.2
3	0	1.0	0.25
4	0	0.333	0.167
5	∞	2.0	5.0
6	0.333	0.667	0.25
7	0.333	0.222	0.167
11	0.5	0.5	0.8
12	0.5	1.5	1.2

It is not fair to compare the numbers in a given column with each other, because, if desired, each choice can be equipped with a trivial change of scale. What is significant for each choice is which varactor it predicts is best (that is, has a highest relative number). Selecting between varactors I and II, two choices would indicate that varactor I is better, and six would indicate that varactor II is better, whereas in fact each is capable of being pumped to the same value of $|S_1|/R_s$, since the same range of elastance $S_{max} - S_{min}$ is available. Only choice 11 predicts this. Furthermore, when varactor III is compared with the other two, only choice 11, and possibly 1, would indicate its selection, in spite of the fact that it is capable of a 60 per cent improvement in $|S_1|/R_s$. Choices 6 and 7 would erroneously lead us to believe that varactor III is the worst of the lot! This comparison should show that the differences between the various figures of merit are by no means trivial.

Using choice 11, we propose to characterize varactor quality by the cutoff frequency

$$\omega_c = \frac{S_{max} - S_{min}}{R_s}, \tag{4.47}$$

or by the same quantity not using radian frequency,

$$f_c = \frac{S_{max} - S_{min}}{2\pi R_s}. \tag{4.48}$$

We further define the modulation ratios

$$m_k = \frac{|S_k|}{S_{max} - S_{min}} \tag{4.49}$$

so that the pumped figure of merit, Expression 4.43, is $m_1\omega_c$. The possible ranges of values for the m_k are discussed in Chapter 7.

We have shown that the cutoff frequency of Equation 4.47 is a valid characterization of varactor quality insofar as it indicates how high a pumped figure of merit the varactor can have. This same quantity is useful in describing the fundamental limits of large-signal devices, such as harmonic multipliers, harmonic dividers, and large-signal frequency converters and amplifiers. The formulas in Chapters 8, 9, and 10 all show that among abrupt-junction varactors those with the same ω_c are capable of the same quality of performance. Thus this characterization of varactor quality is one that applies to all of the varactor applications discussed in this book.

Most microwave measurements of varactor cutoff frequency are made by Houlding's method.[24, 16] For a varactor with a finite S_{min}, the frequency measured by this method is our proposed cutoff frequency, Equation 4.48, and not merely Expression 4.45.

For most varactors, $S_{min} \ll S_{max}$, so that our cutoff frequency reduces to Uhlir's. In the remainder of this book we assume that S_{min} can be nonzero, although in certain places (each carefully noted) the analysis is carried out, for simplicity, under the assumption that S_{min} is negligible.

4.7.2. Normalization Power

A quantity that appears in formulas for pump power (in Chapter 7) and power-handling capabilities of large-signal varactor devices (in Chapters 8, 9, and 10) is the normalization power P_{norm}:

$$P_{norm} = \frac{(V_B + \phi)^2}{R_s}. \tag{4.50}$$

The higher the normalization power (for a given cutoff frequency), the higher the power levels that the varactor is inherently capable of handling, but conversely, the higher the power necessary to pump the varactor. The specific importance of this quantity is discussed in detail in Chapters 7, 8, 9, and 10.

A related quantity, found by combining P_{norm} and ω_c so as to eliminate R_s, is the nominal reactive power P_r,

$$P_r = \frac{(V_B + \phi)^2}{2S_{max}} = \frac{1}{2}\frac{P_{norm}}{\omega_c}, \tag{4.51}$$

defined by Uhlir[59] for varactors with $S_{min} \ll S_{max}$. This quantity, which has the dimensions of power per unit frequency, is usually not as convenient to use as P_{norm} (one exception is calculation of power in low-frequency multipliers).

4.7.3. Other Varactor Characteristics

The most important varactor characteristic is the cutoff frequency, Equation 4.47, which indicates the quality of performance obtainable from the varactor. The second most important quantity seems to be the normalization power, Equation 4.50.

Other varactor properties are also of interest. Some indication of the impedance level, such as R_s or S_{max}, is necessary. Also, knowing the doping distribution (that is, "abrupt-junction," "graded-junction," and so on) is important.

In designing the pump circuit, a knowledge of V_B and ϕ is helpful, as is a knowledge of the dissipation rating for the varactor (as explained in Chapter 7, many varactors cannot be fully pumped at their optimum pump frequency, because of excessive dissipation).

Finally, some parasitic elements that might be specified are the lead inductance and the case capacitance. Other obvious properties are mechanical ones, such as size and weight.

4.7.4. Typical Varactors; Diode Space

It is of interest to know the properties of typical varactors. This information is, of course, apt to be out of date very quickly. It is given only as a rough guide.

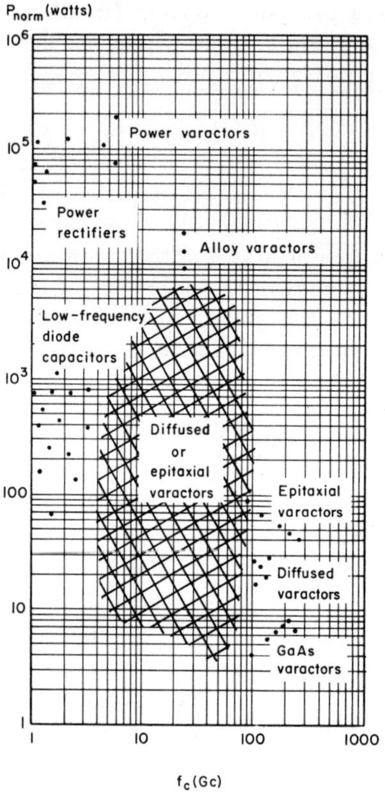

Figure 4.12. Diode space.

Since the most important varactor properties are the cutoff frequency and the normalization power, we restrict our attention to them. We represent varactors by placing them on a cutoff-frequency – normalization-power chart, Figure 4.12. Diodes constructed as high-quality varactors have cutoff frequencies upwards of 10 Gc, whereas other semiconductor diodes, which can be used as varactors, have lower cutoff frequencies. Some typical power rectifiers are indicated in Figure 4.12. The data given in this chart are not complete; only representative diodes have been plotted. For more details on any of these varactors, the reader should get in touch with the manufacturers directly. A list of varactor manufacturers is given in Appendix B.

A plot like Figure 4.12 is often referred to by the nickname, "diode space." In Appendix D, several nomographs for designing parametric amplifiers and harmonic generators include a plot like Figure 4.12, so that varactors can be easily chosen for various applications.

REFERENCES

1. Armstrong, H. L., "Some Reasons for Nonsaturation of Reverse Current in Junction Diodes," IRE Trans. on Electron Devices, ED-5, 66-68 (April, 1958).
2. Biard, J. R., "Low-Frequency Reactance Amplifier," Digest of Technical Papers, 1960 International Solid-State Circuits Conference, Philadelphia, Pa. (February 10-12, 1960), pp. 88-89.

3. Blackwell, L. A., and K. L. Kotzebue, Semiconductor-Diode Parametric Amplifiers, Prentice-Hall, Inc., Englewood Cliffs, N. J. (1961).
4. Boyd, C. R., Jr., "Noise Figure Measurements Relating the Static and Dynamic Cutoff Frequencies of Parametric Diodes," Proc. IRE, 48, 2019-2020 (December, 1960).
5. Boyd, C. R., Jr., "Design Considerations for Parametric Amplifier Low-Noise Performance," IRE Trans. on Military Electronics, MIL-5, 72-80 (April, 1961).
6. Chynoweth, A. G., and K. G. McKay, "Internal Field Emission in Silicon Junctions," Phys. Rev., 106, 418-426 (May 1, 1957).
7. Cohen, J., "Transition Region Properties of Reverse-Biased Diffused p-n Junctions," IRE Trans. on Electron Devices, ED-8, 362-369 (September, 1961).
8. Dunlap, W. C., Jr., An Introduction to Semiconductors, John Wiley & Sons, Inc., New York (1957).
9. Dunlap, W. C., Jr., ibid., Chapter 8.
10. Fletcher, N. H., "The High Current Limit for Semiconductor Junction Devices," Proc. IRE, 45, 862-872 (June, 1957).
11. Giacoletto, L. J., "Junction Capacitance and Related Characteristics Using Graded Impurity Semiconductors," IRE Trans. on Electron Devices, ED-4, 207-215 (July, 1957).
12. Giacoletto, L. J., "Fluctuation Noise in Semiconductor Space-Charge Regions," Proc. IRE, 49, 921-927 (May, 1961).
13. Graphical Symbols for Electrical and Electronic Diagrams, Part I, MIL-STD-15-1, Department of Defense, U. S. Government Printing Office, Washington, D. C. (October 30, 1961).
14. "Graphical Symbols for Electronic Diagrams," Electronics, 35, 33-40 (March 23, 1962).
15. Greene, J. C., and E. W. Sard, "Optimum Noise and Gain-Bandwidth Performance for a Practical One-Port Parametric Amplifier," Proc. IRE, 48, 1583-1590 (September, 1960).
16. Harrison, R. I., "Parametric Diode Q Measurement," Microwave J., 3, 43-46 (May, 1960).
17. Hedderly, D. L., "Parametric Amplification by Charge Storage," Proc. IRE, 49, 966-967 (May, 1961).
18. Heffner, H., "Capacitance Definitions for Parametric Operation," IRE Trans. on Microwave Theory and Techniques, MTT-9, 98-99 (January, 1961).
19. Heffner, H., and G. Wade, "Gain, Band Width, and Noise Characteristics of the Variable-Parameter Amplifier," J. Appl. Phys., 29, 1321-1331 (September, 1958).
20. Hefni, I., "Effect of Minority-Carriers on the Dynamic Characteristic of Parametric Diodes," Electronic Eng., 32, 226-227 (April, 1960).
21. Hefni, I., "The Negative Resistances in Junction Diodes," Proc. IRE, 49, 1427-1428 (September, 1961).
22. Hilibrand, J., and R. D. Gold, "Determination of the Impurity Distribution in Junction Diodes from Capacitance-Voltage Measurements," RCA Rev., 21, 245-252 (June, 1960).
23. Hilibrand, J., and C. F. Stocker, "The Design of Varactor Diodes," RCA Rev., 21, 457-474 (September, 1960).

24. Houlding, N., "Measurement of Varactor Quality," Microwave J., 3, 40-45 (January, 1960).
25. Hunter, L. P., Editor, Handbook of Semiconductor Electronics, McGraw-Hill Book Co., Inc., New York (1956).
26. Hunter, L. P., ibid., Section 3.
27. Knechtli, R. C., and R. D. Weglein, "Diode Capacitors for Parametric Amplification," J. Appl. Phys., 31, 1134-1135 (June, 1960).
28. Knechtli, R. C., and R. D. Weglein, "Low-Noise Parametric Amplifier," Proc. IRE, 48, 1218-1226 (July, 1960).
29. Ko, W. H., and F. E. Brammer, "Semiconductor Diode Amplifiers and Pulse Modulators," IRE Trans. on Electron Devices, ED-6, 341-347 (July, 1959).
30. Kroemer, H., "The Apparent Contact Potential of a Pseudo-Abrupt P-N Junction," RCA Rev., 17, 515-521 (December, 1956).
31. Kurokawa, K., and M. Uenohara, "Minimum Noise Figure of the Variable-Capacitance Amplifier," Bell System Tech. J., 40, 695-722 (May, 1961).
32. Lawrence, H., and R. M. Warner, Jr., "Diffused Junction Depletion Layer Calculations," Bell System Tech. J., 39, 389-403 (March, 1960).
33. McDade, J. C., "RF-Induced Negative Resistance in Junction Diodes," Proc. IRE, 49, 957-958 (May, 1961).
34. McKay, K. G., "Avalanche Breakdown in Silicon," Phys. Rev., 94, 877-884 (May 15, 1954).
35. Middlebrook, R. D., An Introduction to Junction Transistor Theory, John Wiley & Sons, Inc., New York (1957).
36. Misawa, T., "A Note on the Extended Theory of the Junction Transistor," J. Phys. Soc. Japan, 11, 728-739 (July, 1956).
37. Morgan, S. P., and F. M. Smits, "Potential Distribution and Capacitance of a Graded p-n Junction," Bell System Tech. J., 39, 1573-1602 (November, 1960).
38. Mortenson, K. E., "Comments on 'Diode Capacitors for Parametric Amplification' by R. C. Knechtli and R. D. Weglein," J. Appl. Phys., 31, 1135 (June, 1960).
39. Mortenson, K. E., "Parametric Diode Figure of Merit and Optimization," J. Appl. Phys., 31, 1207-1212 (July, 1960).
40. Muss, D. R., "Capacitance Measurements on Alloyed Indium-Germanium Junction Diodes," J. Appl. Phys., 26, 1514-1517 (December, 1955).
41. Penfield, P., Jr., "Noise in Varactor Parametric Amplifiers," First Quarterly Progress Report, Varactor Diodes and Parametric Amplifier for X-Band, Contract NObsr-77621, Microwave Associates, Inc., Burlington, Mass. (September 30, 1959).
42. Rudenberg, H. G., "Optimum Figures of Merit of Varactors," Proc. Natl. Elec. Conf., 15, Chicago, Ill. (1959), pp. 79-91.
43. Sah, C. T., R. N. Noyce, and W. Shockley, "Carrier Generation and Recombination in P-N Junctions and P-N Junction Characteristics," Proc. IRE, 45, 1228-1243 (September, 1957).
44. Sawyer, D. E., "Surface-Dependent Losses in Variable Reactance Diodes," J. Appl. Phys., 30, 1689-1691 (November, 1959).

45. Scarlett, R. M., "Space-Charge Layer Width in Diffused Junctions," IRE Trans. on Electron Devices, ED-6, 405-408 (October, 1959).
46. Shimizu, A., and J. Nishizawa, "Alloy-Diffused Variable Capacitance Diode with Large Figure-of-Merit," IRE Trans. on Electron Devices, ED-8, 370-377 (September, 1961).
47. Shockley, W., "The Theory of p-n Junctions in Semiconductors and p-n Junction Transistors," Bell System Tech. J., 28, 435-489 (July, 1949).
48. Shockley, W., ibid., Section 4.
49. Shockley, W., Electrons and Holes in Semiconductors, D. Van Nostrand Co., Inc., New York (1950).
50. Shockley, W., ibid., pp. 86-95.
51. Siegel, K., "Anomalous Reverse Current in Varactor Diodes," Proc. IRE, 48, 1159-1160 (June, 1960).
52. Siegel, K., "Comparative Figures of Merit for Available Varactor Diodes," Proc. IRE, 49, 809-810 (April, 1961).
53. Spenke, E., Electronic Semiconductors, McGraw-Hill Book Co., Inc., New York (1958).
54. Spenke, E., ibid., Section 4.10.
55. Spenke, E., ibid., Chapter IV, Part 2.
56. Uhlir, A., Jr., "High-Frequency Shot Noise in P-N Junctions," Proc. IRE, 44, 557-558 (April, 1956); and correction, ibid., 1541 (November, 1956).
57. Uhlir, A., Jr., "The Potential of Semiconductor Diodes in High-Frequency Communications," Proc. IRE, 46, 1099-1115 (June, 1958).
58. Uhlir, A., Jr., "Shot Noise in p-n Junction Frequency Converters," Bell System Tech. J., 37, 951-988 (July, 1958).
59. Uhlir, A., Jr., "Similarity Considerations for Varactor Multipliers," Microwave J., 5 (forthcoming, July, 1962).
60. Valdes, L. B., The Physical Theory of Transistors, McGraw-Hill Book Co., Inc., New York (1961).
61. van der Ziel, A., Noise, Prentice-Hall, Inc., New York (1954).
62. Weglein, R. D., "Some Limitations on Parametric Amplifier Noise Performance," IRE Trans. on Microwave Theory and Techniques, MTT-8, 538-544 (September, 1960).

Chapter 5

FREQUENCY CONVERTERS

Although this entire book is about frequency conversion, we reserve the name frequency converter for a device that is pumped at some frequency ω_p (and its harmonics), and in which a signal enters at some other frequency ω_s and leaves at a third frequency $n\omega_p \pm \omega_s$ (or at two or more such frequencies). Examples of frequency converters are modulators, mixers, upconverters, and downconverters. The varactor is useful as a frequency converter because of its good noise properties, and because gain can be achieved, under certain conditions to be described.

Frequency converters are classified (1) according to whether the output frequency is greater than or less than the input frequency (the converters are known, respectively, as upconverters and downconverters); (2) according to the integer n in the formula $n\omega_p + \omega_s$ for the output frequency (if $n = 1$, the device is an upper-sideband device; if $n = -1$, it is a lower-sideband device); and (3) according to whether or not power is dissipated at any other sidebands (idlers).

In Sections 5.2 to 5.5 we discuss, in order, the upper-sideband upconverter, the lower-sideband upconverter, the upper-sideband downconverter, and the lower-sideband downconverter, both without idlers and with a single idler. In each case, we assume that the varactor is open-circuited at all extra sidebands, to prevent power from being dissipated at these frequencies. Note that any other assumption allows sideband power to be dissipated, at least in the varactor series resistance. Then in Section 5.6, we allow current at all sidebands.

As might be expected, the formulas that describe these frequency converters are similar, although the optimization procedures are not; these similarities are discussed in Section 5.7.

The relations between the various frequency converters are conveniently pictured by using the diagrams of Figure 1.2. In Figure 5.1 we show several frequency converters, and indicate, in the caption, the sections of the book in which they are analyzed. The letters s, p, u, and i stand for signal, pump, upper sideband, and lower sideband, respectively.

In Section 3.1 we derived many results about frequency converters that use ideal nonlinear reactances. These would hold for varactor frequency converters if there were no series resistance. The results for varactors with a finite series

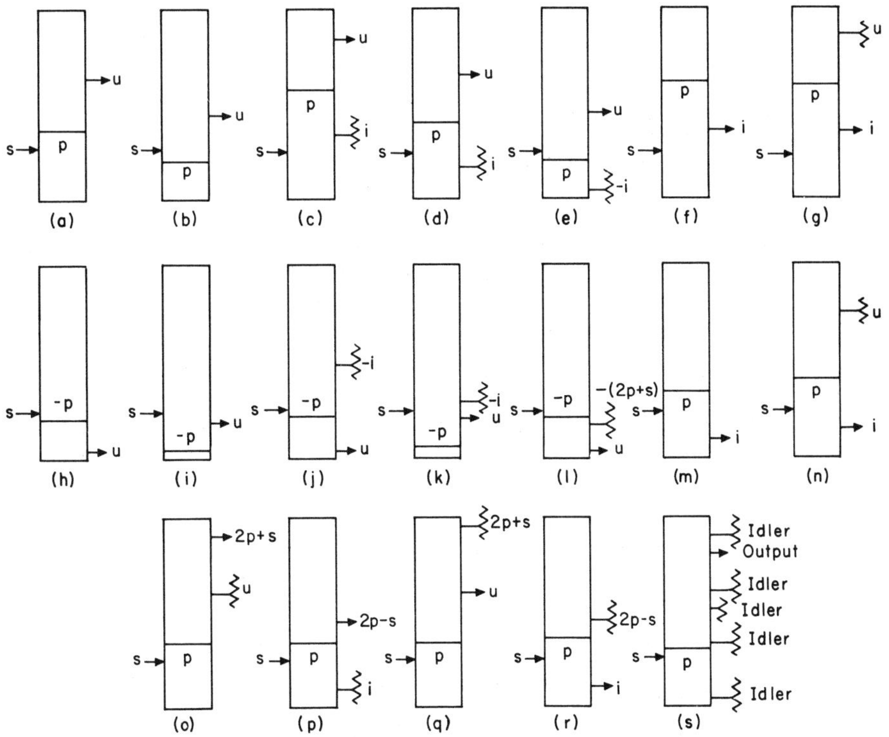

Figure 5.1. Typical frequency converters. The input and output are indicated by arrows; the resistor symbols indicate idlers.

(a) and (b). Upper-sideband upconverters, Section 5.2.

(c), (d), and (e). Upper-sideband upconverters with idlers, Section 5.2.4.

(f). Lower-sideband upconverter, Section 5.3.

(g). Lower-sideband upconverter with idler, Section 5.3.4.

(h) and (i). Upper-sideband downconverters, Section 5.4.

(j), (k), and (l). Upper-sideband downconverters with idlers, Section 5.4.4.

(m). Lower-sideband downconverter, Section 5.5.

(n). Lower-sideband downconverter with idler.

(o) and (p). Upconverters with idlers.

(q). Upper-sideband upconverter with idler.

(r). Lower-sideband downconverter with idler.

(s). Frequency converter with arbitrary idlers, Section 5.6.

Devices (n), (o), (p), (q), and (r) are not discussed in Chapter 5.

resistance R_s are rather different; they lead to optimization and design procedures unlike what one would expect from reading Chapter 3.

5.1. Introduction and Major Results

In the derivations to follow, we make a number of simplifying assumptions:

1. The varactor is described at all frequencies of interest by the model of Figure 4.1, with a constant series resistance.
2. The varactor is pumped at some frequency ω_p, so that its incremental elastance is given by Equation 4.41 (in Sections 5.2.4, 5.3.4, and 5.4.4, we further assume for simplicity that $S_2 = 0$).
3. The voltages and currents at all frequencies except the pump frequency and its harmonics are small, so that the linearized equations, Equations 4.42, hold.
4. The only noise comes from thermal noise in the varactor series resistance and the terminations.
5. Only one varactor is used in the frequency conversion process.
6. We neglect losses in the coupling networks.
7. We assume the varactor is open-circuited at all but certain frequencies (except in Section 5.6).
8. For most of the formulas, we assume the varactor terminations tune out the average elastance S_0.

Assumptions 1, 2, and 3 serve to define our field of interest: small-signal behavior of parametrically pumped varactors. Assumption 4 is discussed in Section 4.3, and is believed to be very reasonable. Assumptions 5, 6, 7, and 8 serve only to simplify the analysis. Similar analyses could be carried through (at least in principle) under more general conditions. Assumption 5 can be relaxed (for example, for traveling-wave devices) by using a matrix approach; see, for example, Section 6.3. Our analysis can be extended to account for losses in the passive coupling networks (that is, those networks between the source and the varactor, and between the varactor and the load) by straightforward methods of circuit theory; such losses serve only to increase the noise and decrease the gain (although they can be helpful in stabilizing lower-sideband devices). Assumption 7 is made strictly for simplicity. Nevertheless, in practical frequency converters the sideband currents are usually small, so that the theory based on open-circuit sidebands should still be pertinent. Furthermore, there is a limit to the improvement in noise temperature obtainable with arbitrary sideband terminations (see Section 5.6); this limit is of the same order of magnitude as the limit for open-circuit sidebands.

Finally, assumption 8, that the terminations tune out the average elastance, is justified insofar as it is usually profitable to do this; see the remarks at the beginning of Section 5.2.2.

The treatments of the various frequency converters are very similar. In each case, we compute the exchangeable gain G_e, the noise temperature T, and the input and output impedances Z_{in} and Z_{out}. (The noise behavior is also expressed in terms of noise equivalent circuits.) These expressions are simplified by assuming the input and output (and idler) terminations are tuned. The source resistance then is adjusted to give, first, optimum gain, and second, optimum noise temperature.

In Section 5.2 we discuss the upper-sideband upconverter. The output is at frequency $\omega_u = \omega_p + \omega_s$, with $\omega_p > 0$. A straightforward derivation leads to Equations 5.33, 5.34, 5.35, and 5.37 for the pertinent quantities. The noise temperature T is independent of the pump frequency, and it has a minimum when R_o is set equal to

$$R_o^T = R_s \sqrt{1 + \left(\frac{m_1 \omega_c}{\omega_s}\right)^2}\,; \tag{5.1}$$

and the resulting minimum noise temperature is

$$T^T = T_d \frac{2\omega_s}{m_1 \omega_c} \left[\frac{\omega_s}{m_1 \omega_c} + \sqrt{1 + \left(\frac{\omega_s}{m_1 \omega_c}\right)^2} \right]. \tag{5.2}$$

In these formulas m_1, defined in Section 4.7.1,

$$m_1 = \frac{|S_1|}{S_{max} - S_{min}}, \tag{5.3}$$

and called the <u>modulation ratio</u>, is discussed extensively in Chapter 7; T_d is the temperature of the varactor, and ω_c is the varactor <u>cutoff frequency</u> discussed in Section 4.7.1,

$$\omega_c = \frac{S_{max} - S_{min}}{R_s}, \tag{5.4}$$

where S_{max} and S_{min} are the varactor maximum and minimum elastances. The superscript T indicates evaluation for the value of R_o that optimizes the noise temperature.

Equation 5.2 is rather general; it is not restricted to the upper-sideband upconverter. It is a fundamental limit on the noise performance of many other varactor amplifying devices, including other upconverters and downconverters, and parametric amplifiers. Note that in Equation 5.2, the only varactor parameter (aside from the temperature T_d) is the quantity $m_1\omega_c = |S_1|/R_s$. Thus two varactors with quite different characteristics have the same limit on noise performance if they have the same value of $m_1\omega_c$. The importance of this quantity leads us to consider it as a figure of merit for the varactor, as pumped (see Sections 4.6 and 4.7); $m_1\omega_c$ appears throughout the analyses of Chapters 5 and 6.

It is shown in Section 5.2 that there is gain for optimum noise temperature (that is, G_e^T is greater than 1) only for signal frequencies smaller than $0.455m_1\omega_c$, and even then only if the pump frequency is high enough.

A different value of R_o serves to make the exchangeable gain (which is always positive) the highest:

$$R_o^G = R_s \sqrt{1 + \frac{m_1^2\omega_c^2}{\omega_s\omega_u}}, \tag{5.5}$$

where the superscript G indicates evaluation for optimum gain. The resulting gain is

$$G_e^G = \left(\frac{\dfrac{m_1\omega_c}{\omega_s}}{1 + \sqrt{1 + \dfrac{m_1^2\omega_c^2}{\omega_s\omega_u}}} \right)^2, \tag{5.6}$$

which evidently becomes higher as ω_u increases. In the limit of very high pump frequency, it is

$$\left(G_e^G\right)_{\omega_u=\infty} = \left(\frac{m_1\omega_c}{2\omega_s}\right)^2, \tag{5.7}$$

and therefore, there is no gain unless ω_s is less than $0.5m_1\omega_c$. When R_o is chosen for maximum G_e, the output resistance becomes equal to R_o^G, and if the output load is matched, the input resistance is also equal to R_o^G. This condition is therefore one of conjugate image matching.

If we want to use the upconverter as a low-noise amplifier, followed by a second stage with finite noise, we might ask about the value of R_o that minimizes the cascade noise figure. If the second-stage noise figure is very small, we should choose R_o so that the first-stage noise temperature is small; on the other hand, if the second-stage noise is large, we should choose R_o to obtain very high gain. Clearly there is a suitable compromise value of R_o, lying somewhere between R_o^T and R_o^G, to minimize the over-all noise figure. Such compromises are discussed in Section 5.2.2(c).

The analysis of Section 5.2 indicates what circuit conditions must be met to achieve the performance indicated. Synthesis of actual circuits is discussed briefly in Section 5.2.3.

The noise and gain formulas reported here were derived earlier by Leenov,[11] ‡ who gave them in somewhat different form. Later, independent work includes that of Kotzebue,[8] Blackwell,[2] and Kurokawa and Uenohara,[10] each of whom also showed that the fundamental limit for upconverters is the same as that for parametric amplifiers.

The effect of an idler on the upper-sideband upconverter is discussed in Section 5.2.4. We analyze the case with the idler at frequency $\omega_i = \omega_p - \omega_s$ and the elastance pumped sinusoidally. This idler can improve the upconverter in two ways. First, the gain can be increased, and in certain frequency ranges, it can become as high as desired. And second, the noise temperature of the upconverter can be reduced. In fact, both beneficial effects can occur simultaneously. The minimum noise temperature has the same form as Equation 5.2, except that m_1 is replaced by $\sqrt{2}\, m_1$. This minimum noise temperature can be achieved with a lossless idler by suitable choice of the pump frequency and the source impedance.

The lower-sideband upconverter is discussed in Section 5.3. The output is at frequency $\omega_i = \omega_p - \omega_s$, with $\omega_p > 2\omega_s$. A derivation quite similar to that of Section 5.2 leads to Equations 5.103 to 5.106 for the pertinent quantities. The noise temperature is the same as that of the upper-sideband upconverter, so that the same optimum value of R_o, given by Equation 5.1, leads to the same minimum of Equation 5.2.

The gain conditions are not as clear, because the exchangeable gain can be negative. When it is, the output impedance has a negative real part, and the device is potentially unstable. This potential instability can occur when

$$\omega_s \omega_i < m_1^2 \omega_c^2; \qquad (5.8)$$

---

‡Superscript numerals denote references listed at the end of each chapter.

in fact, when the source resistance R_o is adjusted to

$$R_o = R_s\left(\frac{m_1^2\omega_c^2}{\omega_s\omega_i} - 1\right), \tag{5.9}$$

the exchangeable gain is infinite. By proper choice of R_o and the pump frequency, we can simultaneously minimize the noise temperature and make the exchangeable gain infinite. This "optimum pump frequency" is

$$\omega_p = \sqrt{m_1^2\omega_c^2 + \omega_s^2}, \tag{5.10}$$

or approximately $m_1\omega_c$ if the signal frequency is small. We shall find that this pump frequency is often desirable (although for different reasons) in other applications of varactors.

If the input and output frequencies are so high that Condition 5.8 is not satisfied, gain is still possible for optimized noise whenever ω_s is smaller than $0.455m_1\omega_c$, and in some cases for higher ω_s. On the other hand, if the source resistance is adjusted to optimize the gain, somewhat larger values of ω_s, up to $0.5m_1\omega_c$, can yield gain.

Synthesis of lower-sideband upconverters is discussed very briefly in Section 5.3.3, and in Section 5.3.4 the effect of an idler at the upper sideband is shown to have little benefit.

The upper-sideband downconverter is discussed in Section 5.4. The final formulas are exactly the same as those for the upper-sideband upconverter. The value of R_o given in Equation 5.1 still minimizes the noise temperature to the value given in Equation 5.2. There is, however, no gain, for any value of R_o. This does not imply, of course, that the device is useless; it is useful, in spite of its loss, for conversion of a signal to a new frequency where in some way it can be operated upon more effectively. A properly chosen idler can lower the loss, but only at the expense of higher noise.

The lower-sideband downconverter is discussed in Section 5.5. The formulas are identical to those for the lower-sideband upconverter, except that the pump frequency is no longer restricted to be above $2\omega_s$, but instead $\omega_s < \omega_p < 2\omega_s$. The noise formulas are the same, and the noise temperature is optimized to the value of Equation 5.2 by the source resistance of Equation 5.1. Only negative-resistance type of gain is possible, occurring only if Condition 5.8 holds. The value of source resistance in Equation 5.9 makes the exchangeable gain infinite; if the input frequency ω_s is greater than $0.578m_1\omega_c$, then the pump frequency of

Equation 5.10 makes the two values of R_o, in Equations 5.1 and 5.9, the same, so that simultaneously the noise figure is optimized and the exchangeable gain is made infinite.

We have seen that the noise performance of the upper-sideband upconverter can be improved by allowing some lower-sideband idling. Suppose more idlers were allowed: Could the noise performance keep improving? The answer is that it cannot improve beyond a new fundamental limit, which we derive in Section 5.6. This new fundamental limit holds for arbitrary idler terminations (even including noiseless negative resistors) and arbitrary elastance waveforms. It is of the same form as Equation 5.2, with m_1 replaced by m_t, where

$$m_t^2 = 2\frac{|S_1|^2 + |S_2|^2 + \cdots}{(S_{max} - S_{min})^2}. \tag{5.11}$$

There is no guarantee that this limit is always achievable, although it sometimes is. Furthermore, this new limit is not much below the old limits, since m_t can never exceed twice an easily obtained value of m_1, and for practical varactors is seldom over $1.5m_1$.

It is clear there is much similarity among the various simple frequency converters. The fundamental limit on noise temperature is identical for each case, and in fact is identical to the fundamental limit for parametric amplifiers, derived in Section 6.2. The optimizations are similar. In Section 5.7, the various formulas and optimizations are compared by listing the pertinent formulas in a table. Anticipating the results of Chapter 6, we list the parametric amplifier formulas, as well as the results from Section 5.6 for frequency converters with arbitrary idlers.

5.2. Upper-Sideband Upconverter

One of the simplest frequency converters is the upper-sideband upconverter. The varactor is pumped at frequency ω_p, and a signal is introduced at frequency ω_s (which may be either higher or lower than ω_p). Of all the sidebands generated, the upper sideband $\omega_u = \omega_p + \omega_s$ is selected for the output.

In Sections 5.2.1, 5.2.2, and 5.2.3, we assume for simplicity that the varactor is open-circuited at all other sidebands, especially the two frequencies $\omega_s - \omega_p$ and $\omega_s + 2\omega_p$. In Section 5.2.4, however, we discuss the advantages that arise from allowing current to flow at the idler frequency $\omega_i = \omega_p - \omega_s$.

Specifically, in Section 5.2.1, we derive formulas for the exchangeable gain, the noise temperature, and the input and output impedances of the upconverter without idlers. Then, in Section 5.2.2, we assume that the source and load impedances are tuned, so that the reactive part of the impedances tunes out the varactor average elastance S_0. Then we adjust the source impedance to optimize the noise or the gain, or to effect some compromise between these two optimizations. In Section 5.2.3, we discuss synthesis of this simple upconverter.

5.2.1. Characteristics

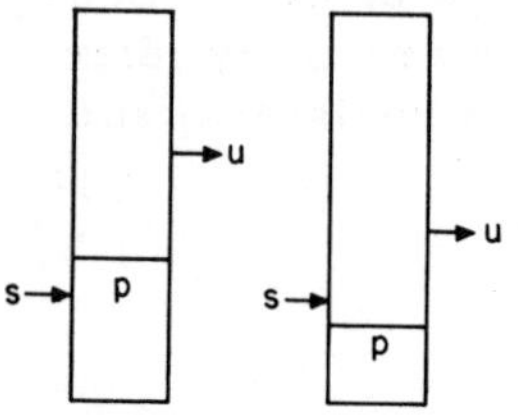

We assume that the varactor is open-circuited at all sidebands except ω_u, so that the only small-signal current is at frequencies ω_s and ω_u. We further assume that the only important noise is thermal noise generated in the varactor series resistance R_s at temperature T_d. We neglect losses in the coupling circuits, since these only reduce the gain and increase the noise temperature. The device we analyze, shown in Figure 5.2, has separate ports for the two different frequencies, in accordance with the ideas of Section 2.2. The small-signal equations are obtained by eliminating from Equations 4.42 all currents except those at frequencies ω_s and ω_u:

$$\begin{bmatrix} V_s \\ \\ V_u \end{bmatrix} = \begin{bmatrix} R_s + \dfrac{S_0}{j\omega_s} & \dfrac{S_1{}^*}{j\omega_u} \\ \\ \dfrac{S_1}{j\omega_s} & R_s + \dfrac{S_0}{j\omega_u} \end{bmatrix} \begin{bmatrix} I_s \\ \\ I_u \end{bmatrix} + \begin{bmatrix} E_{ns} \\ \\ E_{nu} \end{bmatrix}, \quad (5.12)$$

where (as throughout this book) V_s, V_u, I_s, and I_u are half-amplitudes of the components of varactor voltage and current at the appropriate frequencies, and where S_0 and S_1 are the coefficients of varactor incremental elastance, as pumped, according to Equation 4.41, and where E_{ns} and E_{nu} are equiv-

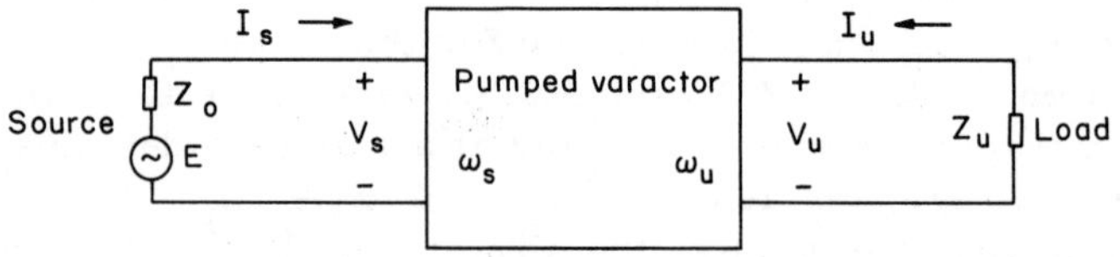

Figure 5.2. Upper-sideband upconverter, without idlers. Although Z_o and Z_u are shown separately, some physical elements may appear in both (see Section 2.2).

alent noise voltages, as discussed in Section 4.3, with mean-squared values‡

$$\overline{|E_{ns}|^2} = \overline{|E_{nu}|^2} = 4kT_dR_s\ \Delta f\ , \tag{5.13}$$

where k is Boltzmann's constant.

We now solve for the upconverter input and output resistance, the exchangeable gain, and the noise temperature. The input constraint is, from Figure 5.2,

$$V_s = -I_sZ_o + E, \tag{5.14}$$

where Z_o is the source impedance. When this constraint is used to eliminate V_s from Equations 5.12, the two remaining equations can be combined so as to eliminate I_s:

$$V_u = \left[R_s + \frac{S_0}{j\omega_u} + \frac{|S_1|^2}{\omega_s\omega_u}\frac{1}{Z_o + R_s + (S_0/j\omega_s)}\right] I_u$$

$$+ E_{nu} + \frac{S_1}{j\omega_s}\frac{E - E_{ns}}{Z_o + R_s + (S_0/j\omega_s)}\ . \tag{5.15}$$

5.2.1(a). Output Impedance. Equation 5.15 implies an output equivalent circuit of a voltage source, represented by the last two terms, in series with an impedance represented by the first term. Thus the output impedance is

$$Z_{out} = R_{out} + jX_{out}$$

$$= R_s + \frac{S_0}{j\omega_u} + \frac{|S_1|^2}{\omega_s\omega_u}\frac{1}{Z_o + R_s + (S_0/j\omega_s)}$$

$$= R_s + \frac{S_0}{j\omega_u} + \frac{m_1^{\,2}\omega_c^{\,2}}{\omega_s\omega_u}\frac{R_s^{\,2}}{Z_o + R_s + (S_0/j\omega_s)}\ , \tag{5.16}$$

where R_{out} and X_{out} are the real and imaginary parts of Z_{out}. We have used the definitions of m_1 and ω_c given by Equations 5.3 and 5.4.

‡The noise voltages are effective rms voltages instead of half-amplitudes. See Section 2.5.1(a).

5.2.1(b). Exchangeable Gain. In calculating the exchangeable gain, we neglect the noise sources. The exchangeable power output is half the square of the magnitude of the output open-circuit voltage, divided by R_{out}; in view of Equation 5.15 this is

$$\frac{1}{2R_{out}} \frac{|S_1|^2}{\omega_s^2} \frac{|E|^2}{|Z_o + R_s + (S_0/j\omega_s)|^2} . \tag{5.17}$$

On the other hand, the input exchangeable power is $|E|^2/2R_o$, where R_o is the real part of Z_o. The ratio of these two exchangeable powers is the exchangeable gain; use of Equation 5.16 for R_{out} leads to the formula for G_e:

$$\frac{1}{G_e} = \left(\frac{\omega_s}{m_1 \omega_c}\right)^2 \frac{|Z_o + R_s + (S_0/j\omega_s)|^2}{R_s R_o} + \frac{\omega_s}{\omega_u} \frac{R_s + R_o}{R_o} . \tag{5.18}$$

If the series resistance vanishes, then G_e approaches ω_u/ω_s, in agreement with Equation 3.16 for the lossless case. Because of the series resistance, G_e is lower than ω_u/ω_s.

5.2.1(c). Noise Temperature. The noise temperature T is (see Section 2.5.4)

$$T = T_0(F - 1), \tag{5.19}$$

where $F - 1$, the exchangeable excess noise figure, is the exchangeable noise output caused by internal sources (that is, caused by E_{ns} and E_{nu}), divided by the exchangeable noise output caused by thermal noise in the source at temperature T_0. Thus $F - 1$ is the ratio of the squares of the open-circuit voltages of each origin, and is therefore

$$F - 1 = \frac{\overline{\left| E_{nu} - \frac{S_1}{j\omega_s} \frac{E_{ns}}{Z_o + R_s + (S_0/j\omega_s)} \right|^2}}{\left| \frac{S_1}{j\omega_s} \right|^2 \frac{4kT_0 R_o \Delta f}{|Z_o + R_s + (S_0/j\omega_s)|^2}} , \tag{5.20}$$

where the bar indicates an average. Thus the noise temperature is (by use of Equation 5.13)

$$T = T_d \left[\frac{R_s}{R_o} + \left(\frac{\omega_s}{m_1 \omega_c}\right)^2 \frac{|Z_o + R_s + (S_0/j\omega_s)|^2}{R_s R_o} \right] . \tag{5.21}$$

A noise equivalent circuit that is valid for any noisy twoport is the Rothe-Dahlke model[13] discussed in Section 2.5.1(d). The excess noise figure for this model is, from Equation 2.26,

$$F - 1 = \frac{R_n}{R_o} + \frac{|Z_o|^2 G_n}{R_o} + \frac{\rho Z_o + \rho^* Z_o^*}{R_o} \sqrt{R_n G_n} . \tag{5.22}$$

To fit this model to the upconverter, we compare Equations 5.21 and 5.22 to find

$$R_n = \frac{T_d}{T_0} \left[1 + \left(\frac{\omega_s}{m_1 \omega_c} \right)^2 + \left(\frac{m_0}{m_1} \right)^2 \right] R_s , \tag{5.23}$$

$$G_n = \frac{T_d}{T_0} \frac{1}{R_s} \left(\frac{\omega_s}{m_1 \omega_c} \right)^2 , \tag{5.24}$$

and

$$\rho = \frac{\dfrac{\omega_s}{m_1 \omega_c} + j \dfrac{m_0}{m_1}}{\sqrt{1 + \left(\dfrac{\omega_s}{m_1 \omega_c} \right)^2 + \left(\dfrac{m_0}{m_1} \right)^2}} , \tag{5.25}$$

where m_0 is

$$m_0 = \frac{S_0}{S_{max} - S_{min}} . \tag{5.26}$$

On the other hand, if the Bauer-Rothe wave model[1,12] discussed in Section 2.5.1(e) is used to represent the upconverter, the temperatures T_a and T_b and the impedance Z_ν are

$$T_a = T_d \frac{2\omega_s}{m_1 \omega_c} \left[\frac{\omega_s}{m_1 \omega_c} + \sqrt{1 + \left(\frac{\omega_s}{m_1 \omega_c} \right)^2} \right] , \tag{5.27}$$

$$T_b = T_d \frac{2\omega_s}{\omega_s + \sqrt{m_1^2 \omega_c^2 + \omega_s^2}} , \tag{5.28}$$

and

$$Z_\nu = R_s\sqrt{1+\left(\frac{m_1\omega_c}{\omega_s}\right)^2} + j\frac{S_0}{\omega_s}. \tag{5.29}$$

5.2.1(d). Input Impedance. The same method used to derive the output impedance Z_{out} can be used to derive the input impedance Z_{in}. We introduce the load constraint

$$V_u = -Z_uI_u \tag{5.30}$$

into Equations 5.12 and eliminate both V_u and I_u. The input impedance, with real and imaginary parts R_{in} and X_{in}, is

$$Z_{in} = R_{in} + jX_{in}$$

$$= R_s + \frac{S_0}{j\omega_s} + \frac{m_1{}^2\omega_c{}^2}{\omega_s\omega_u}\,\frac{R_s{}^2}{Z_u + R_s + (S_0/j\omega_s)}. \tag{5.31}$$

5.2.2. Optimizations

We have derived formulas for the input and output impedances, the noise temperature, and the exchangeable gain of the upper-sideband upconverter. We note from Equations 5.18 and 5.21 that if the input is tuned, so that

$$Z_o + R_s + \frac{S_0}{j\omega_s} \tag{5.32}$$

is real, then simultaneously the noise temperature is reduced and the exchangeable gain is raised. This tuning is therefore beneficial to the upconverter. We henceforth assume that the input is tuned; the output impedance, exchangeable gain, and the noise temperature become

$$Z_{out} = R_{out} + jX_{out}$$

$$= R_s + \frac{S_0}{j\omega_u} + \frac{m_1{}^2\omega_c{}^2}{\omega_s\omega_u}\,\frac{R_s{}^2}{R_s + R_o} \tag{5.33}$$

$$\frac{1}{G_e} = \frac{\omega_s}{\omega_u}\,\frac{R_s + R_o}{R_o} + \left(\frac{\omega_s}{m_1\omega_c}\right)^2\frac{(R_s + R_o)^2}{R_sR_o} \tag{5.34}$$

$$T = T_d \left[\frac{R_s}{R_o} + \left(\frac{\omega_s}{m_1 \omega_c} \right)^2 \frac{(R_s + R_o)^2}{R_s R_o} \right] . \tag{5.35}$$

Note from Equation 5.33 that the reactive part of the output impedance is caused by the average elastance S_0; it is natural, therefore, to tune the output, so that

$$Z_u + R_s + \frac{S_0}{j\omega_u} \tag{5.36}$$

is real; if this is done, the input impedance becomes

$$Z_{in} = R_{in} + jX_{in}$$

$$= R_s + \frac{S_0}{j\omega_s} + \frac{m_1^{\,2} \omega_c^{\,2}}{\omega_s \omega_u} \frac{R_s^{\,2}}{R_s + R_u} . \tag{5.37}$$

The choice of R_o affects the gain and noise temperature. We now discuss optimization of these quantities by choosing R_o appropriately.

5.2.2(a). Gain Optimization. From Equation 5.34, it is clear that if R_o is either very small or very large, the exchangeable gain (which is always positive) is small. As R_o is varied, G_e passes through a maximum, which can be found by differentiating G_e with respect to R_o and setting the result equal to zero. We find that G_e has a maximum equal to

$$G_e^{\,G} = \frac{\left(\dfrac{m_1 \omega_c}{\omega_s} \right)^2}{\left(1 + \sqrt{1 + \dfrac{m_1^{\,2} \omega_c^{\,2}}{\omega_s \omega_u}} \right)^2} , \tag{5.38}$$

which occurs when R_o is set equal to

$$R_o^{\,G} = R_s \sqrt{1 + \frac{m_1^{\,2} \omega_c^{\,2}}{\omega_s \omega_u}} . \tag{5.39}$$

The superscript G indicates evaluation for that value of R_o which optimizes the gain. Under this optimization the output resistance is identical to the source resistance:

$$R_{out}^{G} = R_s \sqrt{1 + \frac{m_1^2 \omega_c^2}{\omega_s \omega_u}}, \tag{5.40}$$

and if the load resistance R_u is chosen to match R_{out}^{G}, then the input resistance is likewise R_o^G, and the upconverter is matched at both input and output. The noise temperature for optimum gain is

$$T^G = T_d \frac{1 + \left(\frac{\omega_s}{m_1 \omega_c}\right)^2 \left(1 + \sqrt{1 + \frac{m_1^2 \omega_c^2}{\omega_s \omega_u}}\right)^2}{\sqrt{1 + \frac{m_1^2 \omega_c^2}{\omega_s \omega_u}}}. \tag{5.41}$$

To obtain some familiarity with these formulas, the reader is referred to Figures 5.3 to 5.6. In Figure 5.3, we plot G_e^G, which is a function of both ω_s and ω_u. Note that the axes of Figure 5.3 (and of many graphs to follow) are calibrated in terms of $m_1 \omega_c$, a fact that indicates the fundamental importance of this figure of merit, which is discussed in Section 4.6. Figure 5.3 shows clearly that only at certain values of ω_s and ω_u can the upconverter operate with gain; the region of possible gain is indicated in Figure 5.4. From Equation 5.38, the gain is greater than 1 only when

$$\omega_u > \frac{m_1 \omega_c}{\frac{m_1 \omega_c}{\omega_s} - 2}, \tag{5.42}$$

and for $\omega_s > 0.5 m_1 \omega_c$, there is no gain. It should be stressed that even though an upconverter does not have gain, it, unlike an ordinary amplifier with no gain, may still be useful.

It is observed from Figure 5.3 (or Equation 5.38) that as ω_u is increased, the gain increases. There is little point, however, in increasing ω_u too far, because the gain G_e^G is within a factor of 2 (3 db) of its value with ω_u very high, whenever ω_u is greater than $0.427 m_1^2 \omega_c^2 / \omega_s$. This is evident from Figure 5.3, where we show this point of diminishing returns.

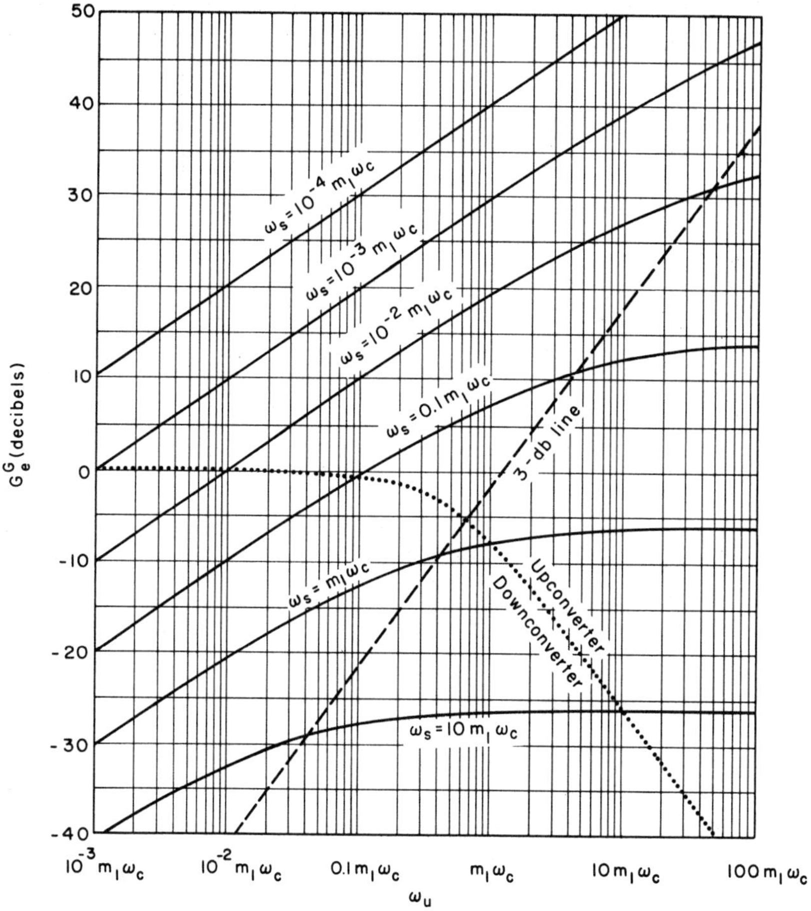

Figure 5.3. Optimum gain G_e^G for an upper-sideband upconverter and downconverter, as a function of ω_s and ω_u, taken from Equation 5.38. The dotted line separates upconverters from downconverters, and the dashed line indicates the output frequency for which the gain is half the gain at infinite pump frequency.

The source resistance for optimum gain R_o^G is plotted in Figure 5.5 (this is also a plot of R_{out}^G). Note that at high frequencies the source and output resistances approach R_s, as might be expected, because the effect of the capacitive part of the varactor disappears.

The noise temperature for optimum gain is plotted in Figure 5.6, under the assumption that the varactor is at room temperature, 290°K. Note that for the upconverter the noise increases as the output frequency increases.

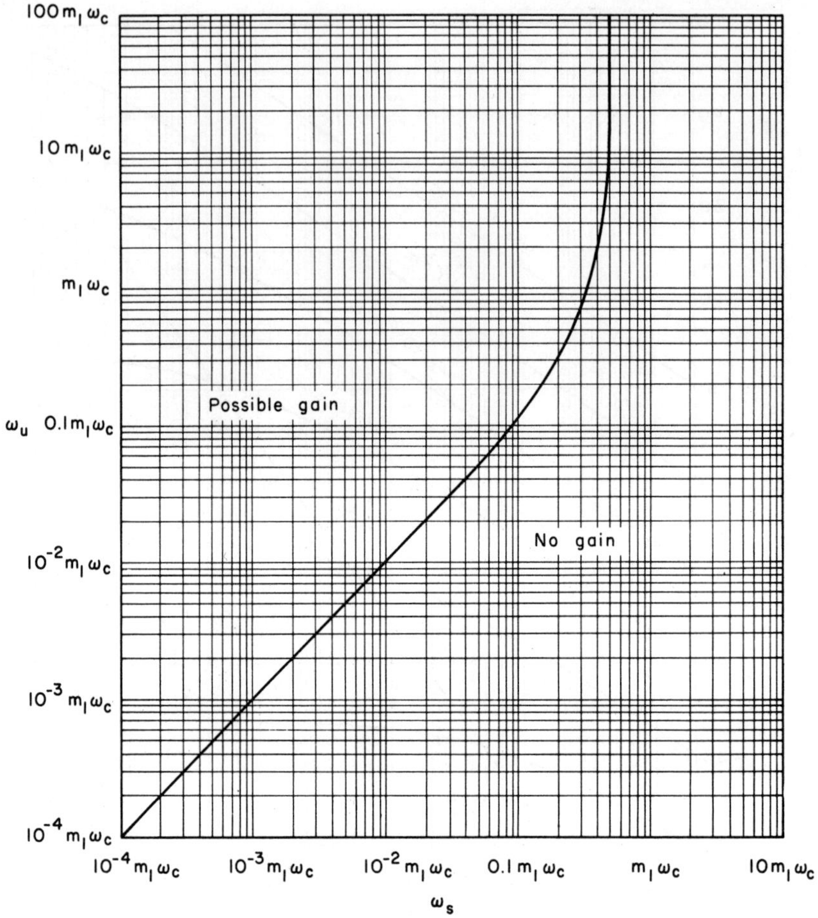

Figure 5.4. Region of possible gain for the upper-sideband upconverter. The portion of the diagram with $\omega_u < \omega_s$ applies to the upper-sideband downconverter. Taken from Equation 5.42.

It is interesting to look at the asymptotic expressions for very large ω_u, especially since the gain is made the greatest for high pump frequencies. Thus, with the subscript $\omega_u = \infty$,

$$(R_o{}^G)_{\omega_u=\infty} = (R_{out}{}^G)_{\omega_u=\infty} = R_s , \tag{5.43}$$

$$(G_e{}^G)_{\omega_u=\infty} = \frac{1}{4}\left(\frac{m_1\omega_c}{\omega_s}\right)^2 , \tag{5.44}$$

and

$$(T^G)_{\omega_u=\infty} = T_d \left[1 + 4\left(\frac{\omega_s}{m_1 \omega_c}\right)^2\right]. \tag{5.45}$$

The resulting noise temperature is seen to be at least T_d, and there is no gain for input frequencies above $0.5m_1 \omega_c$.

On the other hand, if the pump frequency is chosen so that the exchangeable gain is unity (that is, along the curve shown in Figure 5.4), then

$$(R_o^G)_{G=1} = (R_{out}^G)_{G=1} = R_s \left(\frac{m_1 \omega_c}{\omega_s} - 1\right), \tag{5.46}$$

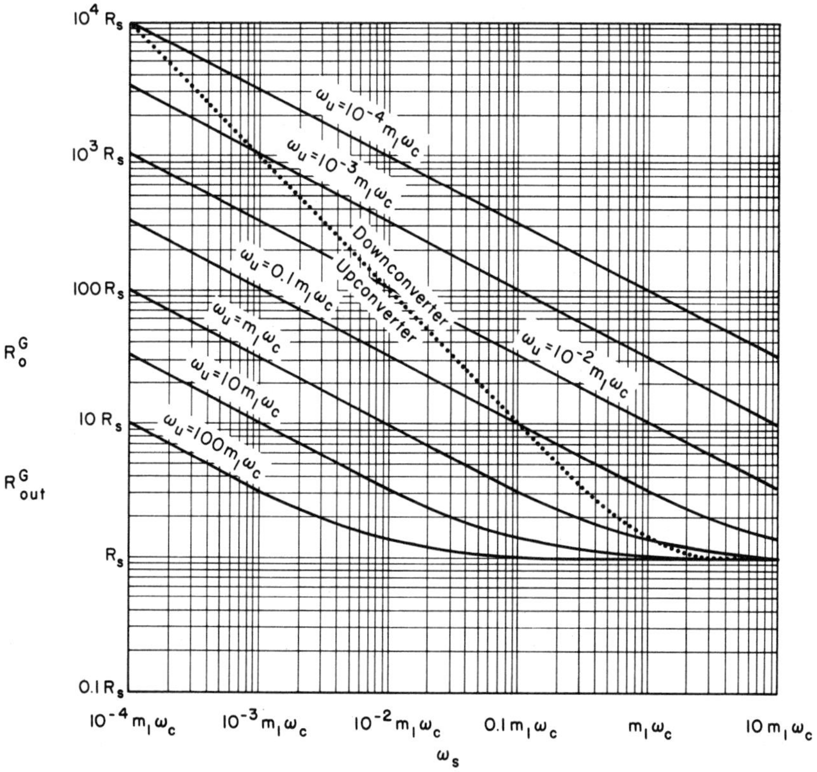

Figure 5.5. Source resistance that makes the upper-side-band upconverter gain the maximum, taken from Equation 5.39. Also, the resulting output resistance (the two are equal). Note that for $\omega_s \omega_u > m_1^2 \omega_c^2$, R_o^G is, for practical purposes, equal to the series resistance R_s. The dotted line distinguishes upconverters from downconverters.

and

$$(T^G)_{G=1} = T_d\left(\frac{2}{\dfrac{m_1\,\omega_c}{\omega_s} - 1}\right), \tag{5.47}$$

where we use the subscript G=1. Since T^G increases as ω_u

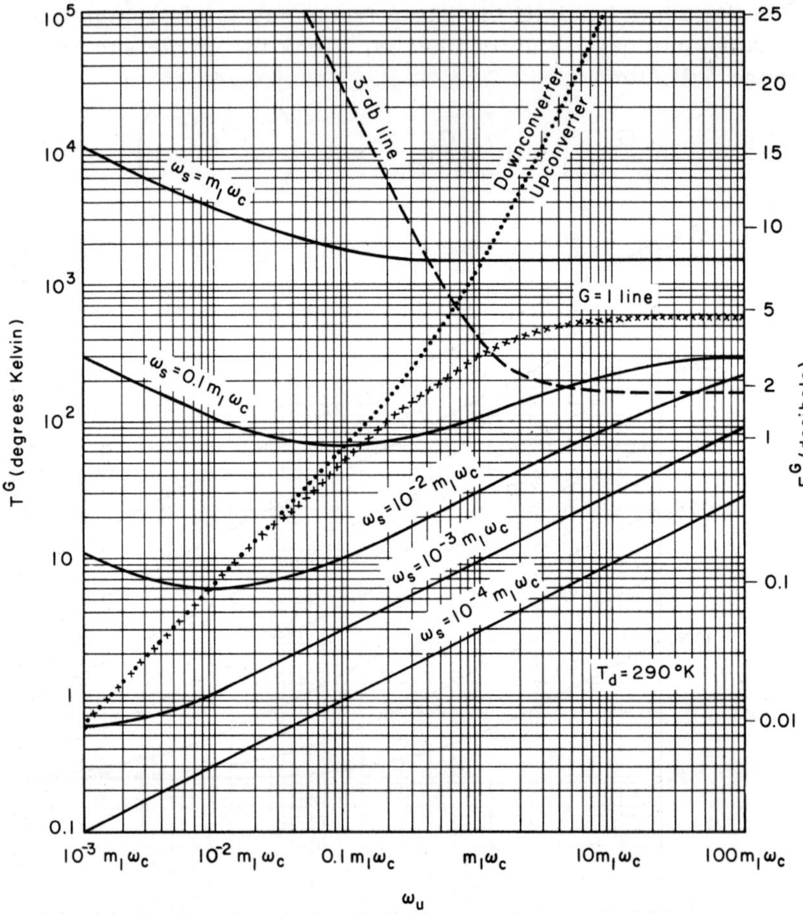

Figure 5.6. Upper-sideband-upconverter (and -downconverter) noise temperature resulting from having the gain optimized, taken from Equation 5.41, for a room-temperature varactor. Note the dotted line separating upconverters from downconverters, the curve denoting unity gain $G_e{}^G$, and the 3-db line indicating the upper-sideband frequency above which the gain can increase only 3 db with increases in pump frequency.

increases, if we want optimum gain greater than 1, the noise temperature will be greater than Equation 5.47.

Often one operates upconverters at signal frequencies low compared with $m_1\omega_c$. In this case, the source resistance for optimum gain is approximately

$$R_o^G \approx \frac{|S_1|}{\sqrt{\omega_s \omega_u}}, \tag{5.48}$$

and the optimum gain is approximately ω_u/ω_s, in accordance with the Manley-Rowe equations (see Section 3.1.1). The noise temperature for maximum gain approaches

$$T \approx T_d \frac{\omega_u + \omega_s}{m_1 \omega_c} \sqrt{\frac{\omega_s}{\omega_u}}. \tag{5.49}$$

5.2.2(b). Noise Optimization. From Equation 5.35, it is clear that R_o can be chosen so as to minimize the noise temperature T. To find the value of R_o that accomplishes this, we differentiate Equation 5.35 by R_o and set the result equal to zero. Letting the superscript T denote evaluation for this value of R_o, we find (consistent with Equation 5.29)

$$R_o^T = R_s \sqrt{1 + \left(\frac{m_1 \omega_c}{\omega_s}\right)^2}, \tag{5.50}$$

and when this value is substituted for R_o in Equations 5.33, 5.34, and 5.35, we find

$$T^T = T_a = T_d \frac{2\omega_s}{m_1 \omega_c} \left[\frac{\omega_s}{m_1 \omega_c} + \sqrt{1 + \left(\frac{\omega_s}{m_1 \omega_c}\right)^2}\right], \tag{5.51}$$

$$G_e^T = \frac{\sqrt{1 + \left(\frac{m_1 \omega_c}{\omega_s}\right)^2}}{\left[\frac{\omega_s}{m_1 \omega_c} + \sqrt{1 + \left(\frac{\omega_s}{m_1 \omega_c}\right)^2}\right]\left[\frac{\omega_s}{m_1 \omega_c} + \sqrt{1 + \left(\frac{\omega_s}{m_1 \omega_c}\right)^2} + \frac{m_1 \omega_c}{\omega_u}\right]}, \tag{5.52}$$

and

$$R_{out}^{T} = R_s \left[\frac{\omega_p}{\omega_u} + \frac{\omega_s}{\omega_u}\sqrt{1 + \left(\frac{m_1 \omega_c}{\omega_s}\right)^2}\right], \qquad (5.53)$$

where T_a is given by Equation 5.27. By comparing Equations 5.39 and 5.50, we see that the same value of R_o cannot optimize the gain and noise temperature simultaneously.

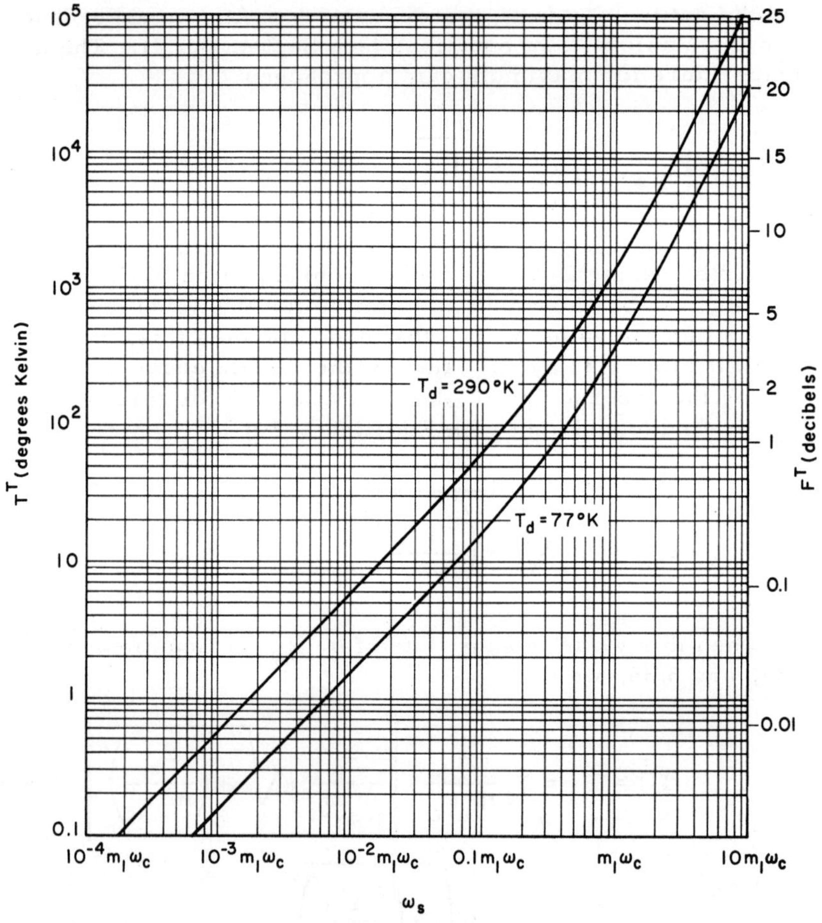

Figure 5.7. Minimum noise temperature T^T of an upper-sideband upconverter or downconverter, taken from Equation 5.51. Note that this minimum is independent of the output frequency ω_u. It is assumed that the varactor is at room temperature or at liquid-nitrogen temperature. Other plots of Equation 5.51 are given in Figures 6.6 and 6.7.

Note that since the original expression for T, Equation 5.35, is independent of ω_u, the value of R_o that optimizes the noise is also independent of ω_u, but the resulting values of exchangeable gain and output resistance do depend on ω_u.

For a given pumped varactor, the value of T^T in Equation 5.51 is a function of only the diode temperature and the signal frequency. This limit, plotted in Figure 5.7, is of fundamental importance, because it applies also to other simple upconverters and downconverters, as well as to parametric amplifiers. A further discussion of this formula is given in Section 6.5.6; for other plots of T^T, see Figures 6.6, 6.7, and 6.18.

The value of source resistance that minimizes the noise temperature, Equation 5.50, is plotted in Figure 5.8, and the resulting exchangeable gain and output resistance are shown in Figures 5.9 and 5.10. Observe that $G_e{}^T$ is a monotonic function of ω_u, but that in order to obtain high gain there is little

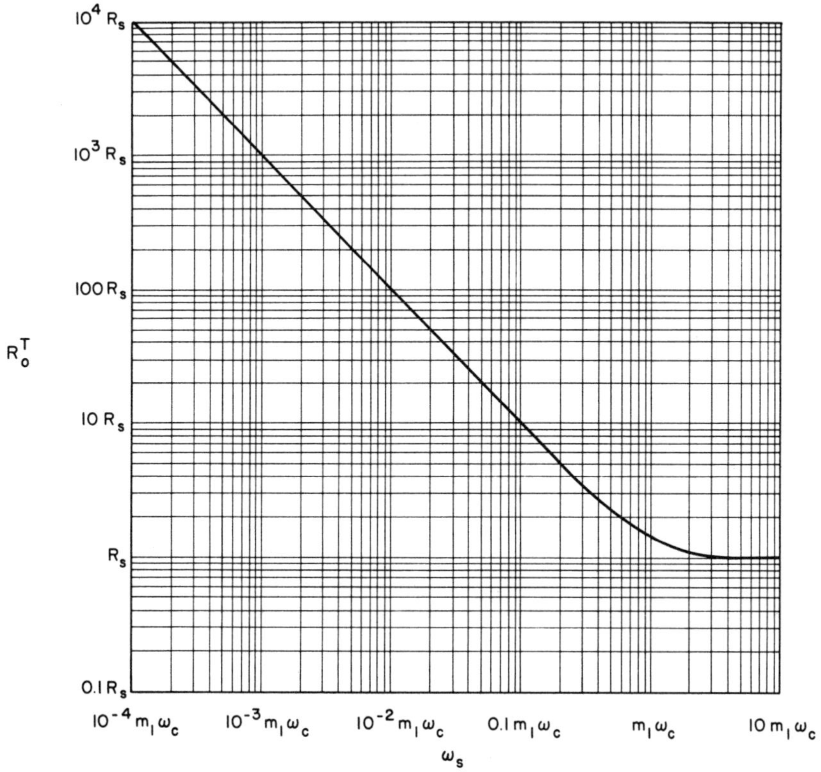

Figure 5.8. Source resistance to realize the minimum noise temperature of an upper-sideband upconverter or downconverter, taken from Equation 5.50. Note that $R_o{}^T$ is independent of the pump frequency.

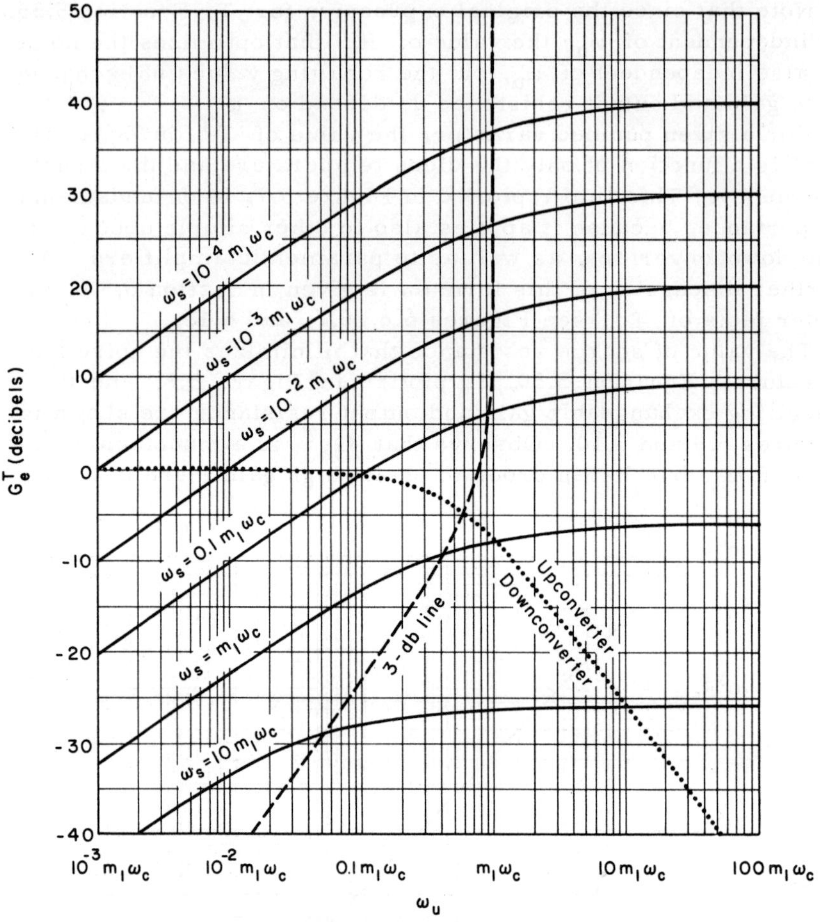

Figure 5.9. Exchangeable gain $G_e{}^T$ for an upper-sideband upconverter or downconverter when the source resistance is chosen to minimize the noise temperature, taken from Equation 5.52. Observe the curve separating upconverters from downconverters, as well as the dashed line indicating a pump frequency high enough so that the gain is within 3 db of the value at very high pump frequencies.

point in choosing ω_u above a point of diminishing returns, as is clear from Figure 5.9. Specifically, the gain $G_e{}^T$ is within a factor of 2 (3 db) of its value for $\omega_u = \infty$ whenever ω_u is greater than

$$\omega_u > \sqrt{m_1{}^2 \omega_c{}^2 + \omega_s{}^2} - \omega_s \tag{5.54}$$

or approximately $m_1 \omega_c$ for small signal frequencies. We shall encounter many instances in which it is wise, for some reason or other, to pump at a frequency close to $m_1\omega_c$; we therefore call $m_1\omega_c$ an optimum pump frequency, although in different circumstances the reasons for selecting this frequency are different.

From Equation 5.52, with ω_u set equal to infinity, we conclude that there is gain only if ω_s is less than $0.455m_1\omega_c$ (not much less than the limit of $0.5m_1\omega_c$ under gain-optimized conditions).

Observe that Figure 5.10 indicates that the output resistance (as well as the source resistance R_o^T) approaches R_s for high pump frequencies; note further that for practical purposes R_{out}^T is independent of ω_s if ω_s is less than about a tenth of $m_1\omega_c$.

Many of the properties of upper-sideband upconverters, under both gain optimization and noise optimization, were given by Leenov,[11] who used a slightly different approach. A similar analysis was reported by Kotzebue[8] and Blackwell.[2] The results

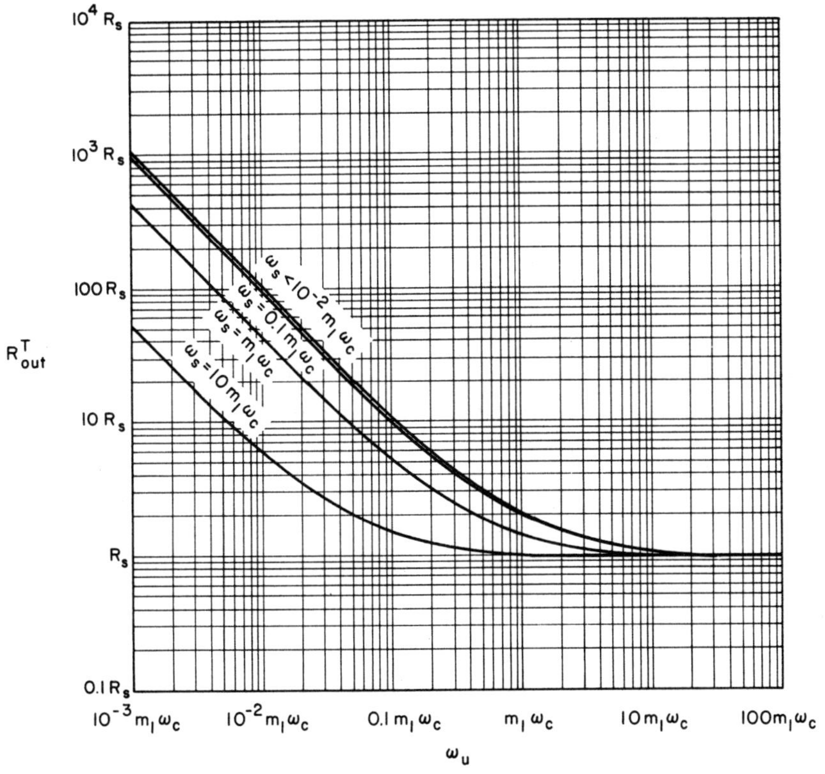

Figure 5.10. Output resistance of upper-sideband upconverter or downconverter when the noise is minimized. Taken from Equation 5.53.

of noise minimization and gain maximization, as well as a discussion of compromises, were given by Kurokawa and Uenohara.[10]

At signal frequencies small compared with $m_1\omega_c$, the source resistance for minimum noise temperature is

$$R_o^T \approx \frac{|S_1|}{\omega_s}, \tag{5.55}$$

merely the "reactance of S_1" at the signal frequency. The minimum noise temperature approaches

$$T^T \approx T_d \frac{2\omega_s}{m_1\omega_c}, \tag{5.56}$$

and the exchangeable gain is approximately

$$G_e^T \approx \frac{\omega_u}{\omega_s} \frac{1}{\dfrac{\omega_u}{m_1\omega_c} + 1}. \tag{5.57}$$

The output resistance becomes

$$R_{out}^T \approx \frac{|S_1|}{\omega_u} + R_s \frac{\omega_p}{\omega_u}. \tag{5.58}$$

5.2.2(c). Compromises. We have seen that the same value of source resistance does not optimize both the gain and the noise temperature of an upper-sideband upconverter. In practice, some compromise optimization is usually desirable: We choose a value of R_o somewhere between R_o^G and R_o^T. This compromise is usually based on the behavior of the amplifier that follows the upconverter (the second stage). If the second stage is extremely noisy, we want all the gain possible in the first stage; therefore, to minimize the over-all noise figure, we choose R_o close to R_o^G. On the other hand, if the second stage is very quiet, we get the best over-all noise performance by adjusting the first stage to have very little noise, even if this means there is no gain. We therefore choose R_o close to R_o^T.

In specific cases this compromise optimization is easy enough to perform. For example, suppose

$$\omega_u = m_1\omega_c \tag{5.59}$$

and

$$\omega_s = 10^{-2} m_1 \omega_c \ . \tag{5.60}$$

The exchangeable gain and noise temperature are plotted against source resistance in Figure 5.11; the conflicting requirements of high gain and low noise temperature are satisfied by values of R_o different by a factor of 10. However, the plot of the cascade noise temperature, under the assumption that the second-stage noise temperature is 3000°K, shows a minimum for an intermediate value of R_o. The principle illustrated by this example holds for any choices of ω_s and ω_u.

Another compromise between noise temperature and gain is the optimization of the noise measure of the upconverter. We define

$$M_e = \frac{F - 1}{1 - \frac{1}{G_e}} = \frac{\frac{T}{T_0}}{1 - \frac{1}{G_e}} \tag{5.61}$$

on a strictly formal basis. The noise measure does not have the

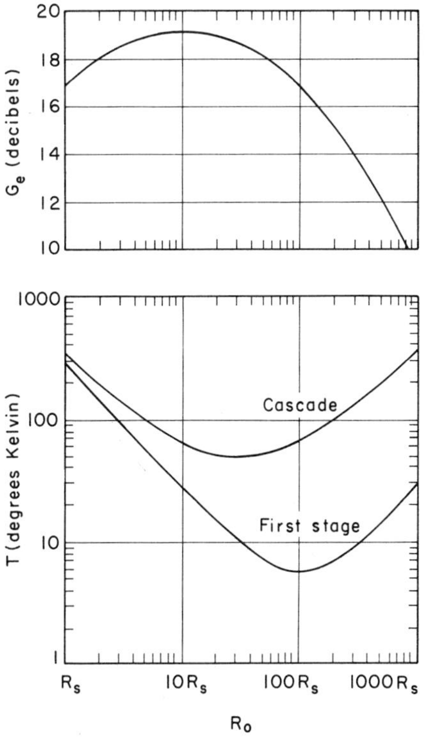

Figure 5.11. Typical behavior of exchangeable gain and noise temperature of an upper-sideband upconverter, as a function of source resistance. Also shown is the cascade noise temperature if the upconverter is followed by a second stage with a noise temperature of 3000°K. Note that the minimum cascade noise temperature is obtained at an intermediate value of R_o. Although this plot is for $\omega_s = 10^{-2} m_1\omega_c$ and $\omega_u = m_1\omega_c$, the qualitative features of these graphs apply generally.

significance for frequency converters that it does for ordinary amplifiers,[6,7] because one cannot meaningfully continue to cascade frequency converters, nor can one imbed them. Moreover, an ordinary amplifier is of no value without gain, whereas a frequency converter may be. Nevertheless, minimizing the noise measure is an interesting compromise between minimizing the noise temperature and maximizing the gain. This compromise might arise if we were to question whether or not to precede a given amplifier by a combination of a varactor upconverter and a noise-free, lossless downconverter.

Evaluating M_e from Equations 5.61, 5.34, and 5.35 and minimizing it with respect to R_o, we find (using a superscript M for evaluation with the value of R_o that minimizes M_e)

$$R_o^M = R_s\left(\frac{m_1\omega_c}{\omega_s} - 1\right) \tag{5.62}$$

$$M_e^M = \frac{T_d}{T_0}\,\frac{2\omega_s\omega_u}{\omega_p m_1\omega_c - 2\omega_s\omega_u}, \tag{5.63}$$

$$G_e^M = \frac{\dfrac{m_1\omega_c}{\omega_s} - 1}{\dfrac{m_1\omega_c}{\omega_u} + 1}, \tag{5.64}$$

$$T^M = \frac{2T_d}{\dfrac{m_1\omega_c}{\omega_s} - 1}, \tag{5.65}$$

$$R_{out}^M = R_s\left(1 + \frac{m_1\omega_c}{\omega_u}\right). \tag{5.66}$$

If Condition 5.42 holds (that is, if there really can be gain), then

$$R_o^G < R_o^M < R_o^T; \tag{5.67}$$

hence minimizing M_e really is a compromise between maximizing G_e and minimizing T, as would be expected.

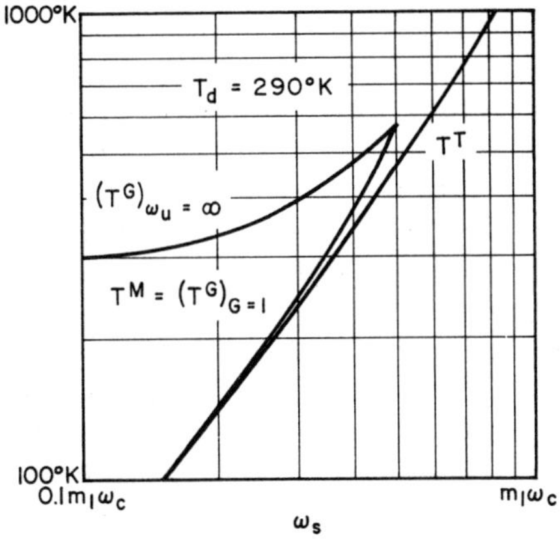

Figure 5.12. Noise temperature for the various optimizations, for a room temperature varactor. This plot shows how much the noise temperature is degraded by optimizing for lowest M_e or highest G_e instead of lowest T. Taken from Equations 5.45, 5.47, 5.51, and 5.65.

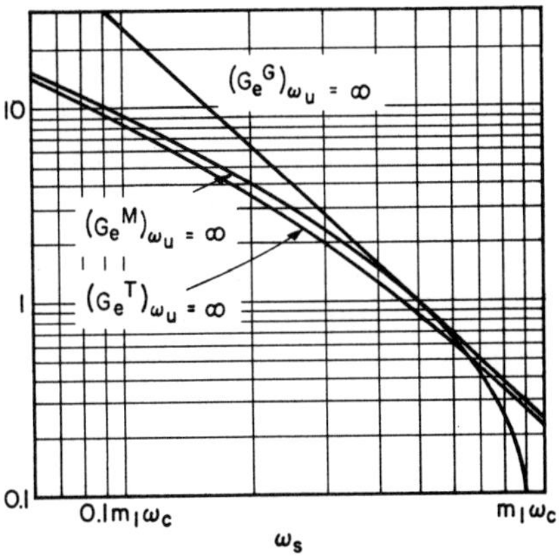

Figure 5.13. Exchangeable gain of an upper-sideband upconverter for the various optimizations, with $\omega_u = \infty$. This figure shows how much the gain is degraded by demanding lowest T or M_e instead of highest G_e. Taken from Equations 5.44, 5.52, and 5.64.

A comparison of the noise temperatures obtained for the various optimizations is given in Figure 5.12, where we plot T^T, T^M, and the range of values of T^G.

A comparison of the upconverter gains obtained for the various optimizations is given in Figure 5.13, where we plot the maximum values of $G_e{}^G$, $G_e{}^T$, and $G_e{}^M$ (that is, those values obtained for $\omega_u = \infty$).

5.2.2(d). Examples. The formulas presented so far can be used in many practical situations, to indicate the quality of the performance obtainable from a given varactor, or to indicate the quality of the varactor necessary for given performance. We indicate several examples of their use.

Observe that all the formulas are functions of the input and output frequency, normalized by the quantity $m_1\omega_c$. In actual calculations, it is usually preferable to use ordinary rather than radian frequency; the formulas can all be rewritten in exactly the same form with the input and output frequencies f_s and f_u, normalized by $m_1 f_c$.

First, suppose we wish to upconvert from 200 Mc to X-band (10 Gc) with a varactor with cutoff frequency f_c = 100 Gc. The discussion of Chapter 7 indicates that a typical value for m_1 might be 0.2, making $m_1 f_c$ = 20 Gc. Thus, $f_s = 10^{-2} m_1 f_c$, and $f_u = 0.5m_1f_c$. If the gain is maximized, Equation 5.38 indicates that it can be as high as 16.4 db, achieved with a noise temperature $T^G = 21°$ if the varactor is at room temperature. From Equation 5.39, the value of source resistance that optimizes the gain is $14R_s$, a quite reasonable number. On the other hand, if the noise temperature is optimized by choosing $R_o{}^T = 100R_s$, we can obtain a gain of only 15.2 db but a noise temperature of 5.9°. Note that the noise temperature is quite a bit lower, at little sacrifice in gain. If the upconverter is followed by a 1000° second stage (6.5-db noise figure), the cascade noise temperature is 44° for optimized gain and 36° for optimized noise. Between these two, of course, there is an even lower optimum compromise. (In practice it is questionable if a noise temperature as low as this is useful at 200 Mc, at least for a radio receiver.)

Second, suppose we wish to upconvert from 3 Gc. How good a varactor is necessary to achieve a 100° noise temperature, for both a room-temperature varactor, and one in liquid nitrogen (say, 100°K, for simplicity)? We transform the equation for lowest noise temperature, Equation 5.51, to the form

$$m_1f_c = 2f_s\sqrt{1 + \frac{T_d}{T^T}}\sqrt{\frac{T_d}{T^T}} \tag{5.68}$$

and find that for a room-temperature varactor, m_1f_c must be at least 20.2 Gc, while for a liquid-air varactor, m_1f_c must be at least 8.5 Gc. Use of the results of Chapter 7 then indicates probable values for m_1 and, therefore, probable values for f_c required.

Third, suppose we are capable of pumping two varactors at 4 Gc, so they have a value of m_1 sufficiently high so that $m_1f_c = 10$ Gc. It is proposed to cascade two upconverters, one from 1 Gc to 5 Gc, and the other to 9 Gc, with an over-all noise temperature of 200°. Can this be done with the varactors at room temperature? From Equation 5.51, the minimum noise temperature for the second stage is 470°. A crude lower bound of the noise temperature of the cascade is found by using the minimum noise temperature of the first stage, along with its maximum exchangeable gain; here we ignore the fact that these cannot be achieved simultaneously. From Equation 5.38, the maximum exchangeable gain is 3.2, while the minimum first-stage noise figure is 70°. Use of the cascade formula then indicates that the cascade noise temperature must be greater than 70° + (470°/3.2) = 217°. The proposed scheme is therefore impossible.

As a final example, suppose we can pump a 140-Gc varactor

so that $m_1 = 0.25$; thus, $m_1 f_c = 35$ Gc. If we are to upconvert an input at 10 Gc, what is the minimum noise temperature, and how high a pump frequency is required to yield gain if the noise is optimized? From Equation 5.51, T^T becomes 219° (if $T_d = 290°$), and from Equation 5.52, $G_e{}^T$ can be greater than 1 only if f_u is greater than $0.70 m_1 f_c = 24.4$ Gc. The pump frequency, therefore, must be greater than 14.4 Gc.

5.2.3. Synthesis Techniques

The performance indicated thus far in Section 5.2 can be achieved if all sideband frequencies except ω_s and ω_u are open-circuited, and provided that the coupling networks have no losses. In Section 7.4, conditions are given for pumping varactors with a sinusoidal current. We now describe some circuits to achieve both of these sets of conditions.

5.2.3(a). Preliminary Design. The synthesis of upconverter circuits should logically proceed in two steps. The first is preliminary design, and consists of selecting a varactor and suitable operating conditions so that the desired performance can be achieved. Since the entire analysis of Section 5.2 and of Chapter 7 is directed toward this preliminary design, we do not discuss it further here.

5.2.3(b). Separation of Frequencies. Once the varactor parameters and the operating conditions are chosen, one must still design a circuit in which to use the varactor. This circuit should separate the various frequencies present and open-circuit the varactor at the unwanted sideband frequencies. In particular, the circuit should fulfill these conditions:

1. The varactor is open-circuited at frequency $2\omega_p$ (see Chapter 7).
2. The varactor is open-circuited at $\omega_p + \omega_u$.
3. The varactor is open-circuited at $\omega_p - \omega_s$.
4. The pump, signal, and upper-sideband frequencies appear at separate terminal pairs, and each only at its own terminal pair.
5. The network is lossless.
6. The pump can be tuned to match the varactor series resistance.

Some of these conditions can be met by using circuits with high symmetry, particularly those with two or more varactors. In addition, many other circuits fulfill these conditions. In practice one would select from among all these circuits those that have desired bandwidth, reliability, simplicity, cost, freedom from spurious oscillations, and the like. To show that such circuits exist (and, in fact, can be drawn almost by inspection to meet the conditions above), we show two of them: one using lumped tuned circuits, and the other using coaxial transmission lines.

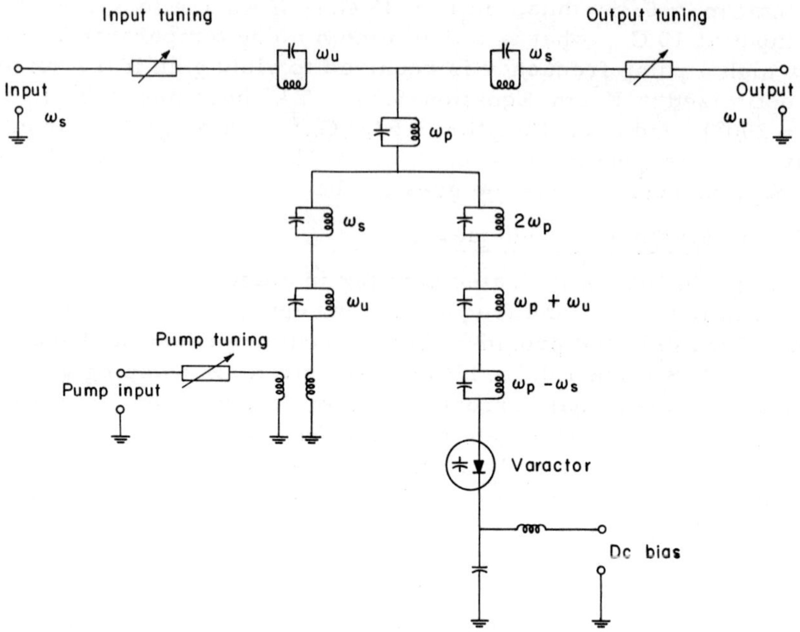

Figure 5.14. A lumped circuit that realizes the frequency separation demanded for the simple upper-sideband upconverter. By proper choies of the impedances of the tuned circuits, the input and output tuning adjustments can be eliminated. The tuned circuits can be adjusted, for example with a grid-dip meter, before assembly.

Figure 5.14 shows the lumped-circuit realization. The varactor is open-circuited at frequencies $2\omega_p$, $\omega_p + \omega_u$, and $\omega_p - \omega_s$ by the tuned circuits shown. The signal and upper-sideband currents are kept away from the pump by the tuned circuits at the lower left, and a pump-frequency matching network is shown leading to the pump terminals.

The pump frequency is excluded from the signal and upper-sideband circuits by the tuned circuit in the middle. The signal and upper sideband are then separated, and brought out to different terminal pairs. The current at the input terminals is exactly what we have called I_s, and the real part of the impedance at the input is actually what we have called R_{in}, because the ω_s tuned circuits appear as open circuits at frequency ω_s, and the remainder of the tuned circuits are purely reactive. Similarly, the output current and resistance are actually what we have called I_u and R_{out}.

A coaxial realization is shown in Figure 5.15. The varactor is mounted as part of the inner conductor, as indicated by Figure 5.16, and not between the inner conductor and the outer

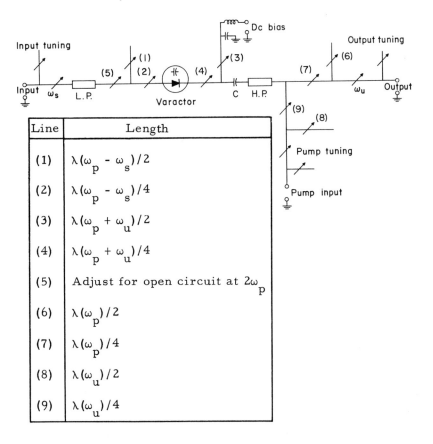

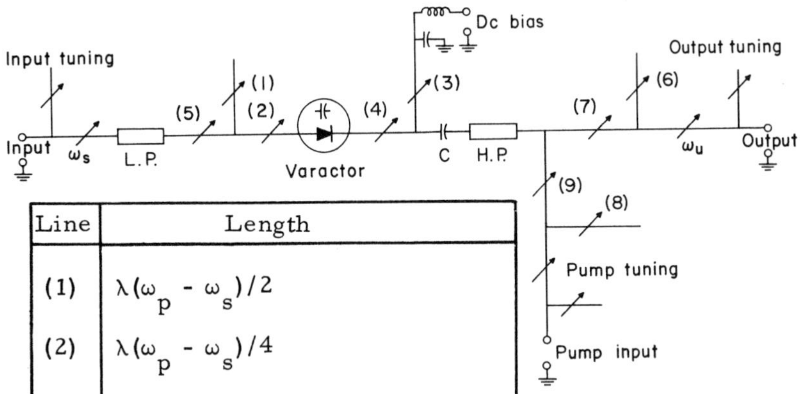

Line	Length
(1)	$\lambda(\omega_p - \omega_s)/2$
(2)	$\lambda(\omega_p - \omega_s)/4$
(3)	$\lambda(\omega_p + \omega_u)/2$
(4)	$\lambda(\omega_p + \omega_u)/4$
(5)	Adjust for open circuit at $2\omega_p$
(6)	$\lambda(\omega_p)/2$
(7)	$\lambda(\omega_p)/4$
(8)	$\lambda(\omega_u)/2$
(9)	$\lambda(\omega_u)/4$

Figure 5.15. A coaxial circuit that realizes the frequency separation demanded for the simple upper-sideband upconverter. The adjustable lines (1) through (9) can be set experimentally before final assembly; their settings depend only on the frequencies and are independent of the varactor parameters.

conductor. The open-circuit conditions are now met by using resonant lengths of coaxial line. Briefly, lines (1) and (2) ensure both that frequency $\omega_p - \omega_s$ cannot propagate to the left of the T-junction and that the varactor is open-circuited at this frequency. Similarly, lines (3) and (4) both open-circuit the varactor at frequency $\omega_p + \omega_u$, and prevent this frequency from propagating to the right. The short-circuited end of line (3) is used to introduce the dc bias voltage to the circuit. Because of the blocking capacitor C and the short circuit at the end of line (1), this dc voltage appears across the varactor.

For simplicity we assume that ω_s is much smaller than ω_p, so that a simple low-pass filter L. P. prevents frequencies ω_p

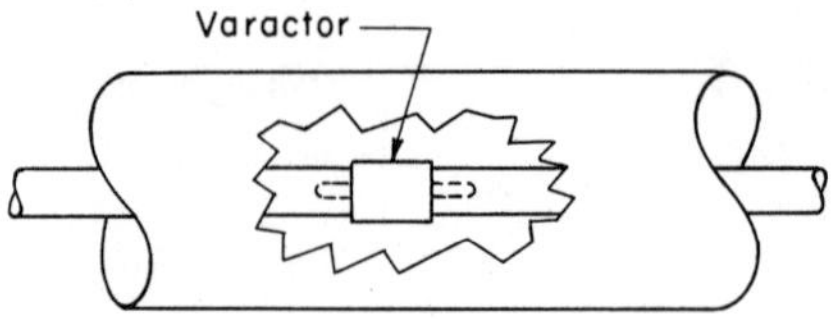

Figure 5.16. A varactor mounted in the inner conductor of a coaxial transmission line.

and ω_u from appearing at the signal-frequency port at the left. Line (5) is adjusted so that the varactor is open-circuited at $2\omega_p$. To the left of the low-pass filter is a signal-frequency tuning adjustment, which can match any source.

Similarly, a high-pass filter H.P. keeps the signal frequency to the left of it. Lines (6), (7), (8), and (9) separate the pump and upper-sideband frequencies, and each is provided with a tuning adjustment to match any loads.

There are several adjustments in the circuit shown, but these adjustments are all independent if performed in the order listed earlier. The adjustments can be made before final assembly; they depend only on the frequencies, and not on the varactor parameters. Both the circuits given here have faults; they are given merely as illustrations of a design procedure based on the conditions of the theory given earlier.

Similar techniques employing waveguide circuitry can be used to realize the conditions listed above. An additional problem, however, is that of coupling from the varactor to the waveguide. Frequency separation can often be accomplished by using waveguides of different sizes.

5.2.4. Effect of Idler

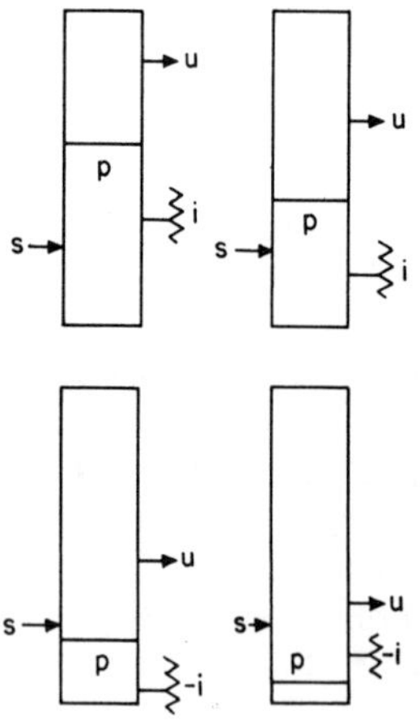

Thus far in Section 5.2, we have assumed that the varactor is open-circuited at all sidebands except the input ω_s and the output ω_u. Now we investigate briefly some of the effects of allowing current to flow at the idler frequency $\omega_i = \omega_p - \omega_s$. Note that ω_i can be either positive or negative, and if positive, it can be either less than or greater than ω_s. If ω_i is negative, its magnitude can be either less than or greater than the pump frequency.

The noise temperature and gain of the simple upper-sideband upconverter are strictly limited, according to the theory given so far. An idler can raise the gain and can lower the noise temperature, and in fact it can do both simultaneously. Thus there may be real advantages in deliberately allowing current to flow at ω_i.

The upconverter with idler is shown in Figure 5.17, where, in accordance with the ideas of Section 2.2, we show separate

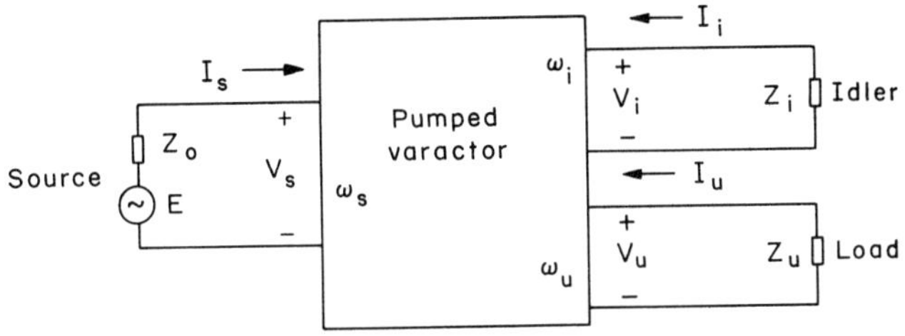

Figure 5.17. Double-sideband varactor upconverter. Although Z_o, Z_i, and Z_u are shown separately, some physical elements may appear in all three; see Section 2.2.

terminations at the various frequencies. The equations that describe the varactor are derived from Equations 4.42, by eliminating all rows and columns except those for frequencies ω_s, ω_u, and ω_i. If, for simplicity, we assume that $S_2 = 0$, then the result is

$$\begin{bmatrix} V_i^* \\ V_s \\ V_u \end{bmatrix} = \begin{bmatrix} R_s - \dfrac{S_0}{j\omega_i} & \dfrac{S_1^*}{j\omega_s} & 0 \\ -\dfrac{S_1}{j\omega_i} & R_s + \dfrac{S_0}{j\omega_s} & \dfrac{S_1^*}{j\omega_u} \\ 0 & \dfrac{S_1}{j\omega_s} & R_s + \dfrac{S_0}{j\omega_u} \end{bmatrix} \begin{bmatrix} I_i^* \\ I_s \\ I_u \end{bmatrix} + \begin{bmatrix} E_{ni}^* \\ E_{ns} \\ E_{nu} \end{bmatrix}, \tag{5.69}$$

where E_{ni}, E_{ns}, and E_{nu}, representing thermal noise of the varactor series resistance,‡ have mean-squared values

$$\overline{|E_{ni}|^2} = \overline{|E_{ns}|^2} = \overline{|E_{nu}|^2} = 4kT_dR_s\,\Delta f. \tag{5.70}$$

The varactor is terminated at the three frequencies with impedances and equivalent sources (representing, for the idler frequency, noise):

$$V_i^* = -Z_i^* I_i^* + E_{nxi}^*, \tag{5.71}$$

‡The noise voltages are effective rms voltages instead of half-amplitudes. See Section 2.5.1(a).

$$V_s = -Z_o I_s + E, \tag{5.72}$$

and

$$V_u = -Z_u I_u, \tag{5.73}$$

where the impedances have real parts, respectively, R_i, R_o, and R_u, and where the idler noise voltage has mean-squared value‡

$$\overline{|E_{nxi}|^2} = 4kT_i R_i \, \Delta f \tag{5.74}$$

with T_i being the temperature of the idler termination.

Equations 5.69 to 5.74 can be solved to find the upconverter input and output impedances, exchangeable gain, and noise temperature. If the source, the load, and the external idler termination are all tuned, these quantities are

$$Z_{in} = R_{in} + jX_{in}$$

$$= R_s + \frac{S_0}{j\omega_s} - \frac{m_1^2 \omega_c^2}{\omega_s \omega_i} \frac{R_s^2}{R_s + R_i} + \frac{m_1^2 \omega_c^2}{\omega_s \omega_u} \frac{R_s^2}{R_s + R_u}, \tag{5.75}$$

$$Z_{out} = R_{out} + jX_{out}$$

$$= R_s + \frac{S_0}{j\omega_u} + \frac{m_1^2 \omega_c^2}{\omega_s \omega_u} \frac{R_s^2}{R_s + R_o - \dfrac{m_1^2 \omega_c^2}{\omega_s \omega_i} \dfrac{R_s^2}{R_s + R_i}}, \tag{5.76}$$

$$\frac{1}{G_e} = \frac{\omega_s^2}{m_1^2 \omega_c^2} \frac{(R_s + R_o)^2}{R_s R_o} + \frac{m_1^2 \omega_c^2}{\omega_i^2} \frac{R_s}{R_o} \frac{R_s^2}{(R_s + R_i)^2}$$

$$- \frac{m_1^2 \omega_c^2}{\omega_i \omega_u} \frac{R_s}{R_o} \frac{R_s}{R_s + R_i} - 2 \frac{\omega_s}{\omega_i} \frac{R_s}{R_o} \frac{R_s + R_o}{R_s + R_i} + \frac{\omega_s}{\omega_u} \frac{R_s + R_o}{R_o}, \tag{5.77}$$

‡The noise voltages are effective rms voltages instead of half-amplitudes. See Section 2.5.1(a).

and

$$T = T_d \frac{R_s}{R_o} \left[1 + \frac{\omega_s^2}{m_1^2 \omega_c^2} \frac{(R_s + R_o)^2}{R_s^2} - 2 \frac{\omega_s}{\omega_i} \frac{R_s + R_o}{R_s + R_i} \right.$$

$$\left. + 2 \frac{m_1^2 \omega_c^2}{\omega_i^2} \frac{R_s^2}{(R_s + R_i)^2} \right] + T_i \frac{R_i}{R_o} \frac{m_1^2 \omega_c^2}{\omega_i^2} \frac{R_s^2}{(R_s + R_i)^2} . \quad (5.78)$$

Note that if R_i is set to infinity, these formulas reduce to Equations 5.37, 5.33, 5.34, and 5.35 for the upconverter without idler. Also, if the series resistance vanishes, then the exchangeable gain and noise temperature are related by Equations 3.29 and 3.30.

Note from Equations 5.75 and 5.76 that the input and output resistances can be negative, because of the action of the idler, but only if the idler frequency ω_i is positive, and even then, only if

$$\omega_i \omega_s < m_1^2 \omega_c^2 . \quad (5.79)$$

This condition is also the condition for gain of a varactor parametric amplifier (see Chapter 6). If either the input or output resistance is negative, the upconverter with idler is apt to be unstable.

Two additional effects are important. First, the exchangeable gain can be increased by the idler. Second, the noise temperature can be reduced by the idler. In fact, in some cases both of these benefits can be achieved together.

Breon[3,4] has shown how the exchangeable gain of the upconverter is increased by the idler. Consider three cases: (1) $\omega_i < 0$; (2) $\omega_i > 0$ but $\omega_s \omega_i > m_1^2 \omega_c^2$; and (3) $\omega_i > 0$ and $\omega_s \omega_i < m_1^2 \omega_c^2$. In the first case, the idler frequency is negative, because the pump frequency is smaller than the input frequency. It is clear from Equation 5.77 that the highest gain is achieved when R_i is infinite, that is, when no idler current is allowed to flow. The analysis then reduces to that in Sections 5.2.1 and 5.2.2.

In the second case, the exchangeable gain, always positive, is made the highest by setting $R_i = 0$. Then, as a function of the source resistance R_o, the gain is highest when

$$R_o^G = R_s \sqrt{\left(1 - \frac{m_1^2 \omega_c^2}{\omega_s \omega_i}\right) \left(1 - \frac{2m_1^2 \omega_c^2}{\omega_i \omega_u}\right)} , \quad (5.80)$$

and the highest gain so achieved is

$$G_e{}^G = \left(\frac{\dfrac{m_1 \omega_c}{\omega_s}}{\sqrt{1 - \dfrac{m_1^2 \omega_c^2}{\omega_s \omega_i}} + \sqrt{1 - \dfrac{2m_1^2 \omega_c^2}{\omega_i \omega_u}}} \right)^2 . \tag{5.81}$$

In the third case, the exchangeable gain can be made as high as desired and, in fact, can even be negative.

The values of ω_s and ω_u for which gain is possible are seen to include the whole region in which $\omega_s \omega_i < m_1^2 \omega_c^2$, as well as the whole region for the upconverter without idler, Figure 5.4, and in addition the region over which Equation 5.81 is greater than 1. These regions are shown in Figure 5.18.

The idler is also useful because the noise temperature can be reduced. Equation 5.78 shows that if ω_i is positive (that is, if $\omega_p > \omega_s$), then it is possible for T to be smaller than it would be without the idler. If we seek that value of R_i which minimizes T (assuming, for simplicity, that $T_i = 0$), we find that it is

$$R_i = \frac{m_1^2 \omega_c^2}{\omega_s \omega_i} \frac{2R_s^2}{R_s + R_o} - R_s . \tag{5.82}$$

This value minimizes the noise temperature to

$$T = T_d \frac{R_s}{R_o} \left[1 + \frac{\omega_s^2}{2m_1^2 \omega_c^2} \frac{(R_s + R_o)^2}{R_s^2} \right], \tag{5.83}$$

which is smaller than the noise temperature achievable without the idler; in fact, it may be obtained from the noise temperature of the upconverter without the idler, by merely replacing m_1 by $\sqrt{2}\, m_1$. Substituting this value of R_i in Equations 5.75, 5.76, and 5.77, we find that the input and output resistances become

$$R_{in} = \frac{R_s - R_o}{2} + \frac{m_1^2 \omega_c^2}{\omega_s \omega_u} \frac{R_s^2}{R_s + R_u} \tag{5.84}$$

and

$$R_{out} = R_s + \frac{2m_1^2 \omega_c^2}{\omega_s \omega_u} \frac{R_s^2}{R_s + R_o} , \tag{5.85}$$

and that the exchangeable gain becomes

$$\frac{1}{G_e} = \frac{1}{2}\left[\frac{\omega_s^2}{2m_1^2\omega_c^2}\,\frac{(R_s + R_o)^2}{R_s R_o} + \frac{\omega_s}{\omega_u}\,\frac{R_s + R_o}{R_o}\right]. \quad (5.86)$$

Note that R_{out} can be calculated from Equation 5.33, for the upper-sideband upconverter without idlers, by merely substituting $\sqrt{2}\,m_1$ for m_1. Similarly, G_e in Equation 5.86 is exactly twice G_e of Equation 5.34 with m_1 replaced by $\sqrt{2}\,m_1$.

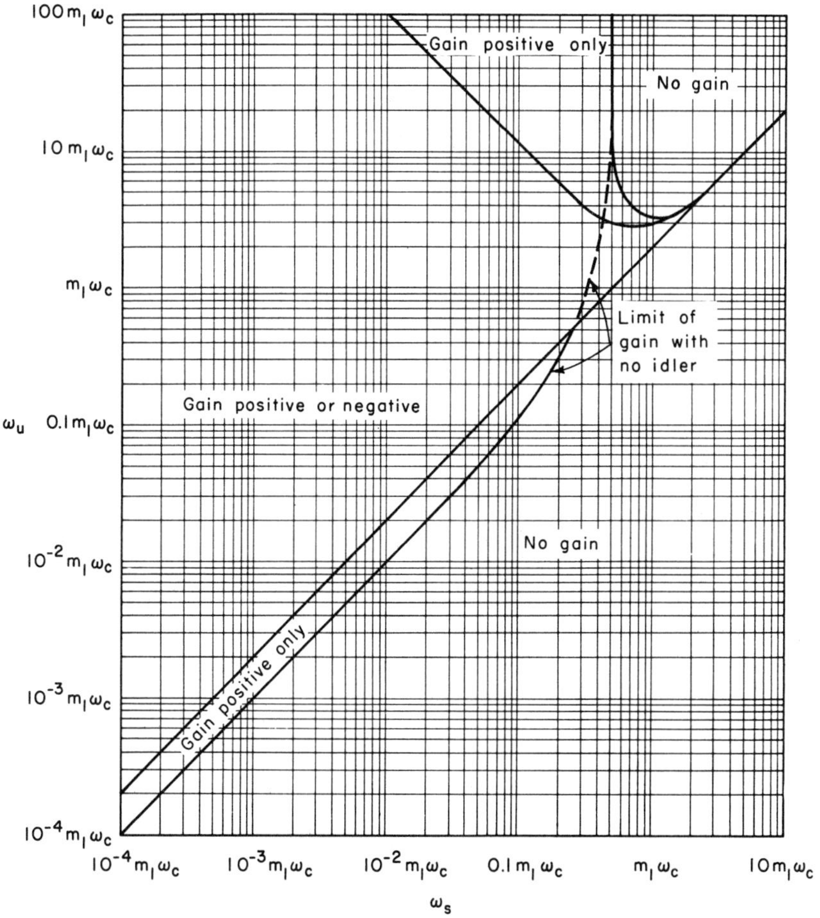

Figure 5.18. Regions in which gain is possible for the upper-sideband upconverter with idler. Note that in many regions indicated, gain is not possible without the idler.

When is this optimization achievable, that is, when is R_i, given in Equation 5.82, positive? It is, provided that

$$2\omega_s < \omega_u < 2\omega_s + \frac{2m_1^2\omega_c^2}{\omega_s} \frac{R_s}{R_s + R_o} \tag{5.87}$$

and thus in particular, for some values of R_o, whenever ω_u is simultaneously greater than $2\omega_s$ and less than $4m_1^2\omega_c^2$.

If we now vary R_o in an attempt to minimize T as given in Equation 5.83, we find that for the value

$$R_o^T = R_s\sqrt{1 + \frac{2m_1^2\omega_c^2}{\omega_s^2}} \tag{5.88}$$

the noise temperature is reduced to the value

$$T^T = T_d \frac{2\omega_s}{\sqrt{2}m_1\omega_c}\left(\frac{\omega_s}{\sqrt{2}m_1\omega_c} + \sqrt{1 + \frac{\omega_s^2}{2m_1^2\omega_c^2}}\right), \tag{5.89}$$

which is of the same form as the fundamental limit derived earlier, but with m_1 replaced by $\sqrt{2}m_1$. This optimization is valid (that is, it produces positive values for R_i, from Equation 5.82) only for certain ranges of ω_s and ω_u. Note that the limit of Equation 5.89 is lower than the limit of Equation 5.51; thus the freedom of having an additional idler allows us to achieve (at least in principle) a better noise temperature. Note also that the gain and noise are improved simultaneously.

The previous optimization is of limited value because the idler temperature was assumed to be zero. This limitation can be overcome, however, by now setting the pump frequency so that R_i, given by Equation 5.82, is zero (thus, the idler temperature is irrelevant). This procedure leads to the following conditions:

$$\omega_p = \sqrt{2m_1^2\omega_c^2 + \omega_s^2}, \tag{5.90}$$

$$R_i = 0, \tag{5.91}$$

and

$$R_o^T = R_s\sqrt{1 + \frac{2m_1^2\omega_c^2}{\omega_s^2}} = R_s\frac{\omega_p}{\omega_s}, \tag{5.92}$$

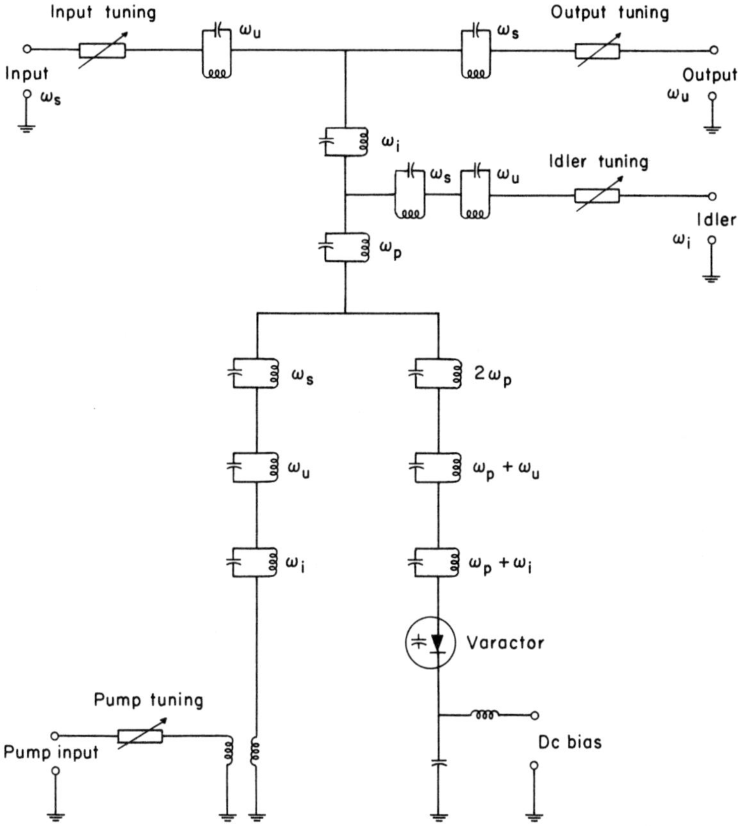

Figure 5.19. A lumped circuit realizing the frequency separation demanded by the theory of Section 5.2.4. Note that separate ports are provided at ω_s, ω_i, and ω_u, so that the circuit can operate with output either at ω_u or ω_i. The tuned circuits can be adjusted, for example with a grid-dip meter, before assembly.

whereupon the noise temperature, exchangeable gain, and output resistance become

$$T^T = T_d \frac{2\omega_s}{\sqrt{2}m_1\omega_c}\left(\frac{\omega_s}{\sqrt{2}m_1\omega_c} + \sqrt{1 + \frac{\omega_s^2}{2m_1^2\omega_c^2}}\right)$$

$$= T_d \frac{\omega_s\omega_u}{m_1^2\omega_c^2}, \tag{5.93}$$

$$G_e^T = \frac{2m_1^2\omega_c^2}{\omega_s\omega_u}, \tag{5.94}$$

and

$$R_{out}^{T} = 2R_o \frac{\omega_s}{\omega_u}$$

$$= 2R_s \frac{\omega_p}{\omega_u}. \quad (5.95)$$

Observe that there is gain ($G_e > 1$) if $\omega_u < \sqrt{6}\, m_1 \omega_c$, that is, if $\omega_s < 0.816 m_1 \omega_c$.

The two optimizations described here are by no means exhaustive. They are given primarily as examples (although important ones) to show the techniques useful for more complete optimizations. The fact that the noise limit of Equation 5.93 is achievable with finite idler temperature is of some importance.

The upper-sideband upconverter with idler can be synthesized in a manner similar to the regular upconverter, Section 5.2.3. The problem is somewhat more difficult, however, because of the extra frequency. A lumped circuit, shown in Figure 5.19, is very similar to the circuit of Figure 5.14 for the regular upconverter.

5.3. Lower-Sideband Upconverter

The lower-sideband upconverter is made by pumping a varactor at frequency ω_p, introducing a signal at frequency ω_s, and using the lower sideband $\omega_i = \omega_p - \omega_s$ for the output. In Sections 5.3.1, 5.3.2, and 5.3.3, we assume for simplicity that all other sidebands are open-circuited. In Section 5.3.4, we allow current to flow in one idler, located at frequency $\omega_u = \omega_p + \omega_s$.

In Section 5.3.1, we derive formulas for the upconverter input and output resistances, exchangeable gain, and noise temperature. Then, in Section 5.3.2, we assume that the source and load are tuned, and discuss gain and noise optimizations. In Section 5.3.3, we briefly treat the synthesis of lower-sideband upconverters. Finally, in Section 5.3.4, we show that an idler at frequency ω_u is not particularly helpful.

The reader should bear in mind that the lower-sideband upconverter may have negative input and output resistances. These imply severe problems of stability or spurious oscillations. All pumped varactors have such problems, but for lower-sideband upconverters and parametric amplifiers they are more severe than for upper-sideband upconverters. The usual concepts of available power, and of the noise temperature based on available power, fail and are replaced by their logical extensions, the exchangeable gain and corresponding noise temperature based thereon [5,6,7] (see Section 2.3). When the exchangeable gain is negative, the output impedance is negative, and one can always obtain gain and, in fact, arbitrarily high transducer gain,‡ but only at the price of possible instabilities.

5.3.1. Characteristics

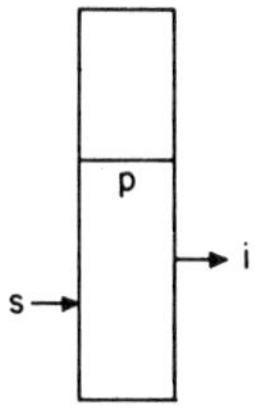

We now assume the varactor is open-circuited at all frequencies except the input ω_s and the output ω_i. We further assume that the only important noise is thermal noise generated in the series resistance R_s, which is at temperature T_d. We neglect losses in the coupling circuits, because these only reduce the gain and increase the noise temperature. The device we analyze is shown in Figure 5.20, with separate ports for the two different frequencies, in accordance with the ideas of Section 2.2. The small-signal equations of the varactor are obtained by eliminating all currents except those at frequencies ω_s and ω_i from Equations 4.42; the result is written in matrix form as

$$\begin{bmatrix} V_i^* \\ V_s \end{bmatrix} = \begin{bmatrix} R_s - \dfrac{S_0}{j\omega_i} & \dfrac{S_1^*}{j\omega_s} \\ -\dfrac{S_1}{j\omega_i} & R_s + \dfrac{S_0}{j\omega_s} \end{bmatrix} \begin{bmatrix} I_i^* \\ I_s \end{bmatrix} + \begin{bmatrix} E_{ni}^* \\ E_{ns} \end{bmatrix}, \quad (5.96)$$

where V_s, V_i, I_s, and I_i are half-amplitudes of the components of varactor voltage and current at the appropriate frequencies, where S_0 and S_1 are the coefficients of varactor incremental elastance, as pumped, according to Equation 4.41, and where E_{ns} and E_{ni} are equivalent noise voltages, as discussed in

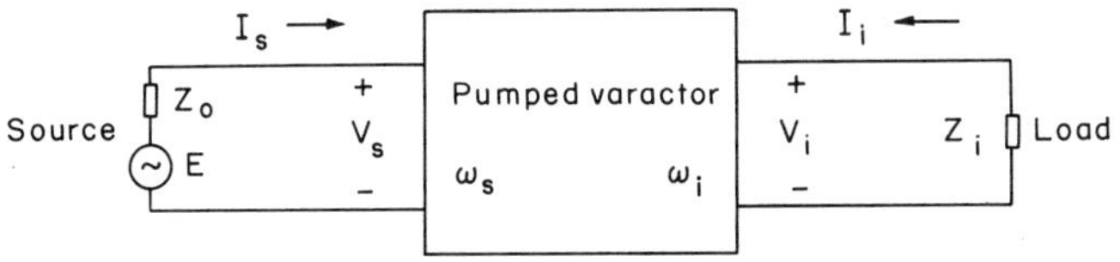

Figure 5.20. Lower-sideband upconverter without idlers. Although Z_o and Z_i are shown separately, some physical elements may appear in both.

‡Transducer gain is actual power output divided by the available power of the source (see Section 2.3).

Section 4.3, with mean-squared values‡

$$\overline{|E_{ns}|^2} = \overline{|E_{ni}|^2} = 4kT_d R_s \, \Delta f \,. \tag{5.97}$$

The input constraint is

$$V_s = -I_s Z_o + E \,, \tag{5.98}$$

and this can be used in Equation 5.96, to eliminate V_s and I_s, leaving one equation, similar to Equation 5.15:

$$V_i^* = \left[R_s - \frac{S_0}{j\omega_i} - \frac{|S_1|^2}{\omega_s \omega_i} \frac{1}{Z_o + R_s + (S_0/j\omega_s)} \right] I_i^*$$

$$+ E_{ni}^* + \frac{S_1}{j\omega_s} \frac{E - E_{ns}}{Z_o + R_s + (S_0/j\omega_s)} \,. \tag{5.99}$$

Equation 5.99 can be obtained from Equation 5.15 (and, in fact, Equations 5.96 can be obtained from Equations 5.12) by substituting V_i^*, I_i^*, $-\omega_i$, and E_{ni}^* for, respectively, V_u, I_u, ω_u, and E_{nu}. The steps used to obtain the output impedance, exchangeable gain, noise temperature, and input impedance are similar to the steps followed in Section 5.2.

5.3.1(a). Output Impedance. Equation 5.99 implies that the output impedance at frequency ω_i is

$$Z_{out} = R_{out} + jX_{out}$$

$$= R_s + \frac{S_0}{j\omega_i} - \frac{|S_1|^2}{\omega_s \omega_i} \frac{1}{Z_o^* + R_s - (S_0/j\omega_s)}$$

$$= R_s + \frac{S_0}{j\omega_i} - \frac{m_1^2 \omega_c^2}{\omega_s \omega_i} \frac{R_s^2}{Z_o^* + R_s - (S_0/j\omega_s)} \,. \tag{5.100}$$

One important point: It is possible to have a negative real part R_{out}, implying that the exchangeable gain is negative. However, this can occur only if $\omega_s\omega_i$ is smaller than $m_1^2 \omega_c^2$.

‡The noise voltages are effective rms values instead of half-amplitudes. See Section 2.5.1(a).

5.3.1(b). Exchangeable Gain. The exchangeable gain can be calculated the same way as for the upper-sideband upconverter; the input exchangeable power is positive, but the output exchangeable power is positive only if R_{out} is positive.

$$\frac{1}{G_e} = \left(\frac{\omega_s}{m_1 \omega_c}\right)^2 \frac{|Z_o^* + R_s - (S_0/j\omega_s)|^2}{R_s R_o} - \frac{\omega_s}{\omega_i} \frac{R_s + R_o}{R_o}. \tag{5.101}$$

If $\omega_s \omega_i < m_1^2 \omega_c^2$, then some choice of Z_o serves to make G_e negative. Observe also that if the series resistance is made vanishingly small, then G_e approaches $-\omega_i/\omega_s$, in agreement with Equation 3.19.

5.3.1(c). Noise Temperature. When the noise temperature of the lower-sideband upconverter is calculated, it is found to be identical to that of the upper-sideband upconverter, Equation 5.21. Because of this, the noise equivalent circuits of Sections 2.5.1(d) and 2.5.1(e), with parameters given by Equations 5.23, 5.24, 5.25, 5.27, 5.28, and 5.29, describe the lower-sideband upconverter. The fact that the noise temperatures are the same could be guessed by observing two facts: (1) The formulas for the lower-sideband device are the same as those for the upper-sideband device if the output frequency is changed from ω_u to $-\omega_i$, and (2), the noise temperature of Equation 5.21 is independent of the output frequency.

5.3.1(d). Input Impedance. The input impedance of the lower-sideband upconverter depends in the following manner on the load impedance Z_i at frequency ω_i:

$$Z_{in} = R_{in} + jX_{in}$$

$$= R_s + \frac{S_0}{j\omega_s} - \frac{m_1^2 \omega_c^2}{\omega_s \omega_i} \frac{R_s^2}{Z_i^* + R_s - (S_0/j\omega_i)}. \tag{5.102}$$

Note that the input impedance can be negative for some choice of Z_i if and only if $\omega_s \omega_i < m_1^2 \omega_c^2$. This negative resistance can produce instabilities at the input, or else in properly designed circuits it can be used as a mechanism for gain at signal frequency (this is the parametric amplifier, discussed in Chapter 6).

5.3.2. Optimizations

We have formulas for the input and output impedances (Equations 5.102 and 5.100), exchangeable gain (Equation 5.101), and noise temperature (Equation 5.21) of the lower-sideband upconverter. If the input is tuned so that $Z_o + R_s + (S_0/j\omega_s)$ is real, then simultaneously the noise temperature is lowered, and $1/G_e$ is made more negative. Since this tuning is usually beneficial, we henceforth assume that it is done. Thus the output impedance, exchangeable gain, and noise temperature become

$$Z_{out} = R_{out} + jX_{out}$$

$$= R_s + \frac{S_0}{j\omega_i} - \frac{m_1^2\omega_c^2}{\omega_s\omega_i}\frac{R_s^2}{R_s + R_o}, \qquad (5.103)$$

$$\frac{1}{G_e} = \left(\frac{\omega_s}{m_1\omega_c}\right)^2 \frac{(R_s + R_o)^2}{R_s R_o} - \frac{\omega_s}{\omega_i}\frac{R_s + R_o}{R_o}, \qquad (5.104)$$

and

$$T = T_d\left[\frac{R_s}{R_o} + \left(\frac{\omega_s}{m_1\omega_c}\right)^2 \frac{(R_s + R_o)^2}{R_s R_o}\right]. \qquad (5.105)$$

Note from Equation 5.103 that the reactive part of the output impedance is simply the reactance of the average elastance S_0; it is natural, therefore, to tune the output, so that $Z_i + R_s + (S_0/j\omega_i)$ is real. If this is done, the input impedance becomes

$$Z_{in} = R_{in} + jX_{in}$$

$$= R_s + \frac{S_0}{j\omega_s} - \frac{m_1^2\omega_c^2}{\omega_s\omega_i}\frac{R_s^2}{R_s + R_i}, \qquad (5.106)$$

where R_i is the real part of the load impedance.

The source resistance R_o can be chosen to optimize the upconverter in various senses; we now turn to a discussion of this.

5.3.2(a). Gain Optimization. If R_o is very small, the exchangeable gain reduces to

$$G_e \approx \frac{R_o}{R_s}\frac{\omega_i}{\omega_s}\frac{m_1^2\omega_c^2}{\omega_s\omega_i - m_1^2\omega_c^2}, \qquad (5.107)$$

whereas for very large R_o, the exchangeable gain approaches

$$G_e = \frac{R_s}{R_o}\left(\frac{m_1\omega_c}{\omega_s}\right)^2 . \tag{5.108}$$

It is clear, therefore, that if $\omega_s\omega_i < m_1^2\omega_c^2$, the exchangeable gain may be made either positive or negative by properly selecting R_o. The exchangeable gain may be made infinite by choosing the source resistance

$$R_o^G = R_s\left(\frac{m_1^2\omega_c^2}{\omega_s\omega_i} - 1\right), \tag{5.109}$$

whereupon the noise temperature becomes

$$T^G = T_d \frac{\omega_s}{\omega_i}\,\frac{m_1^2\omega_c^2 + \omega_i^2}{m_1^2\omega_c^2 - \omega_s\omega_i}, \tag{5.110}$$

and the output resistance becomes zero. (The meaning of the infinite exchangeable gain is that the output impedance is purely reactive, so that the output exchangeable power is infinite.) Although the input resistance R_{in} may be negative, the sum of it and the source resistance is

$$R_{in} + R_o^G = \frac{m_1^2\omega_c^2}{\omega_s\omega_i}\,\frac{R_sR_i}{R_s + R_i} \tag{5.111}$$

and is therefore positive.

The formula for noise temperature is interesting; it is the same as an important noise formula for the parametric amplifier and is plotted in Figures 6.11 and 6.12. Note that if ω_i is very small, then T^G is large; similarly, if ω_i is large enough so that $\omega_s\omega_i$ approaches $m_1^2\omega_c^2$, then T^G becomes large. Therefore, T^G has a minimum, which is

$$(T^G)^T = T_d \frac{2\omega_s}{m_1\omega_c}\left[\frac{\omega_s}{m_1\omega_c} + \sqrt{1 + \left(\frac{\omega_s}{m_1\omega_c}\right)^2}\right]. \tag{5.112}$$

We have encountered this formula before, as the lowest noise temperature T^T of an upper-sideband upconverter. The pump frequency that minimizes T^G to the value of Equation 5.112 is

$$\omega_p = \sqrt{m_1^2\omega_c^2 + \omega_s^2} \tag{5.113}$$

or, for low signal frequencies, simply $m_1 \omega_c$. We have encountered this optimum pump frequency before, under Condition 5.54, although there a different aspect of performance was optimized. When ω_p is chosen according to Equation 5.113, the source resistance R_o^G becomes

$$(R_o^G)^T = R_s \sqrt{1 + \frac{m_1^2 \omega_c^2}{\omega_s^2}} . \tag{5.114}$$

Until now we have assumed that $\omega_s \omega_i < m_1^2 \omega_c^2$; some value of R_o could be found to make the exchangeable gain infinite. Now, what about the case when $\omega_s \omega_i$ is not smaller than $m_1^2 \omega_c^2$? Is gain still possible? Equation 5.104 indicates that the exchangeable gain is always positive; it has a maximum for this value of R_o:

$$R_o^G = R_s \sqrt{1 - \frac{m_1^2 \omega_c^2}{\omega_s \omega_i}} . \tag{5.115}$$

The resulting maximum exchangeable gain is

$$G_e^G = \frac{\left(\dfrac{m_1 \omega_c}{\omega_s}\right)^2}{\left(1 + \sqrt{1 - \dfrac{m_1^2 \omega_c^2}{\omega_s \omega_i}}\right)^2}, \tag{5.116}$$

and the noise temperature is

$$T^G = T_d \frac{1 + \left(\dfrac{\omega_s}{m_1 \omega_c}\right)^2 \left(1 + \sqrt{1 - \dfrac{m_1^2 \omega_c^2}{\omega_s \omega_i}}\right)^2}{\sqrt{1 - \dfrac{m_1^2 \omega_c^2}{\omega_s \omega_i}}} . \tag{5.117}$$

The output resistance is

$$R_{out}^G = R_o^G = R_s \sqrt{1 - \frac{m_1^2 \omega_c^2}{\omega_s \omega_i}} , \tag{5.118}$$

and if the load resistance R_i is equal to this, then so is the input resistance R_{in}.

The region in the ω_s, ω_i -plane for which there is gain is indicated in Figure 5.21. This includes all points for which $\omega_s\omega_i < m_1^2\omega_c^2$, as well as the region in which Equation 5.116 is

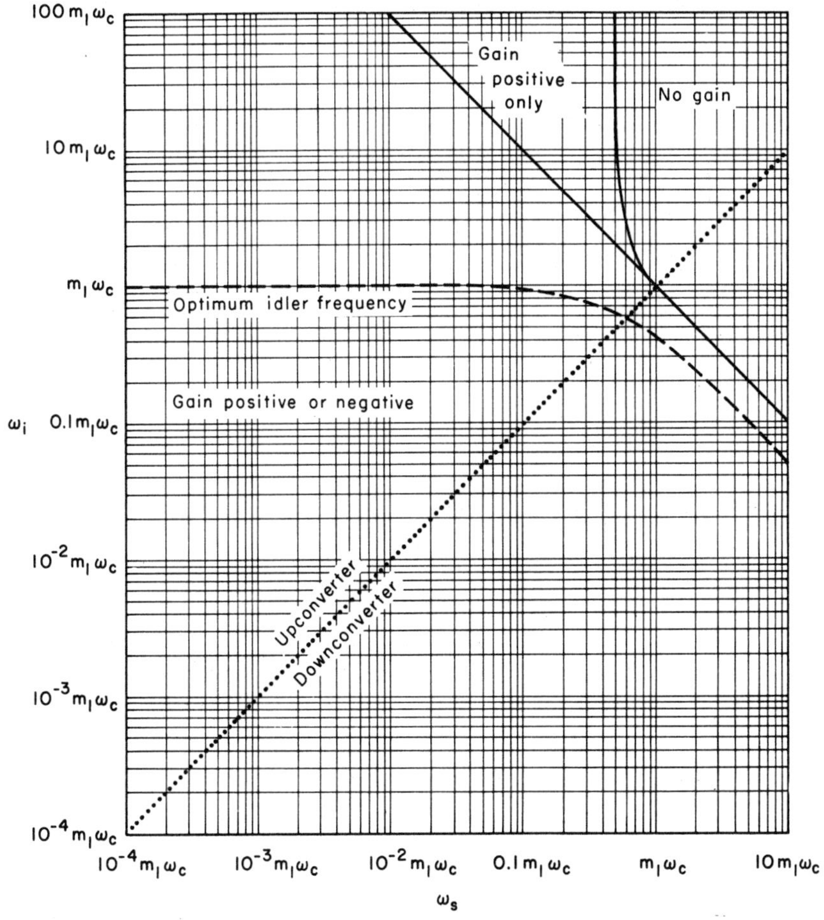

Figure 5.21. Values of ω_s and ω_i for which gain is possible in a lower-sideband upconverter or downconverter. Note the three regions: In one the exchangeable gain can be either positive or negative, depending on the source resistance; in another only positive exchangeable gain is possible, but it can be greater than 1; in the third, the positive exchangeable gain is always less than 1. Also shown is the optimum idler frequency, for which infinite exchangeable gain and lowest noise temperature can be achieved simultaneously.

greater than 1. Also plotted is the idler frequency produced by the optimum pump frequency at which the minimum noise temperature can be realized with infinite gain.

The source resistance that optimizes the gain (that is, makes it infinite if $\omega_s\omega_i < m_1^2\omega_c^2$, and makes it the highest if $\omega_s\omega_i > m_1^2\omega_c^2$) is plotted in Figure 5.22. Finally, the noise temperature T^G, given by Equations 5.110 and 5.117, is plotted in Figure 5.23.

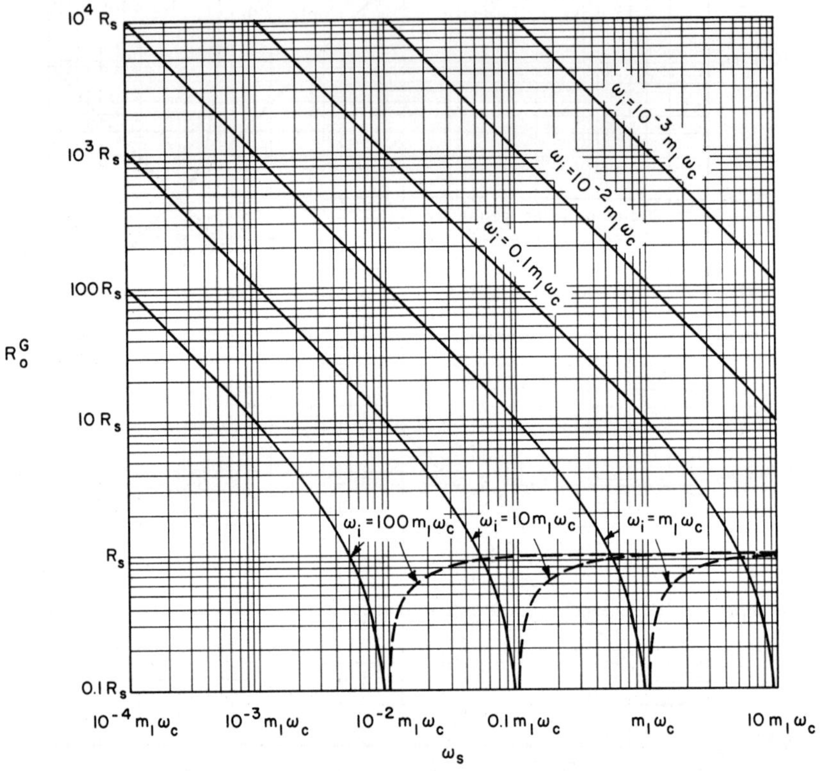

Figure 5.22. Source resistance for optimum gain for lower-sideband upconverter or downconverter. The solid curves are for infinite gain, and the dashed curves are for gain as high as possible.

5.3.2(b). Noise Optimization. Since the noise temperature of Equation 5.105 is the same as for the upper-sideband upconverter, the same value of source resistance, given in Equation 5.50 and plotted in Figure 5.8, optimizes the lower-sideband noise temperature to the same value, given in Equation 5.51 and plotted in Figure 5.7. When this value for R_o is put in Equations 5.103

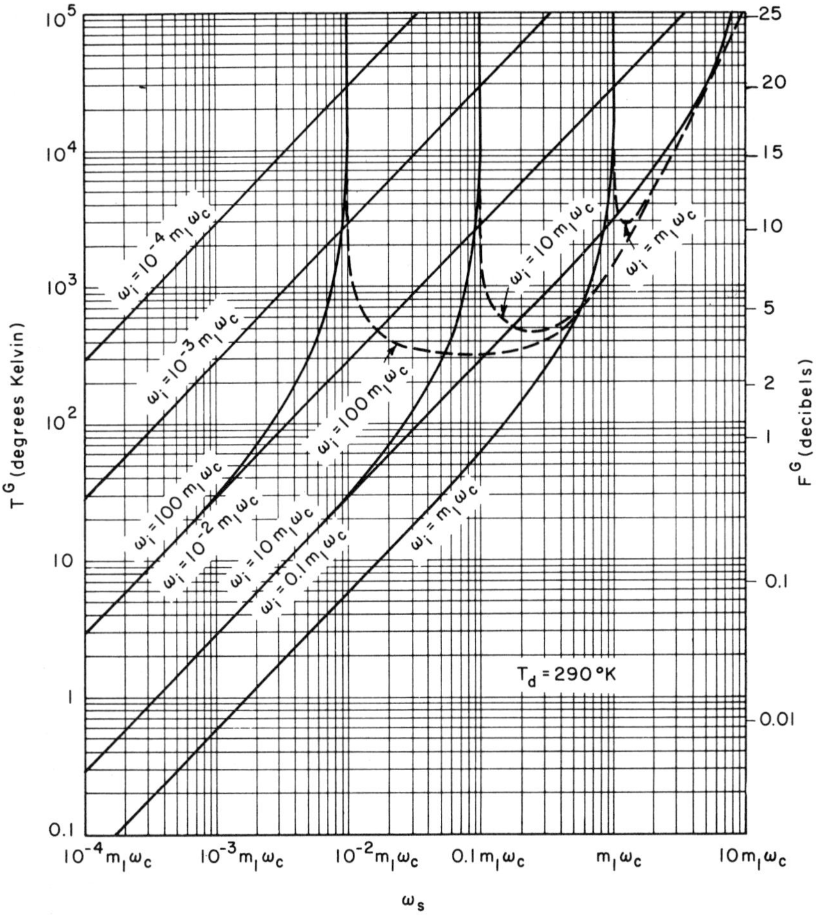

Figure 5.23. Noise temperature for optimum gain of the lower-sideband upconverter or downconverter. The solid curves are for infinite gain, and the dashed curves are for G_e maximum. The varactor is assumed to be at room temperature.

and 5.104, we find that the resulting output resistance and exchangeable gain are

$$R_{out}^{T} = R_s \left(\frac{\omega_p - \sqrt{m_1^2 \omega_c^2 + \omega_s^2}}{\omega_i} \right) \tag{5.119}$$

and

$$G_e^T = \frac{\sqrt{1+\left(\frac{m_1\omega_c}{\omega_s}\right)^2}}{\left[\frac{\omega_s}{m_1\omega_c}+\sqrt{1+\left(\frac{\omega_s}{m_1\omega_c}\right)^2}\right]\left[\frac{\omega_s}{m_1\omega_c}+\sqrt{1+\left(\frac{\omega_s}{m_1\omega_c}\right)^2}-\frac{m_1\omega_c}{\omega_i}\right]} . \tag{5.120}$$

It is clear from Equation 5.120 (and Equation 5.119) that if ω_p is less than its optimum frequency $\sqrt{m_1^2\omega_c^2 + \omega_s^2}$, the output resistance and exchangeable gain are negative, whereas if ω_p is larger, they are positive. Note that if ω_p is equal to its optimum pump frequency, then G_e^T is infinite; we have simultaneously minimized the noise and made the gain infinite. A little thought shows that this condition is the same as setting the gain infinite and then minimizing the resulting noise temperature, Equation 5.110.

5.3.3. Synthesis Techniques

The performance indicated thus far in Section 5.3 can be achieved if all sideband frequencies except ω_s and ω_i are open-circuited, and provided that the coupling networks have no losses. In Section 7.4, conditions are given for pumping varactors with a sinusoidal current. Circuits can be devised deliberately to meet these conditions, and therefore deliberately to achieve the performance indicated in Section 5.3.2.

As with upper-sideband upconverters, the design of such circuits proceeds in two steps. First is the preliminary design, which consists of selecting a varactor and suitable operating frequencies. The entire analysis of Section 5.3 and of Chapter 7 is directed toward preliminary design.

The second step is the actual design of a circuit that simultaneously open-circuits the varactor at the unwanted sidebands and separates the frequencies. The exact conditions to be met are similar to those given in Section 5.2.3(b). Two typical circuits designed specifically to meet these conditions can be obtained from Figures 5.14 and 5.15 by simply replacing everywhere ω_u by ω_i, and $\omega_p - \omega_s$ by $\omega_p + \omega_s$.

An additional problem in lower-sideband upconverters (and downconverters) is possible instability. Such instabilities have, in fact, prevented many lower-sideband frequency converters from operating successfully.

5.3.4. Effect of Idler

We saw in Section 5.2.4 that an idler can greatly improve an upper-sideband upconverter. Simultaneously the noise can be lowered and the gain raised. We now wish to show that similar benefits do not come to the lower-sideband upconverter when we add an idler at frequency $\omega_u = \omega_p + \omega_s$.

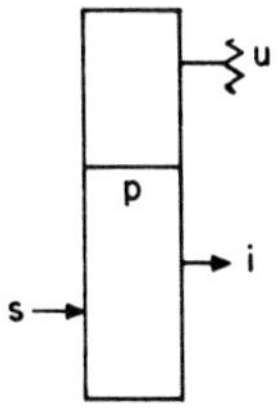

This upconverter with idler is the same as that pictured in Figure 5.17 if the subscripts i and u are interchanged. The basic equations that describe the varactor are the same as Equations 5.69 (we assume for simplicity that $S_2 = 0$), and the load, source, and idler constraints are similar to Equations 5.71 to 5.73. We suppose that the load, source, and idler are tuned. Then, when these equations are solved, we find that the input impedance is given by Equation 5.75, and that the output impedance, exchangeable gain, and noise temperature are

$$Z_{out} = R_s + \frac{S_0}{j\omega_i} - \frac{m_1{}^2\omega_c{}^2}{\omega_s\omega_i}\,\frac{R_s{}^2}{R_s + R_o + \dfrac{m_1{}^2\omega_c{}^2}{\omega_s\omega_u}\,\dfrac{R_s{}^2}{R_s + R_u}}, \tag{5.121}$$

$$\frac{1}{G_e} = \frac{\omega_s{}^2}{m_1{}^2\omega_c{}^2}\,\frac{(R_s + R_o)^2}{R_sR_o} - \frac{m_1{}^2\omega_c{}^2}{\omega_i\omega_u}\,\frac{R_s}{R_o}\,\frac{R_s}{R_s + R_u} + \frac{m_1{}^2\omega_c{}^2}{\omega_u{}^2}\,\frac{R_s}{R_o}\,\frac{R_s{}^2}{(R_s + R_u)^2}$$

$$+ 2\,\frac{R_s}{R_o}\,\frac{\omega_s}{\omega_u}\,\frac{R_s + R_o}{R_s + R_u} - \frac{R_s + R_o}{R_o}\,\frac{\omega_s}{\omega_i}, \tag{5.122}$$

and

$$T = T_d\,\frac{R_s}{R_o}\left[1 + \frac{\omega_s{}^2}{m_1{}^2\omega_c{}^2}\,\frac{(R_s + R_o)^2}{R_s{}^2} + 2\,\frac{\omega_s}{\omega_u}\,\frac{R_s + R_o}{R_s + R_u} + 2\,\frac{m_1{}^2\omega_c{}^2}{\omega_u{}^2}\,\frac{R_s{}^2}{(R_s + R_u)^2}\right]$$

$$+ T_u\,\frac{R_u}{R_o}\,\frac{m_1{}^2\omega_c{}^2}{\omega_u{}^2}\,\frac{R_s{}^2}{(R_s + R_u)^2}. \tag{5.123}$$

It is clear from Equation 5.123 that the idler at frequency ω_u cannot possibly decrease the noise temperature of the lower-sideband upconverter. The lowest noise temperature is obtained with $R_u = \infty$. The idler at ω_u can help to make R_{in} and R_{out} positive and can, in some cases, help to make G_e positive. However, it does so only at the expense of the noise temperature.

In conclusion, an idler at frequency ω_u cannot do much to improve the lower-sideband upconverter, except possibly to prevent spurious oscillations and to improve stability.

5.4. Upper-Sideband Downconverter

The analysis of the upper-sideband downconverter differs from that of the upper-sideband upconverter in only one respect: The upper-sideband frequency ω_u is smaller than, instead of larger than, the input frequency ω_s. The formula

$$\omega_u = \omega_p + \omega_s \tag{5.124}$$

for the upper-sideband frequency seems to imply that ω_u must be larger than ω_s. If, however, we allow the pump frequency to be negative, but larger than $-\omega_s$, Equation 5.124 shows that $\omega_u < \omega_s$. Alternatively, we can consider ourselves to be pumping at a frequency between zero and ω_s and using for output the "upper-sideband" frequency

$$\omega_u = \omega_s - \omega_p . \tag{5.125}$$

From either standpoint, the output frequency is smaller than the input frequency, but the equations of the varactor are not altered (except for possibly replacing S_k by $S_k{}^*$). We can therefore apply the formulas developed for the upper-sideband upconverter to the upper-sideband downconverter. In Sections 5.4.1, 5.4.2, and 5.4.3, we discuss the downconverter without an idler. Then, in Section 5.4.4, we discuss the effect of idling current to increase the gain.

5.4.1. Characteristics

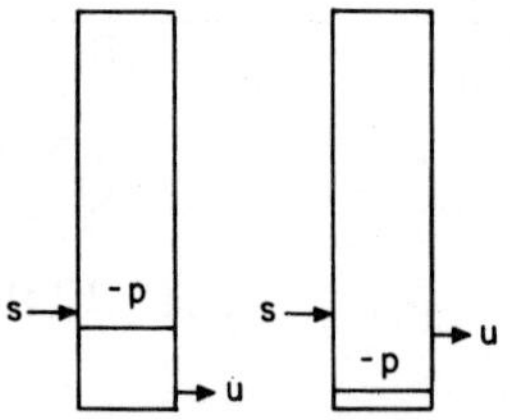

In Section 5.2.1 we derived the input impedance, output impedance, exchangeable gain, and noise temperature of the upper-sideband upconverter. Although the pump frequency was positive, this fact was not used. Therefore, Equations 5.31, 5.16, 5.18, and 5.21 apply also to the upper-sideband downconverter. The noise parameters are given by Equations 5.23, 5.24, 5.25, 5.27, 5.28, and 5.29.

5.4.2. Optimizations

If the input and output are tuned, the input impedance, output impedance, exchangeable gain, and noise temperature reduce to the simplified forms of Equations 5.37, 5.33, 5.34, and 5.35.

5.4.2(a). Gain Optimization. The upper-sideband downconverter cannot give gain but instead always has a loss (see Equation 5.18). If the series resistance vanishes, the loss is a minimum (the "gain" becomes ω_u/ω_s, which is less than unity); effect of the series resistance is to make the loss somewhat greater. For given input and output frequencies, the gain is maximized (that is, the loss is minimized) by choosing the source resistance according to Equation 5.39; the resulting gain, output resistance, and noise temperature are given by the same formulas that describe the upper-sideband upconverter, Equations 5.38, 5.40, and 5.41. These are plotted in Figures 5.3, 5.5, and 5.6 (the same figures are used for both the upconverter and the downconverter).

5.4.2(b). Noise Optimization. Since the noise temperature of Equation 5.21 is independent of output frequency, its optimization clearly cannot be affected by the fact that the output frequency is lower than the input frequency. The same minimum value of Equation 5.51 is obtained when the source resistance is set according to Equation 5.50. The resulting gain, Equation 5.52, cannot exceed 1. In using Equation 5.53 for the output resistance, bear in mind that ω_p appearing there must now be considered as a negative number. The pertinent quantities are plotted in Figures 5.7 to 5.10.

5.4.2(c). Compromises. The concept of wanting a compromise source impedance to minimize a cascade noise figure applies as well to downconverters as it does to upconverters. One source resistance minimizes the loss, and another minimizes the noise. A value of R_o between these two will serve to minimize the cascade noise figure.

5.4.2(d). Examples. We give two examples of the use of the formulas for upper-sideband downconverters. First, suppose we wish to downconvert an incoming signal at 5 Gc so it can be amplified by a parametric amplifier at 500 Mc, with noise temperature 50°. We have a choice of using either a mixer crystal, with 5-db loss and a 6-db noise figure, or a varactor with m_1f_c = 20 Gc. Which is preferable for low noise? With a crystal mixer, the cascade noise formula indicates a contribution of 870° from the crystal (6-db noise figure), and 50° × 3 from the following stage. The over-all noise temperature is 1020°. On the other hand, for a varactor, $f_s/m_1f_c = 0.25$ and $f_u/m_1f_c = 0.025$. If we optimize the first-stage noise figure, the gain $G_e{}^T$ is 0.078, and $T^T = 185°$. The cascade formula would predict an over-all noise temperature of 185° + (50°/0.078) = 825°. This is better noise performance than the crystal mixer. Furthermore, a source resistance lying between $R_o{}^T$ and $R_o{}^G$ can be found to reduce the cascade noise temperature still further. The varactor is seen to be, in theory, capable of better performance in spite of its large loss.

Second, suppose we want a sharp filter at 1 Gc. It is proposed that we down-convert to 10 Mc, use a crystal filter (assumed lossless and noise-free, for simplicity), and upconvert again. What is the minimum loss possible, and when adjusted for minimum loss, what is the noise temperature? Also, suppose each frequency converter is adjusted for minimum noise temperature. What would the cascade noise temperature and loss be? Assume $T_d = 290°$, and $m_1 f_c = 10$ Gc. Under optimum noise conditions, the first stage has a loss of 20.5 db and a noise temperature of 64°; the second stage has a gain of 19.6 db and a noise temperature of 0.58°. The over-all loss is 0.9 db, and noise temperature is 129°. (It must be pointed out that in practice inevitable losses in the circuit at the lower frequency will be extremely important because the noise from them is multiplied about one hundred times when referred to the input.) Under optimum gain conditions, on the other hand, the first stage has a loss of 20.1 db and a noise temperature of 298°; the upconverter has a gain of 19.9 db and a noise temperature of 2.93°. The over-all loss is 0.2 db, and the over-all noise temperature is 598°.

5.4.3. Synthesis Techniques

The synthesis techniques discussed in Section 5.2.3 for the upper-sideband upconverter apply just as well to the downconverter. The lumped realization of Figure 5.14 can be used even if $\omega_u < \omega_s$. The coaxial realization in Figure 5.15 does not, however, because we assumed for simplicity that $\omega_s << \omega_p$, a condition that is of course violated by the downconverter. Instead, let us now assume that $\omega_u << \omega_s$. Then the circuit of Figure 5.15 can be used as a downconverter if the output is used as an input and the left-hand terminal pair is used as the output.

5.4.4. Downconverter with Idler

The upper-sideband upconverter is significantly improved by an idler located at frequency $\omega_i = \omega_p - \omega_s$ when ω_i is positive. For the downconverter, ω_p is inherently negative, so that ω_i cannot be positive, and this idler is of no value.

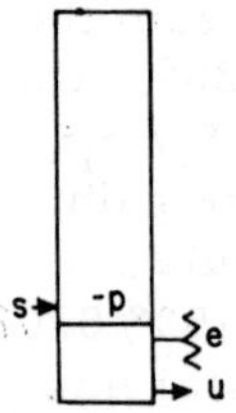

However, consider frequency $\omega_2 = \omega_p + \omega_u$ or, alternatively, $\omega_e = -\omega_2 = |\omega_p| - \omega_u$. Whenever ω_e is positive, that is, whenever ω_u is less than one half of ω_s, then ω_u and ω_e interact through the pump in a fashion to produce negative-resistance type of gain. This idler can help improve the gain of the upper-sideband downconverter, even though it cannot improve the noise temperature.

This downconverter is analyzed the same way the upconverters with idlers are analyzed. We omit the derivation, and simply write down the output impedance, exchangeable gain, and noise temperature, using R_e for the external resistance at frequency ω_e and T_e for its temperature. We assume, as before, that the varactor terminations are tuned.

$$Z_{out} = R_{out} + jX_{out}$$

$$= R_s + \frac{S_0}{j\omega_u} + \frac{m_1^2 \omega_c^2}{\omega_s \omega_u} \frac{R_s^2}{R_s + R_o} - \frac{m_1^2 \omega_c^2}{\omega_u \omega_e} \frac{R_s^2}{R_s + R_e}, \tag{5.126}$$

$$\frac{1}{G_e} = \left(\frac{\omega_s}{m_1 \omega_c}\right)^2 \frac{(R_s + R_o)^2}{R_s R_o} + \frac{\omega_s}{\omega_u} \frac{R_s + R_o}{R_o} - \frac{\omega_s^2}{\omega_u \omega_e} \frac{(R_s + R_o)^2}{R_o (R_s + R_e)}, \tag{5.127}$$

$$T = \frac{T_d R_s}{R_o} + \frac{\omega_s^2}{m_1^2 \omega_c^2} \frac{T_d (R_s + R_o)^2}{R_s R_o} + \frac{T_d R_s + T_e R_e}{R_o} \frac{\omega_s^2}{\omega_e^2} \frac{(R_s + R_o)^2}{(R_s + R_e)^2}. \tag{5.128}$$

These formulas reduce to Equations 5.33, 5.34, and 5.35 when the idler resistance R_e is put infinite. The exchangeable gain can be increased because of the negative term in Equation 5.127; in fact, if ω_e is positive and if $\omega_e \omega_u < m_1^2 \omega_c^2$, this term can predominate, so that the exchangeable gain can become as high as desired. This condition, of course, is precisely the condition that a parametric amplifier with signal and idler frequencies ω_u and ω_e can give gain. The downconverter behaves like an ordinary upper-sideband downconverter with loss, coupled with a parametric-amplification gain mechanism. The same varactor and the same pump are used in both functions.

Observe from Equation 5.128 that the noise temperature is always increased by this idler. Thus, any increases in gain are made at the expense of the noise temperature. This trade between gain and noise temperature is particularly simple if the varactor and the idler termination R_e are at the same temperature. Then R_e can be eliminated from Equations 5.127 and 5.128, to give a relation expressing the trading possible between noise and gain:

$$\frac{T}{T_d} + \frac{\omega_u}{\omega_e} \frac{1}{G_e} = \frac{\omega_s}{\omega_e} + \frac{R_s}{R_o} \frac{2|\omega_p|}{\omega_e} + \left(\frac{\omega_s}{m_1 \omega_c}\right)^2 \frac{(R_s + R_o)^2}{R_s R_o} \frac{|\omega_p|}{\omega_e}. \tag{5.129}$$

In particular, if ω_s is much larger than ω_u,

$$\frac{T}{T_d} + \frac{\omega_u}{\omega_s} \frac{1}{G_e} > 1, \tag{5.130}$$

so that if the noise temperature is low, the exchangeable gain must also be low; similarly, if the gain is high, so that $1/G_e$ is low, then T must be high. Since several inherently positive terms have been omitted from the rather crude Condition 5.130, the noise and gain are actually somewhat worse.

Varactors are not well suited for downconversion, as for example to replace mixers in superheterodyne receivers. If no idler is used, the loss is high, and the i.f. noise appears in the cascade noise formula multiplied at least by the ratio of rf to i.f. frequencies. If an idler is used, more gain can be obtained, so that the contribution to receiver noise temperature from the i.f. stages is reduced, but at the same time the noise from the first stage is increased, as suggested by Condition 5.130. In most practical cases, the over-all noise performance is worse than that from a crystal mixer of good quality.

5.5. Lower-Sideband Downconverter

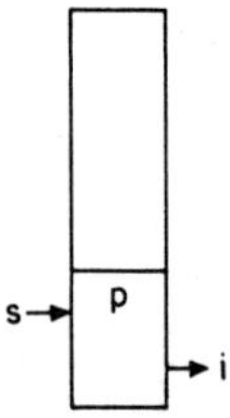

The lower-sideband downconverter is similar to the lower-sideband upconverter. The only change is that the output frequency ω_i is less than ω_s; therefore, the pump frequency, instead of being greater than $2\omega_s$, lies between ω_s and $2\omega_s$.

5.5.1. Characteristics

The input impedance, output impedance, exchangeable gain, and noise temperature are identical in form to those for the lower-sideband upconverter, Equations 5.102, 5.100, 5.101, and 5.21. The input and output impedances can be negative if and only if

$$\omega_s \omega_i < m_1^2 \omega_c^2 , \tag{5.131}$$

and Equation 5.101 predicts that there can be gain only if Condition 5.131 is fulfilled. Note that unless ω_s is greater than $m_1 \omega_c$, Condition 5.131 is automatically satisfied, and for some choice of R_o there can be gain.

5.5.2. Optimizations

If the input and output terminations are tuned, the formulas for input and output impedance, exchangeable gain, and noise temperature are Equations 5.106, 5.103, 5.104, and 5.105.

5.5.2(a). Gain Optimization. If Condition 5.131 holds, then the exchangeable gain can be made negative by choosing R_o small enough. The gain can be made infinite by choosing R_o according to Equation 5.109, whereupon the noise temperature is given by Equation 5.110. If the pump frequency can be varied, the noise temperature can be simultaneously minimized to the value of Equation 5.112 by choosing ω_p according to Equation 5.113.

If Condition 5.131 does not hold, the exchangeable gain is positive and less than 1; it can be maximized (that is, the loss can be minimized) by using the source impedance of Equation 5.115, whereupon the exchangeable gain, noise temperature, and output resistance assume the values of Equations 5.116, 5.117, and 5.118. Note that gain is possible only when Condition 5.131 holds.

5.5.2(b). Noise Optimization. As in the lower-sideband upconverter, the noise temperature is minimized to the value of Equation 5.51 by choosing the source resistance according to Equation 5.50. The resulting formulas for output resistance and exchangeable gain are given in Equations 5.119 and 5.120.

5.5.3. Synthesis Techniques

Synthesis techniques for lower-sideband downconverters are similar to those for upper-sideband downconverters. The circuit of Figure 5.14 can be used with the modifications discussed in Section 5.3.3. The circuit of Figure 5.15 can be used backwards (as discussed in Section 5.4.3) with the changes in frequency given in Section 5.3.3.

5.6. Arbitrary Idler Configurations

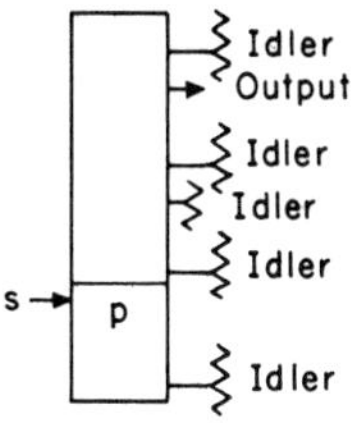

In Section 5.2.4 it was shown that the noise temperature of an upper-sideband upconverter could be lowered by having an idler at frequency $\omega_p - \omega_s$. Is it possible that even better noise performance is obtainable if more idlers are present? The number of possible idler configurations is so large that no useful purpose would be served by analyzing each separately. Instead, we now wish to analyze a very general frequency converter with no restriction on the pumping waveform or the number of idlers present, to show that any such improvement is strictly limited.

We assume that current is allowed to flow at any sideband frequency $n\omega_p + \omega_s$, and that the elastance, as pumped, can have any waveform, that is, any values of the S_k in Equation 4.41, provided only that $S(t)$ is real and lies between S_{min} and S_{max}. The input is at frequency ω_s and output at any sideband, $\ell\omega_p + \omega_s$. This frequency converter is so complicated that we cannot now expect simple formulas for the quantities of interest. Surprisingly, however, there is a simple nontrivial lower limit (probably not achievable) for noise temperature.

In the derivation to follow, we minimize an expression for noise temperature (Equation 5.151) by varying certain parameters (r_k and θ_k). Since the r_k and θ_k parameters do not have a simple physical interpretation, no simple optimization procedure to achieve the limiting noise temperature can be based on the derivation. Furthermore, there is not even any assurance that the r_k and θ_k can physically take on their optimum values; that is, there is no assurance that the minimum to be derived can ever be achieved.

The limiting noise temperature to be derived depends on which sidebands are terminated. If any sidebands are known to be open-circuited, the resulting fundamental limit is somewhat higher. It reduces to Equations 5.51 and 5.93 when the appropriate sidebands are open-circuited. Even if all sidebands are terminated, the new limit is of the same order of magnitude as Equation 5.51; hence, the amount of improvement possible with arbitrary sidebands is definitely limited.

To derive our new fundamental limit on noise temperature, we assume that the varactor is pumped so that $S(t)$ is periodic and constrained to lie between S_{min} and S_{max}:

$$S_{min} \leq S(t) = \sum_{k=-\infty}^{+\infty} S_k e^{jk\omega_p t} \leq S_{max}. \tag{5.132}$$

We suppose that the input to the varactor is at frequency ω_s and the output is at some sideband $\ell\omega_p + \omega_s$, where ℓ is an integer, either positive or negative. We allow the varactor to be terminated at all other sideband frequencies of the form $k\omega_p + \omega_s$ in an impedance Z_k, where Z_k may have either a positive or a negative real part R_k. Since the sideband terminations are in general noisy, we may define a noise "temperature" T_k for each, with the result that the open-circuit noise voltage at each termination is‡

$$\overline{|E_{nxk}|^2} = 4kT_kR_k\,\Delta f\,. \tag{5.133}$$

If any R_k is negative, the corresponding T_k is also.

For simplicity, we set up the problem in matrix form. The equations that govern the parametrically pumped varactor are

$$\underset{\sim}{V} = \underset{\sim}{Z}_c\underset{\sim}{I} + \underset{\sim}{E}_n\,, \tag{5.134}$$

where $\underset{\sim}{V}$ is the column matrix of Fourier coefficients of voltage at the various frequencies $k\omega_p + \omega_s$, in ascending order of k from $-\infty$ to $+\infty$; $\underset{\sim}{I}$ is the similar column matrix of varactor currents; $\underset{\sim}{E}_n$ is a column matrix of noise voltages representing thermal noise of the varactor series resistance, each with mean-squared value‡ $4kT_dR_s\,\Delta f$; $\underset{\sim}{Z}_c$ is the varactor conversion matrix with diagonal elements

$$(\underset{\sim}{Z}_c)_{kk} = R_s + \frac{S_0}{j(k\omega_p + \omega_s)} \tag{5.135}$$

‡The noise voltages are effective rms values instead of half-amplitudes. See Section 2.5.1(a).

and off-diagonal elements

$$(\underset{\sim}{Z}_c)_{nk} = \frac{S_{n-k}}{j(k\omega_p + \omega_s)} , \tag{5.136}$$

and, in particular, elements in the ω_s column

$$(\underset{\sim}{Z}_c)_{k0} = \frac{S_k}{j\omega_s} \qquad (k \neq 0). \tag{5.137}$$

This matrix equation is written out as Equations 4.42.

The sideband terminations are accounted for by another matrix equation,

$$\underset{\sim}{V} = -\underset{\sim}{Z}_{term}\underset{\sim}{I} + \underset{\sim}{E}_{nx} \tag{5.138}$$

where $\underset{\sim}{E}_{nx}$ is a column matrix of the E_{nxk} noise voltages, and $\underset{\sim}{Z}_{term}$ is a diagonal matrix with entries

$$(\underset{\sim}{Z}_{term})_{kk} = Z_k . \tag{5.139}$$

At the outset, we agree that if any sidebands are known to be open-circuited, we set the corresponding current to zero and delete the corresponding voltages and the corresponding rows and columns of both $\underset{\sim}{Z}_c$ and $\underset{\sim}{Z}_{term}$.

To solve for the currents, we eliminate $\underset{\sim}{V}$ from Equations 5.134 and 5.138, to obtain

$$\underset{\sim}{I} = \underset{\sim}{Y}(\underset{\sim}{E}_{nx} - \underset{\sim}{E}_n) , \tag{5.140}$$

where the matrix $\underset{\sim}{Y}$ is

$$\underset{\sim}{Y} = (\underset{\sim}{Z}_c + \underset{\sim}{Z}_{term})^{-1} . \tag{5.141}$$

We now calculate the noise temperature in terms of the matrix $\underset{\sim}{Y}$. We consider all noise from the varactor series resistance and all noise from terminations except the input and the output to be internal to the amplifier. After accounting for all such sources, we find the mean-squared output current caused by internal sources to be

$$\Sigma_k |Y_{\ell k}|^2 4kT_d R_s \, \Delta f + \Sigma_k |Y_{\ell k}|^2 4kT_k R_k \, \Delta f , \tag{5.142}$$

where the first sum, representing noise from the varactor series resistance, is over all k, and the second sum, representing noise from the terminations, is over all k except zero and ℓ. If we make the convention that all sums of this sort are to omit $k = 0$, and that in all such sums we assume $R_\ell = 0$, then Expression 5.142 becomes

$$|Y_{\ell 0}|^2 4kT_d R_s \;\Delta f + \Sigma_k |Y_{\ell k}|^2 (T_d R_s + T_k R_k) 4k \;\Delta f. \quad (5.143)$$

The mean-squared output current arising from a source at temperature T_0 at frequency ω_s is

$$|Y_{\ell 0}|^2 4kT_0 R_0 \;\Delta f\,, \quad (5.144)$$

so that the noise temperature T becomes

$$T = \frac{T_d R_s + \Sigma_k (T_d R_s + T_k R_k) \left|\dfrac{Y_{\ell k}}{Y_{\ell 0}}\right|^2}{R_0}, \quad (5.145)$$

where the sum does not by our convention include $k = 0$.

We wish to eliminate R_0 from this expression. From the definition of $\underset{\sim}{Y}$,

$$\underset{\sim}{Y}(\underset{\sim}{Z}_c + \underset{\sim}{Z}_{term}) = \underset{\sim}{1}, \quad (5.146)$$

where $\underset{\sim}{1}$ is the unit matrix. The $\ell, 0$ entry of this equation is

$$0 = \Sigma_k Y_{\ell k} (\underset{\sim}{Z}_c + \underset{\sim}{Z}_{term})_{k0} + Y_{\ell 0} (\underset{\sim}{Z}_c + \underset{\sim}{Z}_{term})_{00}$$

$$= \Sigma_k Y_{\ell k} \frac{S_k}{j\omega_s} + Y_{\ell 0}\left(Z_0 + R_s + \frac{S_0}{j\omega_s}\right), \quad (5.147)$$

so that

$$R_0 = -R_s - \mathrm{Re}\, \Sigma_k \frac{Y_{\ell k}}{Y_{\ell 0}} \frac{S_k}{j\omega_s}. \quad (5.148)$$

We define the parameters A_k, r_k, and θ_k as follows:

$$A_k = \frac{|S_k|^2}{\omega_s{}^2 R_s{}^2} \frac{T_d R_s}{T_d R_s + T_k R_k}, \quad (5.149)$$

$$r_k e^{j\theta k} = -\frac{Y_{\ell k}}{Y_{\ell 0}} \frac{S_k}{j\omega_s R_s}, \tag{5.150}$$

where r_k and θ_k are the magnitude and phase of the quantity shown. In terms of these parameters, the substitution of Equation 5.148 for R_0 into the expression of Equation 5.145 for T yields

$$T = T_d \frac{\Sigma_k(r_k^2/A_k) + 1}{\Sigma_k r_k \cos\theta_k - 1}. \tag{5.151}$$

This is an expression for the actual noise temperature of the upconverter. It is, of course, much too complicated to be evaluated directly, except in the simplest cases, because the r_k and θ_k parameters depend in a complicated way on the pumping waveform and the sideband terminations. We might suppose that by varying r_k and θ_k in an arbitrary way, the expression for noise temperature, Equation 5.151, could be made small. However, this is not the case: When the r_k and θ_k are varied, Equation 5.151 has a positive minimum. We shall calculate this minimum and call it a lower limit for noise temperature. There is no assurance that this limit can be achieved, because we do not require that the values of r_k and θ_k for the minimum correspond to a physically realizable situation.

It is clear from Equation 5.151 that for the lowest positive T we want all $\theta_k = 0$ (or a multiple of 2π). With this optimization, we have

$$T = T_d \frac{1 + \Sigma_k(r_k^2/A_k)}{\Sigma_k r_k - 1}. \tag{5.152}$$

If we differentiate T by r_k and set the result equal to zero, we obtain a formula that when resubstituted in Equation 5.152 yields for each k

$$\frac{T_{min}}{T_d} = \frac{2r_k}{A_k}. \tag{5.153}$$

If we resubstitute this formula for r_k in Equation 5.152, the result is a quadratic expression for T_{min}:

$$\left(\frac{T_{min}}{T_d}\right)^2 \Sigma_k A_k - \frac{4T_{min}}{T_d} - 4 = 0, \tag{5.154}$$

of which the positive root is

$$T_{min} = \frac{2T_d}{\Sigma_k A_k} \left(1 + \sqrt{1 + \Sigma_k A_k}\right) . \tag{5.155}$$

This minimum is expressed in terms of the A_k parameters, given in Equation 5.149. If the pumping waveform and the terminating resistances and temperatures are known, the result is a function of only the varactor temperature T_d and the input frequency ω_s. In specific cases, it can be evaluated fairly easily. The higher $\Sigma_k A_k$ is, the lower is the fundamental limit T_{min}. It is of interest, therefore, to ask how high $\Sigma_k A_k$ can be under various circumstances.

We define m_t

$$m_t^2 = \sum_{\substack{k=-\infty \\ k\neq 0}}^{\infty} \frac{|S_k|^2}{(S_{max} - S_{min})^2} \tag{5.156}$$

as a kind of total modulation ratio. The summation in Equation 5.156 is over all k from $-\infty$ to $+\infty$, but if some of the sidebands are known to be open-circuited, m_t can be replaced in Equation 5.157 by a similar sum, omitting the open-circuit sidebands.

Clearly, the highest $\Sigma_k A_k$ occurs when the sideband terminations are noise-free. Thus, if we set each $R_k T_k$ to zero, Equation 5.155 becomes, in terms of m_t,

$$T_{min} = 2T_d \frac{\omega_s}{m_t \omega_c} \left[\frac{\omega_s}{m_t \omega_c} + \sqrt{1 + \left(\frac{\omega_s}{m_t \omega_c} \right)^2} \right], \tag{5.157}$$

which is the same form as Equations 5.51 and 5.93, except that m_t is used in place of m_1 and $\sqrt{2}\, m_1$, respectively. Observe that if the summation in Equation 5.156 were to extend over only $k = +1$ or $k = -1$, as would be appropriate for simple upconverters, then the limit of Equation 5.157 reduces to Equation 5.51; similarly for the double-sideband upconverter with finite terminations at $k = +1$ and $k = -1$, m_t can be replaced by a sum like Equation 5.156 over only +1 and -1, and Equation 5.157 reduces to Equation 5.93.

Just as the requirement that $S(t)$ lie between S_{min} and S_{max} restricts m_1 to certain values, m_t is restricted to be less than 0.5, as is shown in Chapter 7 (other restrictions for specific types of pumping are also derived in Chapter 7). For practical varactors, m_t is seldom over 50 per cent higher than m_1.

Noise from the terminations can only increase the noise temperature above what is predicted by Equation 5.157, so that this quantity is a lower limit for noise temperature, regardless of the temperature of the sideband terminations.

Very few assumptions were required to derive the limit of Equation 5.157. Note, in particular,

1. The elastance waveform is not restricted.
2. The output frequency must be different from the input frequency, but it may be either greater or less, so that the restriction applies to both upconverters and downconverters.
3. The output may be at any sideband (ℓ is not specified), and an arbitrary number of idlers can be terminated.
4. The sideband terminations need not be passive.
5. The sideband terminations can be free of noise, and in particular can be noise-free negative resistances, which, if used directly for amplification, would have a zero noise temperature.
6. The termination can even have synchronously pumped frequency converters in it, so that the matrix $\underset{\approx}{Z}_{term}$ is not diagonal, provided only that these do not interact with the signal frequency (so that the elements of $\underset{\approx}{Z}_c + \underset{\approx}{Z}_{term}$ along the ω_s column are given by Equation 5.137.

Thus the limit derived seems to be the minimum penalty for merely having the signal enter the varactor and come back out again, regardless of how the signal is processed at the sidebands. Even if a noise-free amplifier were available at some frequency, we must pay this penalty if we use a varactor to couple to it.

Note particularly that we have not shown that this limit is achievable; in general, it probably is not, although we have just seen some cases in which it is. However, since in practice the minimum temperature given by Equation 5.157 is not much below that of Equation 5.51, the improvement made possible by adding idlers is strictly limited.

5.7. Comparison

It is evident from this chapter that the various frequency converters have much in common. The lowest noise temperatures in the various cases are, if not actually equal, at least identical in form. The choice of a pump frequency of approximately $m_1 \omega_c$ is indicated in several instances to optimize this or that aspect of the frequency converter. The optimization procedures for lowest noise and for highest gain are different but similar.

For comparison of the various frequency converters, we list in Table 5.1 many of the pertinent formulas. The explanation of each row is given in the caption. First, the fundamental noise limits are listed, followed by some of the operating conditions

Row		Upper-Sideband Upconverter	Lower-Sideband Upconverter $\omega_s\omega_i < m_1^2\omega_c^2$	Lower-Sideband Upconverter $\omega_s\omega_i > m_1^2\omega_c^2$
1	$T^T = T_d \frac{2\omega_s}{m\omega_c}\left[\frac{\omega_s}{m\omega_c} + \sqrt{1 + \left(\frac{\omega_s}{m\omega_c}\right)^2}\right]$	$m = m_1$	$m = m_1$	$m = m_1$
2	$R_o^T = R_s\sqrt{1 + \left(\frac{m\omega_c}{\omega_s}\right)^2}$	$m = m_1$	$m = m_1$	$m = m_1$
3	$R_{out}^T = R_s\left(1 + \frac{\sqrt{m_1^2\omega_c^2 + \omega_s^2} - \omega_s}{\omega}\right)$	$\omega = \omega_u$	$\omega = -\omega_i$	$\omega = -\omega_i$
4	$G_e^T = \frac{\sqrt{1 + \left(\frac{m_1\omega_c}{\omega_s}\right)^2}}{\left[\frac{\omega_s}{m_1\omega_c} + \sqrt{1 + \left(\frac{\omega_s}{m_1\omega_c}\right)^2}\right]\left[\frac{\omega_s}{m_1\omega_c} + \sqrt{1 + \left(\frac{\omega_s}{m_1\omega_c}\right)^2} + \frac{m_1\omega_c}{\omega}\right]}$	$\omega = \omega_u$	$\omega = -\omega_i$	$\omega = -\omega_i$
5a	Optimum pump frequency	$\sqrt{m_1^2\omega_c^2 + \omega_s^2} - 2\omega_s$	$\sqrt{m_1^2\omega_c^2 + \omega_s^2}$	
5b	Reason for choosing this optimum pump frequency	G_e^T half-value	$G_e^T = \infty$	
6	Condition of possible gain	$\omega_u > \frac{m_1\omega_c}{\frac{m_1\omega_c}{\omega_s} - 2}$	whole region	$\omega_i < \frac{m_1\omega_c}{2 - \frac{m_1\omega_c}{\omega_s}}$
7	A: $G_e^G = \frac{\left(\frac{m_1\omega_c}{\omega_s}\right)^2}{\left(1 + \sqrt{1 + \frac{m_1^2\omega_c^2}{\omega_s\omega}}\right)^2}$ or B: $G_e^G = \infty$	A; $\omega = \omega_u$	B	A; $\omega = -\omega_i$
8	A: $T^G = T_D \frac{1 + \left(\frac{\omega_s}{m_1\omega_c}\right)^2\left(1 + \sqrt{1 + \frac{m_1^2\omega_c^2}{\omega_s\omega}}\right)^2}{\sqrt{1 + \frac{m_1^2\omega_c^2}{\omega_s\omega}}}$ or B: $T^G = T_d \frac{\omega_s}{\omega_i}\frac{m_1^2\omega_c^2 + \omega_i^2}{m_1^2\omega_c^2 - \omega_s\omega_i}$	A; $\omega = \omega_u$	B	A; $\omega = -\omega_i$
9	A: $R_o^G = R_s\sqrt{1 + \frac{m_1^2\omega_c^2}{\omega_s\omega}}$ or B: $R_o^G = R_s\left(\frac{m_1^2\omega_c^2}{\omega_s\omega_i} - 1\right)$	A; $\omega = \omega_u$	B	A; $\omega = -\omega_i$
10	A: $R_{out}^G = R_s\sqrt{1 + \frac{m_1^2\omega_c^2}{\omega_s\omega}}$ or B: $R_{out}^G = 0$	A; $\omega = \omega_u$	B	A; $\omega = -\omega_i$
11a	Optimum pump frequency	$0.427\frac{m_1^2\omega_c^2}{\omega_s} - \omega_s$	$\sqrt{m_1^2\omega_c^2 + \omega_s^2}$	
11b	Reason for choosing this optimum pump frequency	G_e^G half-value	$T = T^T$	

Table 5.1. Comparison of Various Frequency

Rows 1 to 5, conditions for optimum noise temperature. In rows 1 to 4, use the appropriate values given for m or ω in the formulas.

Row 1: Minimum noise temperature.
Row 2: Source resistance for this minimum.
Row 3: Resulting output resistance.
Row 4: Resulting exchangeable gain.
Row 5a: Optimum pump frequency.
Row 5b: Reason for choosing this frequency.

Upper-Sideband Downconverter	Lower-Sideband Downconverter $\omega_s\omega_i < m_1^2\omega_c^2$	Lower-Sideband Downconverter $\omega_s\omega_i > m_1^2\omega_c^2$	Upper-Sideband Upconverter with Idler (Noise Optimization only)	Upconverter or Downconverter with Arbitrary Idlers	Parametric Amplifier with Single Idler	Parametric Amplifier with Arbitrary Idlers	Row
$m = m_1$	$m = m_1$	$m = m_1$	$m = \sqrt{2}m_1$	$m = m_t$	$m = m_1$	$m = m_t$	1
$m = m_1$	$m = m_1$	$m = m_1$	$m = \sqrt{2}m_1$				2
$\omega = \omega_u$	$\omega = -\omega_i$	$\omega = -\omega_i$	$R_{out}^T = 2R_s\frac{\omega_p}{\omega_u}$				3
$\omega = \omega_u$	$\omega = -\omega_i$	$\omega = -\omega_i$	$G_e^T = \frac{2m_1^2\omega_c^2}{\omega_s\omega_u}$				4
$2\omega_s - \sqrt{m_1^2\omega_c^2 + \omega_s^2}$	$\sqrt{m_1^2\omega_c^2 + \omega_s^2}$		$\sqrt{2m_1^2\omega_c^2 + \omega_s^2}$		$\sqrt{m_1^2\omega_c^2 + \omega_s^2}$		5a
G_e^T half-value	$G_e^T = \infty$		T^T achieved with $R_i = 0$		T^T achieved with $R_i = 0$		5b
no gain	whole region	no gain			$\omega_s\omega_i < m_1^2\omega_c^2$		6
A; $\omega = \omega_u$	B	A; $\omega = -\omega_i$					7
A; $\omega = \omega_u$	B	A; $\omega = -\omega_i$			B (for any gain) if $R_i = 0$		8
A; $\omega = \omega_u$	B	A; $\omega = -\omega_i$					9
A; $\omega = \omega_u$	B	A; $\omega = -\omega_i$					10
$\omega_s - 0.427\frac{m_1^2\omega_c^2}{\omega_s}$	$\sqrt{m_1^2\omega_c^2 + \omega_s^2}$						11a
G_e^G half-value	$T = T^T$						11b

Converters and Parametric Amplifiers

Rows 6 to 11, conditions for optimum gain. In rows 7 to 10, select either formula A or B and use the appropriate value for ω.

Row 6: Frequencies at which gain can be achieved.
Row 7: Highest exchangeable gain.
Row 8: Noise temperature for optimized gain.
Row 9: Source resistance to optimize the gain.
Row 10: Resulting output resistance.
Row 11a: Optimum pump frequency.
Row 11b: Reason for choosing this frequency.

that achieve the minimum noise. Finally, various conditions for optimum gain are compared.

The two most important concepts of this chapter are the fundamental limit on noise performance, caused by the varactor series resistance, and the optimum pump frequencies encountered. These two concepts also apply to parametric amplifiers, as the discussion of Chapter 6 demonstrates. For comparison, some of the formulas relating to parametric amplifiers are given in Table 5.1. Note, in particular, that the fundamental limit on noise performance is identical to that for the upconverters.

5.8. Summary

Several important results were derived, which are the following:

1. Thermal noise of the varactor series resistance provides a fundamental limit on the noise performance obtainable from the frequency converters.
2. A figure of merit for the varactor, as pumped, is the quantity we have called $m_1 \omega_c$.
3. The upper-sideband upconverter is stable, but its gain is limited, and for high signal frequencies no gain is possible.
4. When the upper-sideband upconverter is the first stage in a low-noise system, a compromise optimization is desirable for minimizing the over-all noise figure.
5. An idler at the lower sideband can simultaneously lower the noise temperature and increase the gain of an upper-sideband upconverter.
6. The lower-sideband upconverter is potentially unstable, and can have negative input and output resistances.
7. The exchangeable gain of the lower-sideband upconverter can be made infinite, and simultaneously the noise can be reduced to its fundamental limit (this simultaneous optimization requires that the pump frequency and source resistance both be chosen properly).
8. The upper-sideband downconverter cannot give gain, but the noise temperature can be reduced to the same fundamental limit.
9. The lower-sideband downconverter can give gain for any input frequency, but only if the output frequency is low enough.
10. If the signal frequency is high enough, the exchangeable gain of the lower-sideband downconverter can be made infinite and the noise minimized simultaneously.
11. Providing an idler at the upper sideband cannot improve the lower-sideband upconverter or downconverter, except to help make the input resistance positive.
12. In general, idlers can improve the noise temperature, although only by a strictly limited amount.

Many properties of frequency converters were not discussed. We have not given any fundamental limits on bandwidth, such as those by Kuh,[9] nor have we treated the cascading problem with lower-sideband devices. We have not derived formulas for many of the interesting frequency converters with idlers. We did not study several cascade combinations of frequency converters that are of practical importance. We did not consider the effect of losses in the coupling networks, nor did we discuss multiple-varactor upconverters, such as traveling-wave varieties. Although the theory indicates, for the most part, what must be done to achieve the fundamental limits derived, we did not study exhaustively the practical details involved in separating the frequencies and providing suitable terminations. Pumping is covered in Chapter 7.

This does not, of course, indicate that the theory is not "practical." We believe that it is practical, in the best sense of the word. It provides the basis for a rational design of upconverters; it indicates the quality of varactor necessary to achieve given performance; it indicates, although in somewhat unorthodox terms, how to attain the ultimate performance.

REFERENCES

1. Bauer, H., and H. Rothe, "Der äquivalente Rauschvierpol als Wellenvierpol," Arch. der Elektrischen Übertragung, 10, 241-252 (1956).
2. Blackwell, L. A., and K. L. Kotzebue, Semiconductor-Diode Parametric Amplifiers, Prentice-Hall, Inc., Englewood Cliffs, N. J. (1961).
3. Breon, R. K., "The Effects of Idlers on Varactor Up-Converters," S. B. Thesis, Department of Electrical Engineering, M. I. T., Cambridge (June, 1961).
4. Breon, R. K., The Effects of Idlers on Varactor Up-Converters, Energy Conversion Group Unpublished Internal Memorandum No. 52, Department of Electrical Engineering, M. I. T. (May, 1961).
5. Haus, H. A., and R. B. Adler, "An Extension of the Noise Figure Definition," Proc. IRE, 45, 690-691 (May, 1957).
6. Haus, H. A., and R. B. Adler, "Optimum Noise Performance of Linear Amplifiers," Proc. IRE, 46, 1517-1533 (August, 1958).
7. Haus, H. A., and R. B. Adler, Circuit Theory of Linear Noisy Networks, The Technology Press and John Wiley & Sons, Inc., New York (1959).
8. Kotzebue, K. L., "Optimum Noise Performance of Parametric Amplifiers," Proc. IRE, 48, 1324-1325 (July, 1960).
9. Kuh, E. S., "Theory and Design of Wide-Band Parametric Converters," Proc. IRE, 50, 31-38 (January, 1962).
10. Kurokawa, K., and M. Uenohara, "Minimum Noise Figure of the Variable-Capacitance Amplifier," Bell System Tech. J., 40, 695-722 (May, 1961).
11. Leenov, D., "Gain and Noise Figure of a Variable-Capacitance Up-Converter," Bell System Tech. J., 37, 989-1008 (July, 1958).
12. Penfield, P., Jr., "Wave Representation of Amplifier Noise," IRE Trans. on Circuit Theory, CT-9, 84-86 (March, 1962).
13. Rothe, H., and W. Dahlke, "Theory of Noisy Fourpoles," Proc. IRE, 44, 811-818 (June, 1956).

Chapter 6

PARAMETRIC AMPLIFIERS

One of the most exotic applications of varactors is in negative-resistance parametric amplifiers (also called reactance amplifiers). The varactor is pumped strongly at some frequency ω_p, and a signal is introduced at some other frequency ω_s. Sidebands of the form $n\omega_p + \omega_s$ are generated automatically. The impedance of the varactor at frequency ω_s depends on the pumping and on the varactor terminations at the sidebands. For some sideband terminations, this impedance has a negative real part, so that power comes out of the varactor at frequency ω_s. This device can, therefore, operate as an amplifier with gain; in Chapter 6 we investigate this sort of operation, especially the limitations caused by the varactor series resistance. The series resistance introduces noise and limits the pump frequencies at which gain is possible.

In Section 6.2 we treat the simplest case, that of a single varactor with only one idler (sideband at which current flows). Use of several varactors, especially in transmission-line systems or systems with high symmetry, is promising; in Section 6.3 we show that the fundamental limits derived in Section 6.2 apply also to many configurations of more than one varactor. The use of extra idlers, investigated in Section 6.4, can lead to a slight improvement in noise, but only with more complicated circuitry.

Because the noise formulas of Section 6.2 are so important, we devote all of Section 6.5 to them. Many special cases and asymptotic limits are discussed.

In some applications (particularly radiometry), the so-called degenerate parametric amplifier, with ω_s and $\omega_p - \omega_s$ very close, can be used. This amplifier is examined in Section 6.6, along with the phase-sensitive amplifier, for which $\omega_p = 2\,\omega_s$. Finally, in Section 6.7, we consider some synthesis procedures for parametric amplifiers.

In Section 6.1 are the major results to be derived in detail later. This discussion will suffice for those who are unwilling or unable to go through the complete derivations, and for others it is an appropriate introduction. Note, however, that Section 6.5 is not summarized in Section 6.1.

The various types of parametric amplifiers can be conveniently represented by the diagrams of Figure 1.2. Figure 6.1 shows several parametric amplifiers, and the caption indicates the

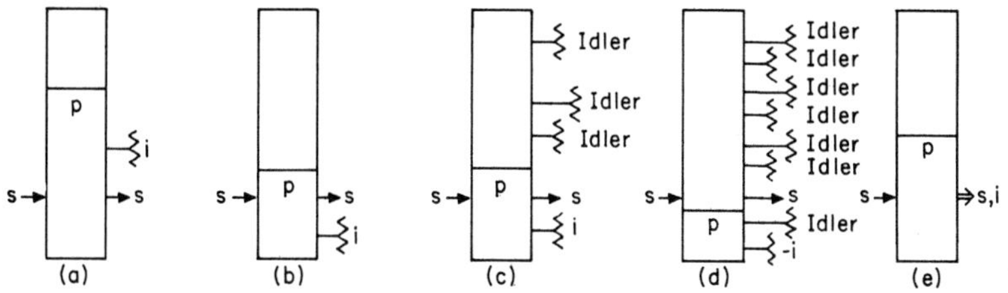

Figure 6.1. Typical parametric amplifiers. Symbols are the following: s = signal frequency, p = pump frequency, i = p - s = idler frequency.
(a) and (b). Simple parametric amplifier, Sections 6.2 and 6.3.
(c) and (d). Parametric amplifier with multiple idlers, Section 6.4.
(e). Degenerate parametric amplifier, Section 6.6.

section of Chapter 6 in which they are treated. Note that for each amplifier the output frequency is the same as the input frequency (otherwise it would be a frequency converter) but that the mechanism for gain necessarily involves flow of current at one or more sideband frequencies.

6.1. Introduction and Major Results

In the derivations to follow, we make a number of assumptions:

1. The varactor is described at all frequencies of interest by the equivalent circuit of Figure 4.1, with a constant series resistance.
2. The varactor is pumped at some frequency ω_p, so that its incremental elastance is given by Equation 4.41. In Section 6.4.1, we further assume, for simplicity, that S is sinusoidal.
3. The voltages and currents at all frequencies except the pump frequency and its harmonics are small, so that the linearized Equations 4.42 hold.
4. A suitable signal-frequency circuit is available, so that the device operates as a linear amplifier.
5. The only noise present is that caused by thermal noise in the varactor series resistance and the terminations.
6. The varactor is open-circuited at all but certain frequencies (except in Section 6.4).
7. For most of the formulas, we assume that the terminations tune out the average varactor elastance S_0.
8. We assume that ω_s is not very close to any of the generated sidebands (except in Section 6.6).

Assumptions 1, 2, 3, and 4 serve to define our field of interest: small-signal behavior of parametrically pumped varactors as amplifiers. Assumption 5, discussed in Section 4. 3, is believed to be very reasonable. Assumption 6 is made for simplicity; any other assumption would allow sideband power to be dissipated, at least in the varactor series resistance. Assumption 7 is justified because it is usually profitable to tune the terminations. Assumption 8 is made to enable us to adjust each of the sideband terminations independently, at least in principle. This assumption is not valid for the degenerate amplifier; see Section 6. 6.

The questions we shall ask about the various parametric amplifiers analyzed here are these: First, for what pump and signal frequencies and for what terminations is gain possible? Second, what is the resulting noise performance? These questions are not always answered in complete detail, because of the complexity of the equations, but enough information is given for preliminary design of an amplifier.

In Section 6. 2 we discuss the simplest parametric amplifier, with just one varactor and with all sidebands open-circuited except the idler frequency $\omega_i = \omega_p - \omega_s$, for which the termination has a real part $R_i \geq 0$. Gain is possible in this amplifier only if the cutoff frequency of the varactor is high enough; in particular, only if

$$0 < \omega_s \omega_i < m_1^2 \omega_c^2, \tag{6.1}$$

where m_1 is the <u>modulation ratio</u>

$$m_1 = \frac{|S_1|}{S_{max} - S_{min}}, \tag{6.2}$$

and ω_c is the cutoff frequency

$$\omega_c = \frac{S_{max} - S_{min}}{R_s}, \tag{6.3}$$

both defined in Section 4. 7. 1; S_{max} and S_{min} are the maximum and minimum varactor elastance, and R_s is the series resistance. If the idler resistance R_i is zero, Condition 6. 1 is sufficient for gain to occur. Otherwise, there can be gain if and only if

$$\omega_s \omega_i (R_s + R_i) < R_s m_1^2 \omega_c^2. \tag{6.4}$$

Note that for a given signal frequency the pump frequency cannot be too high. On the other hand, gain is possible at any signal frequency, no matter how high, provided the pump frequency can be chosen so that the idler frequency is low enough.

The lowest possible noise temperature T at high gain is, as will be shown in the analysis of Section 6.2,

$$T = T_d \frac{\omega_s}{\zeta\omega_i} \frac{m_1^2\omega_c^2 + \zeta^2\omega_i^2}{m_1^2\omega_c^2 - \zeta\omega_s\omega_i} + T_i \frac{R_i}{R_s + R_i} \frac{\omega_s}{\omega_i} \frac{m_1^2\omega_c^2}{m_1^2\omega_c^2 - \zeta\omega_s\omega_i}, \quad (6.5)$$

where ζ is a convenient abbreviation,

$$\zeta = \frac{R_i + R_s}{R_s}, \quad (6.6)$$

and where T_d and T_i are the diode and idler termination temperatures, respectively. Observe that the first term in Equation 6.5 is the contribution from thermal noise originating in the varactor series resistance, and the second term is that caused by the external idler termination. The second term can be made small by terminating the varactor reactively (or at a low temperature). The first term is large when ω_i is small, and again when ω_i is so large that the denominator vanishes. It therefore has a minimum, which is attained at the "optimum" pump frequency (assuming $R_i = 0$)

$$\omega_p = \sqrt{m_1^2\omega_c^2 + \omega_s^2}, \quad (6.7)$$

and is equal to

$$T_{min} = T_d \frac{2\omega_s}{m_1\omega_c} \left[\frac{\omega_s}{m_1\omega_c} + \sqrt{1 + \left(\frac{\omega_s}{m_1\omega_c}\right)^2} \right]. \quad (6.8)$$

This is exactly the minimum noise temperature of the simple frequency converters of Chapter 5 (see Equation 5.2). Furthermore, observe that the optimum pump frequency of Equation 6.7 is approximately equal to $m_1\omega_c$ (for low signal frequencies), just like many of the other optimum pump frequencies discussed in Chapter 5. Thus two of the most important concepts of Chapter 5 (the minimum noise temperature caused by the series resistance, and the optimum pump frequency) apply also to parametric amplifiers. Equation 6.8 has been plotted as Figure 5.7, and also appears later as Figures 6.6 and 6.7 and in Figure 6.18.

The minimum noise temperature and the optimum pump frequency necessary to attain it have been derived by other writers by using an approximate parallel model for the varactor. The first such derivation was performed by Knechtli and Weglein early in 1959 and published‡[20] in 1960. Other similar analyses are those of Greene and Sard,[11] Mortenson,[30] Kotzebue,[22] Blackwell,[2] and Kurokawa and Uenohara.[25]

Parametric-amplifier noise theories that neglect varactor loss[16, 37] indicate that the noise temperature (at high gain) should be (see Equation 3.39)

$$T = T_i \frac{\omega_s}{\omega_i}. \tag{6.9}$$

Such theories imply that for best noise performance the idler termination should be refrigerated, or the amplifier should have a high pump frequency, or both. It is clear that for a varactor parametric amplifier with a finite series resistance the optimization procedure is quite different.

Note that in the fundamental limits derived so far, the only quantity that characterizes the varactor is $m_1\omega_c = |S_1|/R_s$. A higher value of this quantity indicates that gain is more easily obtainable and noise is lower. We are therefore led to consider $|S_1|/R_s$ as a figure of merit for the varactor, as pumped (see Sections 4.6 and 4.7).

We have just seen, for an amplifier consisting of only one varactor, limits on gain and noise caused by the varactor series resistance. However, amplifiers,[1, 21, 35] for example the so-called "traveling-wave parametric amplifiers,"[8] have been built using more than one varactor, and it is important to know whether or not these limits hold for configurations of several varactors. For simplicity, we assume that each of the varactors is identical, pumped with the same value of m_1 (although the varactors may be pumped in different phases). We further assume that each varactor is open-circuited at all sideband frequencies except ω_i but that the idler terminations of the varactors need not be separate, that is, the varactors can "communicate" at idler frequency, The analysis of Section 6.3 shows that under these restrictions, gain is possible if and only if Equation 6.1 is satisfied, and that the noise temperature of such an amplifier is no less than Equation 6.8.

If the use of more than one varactor cannot lead to a lower noise than a single varactor, perhaps the use of more than one

‡Superscript numerals denote references listed at the end of each chapter.

idler can. In Section 6.4 we investigate amplifiers with extra idlers. If the elastance is sinusoidal, extra idlers at frequencies $n\omega_p + \omega_s$ with n positive, cannot help either the gain or the noise. On the other hand, extra lower-sideband idlers, that is, with n negative, can help in two ways. First, if $\omega_p > \omega_s$ (the usual situation), but below the optimum pump frequency of Equation 6.7, the noise can be decreased somewhat by allowing some idling current at frequency $\omega_p + \omega_i = 2\omega_p - \omega_s$. The amount of decrease is not great; in fact, the noise temperature cannot be below Equation 6.8. And second, if the pump frequency is less than the signal frequency, gain may still be possible if other lower-sideband idlers are present.

On the other hand, in Section 6.4.2 we investigate the effects of extra idlers with an elastance waveform that need not be sinusoidal. This problem is so complicated that we do not investigate it in detail but show only a lower limit for the noise temperature. This limit is of the same form as Equation 6.8, but with m_1 replaced by m_t, where

$$m_t^2 = 2\sum_{k=1}^{\infty} m_k^2, \tag{6.10}$$

where the m_k for $k = 1, 2, 3, 4, \ldots$ are defined by formulas similar to Equation 6.2:

$$m_k = \frac{|S_k|}{S_{max} - S_{min}}, \tag{6.11}$$

where the S_k are given by Equation 4.41. In other words, m_t^2 is the variance of the incremental elastance, as pumped. Moreover, if any of the sidebands is known to be open-circuited, then m_t can be replaced by a sum that omits the corresponding m_k. This new lower limit for the noise temperature is probably not achievable in practice (the method of deriving it does not lead to a simple optimization procedure). Nevertheless, it is not much below the limit of Equation 6.8, because, as shown in Chapter 7, m_t cannot be more than twice an easily achievable value of m_1, and for practical varactors is seldom more than 50 per cent greater than m_1. As for extra idlers, the conclusion is that although some improvement in noise is theoretically possible, this improvement is so small that it is probably not worth the increased circuit complexity.

The noise formula for the simple parametric amplifier, Equation 6.5, is one of the most significant single results of this book. Because of its importance, we devote Section 6.5 to its interpretation. This discussion is not summarized here, because such a summary would virtually repeat Section 6.5 in its entirety. Among the special cases discussed, are the forms when $R_s = 0$, when the external idler termination is reactive, when the external idler termination is very cold, and when the external idler termination is at the diode temperature.

We have assumed thus far (see assumption 8 above) that the frequencies are sufficiently separated so that changes in one sideband termination can (at least in principle) be made without affecting the others. Suppose, however, we wish to operate with ω_i close to (but not equal to) ω_s, in order that both frequencies can use the same physical circuit, in the same way. In fact, what if the spectrum of the signal (centered about frequency ω_s) even overlaps the spectrum of the idler? What if the signal frequency is exactly half the pump frequency, so that $\omega_s = \omega_i$?

In case the circuit does not distinguish between ω_s and ω_i, we say that the amplifier is degenerate. The noise performance of degenerate amplifiers is not described by the theory of Sections 6.2 to 6.5. The idler termination cannot be varied independently of the signal-frequency circuit. The ordinary concepts of noise measure, noise figure, and noise temperature do not apply. The noise output from the degenerate amplifier has a spectrum with correlation between the various frequency components, and the signal output has frequency components not included in the input. It is not possible to compare two amplifiers by merely quoting their respective noise temperatures. In general, two amplifiers (one a degenerate amplifier and the other an ordinary amplifier) can be compared only when the statistics of the input signal, the type of detector, and the interpretation of the detector output are all known. Only a few authors[12,7] have recognized this important fact.

In Section 6.6, we describe the noise behavior of degenerate parametric amplifiers, not in complete detail, but thoroughly enough to illustrate the technique of determining the noise performance in specific cases. It is found that two degenerate parametric amplifiers can be compared with each other by looking at a single noise parameter, the "effective noise temperature" T_{eff}. This is not a noise temperature in the usual sense. It is, however, exactly equal to what is called the double-sideband noise temperature, which is what is measured with a wideband noise generator. For the particular case of a varactor degenerate parametric amplifier, T_{eff} is a function of the pumped figure of merit, the varactor temperature, and the pump frequency:

$$T_{eff} = \frac{T_d}{\dfrac{m_1 \omega_c}{\omega_p/2} - 1}. \tag{6.12}$$

The amplifier has gain only when $\omega_p/2 < m_1\omega_c$, a fact that is in agreement with Equation 6.1. Note, however, that T_{eff} is <u>not</u> in agreement with Equation 6.5, and cannot be derived from it merely by setting ω_s equal to ω_i.

For most applications, we cannot compare a degenerate amplifier with an ordinary amplifier with noise temperature T merely by comparing T_{eff} with T. No comparison at all is possible until the detection scheme and the input signal statistics are known. A comparison can be made, however, if the input signal spectrum is located entirely to one side of $\omega_p/2$, and if the degenerate amplifier is followed by a filter or narrow-band amplifier that completely rejects the idler spectrum (located on the other side of $\omega_p/2$). In this case, the output noise (if the output is taken after this filter) is not correlated, and the combination amplifier-plus-filter behaves like an ordinary amplifier with noise described by the ordinary formulas using the <u>single-sideband</u> noise temperature

$$T'_{SSB} = T_i + 2\,T_{eff}, \tag{6.13}$$

where T_i is the temperature of the network seen by the varactor at frequency ω_i. Thus, a single-sideband degenerate amplifier and an ordinary amplifier with noise temperature T can be compared by comparing T with T'_{SSB}.

If the signal and idler spectra overlap, however, the type of detector and the interpretation of the detector output are important in assessing the noise performance of the degenerate amplifier. If the detector is a synchronous detector, phase-locked to the pump of the parametric amplifier, then the degenerate amplifier and an ordinary amplifier can be compared for noise performance by comparing T with T_{eff}. If the detector is of another type, however, no such simple comparison is possible.

In Section 6.7 we briefly discuss the very important problem of the design, or synthesis, of nondegenerate parametric amplifiers. The analysis that leads to the fundamental limits on parametric-amplifier noise also indicates how to achieve this minimum noise temperature.

We rather arbitrarily divide the synthesis into three steps: preliminary design, synthesis of a circuit to satisfy simultaneously conditions at several frequencies, and imbedding the resulting negative resistance to make a two-port amplifier.

The first problem is to select the varactor and the operating frequencies of the amplifier, so that the specified noise figure can be attained. The formulas for regions of gain and noise temperature are given in Section 6.2 and interpreted in Section 6.5. Formulas for pump power are given in Chapter 7, along with a graphical optimization scheme if pump power is limited.

The major purpose of these chapters is to aid in the selection of a varactor, pump frequency, and idler termination that will simultaneously give gain, have a low noise temperature, and require a reasonable pump power. In this sense, then, Chapters 6 and 7 aid preliminary amplifier design.

The second step in the design of the amplifiers is the synthesis of circuitry to fulfill functions at several frequencies simultaneously. We base this design on the analysis of the fundamental limits, and require that the finished amplifier should attain (or at least approach) these limits. In particular, suppose we wish to design a single-idler, single-varactor, nondegenerate parametric amplifier with $R_i = 0$ (this is a desirable amplifier, because it realizes the lowest noise temperature with a room-temperature idler). To obey the conditions in Section 6.2, we wish the frequencies $\omega_p + \omega_s$ and $\omega_p + \omega_i$ to be open-circuited. For pumping with a sinusoidal current (in Chapter 7 it is shown that this is desirable), we wish the varactor to be open-circuited at harmonics of the pump frequency, especially $2\omega_p$. Furthermore, the idler termination should be tuned (see Section 6.2) and lossless. Currents at idler frequency should be excluded from the pump and signal ports, and the signal and pump should be separated to appear at separate ports.

It is a real engineering problem to synthesize circuits to fulfill all these conditions. The usual procedure for designing parametric amplifiers has been to incorporate many adjustments for tuning the device. With enough screws to turn, an adept experimentalist, without much theoretical knowledge of how his adjustments should affect the amplifier gain or noise, can empirically tune up such an amplifier. What we wish to advocate, on the other hand, is a rational procedure for making adjustments independently (at least if done in the proper sequence), in order to satisfy the demands of the theory intentionally. The circuits in Section 6.7.2 were designed with this in mind. We make no claim that these circuits are in any sense optimum; they probably have low bandwidth, use more components than necessary, and use expensive or unreliable components. They are given only to show that the design of parametric amplifiers can be based upon a theory of the fundamental limits.

The second step, outlined above, yields, as far as the signal frequency circuitry is concerned, a negative resistance. The problem of using this negative resistance as an amplifying element is another problem — one we have called step 3. Basically, by use of a three-port network, we must build up a two-port device out of this one-port negative resistance. This problem, which is common to all negative-resistance amplifiers, is treated analytically in Appendix C. Typical three-port circuits discussed there are the circulator, the double-transformer series circuit, the series circuit without transformer, and the general circuit. The problem of cascading negative-resistance amplifiers to noisy second stages is also discussed in Appendix C.

6.2. The Simplest Amplifier

The simplest parametric amplifier, shown in Figure 6.2, uses only one varactor, and has only one idler. To prevent power dissipation at frequencies other than ω_s and ω_i, we agree to open-circuit the varactor at all other sideband frequencies, especially $\omega_p + \omega_s$ and $\omega_p + \omega_i$. Thus the only current (besides the pumping current) flows at frequencies ω_s and ω_i.

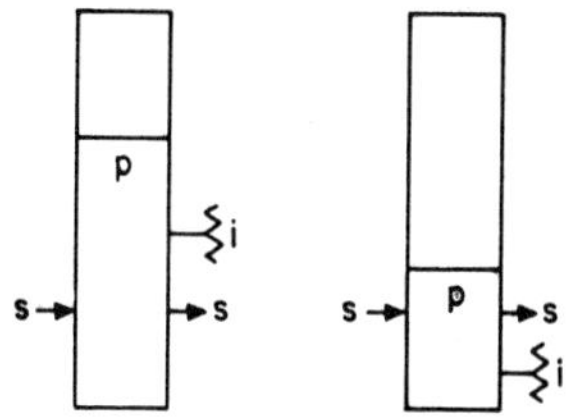

The amplification mechanism is a bit difficult to understand. If current at signal frequency flows through the varactor, the pumping action causes a voltage to be generated at idler frequency. This inducted voltage causes a

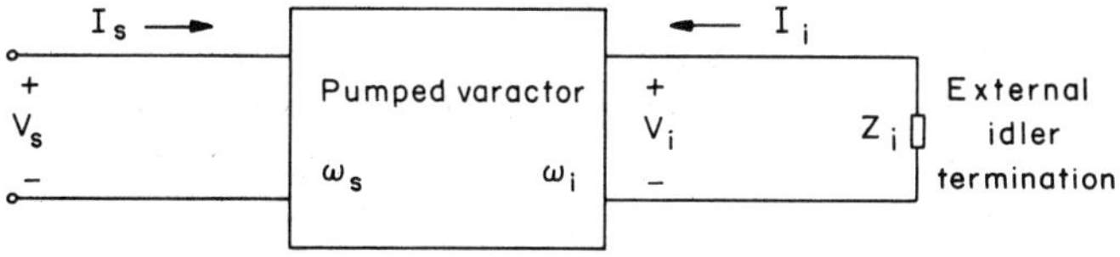

Figure 6.2. Simple single-varactor parametric amplifier with one idler. Although the signal-frequency and idler-frequency terminations are shown separately, some physical elements may appear in both (see Section 2.2).

flow of current at idler frequency; if the varactor is properly terminated, the idler current will be large. This current, in turn, causes an induced voltage at signal frequency. The important point is that this voltage is approximately 180° out of phase with the signal current. The pumped, terminated varactor appears at signal frequency like a negative resistance capable of generating electrical power at this frequency. By the techniques of Appendix C, the negative resistance can be incorporated into an otherwise passive circuit, to make an amplifier.

The Manley-Rowe equations of Section 2.1 predict power generation at frequency ω_s and, therefore, the possibility of gain. The varactor does not obey the Manley-Rowe equations (because of the series resistance), but the capacitive part of the varactor does. Since power is dissipated at only the three frequencies ω_s, ω_i, and ω_p, Equations 2.7 and 2.8 reduce to

$$\frac{P_s}{\omega_s} = \frac{P_i}{\omega_i} = -\frac{P_p}{\omega_p}, \tag{6.14}$$

where P_s, P_i, and P_p are the powers at the corresponding frequencies. Because the termination at ω_i is passive, it must accept power, and therefore, the varactor (at least the nonlinear capacitance part) must put out power at frequency ω_s also (P_s and P_i have the same sign).

Now let us calculate the impedance at frequency ω_s presented by the varactor, as pumped and terminated at frequency ω_i. In addition, we calculate the noise voltages. We start with the varactor equations, Equations 4.42, setting all currents to zero except I_s and I_i:

$$V_i^* = \left(R_s - \frac{S_0}{j\omega_i}\right) I_i^* + \frac{S_1^*}{j\omega_s} I_s + E_{ni}^*, \tag{6.15}$$

$$V_s = \left(R_s + \frac{S_0}{j\omega_s}\right) I_s - \frac{S_1}{j\omega_i} I_i^* + E_{ns}, \tag{6.16}$$

where (as throughout this book) V_i, V_s, I_i, and I_s are half-amplitudes of the components of varactor voltage and current at the corresponding frequencies, where S_0 is the average varactor elastance and S_1 is the half-amplitude of the varactor elastance at frequency ω_p, according to Equation 4.41, and where E_{ns} and E_{ni} are equivalent noise voltages, as discussed in Section 4.3, with mean-squared values‡

$$\overline{|E_{ns}|^2} = \overline{|E_{ni}|^2} = 4kT_d R_s \,\Delta f, \tag{6.17}$$

where k is Boltzmann's constant and T_d the diode temperature.

We need another relation to eliminate V_i^* and I_i^* from Equations 6.15 and 6.16. The passive circuitry surrounding the varactor has some impedance $Z_i = R_i + jX_i$ at idler frequency,

$$V_i = -Z_i I_i + E_{nxi}, \tag{6.18}$$

where $R_i \geq 0$, and E_{nxi} is the noise voltage‡ from this external idler load, at a temperature T_i:

$$\overline{|E_{nxi}|^2} = 4kT_i R_i \,\Delta f. \tag{6.19}$$

By simple algebra V_i^* and I_i^* can be eliminated from Equa-

‡The noise voltages are effective rms voltages, instead of half-amplitudes. See Section 2.5.1(a).

tions 6.15, 6.16, and the complex conjugate of Equation 6.18, with the result

$$V_s = ZI_s + E, \tag{6.20}$$

where the varactor impedance Z at signal frequency is

$$Z = R + jX = R_s + \frac{S_0}{j\omega_s} - \frac{S_1 S_1^*}{\omega_s \omega_i [Z_i^* + R_s - (S_0/j\omega_i)]} \tag{6.21}$$

and the open-circuit noise voltage E is

$$E = E_{ns} + \frac{S_1(E_{ni}^* - E_{nxi}^*)}{j\omega_i [Z_i^* + R_s - (S_0/j\omega_i)]}. \tag{6.22}$$

6.2.1. Conditions for Gain

A one-port device like the varactor can operate as a linear amplifier with gain if and only if its impedance has a negative real part. From Equation 6.21 and the definitions of m_1 and ω_c,

$$R = R_s \left\{ 1 - \frac{m_1^2 \omega_c^2}{\omega_s \omega_i} \; \frac{R_s(R_s + R_i)}{(R_s + R_i)^2 + [X_i - (S_0/\omega_i)]^2} \right\}. \tag{6.23}$$

Therefore, there can be gain only when ω_i is positive, and even then only when the second term in the braces is larger than the first. To help accomplish this, we tune the idler so that

$$X_i = \frac{S_0}{\omega_i}; \tag{6.24}$$

in this case there is gain only when

$$(R_s + R_i)\omega_s \omega_i < R_s m_1^2 \omega_c^2. \tag{6.25}$$

Since R_i must not be negative, the best we can do is make $R_i = 0$, for which

$$\omega_s \omega_i < m_1^2 \omega_c^2. \tag{6.26}$$

No gain is possible if Equation 6.26 does not hold. Whenever

Equation 6.26 holds, gain is possible for some choice of Z_i (in particular $Z_i = jS_0/\omega_i$, the lossless impedance that tunes out the average elastance at frequency ω_i).

Note from Condition 6.26 that gain is possible at any signal frequency,‡ no matter how poor the varactor or how weakly it is pumped, provided only that the pump frequency can be chosen sufficiently close to ω_s. Ordinarily, parametric amplifiers are operated with the idler frequency higher than ω_s (for low noise); this is possible only if $\omega_s < m_1\omega_c$.

It is convenient to rewrite Condition 6.26 in the form

$$\omega_s < \omega_p < \omega_s + \frac{m_1^2\omega_c^2}{\omega_s}, \tag{6.27}$$

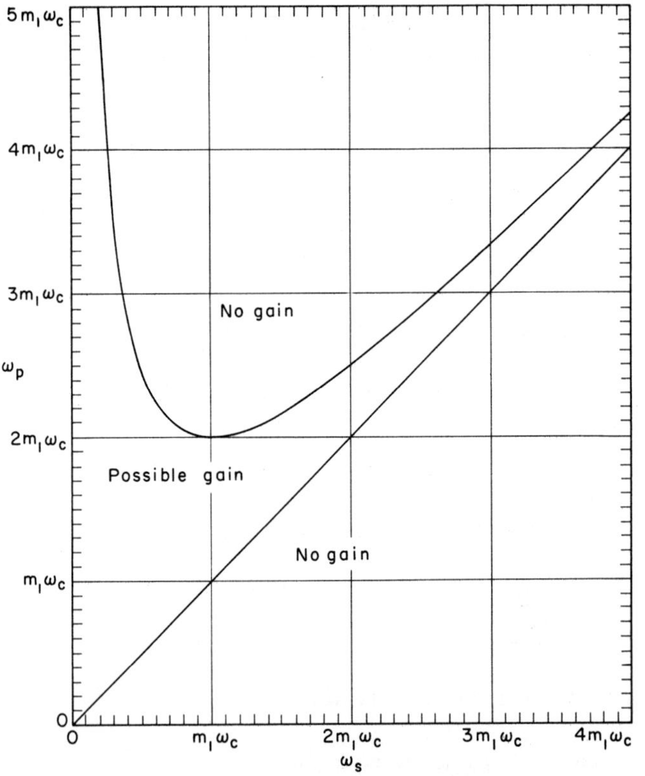

Figure 6.3. For gain with a passive external idler termination, pump and signal frequencies must be chosen according to this chart. Note that the axes in this (and other) graphs are labeled in terms of the varactor pumped figure of merit $m_1\omega_c$. Taken from Condition 6.27.

‡Moreover, in Chapter 7 we show that gain is possible for any signal frequency, even if the pump power is limited.

where the first inequality comes from the remark after Equation 6.23. Graphs of Condition 6.27 are shown in Figures 6.3 and 6.4.

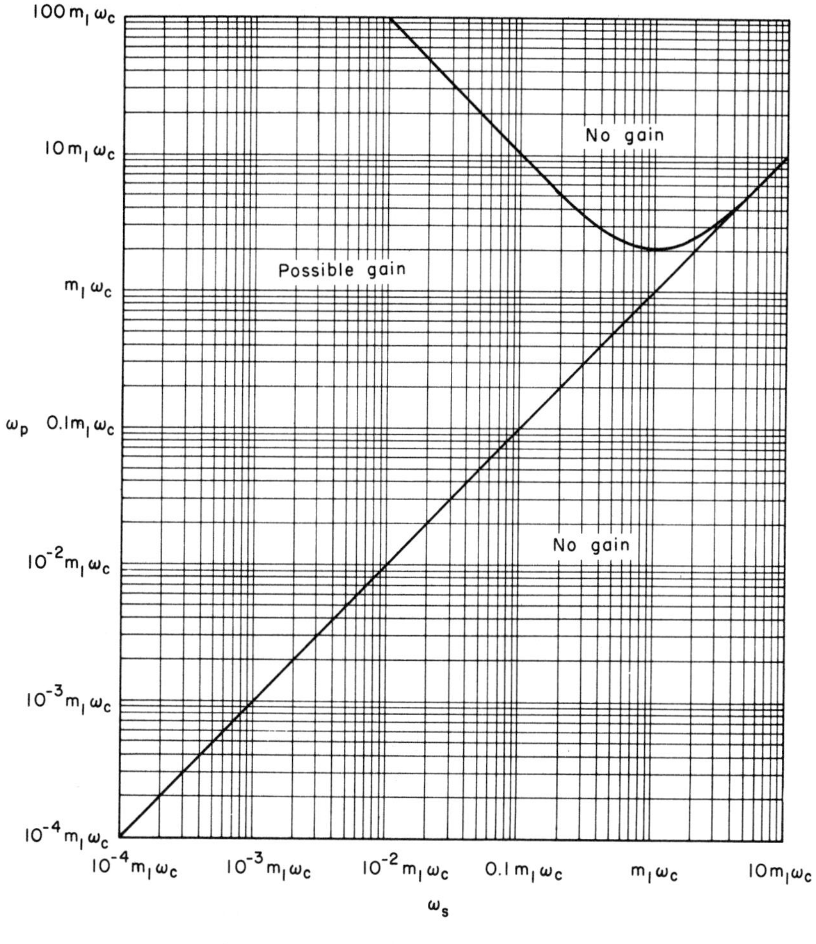

Figure 6.4. The same information given in Figure 6.3, with the frequency axes plotted logarithmically.

(In using Figures 6.3 and 6.4, as well as any of the other diagrams in this book, ordinary, instead of radian, frequencies f_s, f_p, and f_c can be used, since only ratios of frequencies are important.)

6.2.2. Bandwidth

There is a fundamental limit on bandwidth of single-idler parametric amplifiers that arises from Condition 6.27. For a given pump frequency (higher than $2m_1\omega_c$), gain is possible only at small signal frequencies and at signal frequencies close to the pump frequency. The bandwidth of an amplifier certainly cannot

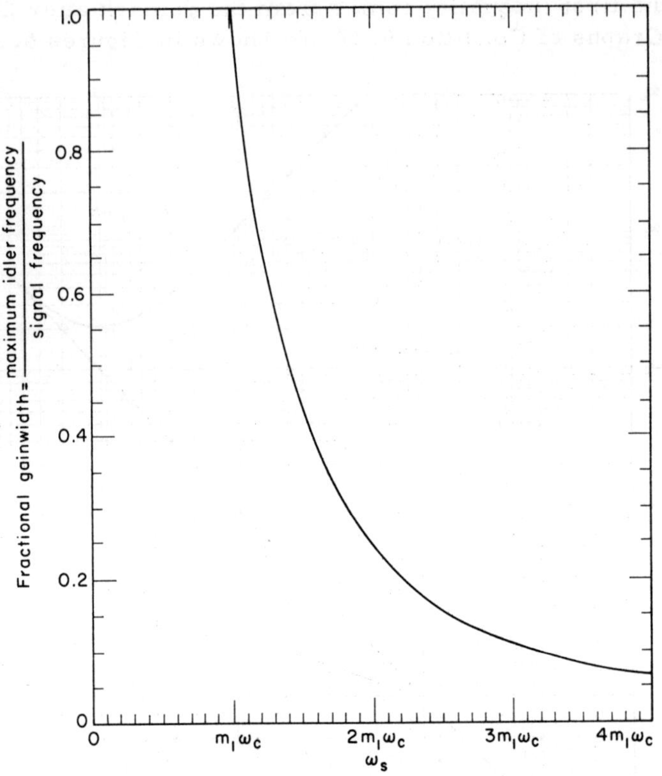

Figure 6.5. Maximum fractional bandwidth of the simple parametric amplifier, taken from Figure 6.3. This limit is probably not achievable.

exceed the range of ω_s over which gain is possible. We call this latter region the gainwidth; it is plotted in Figure 6.5. This limitation is important only at signal frequencies above approximately $m_1\omega_c$.

This bandwidth limit is probably not of much practical value. We have not shown that the limit is achievable; it undoubtedly is not. In most amplifiers, circuitry limits the bandwidth far more. Usually singly tuned resonant circuits are used in both the idler and signal circuits, although more sophisticated circuitry has increased markedly the bandwidths of some amplifiers and frequency converters.[34, 29, 26, 33, 9, 38, 10, 19, 23] It would be of interest to know whether there are achievable fundamental limits on bandwidth. Possibly these could be found from an anlysis similar to that by Youla and Smilen[39] or Kuh and Patterson[24] for tunnel-diode amplifiers.

6.2.3. Noise

If it were not for the series resistance, the noise temperature of our amplifier would be[16,37]

$$T = T_i \frac{\omega_s}{\omega_i}, \tag{6.28}$$

as we show, for example, in Section 3.2.1. The noise could then be reduced arbitrarily by either refrigerating the idler, or raising the pump frequency, or doing both. Because of the series resistance, however, the pump frequency cannot be raised too high (see Condition 6.27). The lowest noise of the varactor amplifier occurs for a finite pump frequency; the minimum noise temperature depends only on the diode temperature T_d and not on T_i. It is clear that Equation 6.28 is misleading for amplifier design.

We now derive the noise temperature of our amplifier and show some of its properties. As an aid to designing optimized amplifiers, a discussion of the noise formula is given in detail in Section 6.5.

It was pointed out in Section 2.5.5 that the noise temperature of a negative-resistance amplifier is given by the simple formula

$$T = -\frac{\text{negative-resistance exchangeable noise power}}{k\,\Delta f}. \tag{6.29}$$

In our case, the exchangeable noise power is $\overline{|E|^2}/4R$, where R is defined by Equation 6.23, and $\overline{|E|^2}$ is calculated from Equations 6.22, 6.17, and 6.19:

$$\overline{|E|^2} = 4k\,\Delta f \left[R_s T_d + \frac{m_1^2 \omega_c^2 R_s^2 (R_s T_d + R_i T_i)}{\omega_i^2 |Z_i + R_s + (S_0/j\omega_i)|^2} \right]. \tag{6.30}$$

Thus,

$$T = \frac{-\overline{|E|^2}}{4Rk\,\Delta f}$$

$$= \frac{T_d + \dfrac{m_1^2 \omega_c^2}{\omega_i^2} \dfrac{R_s(R_s T_d + R_i T_i)}{|Z_i + R_s + (S_0/j\omega_i)|^2}}{\dfrac{m_1^2 \omega_c^2}{\omega_s \omega_i} \dfrac{R_s(R_s + R_i)}{|Z_i + R_s + (S_0/j\omega_i)|^2} - 1}. \tag{6.31}$$

The lowest noise temperature is obtained when the idler termination is tuned so that

$$Z_i + R_s + \frac{S_0}{j\omega_i} \tag{6.32}$$

is real, equal to $R_s + R_i$. If we define as a convenient abbreviation‡

$$\zeta = \frac{R_s + R_i}{R_s}, \tag{6.33}$$

the noise temperature becomes, under tuning,

$$T = T_d \frac{\omega_s}{\zeta\omega_i} \frac{m_1^2\omega_c^2 + \zeta^2\omega_i^2}{m_1^2\omega_c^2 - \zeta\omega_s\omega_i} + T_i \frac{\omega_s}{\omega_i} \frac{R_i}{R_s + R_i} \frac{m_1^2\omega_c^2}{m_1^2\omega_c^2 - \zeta\omega_s\omega_i}. \tag{6.34}$$

This basic noise formula for the parametric amplifier is interpreted extensively in Section 6.5. The second term is the contribution from the idler termination; it can be made small either by refrigerating the idler (putting $T_i = 0$) or by terminating the varactor at idler frequency reactively‡‡ (putting $R_i = 0$). The first term provides a fundamental limit to the noise temperature. It becomes large for very small ω_i, and large for ω_i so great that the denominator vanishes. It has a minimum, therefore, for some value of ω_i (or, equivalently, for some optimum pump frequency). The minimum is determined by differentiating the first term with respect to ω_i:

$$T_{min} = T_d \frac{2\omega_s}{m_1\omega_c}\left[\frac{\omega_s}{m_1\omega_c} + \sqrt{1 + \left(\frac{\omega_s}{m_1\omega_c}\right)^2}\right], \tag{6.35}$$

and the optimum pump frequency to attain this minimum is

$$\omega_p = \frac{R_i\omega_s + R_s\sqrt{m_1^2\omega_c^2 + \omega_s^2}}{R_s + R_i}. \tag{6.36}$$

Equation 6.35 is exactly the minimum found in Chapter 5 for the simple varactor frequency converters. It is plotted in Figures 6.6 and 6.7 as well as in Figure 5.7.

---

‡This quantity can also be interpreted as the ratio of power dissipated at idler frequency, to that portion dissipated in the varactor series resistance.

‡‡The idler-frequency dissipation required by the Manley-Rowe formulas then occurs solely in the varactor series resistance.

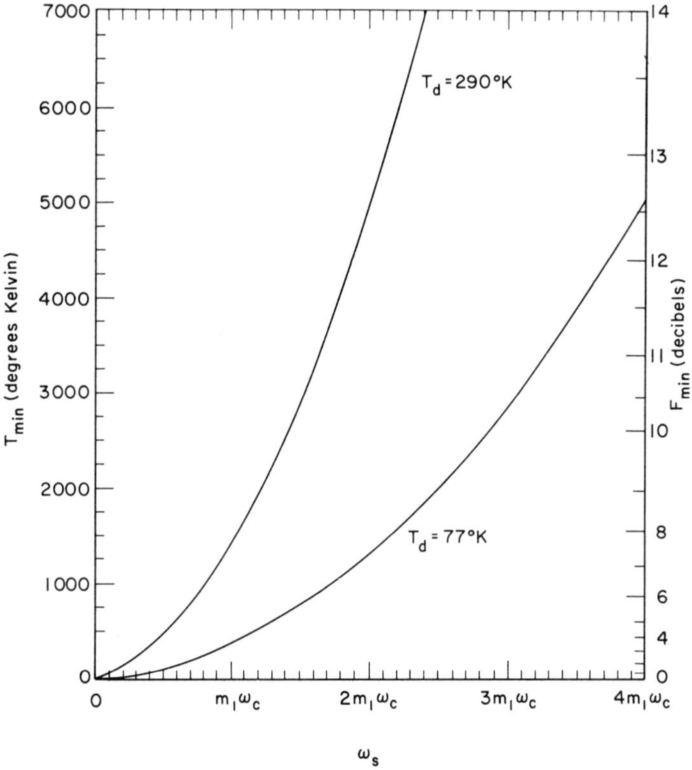

Figure 6.6. Minimum noise temperature of a parametric amplifier at high gain, for two common diode temperatures, taken from Equation 6.35. This quantity is also plotted in Figures 5.7 and 6.7.

Note that two of the most important concepts of Chapter 5 (the minimum noise temperature and the optimum pump frequencies) apply also to the simple parametric amplifier.

The temperature of Equation 6. 35 is a lower limit for the simple parametric amplifier discussed. It can be achieved if and only if

1. The external idler termination is either at zero temperature or is purely reactive.
2. Coupling to all losses (except the series resistance of the varactor) at signal frequency is avoided (see Appendix C).
3. The idler termination is tuned so that $X_i = S_0/\omega_i$.
4. The pump frequency is chosen according to Equation 6. 36.
5. There are no other significant sources of noise (see Section 4. 3).

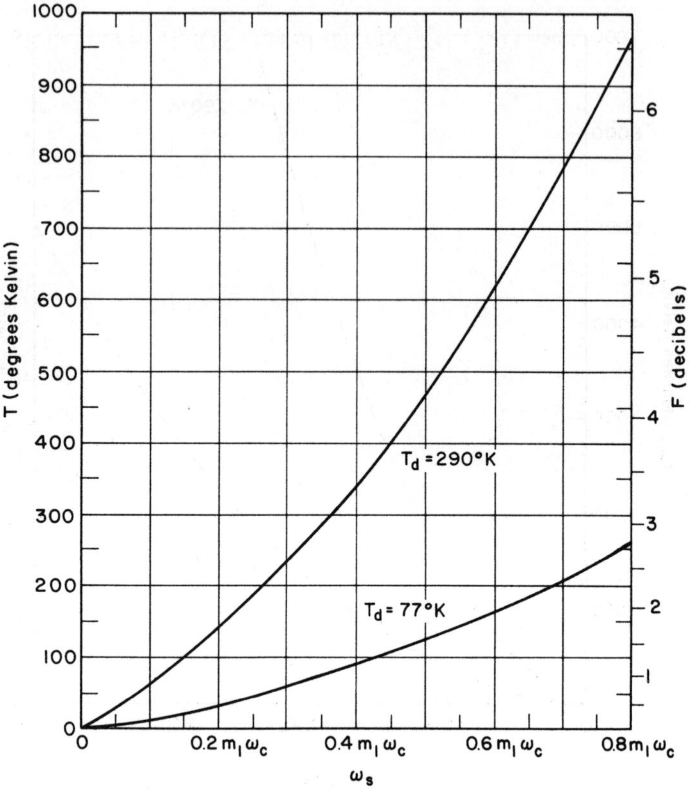

Figure 6.7. Same graph as Figure 6.6, expanded to show detail near the origin.

6.2.4. Examples

We now answer some typical questions by using the formulas developed in Section 6.2. Note that in all the formulas developed, the frequencies appear only divided by the cutoff frequency. It is usually simpler in practice to use ordinary rather than radian frequencies.

First, suppose we have an abrupt-junction varactor with cutoff frequency f_c = 160 Gc, pumped with a sinusoidal current. For a signal frequency of 10 Gc, what is the minimum noise temperature, and what pump frequency can be used to achieve this minimum? Repeat, for a signal frequency of 30 Gc. In Chapter 7 we point out that a current-pumped abrupt-junction varactor can have m_1 only as high as 0.25; thus taking $m_1 f_c$ = 40 Gc, we see from Equations 6.35 and 6.36 (for R_i = 0) that in the first case, $T_{min} = 0.64 T_d$, achieved at ω_p = 41 Gc. In the second case, $T_{min} = 3T_d$, achieved at 50-Gc pump frequency.

Next, suppose we pump a varactor so that $m_1 f_c$ = 30 Gc. What pump frequencies can produce gain at a signal frequency of

1 Gc? From Condition 6.27, f_p can lie anywhere between 1 to 901 Gc. What pump frequencies give gain at a signal frequency of 10 Gc? From Condition 6.27, f_p must lie between 10 and 100 Gc. What pump frequencies for gain at 100 Gc? Now f_p must lie between 100 and 109 Gc.

6.3. Amplifiers with More than One Varactor

The analysis of Section 6.2 indicates that a single pumped varactor can amplify parametrically if and only if Condition 6.26 is obeyed, and that the noise temperature is always greater than Equation 6.35. The analysis of multiple-varactor amplifiers[8, 1, 21, 35] is carried through by extending the analysis of Section 4.2 by matrix algebra. The major results are that gain is possible by the use of several varactors if and only if Condition 6.26 holds, and the minimum noise temperature is again given by Equation 6.35. Thus multiple-varactor amplifiers are subject to the same fundamental limitations as single-varactor amplifiers.

In the discussion to follow, several results of matrix algebra are quoted without proof. The necessary background is available in many textbooks;[18, 31] a good succinct review of the terms and major results is that by Marcus.[28]

For simplicity we assume the varactors (ℓ in number) are all identical and pumped to the same value of m_1 (although possibly at different phases). For simplicity we further assume that each varactor is open-circuited at all sideband frequencies except ω_s and ω_i. All noise sources except thermal noise in the varactor series resistances and the idler termination are ignored.

The network seen by the varactors at the idler frequency has (except in trivial cases) an $\ell \times \ell$ impedance matrix $\underset{\sim}{Z}_i$ and an open-circuit-voltage column matrix $\underset{\sim}{E}_{nxi}$, each element of which is (in general) correlated with every other element.‡ Thus the column matrices of varactor currents and voltages at the idler frequency ω_i are related:

$$\underset{\sim}{V}_i = -\underset{\sim}{Z}_i \underset{\sim}{I}_i + \underset{\sim}{E}_{nxi}. \tag{6.37}$$

Because the voltage sources represent thermal noise in the network[36, 14] (we assume for simplicity that the idler load is all at the same temperature T_i),

$$\overline{\underset{\sim}{E}_{nxi} \underset{\sim}{E}_{nxi}^{\dagger}} = 2kT_i(\underset{\sim}{Z}_i + \underset{\sim}{Z}_i^{\dagger})\,\Delta f, \tag{6.38}$$

where the dagger indicates the Hermitian transpose of a matrix.

‡The noise voltages are effective rms voltages instead of half-amplitudes. See Section 2.5.1(a).

The idler termination can be nonreciprocal (for example, it might contain circulators or isolators), but we assume it is passive, so that the matrix $\underset{\sim}{Z}_i + \underset{\sim}{Z}_i\dagger$ is positive semidefinite.‡ If the idler termination is lossless, $\underset{\sim}{Z}_i = -\underset{\sim}{Z}_i\dagger$; therefore, $\underset{\sim}{E}_{nxi}$ vanishes.

Each varactor individually obeys Equations 6.15 and 6.16. If we wish, however, we may put all these individual equations together into two matrix equations

$$\underset{\sim}{V}_i^* = \left(R_s - \frac{S_0}{j\omega_i}\right)\underset{\sim}{I}_i^* + \frac{|S_1|}{j\omega_s}\underset{\sim}{\Phi}\dagger\underset{\sim}{I}_s + \underset{\sim}{E}_{ni}^* \tag{6.39}$$

and

$$\underset{\sim}{V}_s = \left(R_s + \frac{S_0}{j\omega_s}\right)\underset{\sim}{I}_s - \frac{|S_1|}{j\omega_i}\underset{\sim}{\Phi}\underset{\sim}{I}_i^* + \underset{\sim}{E}_{ns}, \tag{6.40}$$

where $\underset{\sim}{\Phi}$ is the diagonal matrix composed of the phase factors of the various S_1 of the varactors, as pumped, and where $\underset{\sim}{E}_{ni}$ and $\underset{\sim}{E}_{ns}$ are column matrices of noise voltages:‡‡

$$\overline{\underset{\sim}{E}_{ns}\underset{\sim}{E}_{ns}\dagger} = \overline{\underset{\sim}{E}_{ni}\underset{\sim}{E}_{ni}\dagger} = 4kT_dR_s\underset{\sim}{1}\,\Delta f, \tag{6.41}$$

where $\underset{\sim}{1}$ is the unit matrix. Note that the matrix $\underset{\sim}{\Phi}$ obeys the relation

$$\underset{\sim}{\Phi}\underset{\sim}{\Phi}\dagger = \underset{\sim}{1}, \tag{6.42}$$

and therefore $\underset{\sim}{\Phi}$ is unitary.

By vector manipulation, Equations 6.37, 6.39, and 6.40 can be put into the form

$$\underset{\sim}{V}_s = \underset{\sim}{Z}\underset{\sim}{I}_s + \underset{\sim}{E}, \tag{6.43}$$

where the $\ell \times \ell$ impedance matrix $\underset{\sim}{Z}$ of the varactors at signal frequency is

$$\underset{\sim}{Z} = \left(R_s + \frac{S_0}{j\omega_s}\right)\underset{\sim}{1} - \frac{|S_1|^2}{\omega_s\omega_i}\underset{\sim}{\Phi}\left[\underset{\sim}{Z}_i^* + \underset{\sim}{1}\left(R_s - \frac{S_0}{j\omega_i}\right)\right]^{-1}\underset{\sim}{\Phi}\dagger \tag{6.44}$$

‡The power dissipated in the idler is $\underset{\sim}{I}_i\dagger(\underset{\sim}{Z}_i + \underset{\sim}{Z}_i\dagger)\underset{\sim}{I}_i$, which is nonnegative for arbitrary $\underset{\sim}{I}_i$ if and only if the matrix $\underset{\sim}{Z}_i + \underset{\sim}{Z}_i\dagger$ is positive semidefinite. See the discussion leading to Equation 6.46.

‡‡The noise voltages are effective rms voltages instead of half-amplitudes. See Section 2.5.1(a).

and where the column matrix of open-circuit voltages $\underset{\sim}{E}$ is

$$\underset{\sim}{E} = \underset{\sim}{E}_{ns} + \frac{|S_1|}{j\omega_i} \underset{\sim}{\Phi} \left[\underset{\sim}{Z}_i^* + \underset{\sim}{1} \left(R_s - \frac{S_0}{j\omega_i} \right) \right]^{-1} (\underset{\sim}{E}_{ni}^* - \underset{\sim}{E}_{nxi}^*). \quad (6.45)$$

6.3.1. Conditions for Gain

An array of varactors can operate as an amplifier with gain only if (neglecting noise) it can generate power at the signal frequency. The net power put into the array of varactors at frequency ω_s can be expressed in matrix form as

$$P_s = \underset{\sim}{V}_s^\dagger \underset{\sim}{I}_s + \underset{\sim}{I}_s^\dagger \underset{\sim}{V}_s$$

$$= \underset{\sim}{I}_s^\dagger (\underset{\sim}{Z} + \underset{\sim}{Z}^\dagger) \underset{\sim}{I}_s. \quad (6.46)$$

For amplification, the signal-frequency currents at the various varactors must be chosen so as to make P_s negative. It is well known from matrix theory that such a quadratic form as Equation 6.46 can be negative if and only if the matrix $\underset{\sim}{Z} + \underset{\sim}{Z}^\dagger$ has a negative eigenvalue. We shall therefore investigate the eigenvalues of $\underset{\sim}{Z} + \underset{\sim}{Z}^\dagger$.

From Equation 6.44, we find by straightforward matrix algebra

$$\underset{\sim}{Z} + \underset{\sim}{Z}^\dagger = 2R_s \underset{\sim}{1} - \frac{|S_1|^2}{\omega_s \omega_i} (\underset{\sim}{\eta} + \underset{\sim}{\eta}^\dagger) \quad (6.47)$$

or

$$\underset{\sim}{Z} + \underset{\sim}{Z}^\dagger = 2\left(R_s - \frac{|S_1|^2}{R_s \omega_s \omega_i} \right) \underset{\sim}{1}$$

$$+ \frac{|S_1|^2}{\omega_s \omega_i} \underset{\sim}{\eta} \underset{\sim}{\Phi} \left[\frac{2}{R_s} \left(\underset{\sim}{Z}_i^* - \frac{\underset{\sim}{1} S_0}{j\omega_i} \right) \left(\underset{\sim}{Z}_i^* - \frac{\underset{\sim}{1} S_0}{j\omega_i} \right)^\dagger + \underset{\sim}{Z}_i^* + \underset{\sim}{Z}_i^{*\dagger} \right] \underset{\sim}{\Phi}^\dagger \underset{\sim}{\eta}^\dagger, \quad (6.48)$$

where

$$\underset{\sim}{\eta} = \underset{\sim}{\Phi} \left[\underset{\sim}{Z}_i^* + \underset{\sim}{1} \left(R_s - \frac{S_0}{j\omega_i} \right) \right]^{-1} \underset{\sim}{\Phi}^\dagger \quad (6.49)$$

is a convenient abbreviation. Eigenvalues of the matrix $\underset{\sim}{Z} + \underset{\sim}{Z}^\dagger$ are therefore equal to the coefficient of the unit matrix in the first term of Equation 6.48, plus eigenvalues of the second term. But the second term is simply the matrix inside the brackets, conjunctively transformed by the matrix $\underset{\sim}{\eta}\underset{\sim}{\Phi}$. The matrix inside the brackets is positive semidefinite, because the idler impedance is passive, and because any matrix of the form $\underset{\sim}{A}\underset{\sim}{A}\dagger$ is positive semidefinite. Since conjunctive transformations cannot change the definite character of a matrix, the last term in Equation 6.48 cannot have any negative eigenvalues. Therefore, $\underset{\sim}{Z} + \underset{\sim}{Z}\dagger$ can have a negative eigenvalue (and hence the array of varactors can amplify) if and only if the coefficient of the unit matrix in Equation 6.48 is negative, or if (using the definitions of m_1 and ω_c)

$$\omega_s \omega_i < m_1^2 \omega_c^2. \tag{6.50}$$

This, of course, is the same criterion that applies to a single varactor. Equation 6.48 also indicates how to choose $\underset{\sim}{Z}_i$ so as to achieve gain; in particular, if the idler termination is lossless ($\underset{\sim}{Z}_i + \underset{\sim}{Z}_i\dagger = 0$) and if $\underset{\sim}{Z}_i + \underset{\sim}{1}(S_0/j\omega_i)$ is singular, one eigenvalue of the last term in Equation 6.48 is zero.

6.3.2. Noise

It is shown in Section 2.5.6. that the lowest possible noise temperature (at high gain) of an array of varactors is $(1/k\,\Delta f)$ times the lowest positive eigenvalue of the characteristic noise matrix,

$$\underset{\sim}{N} = -\frac{1}{2}(\underset{\sim}{Z} + \underset{\sim}{Z}\dagger)^{-1}\,\overline{\underset{\sim}{E}\underset{\sim}{E}\dagger}. \tag{6.51}$$

From Equations 6.38, 6.41, and 6.45 and the assumption that the thermal noises are not correlated,

$$\overline{\underset{\sim}{E}\underset{\sim}{E}\dagger} = 2k\left\{2T_d R_s \underset{\sim}{1} + \frac{|S_1|^2}{\omega_i^2}\underset{\sim}{\eta}\underset{\sim}{\Phi}[2R_s T_d \underset{\sim}{1} + T_i(\underset{\sim}{Z}_i^* + \underset{\sim}{Z}_i^*\dagger)]\underset{\sim}{\Phi}\dagger\underset{\sim}{\eta}\dagger\right\}\Delta f$$

$$= 4kT_d R_s\left(\underset{\sim}{1} + \frac{m_1^2\omega_c^2}{\omega_i^2} R_s^2 \underset{\sim}{\eta}\underset{\sim}{\eta}\dagger\right)\Delta f$$

$$+ 2kT_i \frac{m_1^2\omega_c^2}{\omega_i^2} R_s^2 \underset{\sim}{\eta}\underset{\sim}{\Phi}(\underset{\sim}{Z}_i^* + \underset{\sim}{Z}_i^*\dagger)\underset{\sim}{\Phi}\dagger\underset{\sim}{\eta}\dagger\;\Delta f, \tag{6.52}$$

where $\underset{\sim}{\eta}$ is given by Equation 6.49. The matrix $\underset{\sim}{Z} + \underset{\sim}{Z}\dagger$ is given in Equations 6.47 and 6.48.

Finding the eigenvalues of the noise matrix $\underset{\sim}{N}$ is not a simple task. We discuss here only a few special cases. First, we suppose that $T_i = T_d$. Next, we suppose that the idler impedance matrix $\underset{\sim}{Z}_i$ is normal. Finally, we show that the noise temperature can never be below the limit of Equation 6.35.

First, suppose that the diode and the idler termination are at the same temperature. Then, by use of Equation 6.47, Equation 6.52 can be rewritten in the form

$$\overline{\underset{\sim}{E}\underset{\sim}{E}^{\dagger}} = 2kT_d \left[\frac{2R_s\omega_p}{\omega_i} \underset{\sim}{1} - \frac{\omega_s}{\omega_i} (\underset{\sim}{Z} + \underset{\sim}{Z}^{\dagger}) \right] \Delta f. \qquad (6.53)$$

Thus the characteristic noise matrix, Equation 6.51, is some function of the matrix $\underset{\sim}{Z} + \underset{\sim}{Z}^{\dagger}$. Since in general the eigenvalues of a function of a matrix are exactly the same function of the eigenvalues of that matrix, we see that any eigenvalue of $\underset{\sim}{N}$ can be written in terms of an eigenvalue of $\underset{\sim}{Z} + \underset{\sim}{Z}^{\dagger}$. Let us denote the most negative eigenvalue of $\underset{\sim}{Z} + \underset{\sim}{Z}^{\dagger}$ by $2R$ (with R hopefully negative). Thus, by Equations 6.51 and 6.53, the lowest noise temperature T obtainable with a given idler termination and given frequencies is

$$T = T_d \left(\frac{\omega_s}{\omega_i} + \frac{\omega_p}{\omega_i} \frac{R_s}{-R} \right). \qquad (6.54)$$

To find how small Equation 6.54 can be, we inquire about the highest possible value for $-R$ as $\underset{\sim}{Z}_i$ is changed. Using the highest value predicted from Equation 6.47,

$$-R \leq R_s \frac{m_1^2\omega_c^2 - \omega_s\omega_i}{\omega_s\omega_i}, \qquad (6.55)$$

we find that

$$T \geq T_d \frac{\omega_s}{\omega_i} \frac{m_1^2\omega_c^2 + \omega_i^2}{m_1^2\omega_c^2 - \omega_s\omega_i}, \qquad (6.56)$$

which is precisely the limit predicted for a single-varactor amplifier, Equation 6.34 with $R_i = 0$. Note that this noise performance can be achieved only if the last term in Equation 6.48 has a zero eigenvalue.

If the idler is at a higher temperature than T_d, then the noise performance is worse than indicated in Equations 6.54 and 6.56.

As a second important special case, suppose that the idler impedance matrix has the property‡ that it commutes with its Hermitian transpose:

$$\underset{\sim}{Z}_i\underset{\sim}{Z}_i^{\dagger} = \underset{\sim}{Z}_i^{\dagger}\underset{\sim}{Z}_i. \qquad (6.57)$$

Such a matrix is called <u>normal</u>, and many practical terminations (as indicated below) have normal impedance matrices. This special case is easily analyzed because two matrices that com-

---

‡This condition is actually more severe than necessary. A weaker condition is that $(\underset{\sim}{Z}_i\underset{\sim}{Z}_i^{\dagger} - \underset{\sim}{Z}_i^{\dagger}\underset{\sim}{Z}_i)$ commutes with $\underset{\sim}{Z}_i$.

mute can be diagonalized by the same transformation. Since $\underset{\sim}{\Phi}$ is a unitary matrix, the eigenvalues of $\underset{\sim}{\Phi}^{\dagger}\underset{\sim}{N}\underset{\sim}{\Phi}$ are identical to the eigenvalues of $\underset{\sim}{N}$. But from Equations 6.44, 6.51, and 6.52, the matrix $\underset{\sim}{\Phi}^{\dagger}\underset{\sim}{N}\underset{\sim}{\Phi}$ is seen to be a function of $\underset{\sim}{Z}_i{}^*$ and $\underset{\sim}{Z}_i{}^{*\dagger}$. Thus any eigenvalue of $\underset{\sim}{\Phi}^{\dagger}\underset{\sim}{N}\underset{\sim}{\Phi}$ is the same function of the eigenvalues of $\underset{\sim}{Z}_i{}^*$ and $\underset{\sim}{Z}_i{}^{*\dagger}$. If we denote some eigenvalue of $\underset{\sim}{Z}_i$ by z_i, with a real part R_i, then the corresponding eigenvalue of $\underset{\sim}{Z}_i{}^{\dagger}$ is $z_i{}^*$. The corresponding eigenvalues of N, if positive, are made lower by tuning the idler so that $z_i + (S_0/j\tilde{\omega}_i)$ is real. If this is done, the lowest eigenvalue of $\underset{\sim}{N}$ corresponds to that z_i with the lowest real part R_i. Since $\underset{\sim}{\tilde{Z}}_i$ represents a passive network, R_i cannot be negative; if the external idler load is lossless, then all R_i vanish. The resulting noise temperature can be written in the form of Equation 6.34, where ζ is given by Equation 6.33. Note, of course, that the meaning of R_i is different — it is no longer merely the idler resistance.

We have shown that if the idler impedance matrix $\underset{\sim}{Z}_i$ is normal, then the limits derived for single-varactor amplifiers hold also for multiple-varactor amplifiers. This analysis with $\underset{\sim}{Z}_i$ normal really covers many cases of practical interest, including, among others, the following networks:

1. Lossless (for which $\underset{\sim}{Z}_i$ is anti-Hermitian).
2. Purely lossy, that is, no reactive power (for which $\underset{\sim}{Z}_i$ is Hermitian).
3. Separate terminations (for which $\underset{\sim}{Z}_i$ is diagonal).
4. All ports connected to a single element or a group of elements with one terminal pair.
5. Any network with a normal impedance matrix, modified by having identical impedances added in series with each port, or placed in shunt across each port.

Finally, let us consider the general problem, for which $\underset{\sim}{Z}_i$ need not be normal and T_i can be arbitrary. We merely wish to show that Equation 6.35 cannot be violated.

Straightforward (but tedious) matrix algebra can be used to write $\overline{\underset{\sim}{E}\underset{\sim}{E}^{\dagger}}$ in a rather lengthy form:

$$\frac{\overline{\underset{\sim}{E}\underset{\sim}{E}^{\dagger}}}{2k\,\Delta f} = -T_{min}(\underset{\sim}{Z} + \underset{\sim}{Z}^{\dagger}) + 2T_d R_s\left[\frac{\omega_s}{m_1\omega_c} + \sqrt{1 + \left(\frac{\omega_s}{m_1\omega_c}\right)^2}\,\right]^2 \underset{\sim}{\eta}\underset{\sim}{\Phi}\underset{\sim}{\xi}\,\underset{\sim}{\xi}^{\dagger}\underset{\sim}{\Phi}$$

$$+\; T_i\,\frac{m_1{}^2\omega_c{}^2}{\omega_i{}^2}\,R_s{}^2\underset{\sim}{\eta}\underset{\sim}{\Phi}(\underset{\sim}{Z}_i{}^* + \underset{\sim}{Z}_i{}^{*\dagger})\underset{\sim}{\Phi}^{\dagger}\underset{\sim}{\eta}^{\dagger}, \qquad (6.58)$$

where

6.4 extra idlers

$$\underset{\sim}{\xi} = \underset{\sim}{Z_i^*} - 1\frac{S_0}{j\omega_i} - R_s \underset{\sim}{1} \left(\frac{\sqrt{m_1^2 \omega_c^2 + \omega_s^2} - \omega_p}{\omega_p - \omega_s} \right), \quad (6.59)$$

and where T_{min} is given by Equation 6.35.

A well-known result of matrix theory is that the eigenvalues of any matrix of the form $\underset{\sim}{A}^{-1}\underset{\sim}{B}$ can be expressed, for some column matrix $\underset{\sim}{x}$, in the form

$$\frac{\underset{\sim}{x}^\dagger \underset{\sim}{B} \underset{\sim}{x}}{\underset{\sim}{x}^\dagger \underset{\sim}{A} \underset{\sim}{x}}. \quad (6.60)$$

Let us write in this way the lowest positive eigenvalue of $\underset{\sim}{N}$:

$$-\frac{1}{2} \frac{\underset{\sim}{x}^\dagger \underset{\sim}{E}\underset{\sim}{E}^\dagger \underset{\sim}{x}}{\underset{\sim}{x}^\dagger (\underset{\sim}{Z} + \underset{\sim}{Z}^\dagger)\underset{\sim}{x}}. \quad (6.61)$$

Since this eigenvalue is positive, the factor in the denominator, $\underset{\sim}{x}^\dagger(\underset{\sim}{Z} + \underset{\sim}{Z}^\dagger)\underset{\sim}{x}$, must be negative. The numerator contains (see Equation 6.58) a term $\underset{\sim}{x}^\dagger(\underset{\sim}{Z} + \underset{\sim}{Z}^\dagger)\underset{\sim}{x}$ multiplied by T_{min}, plus a strictly positive contribution from the last two terms in Equation 6.58. Therefore, the noise temperature of a multiple-varactor amplifier, if positive, can never be below T_{min}. Furthermore, Equation 6.58 indicates how to achieve this ultimate performance. The last term can be eliminated either by making T_i small or by making the idler termination lossless, so that $Z_i^* + Z_i^{*\dagger} = 0$. The contribution from the second term can vanish if the matrix $\underset{\sim}{\xi}$ has a zero eigenvalue or, in other words, if one of the eigenvalues of $\underset{\sim}{Z}_i$ is

$$j\frac{S_0}{\omega_i} + R_s \left(\frac{\sqrt{m_1^2 \omega_c^2 + \omega_s^2} - \omega_p}{\omega_p - \omega_s} \right). \quad (6.62)$$

It should be stressed that we have calculated only the fundamental limit on noise, given by the lowest positive eigenvalue of $\underset{\sim}{N}$; we have not shown how to construct a signal-frequency circuit to achieve this noise performance. This difficult engineering problem is outside the scope of this book.

6.4. Amplifiers with Extra Idlers

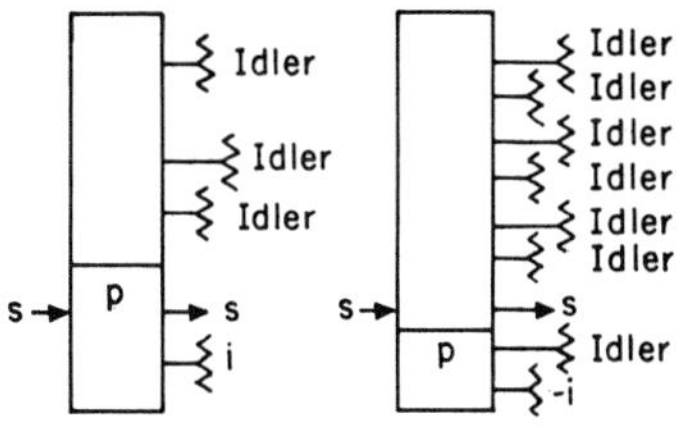

The fundamental limits derived thus far apply when current flows at only one sideband frequency. Is it possible that these limits can be surpassed if extra idling current is allowed? Section 6.4.1 is devoted to this question for the simple case with the elastance sinusoidally pumped. The more general case with an arbitrary elastance waveform is very difficult to treat analytically; in Section 6.4.2 we derive a new fundamental limit for the noise temperature, somewhat lower than Equation 6.35.

6.4.1. Sinusoidal Elastance

If the elastance, as pumped, is a sinusoidal function of time,‡ all S_k are zero, except for $k = 0$ and $k = 1$ (and -1). Therefore, Equations 4.42 show that the voltage at each of the sideband frequencies depends only on the corresponding current and the two currents at the two frequencies removed by ω_p. If we denote the current and voltage half-amplitudes at the sideband frequency $k\omega_p + \omega_s$ by I_k and V_k, then Equations 4.42 predict for each k

$$V_k = \left[R_s + \frac{S_0}{j(k\omega_p + \omega_s)}\right] I_k + \frac{S_1^* I_{k+1}}{j[(k+1)\omega_p + \omega_s]}$$

$$+ \frac{S_1 I_{k-1}}{j[(k-1)\omega_p + \omega_s]} + E_{nk}, \tag{6.63}$$

where E_{nk}, with mean-squared value‡‡

$$\overline{|E_{nk}|^2} = 4kT_d R_s \, \Delta f, \tag{6.64}$$

is the noise generated at frequency $k\omega_p + \omega_s$ by the varactor series resistance. If all currents except I_0 and I_{-1} are set equal to zero, Equation 6.63 reduces to Equation 6.15 (with $k = -1$) and Equation 6.16 (with $k = 0$), provided a slight change in notation is made (V_0, I_0, and E_{n0} to V_s, I_s, and E_{ns}; V_{-1}, I_{-1}, and $E_{n(-1)}$ to V_i^*, I_i^*, and E_{ni}^*).

It is convenient to express Equation 6.63 in equivalent-circuit form. A few of the equations (for small values of k) are shown in Figure 6.8, with dependent sources to account for the coupling between frequencies. The subscripts s and i are used for the signal and idler circuits ($k = 0$ and $k = -1$).

Now let us suppose that the varactor is terminated in a passive load at all frequencies $k\omega_p + \omega_s$ for $k > 0$ and $k < -1$. The effect of the terminations for $k > 0$ is to produce a voltage at

---

‡In practice, this can be achieved by pumping an abrupt-junction varactor with a sinusoidal current.

‡‡These noise voltages are effective rms voltages instead of half-amplitudes. See Section 2.5.1(a).

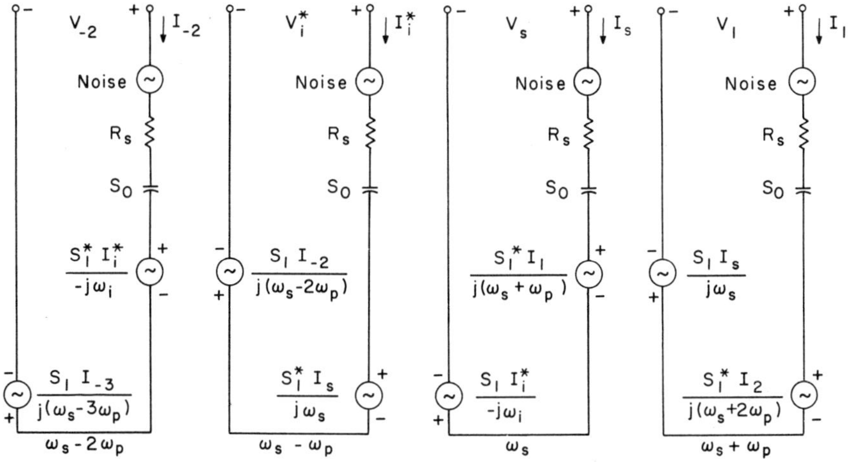

Figure 6.8. Circuits obeying Equation 6.63, for a few small values of k. Note that the idler frequency is $\omega_p - \omega_s$ and not $\omega_s - \omega_p$; therefore, the complex conjugates of V_i and I_i appear.

signal frequency proportional to the signal-frequency current, plus some noise voltage. An equivalent circuit that accounts for this effect would be simply some impedance $(Z_s)_{eq}$ in series with a noise generator. Similarly, the effect of the terminated sidebands for $k < -1$ is the same as some impedance $(Z_i)_{eq}$ in the idler-frequency loop, in series with a noise voltage generator. The resulting equivalent circuit of the pumped varactor as terminated at all frequencies except ω_s and ω_i appears in Figure 6.9.

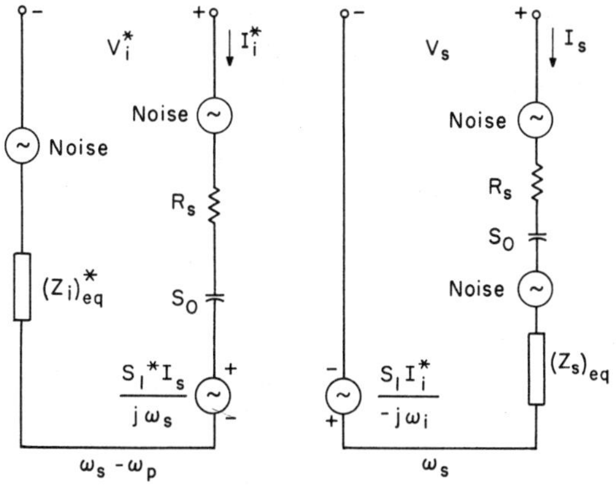

Figure 6.9. Equivalent circuit of the varactor with terminations at all sideband frequencies except the signal and idler.

In other words, the explanation of Figure 6.9 is the following. Since the elastance is sinusoidal, two frequencies can interact directly with each other only if they are separated by ω_p. Therefore, the currents at frequencies $k\omega_p + \omega_s$ for $k > 0$ cannot directly produce a voltage at frequency ω_i (they could, however, if some S_k with $k > 1$ were present). Consequently, their effect is evident only at frequency ω_s. Similarly, the currents that flow at frequencies $k\omega_p + \omega_s$ with $k < -1$ cannot directly produce a voltage at frequency ω_s, so that their effect is evident only at the idler frequency.

6.4.1(a). Gain. Now we wish to investigate the equivalent impedances $(Z_s)_{eq}$ and $(Z_i)_{eq}$. First, we open-circuit the varactor at the idler frequency.‡ The nonlinear-elastance part of the varactor obeys the Manley-Rowe formulas (see Section 2.1). Since the varactor is open-circuited at ω_i and we excite it only at ω_s, power is exchanged only at frequencies $k\omega_p + \omega_s$ with $k \geq 0$. The Manley-Rowe formulas predict

$$\sum_{k=0}^{\infty} \frac{P_k}{k\omega_p + \omega_s} = 0, \tag{6.65}$$

where P_k is the power output of the nonlinear-elastance part of the varactor at frequency $k\omega_p + \omega_s$. Because power cannot flow in at any sideband frequency except ω_s, Equation 6.65 implies that power must not flow out at signal frequency. Referring to Figure 6.9, we see that because $(Z_s)_{eq}$ must absorb power, it has a positive real part.

The effect of the upper sidebands, that is, those with $k > 0$, is to introduce a positive-real impedance at signal frequency; these upper sidebands cannot contribute to gain. Furthermore, the noise they add merely increases the noise temperature of the amplifier. The greatest possibility for gain, and the least noise, therefore occur when the varactor is open-circuited at the upper sidebands, especially the frequency $\omega_p + \omega_s$.

If $\omega_p > \omega_s$, that is, for the region above the straight lines in Figures 6.3 and 6.4, the effect of finite terminations at the lower sidebands (those with $k < -1$) is to introduce into the idler circuit an additional noise and positive-real impedance. The same effect could be achieved by adding to the external idler impedance a passive impedance at some finite temperature. Since any effects of lower-sideband idlers can also be obtained with a passive termination at ω_i, the analysis of Section 6.2 applies: gain is possible only when Condition 6.26 holds, and the noise temperature cannot be lower than Equation 6.35.

‡This of course does not affect the value of $(Z_s)_{eq}$.

According to the discussion of Section 6.5, it is sometimes helpful to use a low-noise idler to reduce the noise figure (although any improvement obtained in this way can also be obtained by a shift in pump frequency). A practical method might be to allow current at frequency $2\omega_p - \omega_s$; the "temperature" of $(Z_i)_{eq}$ can be appreciably below T_d or the ambient temperature.

If $\omega_p < \omega_s$ (that is, for points below the straight line in Figures 6.3 and 6.4), the analysis of Section 6.2 indicates that gain is impossible. This result is modified, however, if lower-sideband currents are allowed to flow. The impedance $(Z_i)_{eq}$ is no longer positive-real. If the frequency $\omega_s - \omega_p$ is terminated in an impedance (to be consistent with previous equations, we call this impedance Z_i^*), then the varactor impedance at signal frequency is, by an analysis similar to that leading to Equation 6.20,

$$Z = R + jX$$

$$= R_s + \frac{S_0}{j\omega_s} + \frac{|S_1|^2}{\omega_s(\omega_s - \omega_p)\{R_s + (Z_i)_{eq}^* + Z_i^* + [S_0/j(\omega_s - \omega_p)]\}} \tag{6.66}$$

and hence can have a negative real part for some passive Z_i if and only if

$$\text{Re}\,(Z_i)_{eq} + R_s < 0. \tag{6.67}$$

But Condition 6.67 is exactly the condition that with the varactor open-circuited at frequency ω_s gain can be achieved at the positive frequency $\omega_s - \omega_p$. The analysis of Section 6.2, therefore, shows that if $\omega_s - 2\omega_p$ is negative, or if

$$\omega_s > \omega_p > \frac{\omega_s}{2}, \tag{6.68}$$

then gain is possible if and only if

$$(\omega_s - \omega_p)(2\omega_p - \omega_s) < m_1^2\omega_c^2. \tag{6.69}$$

Values of ω_p and ω_s consistent with Condition 6.68 but not Condition 6.69 are indicated in Figure 6.10.

The analysis of the case with

$$\frac{\omega_s}{3} < \omega_p < \frac{\omega_s}{2} \tag{6.70}$$

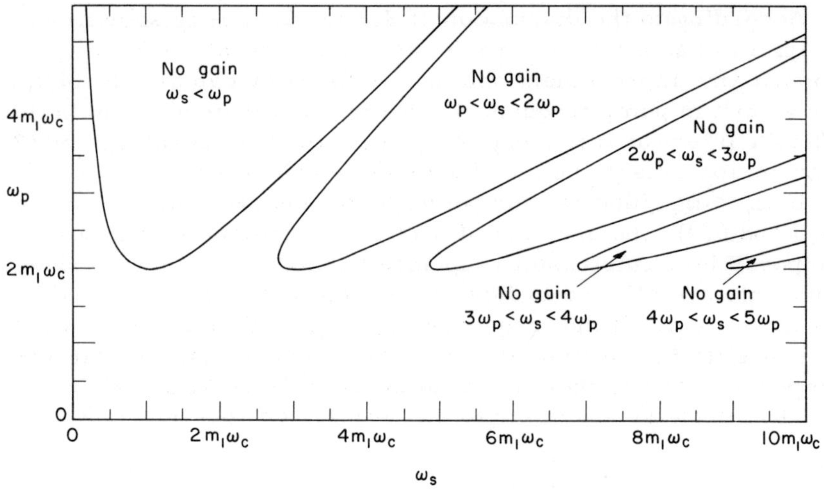

Figure 6.10. Region of possible gain in the ω_s,ω_p-plane if the elastance is pumped sinusoidally but if arbitrary passive sideband terminations are allowed. Note that gain is possible for pump frequencies lower than the signal frequency; in fact, if the pump frequency is less than $2m_1\omega_c$, gain is theoretically possible at any signal frequency. The gain region for $\omega_s < \omega_p$ is the same as predicted by Figure 6.3.

is similar; gain at frequency $\omega_s - 2\omega_p$, and therefore at $\omega_s - \omega_p$, and therefore at ω_s, is possibly for passive terminations at the various frequencies if and only if

$$(\omega_s - 2\omega_p)(3\omega_p - \omega_s) < m_1^2\omega_c^2 . \qquad (6.71)$$

Values of ω_p and ω_s consistent with Condition 6.70 but not Condition 6.71 are indicated in Figure 6.10. For still lower pump frequencies, the analysis is similar. Figure 6.10 shows the values of ω_p and ω_s for which gain can and cannot be achieved if the elastance is pumped sinusoidally. Observe that the limits on gain derived in Section 6.2, and depicted in Figure 6.3, have been modified, but only for the case with ω_p less than ω_s.

6.4.1(b). Noise. The noise analysis of this amplifier is not complete. The upper sidebands can only raise the noise temperature, as mentioned earlier. The lower sidebands help achieve gain with $\omega_p < \omega_s$; with $\omega_p > \omega_s$, they can sometimes lower the noise temperature. In no case can the noise temperature of the amplifier ever be lower than Equation 6.35; the lower-sideband idlers are of practical importance because a noise temperature close to this minimum can be achieved at a lower pump frequency and, therefore, with lower pump power (see Chapter 7).

6.4.1(c). Multiple-Varactor Amplifiers. If an amplifier is made out of several varactors, then (see Section 6.3) its achievable fundamental limits are identical to those of the single-varactor amplifier, provided each varactor is open-circuited at all sideband frequencies except ω_i. An analysis similar to the one presented earlier in Section 6.4.1 shows that if several varactors are pumped to have sinusoidal elastances and if their sideband terminations are arbitrary, then gain is possible from this array of varactors at frequency ω_s for exactly the same values of ω_p as the single-varactor, multiple-idler amplifier. In other words, Figure 6.10 applies to arrays of varactors pumped with sinusoidal elastances, for any sideband terminations.

6.4.2. Arbitrary Elastance Waveform

The most general type of single-varactor amplifier is one with an arbitrary elastance waveform (periodic with frequency ω_p) and unspecified sideband terminations. Needless to say, this amplifier is difficult to analyze; nevertheless, there is a rather simple fundamental limit on the noise temperature, and one that is of some importance, because it is not far below Equation 6.35. The derivation of this limit is quite similar to the derivation of Equation 5.157 in Section 5.6.

The elastance, as pumped, is given by Equation 4.41, and the S_k for all k are assumed nonzero. Furthermore, all I_k and V_k must be assumed nonzero, where I_k and V_k are (as usual) half-amplitudes of varactor current and voltage at sideband frequency $k\omega_p + \omega_s$.

Let the termination at frequency $k\omega_p + \omega_s$ be an impedance Z_k with real part R_k either positive or negative. Since the sideband terminations are in general noisy, we may define a noise "temperature" for each, so that the open-circuit noise voltage at each termination has mean-squared value‡

$$\overline{|E_{nxk}|^2} = 4kT_kR_k\,\Delta f. \tag{6.72}$$

If any R_k is negative, so is the corresponding T_k.

For simplicity, we set the problem up in matrix form. The equations that govern the parametrically pumped varactor are written

$$\underset{\sim}{V} = \underset{\sim}{Z}_c\underset{\sim}{I} + \underset{\sim}{E}_n, \tag{6.73}$$

where $\underset{\sim}{V}$ and $\underset{\sim}{I}$ are column matrices of half-amplitudes of the

‡The noise voltages are effective rms values instead of half-amplitudes. See Section 2.5.1(a).

voltage and current at the various frequencies $k\omega_p + \omega_s$, in order of ascending k from $-\infty$ to $+\infty$; where $\underset{\sim}{E}_n$ is a column matrix of noise voltages representing thermal noise of the varactor series resistance, each voltage with mean-squared value‡ $4kT_dR_s\,\Delta f$; and where $\underset{\sim}{Z}_c$ is the varactor conversion matrix with diagonal elements

$$(\underset{\sim}{Z}_c)_{kk} = R_s + \frac{S_0}{j(k\omega_p + \omega_s)} \tag{6.74}$$

and off-diagonal elements

$$(Z_c)_{k\ell} = \frac{S_{k-\ell}}{j(\ell\omega_p + \omega_s)}, \tag{6.75}$$

and, in particular, elements in the ω_s column

$$(\underset{\sim}{Z}_c)_{k0} = \frac{S_k}{j\omega_s} \qquad (k \neq 0). \tag{6.76}$$

This matrix equation is written out as Equations 4.42.

The sideband terminations are accounted for by another matrix equation,

$$\underset{\sim}{V} = -\underset{\sim}{Z}_{term}\underset{\sim}{I} + \underset{\sim}{E}_{nx}, \tag{6.77}$$

where E_{nx} is a column matrix of the $\underset{\sim}{E}_{nxk}$ noise voltages, and where $\underset{\sim}{\tilde{Z}}_{term}$ is a diagonal matrix with entries

$$(\underset{\sim}{Z}_{term})_{kk} = Z_k. \tag{6.78}$$

At the outset, we agree that if the varactor is known to be open-circuited at any of the sidebands, we set the corresponding current to zero and delete the corresponding voltage and the corresponding row and column of both $\underset{\sim}{Z}_c$ and $\underset{\sim}{Z}_{term}$.

We wish to solve Equations 6.73 and 6.77 for the currents. By eliminating the matrix $\underset{\sim}{V}$, we find

$$\underset{\sim}{I} = \underset{\sim}{Y}(\underset{\sim}{E}_{nx} - \underset{\sim}{E}_n), \tag{6.79}$$

where the matrix $\underset{\sim}{Y}$ is

‡The noise voltages are effective rms values instead of half-amplitudes. See Section 2.5.1(a).

$$\underset{\sim}{Y} = (\underset{\sim}{Z}_c + \underset{\sim}{Z}_{term})^{-1}. \tag{6.80}$$

The noise temperature of the amplifier is the negative of the noise exchangeable power of the pumped, terminated varactor, divided by $k\,\Delta f$ (see Section 2.5.5). The noise exchangeable power is

$$P_e = \frac{\overline{|I_0|^2}}{4 \text{ Re } Y_{00}}, \tag{6.81}$$

where we denote the signal frequency by the subscript 0 (for $k = 0$ in the formula $k\omega_p + \omega_s$). From Equation 6.79 and the formulas for mean-squared noise voltages,

$$\overline{|I_0|^2} = \Sigma_k |Y_{0k}|^2 (4kT_d R_s \ \Delta f) + \Sigma_k |Y_{0k}|^2 (4kT_k R_k \ \Delta f), \tag{6.82}$$

where the first sum, representing noise from the varactor, is over all k from $-\infty$ to $+\infty$, and the second sum is over all k except 0. If we agree to the convention that sums of this sort are to omit the term for $k = 0$, then the noise temperature becomes

$$T = \frac{|Y_{00}|^2 \, T_d R_s + \Sigma_k |Y_{0k}|^2 (T_d R_s + T_k R_k)}{\text{Re } (-Y_{00})}$$

$$= \frac{T_d R_s + \Sigma_k |Y_{0k}/Y_{00}|^2 (T_d R_s + T_k R_k)}{-\text{Re } (1/Y_{00})} \tag{6.83}$$

where by our convention the sum does not include $k = 0$. We now wish to eliminate Re $(1/Y_{00})$ from Equation 6.83. This is easily done, because the definition of $\underset{\sim}{Y}$

$$\underset{\sim}{Y}(\underset{\sim}{Z}_c + \underset{\sim}{Z}_{term}) = \underset{\sim}{1} \tag{6.84}$$

has a 0, 0 element

$$Y_{00}(R_s + jX) + \Sigma_k Y_{0k} \frac{S_k}{j\omega_s} = 1 \tag{6.85}$$

for some real value of X. Thus, dividing by Y_{00} and taking the real part, we obtain

$$\text{Re}\left(\frac{1}{Y_{00}}\right) = R_s - \Sigma_k \text{ Re}\left(\frac{Y_{0k}}{Y_{00}} \frac{jS_k}{\omega_s}\right). \tag{6.86}$$

In substituting Equation 6.86 into Equation 6.83, we can conveniently define the parameters A_k, r_k, and θ_k as follows:

$$A_k = \frac{|S_k|^2}{\omega_s^2 R_s^2} \frac{T_d R_s}{T_d R_s + T_k R_k}, \tag{6.87}$$

$$r_k e^{j\theta_k} = -\frac{Y_{0k}}{Y_{00}} \frac{S_k}{j\omega_s R_s}, \tag{6.88}$$

where r_k and θ_k are the magnitude and phase of the quantity shown. In terms of these parameters, the noise temperature becomes

$$T = T_d \frac{\Sigma_k (r_k^2/A_k) + 1}{\Sigma_k r_k \cos\theta_k - 1}, \tag{6.89}$$

which is identical to Equation 5.151, except for slightly different definitions of r_k and θ_k.

Equation 6.89 is an actual expression for the noise temperature of the parametric amplifier. It is, of course, much too complicated to be evaluated directly, except in the simplest cases. The A_k parameters are easily determined, but the r_k and θ_k parameters depend in a complicated way on the varactor pumping and loading. It might be supposed that by varying the r_k and θ_k parameters in an arbitrary way, the expression for noise temperature could be made small. However, this is not the case: Upon variation of the r_k and θ_k parameters, Equation 6.89 has a positive minimum. The noise temperature obviously cannot be less than this minimum, and probably cannot in general even approach it.

The minimum was calculated in Chapter 5 by differentiating Equation 6.89 with respect to the r_k and θ_k parameters and setting the result equal to zero. Here, we establish this minimum in a slightly different way. If we define the parameters

$$r_k' = \frac{A_k}{\Sigma_k A_k}\left(1 + \sqrt{1 + \Sigma_k A_k}\right), \tag{6.90}$$

then by straightforward algebra Equation 6.89 can be put into the form

$$T = T_{min} + T_d \frac{\Sigma_k |r_k e^{j\theta_k} - r_k'|^2/A_k}{\Sigma_k r_k \cos\theta_k - 1}, \tag{6.91}$$

where

$$T_{min} = \frac{2T_d}{\Sigma_k A_k}\left(1 + \sqrt{1 + \Sigma_k A_k}\right). \tag{6.92}$$

Equation 6.91 shows that T can never be lower than Equation 6.92 if it is to be positive.

This minimum noise temperature is expressed in terms of the A_k parameters defined in Equation 6.87. If the pumping waveform and the terminating resistances and temperatures are known, the result is a function only of the varactor temperature T_d and the signal frequency ω_s. In specific cases it can be evaluated fairly easily.

The higher $\Sigma_k A_k$ is, the lower is the fundamental limit T_{min}. It is of interest, therefore, to ask how high $\Sigma_k A_k$ can be under various circumstances. Let us define m_t

$$m_t^2 = \sum_{\substack{k=-\infty \\ k\neq 0}}^{+\infty} \frac{|S_k|^2}{(S_{max} - S_{min})^2} \tag{6.93}$$

as a kind of total modulation ratio. The summation in Equation 6.93 is over all k (except 0) from $-\infty$ to $+\infty$, but in case some of the sidebands are known to be open-circuited, m_t can be replaced in Equation 6.94 by a similar sum, omitting the open-circuit sidebands.

It is clear from Equation 6.87 that A_k is highest when the sideband terminations are noiseless ($T_k R_k = 0$). In this case, Equation 6.92 becomes

$$T_{min} = T_d \frac{2\omega_s}{m_t \omega_c}\left[\frac{\omega_s}{m_t \omega_c} + \sqrt{1 + \left(\frac{\omega_s}{m_t \omega_c}\right)^2}\right], \tag{6.94}$$

which is of the same form as the fundamental limit for single-idler amplifiers, Equation 6.35, except that m_1 is replaced by m_t. In Chapter 7 it is shown how m_t is restricted by the requirement that S(t) lie between S_{min} and S_{max}. For practical varactors, m_t is seldom more than 50 per cent greater than m_1.

Note particularly that we have not shown that Equation 6.94 is an achievable limit on noise temperature. However, in practical cases it is not much below Equation 6.35, which is known to be achievable by a relatively simple circuit. Therefore, the amount of noise improvement possible by using extra idlers is strictly limited.

Very few assumptions were required to derive the limit of Equation 6.94. In particular:

1. The elastance waveform is not restricted.
2. Any number of idlers can be used.
3. The idler terminations need not be passive, and may in fact include noiseless negative resistors, which if used directly for amplification would have a zero noise temperature.
4. The termination can even have synchronously pumped frequency converters in it, so that the matrix $\underset{\sim}{Z}_{term}$ is not diagonal, provided only that these do not interact with the signal frequency (so that the elements of $\underset{\sim}{Z}_c + \underset{\sim}{Z}_{term}$ along the ω_s column are given by Equation 6.76).

The limit of Equation 6.94 seems to be the minimum penalty for simply having the signal enter the varactor and come back out again, regardless of how it is processed at the sidebands. Even if a noiseless amplifier were available at some frequency, if we couple to it by means of a varactor, we must pay at least this penalty.

6.4.3. Example

The fundamental limit of Equation 6.94 is usually not achievable. For a single-idler amplifier, as in Section 6.2, m_t can be replaced by a summation of the type of Equation 6.93 with only one term $(k = -1)$. The result is exactly the fundamental limit of Equation 6.35, derived in Section 6.2. Therefore, this is one instance in which the fundamental limit predicted in Section 6.4.2 is achievable. Now we describe briefly another such instance.

A simple amplifier with two idlers is one made from a varactor pumped with $S_2 = 0$ but S_1 and S_3 nonzero. Let us open-circuit all sidebands except ω_i and $3\omega_p - \omega_s$ (for simplicity we use the subscript 3 for this sideband). Because $S_2 = 0$, the two idlers do not interact directly, but instead each reacts separately at the signal frequency. Because most of the sidebands are open-circuited, the fundamental limit of Equation 6.94 can be replaced by one of the same form, but with m_t replaced by $m = \sqrt{m_1^2 + m_3^2}$ We shall now show that this fundamental limit is achievable, provided $\omega_s < 0.354 m\omega_c$, provided a noiseless termination at frequency ω_i is available, and provided the pump frequency can be adjusted.

The equations that describe the varactor are obtained from Equations 4.42 by selecting the equations for V_s, V_i^*, and the voltage two rows above V_i^*, and setting all other currents to zero. The two entries in the matrix that relate the two idlers are zero because S_2 is assumed to be zero.

The resistance at signal frequency presented by the varactor is an extension of Equation 6.23:

$$R = R_s\left(1 - \frac{m_1{}^2\omega_c{}^2}{\omega_s\omega_i}\,\frac{R_s}{R_s + R_i} - \frac{m_3{}^2\omega_c{}^2}{\omega_s\omega_3}\,\frac{R_s}{R_s + R_3}\right). \tag{6.95}$$

The open-circuit noise voltage (mean-squared) is likewise an extension of Equation 6. 30:‡

$$\overline{|E|^2} = 4kT_dR_s\,\Delta f\left[1 + \frac{m_1{}^2\omega_c{}^2}{\omega_i{}^2}\,\frac{R_s{}^2}{(R_s + R_i)^2} + \frac{m_3{}^2\omega_c{}^2}{\omega_3{}^2}\,\frac{R_s{}^2}{(R_s + R_3)^2}\right]. \tag{6.96}$$

For simplicity, we have assumed that the terminations at ω_i and ω_3 are tuned and at zero temperature. From Equations 6. 95 and 6. 96, the noise temperature can be written as the negative of the exchangeable power divided by $k\,\Delta f$,

$$T = T_d\,\frac{\dfrac{m_1{}^2\omega_c{}^2}{\omega_i{}^2}\,\dfrac{R_s{}^2}{(R_s + R_i)^2} + \dfrac{m_3{}^2\omega_c{}^2}{\omega_3{}^2}\,\dfrac{R_s{}^2}{(R_s + R_3)^2} + 1}{\dfrac{m_1{}^2\omega_c{}^2}{\omega_s\omega_i}\,\dfrac{R_s}{R_s + R_i} + \dfrac{m_3{}^2\omega_c{}^2}{\omega_s\omega_3}\,\dfrac{R_s}{R_s + R_3} - 1}. \tag{6.97}$$

Note the similarity between Equations 6. 97 and 6. 89. The minimum value of Equation 6. 97 is

$$T_{min} = T_d\,\frac{2\omega_s}{m\omega_c}\left[\frac{\omega_s}{m\omega_c} + \sqrt{1 + \left(\frac{\omega_s}{m\omega_c}\right)^2}\right], \tag{6.98}$$

where m is

$$m^2 = m_1{}^2 + m_3{}^2. \tag{6.99}$$

The minimum noise temperature can be realized with several combinations of ω_p, R_i, and R_3. In particular, if $R_3 = 0$,

$$\omega_p = \frac{1}{3}\sqrt{m^2\omega_c{}^2 + \omega_s{}^2}, \tag{6.100}$$

---

‡ This noise voltage is an effective rms value instead of a half-amplitude. See Section 2.5.1(a).

and

$$R_i = R_s \frac{2\omega_p}{\omega_p - \omega_s}, \tag{6.101}$$

then Equation 6.97 reduces to Equation 6.98. This optimization is valid as long as ω_i is positive, that is, whenever $\omega_s < 0.354m\omega_c$.

Note that an extra idler can help to reduce the noise temperature only by a few per cent, since ordinarily m_3 is less than m_1. If $m_3 = 0.3m_1$, the improvement at low frequencies is about 4.4 per cent.

The properties of this double-idler amplifier are not discussed further, because of the authors' conviction that the extra complexity of the circuitry required for multiple-idler amplifiers is, in general, not justified by the meager improvements possible.

6.5. Discussion of the Noise Formulas

One of the most important single results of Chapter 6 is the expression for noise temperature of the simple single-varactor single-idler amplifier, Equation 6.34:

$$T = T_d \frac{\omega_s}{\zeta\omega_i} \frac{m_1^2\omega_c^2 + \zeta^2\omega_i^2}{m_1^2\omega_c^2 - \zeta\omega_s\omega_i} + T_i \frac{\omega_s}{\omega_i} \frac{R_i}{R_s + R_i} \frac{m_1^2\omega_c^2}{m_1^2\omega_c^2 - \zeta\omega_s\omega_i}. \tag{6.102}$$

We now discuss this formula for noise temperature, explicitly giving its form under a variety of conditions. It is hoped that this discussion will aid in the use of Equation 6.102 in preliminary design of amplifiers. Although specifically derived for the simple amplifier of Section 6.2, Equation 6.102 also describes some more complicated amplifiers, and we note in each instance what amplifiers are covered.

6.5.1. Limit of No Series Resistance

If the series resistance R_s becomes very small, we should expect Equation 6.102 to approach the noise temperature for lossless varactor amplifiers,[16, 37] Equation 3.39 or Equation 6.28. A glance at Equation 6.102 indicates that it does, because ω_c and ζ both approach infinity in the process.

6.5.2. Reactive External Idler Termination

Setting $R_i = 0$ in Equation 6.102, we find

$$T = T_d \frac{\omega_s}{\omega_i} \frac{m_1^2\omega_c^2 + \omega_i^2}{m_1^2\omega_c^2 - \omega_s\omega_i}. \tag{6.103}$$

This is also the formula for noise temperature of a lower-sideband upconverter or downconverter when the source resistance is set to make the exchangeable gain infinite (see Equation 5.110). It is also a lower limit for the noise temperature of an array of many varactors with equal diode and idler temperatures (see Condition 6.56).

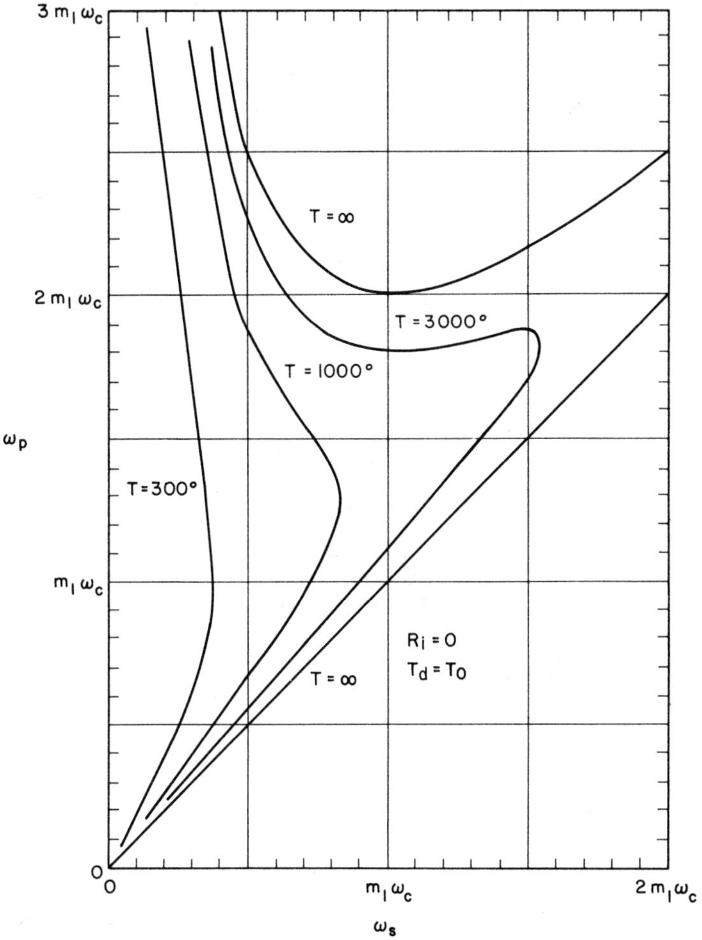

Figure 6.11. Contours of constant noise temperature in the ω_p, ω_s-plane, for a reactive external idler termination, and a room-temperature varactor; taken from Equation 6.103. This figure also applies to a lower-sideband upconverter or downconverter, adjusted to give infinite exchangeable gain. The limits of the region of possible gain, taken from Figure 6.3, are also given. For more detail, see Figure 6.12.

For a given diode temperature, T is now a function of pump and signal frequencies, normalized by $m_1\omega_c$. It is instructive to look at contours of constant noise temperature in the pump-frequency, signal-frequency plane, Figures 6.11 and 6.12. As expected, the noise temperature becomes large near the boundary

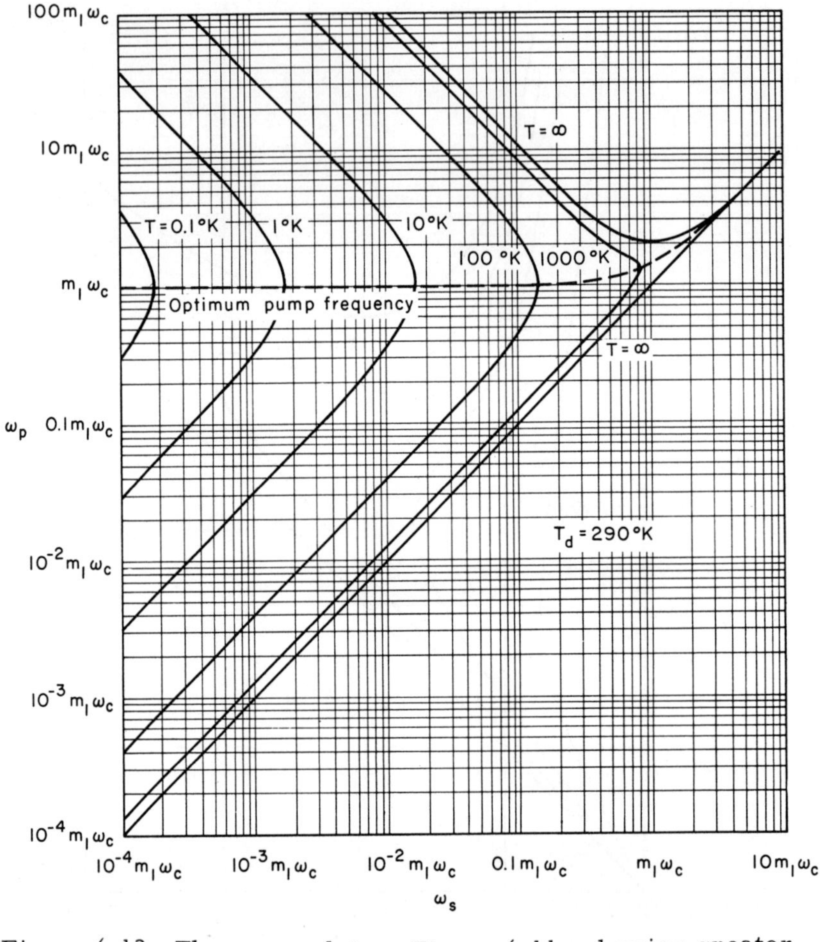

Figure 6.12. The same plot as Figure 6.11, showing greater detail. Note that the frequencies are plotted logarithmically. Also plotted is the optimum pump frequency, given by Equation 6.104.

between regions of gain and no gain. For a given signal frequency, there is a finite pump frequency at which the noise temperature is lowest. This optimum pump frequency is found by differentiating Equation 6.103 with respect to ω_i, and setting the result equal to zero. Thus,

$$\omega_{p,\,opt} = \sqrt{m_1^2\omega_c^2 + \omega_s^2}\,, \tag{6.104}$$

and for this pump frequency the noise temperature is

$$T_{min} = T_d \frac{2\omega_s}{m_1\omega_c}\left[\frac{\omega_s}{m_1\omega_c} + \sqrt{1 + \left(\frac{\omega_s}{m_1\omega_c}\right)^2}\right], \qquad (6.105)$$

which is exactly the fundamental limit of Equation 6.35. Note that T is just twice what would be erroneously predicted from Equation 3.39, using the pump frequency of Equation 6.104. The optimum pump frequency is indicated in Figure 6.13, and the

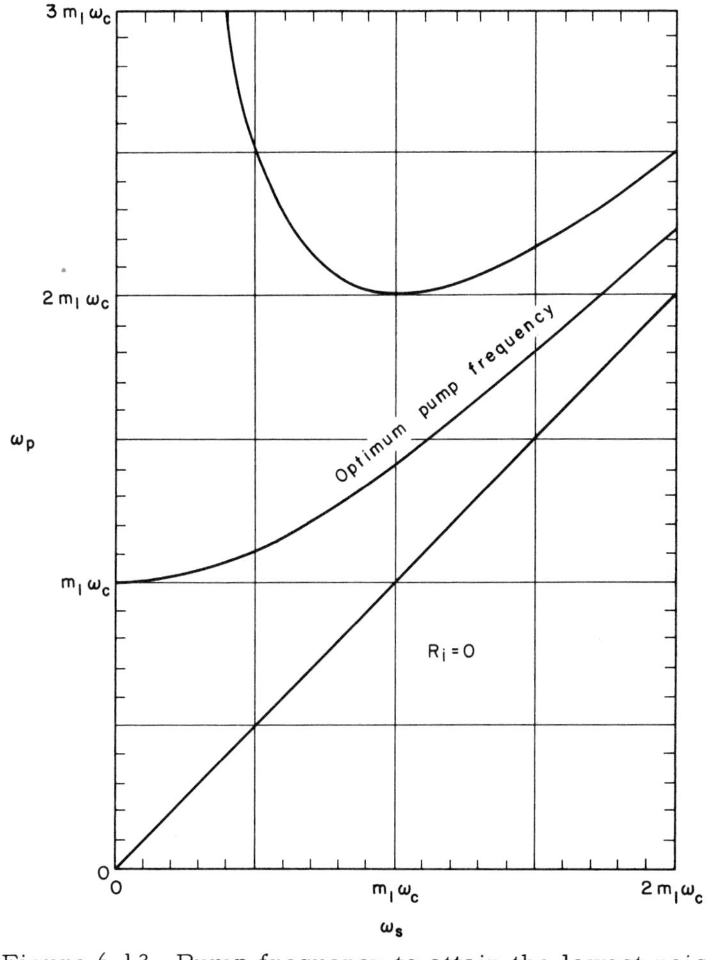

Figure 6.13. Pump frequency to attain the lowest noise figure from amplifiers with reactive external idler terminations, taken from Equation 6.104. The regions of no gain are indicated by the upper and lower lines, from Figure 6.3. This pump frequency is also shown in Figure 6.12.

lowest noise temperature has been given in Figures 6.6 and 6.7. For low signal frequencies, $\omega_{p,\,opt}$ is approximately $m_1\omega_c$.

6.5.3. Zero Idler Temperature

We suspect intuitively that the noise temperature might be improved if we were able to terminate the varactor externally in a noiseless idler impedance, thus "dumping power into a noise-free load." Under some conditions an improvement is possible, but any improvement obtained in this way can also be obtained by a change in pump frequency.

Setting $T_i = 0$, we reduce Equation 6.102 to

$$T = T_d \frac{\omega_s}{\zeta\omega_i} \frac{m_1{}^2\omega_c{}^2 + \zeta^2\omega_i{}^2}{m_1{}^2\omega_c{}^2 - \zeta\omega_s\omega_i} . \tag{6.106}$$

For given R_i, ω_s, and ω_p, this temperature is smaller than that for a finite T_i.

Note that Equation 6.106 depends on R_i and ω_p only in the form of the product $\zeta\omega_i$. If the pump frequency is greater than the optimum frequency of Equation 6.104, the lowest noise temperature is achieved with $R_i = 0$. On the other hand, if ω_p is less than Equation 6.104, T is reduced to Equation 6.105 by setting R_i properly:

$$R_i = R_s \frac{\sqrt{m_1{}^2\omega_c{}^2 + \omega_s{}^2} - \omega_p}{\omega_p - \omega_s} . \tag{6.107}$$

This value of R_i is indicated in Figure 6.14, and the minimum noise temperature attained by adjusting R_i with $T_i = 0$ is shown in Figure 6.15.

On the other hand, if R_i is fixed at some value, the noise-temperature limit of Equation 6.105 can be attained with the following choice of pump frequency (this is lower than Equation 6.104):

$$\omega_p = \frac{R_i\omega_s + R_s\sqrt{m_1{}^2\omega_c{}^2 + \omega_s{}^2}}{R_s + R_i} . \tag{6.108}$$

This is plotted for a few values of R_i in Figure 6.14. Note that Figure 6.14 serves two purposes: For fixed ω_p, it indicates the optimum R_i; for fixed R_i, it indicates the optimum ω_p. At low signal frequencies, the optimum pump frequency is approximately

$$\omega_p = \frac{R_i\omega_s + R_s m_1\omega_c}{R_i + R_s} . \tag{6.109}$$

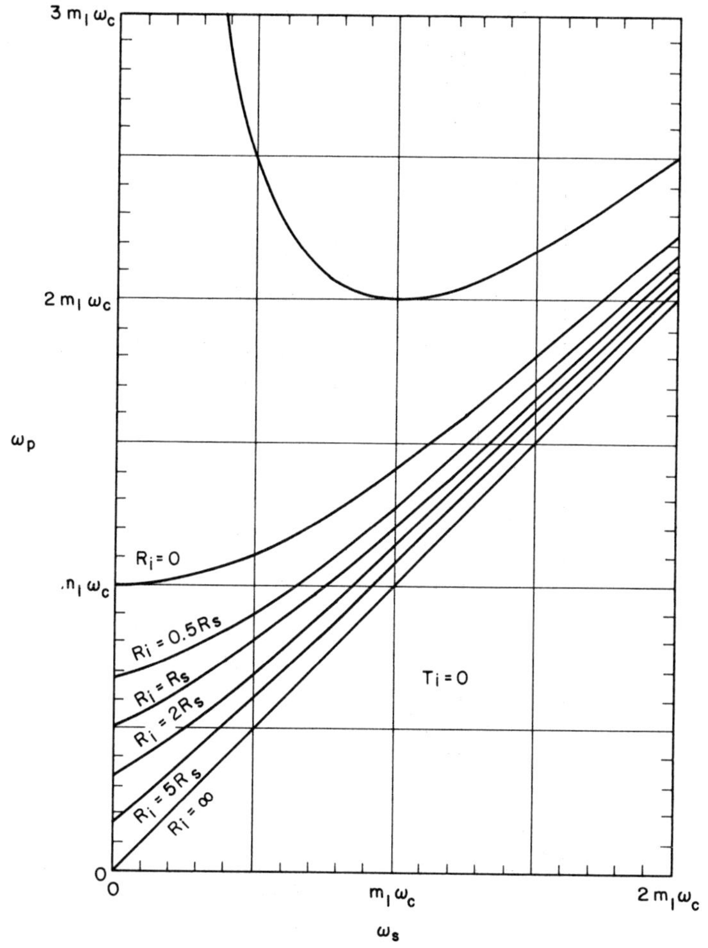

Figure 6.14. Optimum pump frequency for various noise-free external idler terminations. The optimum values are lower than the one of Figure 6.13, which applies to the case with $R_i = 0$. Alternatively, the figure shows, as a function of signal and pump frequencies, the value of R_i necessary to attain the noise performance shown in Figure 6.15; above the line for $R_i = 0$, the optimum R_i is negative, so that the amplifier operates best with $R_i = 0$. Also shown are the limits of the region of gain, taken from Figure 6.3. The curves are taken from Equation 6.107 or Equation 6.108.

If a noise-free idler termination is available, it is evident from Equation 6.108 that we can achieve the minimum noise temperature of Equation 6.105 at a lower pump frequency than would be otherwise possible. We cannot achieve a lower noise temperature.

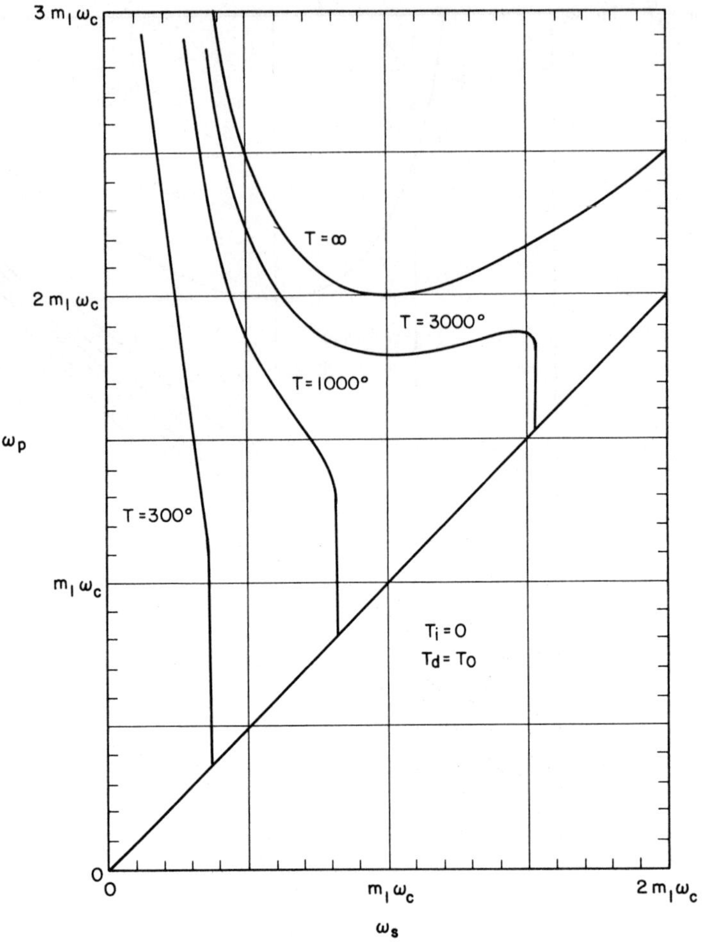

Figure 6.15. Minimum noise temperature, as a function of pump and signal frequencies, for an amplifier with a noiseless external idler termination. For pump frequencies above the optimum shown in Figure 6.13, the curves here coincide with those of Figures 6.11 and 6.12. For lower pump frequencies, a finite R_i can be used (see Figure 6.14) to achieve the minimum noise of Figures 5.7, 6.6, and 6.7. The limits of the region of gain, taken from Figure 6.3, are also shown here.

This is the extent to which a noiseless external idler termination can help lower the noise temperature. Any noise performance achieved with such a termination can also be achieved with a room-temperature idler by a shift in pump frequency. Of course a noise-free idler, if available, is valuable because less pump power is required at the lower pump frequency.

6.5.4. Equal Idler and Diode Temperatures

If $T_i = T_d$ (or, in fact, if $T_i > T_d$), then as R_i is increased from zero, the noise temperature of Equation 6.102 increases. There is therefore no noise advantage in using a lossy external idler termination, regardless of the pump frequency.

If, however, for some reason R_i is nonzero and $T_i = T_d$, then Equation 6.102 becomes

$$T = T_d \frac{\omega_s}{\omega_i} \frac{m_1^2\omega_c^2 + \zeta\omega_i^2}{m_1^2\omega_c^2 - \zeta\omega_s\omega_i} . \qquad (6.110)$$

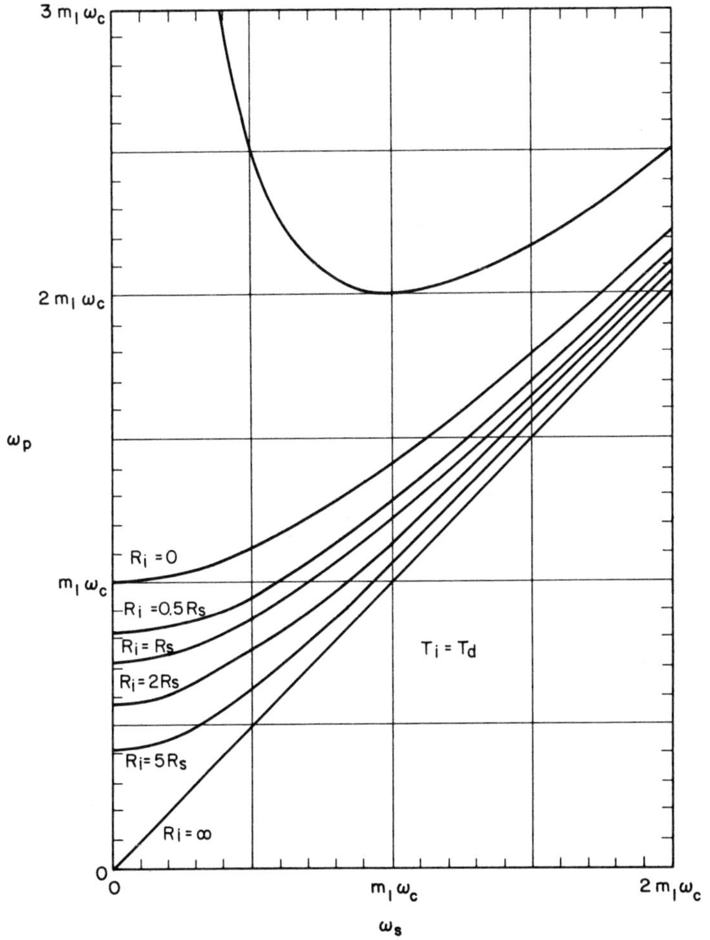

Figure 6.16. Pump frequency for lowest noise when the idler and diode temperatures are the same, for various idler terminations. The curves, taken from Equation 6.111, are different from those of Figure 6.14. Also shown are the boundaries of the region of possible gain, taken from Figure 6.3.

The pump frequency can be adjusted to the value

$$\omega_p = \sqrt{\frac{R_s m_1^2 \omega_c^2}{R_s + R_i} + \omega_s^2} \tag{6.111}$$

to reduce Equation 6.110 to its minimum value:

$$T = T_d \frac{R_s + R_i}{R_s} \frac{2\omega_s}{m_1 \omega_c} \left[\frac{\omega_s}{m_1 \omega_c} + \sqrt{\frac{R_s}{R_s + R_i} + \left(\frac{\omega_s}{m_1 \omega_c}\right)^2} \right]. \tag{6.112}$$

A plot of this optimum pump frequency is given in Figure 6.16 for several values of R_i. The resulting noise figure is shown in Figure 6.17, under the assumption that $T_i = T_d = 290°$ K.

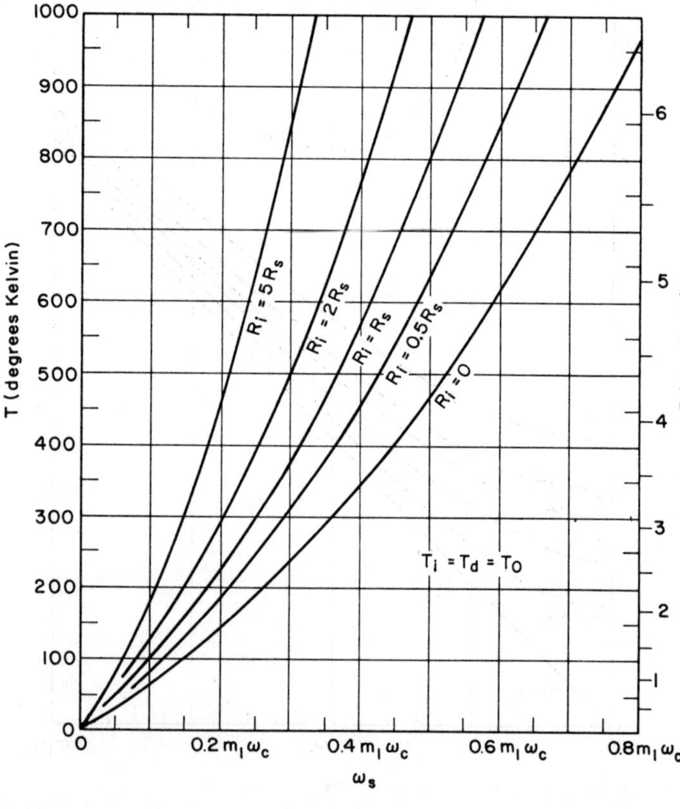

Figure 6.17. Lowest noise temperature at high gain, for various values of R_i, for an assumption that the varactor and the external idler termination are both at room temperature. These noise temperatures, taken from Equation 6.112, are achieved with the pump frequencies indicated in Figure 6.16.

6.5.5. Arbitrary Idler and Diode Temperatures

If T_i and T_d and R_i are arbitrary, the noise temperature of Equation 6.102 can be minimized by proper selection of the pump frequency. By differentiating Equation 6.102 with respect to ω_i and setting the result equal to zero, we find that the pump frequency of

$$\omega_p = \frac{T_d R_s + T_i R_i}{T_d(R_s + R_i)} \sqrt{\omega_s^2 + m_1^2 \omega_c^2 \frac{T_d R_s}{T_d R_s + T_i R_i}}$$

$$+ \frac{T_d - T_i}{T_d} \frac{R_i \omega_s}{R_s + R_i} \qquad (6.113)$$

reduces Equation 6.102 to its minimum value:

$$T = \frac{T_d R_s + T_i R_i}{R_s} \frac{2\omega_s}{m_1 \omega_c} \left[\frac{\omega_s}{m_1 \omega_c} + \sqrt{\frac{T_d R_s}{T_d R_s + T_i R_i} + \left(\frac{\omega_s}{m_1 \omega_c} \right)^2} \right].$$

$$(6.114)$$

Note that Equations 6.113 and 6.114 reduce to Equations 6.104 and 6.105 when $R_i = 0$; they reduce to Equations 6.108 and 6.105 when $T_i = 0$; and they reduce to Equations 6.111 and 6.112 when $T_i = T_d$.

6.5.6. The Fundamental Limit

We have seen several instances in which for some m a fundamental limit on noise temperature is of the form

$$T = T_d \frac{2\omega_s}{m\omega_c} \left[\frac{\omega_s}{m\omega_c} + \sqrt{1 + \left(\frac{\omega_s}{m\omega_c} \right)^2} \right]. \qquad (6.115)$$

For single-varactor, single-idler amplifiers, this is a limit that is achievable with either R_i or T_i zero if m is interpreted as m_1. The simple frequency converters of Sections 5.2, 5.3, 5.4, and 5.5 have an achievable noise limit of this form, with $m = m_1$. The upconverter with idler, Section 5.2.4, has such a limit with $m = \sqrt{2}\, m_1$. Amplifiers and frequency converters with arbitrary pumping waveform and arbitrary sideband terminations have such a limit (probably not achievable) with $m = m_t$.

Equation 6.115 can be used to predict the quality of the varactor necessary to attain a given noise performance at a given signal frequency; we solve Equation 6.115 for $m\omega_c$, obtaining

$$m\omega_c = 2\omega_s \sqrt{\frac{T_d}{T}\left(1 + \frac{T_d}{T}\right)}. \tag{6.116}$$

Equation 6.115 can also be used to predict the largest signal frequency at which a given noise performance can be achieved with a given varactor; we solve Equation 6.115 for ω_s, obtaining

$$\omega_s = \frac{\dfrac{m\omega_c}{2}\dfrac{T}{T_d}}{\sqrt{1 + \dfrac{T}{T_d}}}. \tag{6.117}$$

For low signal frequencies, the noise temperature limit of Equation 6.115 is approximately

$$T = T_d \frac{2\omega_s}{m\omega_c}; \tag{6.118}$$

this approximation is valid within 10 per cent for $\omega_s < 0.1 m_1\omega_c$.

6.5.7. Comparison with Quantum Noise

Recently a quantum-mechanical model for a parametric process has been used to predict the effect of quantization on parametric-amplifier noise performance.[27] The contribution toward noise temperature from this noise source was found to be

$$\frac{\hbar\omega_s}{k \log_e 2}, \tag{6.119}$$

where $h = 2\pi\hbar$ is Planck's constant. To observe this contribution of 0.069° K per Gc of signal frequency, all other noise sources must be lower. From a comparison with Equation 6.118, the quantum contribution cannot be observed unless mf_c is on the order of, or higher than, $29T_d$ Gc. Thus, for a room-temperature varactor, we should need a cutoff frequency above about

30,000 Gc, an unheard-of value. Therefore, for practical purposes, quantum noise can be neglected in varactor parametric amplifiers. This fact is also suggested by Figure 6.18, in which the quantum limit is far below typical plots of Equation 6.115.

6.5.8. Comparison with Other Low-Noise Amplifiers

It is interesting to compare the performance of high-quality varactor parametric amplifiers with other competitive devices

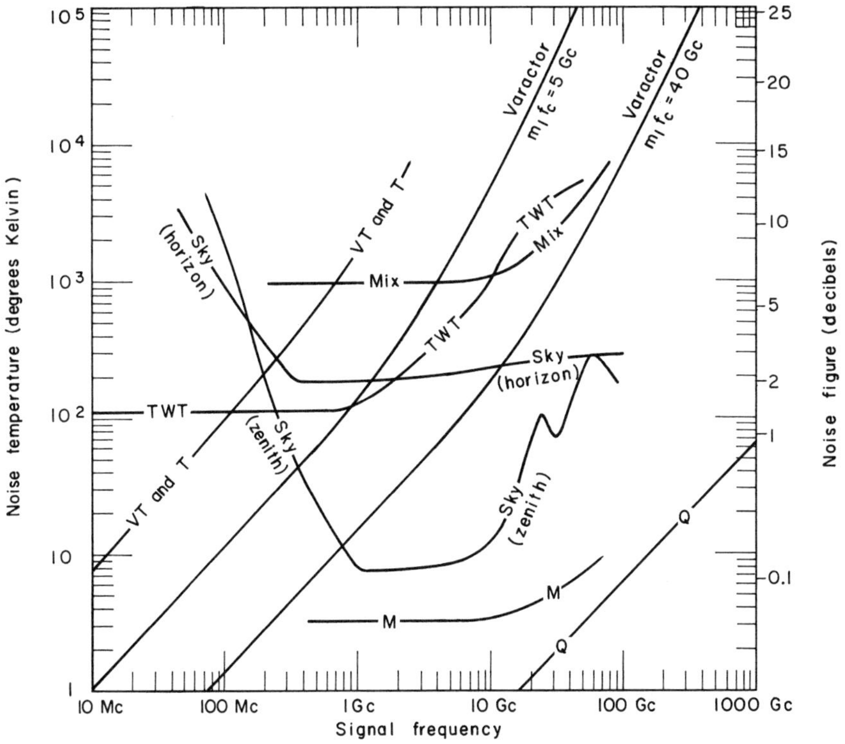

Figure 6.18. Approximate limits on noise temperature, as a function of frequency, of several types of receiver front ends: M = maser, Mix = mixer, T = transistor, TWT = traveling-wave tube, and VT = vacuum tube. Two varactor curves are shown, corresponding to $m_1 f_c = 5$ Gc and 40 Gc, and for $T_d = 290°$K. Also shown are the quantum noise limit Q and two minimum antenna temperatures marked Sky, for vertical (zenith) and horizontal (horizon) orientations. Since all curves are approximate, detailed comparisons cannot be made from this diagram. The curves are either empirical or theoretical limits. Tunnel-diode amplifiers are not indicated because data on them are sparse.

for front ends of receivers. Among the possible devices are vacuum tubes, crystal mixers, transistors, traveling-wave tubes, and masers. In Figure 6.18, we show approximate limits for the noise performance of these devices, either theoretical or empirical limits. Also shown is the limit on varactor noise, taken from Figure 5.7 or Equation 6.105 for two representative values of $m_1 f_c$. We assume that $T_d = 290°$ K.

Also shown are three auxiliary curves of importance. The first is the quantum limit of Equation 6.119. The other two are minimum sky temperatures as seen by horizontal or vertical antennas, taken from the plots by Rafuse.[32] There is little point in designing a receiver with noise temperature much lower than this.

This figure shows, in a rough way, the frequency range over which varactors are useful for receivers. Below about 100 Mc, the minimum sky temperature is great enough so that vacuum tubes and transistors are sufficient. Varactor amplifiers are useful in the approximate range from 100 Mc to 10 Gc and higher. Their noise is, of course, worse than that of masers, but they do not require refrigeration, as masers do.

6.5.9. Examples

We now give a few examples of the use of the noise formulas.

First, suppose a noise temperature of 290° K is desired at 10 Gc from a graded-junction room-temperature varactor. What quality of varactor is required, and for the minimum quality of varactor possible, what pump frequency is appropriate? Assume the idler is at room temperature. The discussion of Section 6.5.4 indicates that R_i should be set equal to zero for best noise performance; thus with $R_i = 0$, Equations 6.116 and 6.104 are used to find that the minimum $m_1 f_c$ is 28.3 Gc and the pump frequency employed should be 30 Gc. If m_1 is restricted to be less than 0.212 (see Chapter 7), then f_c must exceed 133 Gc.

Next, suppose R_i is restricted to be equal to R_s. For a varactor with $m_1 f_c = 20$ Gc and a signal frequency of 15 Gc, what is the range of pump frequencies over which gain is possible, what is the minimum noise temperature (assuming $T_i = T_d = 290°$ K), and what pump frequency realizes this minimum? From Condition 6.25, f_p must be greater than 15 Gc and less than 28.3 Gc. From Equation 6.111, the optimum pump frequency is 20.6 Gc, and from Equation 6.112, the resulting noise temperature is 1550° K. Note that if R_i were zero, a minimum noise temperature of 870° K could be achieved.

6.6. The Degenerate Amplifier

Thus far we have assumed that the signal and idler frequencies are different, so that the varactor termination at ω_i can be adjusted, at least in principle, without affecting the circuit at ω_s.

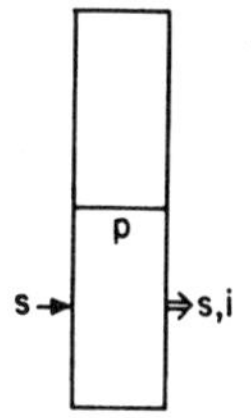

However, for the degenerate amplifier, the signal and idler cannot be physically separated, and the previous theory is not pertinent. In particular, the minimum noise temperature is not T_{min} of Equation 6.35.

The criterion for distinguishing degenerate amplifiers from nondegenerate amplifiers is not merely the relation between ω_s and ω_p. Rather, the criterion is the treatment of the signal and idler by the external amplifier circuitry. Thus, with a single-varactor amplifier, if the signal and idler frequencies are very close or if their spectra overlap, the amplifier circuitry cannot distinguish between them, and the amplifier is degenerate. Even if their spectra do not overlap, the amplifier is degenerate if the varactor termination does not distinguish between them.

On the other hand, with two or more varactors, circuits with special symmetry can be built to separate the signal from the idler, even if they are at the same frequency, because of their relative phases at the varactors. Thus it is possible to build an amplifier in which the signal and idler are separable, even though they may have spectra that overlap. Such an amplifier must be called nondegenerate, in spite of the fact that the signal and idler frequencies are the same. The analysis given thus far in Chapter 6 applies to this amplifier, and the minimum noise temperature is given by Equation 6.34 (or a modification, to account for the fact that several varactors are used).

In summary, then, the essential feature of degenerate amplifiers is not necessarily the relationship between the signal and idler frequencies, but rather is that the amplifier circuitry (exclusive of the varactor) treats the signal and idler identically.

For degenerate amplifiers, the ordinary concepts of noise figure, noise temperature, and noise measure do not apply. Degenerate amplifiers are not amplifiers in the usual sense, because they give output at frequencies not included in the input. Often they are phase-sensitive, and the output noise at various frequencies is generally correlated. The concept of imbedding, upon which the concept of noise measure is based, is not valid, because such imbedding operates on the idler as well as the signal frequency. The noise performance of degenerate amplifiers ordinarily depends upon the type of input signal, the type of detector used, and the interpretation of the detector output. For all these reasons, it is not possible to characterize the noisiness of a degenerate amplifier by a simple noise figure or noise temperature. Although we shall derive a noise figure of merit for the amplifier, the reader is specifically cautioned that this figure of merit cannot, except in specific cases mentioned later, be related to ordinary noise figure. Two amplifiers, one a degenerate parametric amplifier and the other an ordinary amplifier, cannot be compared by means of their respective noise "figures" (except in certain cases discussed later).

Although many of our results hold for any high-gain degenerate parametric amplifier, made from any nonlinear reactance, we are most concerned with varactors. For simplicity, we describe only a varactor amplifier with a circulator, and we assume the gain is high. The modifications for low gain and/or other circuits are straightforward but difficult. In Section 6.6.1 this amplifier is discussed, and in Section 6.6.2 we determine the output caused by input noise and by amplifier noise and find that the output noise at various frequencies is correlated. The noise output of the amplifier defines a figure of merit for the noisiness of degenerate parametric amplifiers. This figure of merit can be used to compare two or more degenerate amplifiers but cannot, in general, be used to compare a degenerate amplifier with a nondegenerate amplifier.

In Section 6.6.3 we consider what happens when a signal is applied to that high-gain circulator amplifier. At least insofar as the spectrum close to $\omega_p/2$ is concerned, the output is the same as would be obtained by merely multiplying the input waveform by $G(1 + 2 \cos \omega_p t)$, where G is the voltage gain, a large number. In fact, this simple procedure also applies to input noise and to noise generated by the varactor series resistance. The discussion of Section 6.6.5 is simplified by the use of this rule.

In Section 6.6.4 we discuss the amplifier with an input whose spectrum does not surround the frequency $\omega_p/2$. Although the circulator does not distinguish between the signal and idler, a sharp filter in a subsequent stage can be used to select ω_s. This single-sideband operation is characterized by a certain degradation in signal-to-noise ratio, calculated in Section 6.6.4. A rule is given for comparing single-sideband degenerate amplifiers with ordinary amplifiers.

If the input spectrum surrounds $\omega_p/2$, we say the operation is double-sideband; we show in Section 6.6.5 that there is no general, simple way to compare the noise performance of a double-sideband degenerate amplifier with that of an ordinary amplifier, because this comparison usually depends on the nature of the input signal, the detector, and the interpretation of the detector output. Several examples are discussed in detail for both synchronous detectors and square-law detectors.

6.6.1. Varactor Circulator Amplifier

Although many of the concepts and formulas of Section 6.6 are valid for other degenerate parametric amplifiers, for example those using electron beams, we shall, for simplicity, base our derivation on the circulator amplifier shown in Figure 6.19. The

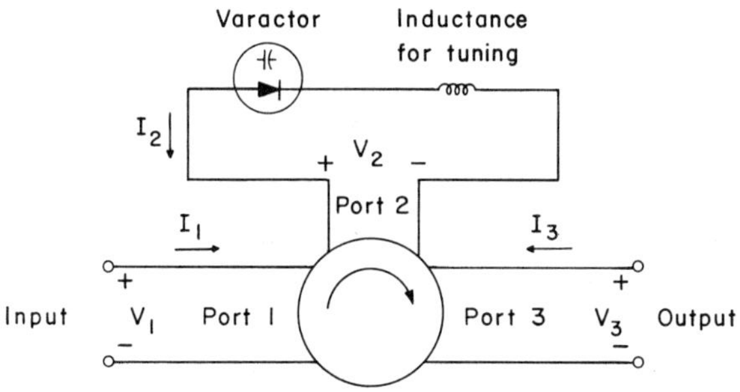

Figure 6.19. Varactor degenerate amplifier using circulator. The pump circuit is not shown; the schematic is valid only for frequencies close to $\omega_p/2$.

three-port ideal circulator is assumed to be lossless over the frequency range of interest. The inductor is the value necessary to tune out the average elastance S_0 close to frequency $\omega_p/2$.

The circulator has the property that a wave incident on port 1 is transmitted to port 2. This wave is reflected by the termination at port 2 and is transmitted to port 3, where a matched load prevents any reflection. We define incoming waves a_k ($k = 1, 2, 3$) and outgoing waves b_k in terms of the circulator characteristic resistance R_0 and the half-amplitude voltages and currents at the three circulator ports:[3]

$$a_k = \frac{V_k + R_0 I_k}{\sqrt{2R_0}}, \tag{6.120}$$

$$b_k = \frac{V_k - R_0 I_k}{\sqrt{2R_0}}. \tag{6.121}$$

In terms of these scattering variables, the power input is

$$\overline{|a_1|^2} = \overline{|b_2|^2}, \tag{6.122}$$

where the horizontal bar is a statistical average necessary only for inputs that have random properties. Similarly, the output into a matched load has power

$$\overline{|a_2|^2} = \overline{|b_3|^2}. \tag{6.123}$$

The varactor obeys Equation 4.40, with $S(t)$ given by Equation 4.41. Without loss of generality, we assume that S_1/j is positive and real, so that the phase of the pumping current is fixed to 180°. Then, the varactor and inductor combined have the equation (for signals close to $\omega_p/2$ in frequency)

$$V_2(\omega) = -R_s I_2(\omega) - \frac{E_n(\omega)}{\sqrt{2}} + \frac{m_1 \omega_c}{\omega_p/2} R_s I_2^*(\omega_p - \omega) \tag{6.124}$$

in the frequency domain. The factor of $\sqrt{2}$ is included because the noise voltage E_n (accounting for thermal noise in the varactor series resistance) is an effective rms value, instead of a half-amplitude [see Section 2.5.1(a)]. Note that the varactor mixes signals at frequency ω with those at frequency $\omega_p - \omega$.

The varactor constraint, Equation 6.124, can be used to relate the input and output waves at frequencies ω and $\omega_p - \omega$. For signals close to frequency $\omega_p/2$, we find in matrix notation

$$\begin{bmatrix} a_2(\omega) \\ a_2^*(\omega_p - \omega) \end{bmatrix} = \frac{1}{\left(\dfrac{m_1\omega_c}{\omega_p/2} R_s\right)^2 - (R_0 + R_s)^2} \left\{ \begin{bmatrix} 2R_0(R_0 + R_s) & 2R_0 R_s \dfrac{m_1\omega_c}{\omega_p/2} \\ 2R_0 R_s \dfrac{m_1\omega_c}{\omega_p/2} & 2R_0(R_0 + R_s) \end{bmatrix} \begin{bmatrix} \dfrac{E_n(\omega)}{\sqrt{4R_0}} \\ \dfrac{E_n^*(\omega_p - \omega)}{\sqrt{4R_0}} \end{bmatrix} \right.$$

$$\left. + \begin{bmatrix} R_0^2 - R_s^2 + \left(\dfrac{m_1\omega_c}{\omega_p/2} R_s\right)^2 & 2R_0 R_s \dfrac{m_1\omega_c}{\omega_p/2} \\ 2R_0 R_s \dfrac{m_1\omega_c}{\omega_p/2} & R_0^2 - R_s^2 + \left(\dfrac{m_1\omega_c}{\omega_p/2} R_s\right)^2 \end{bmatrix} \begin{bmatrix} b_2(\omega) \\ b_2^*(\omega_p - \omega) \end{bmatrix} \right\}. \tag{6.125}$$

Using this formula, we could compute the gain and noise properties of the degenerate amplifier.

The amplifier, however, is of most use when its gain is high. Hence, for simplicity we now assume that the gain is made high by making the denominator of the first factor on the right small. Thus, we set

$$R_0 + R_s = \frac{m_1 \omega_c}{\omega_p/2} R_s - \delta, \tag{6.126}$$

where δ is a small number. This is clearly possible if and only if

$$m_1\omega_c > \frac{\omega_p}{2}. \tag{6.127}$$

Then, Equations 6.125 reduce to

$$\begin{bmatrix} a_2(\omega) \\ \\ a_2^*(\omega_p - \omega) \end{bmatrix} = G \begin{bmatrix} 1 & 1 \\ \\ 1 & 1 \end{bmatrix} \left\{ \begin{bmatrix} b_2(\omega) \\ \\ b_2^*(\omega_p - \omega) \end{bmatrix} + \frac{1}{\sqrt{4R_0}} \begin{bmatrix} E_n(\omega) \\ \\ E_n^*(\omega_p - \omega) \end{bmatrix} \right\}, \tag{6.128}$$

where G is the large number‡

$$G = \frac{R_0}{\frac{m_1\omega_c}{\omega_p/2} R_s - R_s - R_0} = \frac{R_0}{\delta}. \tag{6.129}$$

Thus, using Equation 6.128, we can substitute actual input signals and determine the resulting output a_2. In Section 6.6.2 we investigate noise and in Section 6.6.3, signals.

6.6.2. Noise Outputs

Let us now calculate the output caused by an input noise with temperature T_s and by the noise from the varactor series resistance, with temperature T_d.

First, from the series resistance, the noise voltage $E_n(\omega)$ has a mean-squared value

$$\overline{|E_n(\omega)|^2} = 4kT_dR_s\ \Delta f, \tag{6.130}$$

and the noise generated in different parts of the spectrum is uncorrelated, so that in particular

$$\overline{E_n(\omega)E_n(\omega_p - \omega)} = 0. \tag{6.131}$$

Thus from Equation 6.128, the output power is

---

‡Note that G^2 is the power gain of this amplifier for an input located at one side of frequency $\omega_p/2$.

$$\overline{|a_2(\omega)|^2} = 2G^2 kT_d \frac{R_s}{R_0} \Delta f$$

$$= \frac{2kT_d G^2 \Delta f}{\frac{m_1\omega_c}{\omega_p/2} - 1}, \tag{6.132}$$

and the noise output at frequency ω is perfectly correlated with the output at frequency $\omega_p - \omega$.

Second, we assume for the moment that the input is thermal noise at temperature T_s, so that

$$\overline{|b_2(\omega)|^2} = \overline{|b_2(\omega_p - \omega)|^2}$$

$$= kT_s \Delta f \tag{6.133}$$

and

$$\overline{b_2(\omega)b_2(\omega_p - \omega)} = 0. \tag{6.134}$$

Thus, the output power is

$$\overline{|a_2(\omega)|^2} = 2G^2 kT_s \Delta f \tag{6.135}$$

and the output at frequency ω is perfectly correlated with the output at frequency $\omega_p - \omega$.

Because the two noise sources are uncorrelated, the over-all noise output power is

$$2G^2 k(T_s + T_{eff}) \Delta f, \tag{6.136}$$

where T_{eff} is the effective input noise temperature

$$T_{eff} = \frac{T_d}{\frac{m_1\omega_c}{\omega_p/2} - 1}. \tag{6.137}$$

The reader is specifically cautioned that T_{eff} is not in the usual sense the noise temperature of the degenerate amplifier. Nevertheless, it is a valid figure of merit by which to compare two or more degenerate amplifiers. We shall calculate the noise performance of the degenerate amplifier in terms of T_{eff}. Note also that T_{eff} is what would be measured by a standard noise-temperature measurement if the noise source extended over a region surrounding $\omega_p/2$. The temperature T_{eff} is often called the "double-sideband noise temperature," or the "radiometer noise temperature," and is the number usually quoted to describe the noisiness of a degenerate amplifier.

Although two degenerate amplifiers can be compared for noisiness by comparing their values of T_{eff}, a degenerate amplifier cannot be compared with an ordinary amplifier in a way quite as simple as this.

6.6.3. Signal Outputs

We now calculate the output from the amplifier for a signal input. Two important cases must be distinguished. The input signal has, of course, a spectrum located close to $\omega_p/2$. Either this spectrum is located entirely on one side of $\omega_p/2$, or it is not. In the former case, we say the amplifier is a single-sideband amplifier; in the latter case, a double-sideband amplifier.

In the first case, if noise is momentarily neglected, the input spectrum, lying exclusively on one side of $\omega_p/2$, is centered, say, at some frequency ω_s. To calculate the output, we can let $b_2(\omega_s)$ be the input and $b_2(\omega_p - \omega_s)$ be zero. Then,

$$a_2(\omega_s) = Gb_2(\omega_s), \tag{6.138}$$

and

$$a_2^*(\omega_p - \omega_s) = Gb_2(\omega_s). \tag{6.139}$$

Thus, for a single input, there are two outputs of equal amplitude, located symmetrically about the frequency $\omega_p/2$. The single-sideband amplifier is a combination amplifier and frequency converter; it is discussed more fully in Section 6.6.4.

On the other hand, suppose the spectrum of the input signal surrounds the frequency $\omega_p/2$. Then, neglecting noise, we assume that both $b_2(\omega)$ and $b_2(\omega_p - \omega)$ are nonzero for this double-sideband amplifier. We represent the time function of the signal input as

$$b_2(t) = \text{Re}\, B_2(t)e^{j\omega_p t/2}, \tag{6.140}$$

and the time function of the output as

$$a_2(t) = \text{Re}\, A_2(t)e^{j\omega_p t/2}, \tag{6.141}$$

where, because the signal has a spectrum located close to $\omega_p/2$, the quantities $B_2(t)$ and $A_2(t)$ are slowly varying, possibly random variables (or perhaps fixed amplitudes). Then, putting $\omega = \omega_p/2$ in Equation 6.128, we find

$$\begin{bmatrix} A_2 \\ A_2^* \end{bmatrix} = G \begin{bmatrix} 1 & 1 \\ 1 & 1 \end{bmatrix} \begin{bmatrix} B_2 \\ B_2^* \end{bmatrix}, \tag{6.142}$$

so that

$$A_2 = G(B_2 + B_2^*). \tag{6.143}$$

Note that this result is the same result as would be obtained by multiplying the input time function by $G(1 + 2\cos\omega_p t)$, at least with respect to the portion of the spectrum located near $\omega_p/2$.

This rule is of very great importance. For double-sideband operation of the degenerate parametric amplifier, the input signal can merely be multiplied by $G(1 + 2\cos\omega_p t)$ to obtain the output signal. The same rule holds for input noise and, in fact, holds also for noise generated by the varactor series resistance, provided that it is represented by T_{eff}. In Section 6.6.5, making use of this rule, we discuss double-sideband operation of the degenerate parametric amplifier.

6.6.4. Single-Sideband Operation

Degenerate amplifiers can be used in two essentially different ways. Either the spectrum of the input signal surrounds the frequency $\omega_p/2$, or it does not. The first case is double-sideband operation, discussed in Section 6.6.5; the second case is discussed here. In both types of operation, however, there is generally noise output on both sides of $\omega_p/2$, owing not only to noise input but also to thermal noise generated on both sides of $\omega_p/2$ by the varactor series resistance.

We now wish to describe the noise performance of a single-sideband degenerate parametric amplifier. We shall calculate the degradation in signal-to-noise ratio imposed by the amplifier and give a formula for comparing single-sideband degenerate amplifiers with ordinary amplifiers. The ratio of input signal-to-noise ratio to output signal-to-noise ratio is called F'_{SSB}, and a characteristic temperature T'_{SSB} is calculated by the formula

$$T'_{SSB} = T_s(F'_{SSB} - 1), \tag{6.144}$$

where T_s is the source temperature. Note that if this calculation were applied to an ordinary amplifier, the ratio of signal-to-noise ratios would define the ordinary noise figure F (provided T_s were set equal to standard temperature $T_0 = 290°$ K), and a formula like Equation 6.144 would define the noise temperature T. The newly defined quantities are given primes to stress the fact that they are not ordinary noise temperatures and figures. Obviously, however, their definitions are motivated by similar analyses of ordinary amplifiers.

In the single-sideband case, the input signal has a spectrum located on one side of $\omega_p/2$, and the output is G times as large, so that if the power input is P_i, the power output at the input frequency is $G^2 P_i$. The input also has noise of temperature T_s over the band surrounding $\omega_p/2$, and the output has noise power as given in Expression 6.136. Thus the input signal-to-noise ratio is

$$\left(\frac{S}{N}\right)_{in} = \frac{P_i}{kT_s\,\Delta f}. \tag{6.145}$$

If the output is filtered to reject the idler, so that only a band of width Δf centered about frequency ω_s is passed to the output, then the output signal-to-noise ratio is

$$\left(\frac{S}{N}\right)_{out} = \frac{G^2 P_i}{2G^2 k(T_s + T_{eff})\,\Delta f}, \tag{6.146}$$

so that their ratio is

$$F'_{SSB} = \frac{(S/N)_{in}}{(S/N)_{out}}$$

$$= \frac{2(T_s + T_{eff})}{T_s}. \tag{6.147}$$

Thus

$$T'_{SSB} = T_s(F'_{SSB} - 1)$$

$$= T_s + 2T_{eff}. \tag{6.148}$$

The origin of each term in this expression is clear. The output noise power contains, aside from the contribution of T_s at the input frequency, one contribution of value T_s converted from the idler frequency, plus a contribution of effective value T_{eff} amplified at the input frequency, plus a contribution of effective value T_{eff} converted from the idler frequency. Using Equation 6.137, we find for the varactor amplifier

$$T'_{SSB} = T_s + \frac{2T_d}{\dfrac{m_1 \omega_c}{\omega_s} - 1}. \tag{6.149}$$

The single-sideband degenerate parametric amplifier has two important properties. First, it is not phase-sensitive, because the input and output spectra do not contain $\omega_p/2$. Second, the output noise at the input frequency (that is, the background noise for the output signal) is correlated with the noise at the idler frequency but not with the noise at any frequency on the same side of $\omega_p/2$ as ω_s. Thus since the output is filtered (as, for example, in a subsequent stage of amplification) to reject the idler, the combination of single-sideband amplifier and filter behaves like an ordinary amplifier, and therefore, we can compare the noise performance of ordinary amplifiers with that of single-sideband degenerate amplifiers by comparing T'_{SSB} with the ordinary noise temperature T. Such a comparison is not valid for double-sideband amplifiers.

This point is important. Although there is generally no simple way to compare the noise performance of ordinary amplifiers with that of degenerate parametric amplifiers, the single-sideband amplifier is a special case for which such a comparison is possible.

To obtain T'_{SSB} from T_{eff}, we should, by Equation 6.148, multiply by 2 if T_s is small, or add 3 db to the measured noise figure $F_{eff} = 1 + T_{eff}/T_0$ if $T_s = T_0$. This latter rule is the so-called "3-db rule" often quoted.[15,4] It is limited to cases with $T_s = T_0$, whereas Equation 6.148 holds more generally.

The single-sideband noise temperature T'_{SSB} is always larger than T_{min} for a nondegenerate parametric amplifier even if T_s is small. Therefore, for best noise performance, we should not ordinarily use a single-sideband degenerate amplifier. A comparison between T_{min} and T'_{SSB} is given in Figure 6.20. The two are quite close for low signal frequencies, but for frequencies approaching $m_1\omega_c$, above which the degenerate amplifier does not operate, the nondegenerate amplifier is clearly preferable.

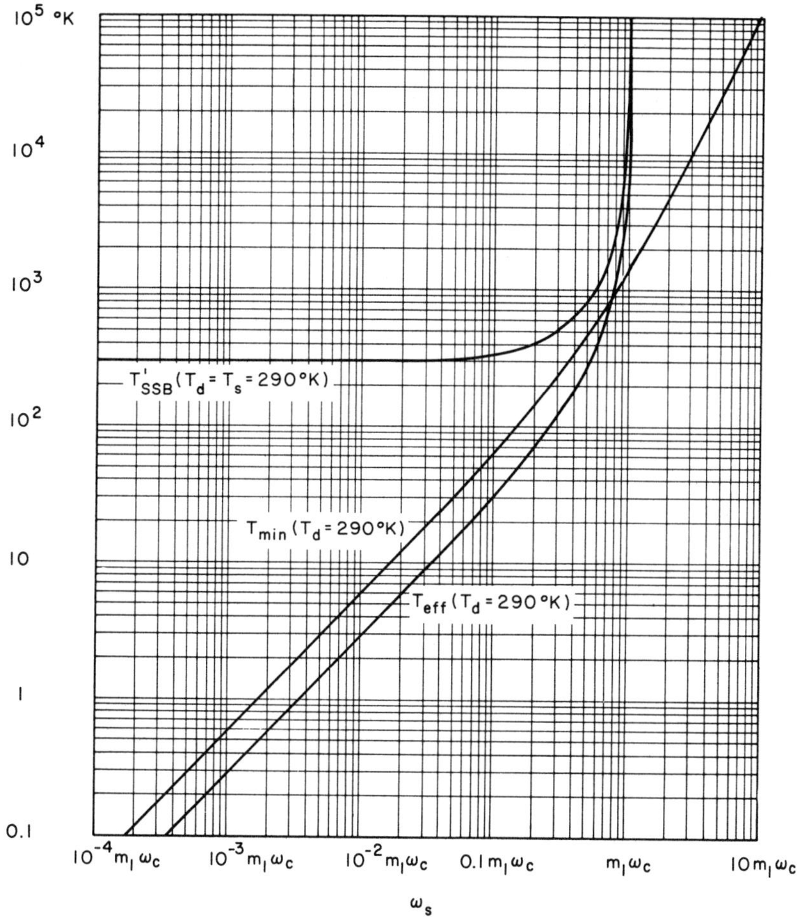

Figure 6.20. Comparison between T_{min}, T_{eff}, and T'_{SSB}, taken from Equations 6.35, 6.137, and 6.149.

6.6.5. Double-Sideband Operation

For input signals with spectra surrounding the frequency $\omega_p/2$, it is usually not possible to compare a degenerate amplifier with an ordinary amplifier for noise performance until the exact nature of the signal, the type of detector, and the interpretation of the detector output are known. Only a few authors,[12,7] have appreciated this important fact. We shall work through some typical examples in this section, concentrating on the synchronous detector and the square-law detector.

We assume the input signal is represented by the time function

$$a_1(t) = b_2(t) = \text{Re } B_2(t)e^{j\omega_p t/2}, \qquad (6.150)$$

where $B_2(t)$ is a slowly varying, possibly random, function with known statistics. The signal input power is

$$\frac{1}{2}(\overline{B_r^2} + \overline{B_i^2}), \tag{6.151}$$

where for convenience we write

$$B_2 = B_r + jB_i. \tag{6.152}$$

The noise output is caused partly by input noise at temperature T_s and partly by internally generated noise at effective input temperature T_{eff}. Thus we include as part of the input,

$$n(t) = \mathrm{Re}\, N(t)e^{j\omega_p t/2}, \tag{6.153}$$

where (because this noise is narrow-band, centered about frequency $\omega_p/2$) the quantity $N(t)$ is a slowly varying random variable with real and imaginary parts N_r and N_i of such a form that

$$\overline{N_r^2} = \overline{N_i^2} = k(T_s + T_{eff})\,\Delta f, \tag{6.154}$$

and

$$\overline{N_r N_i} = 0. \tag{6.155}$$

The input signal-to-noise ratio is

$$\left(\frac{S}{N}\right)_{in} = \frac{\overline{B_r^2} + \overline{B_i^2}}{2kT_s\,\Delta f}. \tag{6.156}$$

The output from the amplifier is approximately

$$[b_2(t) + n(t)]\,G(1 + 2\cos\omega_p t), \tag{6.157}$$

which is for frequencies near $\omega_p/2$

$$G(B_r + N_r)\cos\frac{\omega_p t}{2}. \tag{6.158}$$

At this point we could note that the output signal-to-noise ratio is

$$\left(\frac{S}{N}\right)_{out} = \frac{\overline{B_r^2}}{\overline{N_r^2}}$$

$$= \frac{\overline{B_r^2}}{k(T_s + T_{eff})\,\Delta f}\,. \qquad (6.159)$$

Then we could, if we wished, use the ratio of these two signal-to-noise ratios to define a quantity F'_{DSB} similar to F'_{SSB}. Such a definition, however, would be misleading, because it would not in general indicate the noise performance of the degenerate parametric amplifier. The output noise is correlated, so that the following detector treats this noise in a way different from the way uncorrelated noise is treated. To go further, we must consider the detector that is to be used. In Section 6.6.5(a) we discuss the synchronous detector and in Section 6.6.5(b), the square-law detector.

6.6.5(a). Synchronous Detector. To compare a double-sideband degenerate parametric amplifier with an ordinary amplifier, we must know the input signal statistics and the nature of the detection process. Consider the case in which the output from the parametric amplifier is fed directly (or with additional amplification but no narrow-band filtering) to a synchronous detector run at frequency $\omega_p/2$, synchronously with the parametric amplifier pump. The effect of the synchronous detector is the same as would be obtained by multiplying the waveform by a sinusoid at frequency $\omega_p/2$, and then averaging the result, or at least filtering it to remove all components except those located close to zero frequency. For simplicity we assume the phase between the synchronous detector and the parametric amplifier pump to be optimum, so that all the signals amplified by the amplifier are detected by the detector. Then the output voltage from the detector is

$$[b_2(t) + n(t)]\,G(1 + 2\cos\omega_p t)\cos\frac{\omega_p t}{2} \qquad (6.160)$$

or a component near zero frequency

$$\frac{1}{2}G[B_r(t) + N_r(t)]. \qquad (6.161)$$

On the other hand, suppose an ordinary amplifier with gain G and noise temperature T were used instead of the double-sideband degenerate amplifier. Then the input to the detector would be $G[b_2(t) + n(t)]$, so that the low-frequency output from the detector would be

$$\frac{1}{2}G[B_r(t) + N_r(t)]. \tag{6.162}$$

Note that Expressions 6.161 and 6.162 look identical. The only difference is that in Expression 6.161 $N_r(t)$ comes from a noise temperature $T_{eff} + T_s$, whereas in Expression 6.162 $N_r(t)$ comes from a noise temperature $T + T_s$. Thus, for all purposes, the output from the detector in the two cases is the same if $T = T_{eff}$. The ordinary amplifier is better if $T < T_{eff}$, and the degenerate amplifier is to be preferred if $T > T_{eff}$. The signal-to-noise ratio at the output of the detector is for the ordinary amplifier

$$\left(\frac{S}{N}\right)_{out-det} = \frac{\overline{B_r^2}}{\overline{N_r^2}} = \frac{\overline{B_r^2}}{k(T_s + T)\,\Delta f}, \tag{6.163}$$

whereas the signal-to-noise ratio for the degenerate amplifier coupled to the synchronous detector would merely be Equation 6.159. In either case, the degradation in signal-to-noise ratio of the amplifier and detector combined defines a figure of merit F':

$$F' = \frac{(S/N)_{in}}{(S/N)_{out-det}} = \frac{\overline{B_r^2} + \overline{B_i^2}}{2T_s}\,\frac{T_s + T}{\overline{B_r^2}}; \tag{6.164}$$

an effective temperature can be defined as

$$T' = T_s(F' - 1) = T_s\frac{\overline{B_i^2} - \overline{B_r^2}}{2\,\overline{B_r^2}} + T\frac{\overline{B_r^2} + \overline{B_i^2}}{2\,\overline{B_r^2}} \tag{6.165}$$

(for the degenerate amplifer, T should be replaced in Equations 6.164 and 6.165 by T_{eff}).

If a synchronous detector (with the proper phase) is used, then a double-sideband degenerate amplifier can be easily compared with an ordinary amplifier for noise performance. Simply compare T_{eff} and T and select the amplifier with the lower temperature. This rule is valid for any input statistics, and it does not matter how the detector output is interpreted.

A comparison between T_{eff} and T_{min} for a nondegenerate amplifier is given in Figure 6.20. Note that T_{eff} can be lower than T_{min} at low frequencies.

6.6.5(b). Square-Law Detector. Since the square-law detector is somewhat more complicated than the synchronous detector, we do not discuss it in complete detail.

This detector gives an output of a constant K times the square of the input. Therefore, close to zero frequency, its output voltage is

$$\frac{KG^2}{2}(B_r + N_r)^2 \tag{6.166}$$

when fed from a double-sideband degenerate parametric amplifier. The spectral density of the output clearly depends upon the statistics of B and N.

For simplicity, we now assume that B and N are uncorrelated, and that the noise obeys Gaussian statistics and has a zero mean. Then the output power is

$$\frac{K^2G^4}{4}\overline{(B_r + N_r)^4}$$

$$= \frac{K^2G^4}{4}\left[3k^2(T_s + T_{eff})^2(\Delta f)^2 + 6\overline{B_r{}^2}k(T_s + T_{eff})\,\Delta f + \overline{B_r{}^4}\right], \tag{6.167}$$

where we have used the identity[6] (valid for the Gaussian variable x)

$$\overline{x^4} = 3[\overline{x^2}]^2. \tag{6.168}$$

If we want a comparison of (1) a double-sideband degenerate amplifier with a square-law detector to (2) an ordinary amplifier with the same detector, we must develop an expression similar to Equation 6.167 for the ordinary amplifier. The output voltage (close to zero frequency) from the detector is now

$$\frac{KG^2}{2}[(B_i + N_i)^2 + (B_r + N_r)^2] \tag{6.169}$$

instead of Expression 6.166. We have assumed that the ordinary amplifier has gain G and noise temperature T, and that N is suitably redefined. The difference between Expressions 6.166 and 6.169 is that the degenerate amplifier has rejected one phase of the signal and the noise. The power output from the detector is now

$$\frac{K^2G^4}{4}\left[8k^2(T_s + T)^2\ (\Delta f)^2\ + 8(\overline{B_r{}^2} + \overline{B_i{}^2})k(T_s + T)\ \Delta f\right.$$

$$\left. + \overline{B_r{}^4} + \overline{B_i{}^4} + 2\ \overline{B_r{}^2B_i{}^2}\right], \qquad (6.170)$$

where we have again assumed that the noises are Gaussian, so that Equation 6.168 could be used.

In order to gain a deeper understanding of the relationship between Expression 6.170 and Equation 6.167, we consider explicitly three typical input statistics. First, let us assume that the input "signal" is thermal noise of temperature T_{signal}, as for example the noise in radio astronomy. Second, we assume the phase is fixed, but the amplitude of the signal varies. Finally, we assume the phase of the signal is arbitrary, but the amplitude is fixed.

First, for radiometry purposes, B_r and B_i are considered to be random variables with

$$\overline{B_r{}^2} = \overline{B_i{}^2} = kT_{signal}\ \Delta f, \qquad (6.171)$$

$$\overline{B_rB_i} = 0, \qquad (6.172)$$

and

$$\overline{B_r{}^4} = \overline{B_i{}^4} = 3kT_{signal}\ \Delta f. \qquad (6.173)$$

Then the output power from the detector when driven by the degenerate parametric amplifier is

$$\frac{K^2G^4}{4}\ 3(T_s + T_{eff} + T_{signal})^2k^2\ (\Delta f)^2; \qquad (6.174)$$

the output power from the detector when driven by the ordinary amplifier is

$$\frac{K^2G^4}{4}\ 8(T_s + T + T_{signal})^2k^2\ (\Delta f)^2\ . \qquad (6.175)$$

Note that Expressions 6.174 and 6.175 have the same form. It might be supposed that the signal-to-noise ratios would be the same and, therefore, the amplifiers could be compared by comparing T with T_{eff}. Unfortunately, however, Expressions 6.174 and 6.175 give only the total power at the output with frequency close to zero. They do not indicate how this power is distributed between zero frequency and low frequencies. For this knowledge, we can use plots of the spectral densities of the outputs. In each case, the spectral density consists of an impulse at the origin, representing dc power, and a finite contribution from frequencies below approximately Δf. In the degenerate case, one third of the output power is dc, and two thirds is at frequencies between zero and Δf. In the ordinary amplifier, half the output power is dc, and the other half at low ac frequencies. Thus, the ratio of ac power to dc power is twice as high for the degenerate amplifier as for the ordinary amplifier.[7] This fact may or may not be of importance, depending on the interpretation of the detector output. The ac power can be filtered and treated differently from the dc power. The relatively higher ac content of the degenerate amplifier leads to more "post-detector fluctuations" in some applications.[7] The conclusion is that even though Expressions 6.174 and 6.175 indicate the same signal-to-noise ratio, the two types of amplifiers do not always perform equally satisfactorily. To compare an ordinary amplifier with a degenerate amplifier for radio astronomy, it is necessary to take account of the detector and the postdetector circuits.

As a second example, consider an amplitude-modulated signal at $\omega_p/2$. In this case, the amplitude of the input is some unspecified random variable, whereas the phase is fixed. For a fair evaluation of the degenerate parametric amplifier, we assume that the phase of the signal is the particular phase that is most amplified. In our case, then, $B_i = 0$, and the degenerate amplifier followed by the square-law detector gives an output power

$$\frac{K^2G^4}{4}\left[\overline{B_r^{\,4}} + 6\overline{B_r^{\,2}}k(T_s + T_{eff})\,\Delta f + 3k^2(T_s + T_{eff})^2\,(\Delta f)^2\right], \tag{6.176}$$

whereas the ordinary amplifier followed by this detector gives a power output

$$\frac{K^2G^4}{4}\left[\overline{B_r^{\,4}} + 8\overline{B_r^{\,2}}k(T_s + T)\,\Delta f + 8k^2(T_s + T)^2\,(\Delta f)^2\right]. \tag{6.177}$$

There is no simple way to compare the two types of amplifiers unless we know how the detector output is interpreted. For example, which terms in Expressions 6.176 and 6.177 are to be considered noise and which, signal? Only in specific cases, with known statistics for B_r and known values of T_s, T, and T_{eff}, and a given interpretation of the output power, can a decision be made concerning which type of amplifier gives better noise performance and by how much.

As a third example, suppose the phase of the input signal is random, equally likely at any value from zero to 2π, but that the amplitude is fixed. Thus, for random values of ϕ,

$$B_i = |B| \sin\phi \tag{6.178}$$

and

$$B_r = |B| \cos\phi. \tag{6.179}$$

Then, the double-sideband degenerate amplifier followed by the square-law detector gives an output power

$$\frac{K^2G^4}{4}\left[\frac{3}{8}|B|^4 + 3|B|^2k(T_s + T_{eff})\,\Delta f + 3k^2(T_s + T_{eff})^2\,(\Delta f)^2\right], \tag{6.180}$$

and the ordinary amplifier followed by the same detector gives an output power

$$\frac{K^2G^4}{4}\left[|B|^4 + 8|B|^2k(T_s + T)\,\Delta f + 8k^2(T_s + T)^2\,(\Delta f)^2\right]. \tag{6.181}$$

Although the two types of amplifiers give the same signal-to-noise ratios if $T_{eff} = T$, the ac versus dc distribution of power is different, so that the two types of amplifiers cannot be compared unless we know how the detector output is to be interpreted.

6.6.6. Summary

The degenerate amplifier differs from the nondegenerate parametric amplifier or the ordinary amplifier because two outputs are obtained for one input (one directly amplified, and the other converted). The noise is also both amplified and converted, so that the noise output at one frequency is correlated with the noise at another frequency.

The figure of merit T_{eff} is useful for comparing two or more degenerate amplifiers for noise performance, but we cannot generally compare a degenerate amplifier and an ordinary amplifier unless the input signal statistics are known, the type of detector is specified, and the interpretation of the detector output is given. There are a few cases, however, for which such a comparison can be made. First, for single-sideband amplifiers, it is fair to compare T of an ordinary amplifier with $T_s + 2T_{eff}$ of the degenerate amplifier. Second, if a double-sideband amplifier is followed by a synchronous detector, then it is fair to compare an ordinary amplifier T with T_{eff} of the degenerate amplifier. Ordinarily, however, no simple rules can be given to estimate the relative noisiness of the two types of amplifiers.

The varactor degenerate amplifier using a circulator can give gain only when

$$\frac{\omega_p}{2} < m_1\omega_c, \tag{6.182}$$

and T_{eff} is given by the simple formula,

$$T_{eff} = \frac{T_d}{\dfrac{m_1\omega_c}{\omega_p/2} - 1}. \tag{6.183}$$

6.7. Synthesis Techniques

The derivation of the fundamental limits (especially for noise) in Section 6.2 indicated circuit conditions that must be fulfilled to achieve these limits. We now describe the synthesis of amplifiers to meet these conditions deliberately. Note that this method of designing amplifiers is quite different from the method of putting together what seems like a "logical" circuit but of including many adjustments to be made empirically. The adjustments indicated in the circuits to be described here are (at least if done in the proper order) noninteracting, and can be made before the equipment is running. Many of them are independent of the varactor used.

The design of parametric amplifiers is logically accomplished in three steps. First is the selection of the operating conditions and the varactor parameters required, the preliminary design. Second is the synthesis of a circuit that simultaneously terminates the varactor properly at the various frequencies and provides signal and pump frequency ports. Third, a two-port amplifier must be made out of the single signal-frequency port. Ideally the last two steps would be performed together, but for simplicity we discuss them separately.

6.7.1. Preliminary Design

As part of the preliminary design of parametric amplifiers, we should ordinarily choose the appropriate varactor and choose the pump frequency and the idler termination as well as any other sideband terminations. The theory of Chapter 6 is primarily devoted to this task, by showing how these parameters affect the possibility for gain and the noise temperature. Chapter 7 should be consulted in calculating the pumping parameters, especially the pump power. Since the primary purpose of Chapters 6 and 7 is to aid in this preliminary design, we do not discuss it further here.

6.7.2. Separation of Frequencies

As an example of the synthesis of a circuit to fulfill specific conditions deliberately, we consider a nondegenerate amplifier with a reactive idler termination. We wish to meet the conditions of Section 6.2 for attaining the best noise temperature consistent with the given varactor and the given pump and signal frequencies, as well as the conditions of Section 7.4 for pumping with a sinusoidal current. These conditions are the following:

1. The varactor is open-circuited at frequency $2\omega_p$.
2. The varactor is open-circuited at $\omega_p + \omega_s$.
3. The varactor is open-circuited at $\omega_p + \omega_i$.
4. There is no loss at idler frequency.
5. Currents and voltages at the pump and signal frequencies appear at separate ports.
6. The idler frequency does not appear at the pump and signal ports.
7. The idler frequency termination must be tuned.
8. The pump must be tunable to match the varactor series resistance.

In principle, the varactor should be open-circuited at all sidebands and pumping harmonics, but the most important are $\omega_p + \omega_s$, $\omega_p + \omega_i$, and $2\omega_p$.

Some of these conditions can be met by using circuits with high symmetry, particularly those with two or more varactors. In addition, there are many other circuits that fulfill the conditions. In practice we should select from among these circuits those that have desired bandwidth, reliability, simplicity, cost, freedom from spurious oscillations, and so forth. To show that such circuits exist (and in fact can be drawn almost by inspection to meet the conditions above), we show two of them: one using lumped tuned circuits, and the other using coaxial transmission lines.

6.7.2(a). Lumped Realization. Figure 6.21 shows a lumped circuit that can satisfy the conditions just listed. The bias capacitor and inductor form a low-pass network to admit direct current but to exclude radio frequencies from the bias network. The varactor is open-circuited at $2\omega_p$, $\omega_p + \omega_s$, and $\omega_p + \omega_i$ by the tuned circuits shown. The signal and idler frequencies are kept away from the pump by the tuned circuits at the lower left. The resistance to be matched by the pump is the varactor resistance R_s, and a matching network, consisting of a transformer and a reactance, is shown leading to the pump terminals.

The pump frequency is excluded from the signal and idler circuits by the tuned circuit at the top. The signal and idler frequencies are then separated and tuned individually. The impedance seen looking into the signal-frequency terminals should have a negative real part.

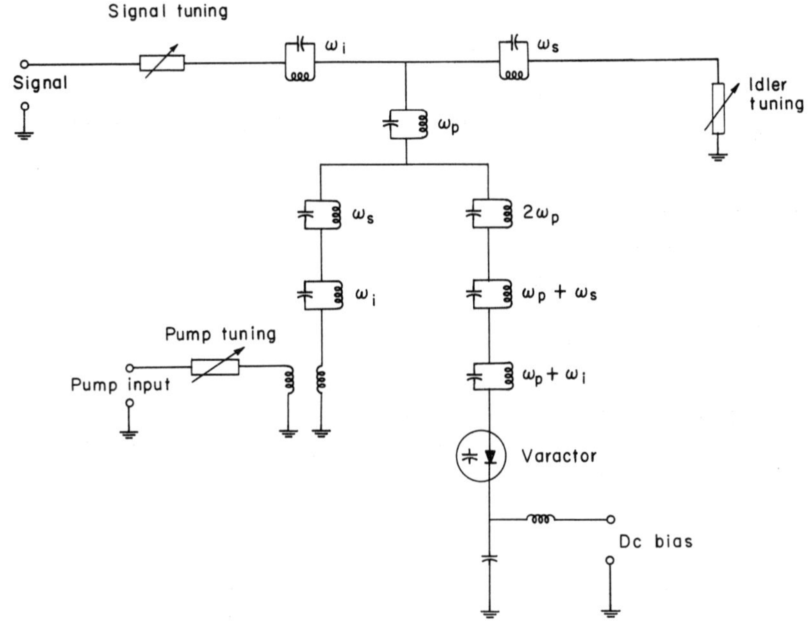

Figure 6.21. A lumped circuit for separating frequencies for a simple parametric amplifier. The tuned circuits can be adjusted, for example with a grid-dip meter, before assembly.

The circuit can probably be simplified or improved in a number of ways. We do not claim that this circuit is in any sense optimum, but only that it does satisfy the conditions just listed.

6.7.2(b). Coaxial Realization. A possible circuit for meeting the conditions listed in Section 6.7.2, using coaxial transmission lines, is sketched in Figure 6.22. The varactor is mounted as part of the inner conductor, as indicated by Figure 6.23, and not between the inner conductor and the outer conductor. The open-circuit conditions are now easily met by using resonant lengths of coaxial line. Lengths (1) and (2) are adjusted to be a half-wavelength and a quarter-wavelength, respectively, at frequency $\omega_p + \omega_s$. Thus the horizontal transmission line is short-circuited at that frequency, and no currents or voltages can appear at that frequency to the left of the T-junction. Since this point is a quarter wavelength from the varactor, no current can flow through the varactor at this frequency. Similarly, lines (3) and (4) open-circuit the varactor at $\omega_p + \omega_i$ and prevent this frequency from appearing to the right of the T-junction. Note that the short circuit at the end of line (3) is at zero frequency an open circuit to allow a bias voltage to be introduced. Because of the blocking capacitor C and the short circuit on the end of line (1), this voltage appears across the varactor.

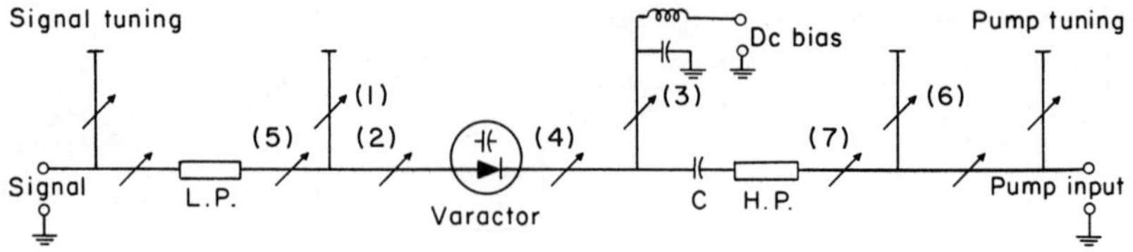

Line	Length
(1)	$\lambda(\omega_p + \omega_s)/2$
(2)	$\lambda(\omega_p + \omega_s)/4$
(3)	$\lambda(\omega_p + \omega_i)/2$
(4)	$\lambda(\omega_p + \omega_i)/4$
(5)	Adjust for open circuit at $2\omega_p$
(6)	$\lambda(\omega_i)/2$
(7)	Adjust to tune at ω_i

Figure 6.22. A coaxial circuit that realizes the frequency separation demanded for the simple parametric amplifier. The adjustable lines (1) through (6) can be set experimentally before final assembly; their settings depend only on the frequencies and are independent of the varactor parameters.

For simplicity we assume that ω_s is much smaller than ω_p, so that a simple low-pass filter L. P. prevents frequencies ω_p and ω_i from appearing at the signal-frequency port at the left. Line (5) is adjusted so that the varactor sees an open circuit at $2\omega_p$, and a signal-frequency tuner is provided.

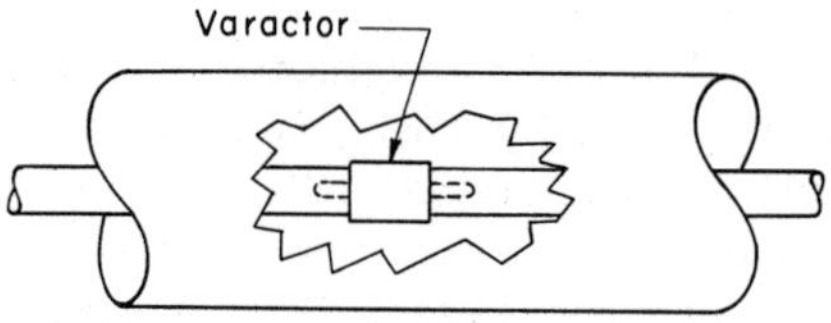

Figure 6.23. A varactor mounted in the inner conductor of a coaxial transmission line.

Similarly, a high-pass filter H. P. keeps signal-frequency currents and voltages to the left of it. Line (6) is a half wavelength at the idler frequency; it places a short circuit across the horizontal line to keep this frequency from the inevitable loss at the pump port. Line (7) can then be adjusted to tune the idler. A matching network for the pump is shown.

Note that there are several adjustments in the circuit shown, but these adjustments are all independent if performed in the order outlined above, except for line (7) and the pump tuning, which interact. Most of the adjustments can be made before assembly, because they depend only on the frequencies and not on the parameters of the varactor.

This circuit probably uses more components than necessary; it is given only as an indication that circuits can be designed to obey deliberately the conditions listed in Section 6.7.2.

Similar techniques can be used to realize the same conditions by means of waveguide circuitry. An additional problem, however, is that of the varactor mounting, that is, the problem of coupling the varactor to the waveguide in an efficient way. Frequency separation in waveguide circuits can often be accomplished by using guides of different sizes.

6.7.3. Signal-Frequency Circuitry

The techniques of Section 6.7.2 yield a negative resistance at the signal frequency. The next step in synthesis is to use the negative resistance, which can deliver power at the signal frequency, in building an amplifier. This problem, common to all negative-resistance amplifiers, usually is solved by the use of a three-port imbedding network, as indicated in Figure 6.24.

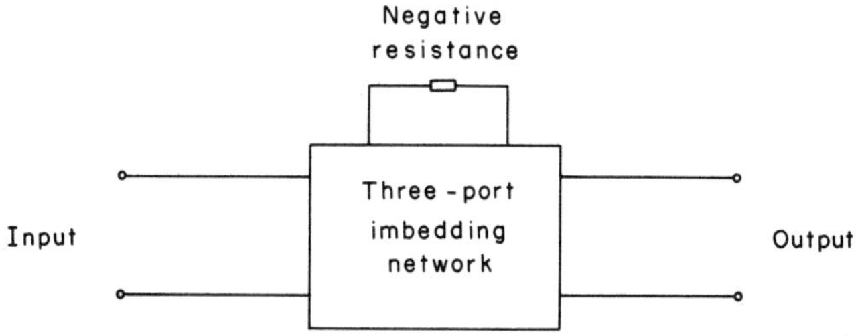

Figure 6.24. Use of a three-port imbedding network converts a negative resistance with a single port into an amplifier with separate input and output. All negative-resistance amplifiers use an imbedding of this type. Often the imbedding network is lossless.

The imbedding network is chosen so that the resulting two-port amplifier has, broadly speaking, desirable characteristics. As this problem is considered in some detail in Appendix C, we do not discuss it further here. In Appendix C, we treat several possible imbedding networks and discuss the problem of cascading to a noisy second stage.

6.8. Summary

Several important results were derived. Among the most significant are the following:

1. The varactor series resistance limits the pump and signal frequencies for which gain can be achieved.
2. Thermal noise of the varactor series resistance provides a fundamental limit on the noise performance obtainable from a parametric amplifier.
3. This minimum noise temperature is the same as that derived for the frequency converters in Chapter 5.
4. If the external idler termination has a finite temperature, this noise performance can be achieved only if the pump frequency is chosen properly and if the external idler termination is made reactive.
5. If a noise-free external idler termination is available, the minimum noise performance can be achieved at a somewhat lower pump frequency. Such an idler termination cannot be used to lower the noise temperature of the amplifier further.
6. Amplifiers with more than one varactor can exhibit nonreciprocal behavior but cannot have gain for pump and signal frequencies outside the range for single-varactor amplifiers, and the noise temperature is likewise governed by the same fundamental limit.
7. Amplifiers with extra idlers can exhibit gain for pump frequencies lower than the signal frequency.
8. If the elastance is pumped sinusoidally, then extra idlers cannot be used to lower the noise temperature below the fundamental limit.
9. If the elastance is not pumped sinusoidally, then a noise temperature below the fundamental limit can in principle be obtained, but any such improvement is slight.
10. For practical varactor amplifiers, the effect of quantum noise is negligible.
11. The degenerate amplifier has a circuit that treats the signal and the idler in an identical fashion. This amplifier can give gain only if the signal frequency is less than $m_1\omega_c$.
12. The degenerate amplifier is not characterized by the ordinary concepts of noise figure, noise measure, and noise temperature. The noise performance depends upon the statistics of the signal, the type of detector, and the interpretation of the detector output.

REFERENCES

1. Baldwin, L. D., "Nonreciprocal Parametric Amplifier Circuits," Proc. IRE, 49, 1075 (June, 1961).
2. Blackwell, L. A., and K. L. Kotzebue, Semiconductor-Diode Parametric Amplifiers, Prentice-Hall, Inc., Englewood Cliffs, N. J. (1961).
3. Carlin, H. J., "The Scattering Matrix in Network Theory," IRE Trans. on Circuit Theory, CT-3, 88-97 (June, 1956).
4. Cohn, S. B., "The Noise Figure Muddle," Microwave J., 2, 7-11 (March, 1959).
5. Davenport, W. B., Jr., and W. L. Root, An Introduction to the Theory of Random Signals and Noise, McGraw-Hill Book Co., Inc., New York (1958).
6. Davenport, W. B., Jr., and W. L. Root, ibid., Equation 8-121.
7. de Jager, J. T., and B. J. Robinson, "Sensitivity of the Degenerate Parametric Amplifier," Proc. IRE, 49, 1205-1206 (July, 1961).
8. Engelbrecht, R. S., "A Low-Noise Nonlinear Reactance Traveling-Wave Amplifier," Proc. IRE, 46, 1655 (September, 1958).
9. Gilden, M., and G. L. Matthaei, "A Nearly Optimum Wide-Band Degenerate Parametric Amplifier," Proc. IRE, 49, 833-834 (April, 1961).
10. Gilden, M., and G. L. Matthaei, "Practical Design and Performance of Nearly Optimum Wide-Band Degenerate Parametric Amplifiers," IRE Trans. on Microwave Theory and Techniques, MTT-9, 484-490 (November, 1961).
11. Greene, J. C., and E. W. Sard, "Optimum Noise and Gain-Bandwidth Performance for a Practical One-Port Parametric Amplifier," Proc. IRE, 48, 1583-1590 (September, 1960).
12. Greene, J. C., W. D. White, and R. B. Adler, "Radar Sensitivity with Degenerate Parametric Amplifier Front End," Proc. IRE, 49, 804-807 (April, 1961).
13. Haus, H. A., and R. B. Adler, Circuit Theory of Linear Noisy Networks, The Technology Press and John Wiley & Sons, Inc., New York (1959).
14. Haus, H. A., and R. B. Adler, ibid., p. 26.
15. Heffner, H., "Masers and Parametric Amplifiers," Microwave J., 2, 33-40 (March, 1959).
16. Heffner, H., and G. Wade, "Minimum Noise Figure of a Parametric Amplifier," J. Appl. Phys., 29, 1262 (August, 1958).
17. Hildebrand, F. B., Methods of Applied Mathematics, Prentice-Hall, Inc., Englewood Cliffs, N. J. (1952).
18. Hildebrand, F. B., ibid., Chapter 1.
19. Johnson, K. M., "Broad-Band S-Band Parametric Amplifier," Proc. IRE, 49, 1943 (December, 1961).

20. Knechtli, R. C., and R. D. Weglein, "Low-Noise Parametric Amplifier," Proc. IRE, 48, 1218-1226 (July, 1960).
21. Korpel, A., and P. Desmares, "Experiments with Nonreciprocal Parametric Devices," Proc. IRE, 49, 1582 (October, 1961).
22. Kotzebue, K. L., "Optimum Noise Performance of Parametric Amplifiers," Proc. IRE, 48, 1324-1325 (July, 1960).
23. Kuh, E. S., "Theory and Design of Wide-Band Parametric Converters," Proc. IRE, 50, 31-38 (January, 1962).
24. Kuh, E. S., and J. D. Patterson, "Design Theory of Optimum Negative-Resistance Amplifiers," Proc. IRE, 49, 1043-1050 (June, 1961).
25. Kurokawa, K., and M. Uenohara, "Minimum Noise Figure of the Variable-Capacitance Amplifier," Bell System Tech. J., 40, 695-722 (May, 1961).
26. Little, A. G., "A Wide-Band Single-Diode Parametric Amplifier Using Filter Techniques," Proc. IRE, 49, 821-822 (April, 1961).
27. Louisell, W. H., A. Yariv, and A. E. Siegman, "Quantum Fluctuations and Noise in Parametric Processes. I." Phys. Rev., 124, 1646-1654 (December, 1961).
28. Marcus, M., Basic Theorems in Matrix Theory, Natl. Bur. Standards (U. S.), Applied Mathematics Series, No. 57, Washington, D. C. (January 22, 1960).
29. Matthaei, G. L., "A Study of the Optimum Design of Wide-Band Parametric Amplifiers and Up-Converters," IRE Trans. on Microwave Theory and Techniques, MTT-9, 23-38 (January, 1961).
30. Mortenson, K. E., "Parametric Diode Figure of Merit and Optimization," J. Appl. Phys., 31, 1207-1212 (July, 1960).
31. Perlis, S., Theory of Matrices, Addison-Wesley Publishing Co., Inc., Cambridge, Mass. (1952).
32. Rafuse, R. P., "Characterization of Noise in Receiving Systems," in Lectures on Communication System Theory, E. J. Baghdady, Editor, McGraw-Hill Book Co., Inc., New York (1961), pp. 369-399.
33. Schaffner, G., and F. Voorhaar, "A Nondegenerate S-Band Parametric Amplifier with Wide Bandwidth," Proc. IRE, 49, 824-825 (April, 1961).
34. Seidel, H., and G. F. Herrmann, "Circuit Aspects of Parametric Amplifiers," IRE Wescon Convention Record, 3, Part 2 (Circuit Theory), 83-90 (1959).
35. Thompson, G. H. B., "Unidirectional Lower Sideband Parametric Amplifier Without Circulator," Proc IRE, 49, 1684-1685 (November, 1961).
36. Twiss, R. Q., "Nyquist's and Thevenin's Theorems Generalized for Nonreciprocal Linear Networks," J. Appl. Phys., 26, 599-602 (May, 1955).
37. van der Ziel, A., "Noise Figure of Reactance Converters and Parametric Amplifiers," J. Appl. Phys., 30, 1449 (September, 1959).
38. Vincent, B. T., Jr., "A C-Band Parametric Amplifier with Large Bandwidth," Proc. IRE, 49, 1682 (November, 1961).
39. Youla, D. C., and L. I. Smilen, "Optimum Negative-Resistance Amplifiers," Proceedings of the Polytechnic Institute of Brooklyn Symposia, Vol. 10, Polytechnic Institute of Brooklyn, Brooklyn, N. Y. (1960), pp. 241-318.

Chapter 7

PUMPING

In the discussions of Chapters 5 and 6, it was assumed that the varactor was pumped at frequency ω_p, so that the elastance, as pumped, could be expressed in a Fourier series

$$S(t) = \sum_{k=-\infty}^{+\infty} S_k e^{jk\omega_p t}, \tag{7.1}$$

where each Fourier coefficient S_k is

$$S_k = \left\langle S(t) e^{-jk\omega_p t} \right\rangle$$

$$= \frac{1}{2\pi} \int_0^{2\pi} S(t) e^{-jk\omega_p t} \, d(\omega_p t), \tag{7.2}$$

where the brackets denote a time average. In terms of the varactor maximum and minimum elastances S_{max} and S_{min}, we defined (in Chapter 4) and used (in Chapters 5 and 6) the modulation ratios

$$m_k = \frac{|S_k|}{S_{max} - S_{min}}, \tag{7.3}$$

especially m_1. We also used the <u>total modulation ratio</u> m_t, which is best expressed in squared form as

$$m_t^2 = 2 \sum_{k=1}^{\infty} \frac{|S_k|^2}{(S_{max} - S_{min})^2}$$

$$= 2 \sum_{k=1}^{\infty} m_k^2. \tag{7.4}$$

In view of Equation 7.1, m_t^2 has the alternate formula

$$m_t^2 = \frac{\langle [S(t)]^2 \rangle - S_0^2}{(S_{max} - S_{min})^2}$$

$$= \frac{\langle [S(t)]^2 \rangle}{(S_{max} - S_{min})^2} - m_0^2; \qquad (7.5)$$

m_t^2 is related to the variance of $S(t)$. However, as yet we have not indicated techniques of pumping varactors or the range of possible values of the m_k. Pumping is the topic of the present chapter. The results are pertinent to all applications of pumped varactors, particularly frequency converters (Chapter 5), parametric amplifiers (Chapter 6), and small-signal subharmonic oscillators (Section 9.3).

7.1. Introduction and Major Results

In Sections 7.2 through 7.5 we discuss the maximum achievable values of m_1 and m_t under various operating conditions. We also calculate the power necessary to pump graded-junction and abrupt-junction varactors. The results indicate that a sinusoidal pumping current is desirable. In Section 7.6 we discuss the circuitry necessary to achieve this "current pumping." Finally, in Section 7.7 we discuss operation of frequency converters and parametric amplifiers with limited pump power.

In Section 7.2 it is shown that if the elastance $S(t)$ is a sinusoidal function of time, then m_1 cannot exceed 0.25. On the other hand, if the capacitance is a sinusoidal function of time, then m_1 cannot exceed 0.172 if $S_{min} = 0$ or a number between 0.172 and 0.25 if $S_{min} \neq 0$. If $S(t)$ is an arbitrary function of time, then m_1 cannot exceed 0.318.

Because of the series resistance, a finite amount of power is required to pump varactors. Realistic estimates of pump power can be made (Section 7.2.4) merely by calculating the power lost in the varactor series resistance.

In Sections 7.3 and 7.4 we discuss practical graded-junction and abrupt-junction varactors, pumped with either a sinusoidal voltage or a sinusoidal current. Under "voltage pumping," m_1 is limited to 0.178 for graded-junction and to 0.212 for abrupt-junction varactors. Similarly, under current pumping (that is, with the current a sinusoidal function of time), m_1 is limited to 0.212 and 0.25 for graded-junction and abrupt-junction varactors, respectively. As the derivations in Section 7.3 and 7.4 make clear, the numbers 0.178 and 0.212 are valid only if $S_{min} = 0$.

The maximum values of m_1 under these various conditions are listed in Table 7.1.

Table 7.1. Maximum Achievable Values of m_1 for Various Pumping Schemes

The values without double daggers hold only if $S_{min} = 0$. If $S_{min} \neq 0$, each of these should be replaced by a number lying somewhere between the number shown and 0.25.

	Graded-Junction Varactor	Abrupt-Junction Varactor	Any Varactor
S Sinusoidal	0.250‡	0.250‡	0.250‡
C Sinusoidal	0.172	0.172	0.172
Voltage Sinusoidal	0.178	0.212	–
Current Sinusoidal	0.212	0.250‡	–
S Arbitrary	0.318‡	0.318‡	0.318‡

The limits shown for graded-junction and abrupt-junction varactors in Table 7.1 can be achieved if the varactor is pumped fully, from a forward voltage ϕ to the reverse voltage V_B, where ϕ is the contact potential and V_B the breakdown voltage. Under this full pumping, the other m_k, the bias voltage V_0, and the pump power P have the values given in Table 7.2.

If the varactor is not fully pumped, a bias voltage different from the value in Table 7.2 can be chosen. If the voltage is equal to the value in Table 7.2 (a condition we call fixed-bias), the values of pump power and m_k for lower pumping levels can be calculated. Similarly, if the bias voltage is placed as low as possible (a condition we call self-bias), P and m_k can also be calculated as a function of the drive level. Plots of these calculations are given in Sections 7.3 and 7.4. In addition, directions are given for finding P and m_k for any bias and drive level.

For a given type of varactor (either graded-junction or abrupt-junction), it is interesting to compare current pumping to voltage pumping and to ask which is "better." Strictly speaking, voltage pumping is very difficult, because the voltage that is to be sinusoidal is that across the nonlinear capacitance only, that is, inside the series resistance. On the other hand, current pumping is quite possible. Table 7.1 shows that larger values of m_1 are possible under current pumping. Furthermore, the calculations

Table 7.2. Properties of Fully Pumped Varactors

We assume that $S_{min} = 0$; the values with double daggers hold also for $S_{min} \neq 0$. Here Γ is the gamma function. Values of m_k for higher values of k can be calculated from the expressions given; alternatively refer to Tables 7.3 through 7.6.

Full Pumping	Voltage Sinusoidal		Current Sinusoidal	
	Graded-Junction Varactor	Abrupt-Junction Varactor	Graded-Junction Varactor	Abrupt-Junction Varactor
m_0	0.713	0.637	0.637	0.500
m_1	0.178	0.212	0.212	0.250‡
m_2	0.0509	0.0424	0.0424	0‡
m_3	0.0255	0.0182	0.0182	0‡
m_k	$\dfrac{\Gamma(5/3)}{\sqrt[3]{4}\left\|\Gamma\left(\frac{4}{3}+k\right)\Gamma\left(\frac{4}{3}-k\right)\right\|}$	$\dfrac{2}{\pi\|4k^2-1\|}$	$\dfrac{2}{\pi\|4k^2-1\|}$	$m_k = 0$‡ $(k \geq 2)$
m_t	0.267	0.308	0.308	0.354‡
V_0	$0.500(V_B+\phi)-\phi$	$0.500(V_B+\phi)-\phi$	$0.424(V_B+\phi)-\phi$	$0.375(V_B+\phi)-\phi$
P	$0.267\,P_{norm}\left(\frac{\omega_p}{\omega_c}\right)^2$	$0.500\,P_{norm}\left(\frac{\omega_p}{\omega_c}\right)^2$	$0.281\,P_{norm}\left(\frac{\omega_p}{\omega_c}\right)^2$	$0.500\,P_{norm}\left(\frac{\omega_p}{\omega_c}\right)^2$

in Sections 7.3 and 7.4 show that to obtain a given value of m_1, more power is required using voltage pumping than using current pumping.

In Section 7.5 we discuss harmonic pumping, in which pump power is introduced at some submultiple of the desired pump frequency. The varactor nonlinear capacitance serves both to generate the desired harmonic and to act as the parametric amplifier or frequency converter. For the most important case, pump power input at half the desired pump frequency, some simple limits on modulation ratios and pump power are derived.

Since current pumping is desirable, the question that arises is how to achieve it. In Section 7.6 we discuss pump circuitry; the major point is that the impedance seen by the varactor should be very high at harmonics of the pump frequency, especially the second. The pump current that flows is therefore constrained to be sinusoidal. Standard matching techniques using lumped circuits, transmission lines, or waveguides can then be used to match any source.

The pump power formulas given in this chapter are important, because power is often very precious at frequencies where varactors are likely to be used. The analyses of Chapters 5 and 6 indicate that it is often desirable to pump at an optimum pump frequency $m_1 \omega_c$. If an abrupt-junction varactor is pumped (with a sinusoidal current) at this frequency so that m_1 is as high as it can be (0.250), the power required is $0.031 P_{norm}$. A 100-Gc varactor with normalization power of 10 watts would then require over 300 mw of power at 25 Gc. This is difficult to obtain from present sources, and even if it were available, it probably could not be dissipated safely in the varactor. The point we are trying to make is that it is important to know how to design frequency converters and amplifiers when the pump power is limited either by the available sources or by the varactor dissipation rating.

A general graphical optimization for parametric amplifiers is given in Section 7.7. This technique allows the designer to choose pump frequency and pump power, and read from an appropriate graph the resulting noise temperature. If an empirical curve of power available, as a function of frequency, is superimposed over the graphs given, the lowest noise temperature achievable can be read by inspection, together with the pump frequency and pump power to be used. These graphs are discussed in detail in Section 7.7.

7.2. General Pumping

The analyses of Chapters 5 and 6 indicate that the quantity $m_1 \omega_c$ is a good figure of merit for a varactor, as pumped. Since ω_c is a parameter of the varactor, it is of interest to know how high a value of m_1 can be achieved. For a varactor of given quality, how high a pumped figure of merit is possible? The precise definition of ω_c, Equation 4.47, was based on the desire to have the corresponding definition of m_1, Equation 7.3, restricted by fairly definite bounds. We now derive these bounds for a variety of pumping situations and then derive an expression for pump power.

It would seem desirable to make the elastance $S(t)$ sinusoidal, because the higher harmonics of $S(t)$ do not enter into most of the fundamental limits derived earlier, and are therefore superfluous. However, in practice, $S(t)$ is not usually a sinusoidal function, nor is the capacitance $C(t)$. Similarly, neither the varactor pumping voltage $v_p(t)$ nor the pumping current $i_p(t)$ is actually sinusoidal, and even if one of them were, it would not ordinarily produce either a sinusoidal elastance or a sinusoidal capacitance.‡ Nevertheless, in many cases the higher harmonics of each of these — capacitance, elastance, current, and voltage — are small, and the approximation is often made that one or the other is actually sinusoidal. The manner in which

‡An exception is the abrupt-junction varactor, in which a sinusoidal current produces a sinusoidal elastance.

these approximations constrain the m_k is discussed in Sections 7.2, 7.3, and 7.4.

7.2.1. Sinusoidal Elastance

Suppose $S(t)$, as pumped, is a sinusoidal function of time. A typical plot, Figure 7.1, shows S_{min} and S_{max}, which are

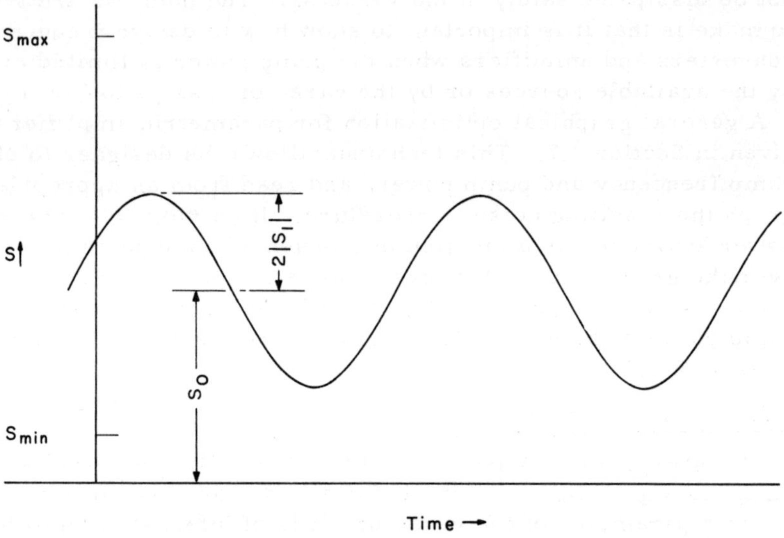

Figure 7.1. Sinusoidal elastance, as pumped. The highest value of m_1 is achieved when $S(t)$ varies between S_{min} and S_{max}. Remember that S_{min} and S_{max} are parameters of the varactor, and are not necessarily the lowest and highest elastances attained during the pumping.

parameters of the varactor independent of the pumping, and S_0 and $|S_1|$, which depend upon the varactor and the pumping. Since S_1 is a half-amplitude of the elastance, the peak-to-peak excursion of $S(t)$ is $4|S_1|$, so that m_1 clearly cannot be greater than one quarter. All other m_k are zero, except for m_0, which is at least twice m_1.

Also, from Equation 7.4, the total modulation ratio m_t is equal to $1.414m_1$, and, therefore, cannot exceed 0.354.

7.2.2. Sinusoidal Capacitance

If the assumption is made that the capacitance $C(t)$ is sinusoidal, then m_1 is similarly restricted. To find this restriction, we write

$$C(t) = C_0 + C_1 e^{j\omega_p t} + C_1^* e^{-j\omega_p t}, \tag{7.6}$$

where the asterisk indicates a complex conjugate; C_1 is a complex number, and $C(t)$ is restricted to lie between $C_{min} = 1/S_{max}$ and $C_{max} = 1/S_{min}$. A method of finding the Fourier coefficients of elastance has been given by Leenov,[1‡] who ascribed it to Uhlir. We define the complex parameter

$$\xi = \frac{2C_1}{C_0 + \sqrt{C_0^2 - 4|C_1|^2}} \tag{7.7}$$

so that

$$\frac{C_1}{C_0} = \frac{\xi}{1 + |\xi|^2}. \tag{7.8}$$

Then the elastance is

$$S(t) = \frac{1}{C_0 + C_1 e^{j\omega_p t} + C_1^* e^{-j\omega_p t}}$$

$$= \frac{1}{C_0} \frac{1 + |\xi|^2}{1 + |\xi|^2 + \xi e^{j\omega_p t} + \xi^* e^{-j\omega_p t}}$$

$$= \frac{1}{C_0} \frac{1 + |\xi|^2}{1 - |\xi|^2} \left(\frac{1}{1 + \xi e^{j\omega_p t}} + \frac{1}{1 + \xi^* e^{-j\omega_p t}} - 1 \right)$$

$$= \frac{1}{C_0} \frac{1 + |\xi|^2}{1 - |\xi|^2} \left(1 - \xi e^{j\omega_p t} - \xi^* e^{-j\omega_p t} + \xi^2 e^{j2\omega_p t} + \xi^{*2} e^{-j2\omega_p t} \right.$$

$$\left. - \xi^3 e^{j3\omega_p t} - \cdots \right), \tag{7.9}$$

and, therefore, the Fourier coefficients of elastance are

$$S_k = \frac{1}{C_0} \frac{1 + |\xi|^2}{1 - |\xi|^2} (-\xi)^k. \tag{7.10}$$

We now ask how to pump to achieve the largest value of m_1,

‡Superscript numerals denote references listed at the end of each chapter.

assuming for the moment that $S_{min} = 0$. If $|\xi|$ is low, then from Equation 7.10 S_1 and therefore m_1 are low; on the other hand if $|\xi|$ approaches its upper limit of unity, then C_0 is high, and m_1 again is low (this fact is not obvious from Equation 7.10). The maximum value of m_1 is found to occur when $C(t)$ varies between C_{min} and $5.83C_{min}$. The resulting value of $|\xi|$ is $\sqrt{2} - 1 = 0.414$, and m_1 is $3 - 2\sqrt{2} = 0.172$; the other m_k can be calculated from the formula

$$m_k = (\sqrt{2} - 1)^{k+1}, \tag{7.11}$$

and m_t under these conditions is, from Equation 7.4, equal to 0.267.

The numbers given in the previous paragraph are correct for varactors in which $S_{min} = 0$. If $S_{min} \neq 0$, then, by Equation 7.3, a different denominator, $(S_{max} - S_{min})$, must be used in calculating m_k. Furthermore, if S_{min} is greater than $0.172S_{max}$, the best way to pump the varactor is between C_{min} and C_{max}. The resulting maximum values of m_1 are plotted in Figure 7.2, as a function of S_{min}. Note that if S_{min} is

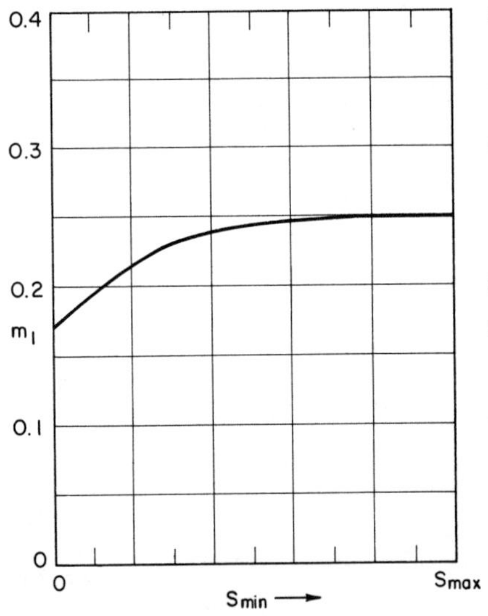

Figure 7.2. Maximum achievable modulation ratio m_1 if the capacitance is sinusoidal. For $S_{min} \geq 0.172S_{max}$, pump between S_{min} and S_{max}; for $S_{min} < 0.172S_{max}$, pump between $0.172S_{max}$ and S_{max}, to achieve the values of m_1 indicated here.

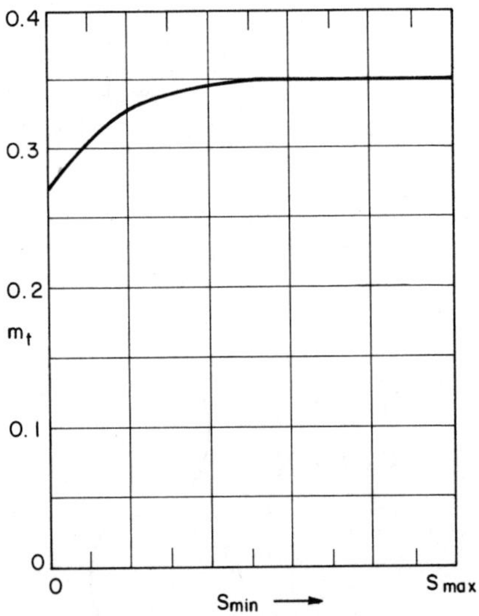

Figure 7.3. Maximum achievable total modulation ratio m_t if the capacitance is sinusoidal. For $S_{min} < 0.111S_{max}$, pump between $0.111S_{max}$ and S_{max}; for $S_{min} \geq 0.111S_{max}$, pump between S_{min} and S_{max}, to achieve the values of m_t indicated here.

close to S_{max}, the harmonics of elastance are small, and we can achieve values of m_1 close to those achievable under sinusoidal-elastance pumping.

On the other hand, suppose we want to achieve the largest possible value of m_t under sinusoidal-capacitance pumping. A similar analysis shows that if $S_{min} = 0$, we should pump between C_{min} and $9C_{min}$ to obtain $m_t = 0.272$. If $S_{min} \neq 0$, then the achievable value of m_t is larger, as shown in Figure 7.3. Note that if S_{min} is below $0.172 S_{max}$, then the highest values of m_t and m_1 are not achieved simultaneously.

7.2.3. Arbitrary Elastance

If neither the elastance nor the capacitance is required to be sinusoidal, there is a possibility that m_1 and m_t can be greater than the limits given earlier. The waveform for $S(t)$ that realizes the largest possible value for m_1 also realizes the largest possible value for m_t; it is a square wave in elastance, with $S(t)$ equal to S_{min} half the time and equal to S_{max} half the time. The resulting value for m_1 is $1/\pi = 0.318$, and the resulting value for m_t is 0.500. The other m_k are given by the formulas

$$m_0 = \frac{1}{2} \frac{S_{max} + S_{min}}{S_{max} - S_{min}}, \tag{7.12}$$

$$m_k = \frac{1}{k\pi}, \qquad (k = 1, 3, 5, 7, \cdots), \tag{7.13}$$

$$m_k = 0, \qquad (k = 2, 4, 6, 8, \cdots). \tag{7.14}$$

In practical cases, of course, it is very difficult to pump a varactor to obtain a square wave of elastance; this would require a square wave of charge and, as shown in the next section, infinite pump power. It is doubtful whether in practice the amplitudes and phases of pumping harmonics can be controlled to obtain any substantial increase in m_1 by square-wave pumping.

7.2.4. Pump Power

It is important to calculate the power required to pump a varactor, because in practice this may exceed the capabilities of sources at hand, or it may exceed the varactor dissipation rating. Also, as we show in Chapter 10, it gives a comparative index of performance for dynamic range. A reasonable estimate of the pump power is made by calculating the power dissipated in the series resistance by the pumping current.

To calculate this power, we note that the pumping charge (integral of the pumping current) $q_p(t)$ is periodic, and may be written in a Fourier series

$$q_p(t) = \sum_{k=-\infty}^{+\infty} Q_k e^{jk\omega_p t}. \tag{7.15}$$

If the pumping current is sinusoidal, only Q_0, Q_{-1}, and Q_1 are present; the remaining Q_k vanish. In general, however, all Q_k are present and contribute toward dissipation in the series resistance.

The time-averaged power dissipated in the series resistance R_s is

$$P = R_s \left\langle \left(\frac{dq_p(t)}{dt} \right)^2 \right\rangle \tag{7.16}$$

or, from Equation 7.15,

$$P = 2R_s\omega_p^2(|Q_1|^2 + 4|Q_2|^2 \cdots + k^2|Q_k|^2 + \cdots). \tag{7.17}$$

For a given elastance waveform, that is, a given charge waveform, the pump power required is proportional to the square of the pump frequency. For square-wave pumping, the summation in Equation 7.17 diverges, indicating that a very large (in fact, theoretically infinite) pump power is required.

7.3. Voltage Pumping

For most varactors it is difficult (and not necessarily desirable) to make the capacitance or the elastance exactly sinusoidal. It is far easier to make the current or voltage across the varactor sinusoidal. Here in Section 7.3, we discuss the implications of having a sinusoidal pumping voltage, and in Section 7.4, a sinusoidal pumping current, for both graded-junction and abrupt-junction varactors.

The voltage that is to be sinusoidal is not the voltage across the entire varactor, but rather the voltage across the nonlinear capacitance, that is, inside the series resistance. This fact does no harm at low pump frequencies, where the effect of the series resistance is negligible, but it does cause trouble at high frequencies. If we wish to pump at the optimum pump frequency $m_1\omega_c$ for any reasonable value of m_1, the series resistance is an appreciable fraction of the impedance of the capacitive part; even if the applied voltage were sinusoidal, the voltage across the capacitive part would not be.

Nevertheless, we wish to discuss voltage pumping for two reasons. First, it is often assumed that this occurs in practice (probably this assumption is unwarranted). But more important, we wish to show that voltage pumping is generally less desirable than current pumping.

For simplicity, we assume that $S_{min} = 0$, so that the voltage may be run over the entire range from the contact potential $-\phi$ to the breakdown voltage V_B. Since we are interested in only graded-junction and abrupt-junction varactors, we use the elastance law

$$S = S_{max}\left(\frac{v + \phi}{V_B + \phi}\right)^{\gamma}, \tag{7.18}$$

where the exponent γ is one third for graded-junction varactors and one half for abrupt-junction varactors (see Section 4.2.4). In addition, we shall use Equations 4.11 to 4.14, which are all based on Equation 7.18.

The pumping voltage $v_p(t)$ is sinusoidal:

$$v_p(t) = V_0 + 2V_1 \cos \omega_p t, \tag{7.19}$$

where, without loss of generality, we have chosen the pumping phase. The average voltage, or bias voltage is V_0. Using Equations 7.2, 7.3, and 4.11, we have

$$m_k = \frac{|S_k|}{S_{max}} = \left| \frac{1}{2\pi} \int_0^{2\pi} \left(\frac{\phi + V_0 + 2V_1 \cos\psi}{V_B + \phi} \right)^{\gamma} \cos k\psi \, d\psi \right|. \tag{7.20}$$

It is convenient to have a compact notation for integrals such as that in Equation 7.20. Let us define

$$I(a,\ n;\ \beta) = \frac{1}{2\pi}\int_0^{2\pi} (1 + \beta \cos\psi)^{a} \cos(n\psi)\, d\psi \tag{7.21}$$

as a convenient abbreviation.[3] Then, Equation 7.20 is the same as

$$m_k = \left(\frac{V_0 + \phi}{V_B + \phi}\right)^{\gamma} \left| I\left(\gamma,\ k;\ \frac{2V_1}{V_0 + \phi}\right)\right|. \tag{7.22}$$

Using this same notation, we can express m_t as the square root of the average of the elastance squared, less the square of the average elastance, as in Equation 7.5:

$$m_t^2 = \left(\frac{V_0 + \phi}{V_B + \phi}\right)^{2\gamma} \left[I\left(2\gamma,\ 0;\ \frac{2V_1}{V_0 + \phi}\right) - I^2\left(\gamma,\ 0;\ \frac{2V_1}{V_0 + \phi}\right)\right]. \tag{7.23}$$

In a similar way, we can calculate the pump power P from Equation 7.16. Thus, using Equation 4.13, we find

$$P = R_s(Q_B + q_\phi)^2 \frac{1}{2\pi} \int_0^{2\pi} \left[\frac{d}{dt}\left(\frac{V_0 + \phi + 2V_1 \cos(\omega_p t)}{V_B + \phi}\right)^{(1-\gamma)}\right]^2 d(\omega_p t). \tag{7.24}$$

We now use the definitions of normalization power P_{norm},

$$P_{norm} = \frac{(V_B + \phi)^2}{R_s}, \tag{7.25}$$

and cutoff frequency,‡

$$\omega_c = \frac{S_{max}}{R_s}, \tag{7.26}$$

along with Equation 4.14, evaluated at breakdown, to determine $(Q_B + q_\phi)$. Then, the integral of Equation 7.24 can be evaluated from the identity[4]

$$\frac{1}{2\pi} \int_0^{2\pi} \left[\frac{d}{dt}(1 + \beta \cos \omega_p t)^a\right]^2 d(\omega_p t) = \frac{a^2 \omega_p^2 \beta}{2a - 1} I(2a - 1,\ 1;\ \beta), \tag{7.27}$$

where the I function is given by Equation 7.21. Thus,

$$P = P_{norm} \frac{\omega_p^2}{\omega_c^2} \left(\frac{V_0 + \phi}{V_B + \phi}\right)^{2(1-\gamma)} \frac{2V_1}{V_0 + \phi} \frac{1}{1 - 2\gamma} I\left(1 - 2\gamma,\ 1;\ \frac{2V_1}{V_0 + \phi}\right). \tag{7.28}$$

---

‡Recall that for simplicity we have assumed in Section 7.3 that $S_{min} = 0$.

We have derived formulas for the m_k, m_t, and P, in terms of the integral notation, Equation 7.21. For given relative values of V_1, $V_0 + \phi$, and $V_B + \phi$, these integrals can be evaluated in terms of the rapidly converging hypergeometric series.[5,3] In the remainder of Section 7.3, we describe the results of these calculations. First, in Section 7.3.1, we assume that the varactor is pumped fully, so that $v_p(t)$ goes all the way from $-\phi$ to V_B during each cycle. Then, in Sections 7.3.2, 7.3.3, and 7.3.4, we assume that the varactor is only partially pumped. Different descriptions are appropriate, depending upon the bias voltage V_0.

7.3.1. Full Pumping

If the varactor is pumped fully, then $v_p(t)$ goes between $-\phi$ and V_B, so that

$$\phi + V_0 - 2V_1 = 0 \tag{7.29}$$

and

$$V_0 + 2V_1 = V_B. \tag{7.30}$$

Thus the bias voltage is

$$V_0 = \frac{V_B - \phi}{2}. \tag{7.31}$$

Use of Equations 7.29 and 7.30 reduces Equation 7.20 to the form

$$m_k = \frac{1}{2^\gamma} \left| \frac{1}{2\pi} \int_0^{2\pi} (1 + \cos\psi)^\gamma \cos k\psi \, d\psi \right|$$

$$= \frac{1}{2^\gamma} \left| I(\gamma, k; 1) \right|, \tag{7.32}$$

which is a function of k and γ only. This can be calculated in closed form in terms of gamma functions:[3]

$$m_k = \frac{1}{4^\gamma} \left| \frac{\Gamma(1 + 2\gamma)}{\Gamma(1 + \gamma + k)\Gamma(1 + \gamma - k)} \right|. \tag{7.33}$$

In a similar way, m_t can be calculated from Equation 7.23 as

$$m_t^2 = \frac{1}{16^\gamma} \frac{\Gamma(1 + 4\gamma)}{[\Gamma(1 + 2\gamma)]^2} - m_0^2, \tag{7.34}$$

and the pump power is

$$P = P_{norm} \frac{\omega_p^2}{\omega_c^2} \frac{16^{(\gamma-1)}}{(1 - \gamma)} \frac{\Gamma(3 - 4\gamma)}{[\Gamma(2 - 2\gamma)]^2}. \tag{7.35}$$

We now give the results of these calculations for graded-junction and abrupt-junction varactors.

7.3.1(a). Graded-Junction Varactors. If we set $\gamma = \frac{1}{3}$, to describe graded-junction varactors, we can evaluate all the parameters of interest. The various m_k values are listed in Table 7.3, from the formula

$$m_k = \frac{1}{\sqrt[3]{4}} \frac{\Gamma\left(\frac{5}{3}\right)}{\left|\Gamma\left(\frac{4}{3} + k\right)\Gamma\left(\frac{4}{3} - k\right)\right|}. \tag{7.36}$$

Also listed in Table 7.3 are the bias voltage, the total modulation ratio m_t, and the pump power.

Table 7.3. Properties of a Graded-Junction Varactor Fully Pumped with a Sinusoidal Voltage

$m_0 = 0.713$	$m_5 = 0.0108$	$m_{10} = 0.0034$
$m_1 = 0.178$	$m_6 = 0.0079$	$m_{11} = 0.0029$
$m_2 = 0.0509$	$m_7 = 0.0061$	$m_{12} = 0.0025$
$m_3 = 0.0255$	$m_8 = 0.0049$	
$m_4 = 0.0157$	$m_9 = 0.0040$	
$m_t = 0.267$	$V_0 + \phi = 0.500(V_B + \phi)$	
$m_0 = 4m_1$	$P = 0.267P_{norm}(\omega_p/\omega_c)^2$	

7.3.1(b). Abrupt-Junction Varactors. If we set $\gamma = \frac{1}{2}$, to describe abrupt-junction varactors, we find from Equation 7.33 that the various m_k are

$$m_k = \frac{1}{2} \frac{1}{\left|\Gamma\left(\frac{3}{2} + k\right) \Gamma\left(\frac{3}{2} - k\right)\right|}$$

$$= \frac{2}{\pi |4k^2 - 1|}. \tag{7.37}$$

Furthermore, m_t^2 is easily calculated from Equation 7.34 to be $0.50 - m_0^2$, so that m_t is 0.308. These numbers are listed in Table 7.4.

Table 7.4. Properties of an Abrupt-Junction Varactor Fully Pumped with a Sinusoidal Voltage

$m_0 = 0.637$	$m_5 = 0.0064$	$m_{10} = 0.0016$
$m_1 = 0.212$	$m_6 = 0.0045$	$m_{11} = 0.0013$
$m_2 = 0.0424$	$m_7 = 0.0033$	$m_{12} = 0.0011$
$m_3 = 0.0182$	$m_8 = 0.0025$	
$m_4 = 0.0101$	$m_9 = 0.0020$	
$m_t = 0.308$	$V_0 + \phi = 0.500(V_B + \phi)$	
$m_0 = 3m_1$	$P = 0.500 P_{norm}(\omega_p/\omega_c)^2$	

Note that the maximum value of m_1 for the abrupt-junction varactor is some 18 per cent larger than that for the graded-junction varactor.

7.3.2. Fixed-Bias Partial Pumping

If the varactor is not pumped fully, we have some freedom in the choice of the bias voltage. If we choose the bias the same as for full pumping, we have fixed-bias operation; if we choose the bias voltage as low as possible, we have self-bias operation. These two types of operation are discussed here and in Section 7.3.3; arbitrary partial pumping is discussed in Section 7.3.4.

The names "fixed-bias" and "self-bias" may be misleading. The bias can be "fixed" at any value, not just the value necessary for full pumping. The term "self-bias" was chosen because if the varactor is open-circuited at zero frequency, the rectification mechanism clamps the bias voltage to be (approximately) the smallest value possible, so that the voltage swing just passes through $-\phi$. However, rectification failure and the resulting avalanche multiplication (see Sections 4.1.4 and 4.5.2) may prevent the practical use of this technique to attain self-bias. It should be stressed that the terms "fixed-bias" and "self-bias" do not necessarily reflect the technique used to attain the corresponding bias. Typical self-bias and fixed-bias pumpings are illustrated in Figure 7.4.

To characterize fixed-bias operation we define the amplitude parameter a as

$$a = \frac{4V_1}{V_B + \phi}, \tag{7.38}$$

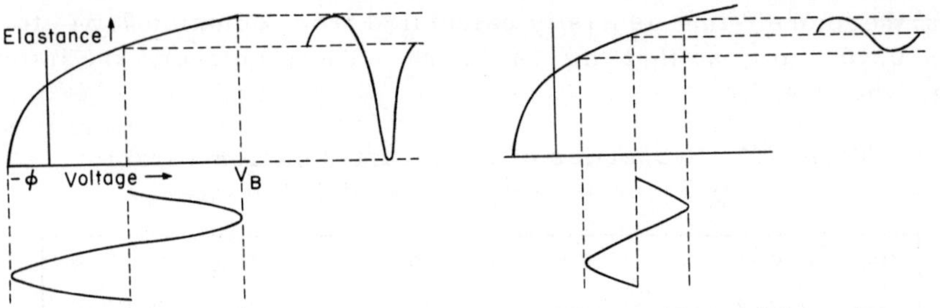

(a) Full pumping

(b) Fixed-bias partial pumping

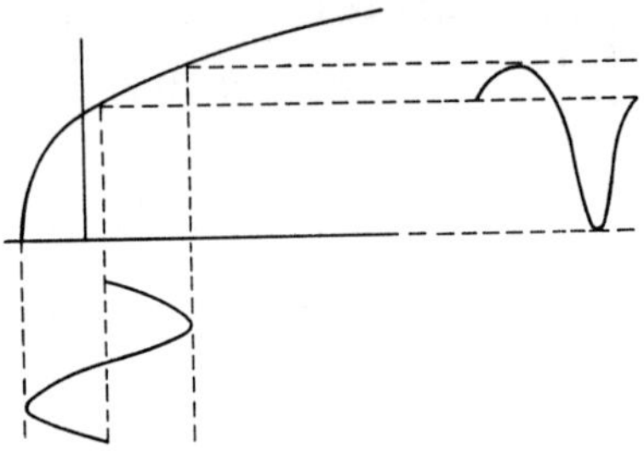

(c) Self - bias partial pumping

Figure 7.4. Illustration of full pumping and two types of partial pumping. Note that in the fixed-bias case the elastance is nearly sinusoidal, so that the values of m_k for $k \geq 2$ are small. The elastance curves are taken from Figure 4.9.

which lies between 0 (no pumping) and 1 (full pumping). For fixed-bias operation, $V_0 = 0.5(V_B - \phi)$, so that

$$m_k = \frac{1}{2^\gamma} \left| \frac{1}{2\pi} \int_0^{2\pi} (1 + a \cos\psi)^\gamma \cos k\psi \, d\psi \right|$$

$$= \frac{1}{2^\gamma} \left| I(\gamma, k; a) \right| . \tag{7.39}$$

In general the integral in Equation 7.39 cannot be expressed in closed form, but for specific values of a, γ, and k it can be integrated numerically or expressed in terms of a rapidly converging hypergeometric series.[5,3]

Similarly, for fixed bias, m_t and P are, from Equations 7.23 and 7.28,

$$m_t^2 = \frac{1}{4^\gamma} I(2\gamma, 0; a) - m_0^2 \tag{7.40}$$

and

$$P = P_{norm} \frac{\omega_p^2}{\omega_c^2} \frac{a}{4^{(1-\gamma)}} \frac{1}{1 - 2\gamma} I(1 - 2\gamma, 1; a). \tag{7.41}$$

7.3.2(a). Graded-Junction Varactors. The results of numerical calculations with $\gamma = \frac{1}{3}$ are given in Figure 7.5. Shown are m_0, m_1, m_2, m_3, m_t, and P, as well as the trivial plot of V_0, which is constant, as a function of the amplitude parameter a. Observe that when a = 1, all the results agree with the results given in Table 7.3 for full pumping. Also, observe that since $1 - 2\gamma = \gamma$, the quantities P, m_1, and a are related by Equations 7.39 and 7.41:

$$P = P_{norm} \frac{\omega_p^2}{\omega_c^2} \frac{3}{2} a m_1. \tag{7.42}$$

7.3.2(b). Abrupt-Junction Varactors. If we set $\gamma = \frac{1}{2}$, we can make similar numerical calculations‡ of the important parameters. These calculations are exhibited in Figure 7.6 and are the same quantities pictured in Figure 7.5 for the graded-junction varactor, namely, the bias voltage, m_0, m_1, m_2, m_3, m_t, and the pump power. Observe that when the amplitude parameter a is unity, then all results reduce to those pertinent to full pumping, given in Table 7.4.

Unfortunately, Equation 7.41 for the pump power cannot be used, because for $\gamma = \frac{1}{2}$ it is an indeterminate form 0/0. However, an alternate formula[4] for Equation 7.27 shows that under fixed-bias conditions P has the closed-form solution,

$$P = P_{norm} \frac{\omega_p^2}{\omega_c^2} \frac{a^2}{2\left(1 + \sqrt{1 - a^2}\right)}, \tag{7.43}$$

which is plotted in Figure 7.6.

7.3.3. Self-Bias Partial Pumping

We have just derived properties of partially pumped varactors for bias voltage fixed at that value necessary for full pumping. Now we derive similar properties for bias voltage as low as possible.

Let us define, as before, the amplitude parameter a by Equation 7.38. The condition that the bias voltage V_0 be as low as possible is that $v_p(t)$ attains the value $-\phi$, regardless of the magnitude of a. Thus,

$$V_0 - 2V_1 = -\phi, \tag{7.44}$$

‡ These calculations, in fact, can be eliminated by expressing the integrals in terms of tabulated complete elliptic integrals.[5,3]

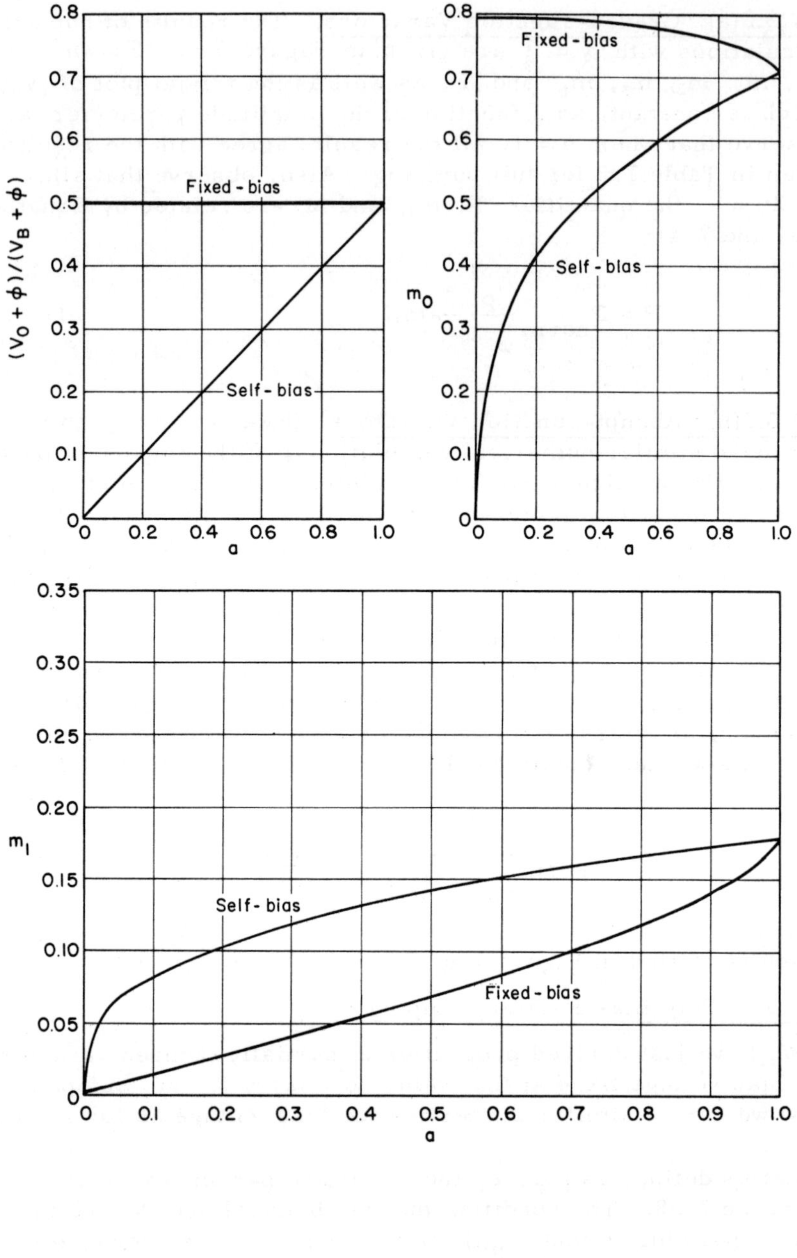

Figure 7.5. Properties of voltage-pumped graded-junction varactors, the amplitude parameter a, which is defined in Equation 7.38.

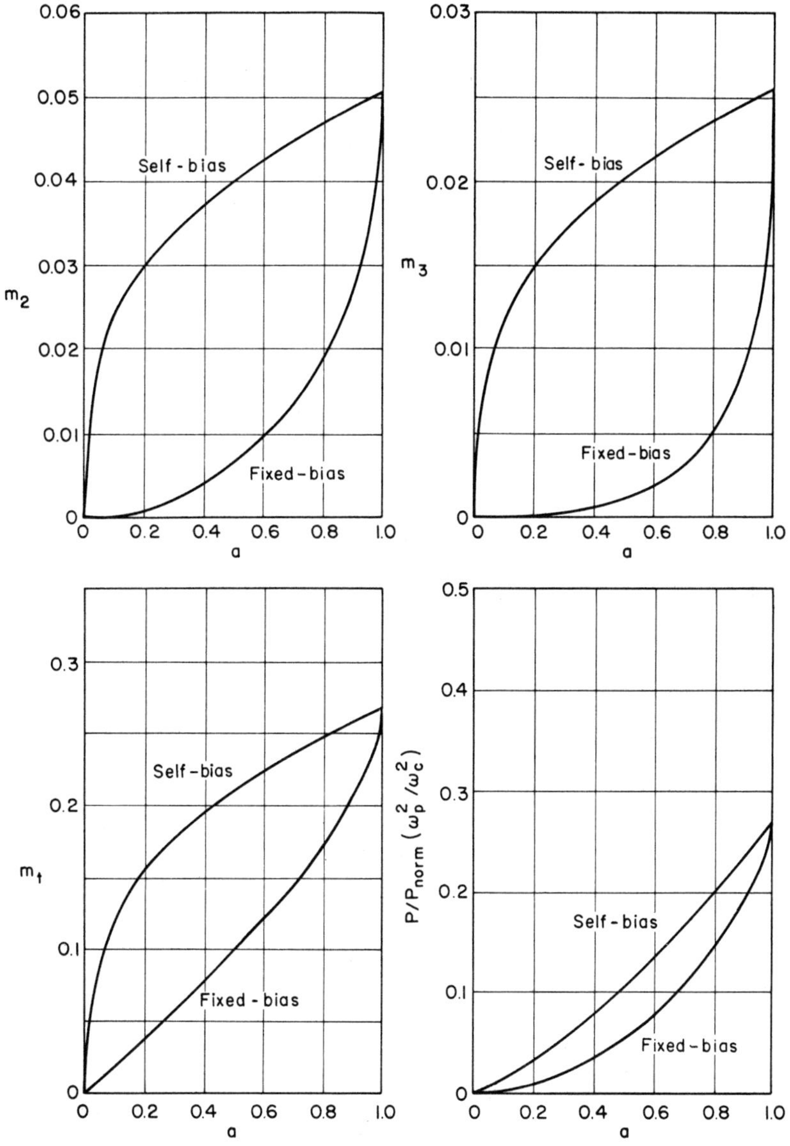

both for fixed-bias and self-bias operation, as a function of

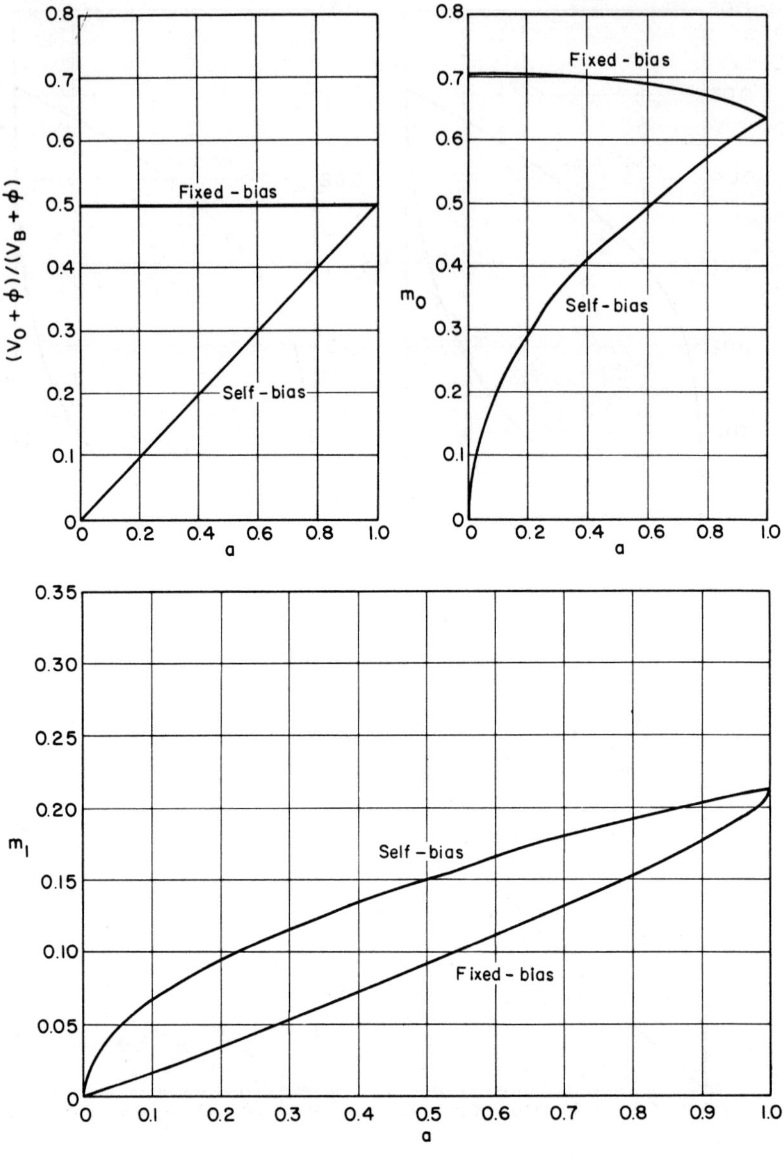

Figure 7.6. Properties of voltage-pumped abrupt-junction varactors, the amplitude parameter a, which is defined in Equation 7.38.

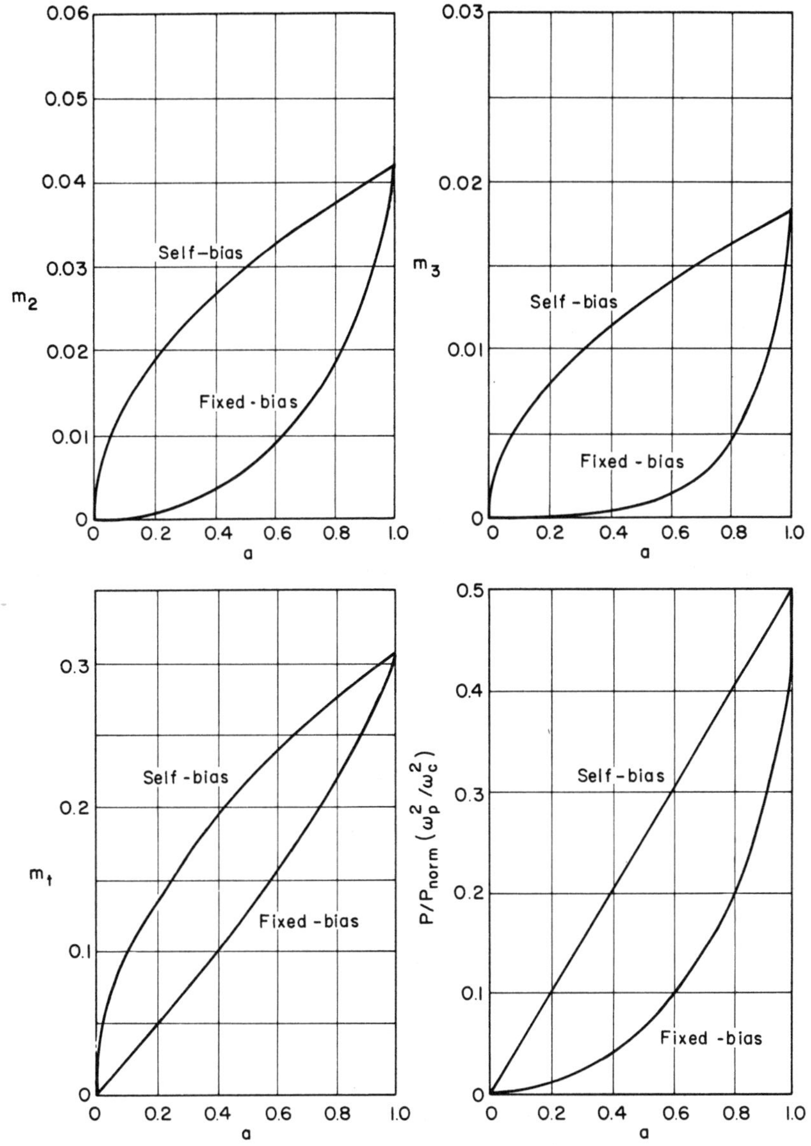

both for fixed-bias and self-bias operation, as a function of

so that m_k becomes, from Equations 7.20 and 7.22,

$$m_k = \frac{a^\gamma}{2^\gamma} \left| \frac{1}{2\pi} \int_0^{2\pi} (1 + \cos\psi)^\gamma \cos k\psi \, d\psi \right|$$

$$= \frac{a^\gamma}{2^\gamma} \left| I(\gamma, k; 1) \right|. \qquad (7.45)$$

Equation 7.45 is a very simple relationship showing that as a function of a all the m_k vary as a^γ. If a is set equal to 1 in Equation 7.45, all the m_k reduce to the values pertinent to full pumping.

Note that m_t^2 can be expressed (Equation 7.4) as a sum of m_k^2; since all the m_k vary as a^γ, m_t must also. On the other hand, Equation 7.28 shows that the pump power varies as $a^{2(1-\gamma)}$. The only remaining problem is the calculation of the bias voltage. From Equation 7.44 and the definition of a, Equation 7.38, it is clear that

$$\frac{V_0 + \phi}{V_B + \phi} = \frac{a}{2}. \qquad (7.46)$$

7.3.3(a). Graded-Junction Varactors. Setting γ to one third, we see that m_t and all m_k vary between zero and the values in Table 7.3 as $a^{1/3}$. The bias voltage is given by Equation 7.46, and the pump power is proportional to $a^{4/3}$. The appropriate calculations, based on these observations, are included in Figure 7.5, along with the fixed-bias calculations, for comparison. Note that the simple relation between P, a, and m_1, Equation 7.42, is still valid.

7.3.3(b). Abrupt-Junction Varactors. Setting γ to one half, we see that m_t and all m_k vary between zero and the values in Table 7.4 as the square root of a. The bias voltage is given in Equation 7.46, and the pump power is proportional to a. The self-bias results are plotted in Figure 7.6, along with the fixed-bias calculations, for comparison.

7.3.4. Arbitrary Partial Pumping

We have just derived the properties of partially pumped varactors in which the bias voltage V_0 was held at that value necessary for full pumping or was as low as possible. What if V_0 is neither? We now show for each value of the amplitude parameter a what values of V_0 are allowable, and give directions for determining the properties of pumped varactors for any allowable bias.

The amplitude parameter a has been defined by Equation 7.38. Now let us define a related symbol a' by the formula

$$a' = \frac{2V_1}{V_0 + \phi} = a\,\frac{V_B + \phi}{2(V_0 + \phi)}. \qquad (7.47)$$

For self-bias, $a' = 1$, and for fixed bias, $a' = a$.

For any value of a, the bias voltage V_0 is restricted, because we cannot allow $v_p(t)$, given in Equation 7.19, to exceed V_B or be less than $-\phi$. This restriction is indicated in Figure 7.7,

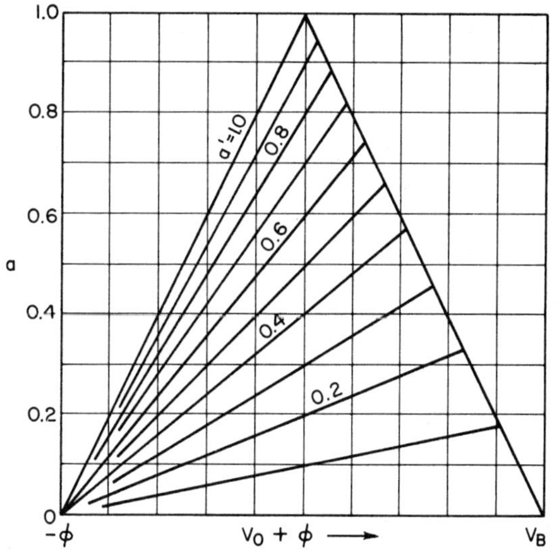

Figure 7.7. Highest achievable value of the pumping parameter $a = 4V_1/(V_B + \phi)$ for any dc bias voltage V_0. The restriction arises because $V_0 + 2V_1 \cos \omega_p t$ must lie between $-\phi$ and V_B. Also shown are lines corresponding to various values of a'.

where we show the maximum value of a that can be achieved with any given bias voltage. Also plotted in Figure 7.7 are lines corresponding to constant values of the parameter a'.

<u>7.3.4(a). Graded-Junction Varactors.</u> A simple set of rules can be given to find the properties of partially pumped graded-junction varactors for any values of a and V_0 falling within the allowed region in Figure 7.7. To find any m_k or m_t, simply follow these steps:

1. Compute a' from Equation 7.47, or from Figure 7.7.

2. Look up m_k (or m_t) using the fixed-bias curves of Figure 7.5, corresponding to a', not a.
3. Multiply the result by $\sqrt[3]{a/a'}$ to obtain m_k or m_t pertinent to the original values of V_0 and a.

The reasoning behind this procedure is clear from Equation 7.22:

$$m_k = \sqrt[3]{\frac{a}{2a'}}\left|I\left(\frac{1}{3}, k; a'\right)\right| . \qquad (7.48)$$

A similar rule can be used to calculate the pump power. Follow the same three steps, except in the last step multiply by $(a/a')^{4/3}$ instead of $(a/a')^{1/3}$. This reasoning follows from Equation 7.28:

$$P = P_{norm}\frac{\omega_p^2}{\omega_c^2}\frac{3a}{2}\sqrt[3]{\frac{a}{2a'}}\, I\left(\frac{1}{3}, 1; a'\right) . \qquad (7.49)$$

Alternatively, the pump power can be found in terms of m_1 and a by the formula,

$$P = P_{norm}\frac{\omega_p^2}{\omega_c^2}\frac{3am_1}{2}, \qquad (7.50)$$

which holds for any bias. Note that Equations 7.48 and 7.49 reduce to the proper formulas for fixed bias when a = a', and to the proper formulas for self-bias when a' = 1.

It is clear from these rules that the lowest value of m_0/m_1 is 4, and this value is achieved under self-bias.

As an example of the use of these rules, suppose a graded-junction varactor is to be voltage-pumped with V_1 equal to one eighth of $(V_B + \phi)$, so that a = 0.5. What is the highest bias voltage that can be applied, and what is the resulting pump power and value of m_1? From Figure 7.7, we see that $V_0 + \phi$ must lie between 0.25 and 0.75 of $(V_B + \phi)$. Thus the largest bias voltage is $V_0 = 0.75V_B - 0.25\phi$. With this bias, Equation 7.47 yields a' = 0.33. From Figure 7.5, we see that m_1 corresponding to an amplitude parameter of 0.33 is 0.045, and multiplying this by $\sqrt[3]{a/a'} = \sqrt[3]{1.5} = 1.14$, we find that $m_1 = 0.051$. Similarly, the pump power is $(1.5)^{4/3} = 1.71$ times the value in Figure 7.5 corresponding to 0.33, or $0.039P_{norm}(\omega_p/\omega_c)^2$. Incidentally, we observe from Figure 7.5 that in self-bias this same value of m_1 can be obtained for only about 6 per cent as much pump power; thus the proposed bias is wasteful of pump power.

7.3.4(b). Abrupt-Junction Varactors. A similar set of rules can be found for the abrupt-junction varactor. For any values of V_0 and a falling within the allowable region of Figure 7.7, we may find the appropriate m_k (and m_t) by following these steps:

1. Compute a' from Equation 7.47 or from Figure 7.7.
2. Look up m_k (or m_t) using the fixed-bias curves of Figure 7.6, corresponding to a', not a.
3. Multiply the result by $\sqrt{a/a'}$ to obtain m_k or m_t pertinent to the original values of V_0 and a.

The reasoning behind these steps is evident from Equation 7.22:

$$m_k = \sqrt{\frac{a}{2a'}} \left| I\left(\frac{1}{2}, k; a'\right)\right| . \tag{7.51}$$

A similar set of rules applies to the pump power. The only difference is that the multiplication in the third step should be by a/a' instead of $\sqrt{a/a'}$. Alternatively, if a and a' are known, the pump power can be calculated from the closed-form expression

$$P = P_{norm} \frac{\omega_p^2}{\omega_c^2} \frac{a}{2} \frac{a'}{1 + \sqrt{1 - (a')^2}} . \tag{7.52}$$

Note that Equations 7.51 and 7.52 reduce to the proper formulas for self-bias when $a' = 1$, and for fixed bias when $a = a'$.

It is clear from these rules that the lowest value of m_0/m_1 is 3, and this value is achieved under self-bias.

As an example of the use of these rules, let us suppose an abrupt-junction varactor is capable of being pumped only between S_{max} and a minimum elastance $S_{min} = 0.5S_{max}$ (for example, because of the pumping-bias anomaly discussed in Section 4.5.2). Can we calculate the maximum value of m_1? Certainly. We proceed in three steps. First, we calculate the minimum voltage, and therefore the maximum value of $4V_1/(V_B + \phi)$ achievable, and then the necessary bias so that $v_p(t)$ extends between the minimum voltage and V_B. Second, we calculate, by the preceding rules, $|S_1|/S_{max}$. Third, we multiply this by $S_{max}/(S_{max} - S_{min})$ to find the value of m_1 achieved.‡ Carrying out the first step, we find from Equation 4.12 that the minimum

‡Note that we are defining m_1 by normalizing with $S_{max} - S_{min}$ instead of S_{max}, so that the cutoff frequency is $(S_{max} - S_{min})/R_s$ instead of S_{max}/R_s.

voltage is

$$\frac{V_{min} + \phi}{V_B + \phi} = \left(\frac{S_{min}}{S_{max}}\right)^{1/\gamma} = 0.25, \tag{7.53}$$

so that the maximum value of $4V_1/(V_B + \phi)$ is 0.75. If we wish to pump with this maximum V_1, with the result that $v_p(t)$ becomes as high as V_B during each cycle, we must use the bias voltage indicated on the right-hand side of the allowed region in Figure 7.7. Thus, $(V_0 + \phi)/(V_B + \phi) = 0.625$, and $a' = 0.6$. Carrying out the second step, we find m_1 corresponding to 0.6 in Figure 7.6 is 0.11, and multiplying by $\sqrt{a/a'} = 1.10$ we discover that $|S_1|/S_{max}$ is 0.12. Thus, multiplying by $S_{max}/(S_{max} - S_{min}) = 2$, we find that $m_1 = 0.24$. Note that this number lies between the maximum m_1 achieved with $S_{min} = 0$, which is 0.212, and the maximum m_1 achieved when S is sinusoidal, which is 0.25.

7.4. Current Pumping

In Section 7.3, we investigated pumping with a sinusoidal voltage. This type of pumping is difficult in practice because of the series resistance. It is more reasonable to suppose that the varactor is pumped with a sinusoidal current, or what is equivalent, a sinusoidal charge

$$q_p(t) = Q_0 + 2Q_1 \cos \omega_p t, \tag{7.54}$$

where without loss of generality we have set the time origin by selecting the phase of the pumping charge. Using Equation 4.11, we can calculate the Fourier coefficients of elastance and, therefore, the various m_k; using Equation 4.12, we can evaluate the bias voltage; using Equation 7.17, we can determine the pump power.

First we discuss the graded-junction varactor, and then the abrupt-junction varactor. The latter treatment is considerably simplified by the fact that the elastance is proportional to the charge, so that a sinusoidal charge produces a sinusoidal elastance. For simplicity, we assume that the graded-junction varactor has a minimum elastance of zero; no such assumption is made for the abrupt-junction varactor.

7.4.1. Graded-Junction Varactors

We suppose that $S_{min} = 0$; Equation 4.11 with $\gamma = \frac{1}{3}$ implies that the elastance varies as the square root of the charge; therefore

$$m_k = \frac{|S_k|}{S_{max}} = \frac{1}{2\pi} \int_0^{2\pi} \left(\frac{q_\phi + Q_0 + 2Q_1 \cos\psi}{q_\phi + Q_B} \right)^{1/2} \cos k\psi \, d\psi$$

$$= \sqrt{\frac{Q_0 + q_\phi}{Q_B + q_\phi}} \left| I\left(\frac{1}{2}, k; \frac{2Q_1}{Q_0 + q_\phi}\right) \right|, \qquad (7.55)$$

where the abbreviation for the integral[3] is defined in Equation 7.21. Furthermore, the bias voltage V_0 is evaluated, as the average value of Equation 4.12,

$$\frac{V_0 + \phi}{V_B + \phi} = \frac{1}{2\pi} \int_0^{2\pi} \left(\frac{q_\phi + Q_0 + 2Q_1 \cos\psi}{q_\phi + Q_B} \right)^{3/2} d\psi,$$

$$= \left(\frac{Q_0 + q_\phi}{Q_B + q_\phi} \right)^{3/2} I\left(\frac{3}{2}, 0; \frac{2Q_1}{Q_0 + q_\phi}\right). \qquad (7.56)$$

The pump power is easily obtained from Equation 7.17 upon setting all Q_k except Q_1 to zero:

$$P = 2R_s Q_1^2 \omega_p^2$$

$$= 4.5 P_{norm} \frac{\omega_p^2}{\omega_c^2} \left(\frac{Q_1}{Q_B + q_\phi} \right)^2, \qquad (7.57)$$

where we have used Equations 4.14, 7.25, and 7.26 in deriving the second half of Equation 7.57. We now calculate these quantities for full pumping, and for partial pumping with self-bias, fixed bias, and arbitrary bias.

7.4.1(a). Full Pumping. For full pumping, during each cycle the charge varies between $-q_\phi$ and Q_B, so that

$$Q_0 + 2Q_1 = Q_B \qquad (7.58)$$

and

$$Q_0 - 2Q_1 = -q_\phi. \qquad (7.59)$$

Thus the expression for m_k reduces from Equation 7.55 to an expression that is identical to the formula for m_k for voltage-pumped abrupt-junction varactors, Equation 7.37. Therefore,

$$m_k = \frac{2}{\pi |4k^2 - 1|}. \tag{7.60}$$

The bias voltage for full pumping is easily evaluated from Equation 7.56:

$$\frac{V_0 + \phi}{V_B + \phi} = \frac{1}{2\pi} \int_0^{2\pi} \left(\frac{1 + \cos \psi}{2} \right)^{3/2} d\psi$$

$$= \frac{4}{3\pi} = 0.424. \tag{7.61}$$

Equations 7.58 and 7.59 predict that Q_1 is $0.25(Q_B + q_\phi)$, so that from Equation 7.57 the pump power is,

$$P = 0.281 P_{norm} \frac{\omega_p^2}{\omega_c^2}. \tag{7.62}$$

These results are summarized in Table 7.5.

Table 7.5. Properties of a Graded-Junction Varactor Fully Pumped with a Sinusoidal Current

$m_0 = 0.637$	$m_5 = 0.0064$	$m_{10} = 0.0016$
$m_1 = 0.212$	$m_6 = 0.0045$	$m_{11} = 0.0013$
$m_2 = 0.0424$	$m_7 = 0.0033$	$m_{12} = 0.0011$
$m_3 = 0.0182$	$m_8 = 0.0025$	
$m_4 = 0.0101$	$m_9 = 0.0020$	
$m_t = 0.308$		$V_0 + \phi = 0.424(V_B + \phi)$
$m_0 = 3m_1$		$P = 0.281 P_{norm}(\omega_p/\omega_c)^2$

7.4.1(b). Fixed-Bias Partial Pumping. If the varactor is not pumped fully, we have to make a choice of bias voltage. As in Section 7.3, we denote operation at the bias voltage necessary for full pumping by "fixed bias," and operation at the smallest possible bias voltage as "self-bias," without intending to imply by these names the manner in which one would actually bias the varactor (see the discussion in Section 7. 3. 2). Note, in particular, that the average voltage V_0 is used to differentiate these types of operation, even when the ac pumping current is sinusoidal.

We define an amplitude parameter a, similar to that for the voltage-pumped cases, by the formula

$$a = \frac{4Q_1}{Q_B + q_\phi}. \tag{7.63}$$

We note first that from Equation 7.57 the pump power is

$$P = 0.281a^2 P_{norm} \frac{\omega_p^2}{\omega_c^2}; \tag{7.64}$$

this formula holds for any type of partial pumping.

The calculation of the m_k parameters is quite difficult, because Equation 7.55 is written in terms of Q_0, whereas the condition of fixed bias is a constraint on V_0. For a given value of a, the relationship between Q_0 and V_0 is very complicated, given by Equation 7.56. The method of calculating the m_k is to use Equation 7.56, for each value of a, to relate Q_0 to V_0, and therefore find that value of Q_0 which makes V_0 satisfy Equation 7.61. This value of Q_0 is then used in Equation 7.55, and the m_k are calculated either by numerical integration or by expressing the integral in terms of the tabulated complete elliptic integrals.[5,3] The values of m_t are calculated in a similar way.

The results of these calculations are given in Figure 7.8, where we plot m_0, m_1, m_2, m_3, m_t, and pump power, as well as the trivial plot of V_0 (which is constant) as a function of the amplitude parameter a. Observe that if a is set equal to 1, then all the quantities plotted take on the values given in Table 7.5.

7.4.1(c). Self-Bias Partial Pumping. Now we ask about partial pumping when the bias voltage is as low as possible. At one time during each pumping period, the charge $q_p(t)$ attains the value $-q_\phi$, so that

$$Q_0 - 2Q_1 = -q_\phi. \tag{7.65}$$

If the amplitude parameter a is defined by Equation 7.63, then the formula for m_k reduces from Equation 7.55 to

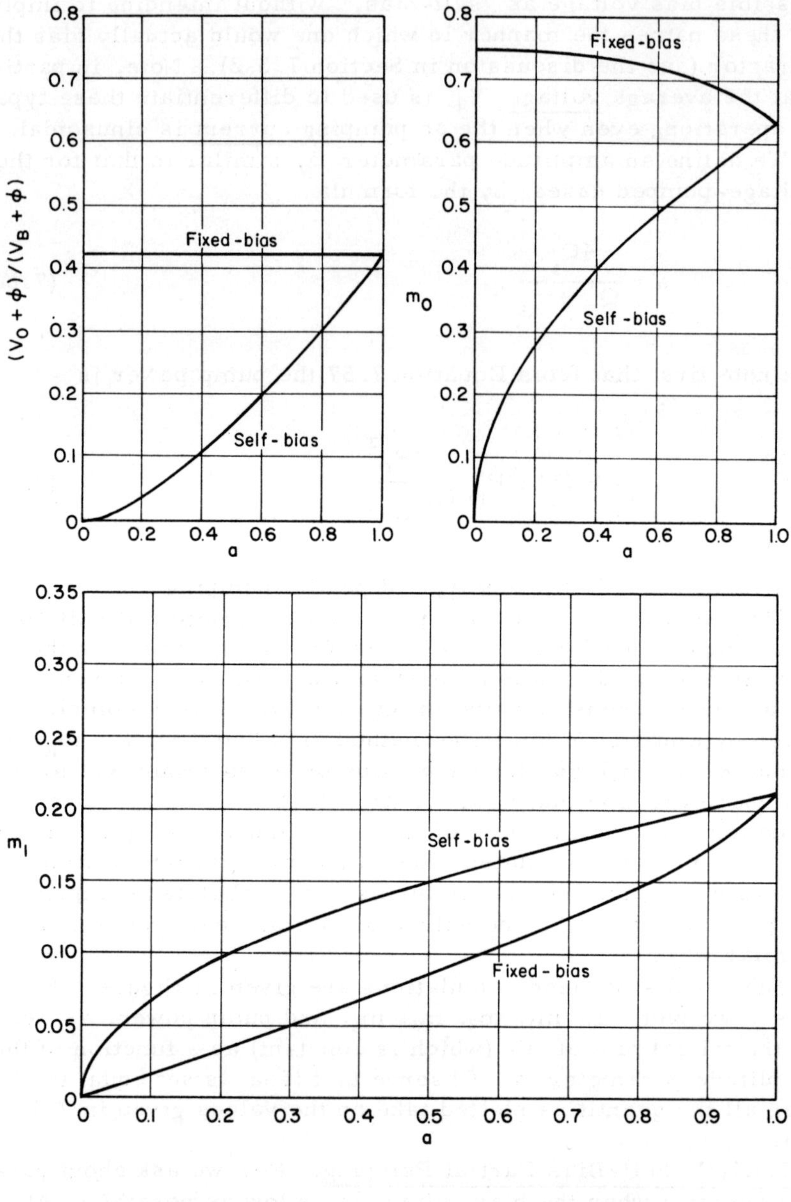

Figure 7.8. Properties of current-pumped graded-junction varactors, the amplitude parameter a.

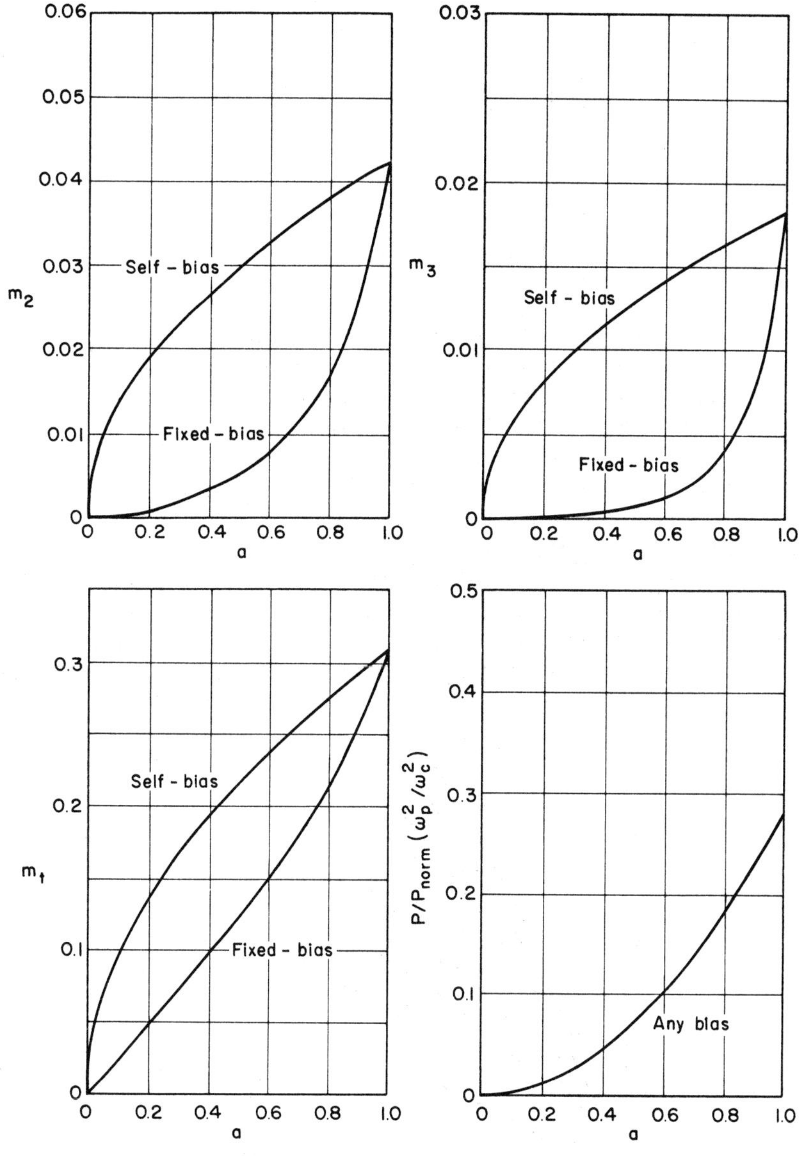

for both fixed-bias and self-bias operation, as a function of

$$m_k = \frac{\sqrt{a}}{\sqrt{2}} \left| \frac{1}{2\pi} \int_0^{2\pi} (1 + \cos\psi)^{1/2} \cos k\psi \, d\psi \right|$$

$$= \frac{\sqrt{a}}{\sqrt{2}} \left| I\left(\frac{1}{2}, k; 1\right) \right| ; \qquad (7.66)$$

this formula shows that all m_k (and therefore m_t also) vary from zero to the values in Table 7.5 as the square root of a. Similarly, the bias voltage $V_0 + \phi$ varies between zero and $0.424(V_B + \phi)$ as $a^{3/2}$. The pump power is given by Equation 7.64, as in the fixed-bias case.

The results for self-bias partial pumping of graded-junction varactors are included for comparison in Figure 7.8.

7.4.1(d). Arbitrary Partial Pumping. We have just derived values of m_k, m_t, and P for bias voltages either equal to that value necessary for full pumping or else a value as low as possible. We now show for any given values of a what other values of bias can be used and then give a set of rules for determining the properties of varactors pumped with arbitrary bias.

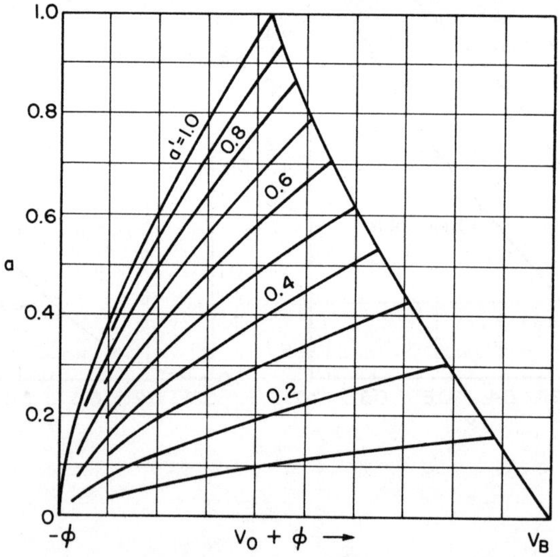

Figure 7.9. Highest achievable value of the pumping parameter $a = 4Q_1/(Q_B + q_\phi)$ for any dc bias voltage V_0 for current-pumped graded-junction varactors. Also shown are curves corresponding to constant values of a', which is defined in Equation 7.67.

For a given amplitude parameter a, the bias voltage is somewhat constrained, because the pumping charge cannot be allowed to be less than $-q_\phi$ or larger than Q_B. This constraint is given for graded-junction varactors in Figure 7.9, which shows, as a function of the bias voltage, the maximum possible value for a.

For values of a and bias voltage that fall within the allowed region of Figure 7.9, let us define the parameter a' by the formula

$$a' = a\left(\frac{4}{3\pi}\,\frac{V_B + \phi}{V_0 + \phi}\right)^{2/3} \tag{7.67}$$

so that

$$a' = \frac{2Q_1}{Q_0 + q_\phi}\left[\frac{I(\frac{3}{2},\,0;\,1)}{I\left(\frac{3}{2},\,0;\,\frac{2Q_1}{Q_0 + q_\phi}\right)}\right]^{2/3} \tag{7.68}$$

Curves corresponding to various values of a' are given in Figure 7.9. Note that for fixed bias, a' = a, and for self-bias, a' = 1. By Equations 7.68 and 7.56, a' is related to the quantity

$$\frac{2Q_1}{Q_0 + q_\phi} \tag{7.69}$$

in the same way that a is for fixed bias.

To find values of m_k or m_t corresponding to specific values of V_0 and a within the allowable region of Figure 7.9, simply follow these steps:

1. Determine a', either from Figure 7.9 or Equation 7.67.
2. Using the fixed-bias curves of Figure 7.8, determine m_k (or m_t) corresponding to a', not a.
3. Multiply the number obtained by $\sqrt{a/a'}$ to obtain the value appropriate to the original values of a and V_0.

For a given value of a, the pump power is independent of the bias, and is given by Equation 7.64.

Clearly the minimum value of m_0/m_1 is 3, and this value is achieved under self-bias.

7.4.1(e). Comparisons. It is interesting to compare the four methods of pumping graded-junction varactors: current or voltage pumping, either fixed bias or self-bias.‡ There are three significant results of such a comparison:

1. Current pumping can give a value of m_1 (0.212) approximately 18 per cent higher than voltage pumping (0.178).
2. The lowest‡‡ value for m_0/m_1 is achieved with self-bias current pumping; this value is 3.

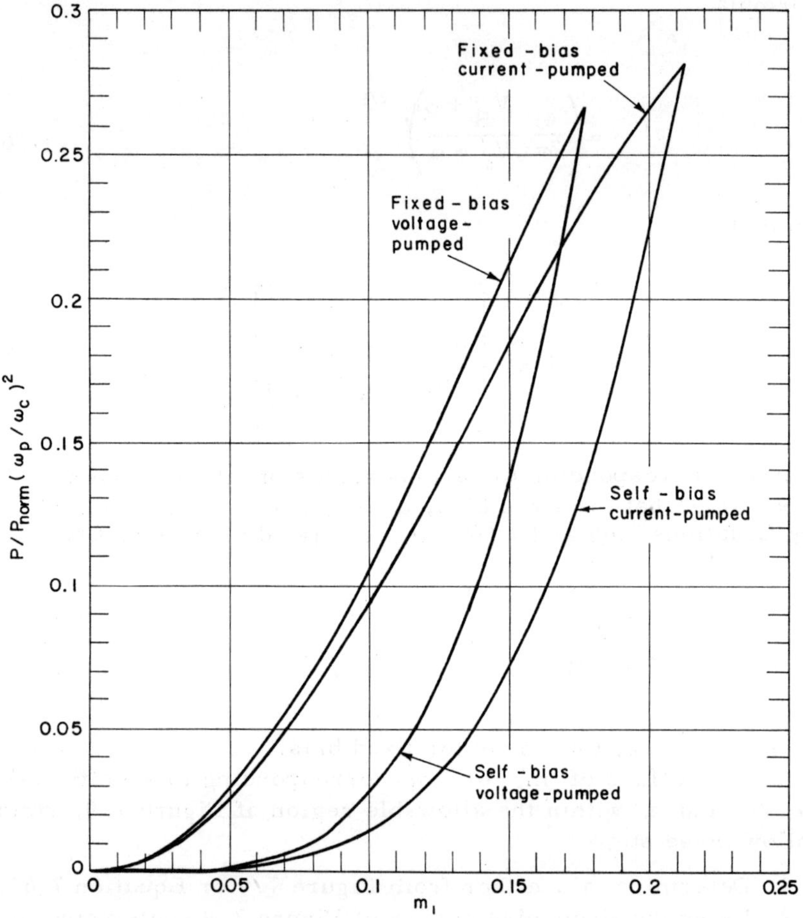

Figure 7.10. Comparison of the pump power required for a graded-junction varactor to attain a given value of m_1.

‡There is, of course, no inherent reason why varactors must be pumped with either a sinusoidal current or voltage; we cannot compare other methods of pumping because we have not discussed them.

‡‡A low value of m_0/m_1 leads to a faster rise time for subharmonic generators, as discussed in Section 9.3.

3. Self-bias current pumping requires the least pump power for a given value of m_1.

By way of emphasing the important third result, we show in Figure 7.10, as a function of pump power, the value of m_1 achieved by each type of operation. The desirability of self-bias current pumping is evident from this figure. The pump power for self-bias operation varies as the fourth power of m_1; under current pumping,

$$P = 139 m_1^4 P_{norm} \frac{\omega_p^2}{\omega_c^2}, \tag{7.70}$$

and under voltage pumping,

$$P = 264 m_1^4 P_{norm} \frac{\omega_p^2}{\omega_c^2}. \tag{7.71}$$

(For a graph of Equation 7.70, see Figure 7.17.)

7.4.2. Abrupt-Junction Varactors

The analysis in Section 7.4.1 was for graded-junction varactors; abrupt-junction varactors are discussed now. Since $\gamma = \frac{1}{2}$, we see that $\gamma/(1 - \gamma)$ is 1, and Equation 4.11 shows that the elastance is proportional to the charge. The results for current pumping are therefore simpler than the previous results.

For simplicity, we have assumed thus far that $S_{min} = 0$, but in the present analysis we need not. We suppose that the varactor has a minimum elastance S_{min}, a corresponding charge Q_{min}, and a voltage V_{min}. Then, Equation 4.11 can be put in the form,

$$\frac{S - S_{min}}{S_{max} - S_{min}} = \frac{q - Q_{min}}{Q_B - Q_{min}}. \tag{7.72}$$

Because the pumping charge is sinusoidal (given by Equation 7.54), the elastance is also sinusoidal, and the m_k values are given by inspection as

$$m_0 = \frac{Q_0 - Q_{min}}{Q_B - Q_{min}} + \frac{S_{min}}{S_{max} - S_{min}}, \tag{7.73}$$

$$m_1 = \frac{Q_1}{Q_B - Q_{min}}, \tag{7.74}$$

and

$$m_k = 0, \qquad k = 2, 3, 4, \cdot \cdot \cdot. \tag{7.75}$$

Furthermore, by Equation 7.4,

$$m_t = \frac{\sqrt{2}\, Q_1}{Q_B - Q_{min}}. \tag{7.76}$$

The bias voltage is obtained from Equation 4.12, which states that $v + \phi$ is proportional to the square of the elastance. The average voltage, therefore, is

$$\frac{V_0 + \phi}{V_B + \phi} = \left(\frac{S_{max} - S_{min}}{S_{max}}\right)^2 (m_0^2 + 2m_1^2). \tag{7.77}$$

From Equation 7.17 the pump power is easily evaluated in terms of the normalization power

$$P_{norm} = \frac{(V_B - V_{min})^2}{R_s} \tag{7.78}$$

as

$$P = 8P_{norm}\left(\frac{S_{max} - S_{min}}{S_{max} + S_{min}}\right)^2 \frac{Q_1^2}{(Q_B - Q_{min})^2} \frac{\omega_p^2}{\omega_c^2}. \tag{7.79}$$

We now evaluate these expressions for full pumping and for partial pumping. Finally, we compare the various ways to pump abrupt-junction varactors.

7.4.2(a). Full Pumping. Under full pumping, the varactor charge is varied between Q_{min} and Q_B, so that

$$Q_0 + 2Q_1 = Q_B \tag{7.80}$$

and

$$Q_0 - 2Q_1 = Q_{min}. \tag{7.81}$$

The parameters of fully pumped varactors are, from Equations 7.73, 7.74, 7.75, 7.76, 7.77, and 7.79,

$$m_0 = \frac{1}{2} + \frac{S_{min}}{S_{max} - S_{min}}, \tag{7.82}$$

$$m_1 = 0.25, \tag{7.83}$$

$$m_k = 0, \qquad k = 2, 3, 4, \cdots, \tag{7.84}$$

$$\frac{V_0 + \phi}{V_B + \phi} = \frac{3}{8}\left(\frac{S_{max} - S_{min}}{S_{max}}\right)^2 + \frac{S_{min}}{S_{max}}, \tag{7.85}$$

and

$$P = \frac{P_{norm}}{2}\frac{\omega_p^2}{\omega_c^2}\left(\frac{S_{max} - S_{min}}{S_{max} + S_{min}}\right)^2. \tag{7.86}$$

These properties are given in Table 7.6 for the important case where $S_{min} = 0$. Note that with current pumping the abrupt-junction varactor can have a value of m_1 18 per cent above that of the graded-junction varactor.

Table 7.6. Properties of an Abrupt-Junction Varactor Fully Pumped with a Sinusoidal Current, with $S_{min} = 0$

$m_0 = 0.500$	$m_1 = 0.250$	$m_k = 0, \; k \geq 2$
$m_t = 0.354$		$V_0 + \phi = 0.375(V_B + \phi)$
$m_0 = 2m_1$		$P = 0.500 P_{norm}(\omega_p/\omega_c)^2$

7.4.2(b). Partial Pumping. Let us define the amplitude parameter

$$a = \frac{4Q_1}{Q_B - Q_{min}}, \tag{7.87}$$

which when $S_{min} = 0$ reduces to the definition of Equation 7.63. Then, regardless of the bias voltage,

$$m_1 = 0.250a, \tag{7.88}$$

$$m_t = 0.354a, \tag{7.89}$$

and

$$P = \frac{P_{norm}}{2}\frac{\omega_p^2}{\omega_c^2}\left(\frac{S_{max} - S_{min}}{S_{max} + S_{min}}\right)^2 a^2. \tag{7.90}$$

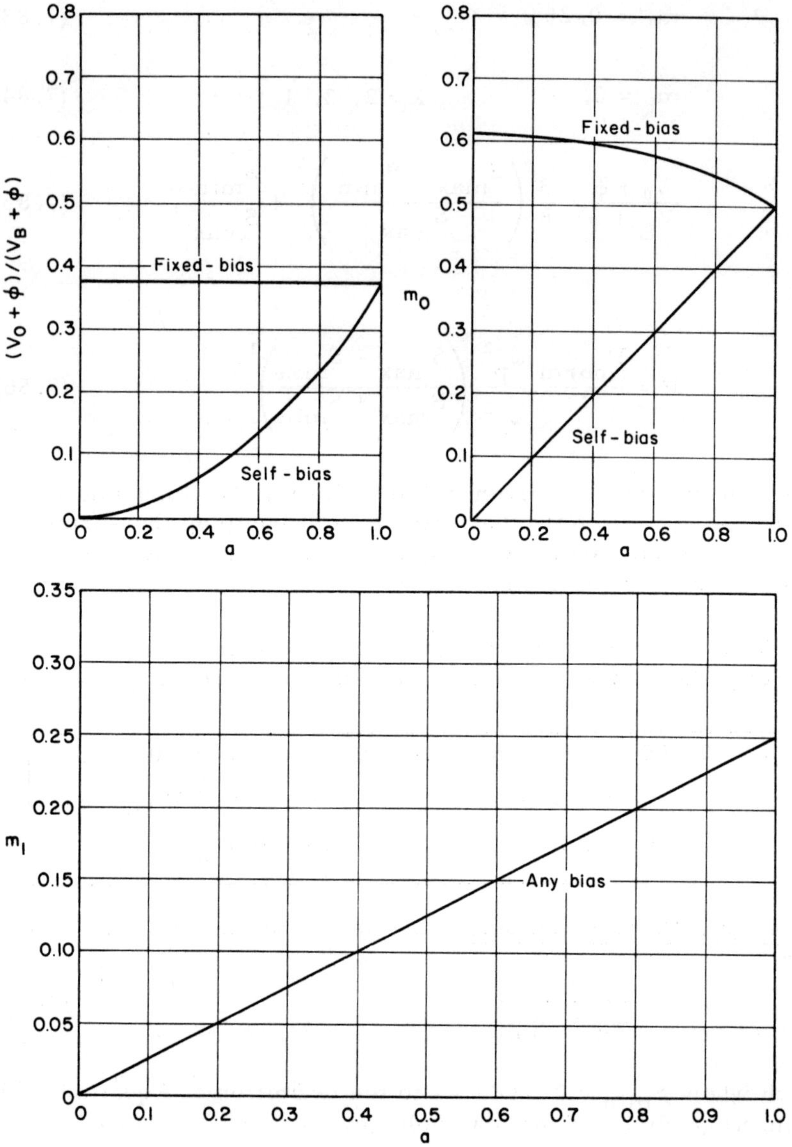

Figure 7.11. Properties of current-pumped abrupt-junction varactors, plitude parameter $a = 4Q_1/(Q_B + q_\phi)$. We have assumed $S_{min} = 0$.

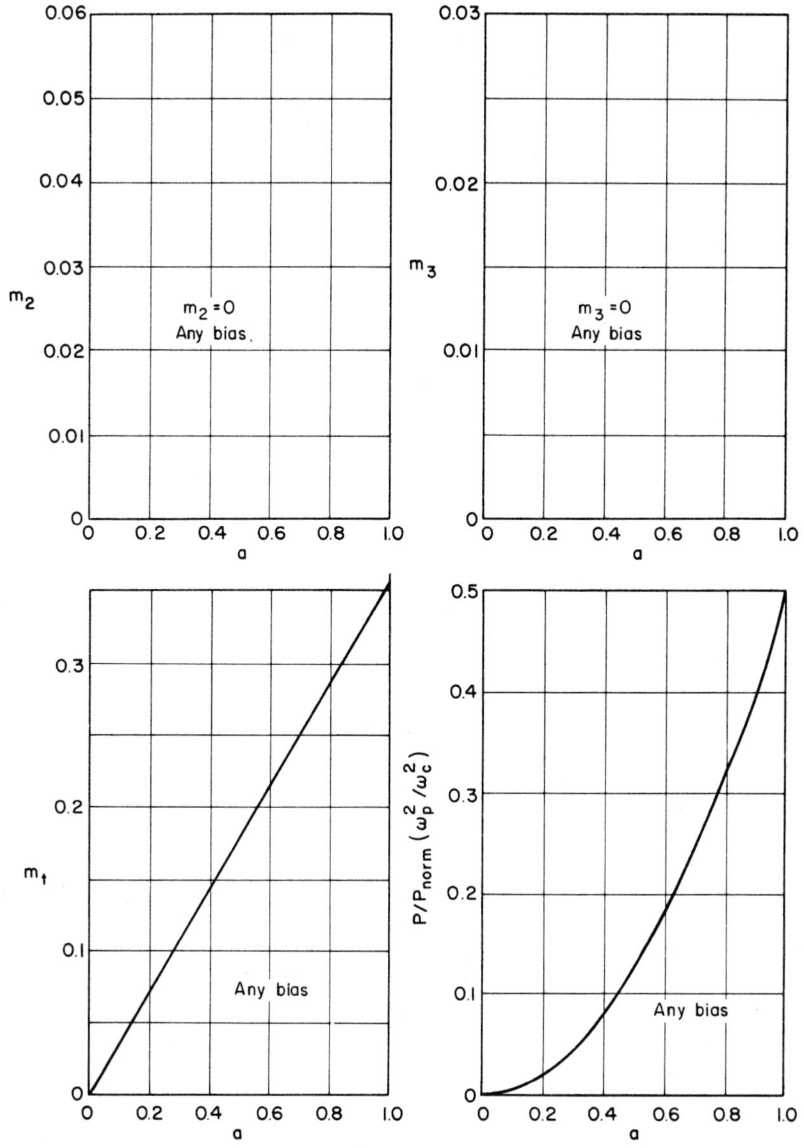

both for fixed-bias and self-bias operation, as a function of the am-

The only parameter left that depends on the bias voltage is m_0, which can be obtained in closed form from Equation 7.77 as

$$m_0^2 = \left(\frac{V_0 + \phi}{V_B + \phi}\right)\left(\frac{S_{max}}{S_{max} - S_{min}}\right)^2 - \frac{a^2}{8}. \qquad (7.91)$$

For <u>fixed-bias</u> operation, we set V_0 equal to the value in Equation 7.85 and obtain

$$m_0^2 = \frac{3 - a^2}{8} + \frac{S_{min} S_{max}}{(S_{max} - S_{min})^2}. \qquad (7.92)$$

On the other hand, for <u>self-bias</u> operation, $Q_0 - 2Q_1 = Q_{min}$, so that

$$m_0 = \frac{a}{2} + \frac{S_{min}}{S_{max} - S_{min}}, \qquad (7.93)$$

and the bias voltage becomes

$$\frac{V_0 + \phi}{V_B + \phi} = \frac{3a^2}{8}\left(\frac{S_{max} - S_{min}}{S_{max}}\right)^2 + \frac{aS_{min}}{S_{max}}\,\frac{S_{max} - S_{min}}{S_{max}} + \frac{S_{min}^2}{S_{max}^2}. \qquad (7.94)$$

The properties of partially pumped abrupt-junction varactors are plotted in Figure 7.11 for the important case with $S_{min} = 0$. Both the fixed-bias and self-bias curves are shown.

For arbitrary bias, the only parameter that depends on bias voltage is m_0, and an explicit formula for arbitrary values of a and V_0 already has been given, as Equation 7.91. The allowable values of bias are restricted by the requirement that $q_p(t)$ be greater than Q_{min} and less than Q_B at all times. For $S_{min} = 0$, this requirement is expressed in Figure 7.12, which shows the maximum value of a that can be achieved for any bias voltage.

It is evident from Equations 7.73 and 7.74 that the smallest value of m_0/m_1 is 2, and that this value is achieved only under self-bias, and only by varactors with $S_{min} = 0$.

<u>7.4.2(c). Comparisons</u>. It is interesting to compare the four types of pumping abrupt-junction varactors: current and volt-

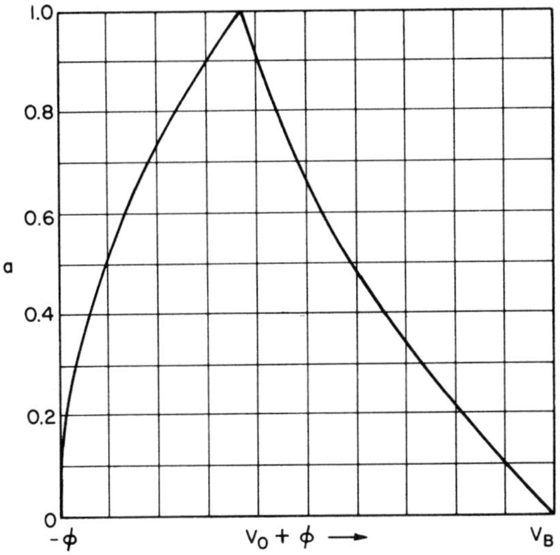

Figure 7.12. Highest achievable value of the pumping parameter $a = 4Q_1/(Q_B + q_\phi)$ for any dc bias voltage V_0 for current-pumped abrupt-junction varactors with $S_{min} = 0$.

age pumping, either fixed-bias or self-bias.‡ There are three significant results of such a comparison:

1. Current pumping can give a value of m_1 (0.250) approximately 18 per cent higher than voltage pumping (0.212).
2. The lowest value‡‡ for m_0/m_1 is achieved with self-bias current pumping; this value is 2.
3. Current pumping requires the least pump power for a given value of m_1.

By way of emphasizing the important third result, we show in Figure 7.13, as a function of pump power, the value of m_1 achieved by each type of pumping. It is clear that, of the four discussed, the most desirable is current pumping. Also note that no other type of pumping gives a lower pump power for a given value of m_1 than current pumping. This is because S is proportional to q; therefore, harmonics of the pumping cur-

‡There is, of course, no inherent reason why varactors must be pumped with either a sinusoidal current or voltage; we cannot compare other types of pumping because we have not discussed them.

‡‡A low value of m_0/m_1 leads to a fast rise time for subharmonic generators, as discussed in Section 9.3.

rent do not contribute toward m_1, although they do contribute toward P.

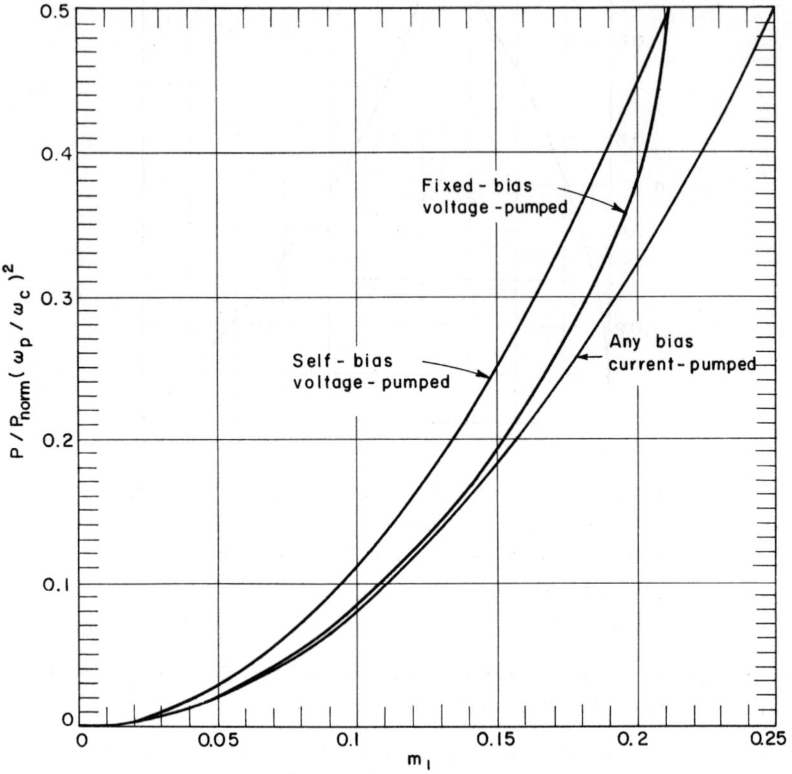

Figure 7.13. Comparison of the pump power required to attain a given value of m_1 under several types of pumping. We have assumed that $S_{min} = 0$.

Under self-bias operation, the pump power varies as the square of the modulation ratio m_1; for current pumping (with $S_{min} = 0$),

$$P = 8 m_1^2 P_{norm} \frac{\omega_p^2}{\omega_c^2}, \tag{7.95}$$

and for voltage pumping (with $S_{min} = 0$),

$$P = 11.11 m_1^2 P_{norm} \frac{\omega_p^2}{\omega_c^2}. \tag{7.96}$$

(For a graph of Equation 7.95, see Figure 7.18.)

7.5. Harmonic Pumping

Sometimes power is available not at the desired pump frequency ω_p but instead at some fraction of it, especially $\omega_p/2$. Is it possible to use this power to pump the varactor directly, or is it necessary to use a frequency multiplier to supply pump power at the correct frequency?

This power can be used directly by means of harmonic pumping.[2] There are three major limitations on the effectiveness of harmonic pumping. In the first place, the circuit is made more complicated because the varactor must be properly terminated at more frequencies. Second, the maximum value of the pumped figure of merit is somewhat smaller than what can be achieved by pumping directly; this effect becomes very important at high frequencies. And third, more power is usually required than would be to pump the varactor directly.

On the other hand, harmonic pumping can be quite useful in certain cases because no separate frequency multiplier is necessary. Only one varactor is used, instead of two. Furthermore, the power required for harmonic pumping, although greater than the power required to pump to the same value of m directly at the higher frequency, is frequently less than the lower-frequency power necessary when a separate frequency multiplier is used, because of losses in the frequency multiplier. In essence, for harmonic pumping the varactor is used both as a frequency multiplier and as the amplifier or frequency converter. Such an arrangement is frequently more efficient of power than use of separate varactors, of equal quality, for the two functions.

One scheme for harmonic pumping that comes to mind immediately is to pump the varactor either with a sinusoidal voltage or a sinusoidal current at some input frequency ω_p and to use the variation of elastance at the kth harmonic $k\omega_p$ as the basis for an amplifier or frequency converter. The pumped figure of merit for such amplifier is $m_k\omega_c$ rather than $m_1\omega_c$. When relatively small values of m_k are required, this technique can be used. Tables 7.3, 7.4, 7.5, and 7.6 show the values of m_k for full pumping, and, except for abrupt-junction current-pumped varactors, finite though small values are achieved. On the other hand, for larger, more useful values of m_k, it is necessary to use currents at the harmonics, especially the desired pump frequency $k\omega_p$.

As an example of the inherent limits of harmonic pumping, we examine in more detail an abrupt-junction varactor with desired pumping at frequency $2\omega_p$ and pump power input at frequency ω_p. For simplicity, we assume that no other pumping currents are present, aside from those at ω_p and $2\omega_p$. We ask for two limitations. First, how high can m_2 be, and therefore how high can the pumped figure of merit $m_2\omega_c$ be? Second, how much pump power is required, and in particular, how much more is required than for pumping at $2\omega_p$ in the first place?

This pumping scheme is easy to analyze because it is similar to the abrupt-junction doubler, which is treated in detail in Chapter 8. In fact, rather than derive any new formulas here, we shall merely borrow those results we need from Chapter 8.

If we do not allow current to flow at $2\omega_p$, we see from Table 7.6 that $m_2 = 0$. In fact, m_2 is proportional to the current at frequency $2\omega_p$, so that to make m_2 high, we should try to have a large second-harmonic current. The largest current occurs when the varactor has a lossless, tuned load at $2\omega_p$ or, in the notation of Chapter 8, when $R_2 = 0$. The detailed charts in Appendix D give the largest possible value of m_2 for each value of frequency $2\omega_p$; this limitation is shown in Figure 7.14. At low frequencies, values very close to 0.25 are possible, whereas for high frequencies, m_2 is severely constrained.

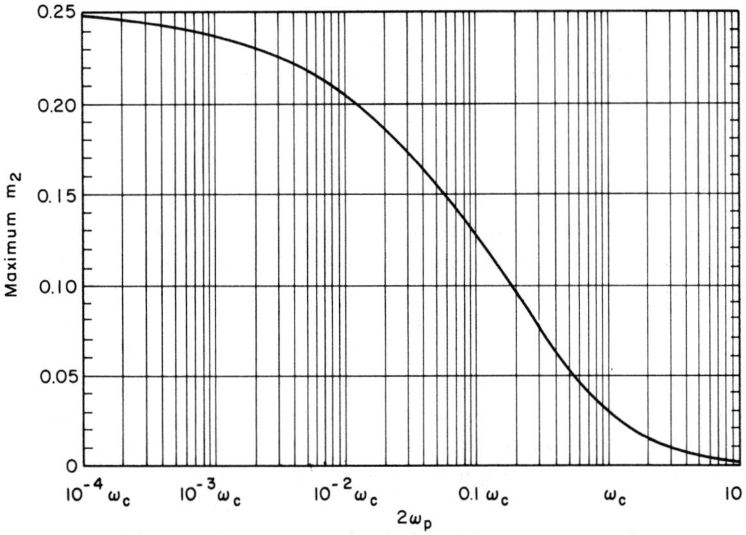

Figure 7.14. Maximum value of m_2 for harmonic pumping, achieved by having a lossless tuned load for the varactor at $2\omega_p$. Pump power input is at ω_p.

The power necessary for harmonic pumping can also be calculated from the results of Chapter 8. For simplicity, we suppose that $S_{min} \ll S_{max}$. The power input is all dissipated and is given by Equation 8.61:

$$P = 8P_{norm} \frac{\omega_p^2}{\omega_c^2} (4m_2^2 + m_1^2). \qquad (7.97)$$

Furthermore, the values of m_1 and m_2 are related by Equation 8.57 (with $R_2 = 0$):

$$m_1^2 = \frac{4m_2\omega_p}{\omega_c}, \tag{7.98}$$

so that the pump power is

$$P = 8P_{norm}\left(\frac{2\omega_p}{\omega_c}\right)^2 m_2^2 \left(1 + \frac{\omega_p}{m_2\omega_c}\right). \tag{7.99}$$

This expression for the pump power is convenient because without the last factor in parentheses, it is the power that would be required to pump that varactor directly at frequency $2\omega_p$ to the same m-value (compare with Equation 7.95). The quantity in the second parentheses determines the extra penalty that must be paid for harmonic pumping, because of the dissipation in the series resistance by current at frequency ω_p. Once $m_2\omega_c$ is known (this can, of course, be related to the desired noise and gain properties of the amplifier or frequency converter by means of the formulas of Chapters 5 or 6), this extra factor can be easily evaluated.

We have discussed very briefly only one example of harmonic pumping. There are many cases in which harmonic pumping is useful and, in particular, is less wasteful of pump power.

7.6. Pump Circuitry

The comparisons of Sections 7.4.1(e) and 7.4.2(c) showed that, at least for graded-junction and abrupt-junction varactors, current pumping is more desirable than voltage pumping for three reasons:

1. A higher value of m_1 is achievable.
2. A lower value of m_0/m_1 is achievable.
3. A given pump power produces a higher m_1.

It was not, of course, shown that current pumping is more desirable than any other type of pumping, only that current pumping is more desirable than voltage pumping. An important question is how to achieve current pumping. We shall see that the requirements for current pumping are actually quite simple, and that all the available power from a source can be used to current-pump the varactor.

A typical pumping network is shown in Figure 7.15. For generality, we do not specify the network that connects the sources (dc for bias, and ac for pumping) to the varactor. The impedance seen by the varactor is denoted by Z_{eq}.

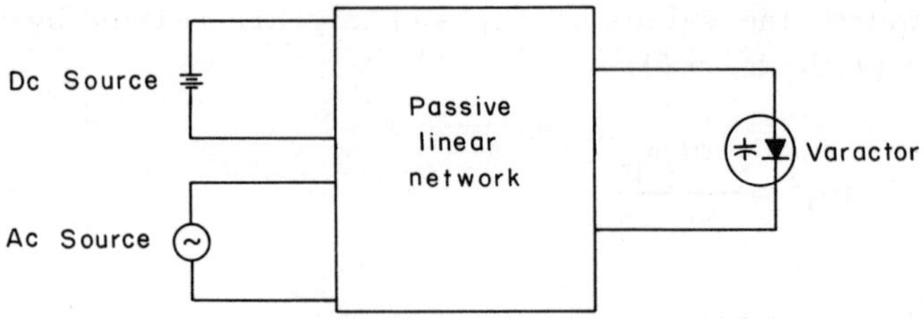

Figure 7.15. General biasing and pumping network for a varactor.

A Thévenin equivalent circuit for the network of Figure 7.15 is shown in Figure 7.16. Notice that both dc and ac sources appear, with voltages different, in general, from those in Figure 7.15.

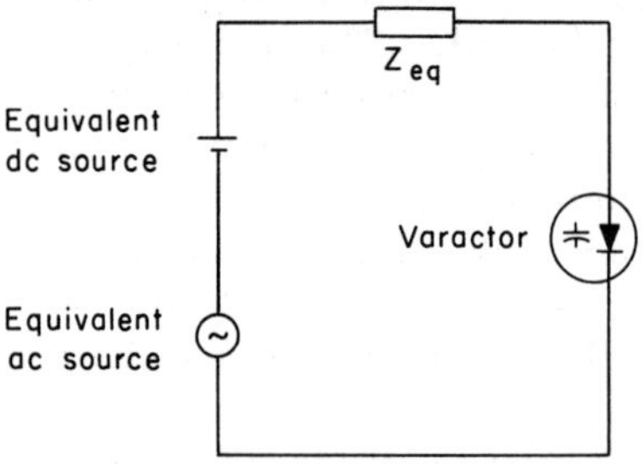

Figure 7.16. Thévenin equivalent circuit for biasing and pumping a varactor; the voltages shown are open-circuit voltages caused by the dc and ac sources of Figure 7.15.

7.6.1. Biasing

The dc bias voltage should be chosen so that the open-circuit dc voltage at the varactor terminals is V_0, the desired bias. No source is necessary if the varactor is to operate with zero average voltage; merely short-circuit the varactor at zero frequency, that is, make $Z_{eq}(\omega = 0) = 0$. If self-bias operation is attempted by using the clamping mechanism of forward conduction, simply open-circuit the varactor at zero frequency, that is, make $Z_{eq}(\omega = 0) = \infty$.

7.6.2. Pumping

To achieve current pumping, we must make sure that the only current through the varactor is at the pump frequency. Since the varactor is a nonlinear device, harmonics (and subharmonics) of the driving frequency may be generated. For current pumping, we wish to block current flow at these harmonics; this is done by open-circuiting the varactor at these frequencies. Thus, to ensure current pumping, simply design the pump circuit so that Z_{eq} is very high at all harmonics‡ of the pump frequency (especially the second).

‡An abrupt-junction varactor needs to be open-circuited at only the second harmonic.

Unfortunately, even if the harmonics of the pump frequency are open-circuited, subharmonic oscillations and spurious oscillations whose frequencies are not related to the pump frequency can easily occur. There does not seem to be any simple, general criterion to suggest when such oscillations can occur.

The impedance presented by the varactor to the pumping source is easily determined from the large-signal varactor equations, Equations 4.31. Since the only pumping current that flows is at the fundamental pump frequency ω_p, the varactor voltage at pump frequency is, from Equations 4.31,

$$V_p = \left(R_s + \frac{S_0}{j\omega_p}\right) I_p + \frac{S_2}{j\omega_p} I_p^{*}, \tag{7.100}$$

where the subscript p indicates the component at pump frequency. The voltage and current are not linearly related (because the varactor is not a linear device), so that the varactor "impedance" as usually defined does not exist. However, the ratio of voltage to current is a well-defined concept, even though it depends on the drive level. Thus,

$$\frac{V_p}{I_p} = R_s + \frac{S_0}{j\omega_p} + \frac{S_2 I_p^{*}/I_p}{j\omega_p}. \tag{7.101}$$

Now let us determine the phase of the last term in Equation 7.101. The elastance S is a function of the charge. If the charge waveform is an even function of time,‡ for example that of Equation 7.54, then the elastance is also an even function of time, and all S_k are real. Since the current amplitude I_p is imaginary, Equation 7.101 predicts

$$\frac{V_p}{I_p} = R_s + \frac{S_0 \pm |S_2|}{j\omega_p}. \tag{7.102}$$

The varactor, although nonlinear, is of course time-invariant, so that the phase of the S_2 term does not depend on the pump phase chosen; Equation 7.102 then holds for any pump phase.

For graded-junction varactors, the plus sign in Equation 7.102 is appropriate; therefore,

‡That is, if $q_p(-t) = q_p(t)$.

$$\frac{V_p}{I_p} = R_s + \frac{S_0 + |S_2|}{j\omega_p}$$

$$= R_s \left[1 - j(m_0 + m_2) \frac{\omega_c}{\omega_p} \right]. \qquad (7.103)$$

Standard matching techniques can be used to match this "impedance" to any source, with the result that all the available power of the source reaches the varactor.

For abrupt-junction varactors, S_2 vanishes for current pumping. Thus

$$\frac{V_p}{I_p} = R_s + \frac{S_0}{j\omega_p}$$

$$= R_s \left(1 - \frac{jm_0\omega_c}{\omega_p} \right), \qquad (7.104)$$

and standard matching techniques can be used to achieve complete transfer of a source's available power to the varactor.

Note that under fixed bias, m_0 and m_2 depend on the drive level; therefore, the reactive part of the input "impedance" changes some 18 per cent for the abrupt-junction and some 10 per cent for the graded-junction varactor, from no drive to full drive.

In summary, to achieve current pumping, simultaneously:

1. Adjust the dc bias properly.
2. Open-circuit the varactor at the harmonics of the pump frequency (especially the second harmonic).
3. Match the ac source to the "impedance" of Equation 7.103 or 7.104.

7.7. Operation with Limited Pump Power

After mentioning the reasons why pump power is in practice often limited, we briefly discuss the effect of insufficient pump power on the performance of upconverters, parametric amplifiers, and degenerate parametric amplifiers.

7.7.1. Pump Power Limits

If pump power is limited, a sinusoidal pumping current should be used rather than a sinusoidal pumping voltage. For graded-junction varactors, it is wise to select the bias voltage as low as possible. The resulting pump power is for graded-junction varactors

$$P = 139 m_1^4 P_{norm} \frac{\omega_p^2}{\omega_c^2}, \tag{7.105}$$

and for abrupt-junction varactors

$$P = 8 m_1^2 P_{norm} \frac{\omega_p^2}{\omega_c^2}. \tag{7.106}$$

These formulas are plotted in Figures 7.17 and 7.18.

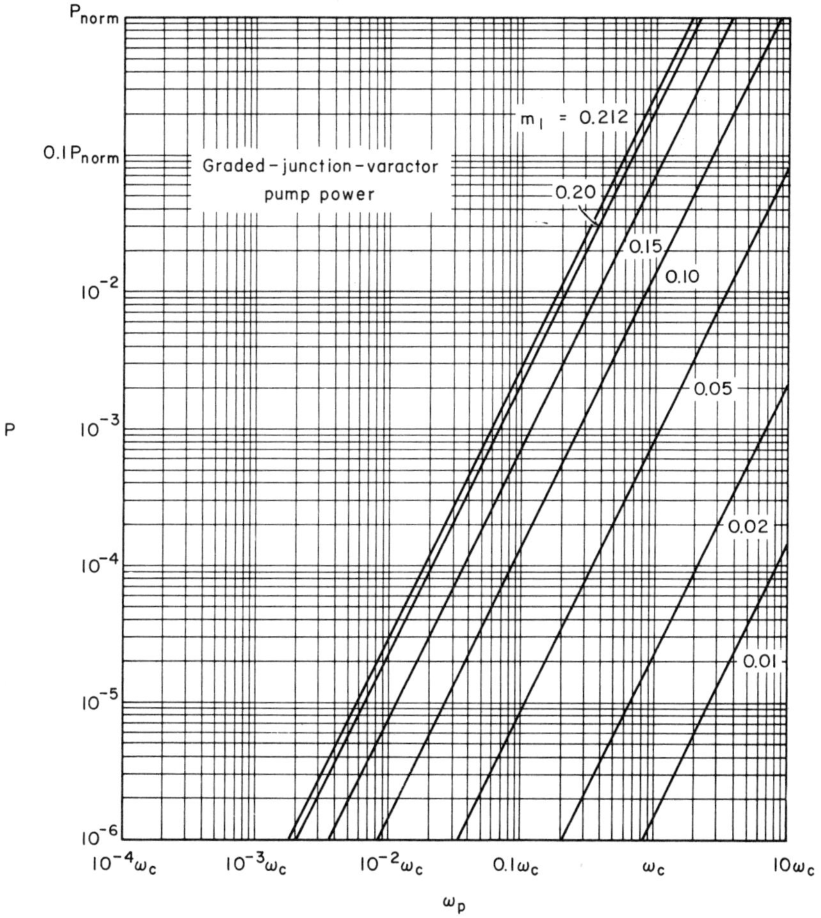

Figure 7.17. Power required to pump a graded-junction varactor, as a function of pump frequency and modulation ratio. Taken from Equation 7.105.

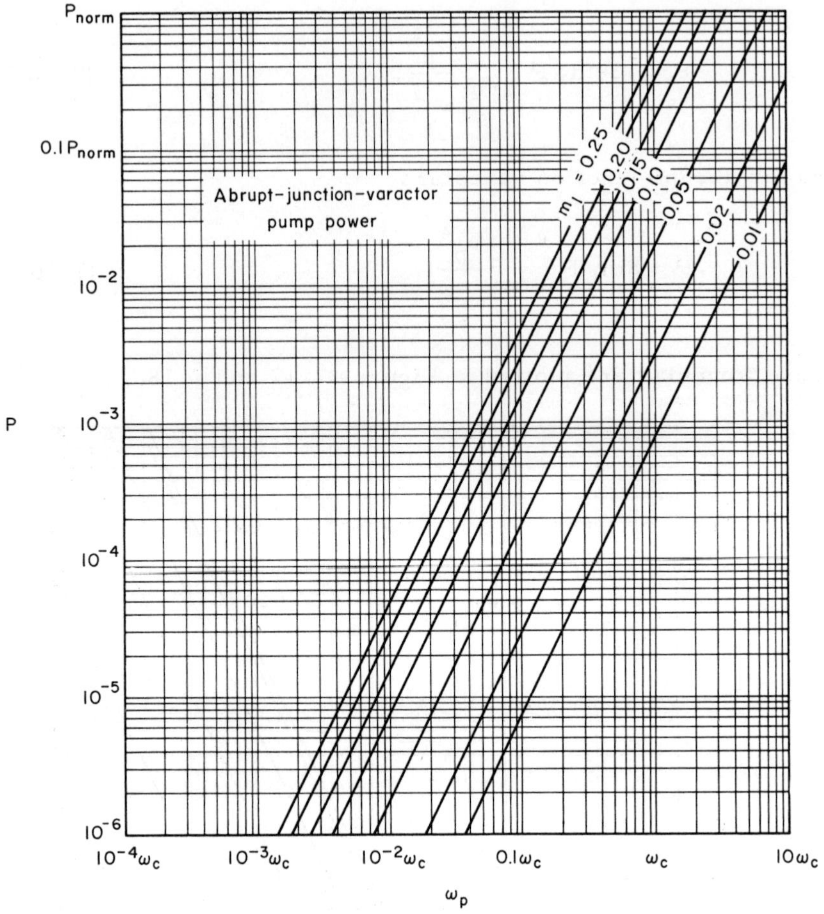

Figure 7.18. Power required to pump an abrupt-junction varactor, as a function of pump frequency and modulation ratio. Taken from Equation 7.106.

It is clear from Figures 7.17 and 7.18 that as the pump frequency is raised, the pump power rises as the square of ω_p. To pump a varactor fully at its optimum pump frequency $m_1\omega_c$ requires $0.013P_{norm}$ for graded-junction and $0.031P_{norm}$ for abrupt-junction varactors. Thus, typical varactors with cutoff frequency of 100 Gc and normalization power of 15 watts would require 190 or 470 mw at 21 or 25 Gc. There are two reasons why this requirement is difficult to meet. Sources may not be available to supply this power at this frequency, and even if such power were available, the varactor could probably not dissipate it safely.

Thus one limitation on pump power is the inability of the varactor to dissipate much power safely. The varactor dissipation rating is not necessarily related to its normalization power. The majority of varactors cannot be pumped fully at their optimum pump frequencies, because the normalization power is greater than 79 (or 32) times the dissipation rating. In the future, efforts probably will be made to produce varactors with lower normalization powers, or larger dissipation ratings, or both.

The second restriction on pump power is caused by the sources that are available. It is well known that high-power sources in the microwave region are not generally available, and even if available may not be desirable because of high cost, high noise, low efficiency, large weight or size, and so on. Probably the best way to describe this very practical restriction is to plot the power available versus the frequency for all sources that are otherwise acceptable. A typical plot of this sort might look like Figure 7.19. If for some reason one of these generators had to

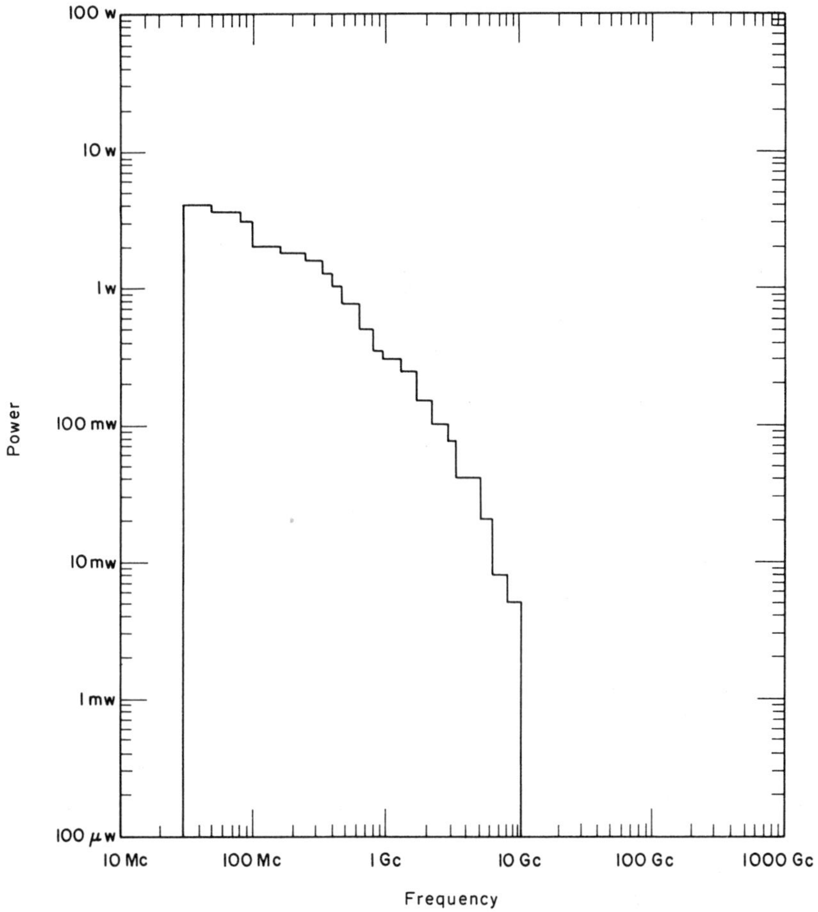

Figure 7.19. Typical plot of power available versus frequency. The selection of sources to be included in this plot is based on factors like availability, cost, weight, and size.

be used as a pump, the curve of Figure 7.19 would be an accurate representation of what is "available." In other circumstances, other curves would be appropriate.

The process of determining what values of m_1 can be achieved at what pump frequencies is now very simple. Merely superimpose the availability curve (Figure 7.19 or its equivalent) over the pump power curve (Figure 7.17 or 7.18), and simultaneously draw in the varactor dissipation rating line. To line up the curves properly, it is necessary to know only the varactor cutoff frequency and normalization power.

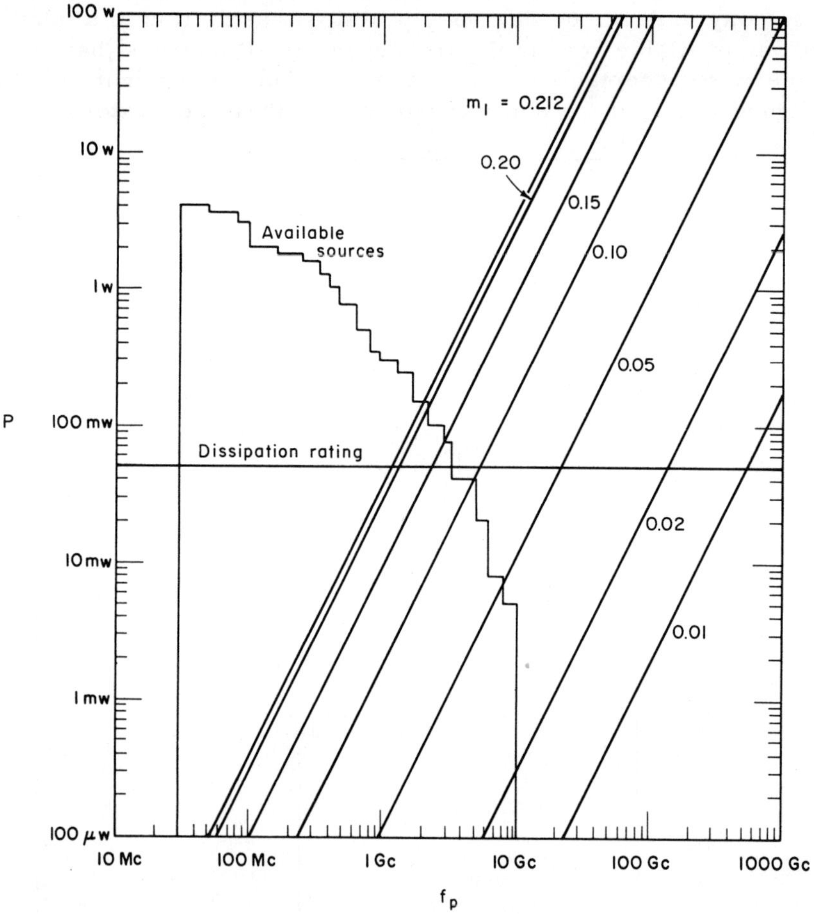

Figure 7.20. Superimposition of Figures 7.17 and 7.19, along with a line corresponding to the varactor dissipation limit. Once the varactor cutoff frequency and normalization power are known, merely slide one curve over the other, both vertically and horizontally, until both cutoff frequency and normalization power line up on the two graphs. In this figure, it was assumed f_c = 20 Gc and P_{norm} = 50 watts.

An example of this procedure, for a 20-Gc graded-junction varactor with a normalization power of 50 watts and a dissipation rating of 50 mw, is shown in Figure 7.20. The allowed region of operation is below the 50-mw line, below the $m_1 = 0.212$ line, and below the available source curve. Note that for some pump frequencies full pumping can occur; for others the varactor is limited by its dissipation rating; and for others pumping is limited by the source available.

In Sections 7.7.3 and 7.7.4, this technique of superimposing curves is extended into a method to optimize the noise performance of insufficiently pumped parametric amplifiers. In Section 7.7.2, we briefly discuss insufficiently pumped frequency converters.

7.7.2. Frequency Converters

We do not discuss in any detail inadequately pumped frequency converters but only point out a few important facts.

For the upper-sideband upconverter, the optimum noise temperature, Equation 5.51, is independent of the pump frequency. For its lowest value, m_1 should be as high as possible. The lowest noise temperature, then, can be achieved for any signal frequency simply by setting the pump frequency low enough so that the varactor can be fully pumped. A graphical technique like that leading to Figure 7.20 is valuable in choosing the pump frequency. There is no assurance that the resulting gain will be adequate.

For the lower-sideband upconverter or downconverter, the lowest noise temperature always results from having m_1 as high as possible. If the exchangeable gain is made infinite by setting the source resistance properly, the noise temperature becomes a function of both input and output frequencies, Equation 5.110. If the output frequency (and therefore the pump frequency) is fixed, the lowest noise results from pumping so as to have as high a value of m_1 as possible. On the other hand, if the pump frequency can be chosen independently, then the lowest noise is not always achieved with the highest value of m_1. Because the formulas are identical, the optimization of the infinite-gain lower-sideband upconverter is the same as the optimization of the parametric amplifier, which is given in detail in Section 7.7.3.

7.7.3. Nondegenerate Parametric Amplifiers

The noise temperature of a parametric amplifier is given by Equation 6.34. We now discuss the minimization of this expression when pump power is inadequate.

If the external idler is cooler than the diode, there may be considerable advantage in simultaneously (1) introducing external idler loss, and (2) pumping at a frequency lower than the ordinary optimum pump frequency. If the external idler is at zero temperature, in fact, this scheme can give as low a noise temperature as the fundamental limit of Equation 6.35, although for finite-temperature external idlers the minimum noise temperature obtained in this way is somewhat larger. As discussed in Section 6.5, this is the extent to which a noise-free idler can help the noise temperature of an amplifier: The pump frequency at which the lowest noise temperature is achieved is reduced. Note that with limited pump power available, this fact can be of very great importance.

If the external idler is as hot as, or hotter than, the diode, it is shown in Section 6.5 that the lowest noise always results when the external idler is lossless. It is this important case that we now wish to discuss in detail.

With a tuned reactive idler termination, the noise temperature‡ is, from Equation 6.103,

$$T = T_d \frac{\omega_s}{\omega_i} \frac{m_1^2 \omega_c^2 + \omega_i^2}{m_1^2 \omega_c^2 - \omega_s \omega_i}. \tag{7.107}$$

Suppose the pump power is unlimited. Then the noise is minimized by pumping to the largest value of m_1 at the optimum pump frequency of Equation 6.104. For limited pump power, we want to lower the pump frequency and lower the value of m_1 somewhat. To choose values of ω_p and m_1, we refer to a plot of the form of Figure 7.20 and see, with our available sources, what combinations of ω_p and m_1 can be achieved. These values are put into Equation 7.107, one after another, and we look for the lowest resulting value of T and choose the combination of ω_p and m_1 that achieves this value of T.

This is a straightforward process but probably too tedious to use in practice. We now describe a technique for graphically selecting ω_p, m_1, and P corresponding to the minimum T. For a fixed signal frequency ω_s, we plot as a function of the pump frequency the pump power necessary to attain a given noise temperature. Analytically, this is done by eliminating m_1 from Equations 7.107 and either 7.105 or 7.106, in order to relate P, T, and ω_p.

---

‡This is identical to the noise temperature of a lower-sideband upconverter (or downconverter) with the source resistance adjusted to make the exchangeable gain infinite. Therefore, the optimization procedure to be described holds equally well for the upconverter.

From this relation, when we plot P vs. ω_p for various values of T, we get a set of curves like those in Figure 7.21 (which apply to graded-junction varactors, with $\omega_s = 10^{-2}\omega_c$). It appears that for sufficient pump power we can obtain very low noise temperatures. However, some combinations of ω_p and P are physically impossible, because they would result in values of m_1 higher than theoretically possible. The line corresponding to the highest possible value of m_1 is shown in Figure 7.21, taken from Figure 7.17; the region to the right of this line is achievable and the region to the left is not.

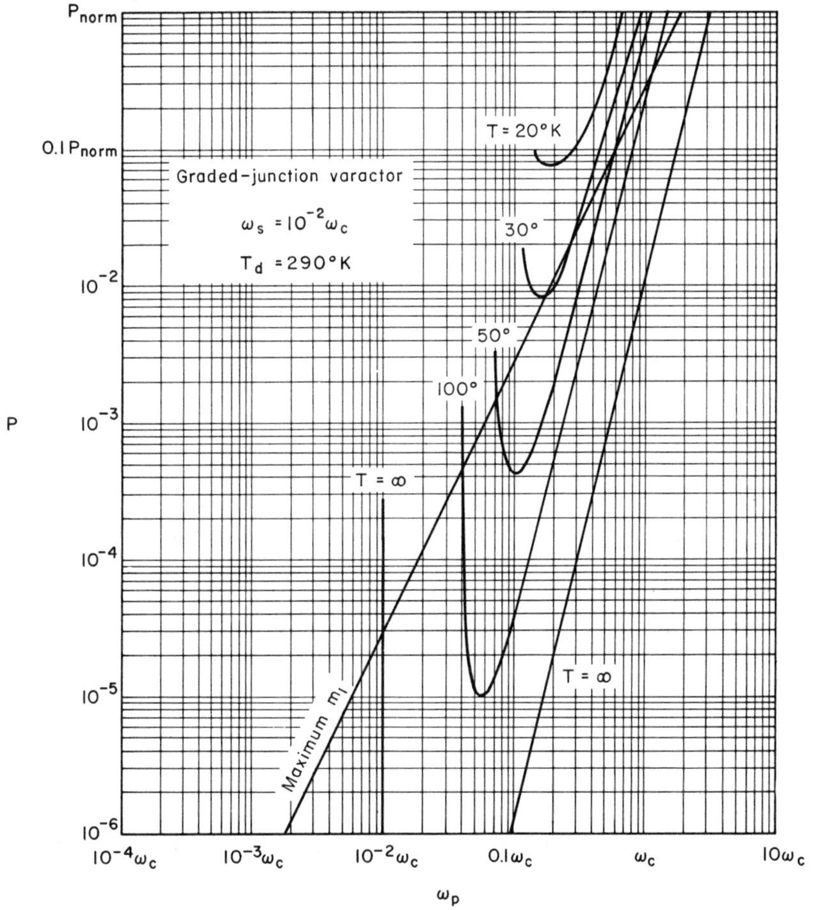

Figure 7.21. Pump power necessary to achieve a given noise performance at a given pump frequency for a graded-junction varactor with $\omega_s = 10^{-2}\omega_c$. The region to the left of the slanted straight line corresponds to a value of m_1 greater than 0.212; only the region to the right of this line represents physically achievable pumping.

The curves for $T = \infty$ are significant because gain is not possible except between them. This fact is consistent with Figure 6.11, which shows that the noise temperature approaches infinity on the lines that separate the region of gain from the two regions of no gain. The left-hand line for $T = \infty$ corresponds to $\omega_p = \omega_s$.

If the pump power is unlimited, then we can optimize the noise temperature by looking for the lowest value of T that falls to the right of the maximum m_1 line. From Figure 7.21, it is clear that for these numbers this lowest value lies between 20° and 30°, and is achieved at a pump frequency approximately $0.2\omega_c$. This optimization is, of course, precisely that of Chapter 6; the pump frequency determined in this way is the optimum pump frequency of Equation 6.104, and the lowest noise temperature is given by Equation 6.105.

The real importance of a plot like Figure 7.21, however, is for limited pump power. Let us suppose the pump power is limited, as a function of pump frequency, in some way (such as that pictured in Figure 7.19). The allowable region of operation in the pump-power, pump-frequency plane is then limited by four factors:

1. m_1 must be less than the theoretical maximum.
2. P must be less than the dissipation rating.
3. The noise temperature must be finite.
4. A source must be available.

The action of these four restrictions is illustrated in Figure 7.22, which shows curves corresponding to $T = \infty$, and to m_1 a maximum value, from Figure 7.21, and a typical dissipation rating and source curve, from Figure 7.19. The region of possible operation is shown shaded.

Now the question is, "Where, within the shaded region of Figure 7.22, is the lowest noise temperature achieved?" The answer is found very easily by superimposing the power-frequency curve for the available sources on the noise-temperature contours of Figure 7.21. It is a simple matter to find the point corresponding to lowest value of T that simultaneously lies below the dissipation rating line and the source power curve and to the right of the curve for maximum m_1 (and, of course, between the two $T = \infty$ curves).

To make this graphical procedure a useful tool, we give in Appendix D accurate plots of curves like those in Figure 7.21, for a considerable range of signal frequencies, for both self-bias graded-junction varactors and abrupt-junction varactors. These graphs are intended to be used as design curves; the accuracy is probably sufficient for practical purposes. For signal frequencies between those for which graphs are provided, interpolation can be used.

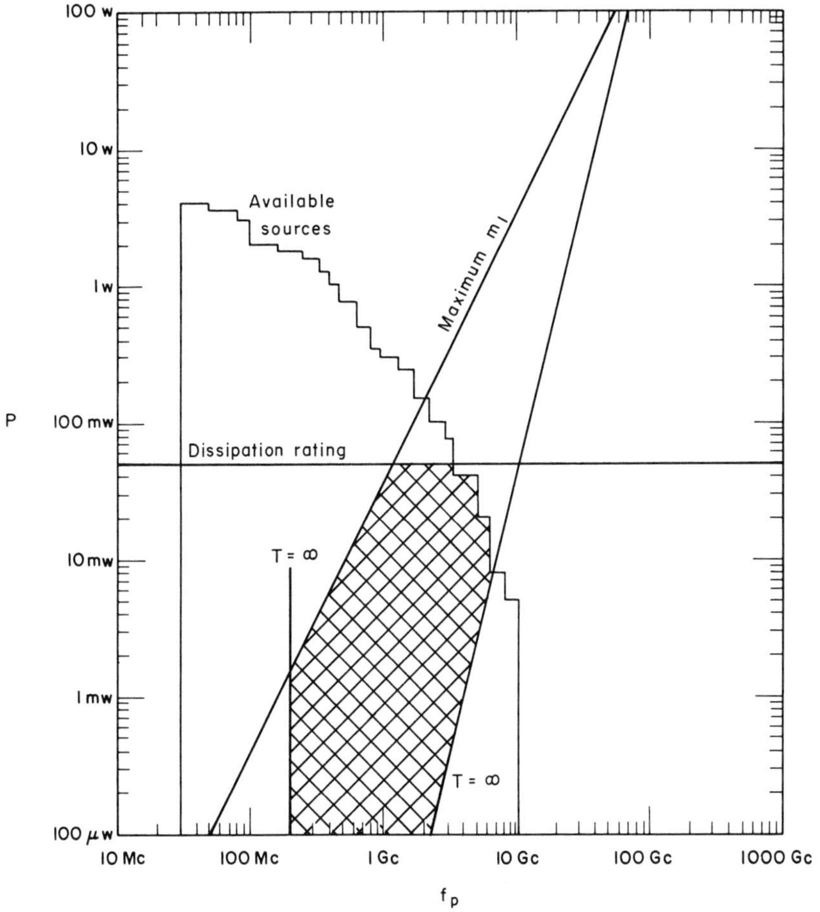

Figure 7.22. The operation of the parametric amplifier is restricted to the shaded region, for which, simultaneously, gain is possible (that is, T is finite and positive), m_1 is less than its maximum value, the pump power is less than the dissipation rating, and there is a source available to supply the power at the pump frequency indicated. This plot happens to be pertinent to a 200-Mc amplifier using a graded-junction varactor with f_c = 20 Gc, P_{norm} = 50 watts, and a dissipation rating of 50 mw.

It should be mentioned, in connection with these graphs, that a single plot of available source power versus frequency is sufficient for all varactors and all signal frequencies. For convenience, this plot can be made by the reader on a sheet of clear plastic, which can be placed over the appropriate graphs and positioned according to the particular values of cutoff frequency and normalization power.

By the use of these graphs, preliminary design of parametric amplifiers with limited pump power can become a routine matter.

7.7.4. Degenerate Parametric Amplifiers

The degenerate parametric amplifier is affected by a limit on pump power more severly than the nondegenerate amplifier. This is because the pump frequency cannot be chosen at will; it must be twice the signal frequency.

It is shown in Section 6.6 that a suitable noise figure of merit for the degenerate amplifier is what we have called T_{eff}. From this number, the noise performance of the amplifier for a variety of situations can be calculated. Accordingly, we work with T_{eff}, Equation 6.137,

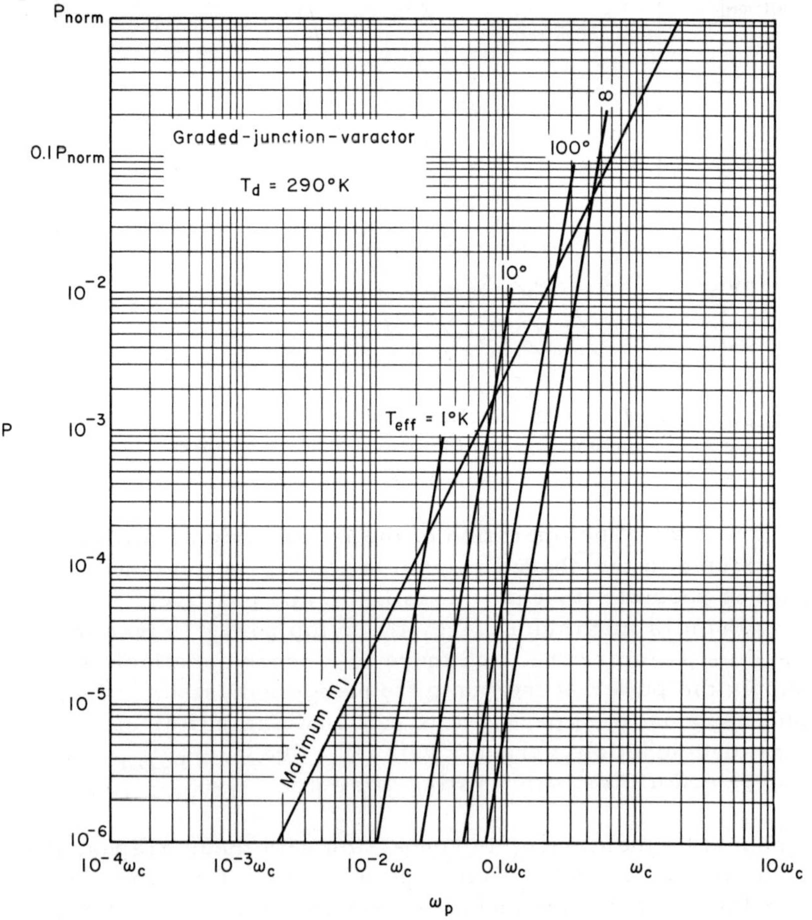

Figure 7.23. Pumping requirement for a graded-junction-varactor degenerate parametric amplifier, in terms of the noise figure of merit T_{eff}.

$$T_{eff} = \frac{T_d}{\dfrac{m_1 \omega_c}{\omega_p/2} - 1}. \qquad (7.108)$$

As in Section 7.7.3, we plot contours of constant noise temperature in the pump-power, pump-frequency plane. These plots are Figures 7.23 and 7.24, for graded-junction and abrupt-junction varactors, respectively (the procedure in making these plots is to eliminate m_1 from Equation 7.108 and either Equation 7.105 or 7.106, obtaining relations between ω_p, P, and T). In Figures 7.23 and 7.24, we have included the line corresponding to the maximum value of m_1. These plots can be used, just as the

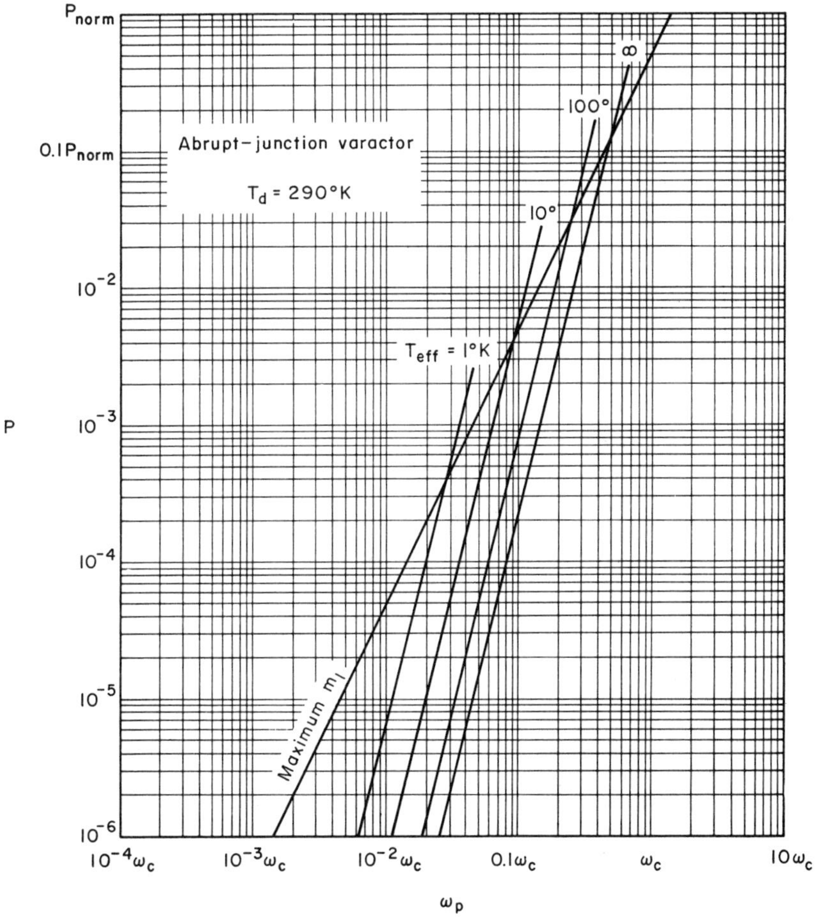

Figure 7.24. Pumping requirement for an abrupt-junction-varactor degenerate parametric amplifier, in terms of the noise figure of merit T_{eff}.

graphs of Appendix D, for preliminary amplifier design. If desired, the plots of source power versus frequency, used in the previous sections, can be superimposed on Figures 7.23 and 7.24.

REFERENCES

1. Leenov, D., "Gain and Noise Figure of a Variable-Capacitance Up-Converter," Bell System Tech. J., 37, 989-1008 (July, 1958).
2. Mortenson, K. E., "An Analysis of Subharmonic Pumping of Parametric Amplifiers," Digest of Technical Papers, 1960 International Solid-State Circuits Conference, Philadelphia, Pa. (February 10-12, 1960), pp. 93-94.
3. Penfield, P., Jr., "Fourier Coefficients of Power-Law Devices," J. Franklin Inst., 273, 107-122 (February, 1962).
4. Penfield, P., Jr., ibid., Eq. (112).
5. Sensiper, S., and R. D. Weglein, "Capacitance and Charge Coefficients for Parametric Diode Devices," Proc. IRE, 48, 1482-1483 (August, 1960).

Chapter 8

HARMONIC MULTIPLIERS

Nonlinear devices can be used to generate harmonics of an input frequency, either for power conversion or for frequency control and comparison. Since in the past most passive harmonic generators have had quite low efficiency, their use in generating power was limited mainly to frequencies, such as the millimeter range, where no other sources were available. A few multipliers, primarily those using electron-beam devices, have obtained higher efficiencies, even greater than 1, but the cause of these high efficiencies is the same as the cause of gain when the same devices are used as amplifiers; that is, these are active, not passive, multipliers.

High efficiency in passive multipliers was not achieved until the development of varactors. According to Page's formulas,‡[8,9] multipliers that depend upon a (passive) nonlinear resistance have at most an efficiency of $1/n^2$, where n is the ratio of output to input frequencies. On the other hand, the Manley-Rowe formulas[7,9] predict no limit below 100 per cent efficiency for nonlinear reactance multipliers. The high efficiency of varactor multipliers, even with the series resistance accounted for, makes them attractive for use in solid-state microwave power sources at frequencies up to several Gc and at power levels up to a few watts. This range of power and frequency has been dominated by low-power klystrons. There is no doubt that, especially in new equipment, solid-state sources consisting of VHF transistor oscillators and varactor multiplier chains will often be used in place of klystrons. The over-all efficiencies of the two types of power sources are comparable (although present trends favor the solid-state sources), and weight and size requirements favor the use of varactors. Probably power applications of varactors, including harmonic generation, will become the major applications in the future years, especially if varactors with large normalization powers become available.

In Chapter 8 we analyze harmonic multipliers using abrupt-junction varactors, giving explicit solutions for the multiply-by-two, -three, -four, -five, -six, and -eight circuits in Sections 8.4 to 8.10. The problem is set up in Section 8.2, and the idler

‡Superscript numerals denote references listed at the end of each chapter.

requirements are discussed in Section 8.3. In Section 8.11 we derive a fundamental limit on efficiency that applies to any type of varactor, any order of multiplication, and any idler configuration. Finally, in Section 8.12 all the results are compared. First, however, we discuss the major results to be derived, not only to introduce the more detailed derivations, but also to help the reader who is unable or unwilling to follow all the mathematical derivations.

Varactor multipliers can be represented schematically by diagrams like those of Figure 1.2. In Figure 8.1 we show typical multipliers, including many covered in Chapter 8. Figures like these are placed in Chapter 8 to identify the device under discussion. In Figure 8.2 we show a few dividers and rational-fraction generators.

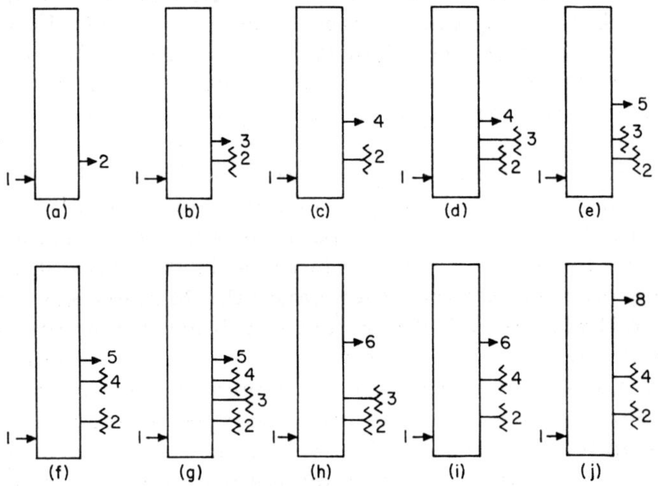

Figure 8.1. Simple representation of some varactor multipliers. The input is at the left, and the output at the right. Idlers are denoted by resistor symbols. The numerals indicate the relative values of the frequencies.

(a) Doubler, Section 8.4.
(b) 1-2-3 tripler, Section 8.6.
(c) 1-2-4 quadrupler, Section 8.7.
(d) 1-2-3-4 quadrupler.
(e) 1-2-3-5 quintupler, Section 8.8.
(f) 1-2-4-5 quintupler, Section 8.8.
(g) 1-2-3-4-5 quintupler.
(h) 1-2-3-6 sextupler, Section 8.9.
(i) 1-2-4-6 sextupler, Section 8.9.
(j) 1-2-4-8 octupler, Section 8.10.

Devices (d) and (g) are not discussed in this book.

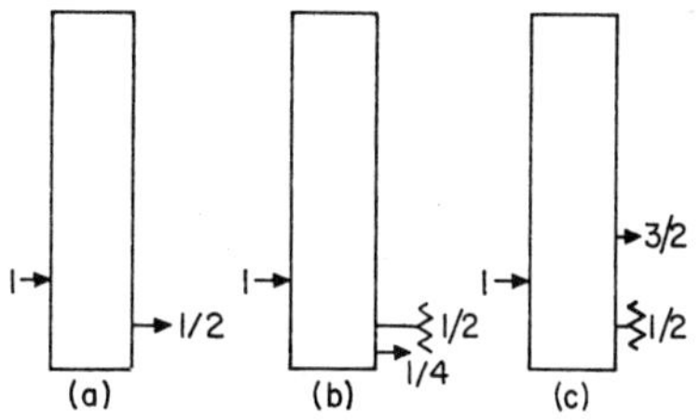

Figure 8.2. Simple representation of a few phase-sensitive varactor frequency dividers and rational-fraction generators.

(a) Divide-by-two circuit, Section 9.4.
(b) Divide-by-four circuit, Section 9.5.
(c) Multiply-by-3/2 circuit.

8.1. Introduction

Chapter 8 is concerned with harmonic multiplication, in which power is put into a varactor at frequency ω_0 and is delivered into a load at an integer-multiple frequency $\ell\omega_0$. However, for generality we sometimes allow the input to be at a frequency $n\omega_0$, for an integer n, so that dividers and rational-fraction multipliers are covered. The analyses are restricted to abrupt-junction varactors (except in Sections 8.2 and 8.11), but they are probably reasonably close for graded-junction varactors.‡

In varactor multipliers, the current and voltage (and therefore the elastance also) are periodic in time, with a fundamental frequency ω_0. In Section 8.2 we give the basic equation (including the frequency-domain form) for varactor multipliers.

In Section 8.3 it is shown that among the abrupt-junction varactor multipliers only the doubler can operate without current flow at one or more frequencies besides the input and output frequencies. These idler currents are essential if any output is to be obtained from the multiply-by-three, -four, -five, and so on, circuits. Although the argument is not strictly true for graded-junction varactors, it is still pertinent in the sense that without these idlers the efficiency of high-order multipliers is unnecessarily small.

It is easy to determine which idler frequencies are required. Consider first the varactor driven only by the input current at frequency ω_0. The voltage is proportional to the square of the charge,‡‡ so that ac voltage is generated at frequency $2\omega_0$. If the varactor is terminated at this frequency, current can flow and power can be delivered to a load. Also, spontaneous phase-locked oscillations can be set up at frequency $\omega_0/2$ by this current (see Chapter 9). A current at any other frequency cannot interact with the current at frequency ω_0, and no power can be delivered by the varactor. Thus only the doubler and the divide-by-two circuit can operate without idlers.

---

‡See Appendix F for the results of an analysis of a graded-junction doubler.

‡‡This is a property of abrupt-junction varactors.

Now suppose that current is known to flow at two or more frequencies. Current at another frequency will interact so as to deliver power to a load only if the new frequency is the sum of two of the previous frequencies, or is the difference between two of the previous frequencies, or is either twice or one half one of the previous frequencies. Thus, for example, if current is known to flow at ω_o and $2\omega_o$, power can be delivered at any of three new frequencies, namely, $\omega_o/2$, $3\omega_o$, and $4\omega_o$. Similarly, if current flows at ω_o and $\omega_o/2$, then power can be delivered by the varactor to a load at any of the three frequencies $\omega_o/4$, $3\omega_o/2$, or $2\omega_o$.

By extending this reasoning to apply to the simultaneous introduction of two or more new frequencies, and by using it repeatedly, we can tabulate possible idler configurations. In Section 8.3.2 we list all possible configurations of two frequencies (two in number), three frequencies (five in number), and four frequencies (nineteen in number).

If an output at a particular frequency is desired, these lists can be searched to find idler configurations that work. With no idlers, power output can be obtained at $\omega_o/2$ and $2\omega_o$; with one idler, at four additional frequencies: $\omega_o/4$, $3\omega_o/2$, $3\omega_o$, and $4\omega_o$. With two idlers, output can be obtained at, besides these frequencies, the seven frequencies $\omega_o/8$, $3\omega_o/4$, $5\omega_o/4$, $5\omega_o/2$, $5\omega_o$, $6\omega_o$, and $8\omega_o$. Fourteen additional frequencies (see Table 8.2) can be reached by use of three idlers. Typical idler configurations are given in Table 8.3.

In several cases there is a choice of which idlers to use. For example, the quintupler can have idlers at $2\omega_o$ and $3\omega_o$, or at $2\omega_o$ and $4\omega_o$, or at $2\omega_o$, $3\omega_o$, and $4\omega_o$ simultaneously. In general, an intelligent choice of idlers cannot be made without actually solving the pertinent equations and comparing the performance.

In Section 8.4 we cover in detail the doubler (multiply-by-two circuit). For simplicity, the discussion is restricted to abrupt-junction varactors, for which most of the expressions can be given in closed form.‡ In addition, we assume that current flows at only two frequencies, the input and the output. In spite of these restrictions, the theory is useful because it probably holds approximately for other varactors, and because the results are probably not severely affected by small currents at $3\omega_o$, $4\omega_o$, and so on. Expressions are derived (Equations 8.33 and 8.29) for the input and load resistances of the multiplier in terms of the modulation ratios m_1 and m_2, defined as

$$m_k = \frac{|S_k|}{S_{max} - S_{min}}, \tag{8.1}$$

‡The graded-junction-varactor doubler is discussed in Appendix

where S_k is the Fourier coefficient of elastance, in the formula

$$S(t) = \sum_{k=-\infty}^{+\infty} S_k e^{jk\omega_o t}, \tag{8.2}$$

and where S_{min} and S_{max} are the varactor minimum and maximum elastances. In addition, expressions are derived for input, output, and dissipated power, for efficiency, and for bias voltage. If the output load is tuned, the efficiency is the greatest (for given values of m_1 and m_2), so that for simplicity we assume such tuning. The formulas can be solved fairly easily by a graphical procedure. We plot contours of constant efficiency in the m_1,m_2-plane for a specified frequency of operation. These contours can then be read like a topographic map to find regions of high efficiency and regions of low efficiency. On the same plane, the various powers and the input and load resistances can also be plotted. Design of a doubler then consists in part of searching this plane for the desired efficiency, powers, and the like.

It is found that as m_1 and m_2 increase, the efficiency and the powers increase. Thus, high values of m_1 and m_2 seem to be desirable. However, the varactor elastance is restricted to lie between S_{min} and S_{max} at all times, so that arbitrarily high values of m_1 and m_2 cannot be achieved; the restriction, more precisely, is that values of m_1 and m_2 to the left of a certain curve in the m_1,m_2-plane are achievable, and points to the right are not. Therefore, for practical purposes only a portion of the m_1,m_2-plane need be consulted.

In Appendix D many plots like this are given, in an effort to make this technique practical.‡ These plots hold for values of ω_o, the input frequency, from $10^{-4}\omega_c$ to $0.1\omega_c$. From the plots the maximum possible efficiency and the maximum possible power output for each frequency can easily be determined. In addition, other questions can be answered. For example, for a given power input, what is the maximum efficiency? For a given value of load resistance, what range of efficiency can be achieved?

The operating points corresponding to maximum efficiency are of particular importance. It is shown that for practical purposes efficiency and power output are optimized simultaneously. As a function of frequency ω_o, we can determine the maximum efficiency and related operating characteristics; see Figures 8.6 to 8.9. For low ω_o, the efficiency is high, close to 100 per

---

‡Also given in Appendix D are similar plots for the divide-by-two circuit, which is discussed in Section 9.4.

cent, and for high values of ω_o it approaches zero. For low frequencies, the maximum efficiency is approximately

$$\epsilon \approx 1 - \frac{20\omega_o}{\omega_c}; \tag{8.3}$$

this formula is valid for efficiencies higher than about 90 per cent. For high frequencies, the efficiency becomes asymptotic to the curve,

$$\epsilon \approx 0.0039 \frac{\omega_c^2}{\omega_o^2}; \tag{8.4}$$

this formula is valid for efficiencies below about 10 per cent. The operating point corresponding to maximum efficiency is very important, and probably the majority of doublers should be designed to operate at or close to this operating point. Nomographs to facilitate the design of optimum-efficiency doublers (and higher-order multipliers) are given in Appendix D.

Like most of the analyses in this book, the treatment of the doubler leads primarily to fundamental limits on operation. No specific circuit is assumed; however, the analysis does indicate those properties of the circuit that are important in achieving the limits predicted. Synthesis of doublers, based on these conditions, is discussed in Section 8.4.4.

The doubler is the simplest multiplier because it is the only one with no idler. The doubler is therefore treated in much more detail than any of the higher-order multipliers. The tripler and quadrupler require one idler, and the quintupler, sextupler, and octupler require two. These five multipliers are treated (in Sections 8.6 to 8.10) in less detail; only the conditions for maximum efficiency and maximum power output are stressed.

In Section 8.5 we derive general formulas for idler and load resistances, input resistance, efficiency, bias voltage, and input, output, and dissipated powers for abrupt-junction-varactor multipliers, without specifying the idler configuration. By merely specifying which idler currents are present, this analysis can easily be made to apply to specific multipliers, such as the doubler, the tripler, and so on. The analysis in Section 8.5 is, in fact, so general that it also applies to dividers and to rational-fraction generators. The formulas for specific multipliers, used in Sections 8.6 to 8.10, are obtained from these more general formulas.

Two multipliers, the tripler and the quadrupler, require only one idler, at frequency $2\omega_o$. These multipliers are discussed in detail in Sections 8.6 and 8.7. Suppose the load and idler resistances are given. Then the efficiency and all the powers depend on the drive level. If the drive level is quite low, the nonlinearities of the elastance are not manifest, and conversion of power is quite limited. As the drive level is increased, the efficiency increases; therefore, it is of interest to drive as hard as possible, that is, to set the drive level so that the elastance traverses the full range from S_{min} to S_{max}. The calculation of the resulting efficiency and powers is very difficult, but it can be done with digital computers. Such a computer program has been developed by B. L. Diamond of M. I. T. Lincoln Laboratory, and Diamond's results[3] are given‡ in Sections 8.6 and 8.7. Highest efficiency is found when the external idler resistance is zero, as would be expected intuitively. In Sections 8.6 and 8.7 are plots of the maximum efficiency and power output, and plots of input and load resistances necessary to attain these maxima. Enough information is given, in graphical form, to make possible the design of optimum triplers and quadruplers, at least to the extent of selecting the necessary varactor and the necessary operating conditions. Nomographs are given in Appendix D to aid in this process. Synthesis of triplers and quadruplers is discussed briefly in Sections 8.6.4 and 8.7.4.

A startling result is that the efficiency of varactor triplers and quadruplers can be quite high — almost as high, in fact, as that of the doubler, when compared on the basis of the same ratio of output frequency to cutoff frequency. At low frequencies, the maximum efficiencies of the tripler and quadrupler are approximately

$$\epsilon \approx 1 - 35 \frac{\omega_o}{\omega_c} \tag{8.5}$$

and

$$\epsilon \approx 1 - 62 \frac{\omega_o}{\omega_c}, \tag{8.6}$$

respectively, where ω_o is in each case the input frequency.

The quadrupler as treated in Section 8.7 has no idler at frequency $3\omega_o$. None is required, but on the other hand there is no evidence that it will not improve the performance. We have analyzed the simple quadrupler with only one idler merely because it is less complicated.

‡With his kind permission.

Three additional multipliers (the quintupler, sextupler, and octupler) can be built with only two idlers; these multipliers are discussed in Sections 8.8, 8.9, and 8.10. In each case, the formulas of Section 8.5 are simplified by assuming specific idler configurations (both 1-2-3-5 and 1-2-4-5 for the quintupler, 1-2-3-6 and 1-2-4-6 for the sextupler, and 1-2-4-8 for the octupler), and then solving numerically at specific frequencies for a variety of idler resistances and load resistances. Qualitatively, the results are not vastly different from the tripler and quadrupler results,‡ except that now there are two idler resistances to specify. In all cases,‡ the best efficiency is found when each idler has no loss (aside from the varactor series resistance); the efficiency is not far below the maximum achievable efficiency for the lower-order multipliers. At low frequencies the maximum efficiencies of the 1-2-4-5 quintupler, the 1-2-4-6 sextupler, and the octupler are approximately

$$\epsilon \approx 1 - 93 \frac{\omega_o}{\omega_c}, \tag{8.7}$$

$$\epsilon \approx 1 - 99 \frac{\omega_o}{\omega_c}, \tag{8.8}$$

and

$$\epsilon \approx 1 - 168 \frac{\omega_o}{\omega_c}, \tag{8.9}$$

respectively, where in each case ω_o is the input frequency.

There is no a priori way of telling what idler configuration is best for a given ratio of output to input frequencies. Both the quintupler and the sextupler can be built with two idlers, in two ways. The results in Sections 8.8 and 8.9 indicate that except in the high-frequency limit, the 1-2-4-5 quintupler is better, and that for the sextuplers higher efficiency and power output are usually achieved with idlers located at $2\omega_o$ and $4\omega_o$ rather than at $2\omega_o$ and $3\omega_o$. At low frequencies the difference between the two types of sextuplers is not great, but in the mid-frequency range the difference in efficiency is as much as 10 per cent. As yet there is no evidence that these configurations would not be improved by allowing current to flow at both $3\omega_o$ and $4\omega_o$; these more complicated three-idler multipliers have not been analyzed.

---

‡An exception is the 1-2-3-5 quintupler, for which more study is required.

Nomographs are given in Appendix D to aid in the design of quintuplers, sextuplers, and octuplers with maximum efficiency.

The six abrupt-junction multipliers discussed here (the doubler, tripler, quadrupler, quintupler, sextupler, and octupler) are the only multipliers we have solved explicitly; they are the only multipliers that require less than three idlers. More complicated multipliers are straightforward but quite tedious to analyze. For that reason, it is important to have a limitation on efficiency that holds regardless of what harmonic is being generated. In Section 8.11 we derive a fundamental limit on multiplier efficiency that holds for:

1. Any ratio of output to input frequencies.
2. Any elastance-charge characteristic.
3. Any number of idlers.
4. Any terminations (not necessarily open circuits) at the unwanted harmonics.

This limit also holds for dividers and rational-fraction multipliers. It holds for graded-junction and abrupt-junction varactors, as well as for all types to be developed in the future (provided that the model of Figure 4.1 is accurate).

Needless to say, this general limit, which depends only on the ratio of output frequency to cutoff frequency, is of wide applicability. Unfortunately, however, it is somewhat higher than the maximum achievable efficiencies of the multipliers that have been analyzed explicitly.‡ Thus for practical purposes the new limit, which is in general not achievable, is not as useful as the actual limits for the specific multipliers that have been analyzed. The new limit is of value only for those multipliers that have not been solved, and even then it is of interest only as an upper bound. The derivation of this fundamental limit is quite similar to the derivations in Sections 5.6 and 6.4.2 for the lowest noise figures of varactor upconverters and parametric amplifiers with arbitrary sideband terminations and arbitrary pumping waveforms.

Three other limits on varactor multipliers can be derived in a similar way. The first of these applies if there is no current at three times the output frequency; the limit on efficiency is slightly less than the previous limit. The second special case is for multipliers with no variation of the elastance at twice the

‡For example, let us compare this new limit with the maximum abrupt-junction doubler efficiency. For a given efficiency, the new limit indicates a certain maximum output frequency above which the given efficiency cannot be achieved. On the other hand, the exact analysis of the doubler indicates that this efficiency can really be achieved only below a frequency approximately 0.2 (0.35, at low efficiencies) times the frequency predicted by the new limit.

output frequency. The efficiency limit in this case is somewhat lower than the previous one. The third limit is appropriate for abrupt-junction multipliers with open-circuited harmonics higher than the output frequency. This new efficiency limit is much lower than the other three, and, in fact, with reasonable assumptions about the operation of the multiplier, the limit approaches very closely the achievable efficiencies found by actual calculation. These four limits are derived in Section 8.11.

Many specific multipliers are discussed in this chapter; it is interesting to compare them. All these multipliers are compared with respect to their maximum efficiencies and maximum power outputs in Section 8.12. Figure 8.60 shows all the efficiency limits, including those of the dividers analyzed in Chapter 9. It is evident that all the efficiencies approach 100 per cent at low frequencies and fall off rapidly at high frequencies, as of course would be expected. What is unexpected, however, is that the higher-order multipliers are capable of almost as high an efficiency as the simple doubler when the multipliers are compared on the basis of the same output frequency.

In a similar way, all the maximum power outputs are shown in Figure 8.61. Note that at low frequencies the power outputs of all the multipliers are very close when compared on the basis of the same input frequency.

There are two topics of great importance that we do not discuss in Chapter 8. These are bandwidth and noise. However, the analyses given can probably be used as a foundation upon which to build a description of the noise and bandwidth properties of varactor multipliers.

8.2. Basic Multiplier Equations

Varactor frequency multipliers (and dividers and rational-fraction generators) have voltages, currents, and elastances that are periodic functions of time and, therefore, have frequency components at multiples of some fundamental frequency ω_0. We suppose that the input is at frequency $n\omega_0$ and the output is at frequency $\ell\omega_0$. In general, there are components of current, voltage, and elastance at other harmonics (idlers) $k\omega_0$, although in some special cases some frequency components may, by design or by accident, vanish.

In the frequency domain, the large-signal varactor equation

$$v(t) = R_s i(t) + \int S(t) i(t)\, dt + e_n \tag{8.10}$$

becomes Equations 4.31:

$$
\begin{bmatrix} \vdots \\ V_k^* \\ \vdots \\ V_2^* \\ V_1^* \\ V_1 \\ V_2 \\ \vdots \\ V_k \\ \vdots \end{bmatrix}
=
\begin{bmatrix}
\ddots & \vdots & & \vdots & \vdots & \vdots & \vdots & & \vdots & \\
\cdots & R_s - \frac{S_0}{jk\omega_o} & \cdots & -\frac{S_{k-2}^*}{jk\omega_o} & -\frac{S_{k-1}^*}{jk\omega_o} & -\frac{S_{k+1}^*}{jk\omega_o} & -\frac{S_{k+2}^*}{jk\omega_o} & \cdots & -\frac{S_{2k}^*}{jk\omega_o} & \cdots \\
 & \vdots & \ddots & \vdots & \vdots & \vdots & \vdots & & \vdots & \\
\cdots & -\frac{S_{k-2}}{j2\omega_o} & \cdots & R_s - \frac{S_0}{j2\omega_o} & -\frac{S_1^*}{j2\omega_o} & -\frac{S_3^*}{j2\omega_o} & -\frac{S_4^*}{j2\omega_o} & \cdots & -\frac{S_{k+2}^*}{j2\omega_o} & \cdots \\
\cdots & -\frac{S_{k-1}}{j\omega_o} & \cdots & -\frac{S_1}{j\omega_o} & R_s - \frac{S_0}{j\omega_o} & -\frac{S_2^*}{j\omega_o} & -\frac{S_3^*}{j\omega_o} & \cdots & -\frac{S_{k+1}^*}{j\omega_o} & \cdots \\
\cdots & \frac{S_{k+1}}{j\omega_o} & \cdots & \frac{S_3}{j\omega_o} & \frac{S_2}{j\omega_o} & R_s + \frac{S_0}{j\omega_o} & \frac{S_1^*}{j\omega_o} & \cdots & \frac{S_{k-1}^*}{j\omega_o} & \cdots \\
\cdots & \frac{S_{k+2}}{j2\omega_o} & \cdots & \frac{S_4}{j2\omega_o} & \frac{S_3}{j2\omega_o} & \frac{S_1}{j2\omega_o} & R_s + \frac{S_0}{j2\omega_o} & \cdots & \frac{S_{k-2}^*}{j2\omega_o} & \cdots \\
 & \vdots & & \vdots & \vdots & \vdots & \vdots & \ddots & \vdots & \\
\cdots & \frac{S_{2k}}{jk\omega_o} & \cdots & \frac{S_{k+2}}{jk\omega_o} & \frac{S_{k+1}}{jk\omega_o} & \frac{S_{k-1}}{jk\omega_o} & \frac{S_{k-2}}{jk\omega_o} & \cdots & R_s + \frac{S_0}{jk\omega_o} & \cdots \\
 & \vdots & & \vdots & \vdots & \vdots & \vdots & & \vdots & \ddots
\end{bmatrix}
\begin{bmatrix} \vdots \\ I_k^* \\ \vdots \\ I_2^* \\ I_1^* \\ I_1 \\ I_2 \\ \vdots \\ I_k \\ \vdots \end{bmatrix}
+
\begin{bmatrix} \vdots \\ E_{nk}^* \\ \vdots \\ E_{n2}^* \\ E_{n1}^* \\ E_{n1} \\ E_{n2} \\ \vdots \\ E_{nk} \\ \vdots \end{bmatrix}
.
\tag{8.11}
$$

For generality we have included the noise terms in Equations 8.10 and 8.11, although from now on in Chapter 8 we ignore noise. In Equations 8.11, we have omitted the formula for the bias voltage V_0. We have assumed that all S_k and all I_k are nonzero; in case some currents are missing, Equations 8.11 are simpler.

In general, any multiplier analysis begins with Equations 8.11. The solution of these equations is very difficult because they are nonlinear (the S_k depend on the I_k in a complicated way). In Chapter 8, however, we solve only abrupt-junction multipliers. Since the elastance is proportional to the charge, the relationship between the S_k and the I_k is fairly simple, so that Equations 8.11 are simplified at the outset.

In Section 8.4 we use Equations 8.11 (without the noise terms) to describe the doubler. However, in Section 8.5, where we derive formulas for higher-order multipliers, we use instead the voltage-charge relationship directly.

8.3. Idlers

We treat in Chapter 8 only abrupt-junction multipliers because of the simplicity of the charge-elastance relationship. For most such multipliers idlers are required, as explained in Section 8.3.1. A tabulation of possible idler configurations is given in Section 8.3.2.

8.3.1. Idler Requirement

The abrupt-junction varactor is a square-law device over its whole range, because the voltage is proportional to the square of the charge.‡ Thus, if the only current flowing is at frequency ω_0, a dc component of voltage is produced, and components at frequencies ω_0 and $2\omega_0$. No other frequencies are generated by this current‡‡ (except for subharmonic generation at frequency $\omega_0/2$).

To generalize this result somewhat, suppose that current at only two frequencies, the input $n\omega_0$ and the output $\ell\omega_0$, flows through the varactor. Then, Equations 8.11 reduce to

$$\begin{bmatrix} E_\ell^* \\ E_n^* \\ E_n \\ E_\ell \end{bmatrix} = \begin{bmatrix} R_s - \dfrac{S_0}{j\ell\omega_0} & -\dfrac{S_{\ell-n}^*}{j\ell\omega_0} & -\dfrac{S_{\ell+n}^*}{j\ell\omega_0} & -\dfrac{S_{2\ell}^*}{j\ell\omega_0} \\ -\dfrac{S_{\ell-n}}{jn\omega_0} & R_s - \dfrac{S_0}{jn\omega_0} & -\dfrac{S_{2n}^*}{jn\omega_0} & -\dfrac{S_{\ell+n}^*}{jn\omega_0} \\ \dfrac{S_{\ell+n}}{jn\omega_0} & \dfrac{S_{2n}}{jn\omega_0} & R_s + \dfrac{S_0}{jn\omega_0} & \dfrac{S_{\ell-n}^*}{jn\omega_0} \\ \dfrac{S_{2\ell}}{j\ell\omega_0} & \dfrac{S_{\ell+n}}{j\ell\omega_0} & \dfrac{S_{\ell-n}}{j\ell\omega_0} & R_s + \dfrac{S_0}{j\ell\omega_0} \end{bmatrix} \begin{bmatrix} I_\ell^* \\ I_n^* \\ I_n \\ I_\ell \end{bmatrix}, \tag{8.12}$$

where ℓ is not necessarily greater than n. Because the varactor is an abrupt-junction type, the only elastance coefficients present are $S_{-\ell}$, S_{-n}, S_0, S_n, and S_ℓ. If both of the coefficients $S_{\ell+n}$ and $S_{\ell-n}$ are zero, then Equations 8.12 reduce to two separate sets of equations for frequencies $n\omega_0$ and $\ell\omega_0$, without coupling. Hence either $S_{\ell+n}$ or $S_{\ell-n}$ must be nonzero, so that either $\ell + n$ or $\ell - n$ must be equal to $-\ell$, $-n$, ℓ, or n. The only possibility (since $\ell \neq 0$ and $n \neq 0$) is that $\ell = 2n$ or $n = 2\ell$. We have established the principle that with no idler an abrupt-junction varactor can generate only twice or one half the input frequency. Abrupt-junction varactors will not operate in

‡We assume (as usual) that the varactor is not driven outside the region in which this is true. That is, the varactor is not overdriven to produce elastance harmonics.

‡‡Therefore, for example, an abrupt-junction tripler with current at only frequencies ω_0 and $3\omega_0$ is impossible. To make a tripler, current must be allowed to flow at some other frequency, such as $2\omega_0$.

tripler or higher-order multipliers without idling currents flowing at one or more frequencies other than the input and output.

This type of reasoning can be easily extended to more complicated idler configurations. If current flows at a number of frequencies, we may ask what additional frequencies can be generated. First, suppose we wish to add one more frequency. Then similar reasoning shows that this frequency must be related in at least one of the following ways to the frequencies of the currents already present:

1. Sum of two frequencies.
2. Difference between two frequencies.
3. Twice some frequency.
4. Half some frequency.

Otherwise, the matrix equation similar to Equations 8.12 breaks up into two unconnected sets of equations, one of which includes only the new frequency. For example, if current flows at ω_o and $2\omega_o$, current can be made to flow at $\omega_o/2$, $3\omega_o$, or $4\omega_o$ but at no other single frequencies.

On the other hand, suppose we wish to add simultaneously two frequencies to a set of frequencies at which current flows through the varactor. And suppose that neither of these frequencies, $\alpha\omega_o$ and $\beta\omega_o$, is the sum of or difference between two of the existing frequencies, or is half or double any of the existing frequencies. Under what conditions can the two new frequencies be introduced simultaneously, when each individually is not acceptable? We shall look at the power delivered by the varactor at these two new frequencies, and we shall require that it be positive. We shall also consider the phase coherence of the two newly generated frequencies and require that each new output shall have a phase that cannot continuously vary; that is, each new output shall have a frequency accurately related to the existing frequencies.

Consider the power output at frequency $\alpha\omega_o$. It is

$$P_\alpha = -2R_s |I_\alpha|^2 - 2\,\mathrm{Re} \sum_{k=-\infty}^{\infty} \frac{S_{\alpha-k} I_\alpha^* I_k}{j\alpha\omega_o}, \tag{8.13}$$

where we have used Equations 8.11. A similar formula holds for the power output at frequency $\beta\omega_o$. For nonzero output power, at least one term in the summation of Equation 8.13 must be present; since S_k is nonzero if and only if I_k is nonzero, the requirement is that, for some k, both $I_{\alpha-k}$ and I_k be present. Thus $\alpha\omega_o$ is the sum or difference of two other frequencies. Since by assumption $\alpha\omega_o$ alone cannot be generated, one of these is $\beta\omega_o$. Therefore, either

$$\alpha - \beta = k, \tag{8.14}$$

or

$$\alpha + \beta = r, \tag{8.15}$$

or both,‡ where current already flows at frequencies $k\omega_o$ or $r\omega_o$.

First, suppose that Equation 8.14 holds but that Equation 8.15 does not. (For example, we might try to build a quadrupler with just an idler at $3\omega_o$.) The summation in Equation 8.13 reduces to two terms, which can be combined to yield

$$P_\alpha = -2R_s|I_\alpha|^2 - 2\,\mathrm{Re}\,\frac{I_\beta S_k I_\alpha^*}{j\beta\omega_o}; \tag{8.16}$$

similarly,

$$P_\beta = -2R_s|I_\beta|^2 + 2\,\mathrm{Re}\,\frac{I_\beta S_k I_\alpha^*}{j\alpha\omega_o}, \tag{8.17}$$

where we have used the fact that for abrupt-junction varactors all I_k are proportional to $jk\omega_o S_k$. Equations 8.16 and 8.17 show that if Equation 8.15 does not hold, P_α and P_β can never be positive together.

Now suppose that Equation 8.14 does not hold but that Equation 8.15 does. An example of this sort is generation of power at frequencies $\omega_o/3$ and $2\omega_o/3$ from an input at frequency ω_o. The powers are

$$P_\alpha = -2R_s|I_\alpha|^2 + 2\,\mathrm{Re}\,\frac{jI_\alpha I_\beta S_r^*}{\beta\omega_o} \tag{8.18}$$

and

$$P_\beta = -2R_s|I_\beta|^2 + 2\,\mathrm{Re}\,\frac{jI_\alpha I_\beta S_r^*}{\alpha\omega_o}. \tag{8.19}$$

It is now possible to get power out at each of the two new frequencies, but the phases of I_α and I_β are not individually constrained in the power relations or in any of the other relations. Only the sum of the two phases appears. There is no physical mechanism to prevent the frequency $\alpha\omega_o$ from dropping a little

‡Without loss of generality we suppose that α, β, k, and r are positive.

bit and the frequency $\beta\omega_0$ from simultaneously rising the same amount. Since the output phase is not coherent with the input phase, the new frequencies are not stable. This may be an important type of oscillation, but it is not useful for frequency control and comparison, and we do not consider it to be frequency multiplication, division, or rational-fraction multiplication, because the ratio of the new frequency to the existing one(s) is not forced to be rational. Thus, for obtaining possible idler configurations, we do not consider the simultaneous introduction of two new frequencies related only by Equation 8.15.

Finally, let us look at the case where the two new frequencies obey both Equations 8.14 and 8.15. An example of this sort is generation of frequencies $3\omega_0/2$ and $5\omega_0/2$ from existing currents at frequencies ω_0, $2\omega_0$, and $4\omega_0$. Neither of the two new frequencies could be generated individually, but together they can be, in a phase-sensitive way. Since both the sum and the difference of the two new phases are constrained by the existing currents, the new oscillations grow with phases that are at least partially fixed by phases of the existing currents. The power to sustain these oscillations comes, crudely speaking, from the fourth harmonic, while the fundamental supplies additional phase information. In summary then, currents at two additional frequencies can be simultaneously introduced into an existing idler configuration only if both Equations 8.14 and 8.15 are obeyed.

This reasoning can be extended to cover more complicated generation schemes. Using the rules obtained thus far (which are adequate for our purposes), we list in Section 8.3.2:

1. All idler configurations with four or less frequencies.
2. All possible output frequencies with 0, 1, 2, or 3 idlers.
3. Typical idler configurations to generate specific desired harmonics, subharmonics, and rational fractions.

For generality these lists, in Tables 8.1, 8.2, and 8.3, apply to dividers and rational-fraction generators as well as to multipliers. We exclude any generators for which the output is not coherent in phase with the input.

8.3.2. Typical Idler Configurations

Using the rules developed in Section 8.3.1, we have found all possible idler configurations for abrupt-junction varactors, with power input at frequency ω_0 and current at either two, three, or four frequencies. These are listed in Table 8.1. The notation used is that of multiples of ω_0; thus, for example, 1 - 2 means currents at only frequencies ω_0 and $2\omega_0$, and $\frac{1}{2}$ - 1 - 2 - 3 means currents only at frequencies $\omega_0/2$, ω_0, $2\omega_0$, and $3\omega_0$.

Table 8.1 indicates the frequencies that can be produced with a limited number of idler currents. If there is no idler, then

Table 8.1. Possible Idler Configurations

Listed are all possible idler configurations for phase-coherent abrupt-junction varactor multipliers with four or less frequencies at which current flows. The configurations with double daggers are those that are analyzed in Chapters 8 and 9 of this book.

Two Frequencies	1 - 2‡	$\frac{1}{2}$ - 1‡	
Three Frequencies	1 - 2 - 3‡	$\frac{1}{2}$ - 1 - 2	$\frac{1}{4}$ - $\frac{1}{2}$ - 1‡
	1 - 2 - 4‡	$\frac{1}{2}$ - 1 - $\frac{3}{2}$	
Four Frequencies	1 - 2 - 3 - 4	$\frac{1}{2}$ - 1 - 2 - 3	$\frac{1}{4}$ - $\frac{1}{2}$ - 1 - 2
	1 - 2 - 3 - 5‡	$\frac{1}{2}$ - 1 - 2 - 4	$\frac{1}{4}$ - $\frac{1}{2}$ - 1 - $\frac{3}{2}$
	1 - 2 - 3 - 6‡	$\frac{1}{2}$ - 1 - $\frac{3}{2}$ - 2	$\frac{1}{4}$ - $\frac{1}{2}$ - $\frac{3}{4}$ - 1
	1 - 2 - 4 - 5‡	$\frac{1}{2}$ - 1 - $\frac{3}{2}$ - $\frac{5}{2}$	$\frac{1}{4}$ - $\frac{1}{2}$ - 1 - $\frac{5}{4}$
	1 - 2 - 4 - 6‡	$\frac{1}{2}$ - 1 - $\frac{3}{2}$ - 3	$\frac{1}{8}$ - $\frac{1}{4}$ - $\frac{1}{2}$ - 1
	1 - 2 - 4 - 8‡	$\frac{1}{2}$ - 1 - 2 - $\frac{5}{2}$	
	1 - $\frac{3}{2}$ - 2 - 3	$\frac{1}{2}$ - $\frac{3}{4}$ - 1 - $\frac{3}{2}$	

(with input at frequency ω_0) power can be delivered only at the two frequencies $\omega_0/2$ and $2\omega_0$. In a similar way, with one idler four additional frequencies can be reached. A list of those frequencies achievable with 0, 1, 2, and 3 idlers is given in Table 8.2. As in Table 8.1, the notation 2 indicates possible power output at frequency $2\omega_0$, and so forth.

When designing a multiplier with a specific ratio of output to input frequency, it is important to know what idler configurations can be used. Although this information is obtainable from Table 8.1, it is convenient to have a table at hand to show which idler configurations are suitable for each ratio of output to input frequency. Such a list is given in Table 8.3 for many multipliers, dividers, and rational-fraction generators. Included are all integral multipliers that require no, one, two, three, or four idlers In most cases there are several possible idler configurations; we have shown all possible configurations with the minimum number of idlers along with all other configurations (if any) that re-

Table 8.2. All Frequencies Attainable with Abrupt-Junction Varactors with 0, 1, 2, or 3 Idlers

In each case, of course, frequencies attainable using a smaller number of idlers are also possible. The double daggers indicate multipliers (or dividers) treated explicitly in Chapters 8 and 9.

No Idlers	$\frac{1}{2}$‡, 2‡
One Idler	$\frac{1}{4}$‡, $\frac{3}{2}$, 3‡, 4‡
Two Idlers	$\frac{1}{8}$, $\frac{3}{4}$, $\frac{5}{4}$, $\frac{5}{2}$, 5‡, 6‡, 8‡
Three Idlers	$\frac{1}{16}$, $\frac{3}{8}$, $\frac{5}{8}$, $\frac{7}{8}$, $\frac{9}{8}$, $\frac{7}{4}$, $\frac{9}{4}$, $\frac{7}{2}$, $\frac{9}{2}$, 7, 9, 10, 12, 16

quire three or less idlers.‡ There is as yet no a priori rule for determining in each case which of the several configurations listed is best.

For the dividers and rational-fraction generators, we show in Table 8.3 all that require no, one, two, or three idlers. Again, in most cases there are several possible idler configurations; we have shown all possible configurations with the minimum number of idlers, as well as all other configurations that require two or less idlers.‡‡ Again, there is as yet no way of determining which idler configuration is best, without an actual solution of each case in detail.

In Table 8.3 the nomenclature is similar to that in Tables 8.1 and 8.2; thus, for example, 1-2-3 indicates current at frequencies ω_0 (input), $2\omega_0$, and $3\omega_0$ only.

For simplicity we have omitted many configurations that can be formed from the configurations in Table 8.3 by simply putting in additional frequencies. For example, a possible quadrupler configuration is 1-2-3-4, although we have not listed it in Table 8.3 because it can be made from a simpler configuration (1-2-4) by adding one more frequency. We do not mean to imply by not listing these additional configurations that improvement cannot be obtained by having current flow at the additional frequencies.

‡These occur for the ×3, ×4, ×5, ×6, and ×8 circuits.

‡‡The only such case is that of the ×3/2 circuit.

×2	1-2‡		
×3	1 2-3‡	1-$\frac{1}{2}$-$\frac{3}{2}$-3	
×4	1-2-4‡	1-$\frac{1}{2}$-$\frac{3}{2}$-$\frac{5}{2}$-4	1-$\frac{1}{2}$-$\frac{3}{2}$-3-4
×5	1-2-3-5‡ 1-2-4-5‡	1-$\frac{1}{2}$-2-$\frac{5}{2}$-5 1-$\frac{1}{2}$-$\frac{3}{2}$-$\frac{5}{2}$-5	
×6	1-2-3-6‡	1-2-4-6‡	1-$\frac{1}{2}$-$\frac{3}{2}$-3-6
×7	1-2-3-4-7 1-2-3-5-7	1-2-3-6-7 1-2-4-5-7	1-2-4-6-7 1-2-4-8-7
×8	1-2-4-8‡	1-2-3-5-8	1-2-3-6-8
×9	1-2-3-6-9	1-2-4-5-9	1-2-4-8-9
×10	1-2-3-5-10 1-2-4-5-10	1-2-4-6-10 1-2-4-8-10	
×11	1-2-3-4-7-11 1-2-3-4-8-11 1-2-3-5-6-11 1-2-3-5-8-11 1-2-3-5-10-11 1-2-3-6-8-11	1-2-3-6-9-11 1-2-3-6-12-11 1-2-4-5-6-11 1-2-4-5-7-11 1-2-4-5-9-11 1-2-4-5-10-11	1-2-4-6-7-11 1-2-4-6-10-11 1-2-4-6-12-11 1-2-4-8-9-11 1-2-4-8-10-11 1-2-4-8-12-11
×12	1-2-3-6-12	1-2-4-6-12	1-2-4-8-12
×13	1-2-3-5-8-13 1-2-3-5-10-13 1-2-3-6-7-13 1-2-3-6-12-13	1-2-4-5-8-13 1-2-4-5-9-13 1-2-4-6-7-13	1-2-4-6-12-13 1-2-4-8-9-13 1-2-4-8-12-13
×14	1-2-3-4-7-14 1-2-3-5-7-14 1-2-3-6-7-14 1-2-3-6-8-14 1-2-3-6-12-14	1-2-4-5-7-14 1-2-4-5-9-14 1-2-4-5-10-14 1-2-4-6-7-14 1-2-4-6-8-14 1-2-4-6-10-14	1-2-4-6-12-14 1-2-4-8-7-14 1-2-4-8-10-14 1-2-4-8-12-14 1-2-4-8-16-14
×15	1-2-3-5-10-15 1-2-3-6-9-15	1-2-3-6-12-15 1-2-4-5-10-15	1-2-4-8-7-15 1-2-4-8-16-15
×16	1-2-4-8-16		
×17	1-2-4-8-9-17	1-2-4-8-16-17	
×18	1-2-3-6-9-18 1-2-3-6-12-18	1-2-4-5-9-18 1-2-4-6-12-18	1-2-4-8-9-18 1-2-4-8-10-18 1-2-4-8-16-18
×20	1-2-3-5-10-20 1-2-4-5-10-20	1-2-4-6-10-20 1-2-4-8-10-20	1-2-4-8-12-20 1-2-4-8-16-20
×24	1-2-3-6-12-24 1-2-4-6-12-24	1-2-4-8-12-24 1-2-4-8-16-24	
×32	1-2-4-8-16-32		

Table 8.3. Idler Configurations (Left and Below)

Listed are possible idler configurations for multipliers, dividers, and rational-fraction generators with specific ratios of output to input frequency. For a listing of what is included, see text. The idler configurations marked with double daggers are those that are analyzed in Chapters 8 and 9.

$\times\frac{1}{2}$		$1-\frac{1}{2}$ ‡	
$\times\frac{3}{2}$	$1-\frac{1}{2}-\frac{3}{2}$		$1-2-3-\frac{3}{2}$
$\times\frac{5}{2}$	$1-\frac{1}{2}-2-\frac{5}{2}$		$1-\frac{1}{2}-\frac{3}{2}-\frac{5}{2}$
$\times\frac{7}{2}$	$1-2-3-\frac{3}{2}-\frac{7}{2}$	$1-\frac{1}{2}-\frac{3}{2}-2-\frac{7}{2}$ $1-\frac{1}{2}-\frac{3}{2}-\frac{5}{2}-\frac{7}{2}$ $1-\frac{1}{2}-\frac{3}{2}-3-\frac{7}{2}$	$1-\frac{1}{2}-2-\frac{5}{2}-\frac{7}{2}$ $1-\frac{1}{2}-2-3-\frac{7}{2}$ $1-\frac{1}{2}-2-4-\frac{7}{2}$
$\times\frac{9}{2}$	$1-2-3-\frac{3}{2}-\frac{9}{2}$ $1-\frac{1}{2}-\frac{3}{2}-3-\frac{9}{2}$		$1-\frac{1}{2}-2-\frac{5}{2}-\frac{9}{2}$ $1-\frac{1}{2}-2-4-\frac{9}{2}$
$\times\frac{1}{4}$		$1-\frac{1}{2}-\frac{1}{4}$ ‡	
$\times\frac{3}{4}$	$1-\frac{1}{2}-\frac{1}{4}-\frac{3}{4}$		$1-\frac{1}{2}-\frac{3}{2}-\frac{3}{4}$
$\times\frac{5}{4}$		$1-\frac{1}{2}-\frac{1}{4}-\frac{5}{4}$	
$\times\frac{7}{4}$	$1-\frac{1}{2}-\frac{1}{4}-\frac{3}{4}-\frac{7}{4}$ $1-\frac{1}{2}-\frac{1}{4}-\frac{5}{4}-\frac{7}{4}$	$1-\frac{1}{2}-\frac{1}{4}-\frac{3}{2}-\frac{7}{4}$ $1-\frac{1}{2}-\frac{1}{4}-2-\frac{7}{4}$	$1-\frac{1}{2}-\frac{3}{2}-\frac{3}{4}-\frac{7}{4}$
$\times\frac{9}{4}$	$1-\frac{1}{2}-\frac{1}{4}-\frac{5}{4}-\frac{9}{4}$	$1-\frac{1}{2}-\frac{1}{4}-2-\frac{9}{4}$	$1-\frac{1}{2}-\frac{3}{2}-\frac{3}{4}-\frac{9}{4}$
$\times\frac{1}{8}$		$1-\frac{1}{2}-\frac{1}{4}-\frac{1}{8}$	
$\times\frac{3}{8}$	$1-\frac{1}{2}-\frac{1}{4}-\frac{1}{8}-\frac{3}{8}$ $1-\frac{1}{2}-\frac{1}{4}-\frac{3}{4}-\frac{3}{8}$		$1-\frac{1}{2}-\frac{3}{2}-\frac{3}{4}-\frac{3}{8}$ $1-\frac{1}{2}-\frac{1}{4}-\frac{3}{8}-\frac{5}{8}$
$\times\frac{5}{8}$	$1-\frac{1}{2}-\frac{1}{4}-\frac{1}{8}-\frac{5}{8}$	$1-\frac{1}{2}-\frac{1}{4}-\frac{5}{4}-\frac{5}{8}$	$1-\frac{1}{2}-\frac{1}{4}-\frac{3}{8}-\frac{5}{8}$
$\times\frac{7}{8}$		$1-\frac{1}{2}-\frac{1}{4}-\frac{1}{8}-\frac{7}{8}$	
$\times\frac{9}{8}$		$1-\frac{1}{2}-\frac{1}{4}-\frac{1}{8}-\frac{9}{8}$	
$\times\frac{1}{16}$		$1-\frac{1}{2}-\frac{1}{4}-\frac{1}{8}-\frac{1}{16}$	

8.4. Doubler

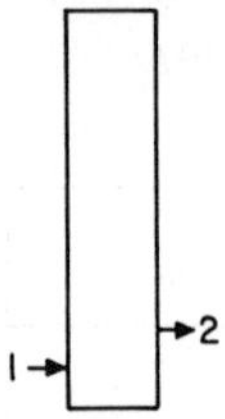

This analysis of the doubler, or multiply-by-two circuit, is restricted to abrupt-junction varactors,‡ for which solutions are obtainable in closed form. We assume further that current flows at only the two frequencies ω_0 (input) and $2\omega_0$ (output). With these restrictions, the nonlinear doubler can be solved exactly. We derive formulas for input and load resistance, input, output, and dissipated power, efficiency, and bias voltage. These formulas are then presented graphically, as an aid in designing doublers with prescribed efficiencies and power capabilities. From these graphs, it is possible to pick out points of highest efficiency and of highest power output; these two points, which are quite close together, are of great practical importance and are discussed in detail. Finally, we take up synthesis of doublers. Some typical (but by no means best possible) circuits are given.

8.4.1. Formulas

The steady-state operation of the doubler is described by the large-signal Equations 4.31 or 8.11. But because we assume the varactor has current flowing at only frequencies ω_0 (input) and $2\omega_0$ (output), all I_k and all S_k vanish except for $k = -2, -1, 1$, and 2 (S_0 is also present). If we neglect noise, Equations 4.31 or 8.11 reduce to

$$V_1 = \left(R_s + \frac{S_0}{j\omega_0}\right) I_1 + \frac{S_2}{j\omega_0} I_1^* + \frac{S_1^*}{j\omega_0} I_2, \tag{8.20}$$

$$V_2 = \left(R_s + \frac{S_0}{j2\omega_0}\right) I_2 + \frac{S_1}{j2\omega_0} I_1. \tag{8.21}$$

But because the elastance is proportional to the charge, which is the time integral of the current,

$$\frac{S_1}{S_2} = \frac{I_1}{j\omega_0}\frac{j2\omega_0}{I_2}$$

$$= 2\frac{I_1}{I_2}, \tag{8.22}$$

and similarly

‡For the graded-junction doubler, see Appendix F.

$$\frac{S_1^*}{S_2} = -2\,\frac{I_1^*}{I_2}. \qquad (8.23)$$

Equations 8.20 and 8.21 then become

$$V_1 = \left(R_s + \frac{S_0}{j\omega_o}\right) I_1 + \frac{S_1^* I_2}{j2\omega_o} \qquad (8.24)$$

and

$$V_2 = \left(R_s + \frac{S_0}{j2\omega_o}\right) I_2 + \frac{S_1^2 I_2}{4jS_2\omega_o}. \qquad (8.25)$$

Let us designate the impedance seen by the varactor at frequency $2\omega_o$ as $Z_2 = R_2 + jX_2$. Thus, if noise is neglected, the output load constraint is

$$V_2 = -Z_2 I_2, \qquad (8.26)$$

and therefore, Equation 8.21 becomes

$$0 = \left(Z_2 + R_s + \frac{S_0}{j2\omega_o}\right) I_2 + \frac{S_1}{j2\omega_o} I_1, \qquad (8.27)$$

and Equation 8.25 is similarly changed. This formula shows the importance of the quantity $Z_2 + R_s + (S_0/j2\omega_o)$. Let us call the phase angle of this quantity θ, so that

$$Z_2 + R_s + \frac{S_0}{j2\omega_o} = \left| Z_2 + R_s + \frac{S_0}{j2\omega_o} \right| e^{j\theta}$$

$$= (R_2 + R_s)\,\frac{e^{j\theta}}{\cos\theta}. \qquad (8.28)$$

We shall see that it is usually wise to tune the load to make $\theta = 0$, but for generality we derive the formulas that describe the doubler with θ unspecified.

8.4.1(a). Load Impedance. Upon substituting Equation 8.26 into Equation 8.25, eliminating I_2, and then taking the absolute values, we find

$$R_2 = R_s\left(\frac{m_1^2}{2m_2}\,\frac{\omega_c}{2\omega_o}\cos\theta - 1\right) \qquad (8.29)$$

by use of Equation 8.28. This formula relates the load impedance to the values of m_1 and m_2. Now, from Equation 8.28, we find a formula for the load impedance,

$$Z_2 = -R_s - \frac{S_0}{j2\omega_o} + R_s \frac{m_1^2}{2m_2} \frac{\omega_c}{2\omega_o} e^{j\theta}. \qquad (8.30)$$

8.4.1(b). Input Impedance. If we substitute the expression for I_2, Equation 8.27, into Equation 8.24, we find

$$V_1 = Z_{in} I_1, \qquad (8.31)$$

where

$$Z_{in} = R_s + \frac{S_0}{j\omega_o} + R_s \frac{\omega_c}{2\omega_o} 2m_2 e^{-j\theta}, \qquad (8.32)$$

in which we have used Equation 8.28. Note that Z_{in} is not the input impedance of the doubler, even though it is the ratio of V_1 to I_1. The circuit is inherently nonlinear, and Z_{in} depends upon the drive level (in particular both S_0 and m_2 change with the drive level). However, the ratio of input voltage to current, which we call Z_{in}, is a useful quantity, and we shall henceforth call it the "input impedance," with the understanding that it depends upon drive level and, therefore, is not an impedance in the usual sense.

The real part of Z_{in} is also useful; we shall call it the "input resistance," again with the understanding that it depends upon drive level. It is

$$R_{in} = \text{Re } Z_{in}$$

$$= R_s \left(1 + 2m_2 \frac{\omega_c}{2\omega_o} \cos\theta\right). \qquad (8.33)$$

8.4.1(c). Input Power. The input power can be calculated from the input resistance:

$$P_{in} = 2|I_1|^2 R_{in}. \qquad (8.34)$$

This can be related to the varactor normalization power. The maximum varactor voltage is V_B, and at breakdown the charge is Q_B and the elastance S_{max}. The minimum varactor voltage is V_{min} (which, if the minimum elastance S_{min} is zero, is the contact potential $-\phi$), and at this minimum voltage the charge is Q_{min} and the elastance S_{min}. The linear relationship between charge and elastance, which is characteristic of abrupt-junction varactors, is

$$\frac{S - S_{min}}{S_{max} - S_{min}} = \frac{q - Q_{min}}{Q_B - Q_{min}}, \tag{8.35}$$

which is the same as Equation 7.72. The range of voltage, $V_B - V_{min}$, is found from integrating the elastance:

$$V_B - V_{min} = \int_{Q_{min}}^{Q_B} S\,dq$$

$$= \frac{1}{2}(Q_B - Q_{min})(S_{max} + S_{min}). \tag{8.36}$$

The normalization power is, by definition,

$$P_{norm} = \frac{(V_B - V_{min})^2}{R_s}. \tag{8.37}$$

From Equation 8.35, the Fourier coefficient of current I_1 can be calculated as

$$I_1 = j\omega_0 Q_1$$

$$= j\omega_0 S_1 \frac{Q_B - Q_{min}}{S_{max} - S_{min}}$$

$$= j\omega_0 m_1 (Q_B - Q_{min}). \tag{8.38}$$

From Equations 8.36 and 8.38 we can derive the useful formula,

$$2|I_1|^2 R_s = 8P_{norm} \left(\frac{S_{max} - S_{min}}{S_{max} + S_{min}}\right)^2 \left(\frac{\omega_0}{\omega_c}\right)^2 m_1^2. \tag{8.39}$$

Similarly, for other values of k,

$$2|I_k|^2 R_s = 8P_{norm} \left(\frac{S_{max} - S_{min}}{S_{max} + S_{min}}\right)^2 \left(\frac{\omega_0}{\omega_c}\right)^2 k^2 m_k^2. \tag{8.40}$$

This quantity has the physical significance of the power dissipated in the varactor series resistance caused by the current at the kth harmonic.

Now Equations 8.33 and 8.39 can be used to help evaluate the input power, Equation 8.34:

$$P_{in} = 2P_{norm}\left(\frac{S_{max} - S_{min}}{S_{max} + S_{min}}\right)^2\left(\frac{2\omega_o}{\omega_c}\right)^2 m_1^2\left(1 + 2m_2\frac{\omega_c}{2\omega_o}\cos\theta\right). \tag{8.41}$$

In case the varactor has a negligible minimum elastance, the awkward factor involving S_{max} and S_{min} is replaced by 1.

8.4.1(d). Output Power. In a similar way, the output power is found to be

$$P_{out} = 2|I_2|^2R_2$$

$$= 8P_{norm}\left(\frac{S_{max} - S_{min}}{S_{max} + S_{min}}\right)^2\left(\frac{2\omega_o}{\omega_c}\right)^2 m_2^2\frac{R_2}{R_s}$$

$$= 8P_{norm}\left(\frac{S_{max} - S_{min}}{S_{max} + S_{min}}\right)^2\left(\frac{2\omega_o}{\omega_c}\right)^2 m_2^2\left(\frac{m_1^2}{2m_2}\frac{\omega_c}{2\omega_o}\cos\theta - 1\right). \tag{8.42}$$

8.4.1(e). Dissipated Power. The dissipated power can be calculated either of two ways. First, it is the I^2R loss in the series resistance,

$$P_{diss} = 2R_s(|I_1|^2 + |I_2|^2); \tag{8.43}$$

second, it is simply $P_{in} - P_{out}$. Calculated either way,

$$P_{diss} = 2P_{norm}\left(\frac{S_{max} - S_{min}}{S_{max} + S_{min}}\right)^2\left(\frac{2\omega_o}{\omega_c}\right)^2(m_1^2 + 4m_2^2). \tag{8.44}$$

8.4.1(f). Efficiency. The efficiency of a doubler can be calculated either of two ways. First, it is the ratio of power output to power input

$$\epsilon = \frac{P_{out}}{P_{in}}. \tag{8.45}$$

Second, consider the input and load resistances. The input resistance, Equation 8.33, consists of R_s plus a term that accounts for the power converted to the second harmonic. Thus, of the power input to the varactor, only a fraction $(R_{in} - R_s)/R_{in}$ is converted to frequency $2\omega_0$. And of the power that is converted, only a fraction $R_2/(R_2 + R_s)$ reaches the load; the remainder is dissipated in the series resistance. Therefore,

$$\epsilon = \frac{R_{in} - R_s}{R_{in}} \frac{R_2}{R_2 + R_s}. \tag{8.46}$$

Calculated either way,

$$\epsilon = \frac{\frac{\omega_c}{2\omega_0} \cos\theta - \frac{2m_2}{m_1^2}}{\frac{\omega_c}{2\omega_0} \cos\theta + \frac{1}{2m_2}}. \tag{8.47}$$

8.4.1(g). Bias Voltage. It is usually necessary to bias the varactor with some average voltage V_0. This can be calculated in terms of m_1 and m_2. For an abrupt-junction varactor the voltage v is proportional to the square of the elastance:

$$\frac{v + \phi}{V_B + \phi} = \left(\frac{S}{S_{max}}\right)^2, \tag{8.48}$$

where ϕ is the contact potential, not necessarily the minimum voltage V_{min}. The average value of Equation 8.48 gives a formula for the bias voltage

$$\frac{V_0 + \phi}{V_B + \phi} = \left(\frac{S_{max} - S_{min}}{S_{max}}\right)^2 (m_0^2 + 2m_1^2 + 2m_2^2). \tag{8.49}$$

8.4.1(h). Phase Condition. The formulas for input and load resistance, input and output power, and efficiency, all depend on the angle θ defined in Equation 8.28. If the load impedance is adjusted to tune out exactly the average elastance, then $\theta = 0$, and for given values of m_1, m_2, and ω_0, the efficiency is the highest. There is some reason then for tuning the load.

However, it is not clear that tuning leads to the highest possible efficiency. A limit on the modulation ratios m_1 and m_2, imposed by the maximum elastance, is derived in Section 8.4.2. This limit depends upon the angle θ. For $\theta \neq 0$, somewhat larger values of m_1 and m_2 are allowable. It is not known whether or not these larger values, together with a nonzero value for θ, can lead to higher efficiencies. Nevertheless, for simplicity we shall assume the load is tuned, so that $\theta = 0$, with the realization that there might be a slight improvement possible in some of the limits to be derived.

8.4.2. Breakdown Limit for the Modulation Ratios

The possible values of m_1 and m_2 are not unlimited. They are restricted because the elastance $S(t)$ must be between S_{min} and S_{max} at all times. We now wish to derive this restriction.

The elastance waveform depends upon the angles of S_1 and S_2, its Fourier coefficients. We decided, in Section 8.4.1(h), to let the angle θ be zero. Then it is clear from Equation 8.27 that if I_1 is chosen to be real and positive (with no loss of generality), then S_1 is imaginary, and therefore I_2 is real and positive. Thus, both S_1 and S_2 are imaginary, so that

$$\frac{S(t)}{S_{max} - S_{min}} = m_0 + 2m_1 \sin \omega_0 t + 2m_2 \sin 2\omega_0 t. \quad (8.50)$$

To fit in the largest possible values of m_1 and m_2 and still have $S_{min} \leq S(t) \leq S_{max}$, we must choose m_0 so that

$$S_0 = \frac{S_{max} + S_{min}}{2}, \quad (8.51)$$

in which case the restriction is that

$$m_1 \sin \omega_0 t + m_2 \sin 2\omega_0 t \leq 0.25, \quad \text{for all time } t. \quad (8.52)$$

To test for this inequality, we set its derivative to zero, finding the time t_0 at which the maximum $S(t)$ occurs:

$$0 = m_1 \cos \omega_0 t_0 + 2m_2 \cos 2\omega_0 t_0, \quad (8.53)$$

or, using trigonometric identities, we solve for $\cos \omega_0 t_0$:

$$\cos \omega_0 t_0 = \frac{m_1}{8m_2} \left(\sqrt{1 + 32 \frac{m_2^2}{m_1^2}} - 1 \right). \quad (8.54)$$

Then, Condition 8.52 becomes

$$4m_1 \left(\sin \omega_0 t_0 + \frac{m_2}{m_1} \sin 2\omega_0 t_0 \right) \leq 1. \quad (8.55)$$

By combining Equation 8.54 and Condition 8.55, we obtain a restriction that relates m_1 and m_1/m_2 in closed form:

$$m_1 < \frac{1}{\left(3 + \sqrt{1 + 32\frac{m_2^2}{m_1^2}}\right)\sqrt{\frac{1}{2} + \frac{m_1^2}{32m_2^2}\left(\sqrt{1 + 32\frac{m_2^2}{m_1^2}} - 1\right)}}. \tag{8.56}$$

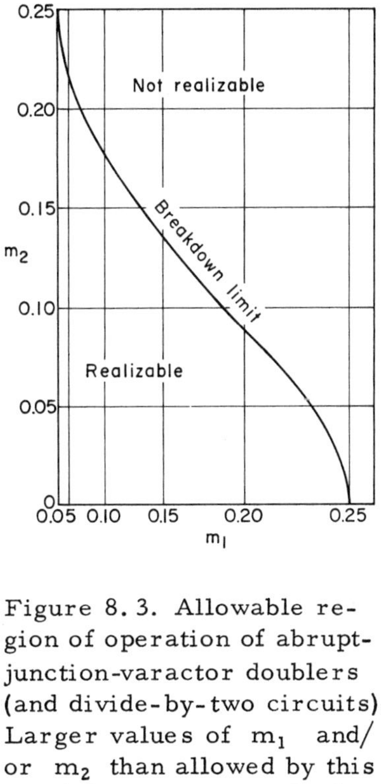

Figure 8.3. Allowable region of operation of abrupt-junction-varactor doublers (and divide-by-two circuits). Larger values of m_1 and/or m_2 than allowed by this breakdown curve imply that the elastance S(t) exceeds S_{max} and/or becomes lower than S_{min} at some times.

This restriction is indicated graphically in Figure 8.3, where we show the m_1, m_2-plane. The region inside the breakdown curve in Figure 8.1 is achievable; the region outside is not. As would be expected, when either m_1 or m_2 is absent, the other is limited to a value of 0.25. The coordinate axes in Figure 8.3 are actually m_1^2 and m_2, but they are labeled in terms of m_1 and m_2.

For many applications, this curve should not be taken too seriously. At high frequencies, the forward conduction mechanism in varactor diodes fails,‡ and the forward charge is no longer limited to the charge at voltage $-\phi$. The varactor elastance is no longer proportional to the charge, and our analyses break down. In practice more power output can be obtained from a varactor than would be calculated according to this theory. On the other hand, this operation is apt to be quite noisy.

8.4.3. Large-Signal Solutions

Upon setting the angle θ to zero, we obtain simpler formulas for the load resistance, input resistance, input, output, and dissipated powers, and efficiency:

‡See Section 4.1.4.

$$R_2 = R_s\left(\frac{m_1^2}{2m_2}\frac{\omega_c}{2\omega_o} - 1\right), \tag{8.57}$$

$$R_{in} = R_s\left(2m_2\frac{\omega_c}{2\omega_o} + 1\right), \tag{8.58}$$

$$P_{in} = 8P_{norm}\left(\frac{S_{max} - S_{min}}{S_{max} + S_{min}}\right)^2\left(\frac{\omega_o}{\omega_c}\right)^2 m_1^2\left(2m_2\frac{\omega_c}{2\omega_o} + 1\right), \tag{8.59}$$

$$P_{out} = 8P_{norm}\left(\frac{S_{max} - S_{min}}{S_{max} + S_{min}}\right)^2\left(\frac{2\omega_o}{\omega_c}\right)^2 m_2^2\left(\frac{m_1^2}{2m_2}\frac{\omega_c}{2\omega_o} - 1\right), \tag{8.60}$$

$$P_{diss} = 8P_{norm}\left(\frac{S_{max} - S_{min}}{S_{max} + S_{min}}\right)^2\left(\frac{\omega_o}{\omega_c}\right)^2 (m_1^2 + 4m_2^2), \tag{8.61}$$

$$\epsilon = \frac{\dfrac{\omega_c}{2\omega_o} - \dfrac{2m_2}{m_1^2}}{\dfrac{\omega_c}{2\omega_o} + \dfrac{1}{2m_2}}. \tag{8.62}$$

The bias voltage is still given by Equation 8.49.

Note that these equations are easily solved for all the parameters of interest once m_1 and m_2 are known. Unfortunately, however, the practical problem is usually posed in the opposite way, with the load resistance, or powers, or efficiency specified, and m_1 and m_2 not known (and, in fact, not even desired). Solution for given values of, for example, R_2 and P_{in}, is a formidable task. By far the simplest technique is graphical inversion of the functional relationships.

The key to this graphical solution is the observation that for each frequency ω_o/ω_c the six quantities in Equations 8.57 to 8.62 depend only on m_1 and m_2. Therefore, curves corresponding to constant values of each of these six parameters can be plotted on the m_1, m_2-plane. It is then a simple matter to see what combinations of powers, resistances, and efficiency are compatible, and to determine the desired operating conditions. The only drawback of this graphical scheme is that a different plot is required for each frequency; in Appendix D we give such plots for a variety of frequencies ranging from $10^{-4}\omega_c$ up to $0.1\omega_c$, in an effort to make this graphical technique a useful one.

We show an example of this technique in Figures 8.4 and 8.5. In Figure 8.4 are the three powers‡ and in Figure 8.5 are the input and load resistances and the efficiency. Both plots are for frequency $\omega_o = 10^{-2}\omega_c$; they should appear superimposed, as in Appendix D. The axes in Figures 8.4 and 8.5 (as in all similar graphs) are not m_1 and m_2 but instead m_1^2 and m_2. However, since the lines of constant m_1 and m_2 are not drawn (their values are usually not important), the user of the graph should not care.

Also shown in Figures 8.4 and 8.5 are two auxiliary curves. The one in Figure 8.5 is the breakdown limit given in Figure 8.3. Only the region under and to the left of this curve can be used. The other curve, in Figure 8.4, denotes a set of operating points that are "optimum" in a sense to be discussed in Section 8.4.3(a). The square and triangle in Figures 8.4 and 8.5 denote, respectively, the points of highest achievable efficiency and power output.

Many preliminary design questions can be answered by referring to these graphs. With a few auxiliary equations and charts (given in Appendix D), the input and output current and the possible range of average elastance corresponding to the various points on these graphs can be found. We now discuss three major features of these graphs.

8.4.3(a). Low-Power Optimum. If the operating point is known to lie inside the breakdown limit of Figure 8.5, then it is often desirable to operate with maximum efficiency or maximum power output or minimum power dissipation, subject to a fixed value for one of the other powers or efficiency. For example, what is the maximum efficiency that can be attained with a given power input? What is the minimum dissipation for a given power output?

Consider a typical question like this. Suppose the varactor dissipation rating restricts the dissipated power to be less than a certain amount. What is the maximum power output that can be achieved with this limited dissipated power? From Figure 8.4, it is clear that such a maximum value exists. The operating point corresponding to this maximum is the point on the dissipated-power curve in Figure 8.4 where this curve is tangent to an output-power curve. Thus, if the varactor dissipation is limited to about $3 \times 10^{-5} P_{norm}$, it is clear from Figure 8.4 that the maximum output power is about $10^{-4}P_{norm}$.

But if for a given dissipated power the output power is the greatest, then the efficiency must also be the highest, and the input power the greatest. Therefore, at this single point on the dissipated power curve, it is tangent not only to a power output

‡We have assumed $S_{min} << S_{max}$. For a finite S_{min}, see the instructions accompanying these graphs in Appendix D.

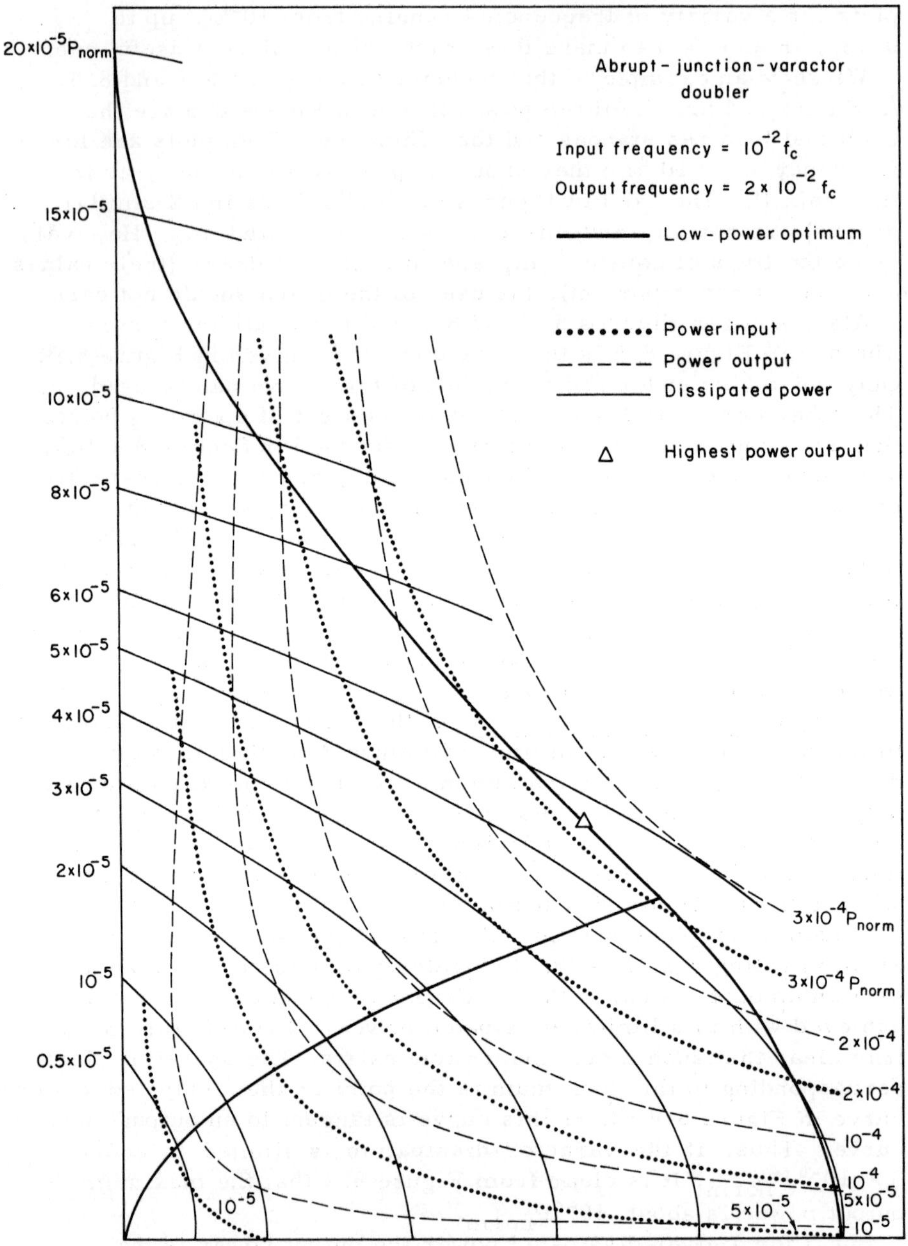

Figure 8.4. Input, output, and dissipated powers of an abrupt-junction-varactor doubler. The heavy solid curve is the low-power-optimum curve, and the triangle is the point of highest power output. This graph should be superimposed on Figure 8.5.

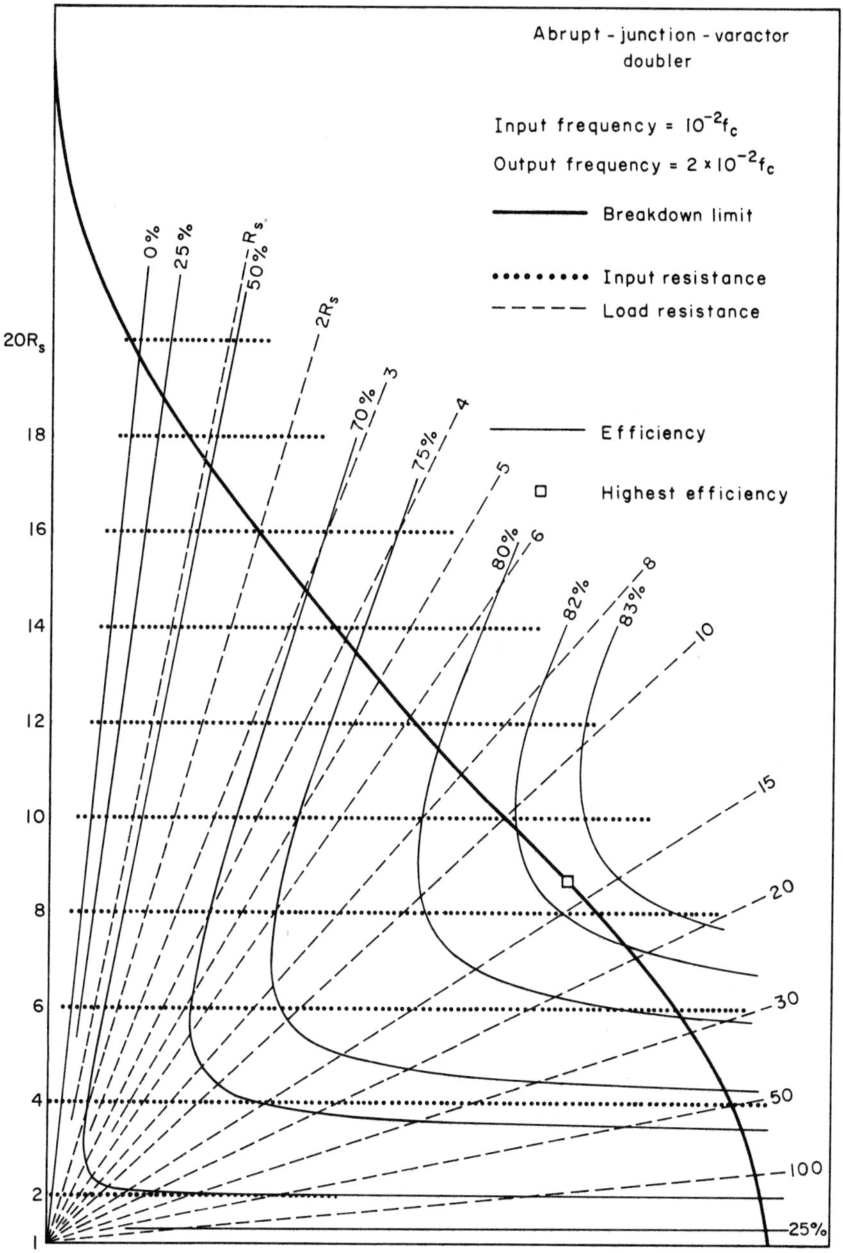

Figure 8.5. Input and load resistances and efficiency of an abrupt-junction-varactor doubler. The heavy solid curve is the breakdown limit of Figure 8.3, and the square is the point of highest achievable efficiency.

curve but also to an efficiency curve and to a power input curve.‡ This point is clearly very special, and the collection of all such points, one for each value of dissipated power, is an important curve in the m_1, m_2-plane. This is the <u>low-power-optimum</u> curve of Figure 8.4. It is given by the formula,

$$m_1^2 = 8m_2^2 \left(1 + \frac{\omega_o}{m_2\omega_c}\right). \tag{8.63}$$

Now consider another problem. Suppose a certain power output is desired; what is the minimum power input? To answer this question, we look for that point on the given output-power curve that is tangent to an input-power curve. But we have just seen that the four sets of curves (input power, output power, dissipated power, and efficiency) are all tangent to each other along the curve of Equation 8.63. Therefore, to answer this question, we need only look for the intersection of the low-power-optimum curve (Equation 8.63) and the desired power-output curve.

Note that the same low-power-optimum curve is used for all such optimizations. The only restriction on the use of this curve is that the operating point must lie within the breakdown-limit curve.

<u>8.4.3(b). Maximum Possible Efficiency</u>. What is the maximum possible efficiency for a given operating frequency? To answer this question, we search the space inside the breakdown limit and select the point with the maximum efficiency. This point lies on the breakdown curve, and is denoted in Figure 8.5 and in the graphs of Appendix D by a square. From the graphs it is easy to determine the operating conditions at this point.

Because of its great importance, we now show, as a function of input frequency ω_0, the maximum possible efficiency and associated operating conditions. In Figure 8.6 we give the maximum efficiency and in Figure 8.7 the input, dissipated, and output power at maximum efficiency. Then, in Figure 8.8 we show the input and load resistances to achieve this efficiency, and in Figure 8.9 the bias voltage.

The curve of efficiency is plotted using probability coordinates for the efficiency axis. Note from Figure 8.6 that at low frequencies the doubler efficiency can be close to 100 per cent, whereas at high frequencies it drops rapidly. This fact is also shown by Figure 8.8. At low frequencies both R_{in} and R_2 are

‡Of course, not all such curves are drawn on the charts given in this book, but only those for a few values of the parameters. The reader must interpolate to draw other curves.

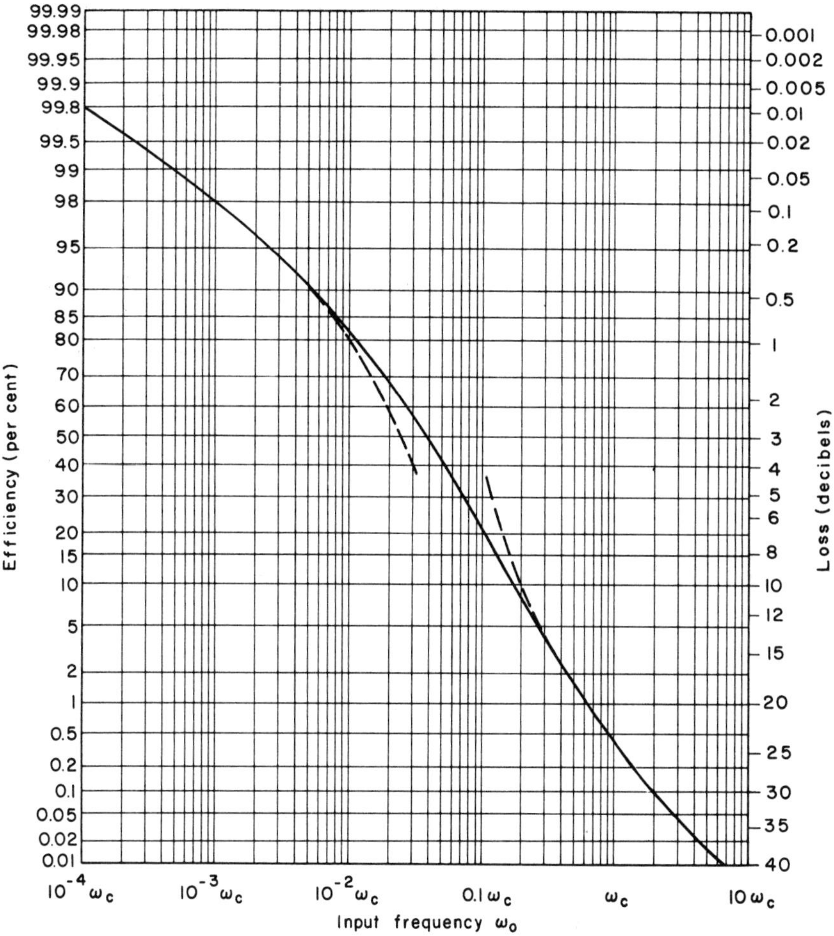

Figure 8.6. Maximum efficiency of an abrupt-junction-varactor doubler. The efficiency for maximum power output is virtually the same. The two dashed curves are the asymptotic expressions given in Table 8.4.

high, so that according to Equation 8.46 the efficiency should be high. At low frequencies both R_{in} and R_2 vary inversely with the frequency, as would be expected from the fact that the non-linear capacitance, with a "reactance" inversely proportional to frequency, dominates. At high frequencies both R_{in} and R_2 approach R_s, as would be expected from the fact that the series resistance now dominates.

Figure 8.7 shows that at low frequencies, the dissipated power is quite small, and the input and output power are approximately the same. At high frequencies the output power approaches $P_{norm}/512$, and the input and dissipated power are practically the same.

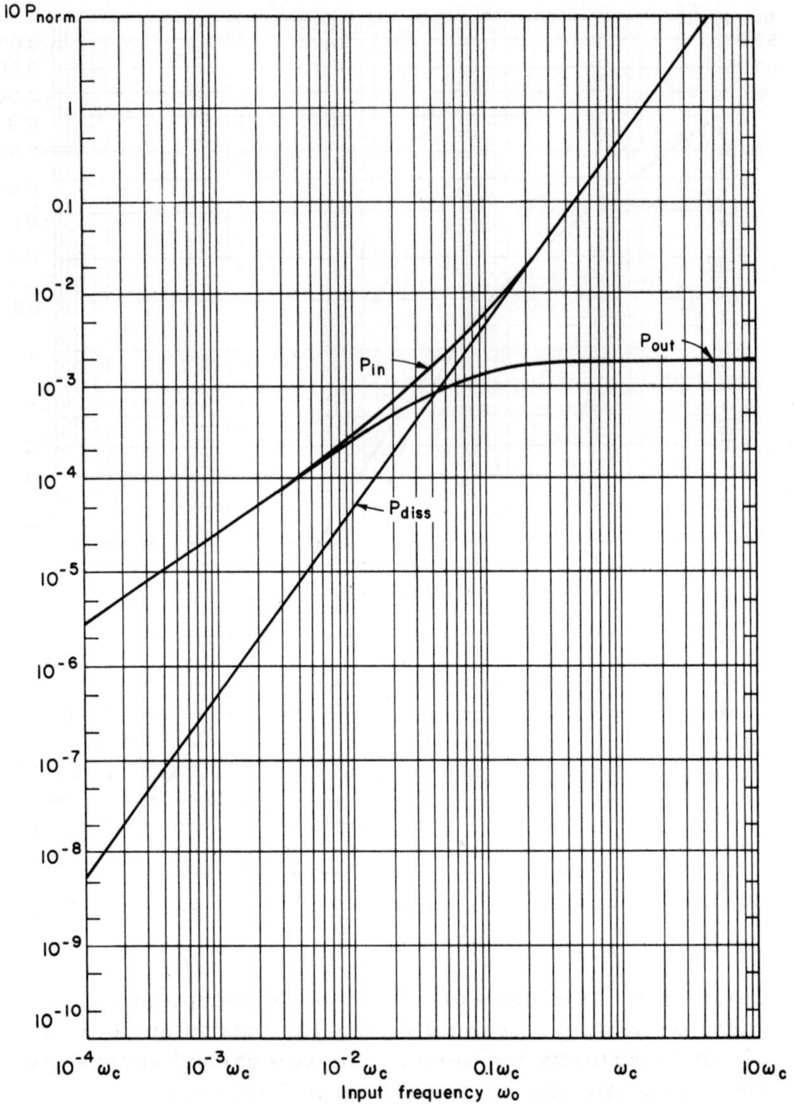

Figure 8.7. Input, output, and dissipated power for an abrupt-junction-varactor doubler operated at maximum efficiency (the curves for operation at maximum power output are virtually identical). We have assumed that $S_{min} \ll S_{max}$. In case S_{min} is finite, all powers should be multiplied by $[(S_{max} - S_{min})/(S_{max} + S_{min})]^2$.

The bias voltage for maximum efficiency (and in fact the bias voltage for any point on the breakdown curve) is found by substituting Equation 8.51 into Equation 8.49:

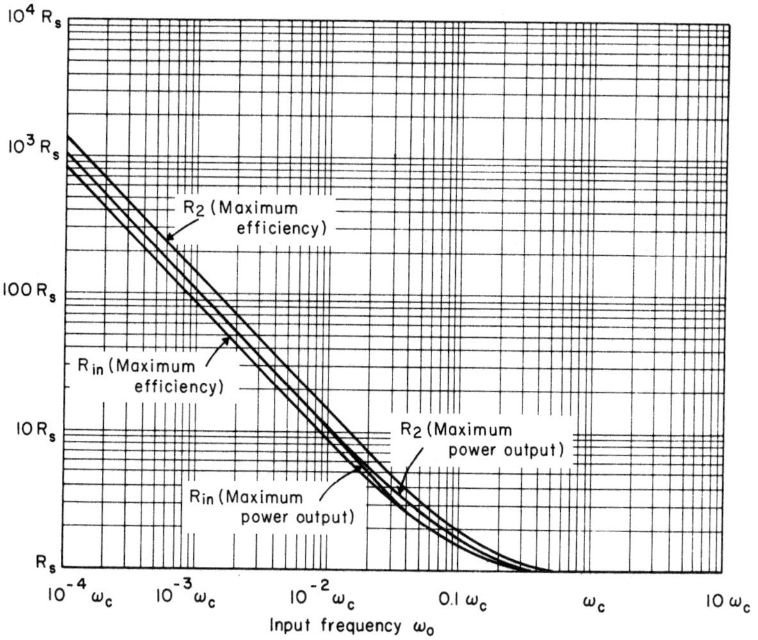

Figure 8.8. Input and load resistances for operation of an abrupt-junction-varactor doubler with either maximum efficiency or maximum power output.

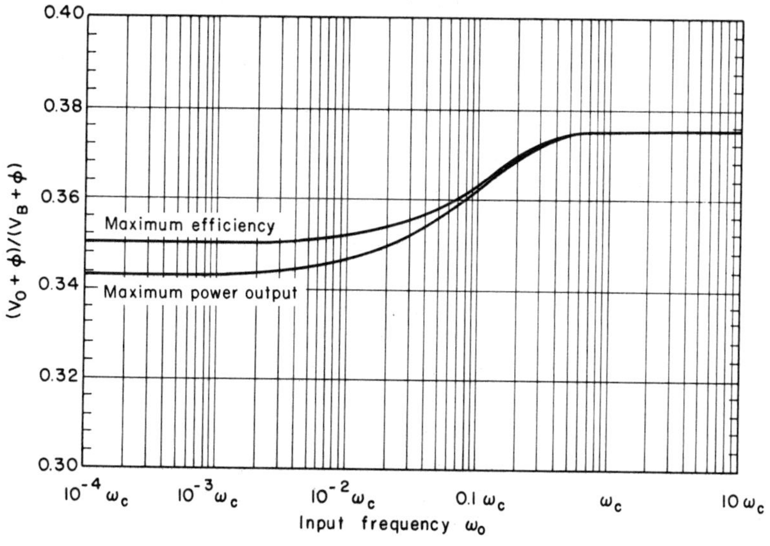

Figure 8.9. Bias voltage V_0 for maximum efficiency or maximum power output of an abrupt-junction-varactor doubler. We have assumed that $S_{min} << S_{max}$. In case S_{min} is finite, $(V_0 + \phi)/(V_B + \phi)$ is $[(S_{max} - S_{min})/S_{max}]^2$ times the quantity plotted vertically, plus (S_{min}/S_{max}).

$$\frac{V_0 + \phi}{V_B + \phi} = \left(\frac{S_{max} - S_{min}}{S_{max}}\right)^2 \left[\frac{1}{4} + 2(m_1^2 + m_2^2)\right] + \frac{S_{min}}{S_{max}}. \qquad (8.64)$$

In case $S_{min} \ll S_{max}$, only the quantity inside the square bracket remains on the right-hand side of Equation 8.64. It is this quantity that is plotted in Figure 8.9.

It is interesting to look at the elastance waveform for the doubler. In Figure 8.10 we show the waveform that occurs for low

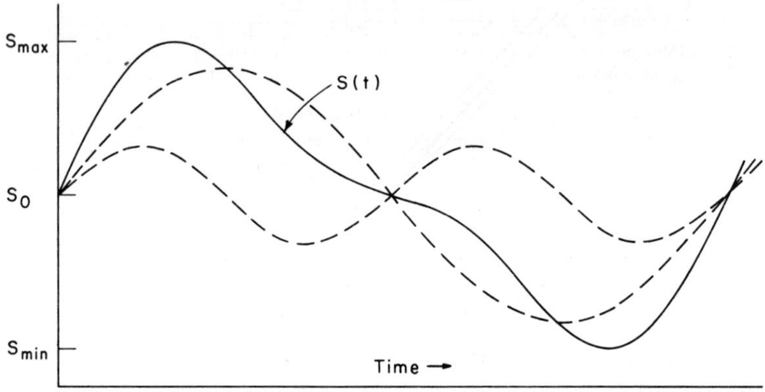

Figure 8.10. Elastance as a function of time for the optimum-efficiency low-frequency abrupt-junction-varactor doubler. Shown are the components of elastance at both the fundamental and the second harmonic, as well as their sum.

frequencies if the efficiency is optimized. Note that the elastance goes through the entire range from S_{min} to S_{max} during each cycle.

The limiting behavior of the doubler at high and low frequencies is quite important. The limiting values of m_1 and m_2 can be used in Equations 8.57 through 8.62 and Equation 8.64 to find expressions for the asymptotic behavior of the parameters of interest. These expressions are given in Table 8.4. The region over which each expression is valid can be determined by comparison with the graphs of Figures 8.6 through 8.9.

Note that at low frequencies the power input and output are proportional to ω_0. The important diode parameter is not the normalization power or the cutoff frequency but rather their ratio, which is independent of R_s:

$$\frac{P_{norm}}{\omega_c} = \frac{(V_B + \phi)^2}{S_{max}}. \qquad (8.65)$$

Table 8.4. Asymptotic Formulas for the Doubler

These asymptotic values hold for the doubler with optimum efficiency. For simplicity we have assumed that $S_{min} = 0$.

	Low-Frequency	High-Frequency
ϵ	$1 - 19.9\frac{\omega_o}{\omega_c}$	$0.0039\left(\frac{\omega_c}{\omega_o}\right)^2$
R_{in}	$0.080\frac{S_{max}}{\omega_o}$	R_s
R_2	$0.271\frac{S_{max}}{2\omega_o}$	R_s
P_{in}	$0.0277P_{norm}\frac{\omega_o}{\omega_c}$	$0.500P_{norm}\left(\frac{\omega_o}{\omega_c}\right)^2$
P_{out}	$0.0277P_{norm}\frac{\omega_o}{\omega_c}$	$0.00195P_{norm}$
P_{diss}	$0.551P_{norm}\left(\frac{\omega_o}{\omega_c}\right)^2$	$0.500P_{norm}\left(\frac{\omega_o}{\omega_c}\right)^2$
$\frac{V_0 + \phi}{V_B + \phi}$	0.349	0.375
m_1	0.208	0.250
m_2	0.080	$0.0078\frac{\omega_c}{\omega_o}$

If the series resistance is really negligible, varactors with a larger value of Equation 8.65 can handle more power than ones with a lower value. Uhlir[10] has called half this quantity the <u>nominal reactive power</u> of varactors. For low-frequency multipliers, it is a convenient parameter by which to judge varactors, since the input and output powers are approximately $0.057\omega_0$ times the nominal reactive power. At higher frequencies, however, it is no more convenient than the normalization power.

The maximum-efficiency operating point is so important that in Appendix D we give nomographs for determining varactor characteristics necessary to attain specified powers and efficiencies. These nomographs are built up around diode space, discussed in Section 4.7.4.

It is interesting to note that the maximum efficiency can be found in closed form. Merely use the expression for the breakdown curve, Equation 8.56, and the expression for efficiency, Equation 8.62, to find the efficiency in terms of m_1/m_2 only. Then, set the derivative of this expression to zero. The result is a set of two equations for the maximum efficiency and the frequency that can be solved parametrically.

8.4.3(c). Maximum Power Output. It is clear from Figure 8.4 that at each frequency the power output is limited by the breakdown limit. The point of greatest possible power output is indicated in Figure 8.4 by a triangle. This point is close to, but not the same as, the point of maximum efficiency. For many practical purposes the points can be considered identical. If we attempted to plot the maximum power output as a function of frequency, we should obtain a curve that coincides with the curve of Figure 8.7 to within the precision of reading the graph. Similarly, if we plotted the efficiency attained at maximum power output, the points would fall right on top of the curve of Figure 8.6. Thus, both the efficiency and the power output are in effect optimized together.

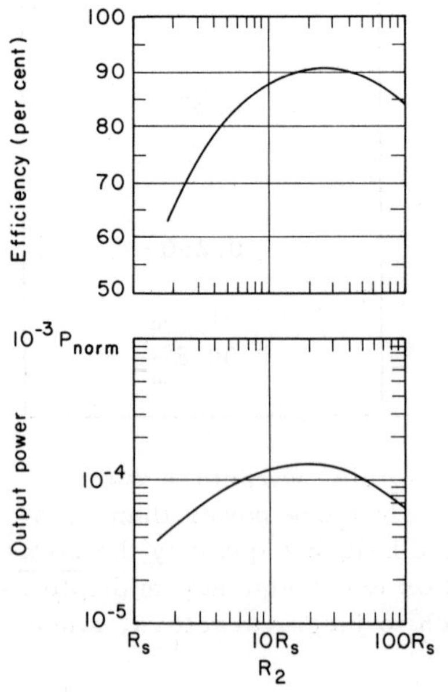

Figure 8.11. Efficiency and output power plotted as a function of load resistance R_2, under the assumption that the elastance waveform goes between S_{min} and S_{max} during each cycle. The input frequency is $0.01\omega_c$. Note that since the two curves have quite broad maxima, the load resistance need not be controlled precisely.

To understand this fact, consider Figure 8.11, which shows efficiency and power output along the breakdown curve as a function of load resistance. It is clear that although the maximum efficiency and the maximum power output do not occur simultaneously, for practical purposes any load resistance relatively close to the resistance that optimizes either will maximize both.

The load and input resistance for optimum power output are slightly different from those for optimum efficiency, and Figure 8.7 shows both sets of resistances. Similarly, the required bias voltage is slightly different, as shown in Figure 8.9.

The low-frequency behavior of the point of maximum power output is indicated in Table 8.5. Note that most of the limits are

Table 8.5. Low-Frequency Optimum-Power-Output Doubler

The only major change from the values in Table 8.4 is the load resistance. Again for simplicity we have assumed that $S_{min} << S_{max}$.

$$\epsilon \approx 1 - 20.8\frac{\omega_o}{\omega_c} \qquad P_{in} \approx 0.0285 P_{norm}\frac{\omega_o}{\omega_c}$$

$$R_{in} \approx 0.096\frac{S_{max}}{\omega_o} \qquad P_{out} \approx 0.0285 P_{norm}\frac{\omega_o}{\omega_c}$$

$$R_2 \approx 0.192\frac{S_{max}}{2\omega_o} \qquad P_{diss} \approx 0.593 P_{norm}\left(\frac{\omega_o}{\omega_c}\right)^2$$

$$\frac{V_0 + \phi}{V_B + \phi} \approx 0.343 \qquad m_1 \approx 0.192 \qquad m_2 \approx 0.096$$

very close to the maximum-efficiency low-frequency behavior indicated in Table 8.4. The major difference is that the load resistance is about 30 per cent lower. At high frequencies the efficiency and power output are optimized together, so that the high-frequency limits are exactly those given in Table 8.4.

8.4.4. Synthesis Techniques

Thus far we have discussed some limits on efficiency and power output and have described a graphical technique for determining operation of abrupt-junction-varactor doublers. If the assumptions of the theory (particularly that no current flows at higher harmonics) can be realized, there is no reason why doublers cannot be synthesized to attain these fundamental limits. The synthesis of doublers consists of two steps. The first is the preliminary design of selecting varactor and operating conditions, and the second is the design of a circuit to fulfill the conditions of the theory.

8.4.4(a). Preliminary Design. Preliminary design involves such tasks as selecting a varactor and determining the operating conditions of the circuit, including the input and load resistances, the bias voltage, and so forth. By use of the formulas and graphs of Section 8.4 and the graphs of Appendix D, this preliminary design can be reduced to a routine matter.

8.4.4(b). Separation of Frequencies. The design of an actual circuit to fulfill the conditions of the theory is difficult. For a doubler these conditions are the following:

1. The varactor should be open-circuited at frequency $3\omega_0$.
2. The varactor should be open-circuited at frequency $4\omega_0$.
3. The input and output frequencies ω_0 and $2\omega_0$ should appear at separate terminals, and these terminals should have tuning adjustments.
4. For highest efficiency the circuit should be lossless.

In general, satisfying all these requirements is a difficult job. Circuits with symmetry can be used to block current flow at the unwanted harmonics, and in some cases even to separate the input and output frequencies. From these and the many other circuits that satisfy the preceding requirements, would be chosen the ones with acceptable bandwidth, cost, reliability, simplicity, freedom from spurious oscillations, and so on. To show that

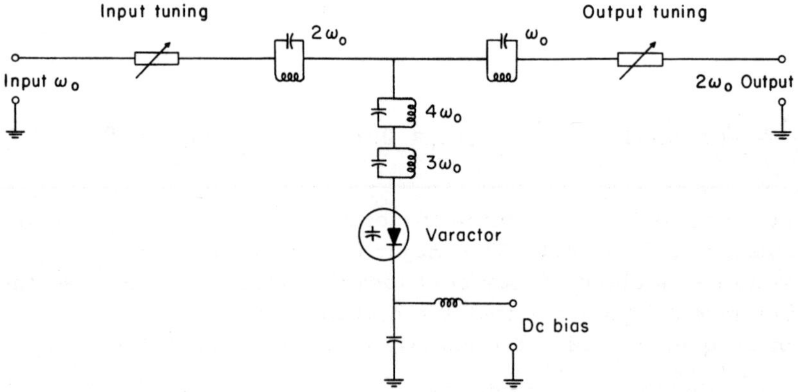

Figure 8.12. Lumped doubler circuit. The tuned circuits separate the input and output frequencies, and the adjustable reactances are used for tuning. This circuit can be used as a divide-by-two circuit if the output is excited, instead of the input.

circuits that satisfy the four stated conditions do exist and, in fact, can be drawn fairly easily, we show three of them. Two use lumped tuned circuits, and the third uses transmission lines.

In Figures 8.12 and 8.13, we show the two lumped circuits. The circuit in Figure 8.12 uses parallel-tuned traps to open-circuit the varactor at the third and fourth harmonic and separate the input and output currents. The input and output are both provided with the reactance necessary to tune the loop reactance to zero, and the bias is introduced with appropriate blocking capacitors and an rf choke. Figure 8.13 illustrates the use of symmetric circuits. The varactor bridge automatically separates the even- and odd-order harmonic currents. Single parallel-resonant traps at the input and output open-circuit the varactor at the third and fourth harmonics and, by appropriate choice of impedance level, tune out the varactor average elastance. The bias is introduced with blocking capacitors and rf chokes.

The coaxial circuit is shown in Figure 8.14. The varactor is mounted within the coaxial line, not across it (see Figure 8.15). The open-circuit conditions are met by using lengths of coaxial line. The input and output are separated by a high-pass filter H. P., and the bias voltage is introduced through one of tuning lines. The adjustments of the lines in Figure 8.14 are independent if carried out in the proper order.

The circuit realizations of Figures 8.12, 8.13, and 8.14 may not be important in themselves as practical circuits. They do indicate, however, that circuits can be synthesized to fulfill

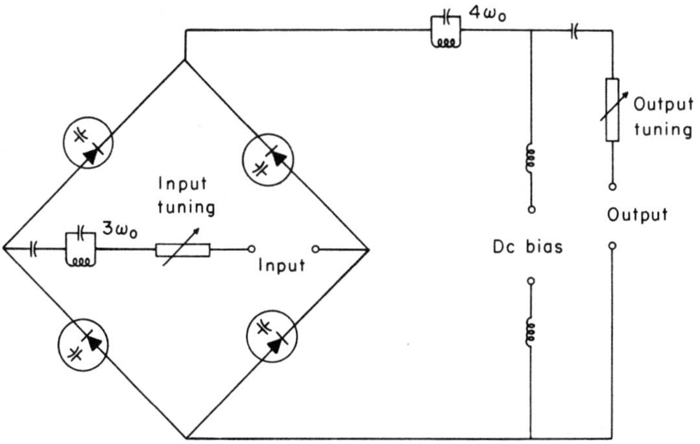

Figure 8.13. Lumped doubler circuit in which symmetry is used to separate the input and output frequencies. This circuit, like that of Figure 8.12, can be used as a divider with input and output reversed. Because four varactors are used, the input, output, and dissipated powers are four times the values calculated for one diode.

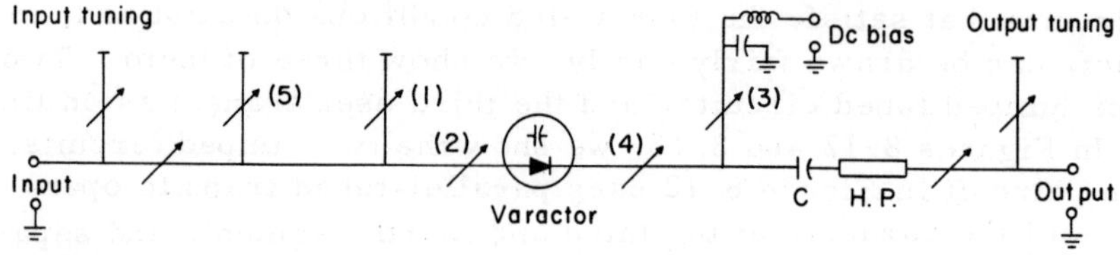

Line	Length
(1)	$\lambda(3\omega_o)/2$
(2)	$\lambda(3\omega_o)/4$
(3)	$\lambda(4\omega_o)/2$
(4)	$\lambda(4\omega_o)/4$
(5)	$\lambda(2\omega_o)/2$

Figure 8.14. Doubler using coaxial transmission lines. The adjustable lines are set according to the conditions shown, before assembly, since their adjustments do not depend upon the varactor used.

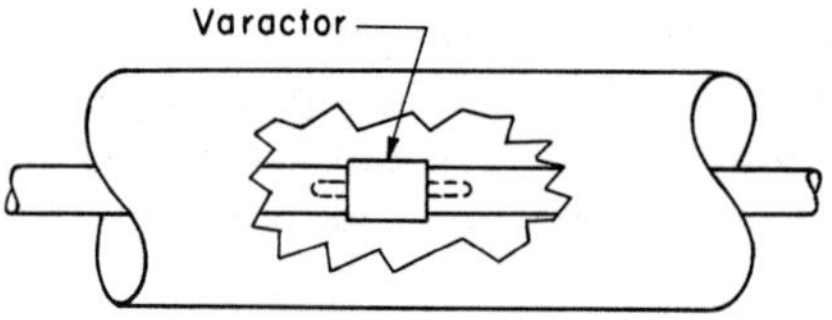

Figure 8.15. A varactor mounted in the inner conductor of a coaxial transmission line.

deliberately the conditions of the theory and thereby attain the performance described in Section 8.4.

8.5. General Multiplier with Idlers

We have seen in Section 8.3 that abrupt-junction multipliers, dividers, and rational-fraction generators, except for the doubler and the divide-by-two circuit,‡ require idler currents. We now derive formulas to describe an abrupt-junction multiplier with any set of idlers. Because the additional complication is so small, we describe at the same time dividers and rational-fraction generators, again with arbitrary idlers present. This analysis will be used in Sections 8.6 to 8.10 to find the equations that describe the tripler, quadrupler, quintupler, sextupler, and octupler, and in Section 9.5, the divide-by-four circuit.

We assume the varactor has an input at frequency $n\omega_o$ and an output to a load at frequency $\ell\omega_o$, where n and ℓ are chosen so that only currents with frequency $k\omega_o$, for integers k, flow through the varactor. For example, if $n = 1$ and $\ell = 3$, we should be describing the tripler. If $n = 2$ and $\ell = 5$, the device would be the multiply-by-5/2 circuit.

‡Although this requirement is not strictly necessary for other varactors, such as the graded-junction, it is helpful in the sense that greater efficiency results from use of idlers. Of course, if the diode is heavily overdriven, idlers may be unnecessary or even detrimental. However, we do not discuss this situation.

For simplicity we restrict this analysis to abrupt-junction varactors, for which

$$\frac{v+\phi}{V_B+\phi} = \left(\frac{S}{S_{max}}\right)^2 + \frac{R_s i}{V_B+\phi}. \tag{8.66}$$

If the minimum elastance S_{min} is different from zero, then the minimum voltage V_{min} is different from $-\phi$, and from Equation 8.66 we can derive the alternate formula,

$$\frac{v - V_{min}}{V_B - V_{min}} = \frac{S^2 - S_{min}^2}{S_{max}^2 - S_{min}^2} + \frac{R_s i}{V_B - V_{min}}. \tag{8.67}$$

In these formulas v, S, and i are functions of time.

8.5.1. Bias Voltage

The bias, or time-averaged, voltage is found by taking the time average of Equation 8.66 or 8.67. Thus,

$$\frac{V_0+\phi}{V_B+\phi} = \left(\frac{S_{max} - S_{min}}{S_{max}}\right)^2 (m_0^2 + 2m_1^2 + 2m_2^2 + \cdots), \tag{8.68}$$

or

$$\frac{V_0 - V_{min}}{V_B - V_{min}} = \frac{S_{max} - S_{min}}{S_{max} + S_{min}} (m_0^2 + 2m_1^2 + 2m_2^2 + \cdots) - \frac{S_{min}^2}{S_{max}^2 - S_{min}^2}, \tag{8.69}$$

where, as usual, the m_k parameters come from the Fourier series expansion for the elastance:

$$\frac{S(t)}{S_{max} - S_{min}} = \sum_{k=-\infty}^{+\infty} M_k e^{jk\omega_o t}, \tag{8.70}$$

with

$$m_k = |M_k|. \tag{8.71}$$

In case the minimum elastance S_{min} is zero, Equation 8.69 reduces to the simple expression,

$$\frac{V_0 + \phi}{V_B + \phi} = m_0^2 + 2m_1^2 + 2m_2^2 + \cdots . \tag{8.72}$$

For the doubler, with all $m_k = 0$ except for $k = 0$, 1, and 2, Equation 8.68 reduces to Equation 8.49.

8.5.2. Input Resistance

Since the multiplier is nonlinear, the input voltage is not proportional to the input current, and strictly speaking, the multiplier does not have an input impedance. However, we shall speak of the ratio of input voltage to input current as the "input impedance," with the understanding that it depends on the drive level.

To find this input impedance, the current must be related to the elastance. For abrupt-junction varactors

$$i(t) = \frac{dq(t)}{dt}$$

$$= 2\,\frac{V_B - V_{min}}{S_{max} + S_{min}}\,\frac{d}{dt}\left(\frac{S(t)}{S_{max} - S_{min}}\right). \tag{8.73}$$

Therefore, each Fourier coefficient of current, I_k, is related to the corresponding Fourier coefficient of elastance:

$$I_k = \frac{V_B - V_{min}}{S_{max} + S_{min}}\,2jk\omega_o M_k, \tag{8.74}$$

or, for the important case with $S_{min} = 0$,

$$I_k = \frac{V_B + \phi}{S_{max}}\,2jk\omega_o M_k. \tag{8.75}$$

An expression for the voltage at each frequency V_k ($k \neq 0$) is obtained from Equation 8.67:

$$V_k = R_s I_k + \frac{S_{max} - S_{min}}{S_{max} + S_{min}}(V_B - V_{min}) \sum_{r=-\infty}^{+\infty} M_r M_{k-r}. \tag{8.76}$$

Note that the Fourier coefficients of $S^2(t)$ are expressed in terms of a summation of products of elastance Fourier coefficients.

In particular, for the input frequency $n\omega_o$, the ratio of Equation 8.76 to Equation 8.74 gives the input impedance

$$Z_{in} = R_{in} + jX_{in}$$

$$= \frac{V_n}{I_n}$$

$$= R_s + \frac{R_s \omega_c}{2n\omega_o} \frac{\sum_{r=-\infty}^{+\infty} M_r M_{n-r}}{jM_n}. \qquad (8.77)$$

Its real part is the "input resistance"

$$R_{in} = R_s \left(1 + \frac{\omega_c}{2n\omega_o} \operatorname{Re} \frac{\sum_{r=-\infty}^{+\infty} M_r M_{n-r}}{jM_n} \right). \qquad (8.78)$$

The input impedance and resistance depend on the drive level, of course. In the summation in Equation 8.78, the terms for $r = 0$ and $r = n$ might as well be omitted, because they contribute only an imaginary part $S_0/jn\omega_o$ corresponding to the time-averaged elastance S_0. Note that if $n = 1$ and if only the M_k for $k = -2$, -1, 0, 1, and 2 are present, then Equation 8.78 reduces to Equation 8.33 for the abrupt-junction doubler.

8.5.3. Idler and Load Resistances

We suppose that the varactor is terminated at each of the other frequencies $k\omega_o$ ($k \neq 0$ or $\pm n$) by a tuned‡ impedance Z_k with real part R_k and imaginary part $S_0/k\omega_o$,

$$V_k = -Z_k I_k$$

$$= -\left(R_k + j\frac{S_0}{k\omega_o}\right) I_k. \qquad (8.79)$$

Then, Equation 8.76 becomes

‡Note particularly that for simplicity we now assume all idlers and the load are tuned. This is analogous to our setting $\theta = 0$ in Section 8.4.1(h).

$$0 = (R_k + R_s)I_k + \frac{jS_0 I_k}{k\omega_0} + \frac{S_{max} - S_{min}}{S_{max} + S_{min}} (V_B - V_{min}) \sum_{r=-\infty}^{+\infty} M_r M_{k-r}, \tag{8.80}$$

and, when we use Equation 8.74 and solve for R_k, we find

$$R_s + R_k = R_s \frac{\omega_c}{2k\omega_0} \frac{\sum_{r=-\infty}^{+\infty} jM_r \, jM_{k-r}}{jM_k}, \tag{8.81}$$

where in the summation the terms for $r = 0$ and $r = k$ are omitted.

Equation 8.81 gives an expression for all the idler resistances in terms of the elastance coefficients M_k. If $k = \ell$, then Equation 8.81 gives a formula for the load resistance R_ℓ. If all M_r are zero except for $r = -2, -1, 0, 1,$ and 2, then Equation 8.81, with $k = 2$, agrees with the expression for load resistance of a doubler, Equation 8.57.

Unfortunately, Equation 8.81 gives the R_k values as a function of the M_k. Usually in practice the problem is put the other way around — we are given or we assume values of R_k, and wish to calculate the multiplier parameters. Thus the set of equations, Equation 8.81, must be "inverted" in some sense. This can be done in specific cases, as will be shown in Sections 8.6 to 8.10. Then, once the M_k are known, the other operating parameters of the multiplier (input resistance, input power, output power, efficiency, dissipated power, bias voltage, and so forth) can be calculated.

8.5.4. Input Power

The input power P_{in} can be calculated two ways. First, in terms of the input current I_n and the input resistance R_{in}, it is

$$P_{in} = 2R_{in}|I_n|^2$$

$$= 8P_{norm} \left(\frac{S_{max} - S_{min}}{S_{max} + S_{min}} \right)^2 \frac{\omega_0^2}{\omega_c^2} n^2 m_n^2 \frac{R_{in}}{R_s}, \tag{8.82}$$

where we have used Equation 8.40. Alternatively, P_{in} is equal to the power dissipated in all idler terminations, in the load termination, and in the varactor series resistance. Thus,

$$P_{in} = 8P_{norm}\left(\frac{S_{max} - S_{min}}{S_{max} + S_{min}}\right)^2 \frac{\omega_o^2}{\omega_c^2} \sum_{k=1}^{\infty} k^2 m_k^2 \frac{R_k + R_s}{R_s}, \qquad (8.83)$$

where, in the summation, for $k = n$, we let "R_n" be zero. Equation 8.82 is usually easier to evaluate, but Equation 8.83 shows clearly how the input power splits up into varactor dissipation, dissipation in the external idlers, and load power. The two expressions can be shown to be equal.

8.5.5. Output Power

It is clear from Equation 8.83 what portion of the input power is dissipated in the load resistance R_ℓ. It is that portion of the term in the summation for $k = \ell$ containing R_ℓ (the portion containing R_s is dissipated in the series resistance at the output frequency). Thus,

$$P_{out} = 8P_{norm}\left(\frac{S_{max} - S_{min}}{S_{max} + S_{min}}\right)^2 \frac{\omega_o^2}{\omega_c^2} \ell^2 m_\ell^2 \frac{R_\ell}{R_s}. \qquad (8.84)$$

8.5.6. Dissipated Power

The dissipated power is simply the difference between Equation 8.83 and Equation 8.84. Thus,

$$P_{diss} = 8P_{norm}\left(\frac{S_{max} - S_{min}}{S_{max} + S_{min}}\right)^2 \frac{\omega_o^2}{\omega_c^2} \left(\sum_{k=1}^{\infty} k^2 m_k^2 + \sum_{\substack{k=1\\k\neq n\\k\neq \ell}}^{\infty} k^2 m_k^2 \frac{R_k}{R_s}\right). \qquad (8.85)$$

Of this dissipated power, a portion

$$P_{diss,v} = 8P_{norm}\left(\frac{S_{max} - S_{min}}{S_{max} + S_{min}}\right)^2 \frac{\omega_o^2}{\omega_c^2} \left(\sum_{k=1}^{\infty} k^2 m_k^2\right) \qquad (8.86)$$

is dissipated as heat in the varactor, and the remainder,

$$P_{diss,i} = 8P_{norm}\left(\frac{S_{max} - S_{min}}{S_{max} + S_{min}}\right)^2 \frac{\omega_o^2}{\omega_c^2}\left(\sum_{\substack{k=1 \\ k\neq n \\ k\neq \ell}}^{\infty} k^2 m_k^2 \frac{R_k}{R_s}\right), \tag{8.87}$$

is dissipated in the external idler terminations. The input power is therefore accounted for by the output power, and the two dissipated powers:

$$\begin{aligned} P_{in} &= P_{out} + P_{diss} \\ &= P_{out} + P_{diss,v} + P_{diss,i}. \end{aligned} \tag{8.88}$$

8.5.7. Efficiency

The efficiency ϵ is the ratio of output power to input power. Thus, from Equations 8.82, 8.83, and 8.84,

$$\begin{aligned} \epsilon &= \frac{P_{out}}{P_{in}} \\ &= \frac{\ell^2 m_\ell^2 R_\ell}{n^2 m_n^2 R_{in}} \\ &= \frac{\ell^2 m_\ell^2 R_\ell}{\sum_{k=1}^{\infty} k^2 m_k^2 (R_k + R_s)}, \end{aligned} \tag{8.89}$$

where in the last expression, when the index k is equal to n, we consider "R_n," which has not been defined as yet, to be zero.

8.5.8. Phase Condition

For the abrupt-junction-varactor doubler, we discovered in Section 8.4.1(h) that if the load impedance has an imaginary part that tunes out exactly the average elastance of the varactor, and if the time origin is set by making the input current I_1 real (with jM_1 real and positive), then the output current is also in phase, so that jM_2 is real and positive. The logical extension of this result would be that, if all the idler and the load terminations are tuned and if the input current I_n is real and positive, then all jM_k are real and positive. This statement, however, is not generally true. Its truth must be tested anew for each different idler configuration. We shall, however, see that it is true for each of the multipliers examined in Sections 8.6 to 8.10.

8.5.9. Summary

We have derived expressions for the bias voltage, input resistance, idler resistances, output resistance, input power, output power, power dissipated in the varactor, power dissipated in the external idler terminations, and efficiency of abrupt-junction-varactor multipliers, dividers, and rational-fraction generators. The idler configuration was not specified. These expressions all reduce to the corresponding expressions for the doubler when all M_k except for $k = -2, -1, 0, 1,$ and 2 are set equal to zero. These formulas are used to describe the abrupt-junction tripler (Section 8.6), quadrupler (Section 8.7), quintupler (Section 8.8), sextupler (Section 8.9), octupler (Section 8.10), and divide-by-four circuit (Section 9.5).

8.6. Tripler

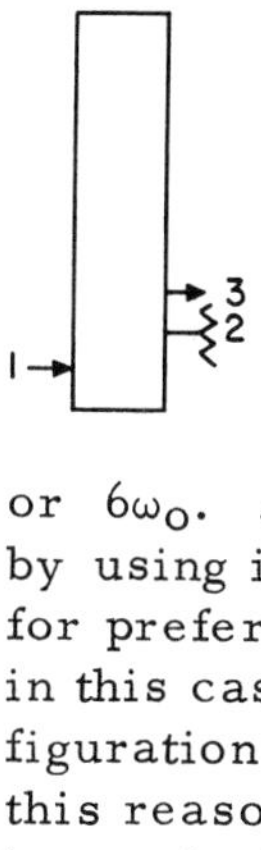

We now solve the equations derived in Section 8.5 for an abrupt-junction varactor tripler.

Without an idler, the tripler will not work. According to Table 8.3, the simplest idler configuration has only one idler, at frequency $2\omega_0$. However, two or more idlers can be used. For example, in addition to frequencies ω_0, $2\omega_0$, and $3\omega_0$, current could be allowed to flow at $\omega_0/2$, or $3\omega_0/2$, or $4\omega_0$, or $5\omega_0$, or $6\omega_0$. Alternatively, the idler at $2\omega_0$ could be eliminated by using idlers at $\omega_0/2$ and $3\omega_0/2$. There is no a priori reason for preferring one idler configuration to another, except that in this case our intuition suggests that the simplest idler configuration might be the easiest to analyze and construct. For this reason we shall analyze only the 1-2-3 tripler. We do not know whether better efficiency, or better power capabilities, or other advantages can or cannot be gained by using different idler configurations.

8.6.1. Formulas

The formulas developed in Section 8.5 apply to the tripler with $n = 1$ and $\ell = 3$. The idler and load equations become, from Equation 8.81,

$$\frac{R_3 + R_s}{R_s} = \frac{\omega_c}{3\omega_0} \frac{(jM_1)(jM_2)}{jM_3} \tag{8.90}$$

and

$$\frac{R_2 + R_s}{R_s} = \frac{\omega_c}{4\omega_o} \frac{(jM_1)^2 - 2(jM_1)^*(jM_3)}{jM_2}. \quad (8.91)$$

Let us choose the time origin so that jM_1 is real and positive, equal to its magnitude m_1. Then, Equation 8.90 shows that the phase of jM_3 is the same as the phase of jM_2, and Equation 8.91 shows that therefore the phase of jM_2 is zero. Thus, jM_2 and jM_3 are both real and positive and equal, respectively, to their magnitudes m_2 and m_3. In terms of the m_k, the formulas for the load resistance R_3, idler resistance R_2, input resistance R_{in}, power input P_{in}, power output P_{out}, dissipated power P_{diss}, varactor dissipation $P_{diss,v}$, external idler dissipation $P_{diss,i}$, efficiency ϵ, and bias voltage V_0 are

$$R_3 = R_s \left(\frac{\omega_c}{3\omega_o} \frac{m_1 m_2}{m_3} - 1 \right), \quad (8.92)$$

$$R_2 = R_s \left[\frac{\omega_c}{4\omega_o} \frac{m_1}{m_2} (m_1 - 2m_3) - 1 \right], \quad (8.93)$$

$$R_{in} = R_s \left[\frac{\omega_c}{\omega_o} \frac{m_2}{m_1} (m_1 + m_3) + 1 \right], \quad (8.94)$$

$$P_{in} = 8P_{norm} \left(\frac{S_{max} - S_{min}}{S_{max} + S_{min}} \right)^2 \frac{\omega_o^2}{\omega_c^2} \left[\frac{\omega_c}{\omega_o} m_1 m_2 (m_1 + m_3) + m_1^2 \right], \quad (8.95)$$

$$P_{out} = 8P_{norm} \left(\frac{S_{max} - S_{min}}{S_{max} + S_{min}} \right)^2 \frac{\omega_o^2}{\omega_c^2} \left(\frac{3\omega_c}{\omega_o} m_1 m_2 m_3 - 9m_3^2 \right), \quad (8.96)$$

$$P_{diss} = 8P_{norm} \left(\frac{S_{max} - S_{min}}{S_{max} + S_{min}} \right)^2 \frac{\omega_o^2}{\omega_c^2} \left[\frac{\omega_c}{\omega_o} (m_1^2 m_2 - 2m_1 m_2 m_3) + m_1^2 + 9m_3^2 \right], \quad (8.97)$$

$$P_{diss,v} = 8P_{norm} \left(\frac{S_{max} - S_{min}}{S_{max} + S_{min}} \right)^2 \frac{\omega_o^2}{\omega_c^2} (m_1^2 + 4m_2^2 + 9m_3^2), \quad (8.98)$$

$$P_{diss,i} = 8P_{norm}\left(\frac{S_{max} - S_{min}}{S_{max} + S_{min}}\right)^2 \frac{\omega_o^{\ 2}}{\omega_c^{\ 2}}\left[\frac{\omega_c}{\omega_o}(m_1^2 m_2 - 2m_1m_2m_3) - 4m_2^2\right], \tag{8.99}$$

$$\epsilon = \frac{3m_3}{m_1}\,\frac{\dfrac{\omega_c}{\omega_o} m_1m_2 - 3m_3}{\dfrac{\omega_c}{\omega_o} m_2(m_1 + m_3) + m_1}, \tag{8.100}$$

$$\frac{V_0 + \phi}{V_B + \phi} = \left(\frac{S_{max} - S_{min}}{S_{max}}\right)^2 (m_0^{\ 2} + 2m_1^{\ 2} + 2m_2^{\ 2} + 2m_3^{\ 2}). \tag{8.101}$$

8.6.2. Technique of Solution

The solution of these equations is by no means a simple task. If m_1, m_2, and m_3 are known, then all quantities of interest can be easily calculated. Often the easiest way to get quick, approximate solutions is to try various combinations of m_1, m_2, and m_3 until the desired solution is obtained. In doing this, however, be careful not to choose m_1, m_2, and m_3 too large. Since the varactor elastance must lie between S_{min} and S_{max},

$$m_1 \sin\omega_o t + m_2 \sin 2\omega_o t + m_3 \sin 3\omega_o t \leq 0.25 \tag{8.102}$$

for all values of t. In addition, the values of R_3 and R_2 in Equations 8.92 and 8.93 must be positive. Aside from these restrictions, any values of m_k are suitable.

More often, however, the values of R_3 and R_2 are known, and the m values are not (nor are they even desired). In general, if R_2, R_3, and some measure of the drive level (such as the input current I_1) are known, then the values of the m_k can be calculated. As an aide to this calculation, we can solve Equations 8.92 and 8.93 for m_2 and m_3 as functions of m_1:

$$m_2 = \frac{m_1}{\dfrac{R_2 + R_s}{R_s}\,\dfrac{4\omega_o}{m_1\omega_c} + \dfrac{2m_1\omega_c}{3\omega_o}\,\dfrac{R_s}{R_3 + R_s}}, \tag{8.103}$$

and

$$m_3 = \frac{m_1}{2 + \dfrac{R_2 + R_s}{R_s}\,\dfrac{R_3 + R_s}{R_s}\,\dfrac{12\omega_o^2}{m_1^2\omega_c^2}}. \tag{8.104}$$

Now if the drive level is known, a value of m_1 can be chosen and m_2 and m_3 calculated. For example, if I_1 is given, then Equation 8.74 can be used to calculate m_1. On the other hand, if the input or output power is specified, then estimate the value of m_1, and, using Equations 8.103 and 8.104 to calculate m_2 and m_3, test this value by comparing the given power with either Equation 8.95 or 8.96. Then, a better approximation can be made; if the calculated power is smaller than the given value, raise the estimate of m_1.

Keep trying until the calculated power is close enough to the given power. In all cases, check to see that Condition 8.102 is not violated.

Perhaps the most important solutions are those for which Condition 8.102 is barely fulfilled, such as those solutions corresponding to maximum power levels and maximum efficiency. Equations 8.103 and 8.104 can be used in conjunction with Condition 8.102 (with the $\leq$ sign replaced by an equality) to find the values of m_1, m_2, and m_3 pertinent to maximum possible drive, that is, with the elastance attaining the values of S_{min} and S_{max} on each cycle.‡

The resulting values of efficiency, powers, and input resistance depend upon the frequency ω_o and the resistances R_2 and R_3. A computer program to solve Equations 8.102, 8.103, and 8.104 simultaneously for specified values of ω_o/ω_c, R_3/R_s, and R_2/R_s has been written by Bliss L. Diamond,[3] of the M. I. T. Lincoln Laboratory. Through the courtesy of Diamond, we describe in Section 8.6.3 the solutions he has obtained. These differ very little from previous approximate solutions for maximum efficiency.[2]

8.6.3. Solutions for Maximum Drive

We now present the highlights of Diamond's computer solutions, for the maximum-drive condition. For a specific frequency and a specific idler resistance R_2, the efficiency and power output as functions of the load resistance go through maxima. Figure 8.16 shows a typical plot of efficiency and power output versus load resistance. Note that the two maxima occur at slightly different values of R_3; but, at least for these values of ω_o and R_2, the efficiency is not much reduced where the power output is a maximum, and vice versa. Thus at least in this case, for

‡These solutions are analogous to the doubler solutions along the breakdown curve of Figure 8.3.

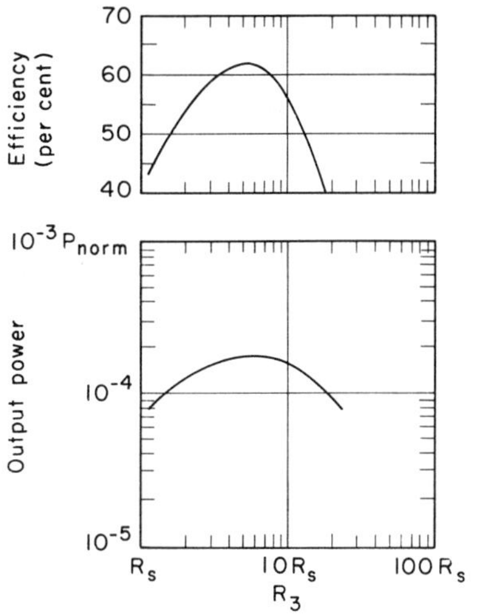

Figure 8.16. Efficiency and power output as functions of load resistance for an abrupt-junction-varactor tripler. Although in this plot we have assumed that $R_2 = 2R_s$ and $\omega_0 = 10^{-2}\omega_c$, the qualitative features are the same for other idler resistances and frequencies.

practical purposes power output and efficiency can be maximized simultaneously.

The point of maximum efficiency is quite important; it is described in detail in Figures 8.17 to 8.23. These figures are plots, for the load resistance that maximizes the efficiency, of the following quantities: efficiency, input resistance, load resistance, power input, power output, dissipated power, and bias voltage. Each appears as a function of frequency, for several values of idler resistance R_2.

As would be expected, Figure 8.17 demonstrates that the highest efficiency is achieved with a lossless idler. The efficiency can be quite close to 100 per cent at low frequencies, even for a finite idler resistance. The input and load resistances become large at small frequencies, in agreement with our intuition about "matching" to a nonlinear capacitance with "reactance" inversely proportional to frequency. At high frequencies, on the other hand, the series resistance of the varactor dominates, and the input and load resistances approach R_s.

At low frequencies the input and output power are approximately the same, and vary linearly with frequency, as shown in Figures 8.20 and 8.21. The dissipated power, Figure 8.22, is much smaller. At higher frequencies, the input and dissipated powers vary as the square of the frequency, and the output power goes through a maximum and then drops off. In these figures we have neglected the factor containing S_{min} in Equations 8.95, 8.96, and 8.97. If S_{min} is not negligible, multiply the values shown by $(S_{max} - S_{min})^2/(S_{max} + S_{min})^2$.

Maximum drive can be achieved only if the average elastance is midway between S_{min} and S_{max}:

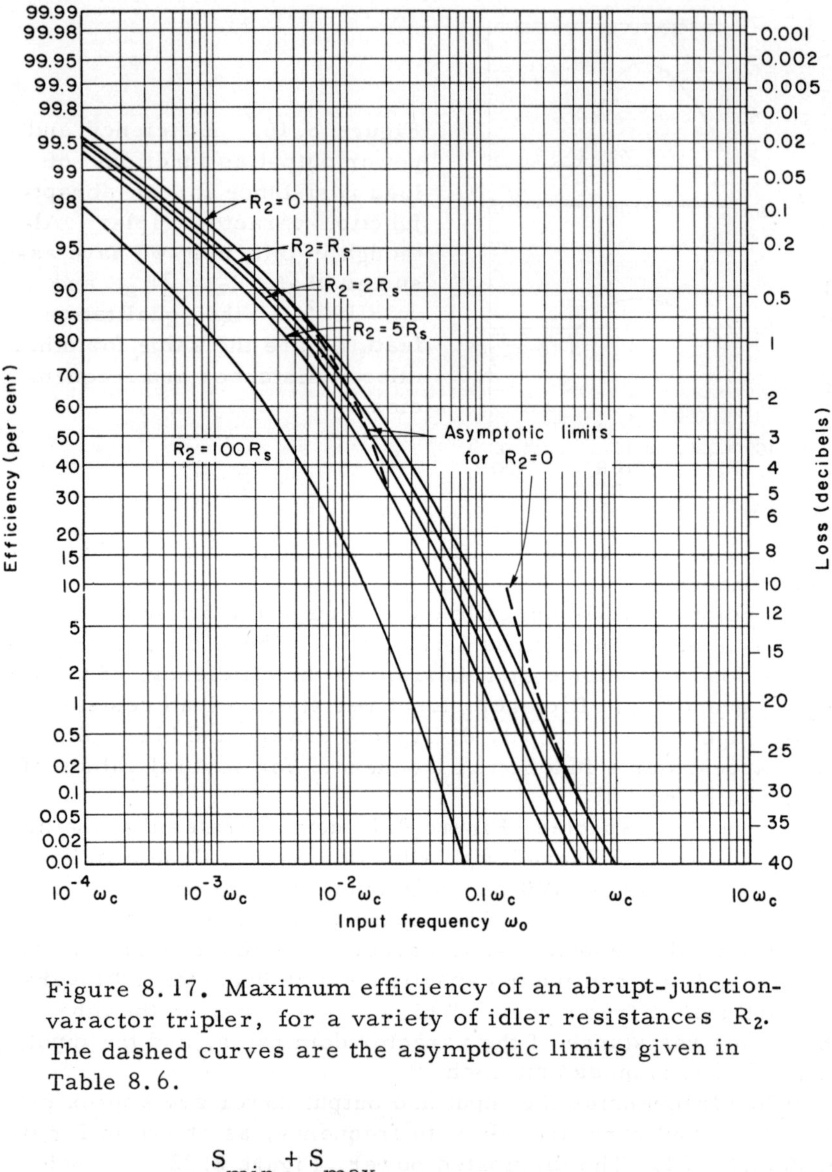

Figure 8.17. Maximum efficiency of an abrupt-junction-varactor tripler, for a variety of idler resistances R_2. The dashed curves are the asymptotic limits given in Table 8.6.

$$S_0 = \frac{S_{min} + S_{max}}{2}. \tag{8.105}$$

Thus the bias voltage is, from Equation 8.101,

$$\frac{V_0 + \phi}{V_B + \phi} = \left(\frac{S_{max} - S_{min}}{S_{max}}\right)^2 \left[\frac{1}{4} + 2(m_1^2 + m_2^2 + m_3^2)\right] + \frac{S_{min}}{S_{max}}. \tag{8.106}$$

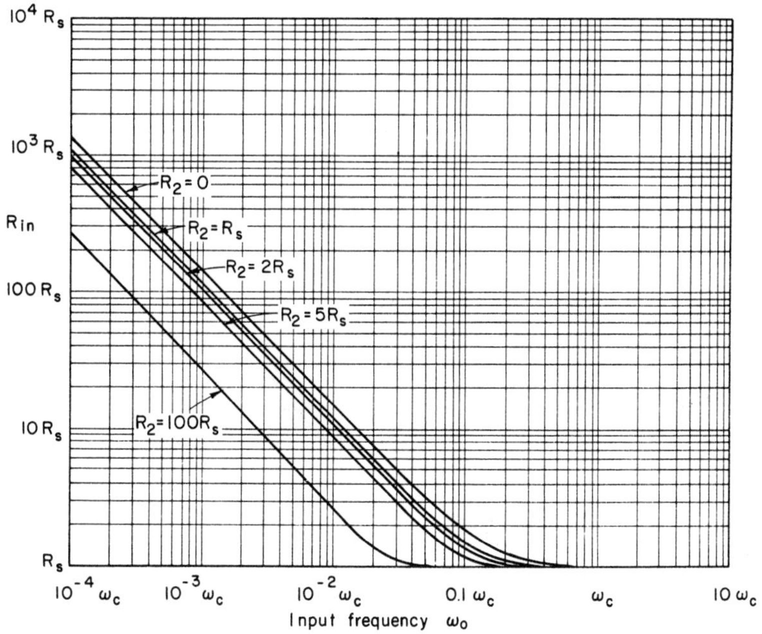

Figure 8.18. Input resistance for maximum efficiency of an abrupt-junction-varactor tripler, for a variety of idler resistances R_2.

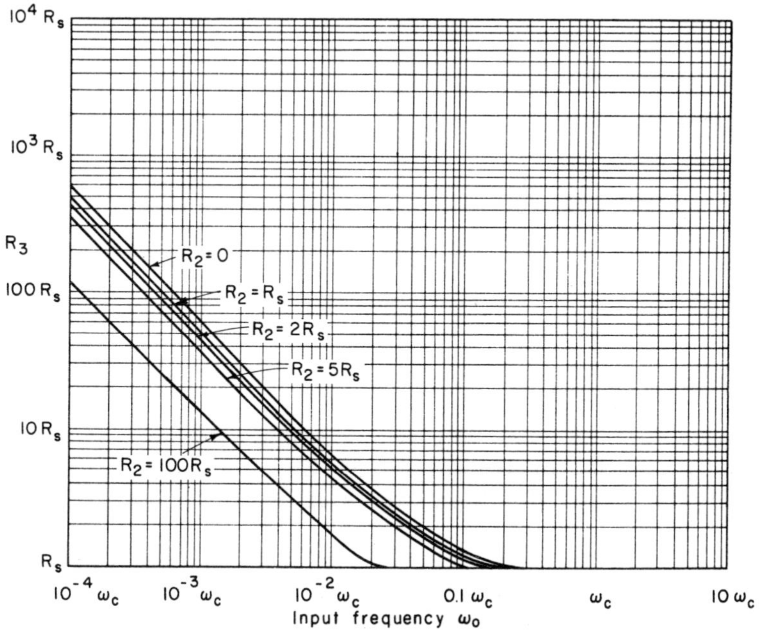

Figure 8.19. Load resistance for an abrupt-junction-varactor tripler with maximum efficiency.

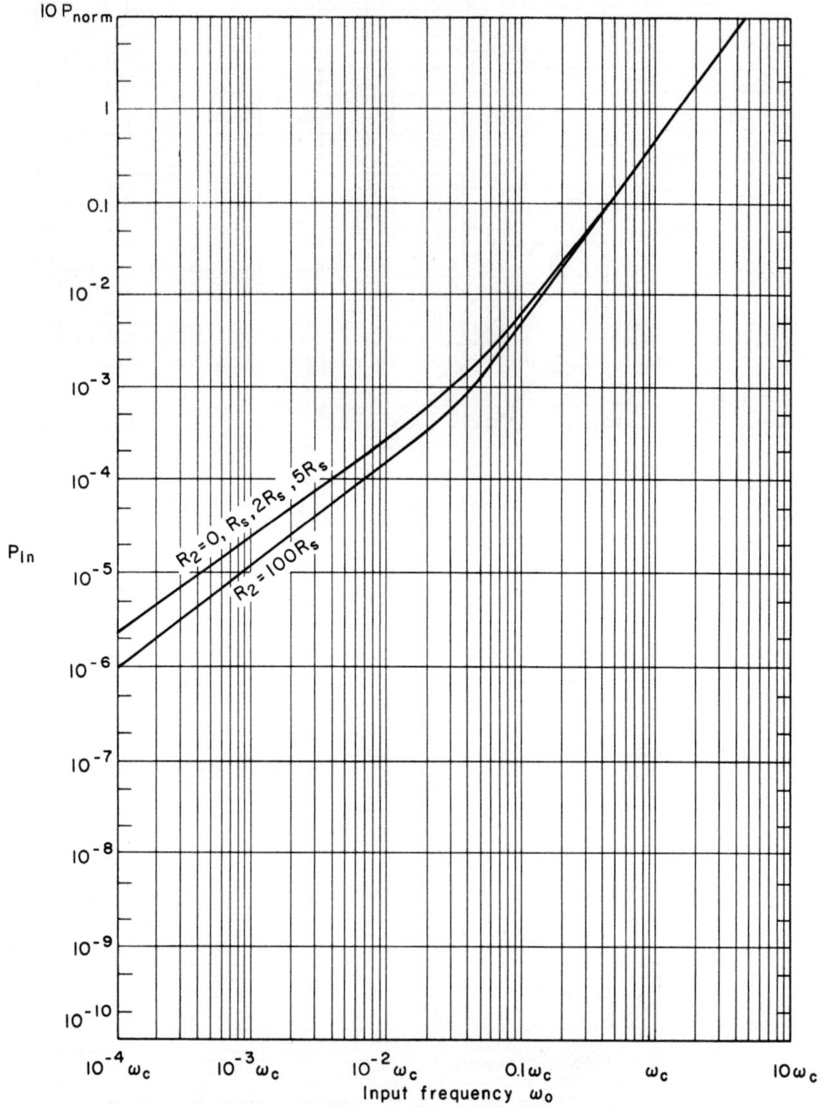

Figure 8.20. Input power of an abrupt-junction-varactor tripler when adjusted to give maximum efficiency. The input power for maximum output power is insignificantly different, except for high values of R_2.

If $S_{min} = 0$, then only the quantity in the square brackets remains; it is this quantity that is plotted in Figure 8.23. If S_{min} is nonzero, the numbers shown in Figure 8.23 must be multiplied by $(S_{max} - S_{min})^2/S_{max}{}^2$ and the result added to S_{min}/S_{max}.

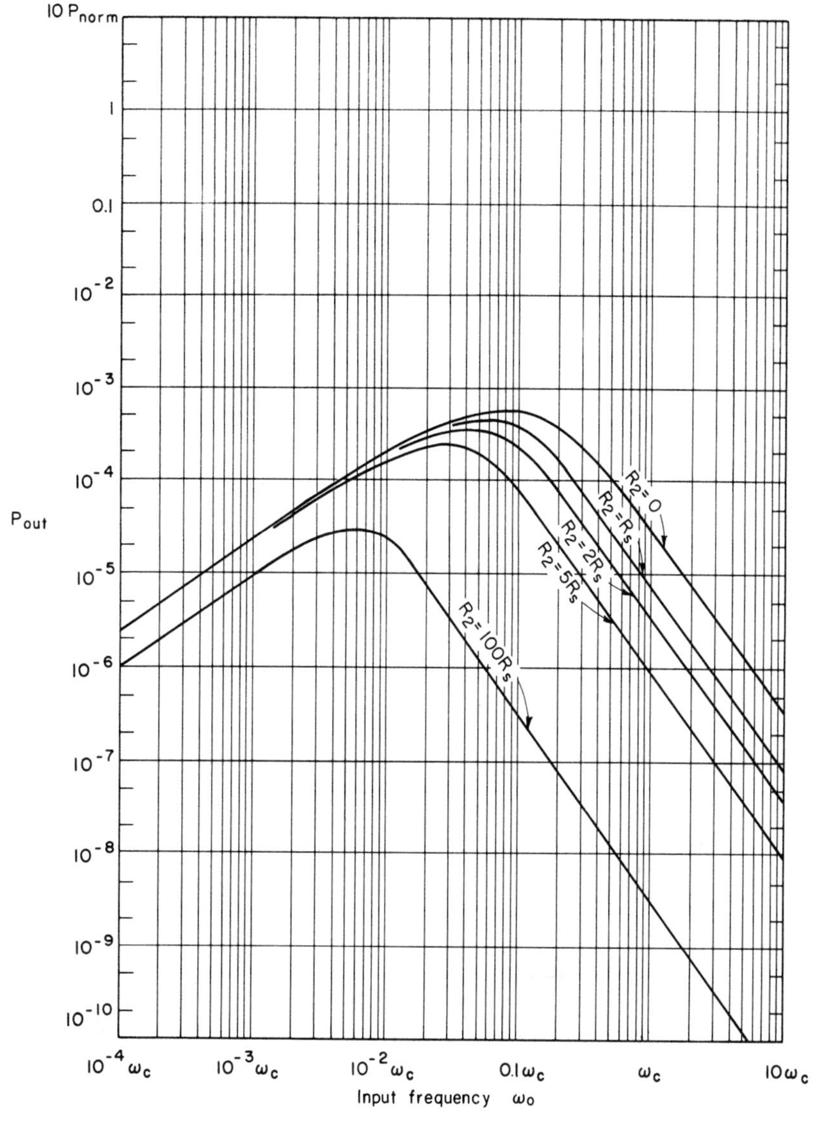

Figure 8.21. Output power of an abrupt-junction-varactor tripler with maximum efficiency. The output power can be only slightly larger than this, except for the case with $R_2 = 100R_s$.

The conditions for maximum power output, rather than maximum efficiency, are qualitatively very similar. For low values of idler resistance R_2, the efficiency, input power, output power, and dissipated power are not significantly different from Figures 8.17, 8.20, 8.21, and 8.22. The load and input resistances are

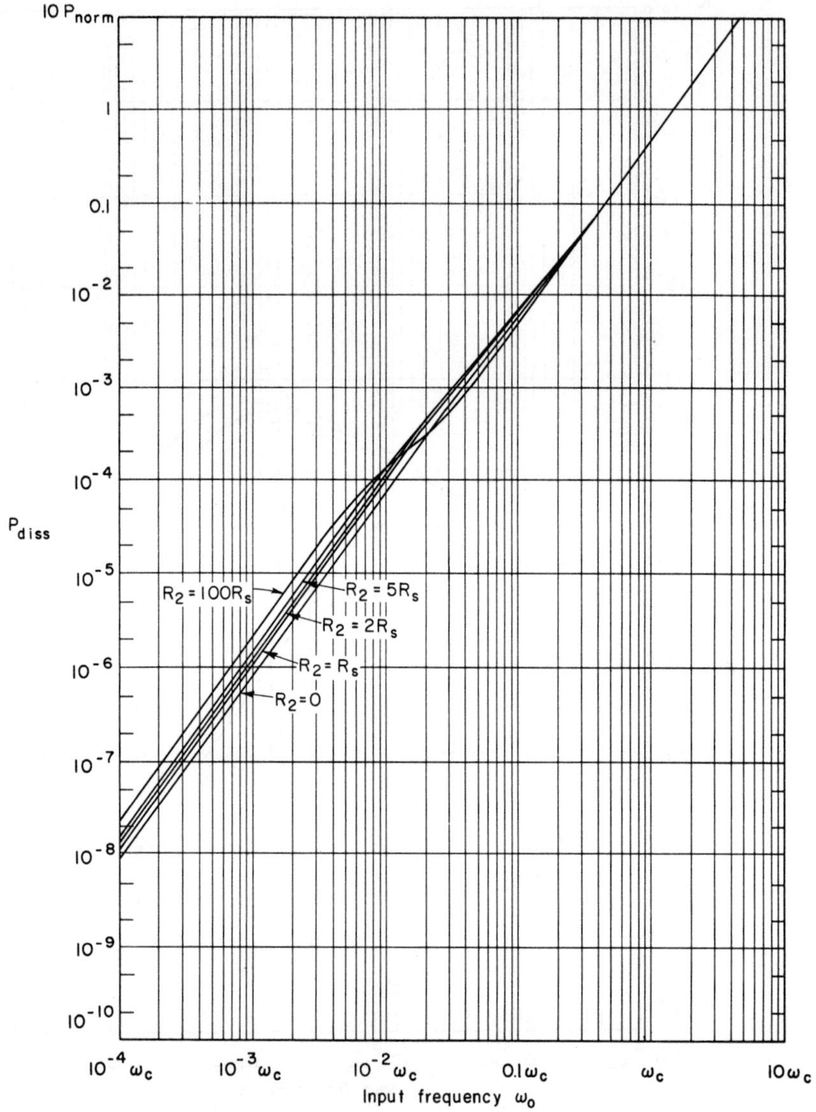

Figure 8.22. Dissipated power of an abrupt-junction-varactor tripler with optimized efficiency. This includes both the power dissipated in the varactor and in the idler resistance R_2.

slightly different, but not enough to make replotting worth while. Similarly, the bias voltage is changed, but not very significantly. For high values of R_2, the two optimizations differ enough to show on our graphs, but for simplicity we do not plot the optimum power-output results with high R_2.

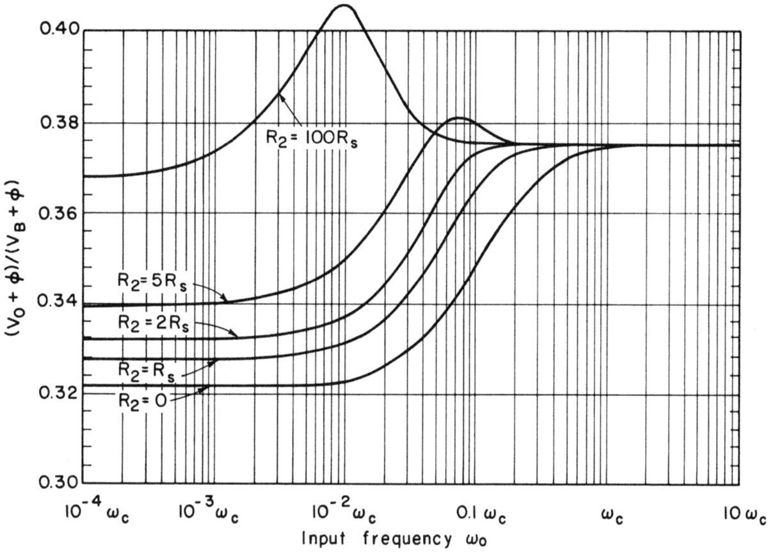

Figure 8.23. Bias voltage for an abrupt-junction-varactor tripler with maximum efficiency.

Note that the power output and the efficiency are always the highest when the idler is lossless, that is, when $R_2 = 0$. Figures 8.24 and 8.25 show the resistances and the powers for maximum

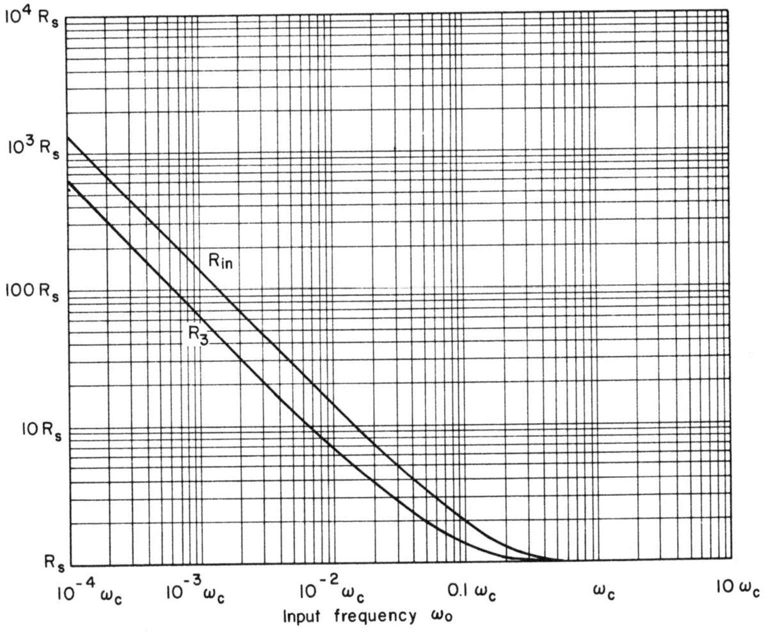

Figure 8.24. Input and load resistance for maximum efficiency of an abrupt-junction-varactor tripler with a lossless idler $(R_2 = 0)$.

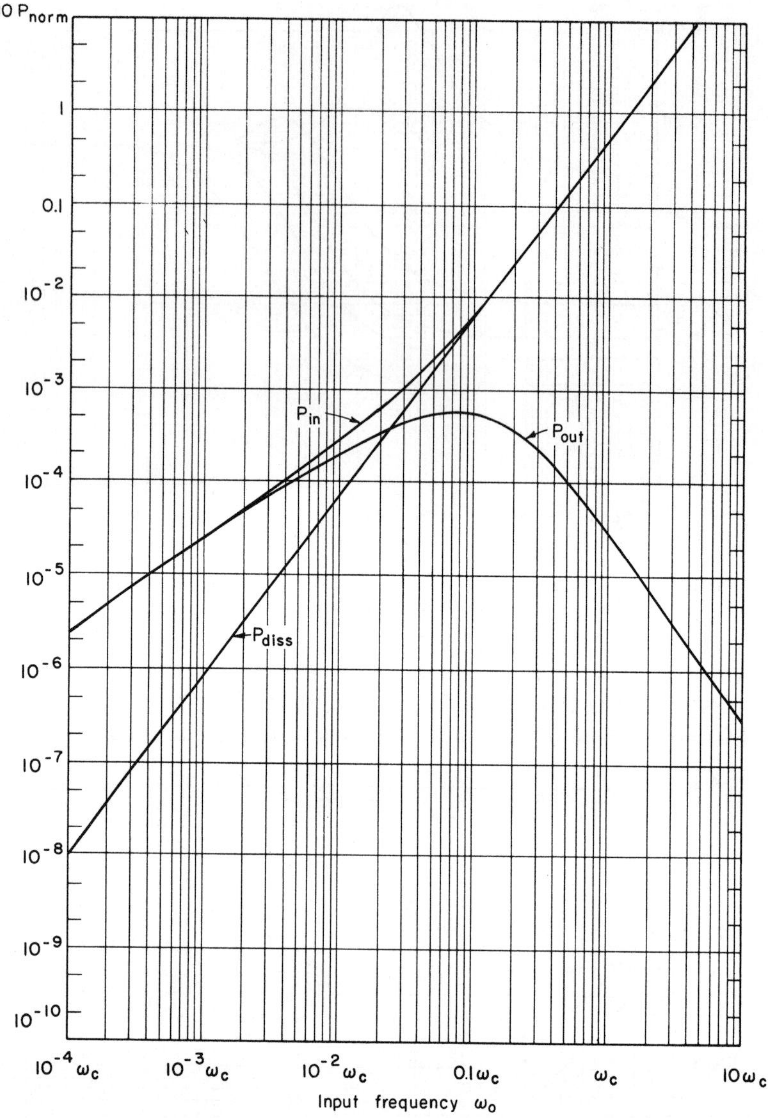

Figure 8.25. Input, output, and dissipated power for an abrupt-junction-varactor tripler with maximum efficiency and a lossless idler.

efficiency for this important case of a lossless idler termination. Nomographs given in Appendix D, based on the figures in this section, can be used to speed the design of abrupt-junction-varactor triplers with maximum efficiency in the case with $R_2 = 0$.

It is interesting to examine the elastance waveform under typical operation. The example shown in Figure 8.26 is for

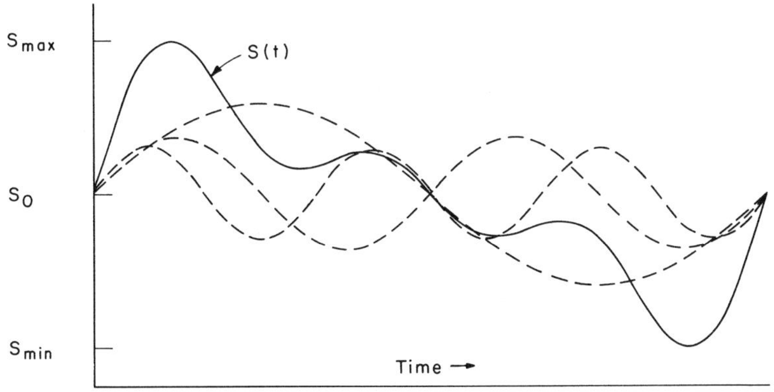

Figure 8.26. Elastance waveform of an abrupt-junction-varactor tripler at low frequencies, adjusted to yield maximum efficiency with a lossless idler termination (m_1 = 0.148, m_2 = 0.092, m_3 = 0.074). We show the fundamental, second harmonic, and third harmonic individually and their sum.

the low-frequency values of m_1, m_2, and m_3 for maximum efficiency. Since the solution is for maximum drive, S(t) attains S_{max} and S_{min} once each cycle.

The behavior of the tripler at low and high frequencies is quite important. The limiting values of m_1, m_2, and m_3 can be found from Diamond's computer solution and used in Equations 8.92 to 8.101 to determine the asymptotic behavior of the tripler. Table 8.6 shows the low- and high-frequency limits, both for highest efficiency and highest power output. At high frequencies the efficiency and power output are maximized simultaneously, but at low frequencies they are maximized for slightly different load resistances.

8.6.4. Synthesis

Synthesis of abrupt-junction-varactor tripler circuits is very similar to synthesis of doublers and other multipliers. The first step is the preliminary design, in which the varactor is selected to meet specific efficiency and power requirements, according to the theory given thus far in Section 8.6. Then, a circuit is designed to separate the frequencies of interest. In particular, for the tripler:

1. The varactor should be open-circuited at frequencies $4\omega_o$, $5\omega_o$, and $6\omega_o$.
2. The varactor should see a tuned load (preferably lossless) at frequency $2\omega_o$.
3. The input and output frequencies should appear at separate terminals, and these terminals should have tuning adjustments.

Table 8.6. Asymptotic Formulas for the Tripler

Low-frequency and high-frequency formulas are given for the abrupt-junction-varactor tripler, with idler at $2\omega_0$. For simplicity we have assumed that $S_{min} \ll S_{max}$. Since maximum power output and efficiency are achieved with a lossless idler, we have assumed that $R_2 = 0$.

	Low-Frequency		High-Frequency
	Maximum ϵ	Maximum P_{out}	Maximum ϵ and P_{out}
ϵ	$1 - 34.8\frac{\omega_0}{\omega_c}$	$1 - 35.1\frac{\omega_0}{\omega_c}$	$6.08 \times 10^{-5}\left(\frac{\omega_c}{\omega_0}\right)^4$
R_{in}	$0.137\frac{S_{max}}{\omega_0}$	$0.126\frac{S_{max}}{\omega_0}$	R_s
R_3	$0.184\frac{S_{max}}{3\omega_0}$	$0.168\frac{S_{max}}{3\omega_0}$	R_s
P_{in}	$0.0241P_{norm}\frac{\omega_0}{\omega_c}$	$0.0242P_{norm}\frac{\omega_0}{\omega_c}$	$0.500P_{norm}\left(\frac{\omega_0}{\omega_c}\right)^2$
P_{out}	$0.0241P_{norm}\frac{\omega_0}{\omega_c}$	$0.0242P_{norm}\frac{\omega_0}{\omega_c}$	$3.04 \times 10^{-5}P_{norm}\left(\frac{\omega_c}{\omega_0}\right)^2$
P_{diss}	$0.837P_{norm}\left(\frac{\omega_0}{\omega_c}\right)^2$	$0.849P_{norm}\left(\frac{\omega_0}{\omega_c}\right)^2$	$0.500P_{norm}\left(\frac{\omega_0}{\omega_c}\right)^2$
$\frac{V_0 + \phi}{V_B + \phi}$	0.321	0.324	0.375
m_1	0.148	0.155	0.250
m_2	0.091	0.0840	$0.0156\frac{\omega_c}{\omega_0}$
m_3	0.074	0.0775	$0.00065\left(\frac{\omega_c}{\omega_0}\right)^2$

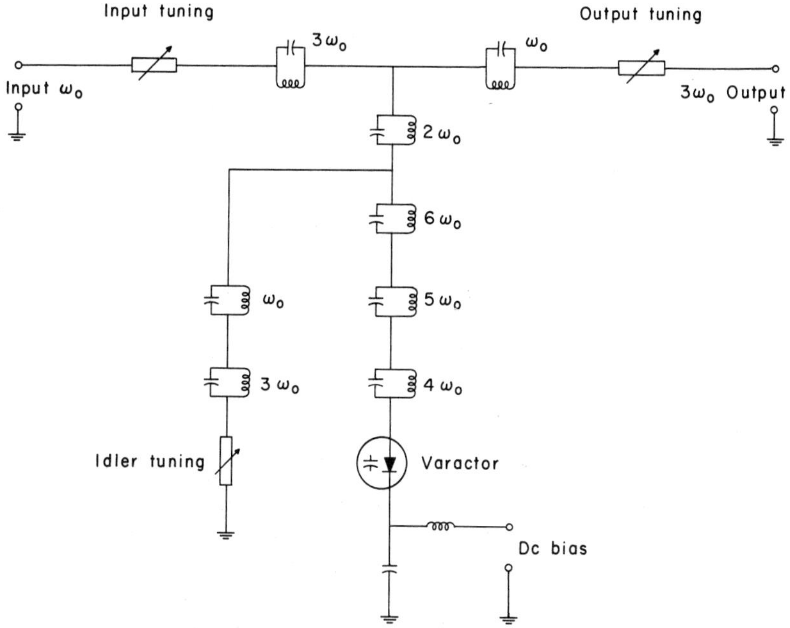

Figure 8.27. Tripler circuit using lumped tuned circuits to separate the frequencies. Separate tuning elements are shown at each of the three frequencies at which current flows.

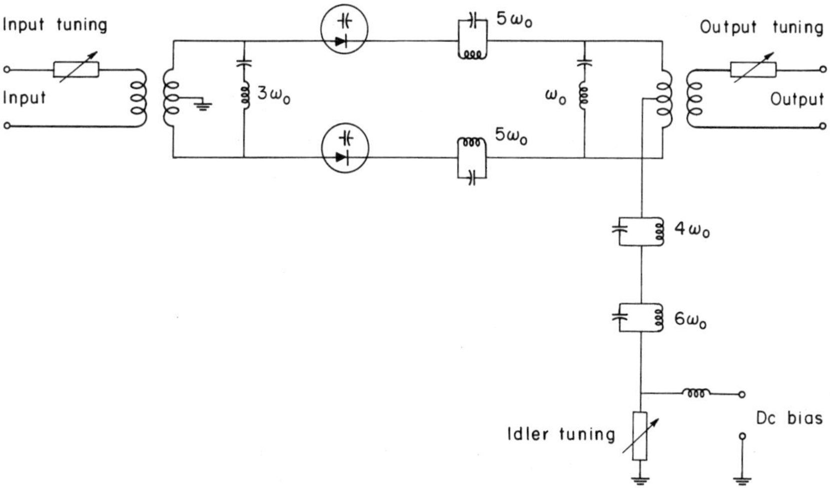

Figure 8.28. Tripler circuit in which the frequency separation is accomplished by circuit symmetries.

To demonstrate that such circuits actually exist, we show two, Figures 8.27 and 8.28. In Figure 8.27, currents at frequencies $4\omega_o$, $5\omega_o$, and $6\omega_o$ are suppressed by the tuned circuits in the middle leg. The idler current is restricted to the left-hand leg, and the other tuned circuits separate the input and output frequencies, which appear at separate terminals. This identical circuit may not be suitable for practical multipliers, but it does indicate that circuits can be designed to achieve deliberately the efficiency, and so forth, predicted earlier in Section 8.6.

In Figure 8.28 we illustrate the use of symmetry in separating the idler from the input and output. All the even harmonics appear in the lower leg, which is open-circuited at the fourth and sixth, while the second is tuned. Traps at $5\omega_o$ are placed in series with the varactors. The input and output circuits are provided with traps for the undesired frequencies. Note that with two varactors present, the power handling is double that of one varactor.

8.7. Quadrupler

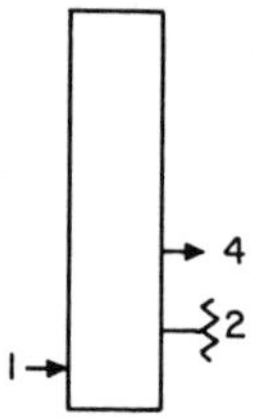

Except for the tripler, the only abrupt-junction multiplier with one idler is the quadrupler, or multiply-by-four circuit. The quadrupler that we analyze here has input at frequency ω_o, output at frequency $4\omega_o$, and a single idler at $2\omega_o$. There is no evidence that current at frequency $3\omega_o$ is harmful to the operation of the quadrupler, but we have picked the simplest idler configuration for our analysis. The analysis is very similar to that of the tripler. The techniques of solution and the general nature of the results are unchanged, although of course the formulas and specific curves are different.

8.7.1. Formulas

Since current flows through the varactor at only frequencies ω_o, $2\omega_o$, and $4\omega_o$, the formulas developed in Section 8.5 apply with $n = 1$, $\ell = 4$, and all I_k and S_k zero except for $k = -4$, -2, -1, 1, 2, and 4 (S_0 is also nonzero). The idler and load resistance equations become, from Equation 8.81,

$$\frac{R_4 + R_s}{R_s} = \frac{\omega_c}{8\omega_o}\frac{(jM_2)^2}{jM_4}, \tag{8.107}$$

$$\frac{R_2 + R_s}{R_s} = \frac{\omega_c}{4\omega_o}\frac{(jM_1)^2 - 2(jM_2)^*(jM_4)}{jM_2}. \tag{8.108}$$

We may, without loss of generality, choose the time origin so that jM_1 is real and positive and, therefore, equal to its magnitude m_1. Then, Equation 8.107 predicts that the phase angle of jM_4 is twice the phase angle of jM_2. This information, used in Equation 8.108, implies that jM_2 (and therefore jM_4) must be real and positive; thus $jM_2 = m_2$ and $jM_4 = m_4$. Then, in terms of the m_k, the formulas for load resistance R_4, idler resistance R_2, input resistance R_{in}, power input P_{in}, output power P_{out}, dissipated power P_{diss}, varactor dissipation $P_{diss,v}$, external idler dissipation $P_{diss,i}$, efficiency ϵ, and bias voltage V_0 are

$$R_4 = R_s\left(\frac{\omega_c}{8\omega_o}\frac{m_2^2}{m_4} - 1\right), \tag{8.109}$$

$$R_2 = R_s\left[\frac{\omega_c}{4\omega_o}\left(\frac{m_1^2}{m_2} - 2m_4\right) - 1\right], \tag{8.110}$$

$$R_{in} = R_s\left(\frac{\omega_c}{\omega_o} m_2 + 1\right), \tag{8.111}$$

$$P_{in} = 8P_{norm}\left(\frac{S_{max} - S_{min}}{S_{max} + S_{min}}\right)^2 \frac{\omega_o^2}{\omega_c^2}\left(\frac{\omega_c}{\omega_o} m_2 m_1^2 + m_1^2\right), \tag{8.112}$$

$$P_{out} = 8P_{norm}\left(\frac{S_{max} - S_{min}}{S_{max} + S_{min}}\right)^2 \frac{\omega_o^2}{\omega_c^2}\left(\frac{2\omega_c}{\omega_o} m_2^2 m_4 - 16m_4^2\right), \tag{8.113}$$

$$P_{diss} = 8P_{norm}\left(\frac{S_{max} - S_{min}}{S_{max} + S_{min}}\right)^2 \frac{\omega_o^2}{\omega_c^2}\left[\frac{\omega_c}{\omega_o}(m_1^2 m_2 - 2m_2^2 m_4) + m_1^2 + 16m_4^2\right], \tag{8.114}$$

$$P_{diss,v} = 8P_{norm}\left(\frac{S_{max} - S_{min}}{S_{max} + S_{min}}\right)^2 \frac{\omega_o^2}{\omega_c^2}(m_1^2 + 4m_2^2 + 16m_4^2), \tag{8.115}$$

$$P_{diss,i} = 8P_{norm}\left(\frac{S_{max} - S_{min}}{S_{max} + S_{min}}\right)^2 \frac{\omega_o^2}{\omega_c^2}\left[\frac{\omega_c}{\omega_o}(m_1^2 m_2 - 2m_2^2 m_4) - 4m_2^2\right], \tag{8.116}$$

$$\epsilon = \frac{2m_4}{m_1} \frac{\frac{\omega_c}{\omega_o} m_2^2 - 8m_4}{\frac{\omega_c}{\omega_o} m_1 m_2 + m_1} \tag{8.117}$$

$$\frac{V_0 + \phi}{V_B + \phi} = \left(\frac{S_{max} - S_{min}}{S_{max}} \right)^2 (m_0^2 + 2m_1^2 + 2m_2^2 + 2m_4^2). \tag{8.118}$$

8.7.2. Technique of Solution

The solution of these equations is similar to the solution of Equations 8.92 to 8.101 for the abrupt-junction tripler. If the values of m_1, m_2, and m_4 are known, then all quantities of interest can be calculated quickly. Often the easiest way to get quick, approximate solutions is to try various combinations of m_1, m_2, and m_4 until the desired solution is obtained. In doing this, however, be careful not to choose m_1, m_2, and m_4 larger than allowable. Since the varactor elastance must lie between S_{min} and S_{max},

$$m_1 \sin \omega_o t + m_2 \sin 2\omega_o t + m_4 \sin 4\omega_o t \leq 0.25 \tag{8.119}$$

for all values of t. In addition, the values used must yield positive R_2 and R_4 from Equations 8.109 and 8.110. Aside from these restrictions, any values of the m_k are suitable.

More often, however, the values of R_2 and R_4 are known, and the m values are not, nor are they usually desired. If R_2, R_4, and some indication of the drive level, such as the magnitude of I_1, are known, then the values of m_k can be calculated. To help in this calculation, Equations 8.109 and 8.110 can be solved for m_2 and m_4 as functions of m_1, the frequency ω_o, the resistances R_2 and R_4, and the varactor parameters. Thus, to find m_2, solve the cubic equation

$$\left(\frac{m_2}{m_1} \right)^3 \frac{R_s}{R_4 + R_s} \frac{m_1 \omega_c}{4\omega_o} + \left(\frac{m_2}{m_1} \right) \frac{R_2 + R_s}{R_s} \frac{4\omega_o}{m_1 \omega_c} - 1 = 0, \tag{8.120}$$

which has only one real root. Then, m_4 can be found from the formula

$$\frac{m_4}{m_1} = \left(\frac{m_2}{m_1} \right)^2 \frac{R_s}{R_4 + R_s} \frac{m_1 \omega_c}{8\omega_o}. \tag{8.121}$$

Alternatively, to avoid this cubic equation, we can solve for m_1 and m_4 in terms of m_2:

$$\left(\frac{m_1}{m_2}\right)^2 = \frac{R_2 + R_s}{R_s} \frac{4\omega_o}{m_2\omega_c} + \frac{R_s}{R_4 + R_s} \frac{m_2\omega_c}{4\omega_o}, \tag{8.122}$$

$$\frac{m_4}{m_2} = \frac{m_2\omega_c}{8\omega_o} \frac{R_s}{R_4 + R_s}. \tag{8.123}$$

The final values of m_1, m_2, and m_4 obey either Equations 8.120 and 8.121, or 8.122 and 8.123, and also are compatible with the prescribed power level. In all cases, check to see that Condition 8.119 is obeyed.

Probably the most important solutions are those for which Condition 8.119 is obeyed with the equality sign; that is, those solutions corresponding to maximum drive. Equations 8.119 and either 8.120 and 8.121, or 8.122 and 8.123 can be used to determine the values of m_1, m_2, and m_4 for specific values of R_2 and R_4. These values are pertinent to cases in which the elastance waveform varies between S_{min} and S_{max} during each cycle.‡

For a given varactor, the resulting values of efficiency, powers, and input resistance depend on the frequency ω_o and the resistances R_2 and R_4. A computer program to solve Equations 8.119, 8.122, and 8.123 simultaneouslv for specified values of ω_o/ω_c, R_2/R_s, and R_4/R_s has been written by B. L. Diamond,[3] of M. I. T. Lincoln Laboratory. In Section 8.7.3, we discuss the solutions he has obtained. These solutions differ very little from approximate maximum-efficiency quadrupler solutions previously given by Diamond.[2]

8.7.3. Solutions for Maximum Drive

We now give solutions for the abrupt-junction quadrupler with maximum drive as obtained by Diamond. The results are qualitatively quite similar to those for the tripler.

For a given frequency and a specific value of R_2, the efficiency and power output at maximum drive as functions of the load resistance look somewhat like Figure 8.16 for the tripler. The two maxima are at slightly different values of R_4, but in most cases the efficiency is not much reduced where the power output is a maximum, and vice versa. Thus for practical purposes both the efficiency and power output can be maximized simul-

‡These solutions are analogous to the doubler solutions along the breakdown curve of Figure 8.3.

taneously, provided the idler circuit is not too lossy.

The point of maximum efficiency, which is quite important, is described in detail in Figures 8.29 to 8.35. In these figures we show, respectively, the efficiency, input resistance, load

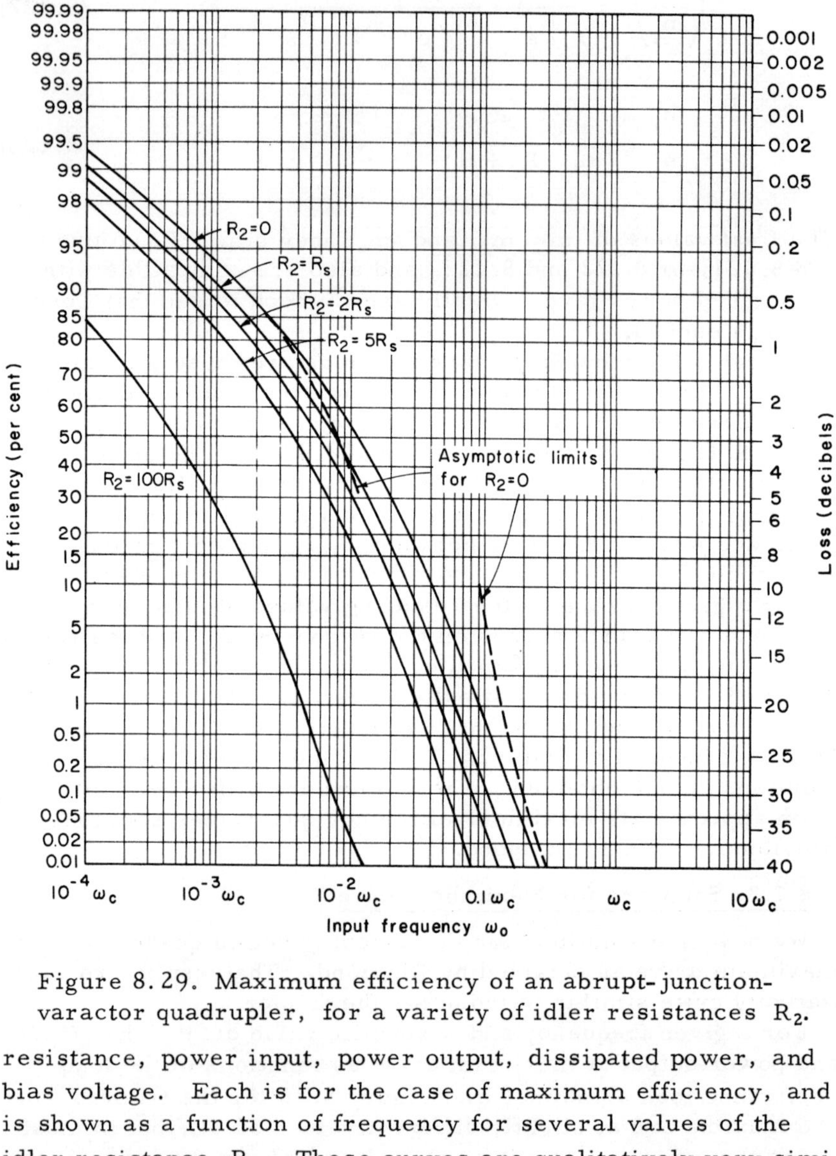

Figure 8.29. Maximum efficiency of an abrupt-junction-varactor quadrupler, for a variety of idler resistances R_2.

resistance, power input, power output, dissipated power, and bias voltage. Each is for the case of maximum efficiency, and is shown as a function of frequency for several values of the idler resistance R_2. These curves are qualitatively very similar to the tripler curves, and are interpreted in the same way.

In Figures 8.32 to 8.35 we have neglected the minimum elastance S_{min}. In case S_{min} is nonzero, the powers should be multiplied by $(S_{max} - S_{min})^2/(S_{max} + S_{min})^2$, and the values

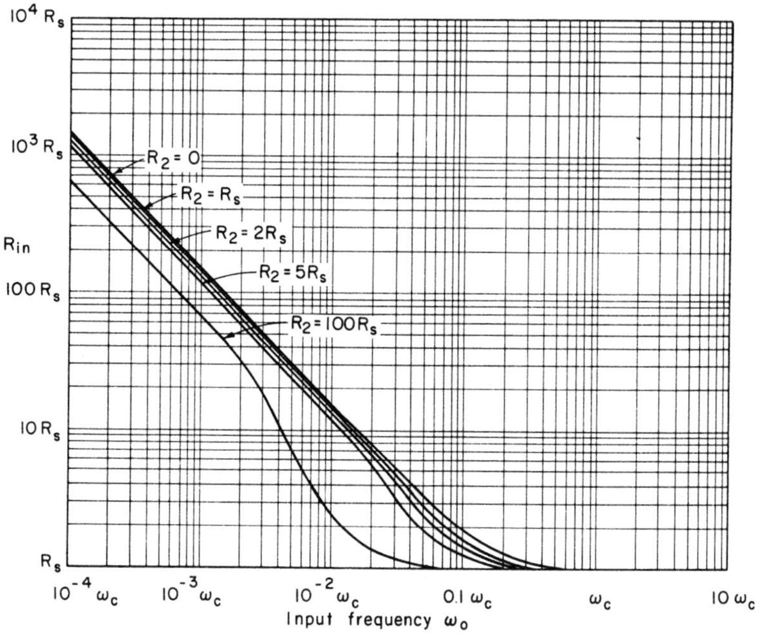

Figure 8.30. Input resistance of an abrupt-junction-varactor quadrupler adjusted for maximum efficiency.

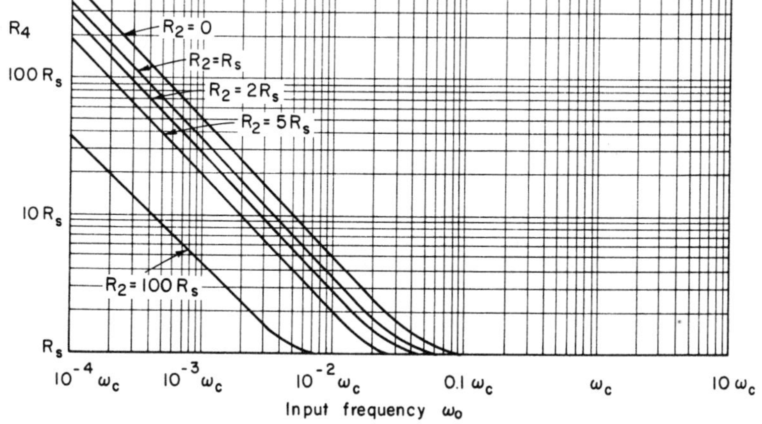

Figure 8.31. Load resistance for maximum efficiency of an abrupt-junction-varactor quadrupler. The load resistance for maximum output power is slightly different.

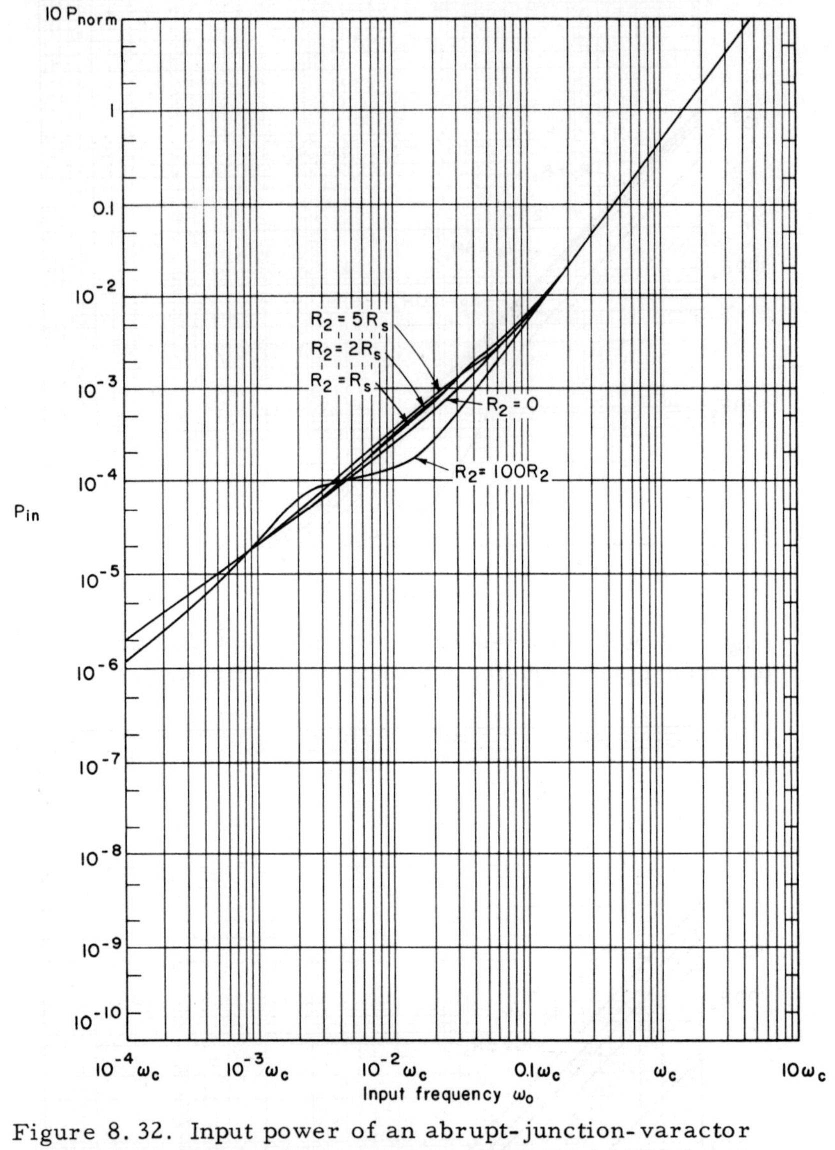

Figure 8.32. Input power of an abrupt-junction-varactor quadrupler adjusted for maximum efficiency. The input power for maximum power output is slightly different.

in Figure 8.35 should be multiplied by $(S_{max} - S_{min})^2/S_{max}^2$ and then added to S_{min}/S_{max}. Since the solution is for maximum drive, the average elastance is given by Equation 8.105, and the bias voltage formula, Equation 8.118, becomes

$$\frac{V_0 + \phi}{V_B + \phi} = \left(\frac{S_{max} - S_{min}}{S_{max}}\right)^2 \left[\frac{1}{4} + 2(m_1^2 + m_2^2 + m_4^2)\right] + \frac{S_{min}}{S_{max}}. \quad (8.124$$

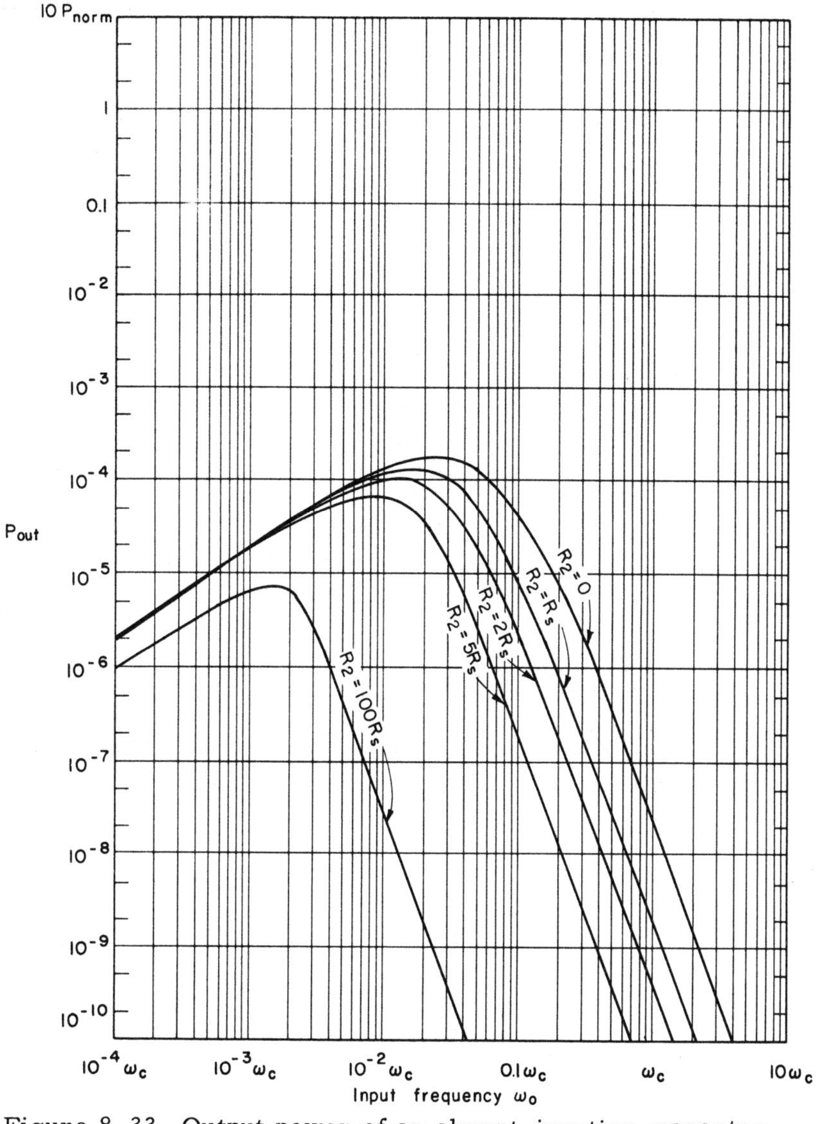

Figure 8.33. Output power of an abrupt-junction-varactor quadrupler with maximum efficiency.

The conditions for maximum power output, rather than maximum efficiency, are very similar. For low values of R_2, the efficiency and the powers are not significantly different from Figures 8.29, 8.32, 8.33, and 8.34. The resistances and bias voltage are slightly different, but not enough to make replotting necessary.

The power output and efficiency are always highest when the idler termination is lossless, that is, when $R_2 = 0$. Figure

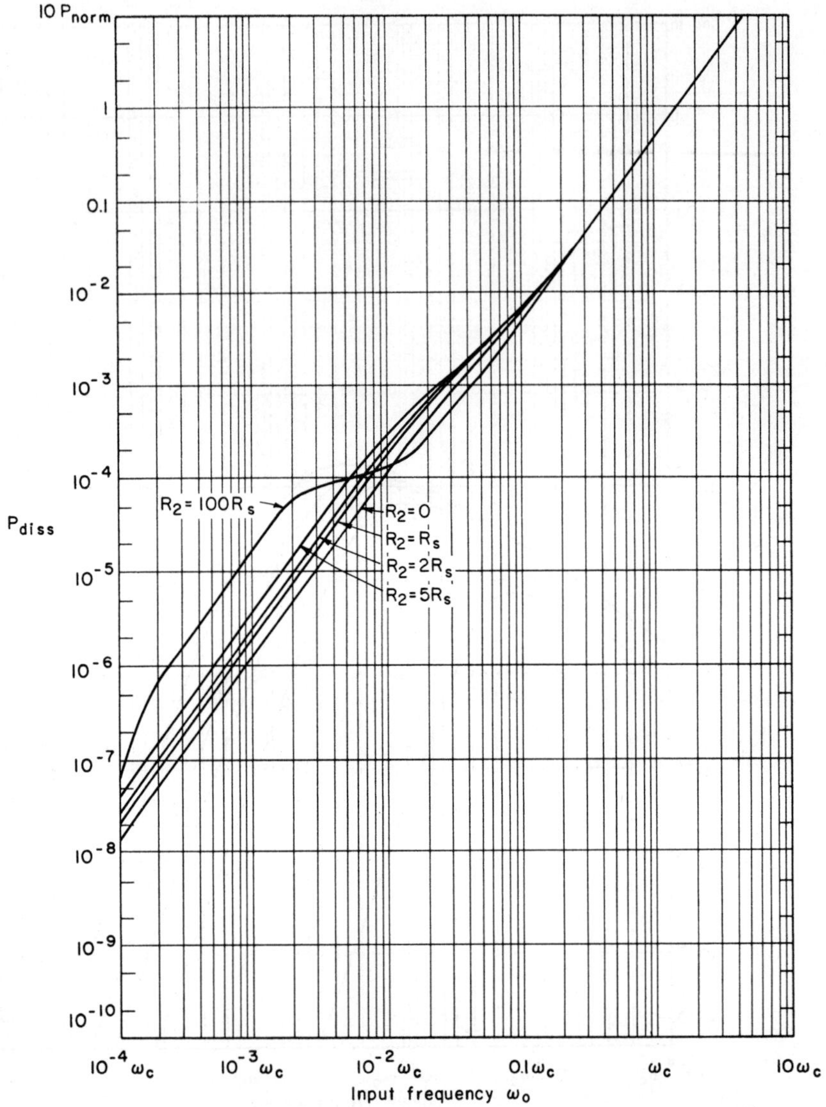

Figure 8.34. Power dissipated (both within the varactor and in the external idler termination) in an abrupt-junction-varactor quadrupler with optimum efficiency.

8.36 shows the input and load resistances for this case, and Figure 8.37 the input, output, and dissipated powers. Some nomographs in Appendix D are based on this important case; these can be used to aid the design of abrupt-junction-varactor quadruplers with maximum efficiency.

It is interesting to look at the elastance waveform of the quadrupler. In Figure 8.38 is the waveform for low frequencies, if

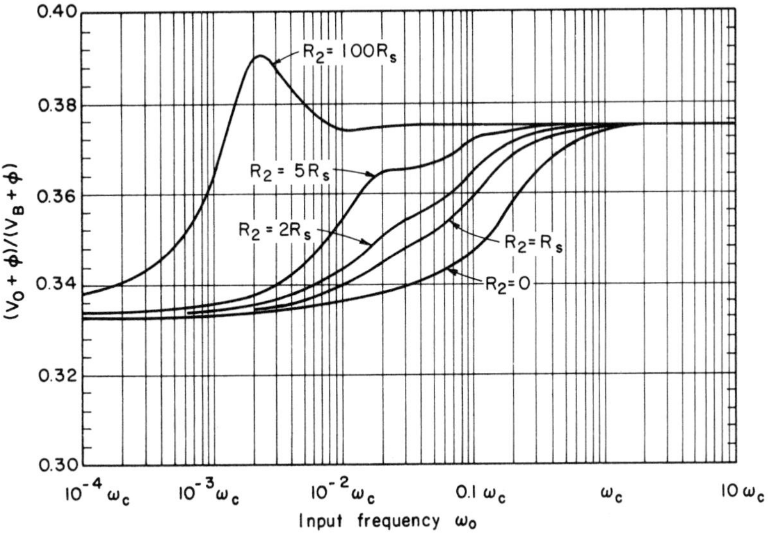

Figure 8. 35. Bias voltage for an abrupt-junction-varactor quadrupler with maximum efficiency. The bias voltage changes slightly when the output power, rather than the efficiency, is maximized.

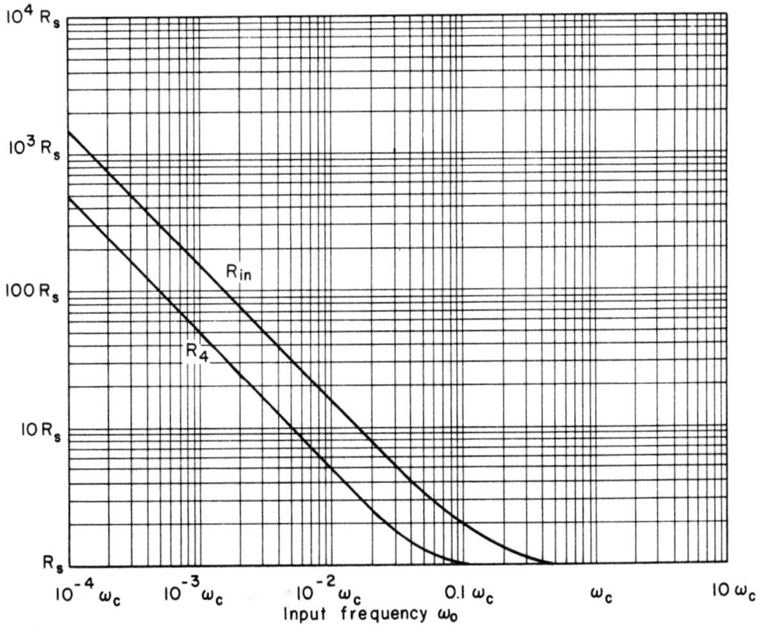

Figure 8. 36. Input and load resistances for an abrupt-junction-varactor quadrupler with maximum efficiency, using a lossless idler termination.

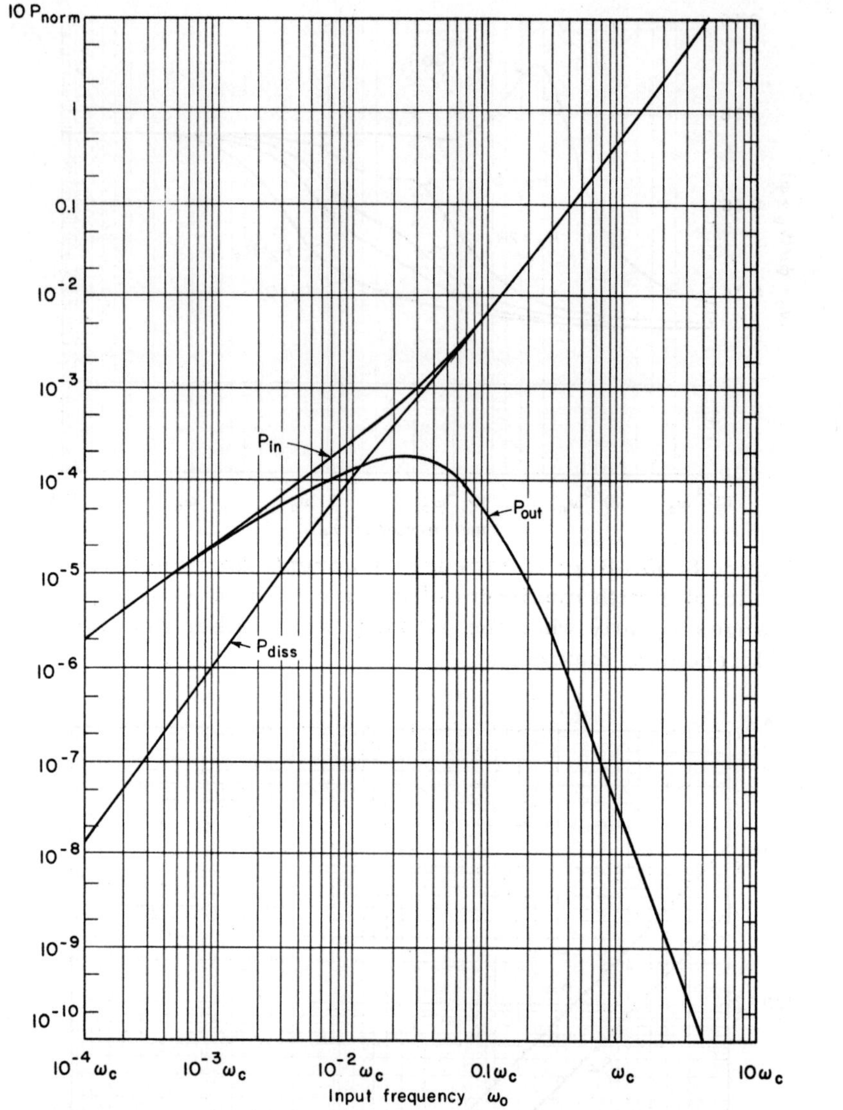

Figure 8.37. Input, output, and dissipated power of an abrupt-junction-varactor quadrupler with maximum efficiency and a lossless idler.

the efficiency is optimized with a lossless idler termination. Since the solution is for maximum drive consistent with Condition 8.119, the elastance varies all the way from S_{min} to S_{max} during each cycle.

The behavior of the quadrupler at high and low frequencies is important. The limiting values of m_1, m_2, and m_4 can be de-

Table 8.7. Asymptotic Formulas for the Quadrupler

Low-frequency and high-frequency formulas are given for the abrupt-junction-varactor quadrupler with idler at $2\omega_0$. For simplicity we have assumed that $S_{min} \ll S_{max}$. Since maximum power output and efficiency are achieved with a lossless idler, we have assumed that $R_2 = 0$.

	Low-Frequency		High-Frequency
	Maximum ϵ	Maximum P_{out}	Maximum ϵ and P_{out}
ϵ	$1 - 62.5\frac{\omega_0}{\omega_c}$	$1 - 66.2\frac{\omega_0}{\omega_c}$	$5.96 \times 10^{-8}\left(\frac{\omega_c}{\omega_0}\right)^6$
R_{in}	$0.150\frac{S_{max}}{\omega_0}$	$0.136\frac{S_{max}}{\omega_0}$	R_s
R_4	$0.205\frac{S_{max}}{4\omega_0}$	$0.136\frac{S_{max}}{4\omega_0}$	R_s
P_{in}	$0.0196P_{norm}\frac{\omega_0}{\omega_c}$	$0.0201P_{norm}\frac{\omega_0}{\omega_c}$	$0.500P_{norm}\left(\frac{\omega_0}{\omega_c}\right)^2$
P_{out}	$0.0196P_{norm}\frac{\omega_0}{\omega_c}$	$0.0201P_{norm}\frac{\omega_0}{\omega_c}$	$2.98 \times 10^{-8}P_{norm}\left(\frac{\omega_c}{\omega_0}\right)^4$
P_{diss}	$1.23P_{norm}\left(\frac{\omega_0}{\omega_c}\right)^2$	$1.33P_{norm}\left(\frac{\omega_0}{\omega_c}\right)^2$	$0.500P_{norm}\left(\frac{\omega_0}{\omega_c}\right)^2$
$\frac{\cdots + \phi}{\cdots 3 + \phi}$	0.334	0.334	0.375
m_1	0.128	0.136	0.250
m_2	0.150	0.136	$0.0156\frac{\omega_c}{\omega_0}$
m_4	0.055	0.068	$1.53 \times 10^{-5}\left(\frac{\omega_c}{\omega_0}\right)^3$

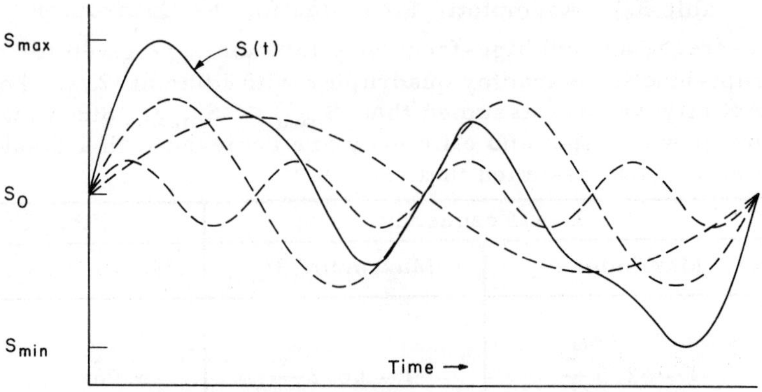

Figure 8.38. Elastance waveform of a low-frequency abrupt-junction-varactor quadrupler adjusted to yield maximum efficiency with a lossless idler termination. The plot is for $m_1 = 0.13$, $m_2 = 0.15$, and $m_4 = 0.053$. The fundamental and second and fourth harmonics are shown individually as well as their sum.

termined from the computer solution and used in Equations 8.109 to 8.118 to find all quantities of interest. In Table 8.7 we show the low- and high-frequency limits, both for highest efficiency and for highest power output. At high frequencies the efficiency and power output are maximized simultaneously, whereas at low frequencies the two points are slightly different.

8.7.4. Synthesis

Synthesis of abrupt-junction quadruplers is quite similar to the synthesis of triplers. The discussion of Section 8.6.4. can be modified very easily to pertain to the quadrupler, by requiring that the varactor be open-circuited at frequencies $3\omega_0$, $5\omega_0$, $6\omega_0$, and $8\omega_0$, instead of the frequencies given there. An obvious modification of Figure 8.27, with additional tuned circuits for the additional frequencies that must be open-circuited, is an example of a circuit designed to fulfill deliberately the conditions of the theory of Section 8.7. Symmetry can also be used to separate the input and output currents.

8.8. Quintupler

According to Table 8.3, there are two possible idler configurations for the abrupt-junction quintupler (or multiply-by-five circuit) with only two idlers. The first has idlers at frequencies $2\omega_0$ and $4\omega_0$, and the other at $2\omega_0$ and $3\omega_0$. There appears to be no a priori way of determining which of these is better, or, in fact, telling whether either of these is better than some other configuration with more than two idlers (such as the 1-2-3-4-5 quintupler). In this section we discuss only the 1-2-3-5 and 1-2-4-5 quintuplers. A direct comparison will not be made because, as we shall see, the 1-2-3-5 quintupler exhibits an unusual behavior about which more study is required.

8.8.1. Formulas for the 1-2-4-5 Quintupler

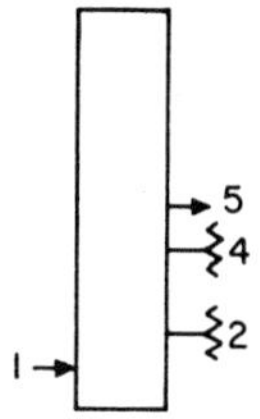

With current flowing only at frequencies ω_o, $2\omega_o$, $4\omega_o$, and $5\omega_o$, the formulas in Section 8.5 apply with $n = 1$ and $\ell = 5$. The idler and load equations become

$$\frac{R_5 + R_s}{R_s} = \frac{\omega_c}{5\omega_o} \frac{(jM_1)(jM_4)}{jM_5}, \tag{8.125}$$

$$\frac{R_4 + R_s}{R_s} = \frac{\omega_c}{8\omega_o} \frac{(jM_2)^2 - 2(jM_1)^*(jM_5)}{jM_4} \tag{8.126}$$

$$\frac{R_2 + R_s}{R_s} = \frac{\omega_c}{4\omega_o} \frac{(jM_1)^2 - 2(jM_2)^*(jM_4)}{jM_2}. \tag{8.127}$$

Let us choose the time origin so that jM_1 is real and positive. Equation 8.125 shows that the angle of jM_5 is equal to the angle of jM_4, and Equation 8.126 indicates that the angle of jM_4 is twice the angle of jM_2. Using this information in Equation 8.127, we see that the angle of jM_2 must be zero, so that $jM_5 = m_5$, $jM_4 = m_4$, $jM_2 = m_2$, and $jM_1 = m_1$.

In terms of m_1, m_2, m_4, and m_5, the equations for load resistance R_5, idler resistances R_2 and R_4, input resistance R_{in}, power input P_{in}, power output P_{out}, dissipated power P_{diss}, varactor dissipation $P_{diss,v}$, external idler dissipation $P_{diss,i}$, efficiency ϵ, and bias voltage V_0 are

$$R_5 = R_s\left(\frac{\omega_c}{5\omega_o}\frac{m_1 m_4}{m_5} - 1\right), \tag{8.128}$$

$$R_4 = R_s\left(\frac{\omega_c}{8\omega_o}\frac{m_2^2 - 2m_1 m_5}{m_4} - 1\right), \tag{8.129}$$

$$R_2 = R_s\left[\frac{\omega_c}{4\omega_o}\left(\frac{m_1^2}{m_2} - 2m_4\right) - 1\right], \tag{8.130}$$

$$R_{in} = R_s \left[\frac{\omega_c}{\omega_o} \left(\frac{m_4 m_5}{m_1} + m_2 \right) + 1 \right], \tag{8.131}$$

$$P_{in} = 8P_{norm} \left(\frac{S_{max} - S_{min}}{S_{max} + S_{min}} \right)^2 \frac{\omega_o^2}{\omega_c^2} \left[\frac{\omega_c}{\omega_o} m_1(m_1 m_2 + m_4 m_5) + m_1^2 \right], \tag{8.132}$$

$$P_{out} = 8P_{norm} \left(\frac{S_{max} - S_{min}}{S_{max} + S_{min}} \right)^2 \frac{\omega_o^2}{\omega_c^2} \left(\frac{5\omega_c}{\omega_o} m_1 m_4 m_5 - 25 m_5^2 \right), \tag{8.133}$$

$$P_{diss} = 8P_{norm} \left(\frac{S_{max} - S_{min}}{S_{max} + S_{min}} \right)^2 \frac{\omega_o^2}{\omega_c^2} \left[\frac{\omega_c}{\omega_o} m_1(m_1 m_2 - 4 m_4 m_5) + m_1^2 + 25 m_5^2 \right] \tag{8.134}$$

$$P_{diss,v} = 8P_{norm} \left(\frac{S_{max} - S_{min}}{S_{max} + S_{min}} \right)^2 \frac{\omega_o^2}{\omega_c^2} (m_1^2 + 4m_2^2 + 16m_4^2 + 25m_5^2), \tag{8.135}$$

$$P_{diss,i} = 8P_{norm} \left(\frac{S_{max} - S_{min}}{S_{max} + S_{min}} \right)^2 \frac{\omega_o^2}{\omega_c^2} \left[\frac{\omega_c}{\omega_o} m_1(m_1 m_2 - 4 m_4 m_5) - 4m_2^2 - 16m_4^2 \right] \tag{8.136}$$

$$\epsilon = \frac{5m_5}{m_1} \frac{\dfrac{\omega_c}{\omega_o} m_1 m_4 - 5m_5}{\dfrac{\omega_c}{\omega_o}(m_1 m_2 + m_4 m_5) + m_1}, \tag{8.137}$$

$$\frac{V_0 + \phi}{V_B + \phi} = \left(\frac{S_{max} - S_{min}}{S_{max}} \right)^2 (m_0^2 + 2m_1^2 + 2m_2^2 + 2m_4^2 + 2m_5^2). \tag{8.138}$$

8.8.2. Technique of Solution of the 1-2-4-5 Quintupler

The solution of these equations is similar to the solution of the other abrupt-junction multiplier equations, such as those for the tripler and the quadrupler. If m_1, m_2, m_4, and m_5 are known, then all quantities of interest can be calculated. On the other hand, if other information is available, the m_k values must be calculated somehow. Sometimes the easiest method of solution is to try values of the m_k until the desired solution is obtained. In choosing values of the m_k, be certain that they are not so large that the elastance is forced below S_{min} or above S_{max}. The values chosen must obey the inequality

$$m_1 \sin \omega_o t + m_2 \sin 2\omega_o t + m_4 \sin 4\omega_o t + m_5 \sin 5\omega_o t \le 0.25 \tag{8.139}$$

for all values of t, and in addition, the values of R_5, R_4, and R_2 predicted from Equations 8.128 to 8.130 must be positive. Aside from these restrictions, any choice of the m_k describes a possible solution for the 1-2-4-5 quintupler.

Often the values of the m_k must be chosen to be compatible with prescribed values of R_5, R_4, and R_2. In this case, it is convenient to solve Equations 8.128 to 8.130 for ratios of the m_k. Thus if it is assumed that m_1 is known, m_2, m_4, and m_5 can be found by solving these equations:

$$\left(\frac{m_2}{m_1}\right)^3 \left(\frac{1}{\dfrac{R_4 + R_s}{R_s}\dfrac{4\omega_o}{m_1\omega_c} + \dfrac{R_s}{R_5 + R_s}\dfrac{m_1\omega_c}{5\omega_o}}\right) + \left(\frac{m_2}{m_1}\right)\left(\frac{R_2 + R_s}{R_s}\frac{4\omega_o}{m_1\omega_c}\right) - 1 = 0, \tag{8.140}$$

$$\frac{m_4}{m_1} = \frac{1/2}{m_2/m_1} - \frac{R_2 + R_s}{R_s}\frac{2\omega_o}{m_1\omega_c}, \tag{8.141}$$

$$\frac{m_5}{m_1} = \frac{R_s}{R_5 + R_s}\frac{m_1\omega_c}{5\omega_o}\frac{m_4}{m_1}. \tag{8.142}$$

The cubic equation for m_2 has only one real root. If, on the other hand, m_2 is assumed to be known, instead of m_1, it is necessary to solve only a quadratic, instead of a cubic equation:

$$\left(\frac{m_4}{m_2}\right)^2 \left(\frac{4}{5}\frac{R_s}{R_5 + R_s}\frac{m_2\omega_c}{\omega_o}\right) + \left(\frac{m_4}{m_2}\right)\left(\frac{8}{5}\frac{R_2 + R_s}{R_5 + R_s} + \frac{R_4 + R_s}{R_s}\frac{8\omega_o}{m_2\omega_c}\right) - 1 = 0, \tag{8.143}$$

$$\left(\frac{m_1}{m_2}\right)^2 = 2\frac{m_4}{m_2} + \frac{R_2 + R_s}{R_s}\frac{4\omega_o}{m_2\omega_c}, \tag{8.144}$$

$$\frac{m_5}{m_2} = \frac{R_s}{R_5 + R_s} \frac{m_2 \omega_c}{5\omega_o} \frac{m_1}{m_2} \frac{m_4}{m_2}. \tag{8.145}$$

If, in addition to the values of R_2, R_4, and R_5, some indication of the power level is given, then the values of m_1, m_2, m_4, and m_5 can be determined by using the specification of the drive level and either Equations 8.140 to 8.142 or 8.143 to 8.145. In no case, however, should Condition 8.139 be violated.

Like the tripler and the quadrupler, the quintupler has particularly important solutions for full drive, that is, for the drive so large that Condition 8.139 is barely obeyed. The largest power capabilities and the highest efficiencies are found for maximum drive. A computer program to solve Equations 8.143 to 8.145 simultaneously for maximum drive (Condition 8.139 with equality attained at one point in each cycle) has been written by Diamond.[3] Diamond's results, which we give in Section 8.8.3, differ qualitatively very little from the results derived so far for the tripler and the quadrupler, except that now there are two separate idler resistances.

8.8.3. Solutions for Maximum Drive, 1-2-4-5 Quintupler

If the load resistance R_5 is varied so as to obtain highest efficiency, this maximum efficiency depends on the idler resistances and the frequency. The maximum efficiency is obtained when the idler terminations are lossless, that is, when $R_2 = R_4 = 0$ Because this case represents the fundamental performance limit, we concentrate on it from now on, and do not present the rest of the results for lossy idlers.

This maximum-efficiency operation is described in Figures 8.39 to 8.42. In these figures we show, respectively, the efficiency, input, output, and dissipated power, input and load resistances, and bias voltage. These plots are very similar to the tripler and quadrupler curves, and are interpreted in the same way. In each case we have set R_2 and R_4 to zero.

In Figures 8.40 and 8.42 we have neglected the minimum elastance S_{min}. In case S_{min} is nonzero, the powers should be multiplied by $(S_{max} - S_{min})^2/(S_{max} + S_{min})^2$, and the values from Figure 8.42 should be multiplied by $(S_{max} - S_{min})^2/S_{max}^2$ and then added to S_{min}/S_{max}. Since the solution here is for maximum drive, the average elastance S_0 is given by Equation 8.105, and the bias voltage V_0 becomes, from Equation 8.138,

$$\frac{V_0 + \phi}{V_B + \phi} = \left(\frac{S_{max} - S_{min}}{S_{max}}\right)^2 \left[\frac{1}{4} + 2(m_1^2 + m_2^2 + m_4^2 + m_5^2)\right] + \frac{S_{min}}{S_{max}}. \tag{8.146}$$

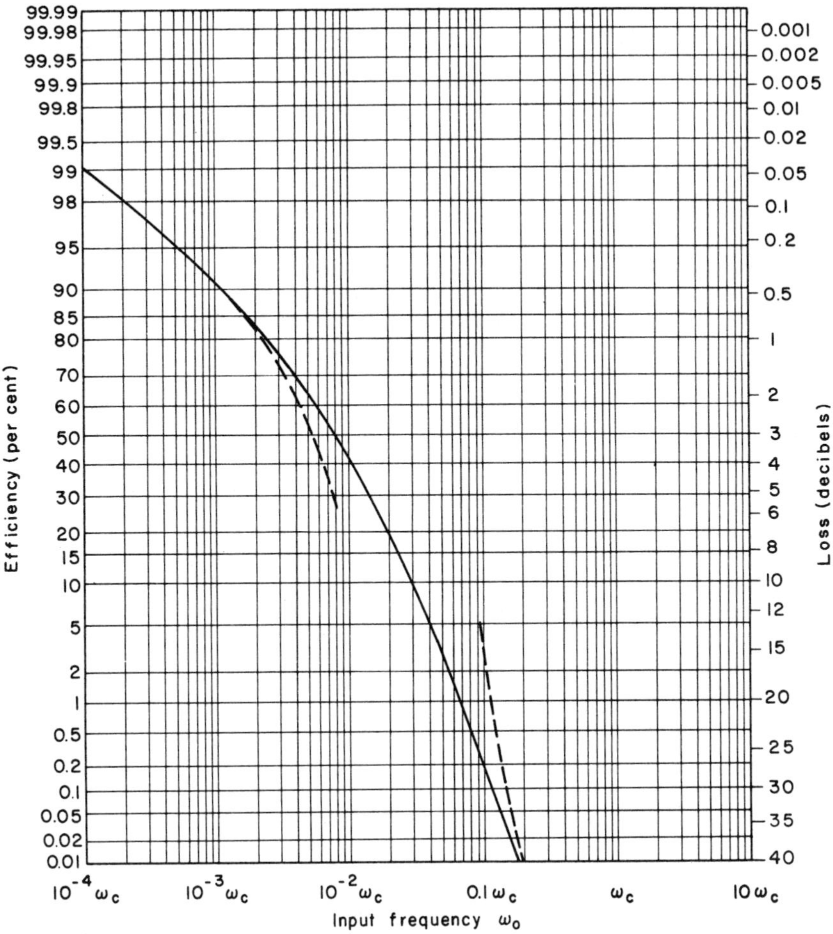

Figure 8.39. Maximum efficiency of an abrupt-junction-varactor 1-2-4-5 quintupler. The dashed curves are the asymptotic curves given in Table 8.8.

If $S_{min} \ll S_{max}$, only the factor in the square brackets remains in Equation 8.146; it is this quantity that is plotted in Figure 8.42.

The solution for maximum power output, rather than maximum efficiency, is so close to Figures 8.39 to 8.42 that no purpose is served by examining it further.

Some nomographs in Appendix D are based on Figures 8.39 and 8.40, these can be used as an aid in preliminary design of maximum-efficiency abrupt-junction-varactor quintuplers with lossless idlers.

It is interesting to consider the elastance waveform. In Figure 8.43 we show the various components of elastance as well as

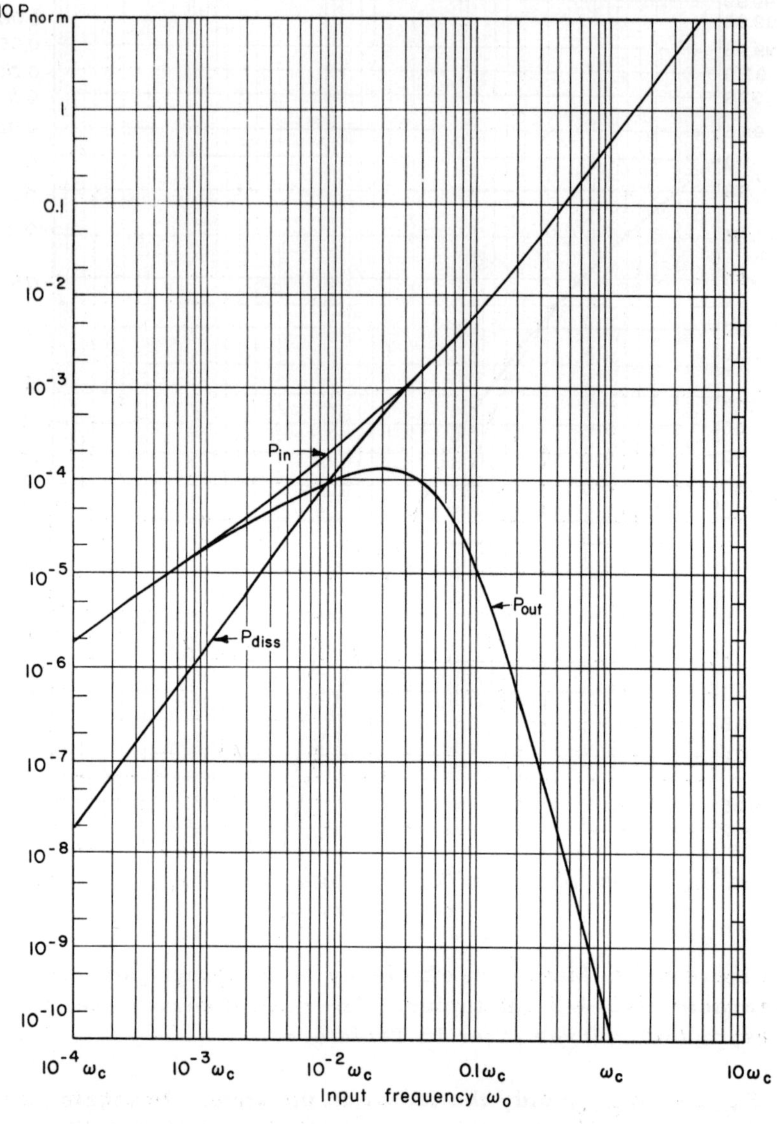

Figure 8.40. Input, output, and dissipated power of an abrupt-junction-varactor 1-2-4-5 quintupler adjusted for maximum efficiency with lossless idlers.

their sum $S(t)$ for the low-frequency quintupler with optimized efficiency. Since the solution pictured is for maximum drive, the elastance varies between S_{min} and S_{max} during each cycle.

The behavior of the quintupler at high and low frequencies is important. The asymptotic behavior is indicated in Table 8.8 for the case of optimized efficiency. It happens that for the

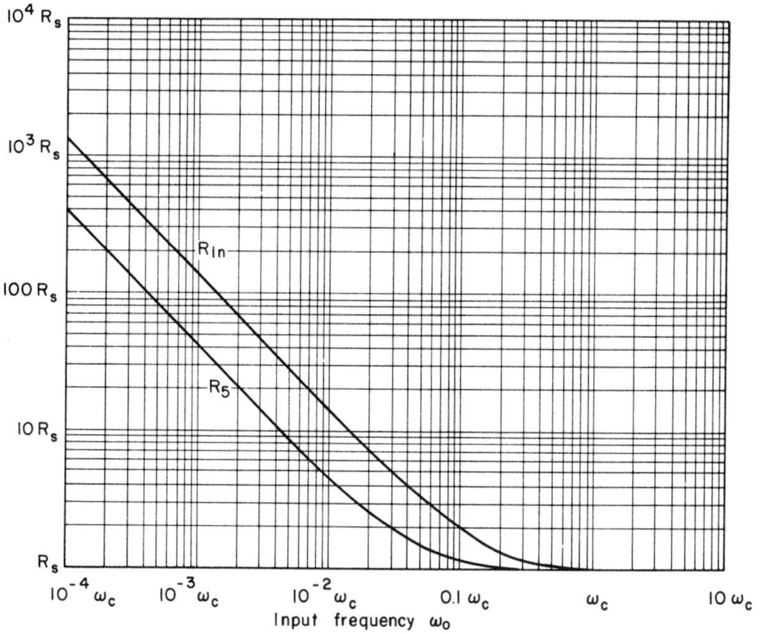

Figure 8.41. Input and load resistances of an abrupt-junction-varactor 1-2-4-5 quintupler for maximum efficiency with lossless idlers.

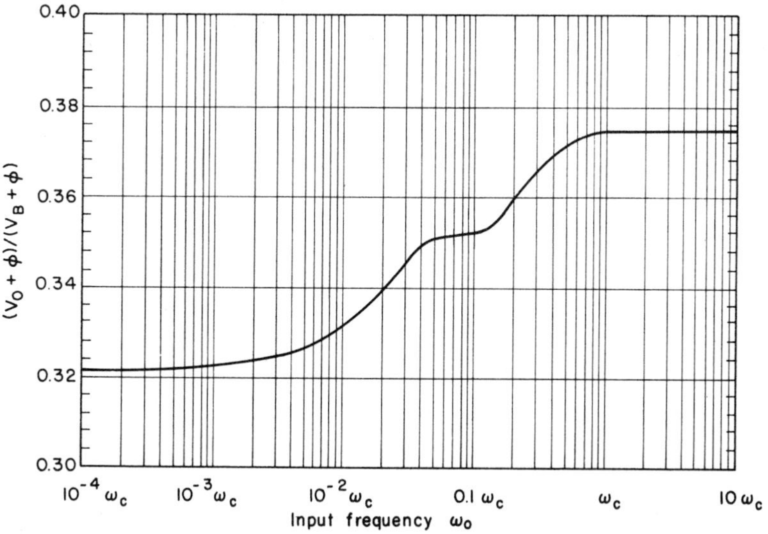

Figure 8.42. Bias voltage of an abrupt-junction-varactor 1-2-4-5 quintupler adjusted for maximum efficiency if the external idler terminations are lossless.

Table 8.8. Asymptotic Formulas for the 1-2-4-5 Quintupler

Low-frequency and high-frequency formulas are given for the abrupt-junction-varactor quintupler with idlers at $2\omega_o$ and $4\omega_o$. We have assumed that $S_{min} \ll S_{max}$ and that $R_2 = R_4 = 0$. To within three decimal places, the values of m_k are the same for maximum power output as for maximum efficiency.

	Low-Frequency Maximum ϵ and P_{out}	High-Frequency Maximum ϵ and P_{out}
ϵ	$1 - 92.9\frac{\omega_o}{\omega_c}$	$2.33 \times 10^{-10}\left(\frac{\omega_c}{\omega_o}\right)^8$
R_{in}	$0.136\frac{S_{max}}{\omega_o}$	R_s
R_5	$0.209\frac{S_{max}}{5\omega_o}$	R_s
P_{in}	$0.0178P_{norm}\frac{\omega_o}{\omega_c}$	$0.500P_{norm}\left(\frac{\omega_o}{\omega_c}\right)^2$
P_{out}	$0.0178P_{norm}\frac{\omega_o}{\omega_c}$	$1.16\times 10^{-10} P_{norm}\left(\frac{\omega_c}{\omega_o}\right)^6$
P_{diss}	$1.65P_{norm}\left(\frac{\omega_o}{\omega_c}\right)^2$	$0.500P_{norm}\left(\frac{\omega_o}{\omega_c}\right)^2$
$\frac{V_0 + \phi}{V_B + \phi}$	0.322	0.375
m_1	0.128	0.250
m_2	0.109	$0.0156\frac{\omega_c}{\omega_o}$
m_4	0.075	$3.05 \times 10^{-5}\left(\frac{\omega_c}{\omega_o}\right)^3$
m_5	0.046	$7.63 \times 10^{-7}\left(\frac{\omega_c}{\omega_o}\right)^4$

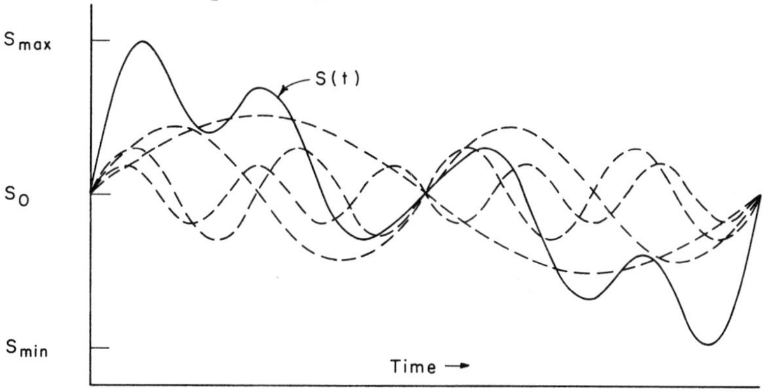

Figure 8.43. Elastance waveform of an abrupt-junction-varactor 1-2-4-5 quintupler, for the values $m_1 = 0.13$, $m_2 = 0.11$, $m_4 = 0.075$, and $m_5 = 0.046$. In addition, each harmonic is shown individually.

1-2-4-5 quintupler the solution with maximum power output, although slightly different, is so close to the maximum-efficiency solution that the values of m_1, m_2, m_4, and m_5 are the same, to within the accuracy shown in Table 8.8. Therefore, only one set of low-frequency limiting formulas is given.

8.8.4. Formulas for the 1-2-3-5 Quintupler

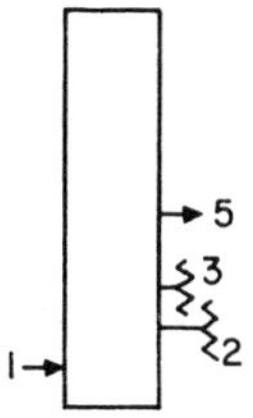

If the second idler of the quintupler is at frequency $3\omega_0$ instead of $4\omega_0$, the formulas are somewhat different from those given in Section 8.8.1. The idler and load equations are

$$\frac{R_5 + R_s}{R_s} = \frac{\omega_c}{5\omega_0} \frac{(jM_2)(jM_3)}{jM_5}, \tag{8.147}$$

$$\frac{R_3 + R_s}{R_s} = \frac{\omega_c}{3\omega_0} \frac{(jM_1)(jM_2) - (jM_2)^*(jM_5)}{jM_3}, \tag{8.148}$$

$$\frac{R_2 + R_s}{R_s} = \frac{\omega_c}{4\omega_0} \frac{(jM_1)^2 - 2(jM_1)^*(jM_3) - 2(jM_3)^*(jM_5)}{jM_2}. \tag{8.149}$$

We may, without loss of generality, choose the time $t = 0$ so that jM_1 is real and positive. Equation 8.147 indicates that the angle of jM_5 is the sum of the angles of jM_2 and jM_3. Then, Equation 8.148 shows that the angle of jM_3 is the same as the angle of jM_2. Finally, from Equation 8.149 we note that these facts imply that jM_2 is positive and real. Thus $jM_1 = m_1$, $jM_2 = m_2$, $jM_3 = m_3$, and $jM_5 = m_5$.

In terms of m_1, m_2, m_3, and m_5, the equations in Section 8.5 for load resistance, idler resistances, input resistance, power input, power output, dissipated power, varactor dissipation, external idler dissipation, efficiency, and bias voltage reduce to

$$R_5 = R_s \left(\frac{\omega_c}{5\omega_o} \frac{m_2 m_3}{m_5} - 1 \right), \tag{8.150}$$

$$R_3 = R_s \left[\frac{\omega_c}{3\omega_o} \frac{m_2(m_1 - m_5)}{m_3} - 1 \right], \tag{8.151}$$

$$R_2 = R_s \left[\frac{\omega_c}{4\omega_o} \frac{m_1^2 - 2m_3(m_1 + m_5)}{m_2} - 1 \right], \tag{8.152}$$

$$R_{in} = R_s \left[\frac{\omega_c}{\omega_o} \left(\frac{m_2 m_3}{m_1} + m_2 \right) + 1 \right], \tag{8.153}$$

$$P_{in} = 8P_{norm} \left(\frac{S_{max} - S_{min}}{S_{max} + S_{min}} \right)^2 \frac{\omega_o^2}{\omega_c^2} \left[\frac{\omega_c}{\omega_o} m_1 m_2 (m_1 + m_3) + m_1^2 \right], \tag{8.154}$$

$$P_{out} = 8P_{norm} \left(\frac{S_{max} - S_{min}}{S_{max} + S_{min}} \right)^2 \frac{\omega_o^2}{\omega_c^2} \left(\frac{5\omega_c}{\omega_o} m_2 m_3 m_5 - 25 m_5^2 \right), \tag{8.155}$$

$$P_{diss} = 8P_{norm} \left(\frac{S_{max} - S_{min}}{S_{max} + S_{min}} \right)^2 \frac{\omega_o^2}{\omega_c^2} \left[\frac{\omega_c}{\omega_o} m_2 (m_1^2 + m_1 m_3 - 5 m_3 m_5) + m_1^2 + 25 m_5^2 \right], \tag{8.156}$$

$$P_{diss,v} = 8P_{norm} \left(\frac{S_{max} - S_{min}}{S_{max} + S_{min}} \right)^2 \frac{\omega_o^2}{\omega_c^2} (m_1^2 + 4m_2^2 + 9m_3^2 + 25m_5^2), \tag{8.157}$$

$$P_{diss,i} = 8P_{norm} \left(\frac{S_{max} - S_{min}}{S_{max} + S_{min}} \right)^2 \frac{\omega_o^2}{\omega_c^2} \left[\frac{\omega_c}{\omega_o} m_2 (m_1^2 + m_1 m_3 - 5 m_3 m_5) - 4m_2^2 - 9m_3^2 \right] \tag{8.158}$$

$$\epsilon = \frac{5m_5}{m_1} \frac{\frac{\omega_c}{\omega_o} m_2 m_3 - 5m_5}{\frac{\omega_c}{\omega_o} m_2(m_1 + m_3) + m_1}, \quad (8.159)$$

$$\frac{V_0 + \phi}{V_B + \phi} = \left(\frac{S_{max} - S_{min}}{S_{max}}\right)^2 (m_0^2 + 2m_1^2 + 2m_2^2 + 2m_3^2 + 2m_5^2).$$

(8.160)

8.8.5. Technique of Solution of the 1-2-3-5 Quintupler

The solution of these equations is similar to the solution of Equations 8.128 to 8.138 for the 1-2-4-5 quintupler. If m_1, m_2, m_3, and m_5 are known, then all quantities of interest can be calculated. On the other hand, if other information is available, the m_k values must be calculated somehow. If values of the m_k are chosen so that the desired solution is obtained, the problem is solved. Often the simplest technique is to try values until a set is found that is satisfactory. The values must be small enough so that at all times the elastance lies between S_{min} and S_{max}:

$$m_1 \sin \omega_0 t + m_2 \sin 2\omega_0 t + m_3 \sin 3\omega_0 t + m_5 \sin 5\omega_0 t \le 0.25$$

(8.161)

for all values of t. In addition, the values of R_5, R_3, and R_2 calculated from Equations 8.150 to 8.152 must be positive. Aside from these conditions, any choice of the m_k describes a possible solution for the 1-2-3-5 quintupler.

Often the values of the m_k must be chosen to be compatible with prescribed values of R_5, R_3, and R_2. In this case, it is helpful to have Equations 8.150 to 8.152 solved for ratios of the m_k. Thus, if m_1 is assumed to be known, then m_2, m_3, and m_5 can be obtained by solving a fifth-degree polynomial. On the other hand, if m_2 is known, the other m_k can be found from these formulas:

$$\left(\frac{m_3}{m_2}\right)^2 = \frac{\frac{4\omega_0}{m_2\omega_c} \frac{R_2 + R_s}{R_s}}{\left(\frac{3\omega_0}{m_2\omega_c} \frac{R_3 + R_s}{R_s} + \frac{m_2\omega_c}{5\omega_0} \frac{R_s}{R_5 + R_s}\right)^2 - \frac{6\omega_0}{m_2\omega_c} \frac{R_3 + R_s}{R_s} - \frac{4}{5} \frac{m_2\omega_c}{\omega_0} \frac{R_s}{R_5 + R_s}},$$

(8.162)

$$\frac{m_1}{m_2} = \frac{m_3}{m_2}\left(\frac{3\omega_o}{m_2\omega_c}\frac{R_3 + R_s}{R_s} + \frac{m_2\omega_c}{5\omega_o}\frac{R_s}{R_5 + R_s}\right), \quad (8.163)$$

$$\frac{m_5}{m_2} = \frac{m_3}{m_2}\left(\frac{m_2\omega_c}{5\omega_o}\frac{R_s}{R_5 + R_s}\right). \quad (8.164)$$

If, in addition to the values of R_2, R_3, and R_5, some indication of the drive level is given, then the values of m_1, m_2, m_3, and m_5 can be determined. In all cases, Condition 8.161 must be obeyed.

8.8.6. Anomaly of the 1-2-3-5 Quintupler

The 1-2-3-5 quintupler has an unusual property‡ that can be appreciated by studying Equation 8.162. If ω_o, m_2, R_3, and R_5 are specified, m_3 is found by taking the square root of this equation. This can be done only if the denominator is positive. If the denominator is negative, the corresponding combination of values of ω_o, m_2, R_3, and R_5 is not possible.‡‡

Exploring this anomaly, we find that the denominator of Equation 8.162 is a function of only two quantities:

$$\frac{m_2\omega_c}{3\omega_o}\frac{R_s}{R_3 + R_s} \quad (8.165)$$

and

$$\frac{m_2\omega_c}{5\omega_o}\frac{R_s}{R_5 + R_s}. \quad (8.166)$$

The denominator is negative for certain combinations of these and positive for other combinations, as indicated in Figure 8.44.

As m_2 is varied for specific values of R_3 and R_5, both the abscissa and the ordinate of Figure 8.44 are increased simultaneously along a straight line going through the origin. Figure 8.45 shows such a line. If m_2 is small, the point describing the operation is close to the origin, in an allowed region. As m_2 increases, this point enters the forbidden region and then comes back out. Thus, if m_2 is high enough (and if Condition 8.161 allows such a high value for the m_k), the operation is again permissible.

---

‡Pointed out to the authors by B. L. Diamond.

‡‡This condition is separate from Condition 8.161, which requires that the elastance be between S_{min} and S_{max}.

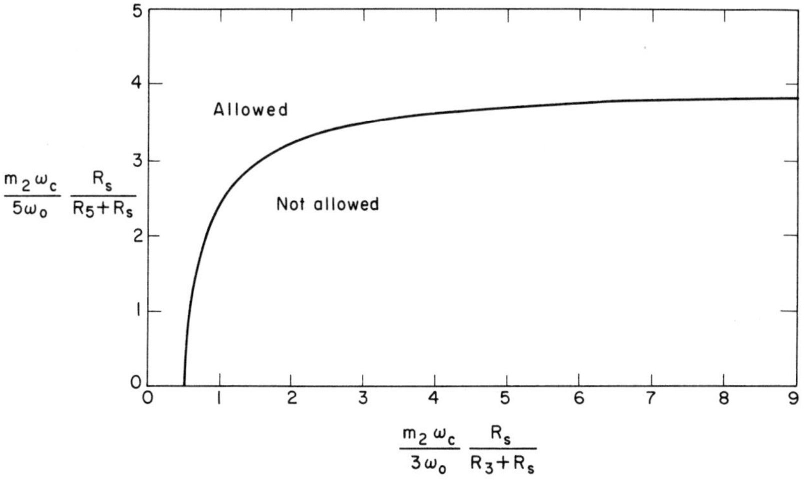

Figure 8.44. Assumed values of R_3, R_5, ω_o, and m_2 must be compatible with the figure for the calculated values of m_1, m_3, and m_5 to be real.

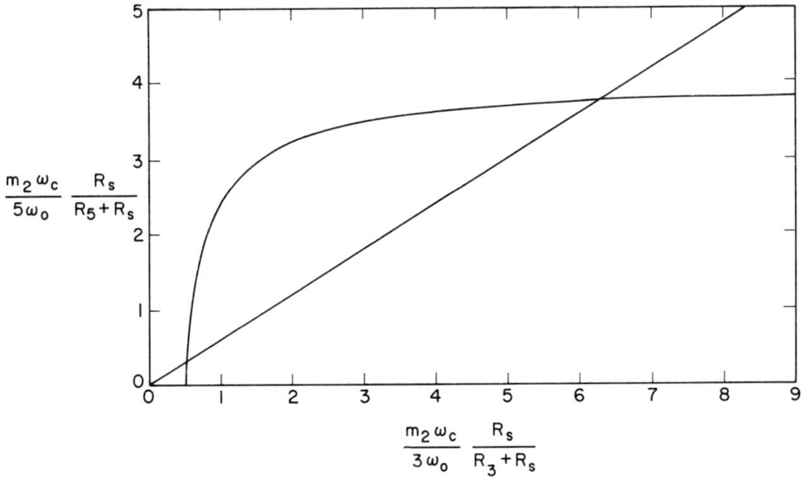

Figure 8.45. Line formed on the plot of Figure 8.44 by varying m_2. This particular line is for $R_3 = 0$, $R_5 = 0$. If $R_3 = 0$ and R_5 is positive, the resulting points fall below the straight line. More generally, the region below the line is for $R_5 > R_3$, and the region above the line is for $R_5 < R_3$.

Consider a quintupler with values of R_3 and R_5 indicated by the line in Figure 8.45. Before the multiplier is excited, $m_2 = 0$, and the point describing the operation is located at the origin of Figure 8.45. If the input power is gradually increased, m_2 increases, and the operation is described in Figure 8.45 by a point that moves along the line indicated. Even if the frequency is low enough to allow operation along this line above the forbidden region, there is no apparent way for the multiplier to reach this allowed region, because it cannot (while it obeys Equations 8.150 to 8.152) go through the forbidden region. Low values of m_2 have, in general, low efficiency, and high values, high efficiency. Although the high-efficiency mode of operation is theoretically possible, it cannot, except possibly by means of some transient effect, be attained by merely turning on the multipler. This effect may make the 1-2-3-5 quintupler impractical as a low-frequency multiplier.

This difficulty always occurs unless the line in Figure 8.45 is steep enough. If

$$\frac{R_3 + R_s}{R_5 + R_s} \geq 4.17, \tag{8.167}$$

then all values of m_2 are allowed; if Condition 8.167 does not hold, then some values of m_2 are forbidden. In particular, some values of m_2 are forbidden by Figure 8.44 whenever $R_3 < 3.17R_s$.

It is clear from Figure 8.44 that if we are far enough to the left, we do not encounter this restriction. Thus, any value of m_2 such that

$$\frac{m_2\omega_c}{\omega_o}\frac{R_s}{R_3 + R_s} \leq 1.5 \tag{8.168}$$

is satisfactory. Similarly, we are above the forbidden region if

$$\frac{m_2\omega_c}{\omega_o}\frac{R_s}{R_5 + R_s} \geq 20.0. \tag{8.169}$$

If any one or more of Conditions 8.167 to 8.169 hold, the corresponding values of ω_o, m_2, R_3, and R_5 are allowable. If none of them holds, the combination of values may or may not be allowable. It should be emphasized that this condition is separate from Condition 8.161; values that are compatible with Figure 8.44 are not necessarily realizable. Both Condition 8.161 and the requirement of Figure 8.44 must hold.

8.8.7. High-Frequency Solution of the 1-2-3-5 Quintupler

At high frequencies the 1-2-3-5 quintupler behaves very well, with an efficiency several times that of the 1-2-4-5 quintupler.

Table 8.9. Asymptotic Formulas for the 1-2-3-5 Quintupler

High-frequency behavior of the abrupt-junction-varactor 1-2-3-5 quintupler with lossless idler terminations and with power output and efficiency simultaneously optimized.

$\epsilon \approx 16.6 \times 10^{-10}\left(\frac{\omega_c}{\omega_o}\right)^8$	$P_{in} \approx 0.500 P_{norm}\left(\frac{\omega_o}{\omega_c}\right)^2$
$R_{in} \approx R_s$	$P_{out} \approx 8.28 \times 10^{-10} P_{norm}\left(\frac{\omega_c}{\omega_o}\right)^6$
$R_5 \approx R_s$	$P_{diss} \approx 0.500 P_{norm}\left(\frac{\omega_o}{\omega_c}\right)^2$
$\frac{V_0 + \phi}{V_B + \phi} \approx 0.375$	
$m_1 \approx 0.250$	$m_3 \approx 0.0013\left(\frac{\omega_c}{\omega_o}\right)^2$
$m_2 \approx 0.0156\frac{\omega_c}{\omega_o}$	$m_5 \approx 2.03 \times 10^{-6}\left(\frac{\omega_c}{\omega_o}\right)^4$

Specifically, if the input frequency is large enough, then Condition 8.168 is satisfied for any positive values of R_3, since m_2 is restricted in value by Condition 8.161. Thus the requirement of Figure 8.44 is automatically satisfied. The high-frequency behavior is shown in Table 8.9. Note particularly that the 1-2-3-5 quintupler has greater power output and efficiency than the 1-2-4-5 quintupler at high frequencies. In making Table 8.9, we have assumed that the idler terminations are lossless, that the varactor is fully pumped, and that the output load is adjusted for maximum power output and efficiency.

8.9. Sextupler

The abrupt-junction sextupler, or multiply-by-six circuit, requires two idlers. Table 8.3 shows that there are two possible idler configurations with only two idlers; the first has the idlers at $2\omega_o$ and $4\omega_o$, and is called the 1-2-4-6 sextupler;

the second has idlers at $2\omega_0$ and $3\omega_0$, and is referred to as the 1-2-3-6 sextupler. There is no a priori way of knowing which of these is better, or in fact of knowing if either is better than a sextupler with more than two idlers. In Sections 8.9.1, 8.9.2, and 8.9.3 we discuss the 1-2-4-6 sextupler, and in Sections 8.9.4, 8.9.5, and 8.9.6, the 1-2-3-6 sextupler. The two are compared in Section 8.9.7.

8.9.1. Formulas for the 1-2-4-6 Sextupler

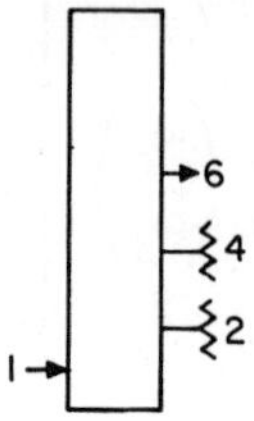

This sextupler has current flowing only at frequencies ω_0, $2\omega_0$, $4\omega_0$, and $6\omega_0$. The formulas in Section 8.5 apply with $n = 1$ and $\ell = 6$. The idler and load equations, Equation 8.81, become

$$\frac{R_6 + R_s}{R_s} = \frac{\omega_c}{6\omega_0} \frac{(jM_2)(jM_4)}{jM_6}, \tag{8.170}$$

$$\frac{R_4 + R_s}{R_s} = \frac{\omega_c}{8\omega_0} \frac{(jM_2)^2 - 2(jM_2)^*(jM_6)}{jM_4}, \tag{8.171}$$

$$\frac{R_2 + R_s}{R_s} = \frac{\omega_c}{4\omega_0} \frac{(jM_1)^2 - 2(jM_2)^*(jM_4) - 2(jM_4)^*(jM_6)}{jM_2}. \tag{8.172}$$

As before, we choose our time origin so that jM_1 is real and positive. Equation 8.170 then shows that the phase angle of jM_6 is the sum of the angles of jM_2 and jM_4. When we use this information in Equation 8.171, we find that the phase angle of jM_4 is twice the angle of jM_2. Finally, this information together with Equation 8.172 shows that the angle of jM_2 is twice the angle of jM_1, or zero. Thus $jM_1 = m_1$, $jM_2 = m_2$, $jM_4 = m_4$, and $jM_6 = m_6$, where, as usual, the m_k are the magnitudes of the M_k.

In terms of m_1, m_2, m_4, and m_6, the formulas from Section 8.5 for load resistance R_6, idler resistances R_4 and R_2, input resistance R_{in}, power input P_{in}, output power P_{out}, dissipated power P_{diss}, varactor dissipation $P_{diss,v}$, idler dissipation $P_{diss,i}$ external to the varactor, efficiency ϵ, and bias voltage V_0 are

$$R_6 = R_s \left(\frac{\omega_c}{6\omega_0} \frac{m_2 m_4}{m_6} - 1 \right), \tag{8.173}$$

$$R_4 = R_s\left[\frac{\omega_c}{8\omega_o}\frac{m_2(m_2 - 2m_6)}{m_4} - 1\right], \qquad (8.174)$$

$$R_2 = R_s\left[\frac{\omega_c}{4\omega_o}\left(\frac{m_1^2 - 2m_4m_6}{m_2} - 2m_4\right) - 1\right], \qquad (8.175)$$

$$R_{in} = R_s\left(\frac{\omega_c}{\omega_o}m_2 + 1\right), \qquad (8.176)$$

$$P_{in} = 8P_{norm}\left(\frac{S_{max} - S_{min}}{S_{max} + S_{min}}\right)^2\frac{\omega_o^2}{\omega_c^2}\left(\frac{\omega_c}{\omega_o}m_1^2m_2 + m_1^2\right), \qquad (8.177)$$

$$P_{out} = 8P_{norm}\left(\frac{S_{max} - S_{min}}{S_{max} + S_{min}}\right)^2\frac{\omega_o^2}{\omega_c^2}\left(\frac{6\omega_c}{\omega_o}m_2m_4m_6 - 36m_6^2\right), \qquad (8.178)$$

$$P_{diss} = 8P_{norm}\left(\frac{S_{max} - S_{min}}{S_{max} + S_{min}}\right)^2\frac{\omega_o^2}{\omega_c^2}\left[\frac{\omega_c}{\omega_o}m_2(m_1^2 - 6m_4m_6) + m_1^2 + 36m_6^2\right], \qquad (8.179)$$

$$P_{diss,v} = 8P_{norm}\left(\frac{S_{max} - S_{min}}{S_{max} + S_{min}}\right)^2\frac{\omega_o^2}{\omega_c^2}\left(m_1^2 + 4m_2^2 + 16m_4^2 + 36m_6^2\right), \qquad (8.180)$$

$$P_{diss,i} = 8P_{norm}\left(\frac{S_{max} - S_{min}}{S_{max} + S_{min}}\right)^2\frac{\omega_o^2}{\omega_c^2}\left[\frac{\omega_c}{\omega_o}m_2(m_1^2 - 6m_4m_6) - 4m_2^2 - 16m_4^2\right], \qquad (8.181)$$

$$\epsilon = \frac{6m_6}{m_1}\,\frac{\dfrac{\omega_c}{\omega_o}m_2m_4 - 6m_6}{\dfrac{\omega_c}{\omega_o}m_1m_2 + m_1}, \qquad (8.182)$$

$$\frac{V_0 + \phi}{V_B + \phi} = \left(\frac{S_{max} - S_{min}}{S_{max}}\right)^2\left(m_0^2 + 2m_1^2 + 2m_2^2 + 2m_4^2 + 2m_6^2\right). \qquad (8.183)$$

8.9.2. Technique of Solution of the 1-2-4-6 Sextupler

The solution of these equations is similar to the solution of the other abrupt-junction multiplier equations, particularly the 1-2-4-5 quintupler. If m_1, m_2, m_4, and m_6 are known, then all quantities of interest can be calculated. On the other hand, if other information is available, the m_k must be calculated. Often the simplest method of calculating the m_k values is to try combinations of m_1, m_2, m_4, and m_6 until the desired solution is obtained. In doing this, however, be careful that the m_k are small enough so that the elastance remains between S_{min} and S_{max} for all time. The values chosen must obey the inequality

$$m_1 \sin \omega_0 t + m_2 \sin 2\omega_0 t + m_4 \sin 4\omega_0 t + m_6 \sin 6\omega_0 t \leq 0.25 \tag{8.184}$$

for all values of t. In addition, the load and idler resistances R_6, R_4, and R_2, as calculated from the m_k by use of Equations 8.173 to 8.175, must be positive. Any set of m_1, m_2, m_4, and m_6 that obeys these restrictions describes a possible solution for the 1-2-4-6 sextupler.

In many cases values of R_6, R_4, and R_2 are given, and the m_k must be selected to be compatible with these. Then, it is helpful to solve Equations 8.173 to 8.175 for the m_k, in terms of one of them. Thus, if m_1 is assumed to be known, the other m_k's are found by solving a fifth-degree polynomial. However, if m_2 is assumed to be known, the values of m_1, m_4, and m_6 can be obtained more easily:

$$\left(\frac{m_4}{m_2}\right) = \frac{1}{\dfrac{8\omega_0}{m_2\omega_c}\dfrac{R_4 + R_s}{R_s} + \dfrac{m_2\omega_c}{3\omega_0}\dfrac{R_s}{R_6 + R_s}}, \tag{8.185}$$

$$\frac{m_6}{m_2} = \frac{m_4}{m_2}\frac{m_2\omega_c}{6\omega_0}\frac{R_s}{R_6 + R_s}, \tag{8.186}$$

$$\left(\frac{m_1}{m_2}\right)^2 = \frac{4\omega_0}{m_2\omega_c}\frac{R_2 + R_s}{R_s} + 2\left(\frac{m_4}{m_2}\right)\left(1 + \frac{m_6}{m_2}\right). \tag{8.187}$$

These three equations, together with some specification of the power level or the drive level, are sufficient to determine m_1, m_2, m_4, and m_6. In all cases Condition 8.184 must not be violated.

The most important solutions for the sextupler are for full drive, that is, for a power level so high that the elastance waveform hits S_{min} and S_{max} periodically but does not go below S_{min} or above S_{max}. The largest efficiency and power-handling capabilities are experienced for maximum drive. Diamond[3] has written a computer program that solves Equations 8.185 to 8.187 simultaneously for maximum drive. The results are qualitatively very similar to the other multipliers.

8.9.3. Solutions for Maximum Drive, 1-2-4-6 Sextupler

We now describe briefly Diamond's results for the 1-2-4-6 sextupler with maximum drive.[3] We suppose the load resistance R_6 is, for each frequency and for each set of idler resistances, varied so as to maximize the efficiency. Since this maximum efficiency is the highest when the idlers are lossless, we assume from now on that $R_2 = R_4 = 0$, and do not even discuss Diamond's results for the case with lossy idlers.

The maximum-efficiency operation results in efficiency, input, output, and dissipated powers, input and load resistances, and bias voltage as shown in Figures 8.46 to 8.49. The plots are qualitatively similar to previous plots for other multipliers, and are interpreted in the same way. In each of these figures we have assumed the idler terminations are lossless. In Figures 8.47 and 8.49 we have assumed that $S_{min} \ll S_{max}$, but if S_{min} is nonzero, the instructions accompanying Figures 8.40 and 8.42 can be used. Since the solution is for maximum drive, the average elastance is

$$S_0 = \frac{S_{max} + S_{min}}{2}, \tag{8.188}$$

and therefore the bias voltage is given by the formula

$$\frac{V_0 + \phi}{V_B + \phi} = \left(\frac{S_{max} - S_{min}}{S_{max}}\right)^2 \left[\frac{1}{4} + 2(m_1^2 + m_2^2 + m_4^2 + m_6^2)\right] + \frac{S_{min}}{S_{max}}. \tag{8.189}$$

The solutions for maximum power output, rather than maximum efficiency, approximate Figures 8.46 and 8.47 so closely that no useful purpose would be served by plotting them. For practical purposes, efficiency and power output are maximized simultaneously, at least for small idler resistances. Some nomographs in Appendix D, based on Figures 8.46 to 8.49, can be used in the practical design of 1-2-4-6 sextuplers with maximum efficiency.

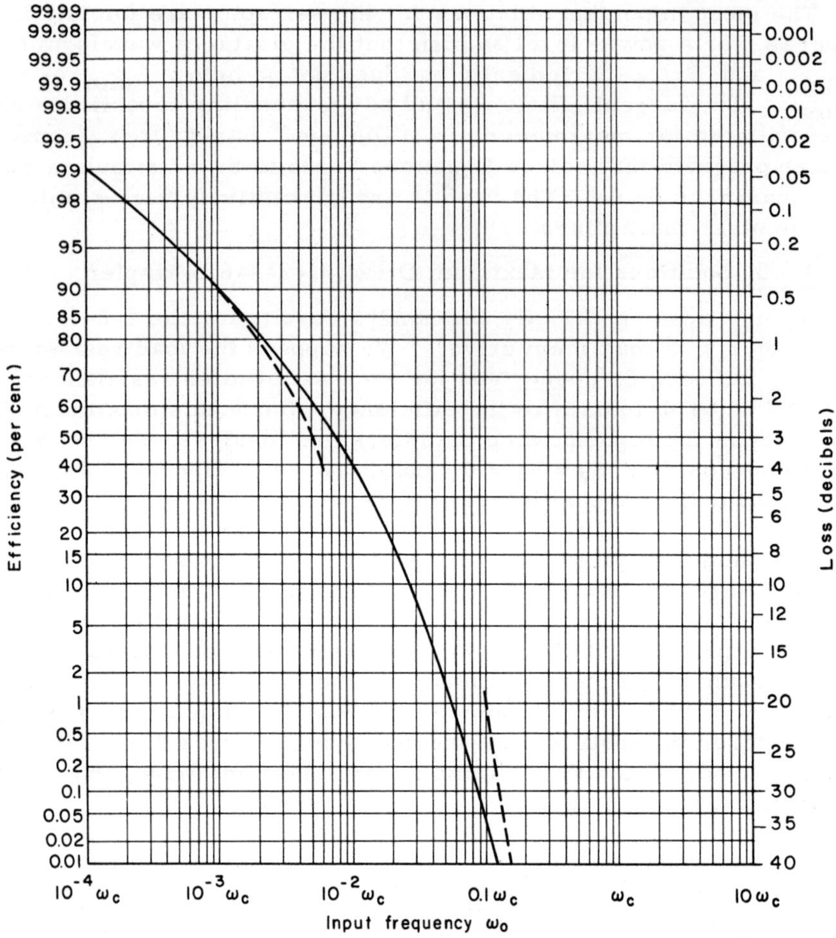

Figure 8.46. Maximum efficiency of a 1-2-4-6 abrupt-junction-varactor sextupler. It is assumed that the idler terminations are tuned and lossless, that the load is tuned and adjusted to give maximum efficiency, and that the varactor is fully driven. The dashed curves are the asymptotic limits given in Table 8.10.

The behavior of the 1-2-4-6 sextupler at high frequencies and low frequencies is important. At low frequencies the efficiency approaches 1, and the input and load resistances vary inversely with the frequency. The input and output power, approximately the same, are proportional to the frequency, whereas the dissipation is proportional to $\omega_0{}^2$. At high frequencies, on the other hand, the input and load resistances approach R_s, and the efficiency drops toward zero. The output power also drops to zero, and the input and dissipated power vary as the square of the fre-

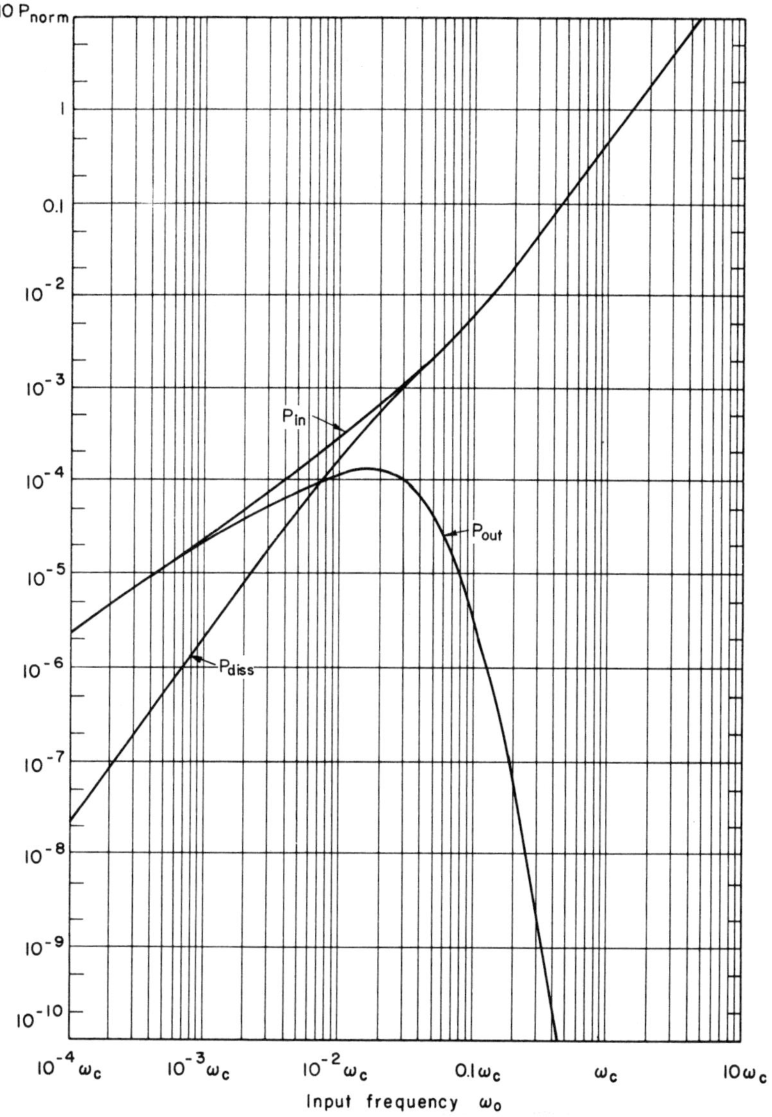

Figure 8.47. Input, output, and dissipated powers for maximum efficiency of an abrupt-junction-varactor 1-2-4-6 sextupler.

quency. This behavior is shown in Table 8.10 for both the maximum-efficiency and maximum-power-output optimizations. Qualitatively, the behavior is similar to that of all the other multipliers. Quantitatively, these limits should be compared with the limits for the 1-2-3-6 sextupler, given in Table 8.11.

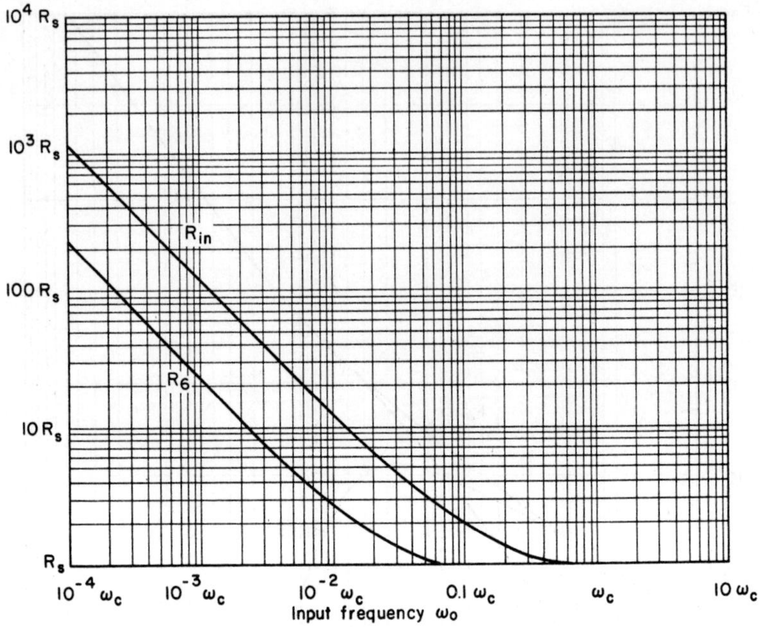

Figure 8.48. Input and load resistances to attain maximum efficiency in an abrupt-junction-varactor 1-2-4-6 sextupler with lossless idlers.

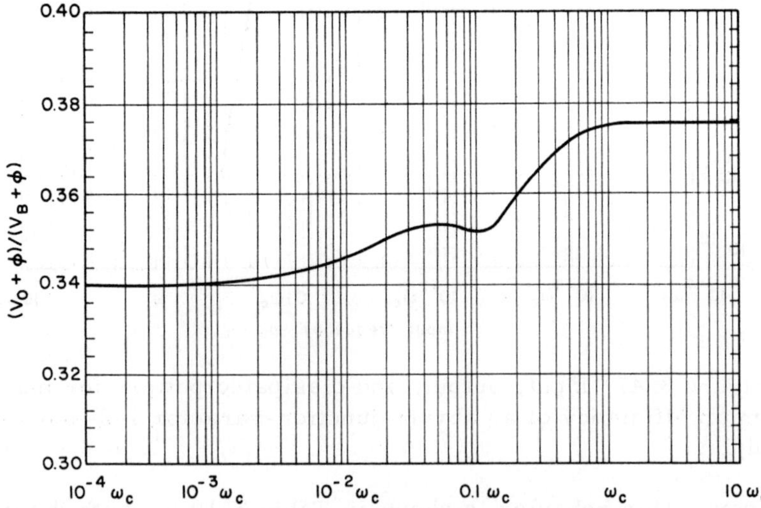

Figure 8.49. Bias voltage for maximum efficiency in an abrupt-junction-varactor 1-2-4-6 sextupler with lossless idlers, if $S_{min} << S_{max}$.

Table 8.10. Asymptotic Formulas for the 1-2-4-6 Sextupler

Low-frequency and high-frequency formulas are given for the abrupt-junction-varactor sextupler with idlers at $2\omega_o$ and $4\omega_o$. We have assumed that S_{min} is negligible in comparison to S_{max}, and that $R_2 = R_4 = 0$.

	Low-Frequency		High-Frequency
	Maximum ϵ	Maximum P_{out}	Maximum ϵ and P_{out}
ϵ	$1 - 99\frac{\omega_o}{\omega_c}$	$1 - 104\frac{\omega_o}{\omega_c}$	$9.10 \times 10^{-13}\left(\frac{\omega_c}{\omega_o}\right)^{10}$
R_{in}	$0.117\frac{S_{max}}{\omega_o}$	$0.130\frac{S_{max}}{\omega_o}$	R_s
R_6	$0.135\frac{S_{max}}{6\omega_o}$	$0.110\frac{S_{max}}{6\omega_o}$	R_s
P_{in}	$0.0219P_{norm}\frac{\omega_o}{\omega_c}$	$0.0225P_{norm}\frac{\omega_o}{\omega_c}$	$0.500P_{norm}\left(\frac{\omega_o}{\omega_c}\right)^2$
P_{out}	$0.0219P_{norm}\frac{\omega_o}{\omega_c}$	$0.0225P_{norm}\frac{\omega_o}{\omega_c}$	$4.55 \times 10^{-13}P_{norm}\left(\frac{\omega_c}{\omega_o}\right)^8$
P_{diss}	$2.18P_{norm}\left(\frac{\omega_o}{\omega_c}\right)^2$	$2.34P_{norm}\left(\frac{\omega_o}{\omega_c}\right)^2$	$0.500P_{norm}\left(\frac{\omega_o}{\omega_c}\right)^2$
$\frac{V_0 + \phi}{V_B + \phi}$	0.340	0.342	0.375
m_1	0.153	0.147	0.250
m_2	0.117	0.131	$0.016\frac{\omega_c}{\omega_o}$
m_4	0.067	0.055	$3.05 \times 10^{-5}\left(\frac{\omega_c}{\omega_o}\right)^3$
m_6	0.058	0.066	$3.97 \times 10^{-8}\left(\frac{\omega_c}{\omega_o}\right)^5$

8.9.4. Formulas for the 1-2-3-6 Sextupler

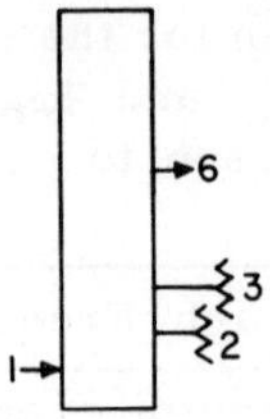

If the second idler of the sextupler is at frequency $3\omega_o$ instead of $4\omega_o$, the formulas are somewhat different from those in Section 8.9.1. The idler and load equations are, from Equation 8.81,

$$\frac{R_6 + R_s}{R_s} = \frac{\omega_c}{12\omega_o} \frac{(jM_3)^2}{jM_6}, \tag{8.190}$$

$$\frac{R_3 + R_s}{R_s} = \frac{\omega_c}{3\omega_o} \frac{(jM_1)(jM_2) - (jM_3)^*(jM_6)}{jM_3}, \tag{8.191}$$

$$\frac{R_2 + R_s}{R_s} = \frac{\omega_c}{4\omega_o} \frac{(jM_1)^2 - 2(jM_1)^*(jM_3)}{jM_2}. \tag{8.192}$$

As before, we find it convenient to choose the time origin so that jM_1 is real and positive. Equation 8.190 implies that the phase angle of jM_6 is twice the phase angle of jM_3. When this information is used in Equation 8.191, we find that the angle of jM_3 is equal to the angle of jM_2. Then, from Equation 8.192, we learn that, since jM_1 is real and positive, jM_2 is also real and positive. Therefore, each jM_k is real and positive; that is, $jM_1 = m_1$, $jM_2 = m_2$, $jM_3 = m_3$, and $jM_6 = m_6$.

In terms of m_1, m_2, m_3, and m_6, the formulas for load resistance, idler resistances R_3 and R_2, input resistance, power input, power output, dissipated power, varactor dissipation, external idler dissipation, efficiency, and bias voltage for the 1-2-3-6 sextupler are

$$R_6 = R_s \left(\frac{\omega_c}{12\omega_o} \frac{m_3^2}{m_6} - 1 \right), \tag{8.193}$$

$$R_3 = R_s \left[\frac{\omega_c}{3\omega_o} \left(\frac{m_1 m_2}{m_3} - m_6 \right) - 1 \right], \tag{8.194}$$

$$R_2 = R_s \left(\frac{\omega_c}{4\omega_o} \frac{m_1^2 - 2m_1 m_3}{m_2} - 1 \right), \tag{8.195}$$

$$R_{in} = R_s \left[\frac{\omega_c}{\omega_o} \left(\frac{m_2 m_3}{m_1} + m_2 \right) + 1 \right], \tag{8.196}$$

$$P_{in} = 8P_{norm}\left(\frac{S_{max} - S_{min}}{S_{max} + S_{min}}\right)^2 \frac{\omega_o^2}{\omega_c^2}\left(\frac{\omega_c}{\omega_o} m_1m_2(m_1 + m_3) + m_1^2\right), \tag{8.197}$$

$$P_{out} = 8P_{norm}\left(\frac{S_{max} - S_{min}}{S_{max} + S_{min}}\right)^2 \frac{\omega_o^2}{\omega_c^2}\left(\frac{3\omega_c}{\omega_o} m_3^2m_6 - 36m_6^2\right), \tag{8.198}$$

$$P_{diss} = 8P_{norm}\left(\frac{S_{max} - S_{min}}{S_{max} + S_{min}}\right)^2 \frac{\omega_o^2}{\omega_c^2}\left[\frac{\omega_c}{\omega_o}(m_1^2m_2 + m_1m_2m_3 - 3m_3^2m_6) + m_1^2 + 36m_6^2\right], \tag{8.199}$$

$$P_{diss,v} = 8P_{norm}\left(\frac{S_{max} - S_{min}}{S_{max} + S_{min}}\right)^2 \frac{\omega_o^2}{\omega_c^2}(m_1^2 + 4m_2^2 + 9m_3^2 + 36m_6^2), \tag{8.200}$$

$$P_{diss,i} = 8P_{norm}\left(\frac{S_{max} - S_{min}}{S_{max} + S_{min}}\right)^2 \frac{\omega_o^2}{\omega_c^2}\left[\frac{\omega_c}{\omega_o}(m_1^2m_2 + m_1m_2m_3 - 3m_3^2m_6) - 4m_2^2 - 9m_3^2\right], \tag{8.201}$$

$$\epsilon = \frac{3m_6}{m_1} \frac{\frac{\omega_c}{\omega_o} m_3^2 - 12m_6}{\frac{\omega_c}{\omega_o} m_2(m_1 + m_3) + m_1}, \tag{8.202}$$

$$\frac{V_0 + \phi}{V_B + \phi} = \left(\frac{S_{max} - S_{min}}{S_{max}}\right)^2 (m_0^2 + 2m_1^2 + 2m_2^2 + 2m_3^2 + 2m_6^2). \tag{8.203}$$

8.9.5. Technique of Solution of the 1-2-3-6 Sextupler

The technique by which these equations are solved is similar to that used for the 1-2-4-6 sextupler. If m_1, m_2, m_3, and m_6 are all known, then all quantities of interest can be calculated. More often, however, the m_k values must be computed from some auxiliary conditions. Sometimes the simplest procedure is to assume values of the m_k until the auxiliary conditions are met. However, the values chosen for each trial must be small enough so that the elastance remains between S_{min} and S_{max}

for all time. As before, the condition on the values of m_1, m_2, m_3, and m_6 is

$$m_1 \sin \omega_0 t + m_2 \sin 2\omega_0 t + m_3 \sin 3\omega_0 t + m_6 \sin 6\omega_0 t \leq 0.25 \tag{8.204}$$

for all values of t. In addition, the values of R_6, R_3, and R_2 calculated from Equations 8.193 to 8.195 must be positive. Except for these restrictions, any set of values for the m_k corresponds to a possible solution for the 1-2-3-6 sextupler.

Often the values of m_1, m_2, m_3, and m_6 must be chosen to be compatible with specified values of R_6, R_3, and R_2. If this is so, it is helpful to be able to solve Equations 8.193 to 8.195 for the remaining m_k when one of the m_k is known. For example, if m_1 is known,

$$\left(\frac{m_3}{m_1}\right)^3 + \frac{m_3}{m_1}\left(\frac{12\omega_0}{m_1\omega_c}\frac{R_6 + R_s}{R_s} - \frac{3\omega_0}{m_1\omega_c}\frac{R_3 + R_s}{R_s} + 6\frac{R_6 + R_s}{R_2 + R_s}\right) - 3\frac{R_6 + R_s}{R_2 + R_s} = 0, \tag{8.205}$$

$$\frac{m_2}{m_1} = \frac{m_1\omega_c}{4\omega_0}\frac{R_s}{R_2 + R_s}\left(1 - 2\frac{m_3}{m_1}\right), \tag{8.206}$$

$$\frac{m_6}{m_1} = \frac{m_1\omega_c}{12\omega_0}\frac{R_s}{R_6 + R_s}\left(\frac{m_3}{m_1}\right)^2. \tag{8.207}$$

The cubic equation for m_3 has only one real root. On the other hand, if m_3 is known, then m_1, m_2, and m_6 can be calculated from the formulas

$$\left(\frac{m_1}{m_3}\right)^3 - 2\left(\frac{m_1}{m_3}\right)^2 - \left(\frac{4\omega_0}{m_3\omega_c}\frac{R_2 + R_s}{R_s} - \frac{3\omega_0}{m_3\omega_c}\frac{R_3 + R_s}{R_s} + \frac{1}{3}\frac{R_2 + R_s}{R_6 + R_s}\right) = 0, \tag{8.208}$$

$$\frac{m_2}{m_3} = \frac{\dfrac{3\omega_0}{m_3\omega_c}\dfrac{R_3 + R_s}{R_s} + \dfrac{m_3\omega_c}{12\omega_0}\dfrac{R_s}{R_6 + R_s}}{\dfrac{m_1}{m_3}}, \tag{8.209}$$

$$\frac{m_6}{m_3} = \frac{m_3\omega_c}{12\omega_o} \frac{R_s}{R_6 + R_s} . \tag{8.210}$$

The cubic equation for m_1 has one real root. In solving in terms of one of the m_k, it is apparently not possible to avoid a cubic equation. If, in addition to the values of R_2, R_3, and R_6, some indication of the drive level, such as the power input, is given, then m_1, m_2, m_3, and m_6 can be determined. In all cases, Condition 8.204 must be obeyed.

The most important solutions for the 1-2-3-6 sextupler are those for full drive, that is, for the drive level so high that in each cycle the elastance goes all the way from S_{min} to S_{max}. The largest efficiencies and power capabilities occur for maximum drive. A computer program to solve Equations 8.205 to 8.207 for maximum drive has been written by Diamond.[3] His results, given now, are compared with the results for the 1-2-4-6 sextupler in Section 8.9.7.

8.9.6. Solutions for Maximum Drive, 1-2-3-6 Sextupler

As a function of the load resistance, the efficiency has a maximum value which depends upon the frequency and the idler resistances R_2 and R_3. The highest efficiency is obtained when the idlers are lossless, that is, when $R_2 = R_3 = 0$. We shall therefore concentrate now on this important case and neglect the less important case of lossy idlers.

The case of maximum efficiency is portrayed in Figures 8.50 to 8.53, where we show, respectively, the efficiency, the input, output, and dissipated powers, the input and load resistances, and the bias voltage. In each plot we have assumed that the idler terminations are lossless, and in Figures 8.51 and 8.53 we have further assumed that the minimum elastance S_{min} is negligible. If S_{min} is not negligible, the instructions accompanying Figures 8.40 and 8.42 can be followed. Since the solution pictured here is for maximum drive, the average elastance is equal to Equation 8.188, so that the bias-voltage formula reduces to

$$\frac{V_0 + \phi}{V_B + \phi} = \left(\frac{S_{max} - S_{min}}{S_{max}}\right)^2 \left[\frac{1}{4} + 2(m_1^2 + m_2^2 + m_3^2 + m_6^2)\right] + \frac{S_{min}}{S_{max}} . \tag{8.211}$$

The solution for maximum power output, rather than maximum efficiency, is so close to this solution that there is little point in replotting the parameters. Some nomographs are given in Appendix D to help the design of 1-2-3-6 sextuplers with maximum efficiency. They are based on Figures 8.50 and 8.51.

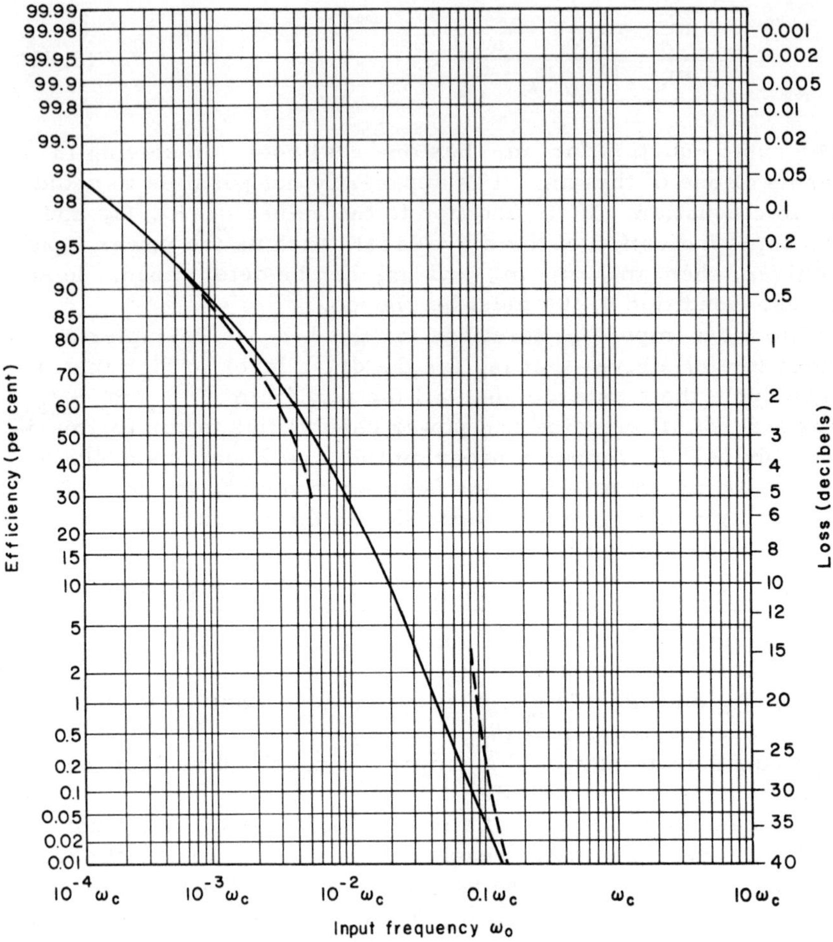

Figure 8.50. Maximum efficiency of an abrupt-junction-varactor 1-2-3-6 sextupler. The dashed curves are the asymptotic limits taken from Table 8.11.

The low-frequency and high-frequency asymptotic limits for the 1-2-3-6 sextupler are indicated in Table 8.11. Shown are the limits both for maximum power output and for maximum efficiency. At low frequencies the input and load resistances are inversely proportional to the frequency, as would be expected because the input and load circuits must "match" an impedance that arises from a capacitance and, therefore, becomes large at low frequencies. The efficiency is close to 100 per cent, and the power input and output are proportional to the frequency. At high frequencies, on the other hand, the input and load resistances approach R_s, as would be expected because at high frequencies the "reactance" of the capacitive part of the varactor becomes

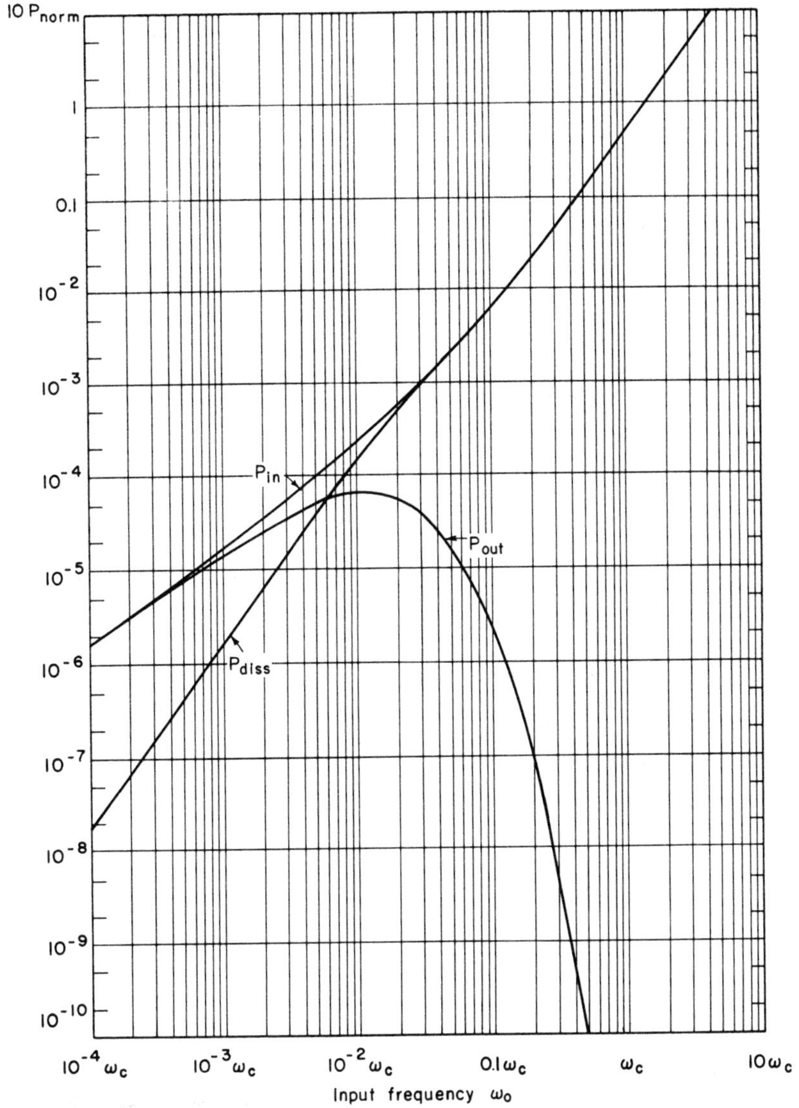

Figure 8.51. Input, output, and dissipated power of an abrupt-junction-varactor 1-2-3-6 sextupler adjusted to have maximum efficiency.

small in comparison with R_s. The efficiency and power output approach zero, and the input and dissipated power rise as the square of the frequency. It is assumed in the formulas for R_{in}, R_6, P_{in}, P_{out}, P_{diss}, and the bias voltage in Table 8.11 that the minimum elastance S_{min} is small in comparison with S_{max}, and that the idler terminations are tuned and lossless.

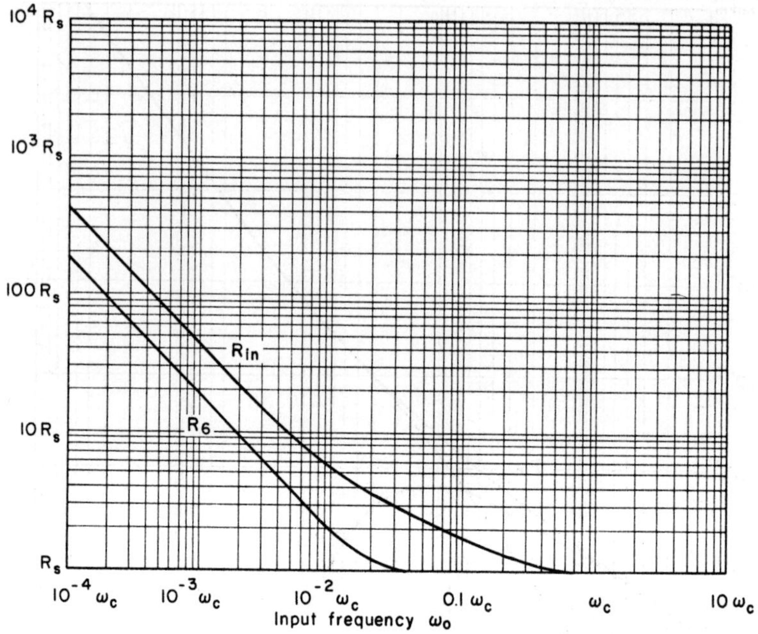

Figure 8.52. Load resistance for maximum efficiency of an abrupt-junction-varactor 1-2-3-6 sextupler with lossless idler terminations. The resulting input resistance is also given.

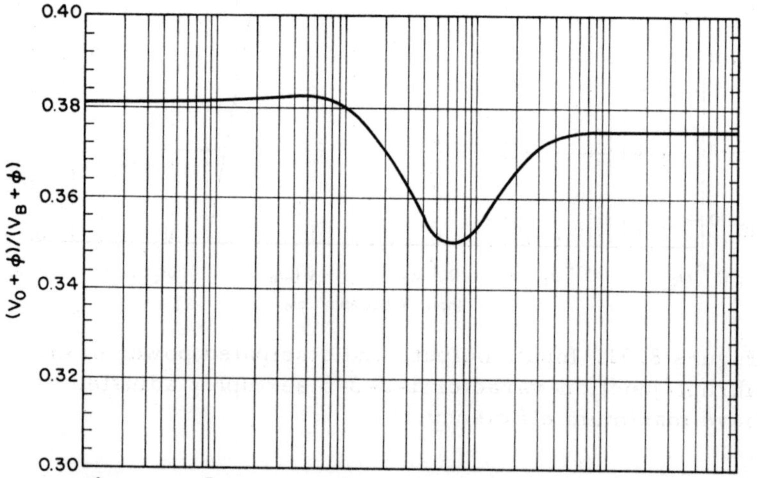

Figure 8.53. Bias voltage for maximum-efficiency operation of an abrupt-junction-varactor 1-2-3-6 sextupler.

Table 8.11. Asymptotic Behavior of the 1-2-3-6 Sextupler

Low-frequency and high-frequency formulas are given for the abrupt-junction-varactor sextupler with idlers at $2\omega_o$ and $3\omega_o$.

	Low-Frequency		High-Frequency
	Maximum ϵ	Maximum P_{out}	Maximum ϵ and P_{out}
ϵ	$1 - 134\frac{\omega_o}{\omega_c}$	$1 - 154\frac{\omega_o}{\omega_c}$	$2.87 \times 10^{-12}\left(\frac{\omega_c}{\omega_o}\right)^{10}$
R_{in}	$0.041\frac{S_{max}}{\omega_o}$	$0.059\frac{S_{max}}{\omega_o}$	R_s
R_6	$0.114\frac{S_{max}}{6\omega_o}$	$0.062\frac{S_{max}}{6\omega_o}$	R_s
P_{in}	$0.0162 P_{norm}\frac{\omega_o}{\omega_c}$	$0.0183 P_{norm}\frac{\omega_o}{\omega_c}$	$0.500 P_{norm}\left(\frac{\omega_o}{\omega_c}\right)^2$
P_{out}	$0.0162 P_{norm}\frac{\omega_o}{\omega_c}$	$0.0183 P_{norm}\frac{\omega_o}{\omega_c}$	$1.44 \times 10^{-12} P_{norm}\left(\frac{\omega_c}{\omega_o}\right)^8$
P_{diss}	$2.17 P_{norm}\left(\frac{\omega_o}{\omega_c}\right)^2$	$2.83 P_{norm}\left(\frac{\omega_o}{\omega_c}\right)^2$	$0.500 P_{norm}\left(\frac{\omega_o}{\omega_c}\right)^2$
$\frac{V_0 + \phi}{V_B + \phi}$	0.381	0.363	0.375
m_1	0.222	0.198	0.250
m_2	0.027	0.039	$0.0156\frac{\omega_c}{\omega_o}$
m_3	0.111	0.099	$0.0013\left(\frac{\omega_c}{\omega_o}\right)^2$
m_6	0.054	0.078	$7.06 \times 10^{-8}\left(\frac{\omega_c}{\omega_o}\right)^5$

8.9.7. Comparison of the 1-2-4-6 and 1-2-3-6 Sextuplers

Since two practical idler configurations exist for the abrupt-junction-varactor sextupler, it is worth-while to compare them. Most of the comparisons can be made by referring either to Figures 8.46 to 8.49 and Table 8.10 or to Figures 8.50 to 8.53 and Table 8.11.

The two maximum efficiencies are shown together in Figure 8.54. Higher efficiency is obtainable from the 1-2-4-6 sextupler, although the difference is not a great deal. The maximum difference in efficiency is about 11 per cent. For very high frequencies, ω_o higher than about $0.1\omega_c$, the 1-2-3-6 sextupler gives

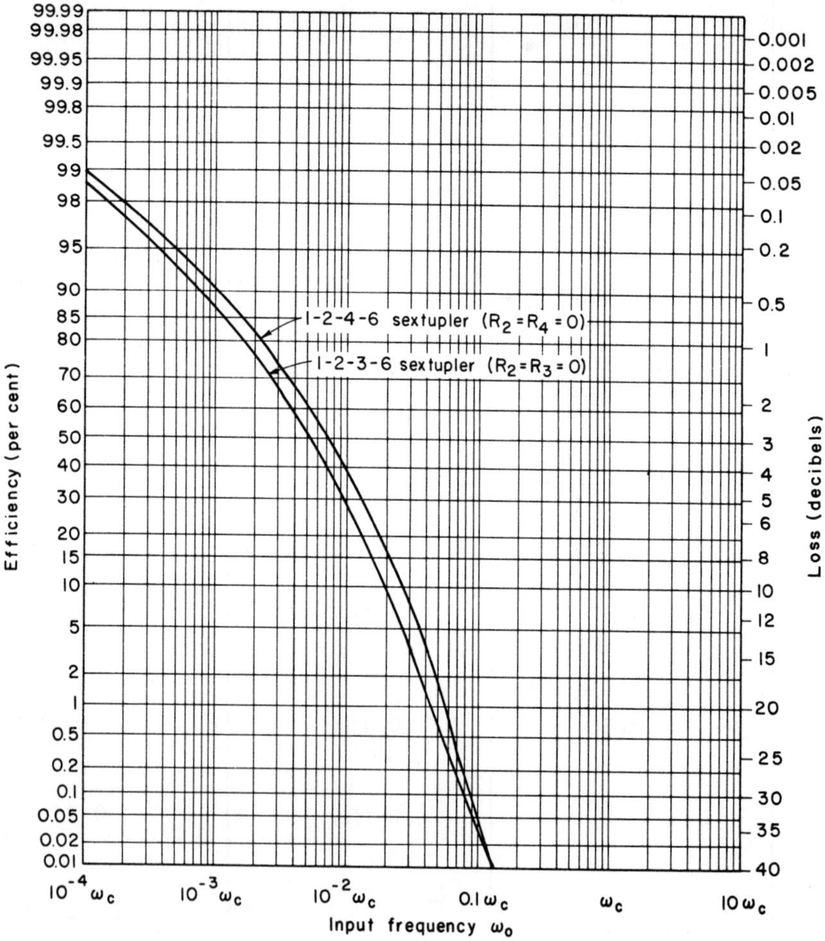

Figure 8.54. Comparison of the maximum efficiency obtainable from the abrupt-junction-varactor 1-2-4-6 and 1-2-3-6 sextuplers. In both cases the idler terminations are lossless.

higher efficiency, although this fact is not evident in Figure 8.54 because both efficiencies are so low.

8.10. Octupler

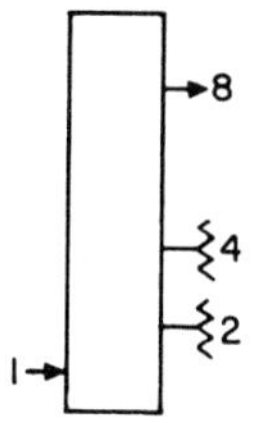

We analyze here the abrupt-junction octupler, or multiply-by-eight circuit, with input at frequency ω_o, output at $8\omega_o$, and current flowing at the two idler frequencies, $2\omega_o$ and $4\omega_o$. This is the only possible two-idler configuration for the octupler.

8.10.1. Formulas

The formulas in Section 8.5 apply to the octupler if $n = 1$ and $\ell = 8$. The idler and load equations are

$$\frac{R_8 + R_s}{R_s} = \frac{\omega_c}{16\omega_o} \frac{(jM_4)^2}{jM_8}, \tag{8.212}$$

$$\frac{R_4 + R_s}{R_s} = \frac{\omega_c}{8\omega_o} \frac{(jM_2)^2 - 2(jM_4)^* (jM_8)}{jM_4}, \tag{8.213}$$

$$\frac{R_2 + R_s}{R_s} = \frac{\omega_c}{4\omega_o} \frac{(jM_1)^2 - 2(jM_2)^* (jM_4)}{jM_2}. \tag{8.214}$$

We see from Equation 8.212 that the phase angle of jM_8 is twice the angle of jM_4. Then, Equation 8.213 indicates that the phase angle of jM_4 is twice the phase angle of jM_2. This information, applied to Equation 8.214, shows that the phase angle of jM_2 is twice the angle of jM_1. We can without loss of generality assume that jM_1 is real and positive; then each jM_k is real and positive and, therefore, equal to its magnitude m_k. Thus, $jM_1 = m_1$, $jM_2 = m_2$, $jM_4 = m_4$, and $jM_8 = m_8$.

In terms of m_1, m_2, m_4, and m_8, the equations for load resistance, idler resistances, input resistance, input power, output power, dissipated power, varactor dissipation, external idler dissipation, efficiency, and bias voltage become, respectively,

$$R_8 = R_s \left(\frac{\omega_c}{16\omega_o} \frac{m_4^2}{m_8} - 1 \right), \tag{8.215}$$

$$R_4 = R_s \left[\frac{\omega_c}{8\omega_o} \left(\frac{m_2^2}{m_4} - 2m_8 \right) - 1 \right], \tag{8.216}$$

$$R_2 = R_s\left[\frac{\omega_c}{4\omega_o}\left(\frac{m_1^2}{m_2} - 2m_4\right) - 1\right], \tag{8.217}$$

$$R_{in} = R_s\left(\frac{\omega_c}{\omega_o} m_2 + 1\right), \tag{8.218}$$

$$P_{in} = 8P_{norm}\left(\frac{S_{max} - S_{min}}{S_{max} + S_{min}}\right)^2 \frac{\omega_o^2}{\omega_c^2}\left(\frac{\omega_c}{\omega_o} m_1^2 m_2 + m_1^2\right), \tag{8.219}$$

$$P_{out} = 8P_{norm}\left(\frac{S_{max} - S_{min}}{S_{max} + S_{min}}\right)^2 \frac{\omega_o^2}{\omega_c^2}\left(\frac{4\omega_c}{\omega_o} m_4^2 m_8 - 64m_8^2\right), \tag{8.220}$$

$$P_{diss} = 8P_{norm}\left(\frac{S_{max} - S_{min}}{S_{max} + S_{min}}\right)^2 \frac{\omega_o^2}{\omega_c^2}\left[\frac{\omega_c}{\omega_o}(m_1^2 m_2 - 4m_4^2 m_8) + m_1^2 + 64m_8^2\right], \tag{8.221}$$

$$P_{diss,v} = 8P_{norm}\left(\frac{S_{max} - S_{min}}{S_{max} + S_{min}}\right)^2 \frac{\omega_o^2}{\omega_c^2}(m_1^2 + 4m_2^2 + 16m_4^2 + 64m_8^2), \tag{8.222}$$

$$P_{diss,i} = 8P_{norm}\left(\frac{S_{max} - S_{min}}{S_{max} + S_{min}}\right)^2 \frac{\omega_o^2}{\omega_c^2}\left[\frac{\omega_c}{\omega_o}(m_1^2 m_2 - 4m_4^2 m_8) - 4m_2^2 - 16m_4^2\right], \tag{8.223}$$

$$\epsilon = \frac{4m_8}{m_1}\,\frac{\frac{\omega_c}{\omega_o} m_4^2 - 16m_8}{\frac{\omega_c}{\omega_o} m_1 m_2 + m_1}, \tag{8.224}$$

$$\frac{V_\upsilon + \phi}{V_B + \phi} = \left(\frac{S_{max} - S_{min}}{S_{max}}\right)^2 (m_0^2 + 2m_1^2 + 2m_2^2 + 2m_4^2 + 2m_8^2). \tag{8.225}$$

8.10.2. Technique of Solution

The technique of solution of these equations is similar to that for the other multipliers. If m_1, m_2, m_4, and m_8 are known, then all quantities of interest can be calculated. On the other hand, if other information is available, the m_k values must be calculated. Often the simplest method of finding them is to try particular values until the desired solution is obtained. The values of the m_k should be small enough so that the elastance waveform cannot be lower than S_{min} or higher than S_{max}. The values must therefore obey the inequality

$$m_1 \sin \omega_0 t + m_2 \sin 2\omega_0 t + m_4 \sin 4\omega_0 t + m_8 \sin 8\omega_0 t \leq 0.25 \tag{8.226}$$

for all values of t. In addition, the values chosen should not yield, from Equations 8.215 to 8.217, negative values of R_8, R_4, or R_2. Aside from these restrictions, any choice of the m_k is permissible.

Often the values of the m_k must be compatible with specific values of R_2, R_4, and R_8. In this case, it is convenient to have Equations 8.215 to 8.217 solved to find, from one of the m_k, all the others. If m_1 is assumed to be known, then the other m_k can be found by solving a fifth-degree polynomial. On the other hand, if m_4 is assumed to be known, then

$$\frac{m_8}{m_4} = \frac{m_4 \omega_c}{16\omega_0} \frac{R_s}{R_8 + R_s}, \tag{8.227}$$

$$\left(\frac{m_2}{m_4}\right)^2 = \frac{8\omega_0}{m_4 \omega_c} \frac{R_4 + R_s}{R_s} + \frac{m_4 \omega_c}{8\omega_0} \frac{R_s}{R_8 + R_s}, \tag{8.228}$$

$$\left(\frac{m_1}{m_4}\right)^2 = \frac{m_2}{m_4} \left(\frac{4\omega_0}{m_4\omega_c} \frac{R_2 + R_s}{R_s} + 2\right). \tag{8.229}$$

These formulas, together with a specification of the drive level, are sufficient to determine the m_k uniquely. In all cases, however, Condition 8.226 must be obeyed.

The most important solutions for the octupler are for maximum drive, in which during each cycle the elastance varies over the whole permissible range from S_{min} to S_{max}. A computer program to solve Equations 8.227 to 8.229 for maximum drive has been written by Diamond.[3]

8.10.3. Solutions for Maximum Drive

Diamond's results[3] for the octupler under maximum drive are given here very briefly. If the load resistance R_8 is varied so as to obtain the highest efficiency for specified values of frequency and idler resistances, this highest efficiency depends (for a given varactor) only upon frequency and idler resistances. The highest efficiency is obtained for lossless idler terminations, that is, when $R_2 = R_4 = 0$. We concentrate on this important case from now on and ignore the solutions for lossy idlers.

Figures 8.55 to 8.58 pertain to the optimum-efficiency case, with lossless idler terminations. Shown, respectively, are the efficiency, the input, output, and dissipated powers, the input and load resistances, and the bias voltage, all as functions of

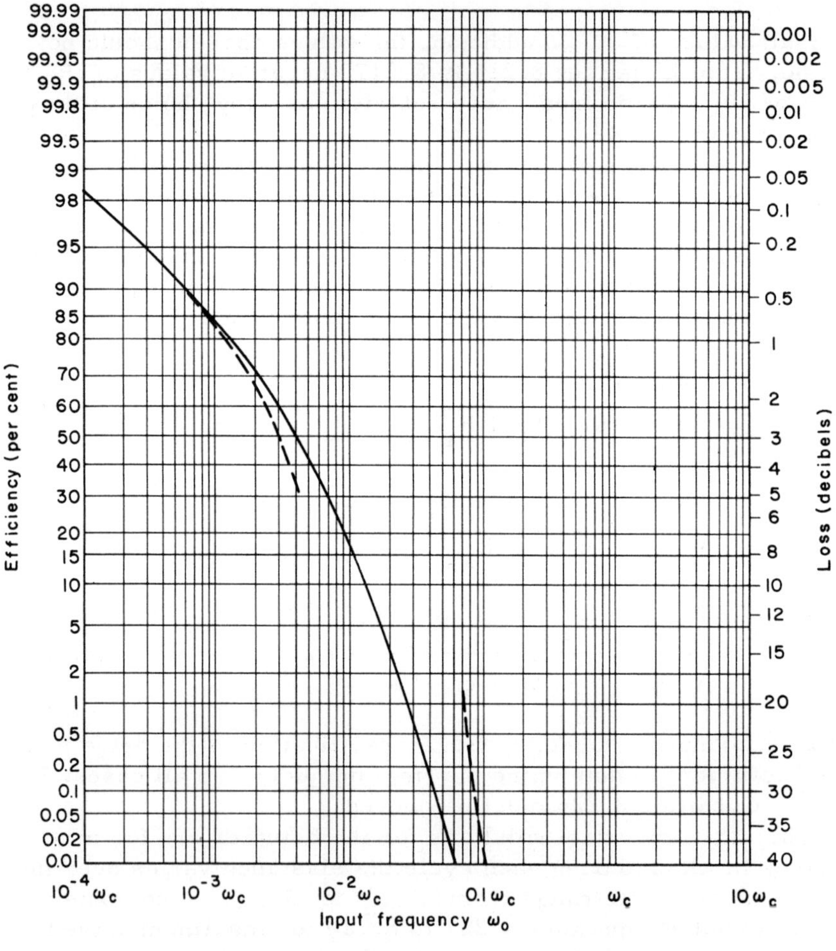

Figure 8.55. Maximum efficiency of an abrupt-junction-varactor octupler. This efficiency is achieved with lossless idler terminations at twice and four times the input frequency. The two dashed curves are the asymptotic formulas given in Table 8.12.

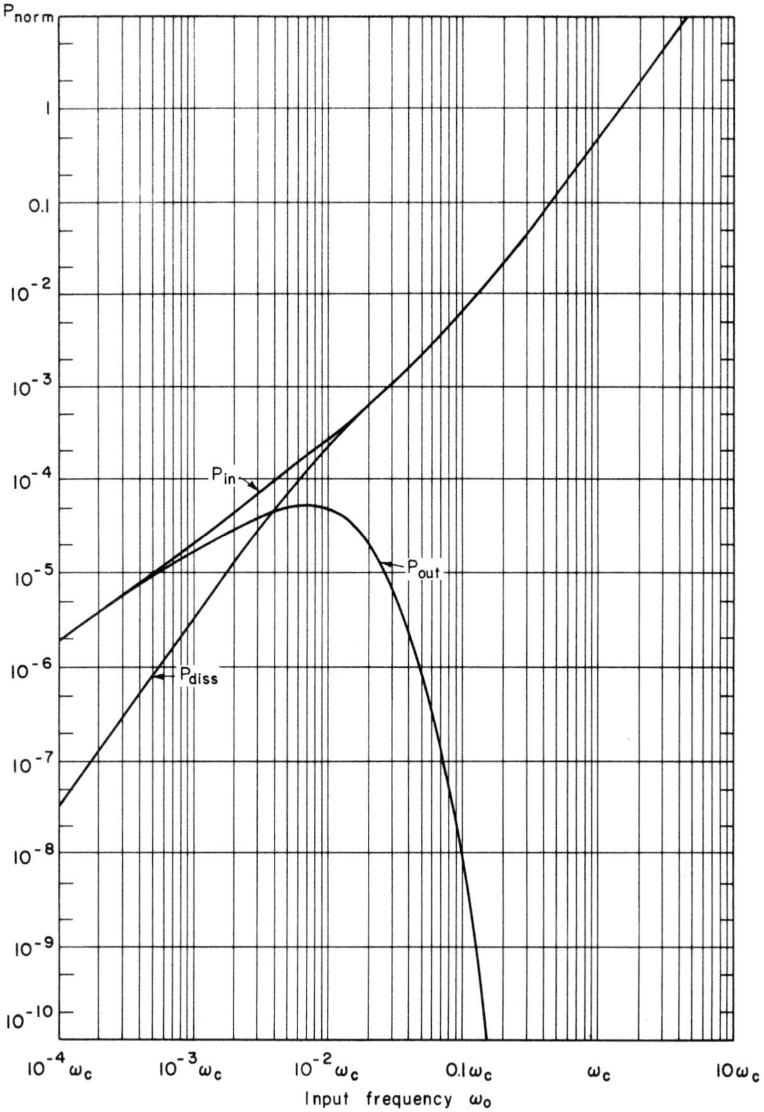

Figure 8.56. Input, output, and dissipated power for the abrupt-junction-varactor octupler adjusted for maximum efficiency.

the input frequency ω_0. The curves are similar to those for the other multipliers and are interpreted in the same way.

In Figures 8.56 and 8.58 we have neglected the minimum elastance S_{min}. If S_{min} is finite, the instructions accompanying Figures 8.40 and 8.42 should be followed. Since the solution is for maximum drive, the average elastance S_0 is given by

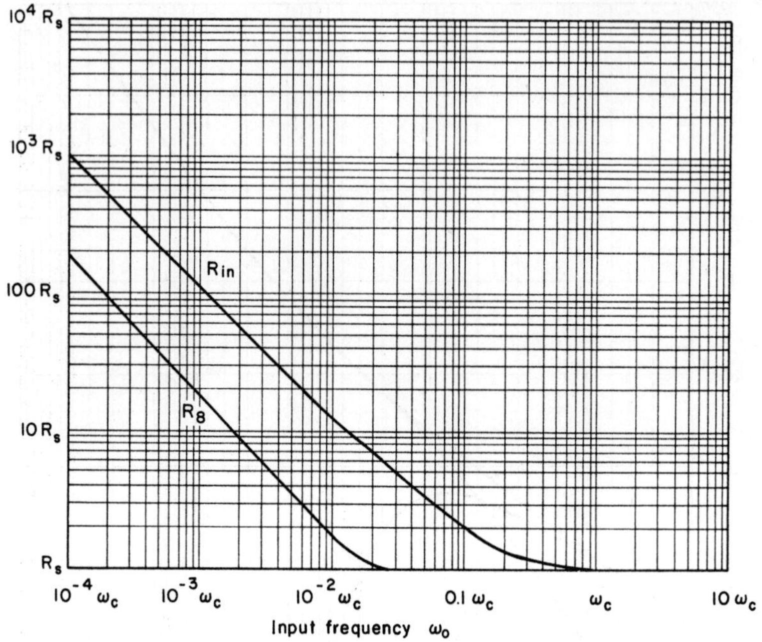

Figure 8.57. Input and load resistances for the abrupt-junction-varactor octupler adjusted for maximum efficiency.

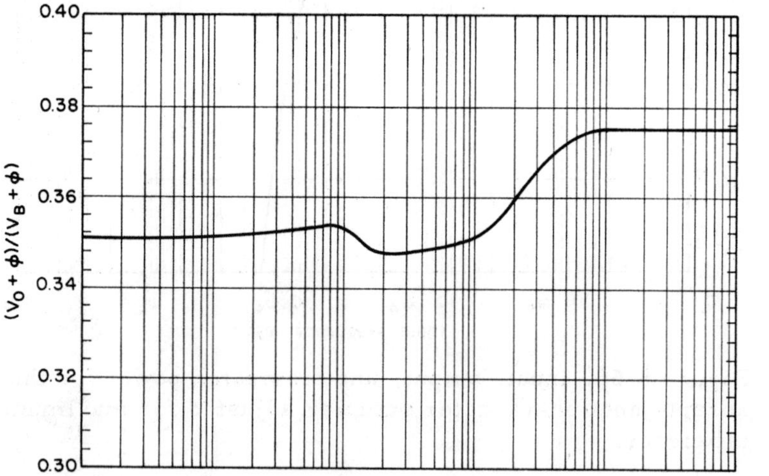

Figure 8.58. Bias voltage for the abrupt-junction-varactor octupler operated with maximum efficiency. We have assumed for this graph that $S_{min} \ll S_{max}$.

Table 8.12. Asymptotic Formulas for the Octupler

Both high-frequency and low-frequency formulas for the abrupt-junction-varactor octupler are shown. It is assumed that the idler terminations at frequencies $2\omega_o$ and $4\omega_o$ are tuned and reactive.

	Low-Frequency		High-Frequency
	Maximum ϵ	Maximum P_{out}	Maximum ϵ and P_{out}
ϵ	$1 - 168\frac{\omega_o}{\omega_c}$	$1 - 193\frac{\omega_o}{\omega_c}$	$8.67 \times 10^{-19}\left(\frac{\omega_c}{\omega_o}\right)^{14}$
R_{in}	$0.103\frac{S_{max}}{\omega_o}$	$0.116\frac{S_{max}}{\omega_o}$	R_s
R_8	$0.150\frac{S_{max}}{8\omega_o}$	$0.076\frac{S_{max}}{8\omega_o}$	R_s
P_{in}	$0.0198P_{norm}\frac{\omega_o}{\omega_c}$	$0.0215P_{norm}\frac{\omega_o}{\omega_c}$	$0.500P_{norm}\left(\frac{\omega_o}{\omega_c}\right)^2$
P_{out}	$0.0198P_{norm}\frac{\omega_o}{\omega_c}$	$0.0215P_{norm}\frac{\omega_o}{\omega_c}$	$4.34 \times 10^{-19}P_{norm}\left(\frac{\omega_c}{\omega_o}\right)^{12}$
P_{diss}	$3.33P_{norm}\left(\frac{\omega_o}{\omega_c}\right)^2$	$4.17P_{norm}\left(\frac{\omega_o}{\omega_c}\right)^2$	$0.500P_{norm}\left(\frac{\omega_o}{\omega_c}\right)^2$
$\frac{V_0 + \phi}{V_B + \phi}$	0.351	0.352	0.375
m_1	0.155	0.153	0.250
m_2	0.103	0.116	$0.0156\frac{\omega_c}{\omega_o}$
m_4	0.117	0.100	$3.05 \times 10^{-5}\left(\frac{\omega_c}{\omega_o}\right)^3$
m_8	0.045	0.067	$2.91 \times 10^{-11}\left(\frac{\omega_c}{\omega_o}\right)^7$

Equation 8.188, and the bias voltage is given by the simple formula,

$$\frac{V_0 + \phi}{V_B + \phi} = \left(\frac{S_{max} - S_{min}}{S_{max}}\right)^2 \left[\frac{1}{4} + 2(m_1^2 + m_2^2 + m_4^2 + m_8^2)\right] + \frac{S_{min}}{S_{max}}. \tag{8.230}$$

Since the solution for maximum power output is close to the maximum-efficiency solution, at least for small values of the idler resistances, there would be little advantage in discussing it. In Appendix D we give some nomographs, based on Figures 8.55 and 8.56, to assist in the practical design of maximum-efficiency octuplers.

At high and low frequencies the octupler is described by the asymptotic formulas given in Table 8.12. It is assumed there that $S_{min} \ll S_{max}$, and that the idler resistances R_4 and R_2 are zero.

8.11. Arbitrary Multipliers

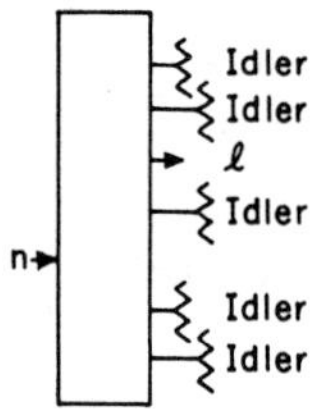

Thus far in Chapter 8 we have analyzed a few abrupt-junction multipliers with very simple idler configurations. Limits on efficiency were found in each case. However, we have as yet no limits on efficiency for multipliers that (1) have additional idlers, (2) use varactors with different elastance-charge characteristics, or (3) have different ratios of output to input frequency.

Thus, for example, we do not know whether or not there is any limit on efficiency less than 100 per cent for a multiplier using a graded-junction varactor, or for a septupler, or for a divide-by-eight circuit, or for a multiply-by-3/2 circuit. Is it possible that with additional idlers the efficiency of a doubler can be much higher than the limit of Figure 8.6? If the varactor is driven into the forward region, and if the forward conduction mechanism fails, our analyses no longer hold. Is there any limit on efficiency? Also, if the load is not tuned, can the efficiency be high?

In Section 8.11 we answer these questions in great generality, by deriving a fundamental limit on efficiency of varactor multipliers (and dividers and rational-fraction generators) that holds for any ratio of output to input frequency, for any number of idlers present, with any passive terminations, and for any elastance-charge characteristic.

This efficiency is in general not achievable, and is decidedly higher than the efficiencies of any of the multipliers we have discussed in detail. However, it is still more useful than no limit at all for those multipliers (and dividers and rational-fraction generators) that have not been solved exactly.

This fundamental limit, given in Section 8.11.1, is analogous to the unachievable fundamental limits on noise in frequency converters and amplifiers derived in Sections 5.6. and 6.4.2. Two simplified multipliers, discussed in Sections 8.11.2 and 8.11.3, have slightly lower fundamental limits on efficiency, also not achievable. For abrupt-junction multipliers with no current at frequencies higher than the output frequency, there is a much tighter bound on efficiency; this bound is derived in Section 8.11.4. This particular limit is not much higher than the known achievable efficiency for the doubler.

The same basic analysis, which we now give, pertains to all these limits. For generality the relationship between charge and capacitance is not specified, so that we are discussing abrupt-junction varactors, graded-junction varactors, epitaxial varactors, and units that have not yet been developed, as well as varactors operated into the forward direction if the forward conduction mechanism fails. Furthermore, we do not specify the number of idlers present.

We suppose that currents flow at frequencies of the form $k\omega_0$. A source is at the input frequency $n\omega_0$ and a load at the output frequency $\ell\omega_0$. The load impedance is $Z_\ell = R_\ell + jX_\ell$, and the terminations of the varactor at the idler frequencies $k\omega_0$ are impedances (or at least ratios of voltage to current) $Z_k = R_k + jX_k$, where R_ℓ and R_k are positive.

The varactor is described by the matrix Equations 8.11. If we neglect noise,

$$\underset{\sim}{V} = \underset{\sim}{Z}_v \underset{\sim}{I}, \tag{8.231}$$

where the matrix $\underset{\sim}{V}$ is the column matrix of varactor voltages, for both positive and negative frequencies, arranged in order, $\underset{\sim}{I}$ is the column matrix of currents, and $\underset{\sim}{Z}_v$ is the square matrix of Equations 8.11. If any of the I_k are known to be zero, we agree at the outset to eliminate the corresponding rows and columns from the matrices $\underset{\sim}{V}$, $\underset{\sim}{I}$, and $\underset{\sim}{Z}_v$. Thus, in particular, the dc voltage is omitted in Equations 8.11 and 8.231; and in all the sums that follow, the term for $k = 0$ is implicitly omitted.

The varactor currents and voltages are also related by the terminal constraints. Thus, for each k except n and $-n$,

$$V_k = -Z_k I_k, \tag{8.232}$$

and for the input frequency,

$$V_n = -Z_n I_n + E, \tag{8.233}$$

where E is the source voltage. The complex conjugate of Equation 8.233 is

$$V_{-n} = -Z_{-n} I_{-n} + E^*. \tag{8.234}$$

Equations 8.232 to 8.234 can be combined into a matrix equation,

$$\underset{\sim}{V} = -\underset{\sim}{Z}_L \underset{\sim}{I} + \underset{\sim}{E}_s, \tag{8.235}$$

where the matrix $\underset{\sim}{Z}_L$ is diagonal, and the matrix $\underset{\sim}{E}_s$ is a column matrix with all zero entries except for E^* and E, corresponding to frequencies $-n\omega_0$ and $+n\omega_0$, respectively.

Equations 8.231 and 8.235 can be combined to eliminate V:

$$\underset{\sim}{I} = \underset{\sim}{Y} \underset{\sim}{E}_s, \tag{8.236}$$

where

$$\underset{\sim}{Y} = (\underset{\sim}{Z}_v + \underset{\sim}{Z}_L)^{-1}. \tag{8.237}$$

For future reference, we note that the row in the matrix $\underset{\sim}{Z}_v + \underset{\sim}{Z}_L$ for frequency $\ell\omega_0$ is

$$(\underset{\sim}{Z}_v + \underset{\sim}{Z}_L)_{\ell k} = \frac{S_{\ell - k}}{j\ell\omega_0} \tag{8.238}$$

for $k \neq \ell$.

The power delivered to the load resistance R_ℓ is

$$P_{out} = 2|I_\ell|^2 R_\ell, \tag{8.239}$$

and the power input is simply the power dissipated in the series resistance of the varactor and in all the idlers and the load. Thus,

$$P_{in} = 2 \sum_{k=1}^{\infty} |I_k|^2 (R_k + R_s), \tag{8.240}$$

where in this summation we let $R_n = 0$. The efficiency is simply the ratio of the power output to the power input:

$$\epsilon = \frac{|I_\ell|^2 R_\ell}{\sum_{k=1}^{\infty} |I_k|^2 (R_k + R_s)}. \tag{8.241}$$

To evaluate the efficiency, we must find the I_k. Without loss of generality, we can choose the time origin so that E, the voltage source at the input, is real. Thus, Equations 8.236 can be written as

$$I_k = (Y_{kn} + Y_{k(-n)})E. \tag{8.242}$$

Now, Equation 8.241 becomes

$$\epsilon = \frac{R_\ell}{\sum_{\substack{k=1 \\ k\neq\ell}}^{\infty} \left| \frac{Y_{kn} + Y_{k(-n)}}{Y_{\ell n} + Y_{\ell(-n)}} \right|^2 (R_k + R_s) + R_\ell + R_s}. \tag{8.243}$$

We wish to eliminate R_ℓ from Equation 8.243. From the definition of $\underset{\sim}{Y}$, we know that

$$(\underset{\sim}{Z}_v + \underset{\sim}{Z}_L)\underset{\sim}{Y} = \underset{\sim}{1}, \tag{8.244}$$

where $\underset{\sim}{1}$ is the unit matrix. The ℓ, n and ℓ,-n entries of Equations 8.244 are

$$\sum_{\substack{k=-\infty \\ k\neq\ell}}^{\infty} (\underset{\sim}{Z}_v + \underset{\sim}{Z}_L)_{\ell k} Y_{kn} + (R_s + R_\ell + jX_{\ell\ell})Y_{\ell n} = 0 \tag{8.245}$$

and

$$\sum_{\substack{k=-\infty \\ k\neq\ell}}^{\infty} (\underset{\sim}{Z}_v + \underset{\sim}{Z}_L)_{\ell k} Y_{k(-n)} + (R_s + R_\ell + jX_{\ell\ell})Y_{\ell(-n)} = 0. \tag{8.246}$$

Taking the sum of these two equations, dividing by $Y_{\ell n} + Y_{\ell(-n)}$, and taking the real part of the result, we find

$$R_\ell = \sum_{\substack{k=-\infty \\ k\neq \ell}}^{\infty} \text{Re}\left[(\underset{\sim}{Z}_v + \underset{\sim}{Z}_L)_{\ell k}\left(-\frac{Y_{kn} + Y_{k(-n)}}{Y_{\ell n} + Y_{\ell(-n)}}\right)\right] - R_s. \tag{8.247}$$

To aid in the substitution of Equation 8.247 into Equation 8.243, we define, with the aid of Equation 8.238, the quantities A_k, r_k, and θ_k, for $k \neq \ell$, as follows:

$$A_k = \frac{|(\underset{\sim}{Z}_v + \underset{\sim}{Z}_L)_{\ell k}|}{\sqrt{R_s(R_k + R_s)}}$$

$$= \frac{|S_{\ell-k}|}{\ell\omega_o R_s}\sqrt{\frac{R_s}{R_k + R_s}}, \tag{8.248}$$

$$r_k e^{j\theta_k} = \frac{jS_{\ell-k}}{|S_{\ell-k}|}\sqrt{\frac{R_k + R_s}{R_s}}\left(\frac{Y_{kn} + Y_{k(-n)}}{Y_{\ell n} + Y_{\ell(-n)}}\right), \tag{8.249}$$

where r_k and θ_k are the amplitude and phase of the quantity shown. (In defining $A_{-\ell}$ and $r_{-\ell}$, we put $R_{-\ell} = 0$ in Equations 8.248 and 8.249.) Note that $r_k = r_{-k}$. The formula for the load resistance, Equation 8.247, becomes,

$$R_\ell = R_s\left[\sum_{\substack{k=-\infty \\ k\neq \ell}}^{\infty} A_k r_k \cos\theta_k - 1\right], \tag{8.250}$$

so that the efficiency is

$$\epsilon = \frac{\displaystyle\sum_{\substack{k=-\infty \\ k\neq \ell}}^{\infty} A_k r_k \cos\theta_k - 1}{\displaystyle\sum_{\substack{k=1 \\ k\neq \ell}}^{\infty} r_k^2 + \sum_{\substack{k=-\infty \\ k\neq \ell}}^{\infty} A_k r_k \cos\theta_k}. \tag{8.251}$$

Equation 8.251 is a valid formula for the efficiency of the multiplier (or divider or rational-fraction generator). It is, however, much too complicated to be evaluated explicitly. The A_k parameters are simple enough, but the r_k and θ_k parameters are complicated functions of the varactor, the excitation, and the terminations. Only certain values of r_k and θ_k are physically realizable. It might be thought that if we substituted values of r_k and θ_k that are not physically realizable into Equation 8.251, then we could make the expression approach as close to 100 per cent as desired. However, this is not so. As r_k and θ_k are varied, Equation 8.251 has a maximum value. The maximum efficiency for achievable values of r_k and θ_k is certainly no greater than this upper limit, which we shall calculate and call ϵ_{max}. This upper limit for efficiency is not, in general, achievable.

We first make the efficiency given by Equation 8.251 as large as possible by varying all the θ_k. Clearly, the highest value is found when all $\cos\theta_k = 1$:

$$\epsilon \leq \frac{\displaystyle\sum_{\substack{k=1\\k\neq\ell}}^{\infty} (A_k + A_{-k})r_k + A_{-\ell} - 1}{\displaystyle\sum_{\substack{k=1\\k\neq\ell}}^{\infty} r_k^2 + \sum_{\substack{k=1\\k\neq\ell}}^{\infty} (A_k + A_{-k})r_k + A_{-\ell}} . \tag{8.252}$$

In deriving Condition 8.252, we have used the fact that $r_{-\ell} = 1$. If we maximize the right-hand side of Condition 8.252 with respect to each r_k, by a procedure similar to that in Section 5.6, we find

$$\epsilon \leq \epsilon_{max}, \tag{8.253}$$

where ϵ_{max} is

$$\epsilon_{max} = \frac{f - 1}{f + 1}, \tag{8.254}$$

with

$$f = A_{-\ell} + \sqrt{(1 - A_{-\ell})^2 + \sum_{\substack{k=1\\k\neq\ell}}^{\infty} (A_k + A_{-k})^2} . \tag{8.255}$$

So far, ϵ_{max} depends upon all the elastance coefficients S_k, the output frequency $\ell\omega_o$, and the idler resistances R_k. Furthermore, it depends on which terms are included in the summation; recall that if any current coefficients I_k are known to vanish, then the corresponding rows and columns in the matrices are eliminated, and therefore the summation in Equation 8.255 is only over those k with $I_k \neq 0$. In the most general case we should not assume any special knowledge about (1) which frequencies have currents flowing, (2) what values of R_k are presented to the varactor, or (3) what values of S_k exist (except that the elastance must remain between S_{min} and S_{max} at all times). If we assume that R_k and S_k are adjusted to maximize ϵ_{max}, the resulting value is quite large. It is derived in Section 8.11.1.

On the other hand, if no current flows at frequency $3\ell\omega_o$, a slightly lower bound, derived in Section 8.11.2, applies. If the elastance coefficient at twice the output frequency $S_{2\ell}$ vanishes, then $A_{-\ell} = 0$, and a still lower bound is derived in Section 8.11.3. These bounds, however, are not much below the bound of Section 8.11.1.

A much lower, and therefore more useful, value of ϵ_{max} is found when all S_k and all I_k vanish for $k > \ell$. These rather drastic assumptions apply to the abrupt-junction multipliers analyzed so far in Chapter 8. The resulting bound for ϵ_{max}, derived in Section 8.11.4, is not much greater than the efficiency that is known to be possible for the doubler.

8.11.1. Most General Multiplier

We wish to interpret Equation 8.254 for ϵ_{max} in the important case where nothing is known about the values of the R_k or the S_k. First, we note that

$$A_k = \frac{m_{\ell-k}\omega_c}{\ell\omega_o}\sqrt{\frac{R_s}{R_k + R_s}} \tag{8.256}$$

has its largest value,‡ $m_{\ell-k}\omega_c/\ell\omega_o$, when $R_k = 0$. If we have no information about R_k, then to find the largest value for f (and therefore the largest value for ϵ_{max}), we must assume that all $R_k = 0$ for $k \neq \ell$.

Next, we note that

$$(A_k + A_{-k})^2 \leq 2(A_k^2 + A_{-k}^2). \tag{8.257}$$

Thus, an upper limit for f is

‡We assume, of course, that the terminations are passive, so that R_k is nonnegative.

$$f \leq \frac{m_{2\ell}\omega_c}{\ell\omega_o} + \sqrt{\left(1 - \frac{m_{2\ell}\omega_c}{\ell\omega_o}\right)^2 + 2\left(\frac{\omega_c}{\ell\omega_o}\right)^2 (m_t^2 - m_\ell^2 - m_{2\ell}^2)}, \tag{8.258}$$

where m_t, the total modulation ratio

$$m_t = \sqrt{2(m_1^2 + m_2^2 + m_3^2 + \cdots)}, \tag{8.259}$$

is always less than 0.5. If one or more of the currents is known to be zero, then m_t^2 can be replaced by a sum omitting one or more of the terms. Thus, if some $I_k = 0$, then one term $m_{\ell-k}^2$ and one term $m_{\ell+k}^2$ can be omitted in a new summation to replace m_t^2 in Conditions 8.258 and 8.261.

Unless we have some information about the value of $m_{2\ell}$, we must assume that it is finite. The right-hand side of Condition 8.258 has a maximum as $m_{2\ell}$ is varied; this maximum is

$$2\sqrt{1 + \left(\frac{\omega_c}{\ell\omega_o}\right)^2 (m_t^2 - m_\ell^2)} - 1. \tag{8.260}$$

When this maximum value for f is used in Equation 8.254 for the efficiency, we find that

$$\epsilon_{max} \leq 1 - \frac{1}{\sqrt{1 + \left(\frac{m_t\omega_c}{\ell\omega_o}\right)^2}}, \tag{8.261}$$

where we have dropped the m_ℓ^2 term because we have no knowledge of its size. Putting $m_t = 0.5$ in Condition 8.261 gives an upper bound for efficiency of a multiplier that cannot be surpassed by any choice of idler configuration, elastance-charge law, or idler terminations. This bound for efficiency is greater than can be actually achieved, because we have made many simplifying estimates. This bound is a function of only the ratio of output frequency to cutoff frequency and is, therefore, useful when nothing whatever is known of the operating details of the multiplier.

This limit is shown as a function of frequency in Figure 8.59. Also given for comparison are the three other limits to be derived now.

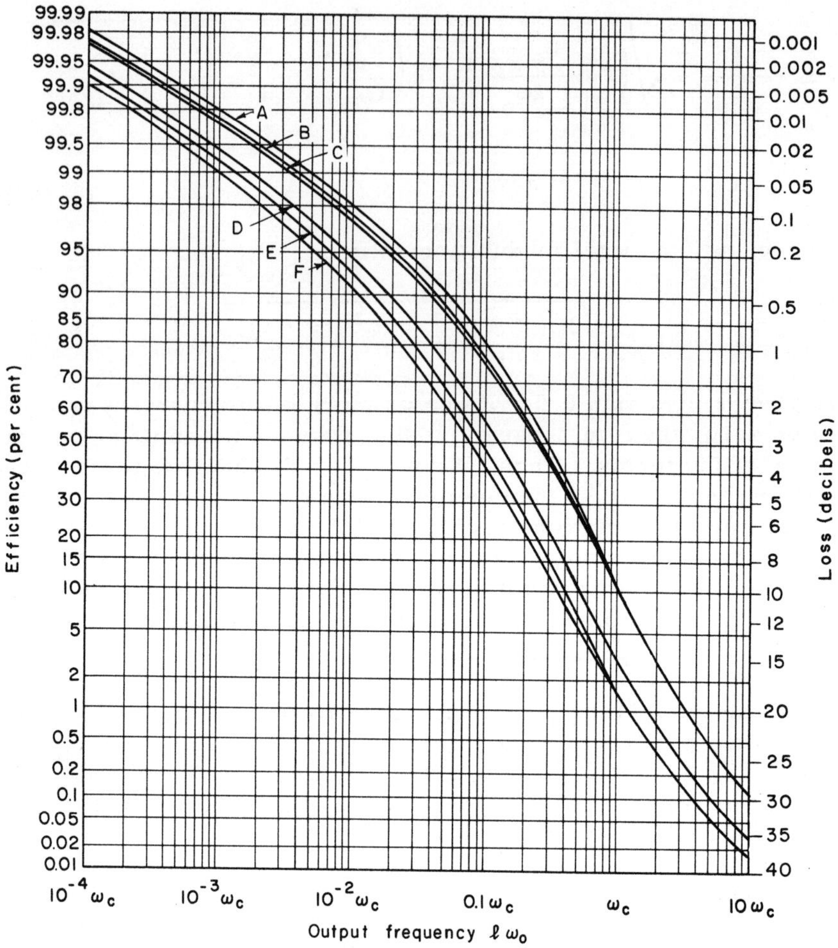

Figure 8.59. Fundamental limits on efficiency derived in Section 8.11, compared with the achievable efficiency of the abrupt-junction doubler.

Curve	Formula	Condition	Value of m_t
A	8.261	General	0.5
B	8.265	$I_{3\ell} = 0$	0.5
C	8.269	$S_{2\ell} = 0$	0.5
D	8.273	$I_k = 0,\ S_k = 0 \quad (k > \ell)$	0.5
E	8.273	$I_k = 0,\ S_k = 0 \quad (k > \ell)$	0.354
F	Abrupt-junction-varactor doubler (Figure 8.6)		

8.11.2. Slightly Lower Limit

If no current flows at three times the output frequency, $I_{3\ell} = 0$, the term for $k = 3\ell$ can be omitted in the summation of Equation 8.255. This information lets us derive a slightly lower limit for ϵ_{max}.

First, we note that each A_k has its largest value when $R_k = 0$. If we have no information about R_k, then to find the largest possible value for f, we must assume that all $R_k = 0$ for $k \neq \ell$. Next, we note that

$$(A_k + A_{-k})^2 \leq 2(A_k^2 + A_{-k}^2). \qquad (8.262)$$

Thus, an upper limit for f is found from Equation 8.255 by evaluating the A_k, setting all R_k to zero, and using Condition 8.262:

$$f \leq \frac{m_{2\ell}\omega_c}{\ell\omega_o} + \sqrt{\left(1 - \frac{m_{2\ell}\omega_c}{\ell\omega_o}\right)^2 + 2\left(\frac{\omega_c}{\ell\omega_o}\right)^2 (m_t^2 - m_\ell^2 - 2m_{2\ell}^2 - m_{4\ell}^2)}. \qquad (8.263)$$

If more currents are known to be zero, then m_t^2 can be replaced by a sum omitting one or more of the terms. Thus, if some $I_k = 0$, then one of the two terms $m_{\ell-k}^2$ and one of the two terms $m_{\ell+k}^2$ can be omitted in a new summation, which replaces m_t^2 in Conditions 8.263 and 8.265.

Condition 8.263 depends upon $m_{2\ell}$, m_ℓ, and $m_{4\ell}$ explicitly. If nothing is known about these values, then we must use those values that make Condition 8.263 the largest. Thus, we assume (in the absence of other information) that $m_\ell = 0$, $m_{4\ell} = 0$, and $m_{2\ell}$ is chosen to maximize Condition 8.263. This maximum is

$$\frac{1}{3}\left[2\sqrt{4 + 6\left(\frac{\omega_c}{\ell\omega_o}\right)^2 m_t^2} - 1\right], \qquad (8.264)$$

and the resulting maximum possible value for efficiency is

$$\epsilon_{max} \leq \frac{\sqrt{4 + 6\left(\frac{m_t\omega_c}{\ell\omega_o}\right)^2} - 2}{\sqrt{4 + 6\left(\frac{m_t\omega_c}{\ell\omega_o}\right)^2} + 1}. \qquad (8.265)$$

By putting m_t equal to 0.5 in Condition 8.265, we find a limit on efficiency that depends only upon the ratio of output frequency to cutoff frequency. This limit, slightly lower than Condition 8.261, is indicated in Figure 8.59.

8.11.3. Still Lower Limit

If it is known that the elastance has no Fourier component at twice the output frequency, then $m_{2\ell} = 0$ and $A_{-\ell} = 0$. Thus, from Equation 8.255, f becomes

$$f = \sqrt{1 + \sum_{\substack{k=1 \\ k\neq\ell}}^{\infty} (A_k + A_{-k})^2}, \tag{8.266}$$

where, as usual, the summation is over those k with $I_k \neq 0$, except $k = \ell$.

First, we note that each A_k has its largest value‡ when $R_k = 0$. If we have no information about R_k, we must assume, to find the largest possible value for f, that all $R_k = 0$ for $k \neq \ell$. Next, we note that

$$(A_k + A_{-k})^2 \leq 2(A_k{}^2 + A_{-k}{}^2). \tag{8.267}$$

An upper limit for f is found from Equation 8.266 by evaluating the A_k, setting all R_k to zero, and using Equation 8.267:

$$f \leq \sqrt{1 + 2\left(\frac{\omega_c}{\ell\omega_o}\right)^2 (m_t{}^2 - m_\ell{}^2)}. \tag{8.268}$$

If some of the currents are known to be zero, then $m_t{}^2$ can be replaced by a sum omitting some of the terms. For example, if a particular $I_k = 0$, then one of the two terms $m_{\ell-k}{}^2$ and one of the two terms $m_{\ell+k}{}^2$ can be omitted in a new summation, to replace $m_t{}^2$ in Conditions 8.268 and 8.269.

The expression for the maximum possible value of f, Condition 8.268, depends upon m_ℓ. If nothing is known about the value of m_ℓ, we must assume that value which maximizes Condition 8.268. Thus,

‡Again, we assume R_k is not negative.

$$\epsilon_{max} \le \frac{\sqrt{1 + 2\left(\frac{m_t \omega_c}{\ell \omega_o}\right)^2} - 1}{\sqrt{1 + 2\left(\frac{m_t \omega_c}{\ell \omega_o}\right)^2} + 1}. \qquad (8.269)$$

By putting m_t equal to its maximum value 0.5, we find a limit on efficiency that depends only upon the ratio of output frequency to cutoff frequency. This limit is slightly lower than Condition 8.265, as indicated in Figure 8.59.

8.11.4. Much Lower Limit

A limit on efficiency that is relatively close to achievable efficiencies can be derived for the case that all $S_k = 0$ for $k > \ell$, and that all $I_k = 0$ for $k > \ell$. These assumptions are pertinent to many abrupt-junction multipliers.

Of the approximations we have used thus far, the crudest are Conditions 8.257, 8.262, and 8.267. The limit now to be derived does not require this approximation, and is therefore much more powerful.

Since all $S_k = 0$ for $k > \ell$, all A_{-k} for k positive are zero. Furthermore, the sum for f in Equation 8.255 can terminate at $\ell - 1$. Therefore, Equation 8.255 becomes

$$f = \sqrt{1 + \sum_{k=1}^{\ell-1} A_k{}^2}$$

$$= \sqrt{1 + \sum_{k=1}^{\ell-1} \left(\frac{m_k \omega_c}{\ell \omega_o}\right)^2 \frac{R_s}{R_{\ell-k} + R_s}}. \qquad (8.270)$$

If we know nothing about the values of R_k, then to obtain the largest possible value for Equation 8.270 we must assume‡ that all $R_k = 0$. Then,

$$f \le \sqrt{1 + \left(\frac{\omega_c}{\ell \omega_o}\right)^2 \sum_{k=1}^{\ell-1} m_k{}^2}. \qquad (8.271)$$

‡As usual, we assume that R_k is not negative.

Since the sum of m_k^2 is $(m_t^2/2) - m_\ell^2$, if we have no knowledge of the value of m_ℓ we must say that

$$f \le \sqrt{1 + \frac{1}{2}\left(\frac{m_t \omega_c}{\ell \omega_o}\right)^2}. \tag{8.272}$$

Thus, the maximum efficiency has an upper limit,

$$\epsilon_{max} \le \frac{\sqrt{1 + \frac{1}{2}\left(\frac{m_t \omega_c}{\ell \omega_o}\right)^2} - 1}{\sqrt{1 + \frac{1}{2}\left(\frac{m_t \omega_c}{\ell \omega_o}\right)^2} + 1}. \tag{8.273}$$

If we set m_t equal to 0.5, its maximum value, the limit of Condition 8.273 is a function of only the ratio of output frequency to cutoff frequency. As indicated in Figure 8.59, this limit is much lower than the other limits and is not much higher than the achievable efficiency found for the doubler in Section 8.4. In Figure 8.59, we also show Condition 8.273 for $m_t = 0.25\sqrt{2}$, because for many abrupt-junction multipliers m_t is known to be less than $0.25\sqrt{2} = 0.354$.

For specific abrupt-junction multipliers, we can use Equation 8.270 to obtain even a tighter bound, provided the idler configuration is specified. Although the limits for the configurations that have been solved are not particularly important, because the actual achievable limits are known, it is interesting to apply Equation 8.270 to some of them. Thus, for example, for the doubler with current at only ω_o and $2\omega_o$,

$$\epsilon \le \frac{\sqrt{1 + \frac{m_1^2 \omega_c^2}{4\omega_o^2}} - 1}{\sqrt{1 + \frac{m_1^2 \omega_c^2}{4\omega_o^2}} + 1}. \tag{8.274}$$

For the quadrupler, with an idler at only $2\omega_o$, we find

$$\epsilon \leq \frac{\sqrt{1 + \left(\frac{m_2\omega_c}{4\omega_o}\right)^2 \frac{R_s}{R_2 + R_s}} - 1}{\sqrt{1 + \left(\frac{m_2\omega_c}{4\omega_o}\right)^2 \frac{R_s}{R_2 + R_s}} + 1}. \tag{8.275}$$

Similarly, for a quintupler with idlers at $2\omega_o$ and $3\omega_o$,

$$\epsilon \leq \frac{\sqrt{1 + \left(\frac{m_2\omega_c}{5\omega_o}\right)^2 \frac{R_s}{R_3 + R_s} + \left(\frac{m_3\omega_c}{5\omega_o}\right)^2 \frac{R_s}{R_2 + R_s}} - 1}{\sqrt{1 + \left(\frac{m_2\omega_c}{5\omega_o}\right)^2 \frac{R_s}{R_3 + R_s} + \left(\frac{m_3\omega_c}{5\omega_o}\right)^2 \frac{R_s}{R_2 + R_s}} + 1}. \tag{8.276}$$

Since we have not actually solved the 1-2-3-5 quintupler, Condition 8.276 may be of value.

In summary, we wish to stress that the four limits on efficiency derived in Section 8.11 are in general not achievable. To achieve these limits, the magnitudes and phases of the currents must be set to obtain the θ_k and r_k parameters that make the efficiency attain Equation 8.254. Then, the idler resistances must be adjusted; and for the first three limits each $m_{\ell+k}$ must be equal to $m_{\ell-k}$. Usually m_ℓ must vanish, and nevertheless m_t should be as high as 0.5. These conditions are, of course, not compatible, and so the limits given are not achievable. They are of value only for multipliers that have not been, or cannot be, solved exactly, and even then they are of limited usefulness. They are, however, better than no limits at all.

8.12. Comparison

It is interesting to compare the many abrupt-junction-varactor multipliers we have described. For generality we also compare their performance with the fundamental limits of Section 8.11, and with the abrupt-junction-varactor dividers discussed in Chapter 9, and with the graded-junction-varactor doubler, Appendix F. First we consider the efficiency, then the power output. Finally, we examine the high-frequency limit of operation.

8.12.1. Efficiency

In Figure 8.60 we plot the maximum efficiency of the abrupt-junction-varactor doubler, tripler, quadrupler, 1-2-4-5 quintupler, 1-2-4-6 sextupler, 1-2-3-6 sextupler, and octupler. These are

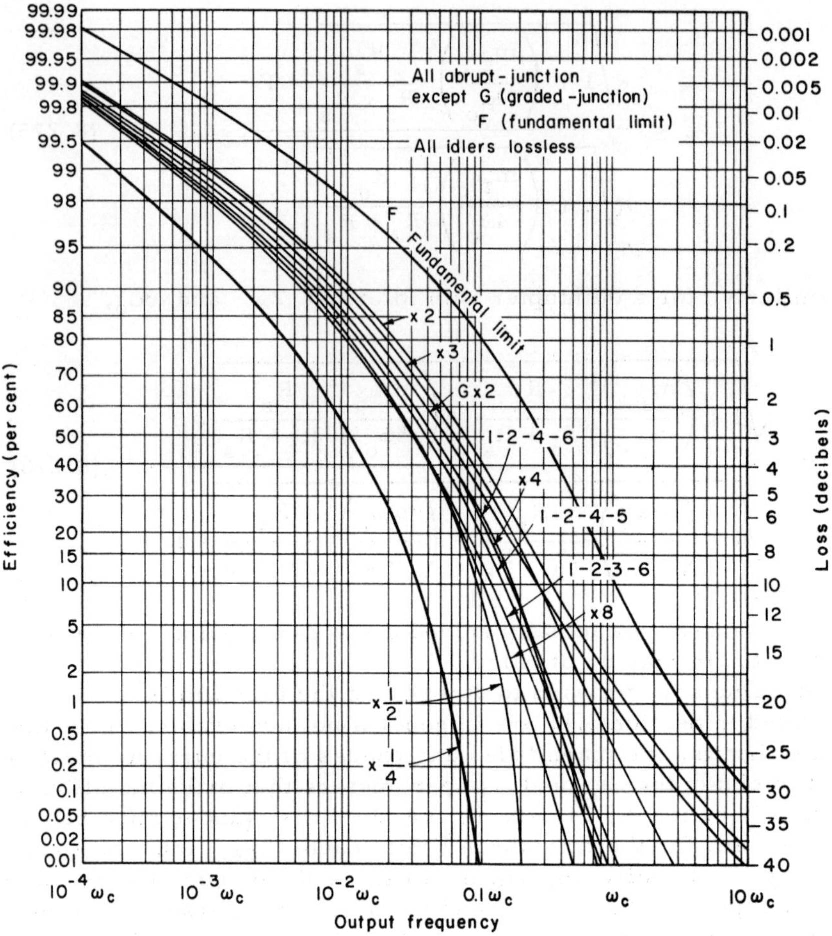

Figure 8.60. Maximum efficiency for several abrupt-junction-varactor multipliers and dividers, as a function of the output frequency. Also shown are the fundamental limit of Condition 8.261 and the efficiency of the graded-junction-varactor doubler.

plotted as a function of output frequency rather than input frequency. Also shown in Figure 8.60 are the maximum efficiencies of the divide-by-two circuit, the divide-by-four circuit, and the graded-junction doubler, as well as the fundamental limit of Condition 8.261.

At low frequencies all of these curves have the asymptotic behavior

$$\epsilon \approx 1 - a \frac{\omega_{out}}{\omega_c}, \qquad (8.277)$$

where α depends upon the particular type of operation. These low-frequency efficiencies are given explicitly in Table 8.13.

A question often asked is, "Which is better, two cascaded multipliers or a single higher-order multiplier?" For example, if we want to generate the sixth harmonic, is it better to use a sextupler, or to use a tripler and a doubler in cascade? As far as efficiency is concerned, it makes very little difference at low or moderate frequencies (but not at higher frequencies). Slightly better efficiency can in theory be attained by cascading two mul-

Table 8.13. Low-Frequency Efficiencies

Low-frequency asymptotes to the curves in Figure 8.60 for a variety of multipliers and dividers.

Abrupt-Junction		Graded-Junction	
$\times 2$	$1 - 9.95\frac{\omega_{out}}{\omega_c}$	$\times 2$	$1 - 13\frac{\omega_{out}}{\omega_c}$
$\times 3$	$1 - 11.6\frac{\omega_{out}}{\omega_c}$	Fundamental Limits	
$\times 4$	$1 - 15.6\frac{\omega_{out}}{\omega_c}$	Equation 8.261	$1 - \frac{1}{m_t}\frac{\omega_{out}}{\omega_c}$
$\times 5$ (1-2-4-5)	$1 - 18.6\frac{\omega_{out}}{\omega_c}$	Equation 8.265	$1 - \frac{1.22}{m_t}\frac{\omega_{out}}{\omega_c}$
$\times 6$ (1-2-4-6)	$1 - 16.6\frac{\omega_{out}}{\omega_c}$	Equation 8.269	$1 - \frac{1.41}{m_t}\frac{\omega_{out}}{\omega_c}$
$\times 6$ (1-2-3-6)	$1 - 22.3\frac{\omega_{out}}{\omega_c}$	Equation 8.273	$1 - \frac{2.83}{m_t}\frac{\omega_{out}}{\omega_c}$
$\times 8$	$1 - 21.0\frac{\omega_{out}}{\omega_c}$		
$\times \frac{1}{2}$	$1 - 19.9\frac{\omega_{out}}{\omega_c}$		
$\times \frac{1}{4}$	$1 - 62.5\frac{\omega_{out}}{\omega_c}$		

tipliers, but in practice this improvement can be wiped out very easily if the coupling network is at all lossy. For example, consider the low-frequency case. Assume that the varactors used in the multipliers have the optimum normalization powers, and that they have the same cutoff frequency. Then the fourth harmonic can be obtained either from a quadrupler or from two cascaded doublers. The maximum theoretical efficiencies in the two cases are

$$\times 4: \qquad \epsilon \approx 1 - 15.6 \frac{4\omega_o}{\omega_c}, \tag{8.278}$$

$$\times 2\text{-}\times 2: \qquad \epsilon \approx 1 - 14.9 \frac{4\omega_o}{\omega_c}. \tag{8.279}$$

Similarly, in order to obtain the sixth harmonic, there are at least four schemes: the 1-2-3-6 sextupler, the 1-2-4-6 sextupler, a tripler followed by a doubler, and a doubler followed by a tripler. The low-frequency efficiencies in these four cases are, respectively,

$$\times 6\ (1\text{-}2\text{-}3\text{-}6): \qquad \epsilon \approx 1 - 22.3 \frac{6\omega_o}{\omega_c}, \tag{8.280}$$

$$\times 6\ (1\text{-}2\text{-}4\text{-}6): \qquad \epsilon \approx 1 - 16.6 \frac{6\omega_o}{\omega_c}, \tag{8.281}$$

$$\times 3\text{-}\times 2: \qquad \epsilon \approx 1 - 15.7 \frac{6\omega_o}{\omega_c}, \tag{8.282}$$

$$\times 2\text{-}\times 3: \qquad \epsilon \approx 1 - 14.9 \frac{6\omega_o}{\omega_c}. \tag{8.283}$$

As a final example, consider the ways the eighth harmonic may be obtained. Four schemes are possible: an octupler, a doubler followed by a quadrupler, a quadrupler followed by a doubler, and finally three doublers in cascade. The low-frequency efficiencies for these four cases are, respectively,

$$\times 8: \qquad \epsilon \approx 1 - 21.0 \frac{8\omega_o}{\omega_c}, \tag{8.284}$$

$$\times 2\text{-}\times 4: \qquad \epsilon \approx 1 - 18.1 \frac{8\omega_o}{\omega_c}, \tag{8.285}$$

$$\times 4\text{-}\times 2: \qquad \epsilon \approx 1 - 17.8 \frac{8\omega_o}{\omega_c}, \tag{8.286}$$

$$\times 2\text{-}\times 2\text{-}\times 2: \qquad \epsilon \approx 1 - 17.4 \frac{8\omega_o}{\omega_c}. \tag{8.287}$$

It is apparent in each of these examples that the efficiencies of the various schemes are about the same. Using many stages has the advantage that each stage is easier to design and that less power is dissipated in each varactor. On the other hand, the coupling networks quite probably will have enough loss to nullify the slight advantage in efficiency, and the use of symmetric circuits to partition the loss among two (or more) varactors can ameliorate the dissipation problem.

At higher frequencies the efficiency can be greatly improved by cascading stages. See the discussion in Section 8.12.3.

Two common misconceptions about multipliers can be disproved by referring to Figure 8.60. The first is that the doubler is at all frequencies inherently much more efficient than higher-order multipliers.[6,5,4,1] This is probably true for multipliers without idlers and is certainly true for nonlinear-resistance multipliers.[8,9] But Figure 8.60 shows that it is not true for varactor multipliers with idlers.

The second, related, misconception is that in order to generate high-order harmonics, a sharply nonlinear characteristic is desirable. Again, without idlers, such a requirement makes sense. A device with a gradual nonlinearity (such as the abrupt-junction or graded-junction varactor) will tend, crudely speaking, to put most of the input power into the second harmonic. Very little is directly converted to the higher harmonics. On the other hand, a "highly nonlinear" device‡ would tend to spread the input power among several harmonics. Thus at least some power can be delivered to a load at a higher harmonic. If idlers are used, however, the situation is different. A device with a gradual nonlinearity would tend to generate mainly the second harmonic. However, if there is an idler at this frequency, the second harmonic can itself generate the fourth or else mix with the fundamental to generate the third. Let us contrast such a device with a sharply nonlinear device that, crudely speaking, spreads the input power over a number of harmonics; at, for example, the eighth harmonic, we might get as much as 10 or 12 per cent efficiency. The smoothly nonlinear device with idlers, on the other hand, might convert greater than 90 per cent of the fundamental power to the second harmonic. This is again converted

‡Whatever that means.

at high efficiency to the fourth and again to the eighth. The over-all efficiency would be perhaps 70 per cent. It is clear that, as least as far as efficiency is concerned, the gentle nonlinearity (provided it is not too gentle, for example, a constant) is preferable.

8.12.2. Power Output

The multipliers can be compared on the basis of power output. The maximum power output is shown in Figure 8.61 for the various multipliers and dividers, as a function of the input frequency. Although the lower-order multipliers have slightly greater power outputs, at low frequencies the difference is slight.

The low-frequency limits are shown explicitly in Table 8.14. Each is proportional to the input frequency. The powers are expressed both in terms of P_{norm} and ω_c and in terms of Uhlir's[10] nominal reactive power P_r:

$$P_r = \frac{(V_B + \phi)^2}{2S_{max}}$$

$$= \frac{1}{2} \frac{P_{norm}}{\omega_c} . \qquad (8.288)$$

Note that P_r has the dimensions of power per unit frequency, and must therefore be multiplied by a frequency, as in Table 8.14.

The limits shown in Figure 8.61 and Table 8.14 are, of course, theoretical powers based on the assumption that the varactor is not driven into the forward region. When the varactor is over-driven, these power outputs can be increased, often greatly, if the forward-conduction mechanism fails.

8.12.3. High-Frequency Limit

The high-frequency limit is inherently less interesting than the low-frequency limit because the efficiency is very small, and the power dissipation is very large. Nevertheless, in some applications these limits are pertinent.

At high frequencies the series resistance dominates, and to transfer maximum power to the load or to obtain maximum efficiency, the load should equal R_s. Similarly, the input resistance is also R_s.

For abrupt-junction varactors, $m_1 \approx 0.25$, and the other m_k become proportional to $(\omega_c/\omega_0)^{k-1}$ and, therefore, approach zero. The output power into a matched load becomes

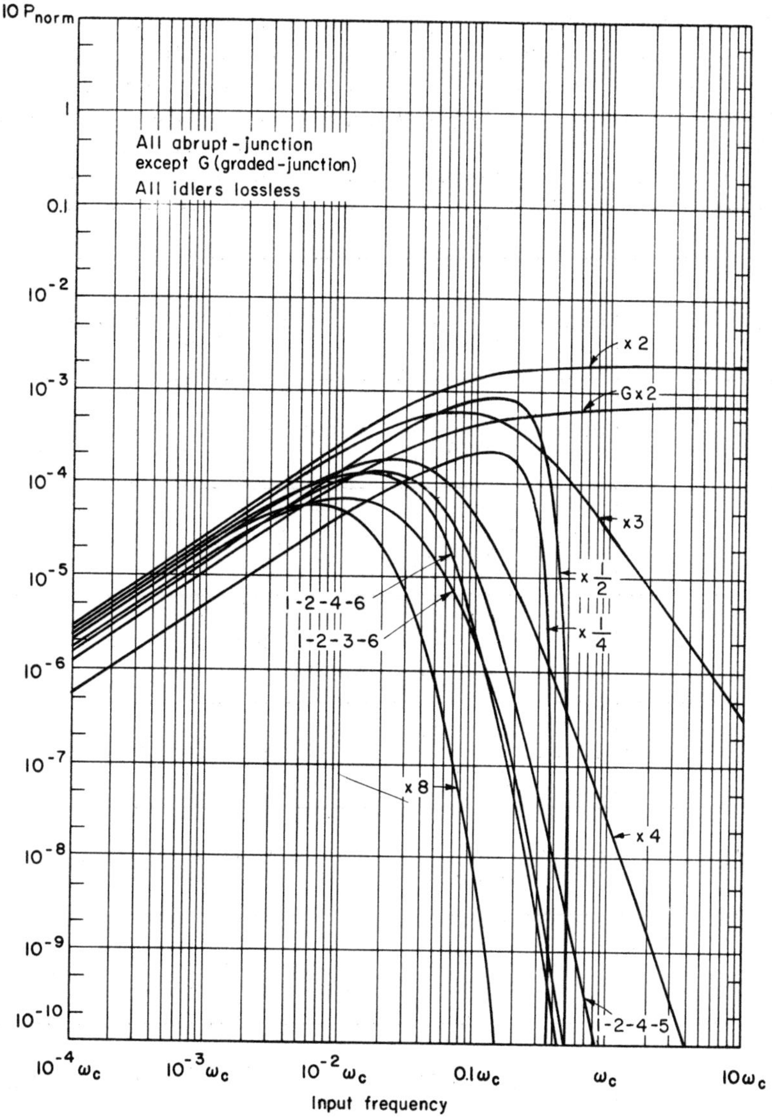

Figure 8. 61. Output power for several abrupt-junction-varactor multipliers and dividers, as a function of the input frequency. Also shown is the power output for a graded-junction-varactor doubler. All are assumed to operate at maximum efficiency with lossless, tuned idlers.

Table 8.14. Maximum Power Output of Multipliers at Low Frequencies

All the power outputs are proportional to frequency. For simplicity, we assume that $S_{min} << S_{max}$. Note that when compared on the basis of the same input frequency, all the abrupt-junction-varactor multipliers have, within a factor of about 1.5, the same power input. The last formula in each case is in terms of ordinary rather than radian frequency.

Abrupt-Junction	
$\times 2$	$0.028 P_{norm} \dfrac{\omega_o}{\omega_c} = 0.057 P_r \omega_o = 0.36 P_r f_o$
$\times 3$	$0.024 P_{norm} \dfrac{\omega_o}{\omega_c} = 0.048 P_r \omega_o = 0.30 P_r f_o$
$\times 4$	$0.020 P_{norm} \dfrac{\omega_o}{\omega_c} = 0.040 P_r \omega_o = 0.25 P_r f_o$
$\times 5$ (1-2-4-5)	$0.018 P_{norm} \dfrac{\omega_o}{\omega_c} = 0.036 P_r \omega_o = 0.22 P_r f_o$
$\times 6$ (1-2-4-6)	$0.022 P_{norm} \dfrac{\omega_o}{\omega_c} = 0.045 P_r \omega_o = 0.28 P_r f_o$
$\times 6$ (1-2-3-6)	$0.018 P_{norm} \dfrac{\omega_o}{\omega_c} = 0.037 P_r \omega_o = 0.23 P_r f_o$
$\times 8$	$0.021 P_{norm} \dfrac{\omega_o}{\omega_c} = 0.043 P_r \omega_o = 0.27 P_r f_o$
$\times \frac{1}{2}$	$0.014 P_{norm} \dfrac{2\omega_o}{\omega_c} = 0.0285 P_r (2\omega_o) = 0.18 P_r (2f_o)$
$\times \frac{1}{4}$	$0.0050 P_{norm} \dfrac{4\omega_o}{\omega_c} = 0.010 P_r (4\omega_o) = 0.063 P_r (4f_o)$
Graded-Junction	
$\times 2$	$0.012 P_{norm} \dfrac{\omega_o}{\omega_c} = 0.024 P_r \omega_o = 0.15 P_r f_o$

$$P_{out} \approx 8\ell^2 m_\ell^{\,2} P_{norm} \frac{\omega_o^{\,2}}{\omega_c^{\,2}}. \tag{8.289}$$

If $\ell = 2$ (doubler), then P_{out} approaches a finite limit. If $\ell > 2$, P_{out} goes to zero as $(\omega_c/\omega_o)^{2(\ell-2)}$. Since the power input is the same value obtained for current pumping of such varactors

$$P_{in} \approx P_{diss} \approx 0.5 P_{norm} \frac{\omega_o^{\,2}}{\omega_c^{\,2}}, \tag{8.290}$$

the efficiency becomes

$$\epsilon \approx 16\ell^2 m_\ell^{\,2} \tag{8.291}$$

and, therefore, approaches zero as $(\omega_c/\omega_o)^{2(\ell-1)}$.

It is apparent that multipliers with idlers have an efficiency that approaches zero very rapidly at high frequencies. There are two ways to improve the efficiency. First, to obtain, say, the fourth harmonic, do not use a quadrupler, with efficiency proportional to $(\omega_c/\omega_o)^6$, but instead use two cascaded doublers. The cascade efficiency goes to zero only as $(\omega_c/\omega_o)^4$ and is, therefore, better at high frequencies. Second, a different type of varactor can be used. Since the efficiency is low anyway, perhaps a device with a sharp nonlinearity could be used to convert a small amount of power directly to the desired harmonic, rather than through the extremely lossy idlers. It is shown in Appendix F that the high-frequency limit of efficiency of a varactor multiplier without idlers is

$$\epsilon \approx 0.25 |m_{\ell-1} \pm m_{\ell+1}|^2 \left(\frac{\omega_c}{\ell\omega_o}\right)^2. \tag{8.292}$$

Thus, unless the m-values vanish identically (as they do for high-order abrupt-junction-varactor multipliers), the efficiency goes to zero only as the square of the frequency. The high-frequency efficiency of graded-junction multipliers without idlers is given in Appendix F.

8.13. Summary

Several important results were derived. Among the most significant are these:

1. Varactors are capable of high efficiency as multipliers.
2. The varactor series resistance limits the attainable efficiency.

3. Abrupt-junction-varactor multipliers of order higher than two require one or more idlers.
4. For the doubler, there is a low-power optimization to obtain the maximum efficiency, maximum power output, or minimum dissipation for a specified value of efficiency, power output, power input, or dissipation.
5. Optimization of a doubler for maximum efficiency yields almost the maximum power output. At low frequencies the efficiency approaches 1, and at high frequencies it goes to zero.
6. For all the multipliers, at low frequencies the power input and output are proportional to the frequency, and the dissipation is proportional to the square of the frequency. They all have approximately the same maximum output power, when compared on the basis of the same input frequency. The power limitations can probably be violated by operating into the forward region if the conduction mechanism fails.
7. For all the multipliers, at low frequencies high resistances must be used to match to the varactor, but at high frequencies the load and input resistances approach R_s.
8. Higher-order multipliers with idlers can be almost as efficient as a doubler, when compared on the basis of the same output frequency.
9. The 1-2-3-5 quintupler has an anomaly that probably prevents it from being a practical multiplier.
10. There is a fundamental limit on efficiency of any varactor multiplier, regardless of the idler terminations, the elastance charge characteristic, or the order of multiplication.

There are many aspects of multipliers that we have not discussed We have not derived any limitations on bandwidth, although surely there must be some. We have not investigated noise in multipliers or the possibility of spurious oscillations or motor boating. Although we gave a few typical circuits, we have not discussed the more practical aspects of circuit design of multipliers; rather, we have tried to give performance limits.

A very important restriction of our analysis is that we have neglected operation into the forward region upon failure of the conduction mechanism. These "charge storage" effects have been found experimentally to allow powers somewhat in excess of the limits calculated here.

REFERENCES

1. Blackwell, L. A., and K. L. Kotzebue, "Semiconductor-Diode Parametric Amplifiers," Prentice-Hall, Inc., Englewood Cliffs, N.J. (1961).
2. Diamond, B. L., "Idler Circuits in Varactor Frequency Multipliers," S.M. Thesis, Department of Electrical Engineering, M.I.T., Cambridge (February, 1961).
3. Diamond, B. L., Some Results for Higher-Order Varactor Frequency Multipliers, M.I.T. Lincoln Laboratory Group Report 47G-5, Lexington, Mass. (forthcoming, 1962).
4. Hyltin, T. M., and K. L. Kotzebue, "A Solid-State Microwave Source from Reactance-Diode Harmonic Generators," IRE Trans. on Microwave Theory and Techniques, MTT-9, 73-78 (January, 1961).
5. Johnson, K. M., "Large Signal Analysis of a Parametric Harmonic Generator," IRE Trans. on Microwave Theory and Techniques, MTT-8, 525-532 (September, 1960).
6. Leeson, D. B., and S. Weinreb, "Frequency Multiplication with Nonlinear Capacitors — A Circuit Analysis," Proc. IRE, 47, 2076-2084 (December, 1959).
7. Manley, J. M., and H. E. Rowe, "Some General Properties of Nonlinear Elements. — Part I. General Energy Relations," Proc. IRE, 44, 904-913 (July, 1956).
8. Page, C. H., "Frequency Conversion With Positive Nonlinear Resistors," J. Res. Nat. Bur. Standards, 56, 179-182 (April, 1956).
9. Penfield, P., Jr., Frequency-Power Formulas, The Technology Press and John Wiley & Sons, Inc., New York (1960).
10. Uhlir, A., Jr., "Similarity Considerations for Varactor Multipliers," Microwave J., 5, (forthcoming, July, 1962).

Chapter 9

HARMONIC DIVIDERS

Like other nonlinear reactances, varactors have the peculiar property that they can, when pumped, generate power not only at harmonics of the pumping frequency but also at subharmonics and at rational fractions of the pumping frequency. Any nonlinear device can generate harmonics of the driving frequency, but many cannot generate subharmonics. In particular, Page's formulas[5,6] ‡ show that nonlinear resistors with positive incremental resistances cannot. The ability to generate subharmonics is therefore not a common one, and ought to be regarded as an oddity.

When generating subharmonics or rational fractions of the driving frequency, varactors behave much like oscillators. The output current builds up with a phase that is at least partially arbitrary, accompanied perhaps by currents at other frequencies. As the output becomes larger, nonlinear effects reduce the rate of rise, and finally a steady state with a constant output amplitude is reached. We discuss for the divide-by-two circuit both the small-signal build-up of oscillations (with an incremental linear theory in Sections 9.2 and 9.3) and the steady state (with a nonlinear theory in Section 9.4). In Section 9.5, we give the large-signal, steady-state theory of the abrupt-junction-varactor divide-by-four circuit.

9.1. Introduction

When power is available to a varactor at frequency ω_p, power can come out at harmonics of that frequency ($2\omega_p$, $3\omega_p$, $4\omega_p$, $\cdots$), or it can under some circumstances come out at subharmonics of that frequency ($\omega_p/2$, $\omega_p/4$, $\cdots$), or at rational fractions of that frequency ($2\omega_p/3$, $3\omega_p/2$, $3\omega_p/4$, $5\omega_p/2$, $\cdots$). We call these three cases <u>harmonic generation</u>, <u>subharmonic generation</u>, and <u>rational-fraction generation</u>. Harmonic generation, or <u>multiplication</u>, was discussed in Chapter 8; now we discuss subharmonic generation, or <u>division</u>; rational-fraction generation is not discussed in detail in this book, although the analyses should be as straightforward as those of Chapters 8 and 9.

‡Superscript numerals denote references listed at the end of each chapter.

Dividers are useful for frequency control and comparison and for use in digital computers as bistable elements, parametrons.[8,3,1,7] In each of these applications, the phase of the output signal must be locked to the driving phase in some way, in order that the frequency of the output be accurately related to the input frequency (in the parametron there is some phase ambiguity, but not enough to enable the output phase to shift in a continuous manner). Among the dividers, only the divide-by-two and the divide-by-2^n circuits have this important property for small signals.‡

In Section 9.2 we discuss the divide-by-n circuit, for small signals, and show that only the divide-by-two circuit can operate without an idler.‡‡ For this reason most other dividers have an arbitrary phase and, in fact, for small signals are neither phase-sensitive nor frequency-stable.

In Section 9.3 we treat the small-signal divide-by-two circuit. The phases of oscillation for which power can be obtained are limited, and indeed there are no such phases unless the output frequency ω_o is

$$\omega_o < m_2\omega_c, \tag{9.1}$$

where m_2 is the modulation ratio of the varactor at the input frequency $2\omega_o$,

$$m_2 = \frac{|S_2|}{S_{max} - S_{min}}, \tag{9.2}$$

and where ω_c is the cutoff frequency of the varactor,

$$\omega_c = \frac{S_{max} - S_{min}}{R_s}, \tag{9.3}$$

where S_{max}, S_{min}, and R_s are the maximum and minimum elastances and the series resistance of the varactor. Thus for $\omega_o > m_2\omega_c$, the device cannot divide by two.

Whenever Condition 9.1 is obeyed, oscillations, which are at one of two possible phases that differ by 180°, can occur. They grow and eventually are limited by nonlinear effects. The fastest

---

‡Similarly, among the rational-fraction generators, only the multiply-by-$k/2^n$ circuits have this important property for small signals.

‡‡This fact is consistent with the results of Section 8.3.

initial rate of growth takes place when the varactor is terminated at frequency ω_0 by an inductance of a value derived in Section 9.3. For typical varactors the resulting rise times are of the order of a few nanoseconds and, therefore, fast enough to make the parametron of interest to computer engineers as a bistable element. Unfortunately, for most varactors the required pump power is high, and even when this power is available, the required circuitry is so complicated that parametrons made from varactors will probably never be competitive bistable computer elements.

The steady-state operation of dividers for power conversion and frequency control is discussed in Sections 9.4 and 9.5. To analyze the divide-by-two-circuit in Section 9.4, we assume that current is allowed to flow at only the driving frequency $2\omega_0$ and the output frequency ω_0. In addition, the analysis is restricted to the abrupt-junction varactor, for which the necessary solutions can be found in closed form. In spite of these restrictions, the theory is a useful one, because it holds approximately for graded-junction varactors, and because the results probably are not much affected by small amounts of current at frequencies $3\omega_0$ and $4\omega_0$. On the other hand, if the circuit allows spontaneous oscillations at other, unrelated frequencies, these oscillations can radically affect the performance.

Our treatment of the divide-by-two circuit is similar to that of the doubler. Expressions are derived in Section 9.4 for the input and load resistances, the input, output, and dissipated powers, and the efficiency. If the load impedance is tuned, these are functions of the operating frequency ω_0 and the modulation ratios m_1 and m_2. For a given frequency ω_0, all this information can be visualized very easily if it is plotted on the m_1, m_2-plane. Contours can be plotted corresponding to various values of input resistance, load resistance, input power, dissipated power, output power, and efficiency. Each point represents an operating point; by searching over this graph we can easily determine possible combinations of efficiency, powers, and the like.

The requirement that the varactor elastance lie between its minimum and maximum values led in Chapter 7 to definite restrictions on the values of m_1 allowed for pumping varactors. In a similar way, this requirement leads to possible combinations of values of m_1 and m_2 for the divide-by-two circuit. This restriction can be expressed by a single curve on the m_1, m_2-plane. Points to the right of this curve (that is, for excessively high values of m_1 and/or m_2) are not physically achievable. Thus, for practical purposes, only a portion of the m_1, m_2-plane must appear in these plots.

In Appendix D we give plots of this type for output frequencies from $10^{-4}\omega_c$ to $0.1\omega_c$. For a given power input, what is the maximum possible power output? For a given dissipation, for example fixed by the dissipation rating of the varactor, what is the highest efficiency achievable? For a given load resistance, what range of efficiency is possible? What load resistance should be used for highest efficiency? All these questions can be answered by referring to the appropriate graph in Appendix D.

For each operating frequency, there is a maximum possible efficiency, which drops to zero as ω_o approaches $0.25\omega_c$ and approaches 100 per cent for small ω_o. The operating point that achieves the maximum efficiency is of great interest, and graphs are given in Section 9.4 to describe it. Many divide-by-two circuits should be designed to operate at this point, or close to it, and design of optimized dividers is facilitated by some nomographs given in Appendix D.

For a given load resistance, there is a certain minimum input power that must be supplied to maintain oscillation. On the other hand, there is a maximum input power that can be supplied and still have the elastance remain between S_{min} and S_{max} (that is, to avoid clipping). The dynamic range of a divide-by-two circuit can be defined as the ratio of (1) the maximum power input before clipping to (2) the minimum power input to maintain oscillation, for a fixed load resistance. Alternatively, the available dynamic range can be defined as the available power input just before clipping divided by the minimum available power to maintain oscillation, at constant generator and load resistances. These two dynamic ranges are discussed in Section 9.4.

Like most of the analyses in this book, our treatment of the divide-by-two circuit determines fundamental limits of operation. No specific circuit is assumed, and our limits cannot be violated by a tricky choice of circuit. However, the analysis does indicate the circuit conditions that must be met to attain the fundamental limits. Synthesis of practical divider circuits, from this point of view, is discussed in Section 9.4.4.

The analysis of the divide-by-two circuit is very similar to that of the doubler, Section 8.4. The plots in the m_1, m_2-plane are quite similar to the doubler plots, especially at very low frequencies, where they are, for practical purposes, the same.

In Section 9.5 we take a brief look at the large-signal properties of the divide-by-four circuit. This device has an input at $4\omega_o$, an idler at $2\omega_o$, and an output at ω_o. For simplicity, we investigate only the case with a lossless, tuned, idler termination, and a tuned load. Only the point of maximum efficiency is discussed in detail; graphs are given for the maximum efficiency, for the input, output, and dissipated powers at maximum efficiency, for the input and load resistances, and for the bias voltage. These curves are qualitatively very similar to the corresponding curves for the divide-by-two circuit.

Suppose the divide-by-four circuit is turned on from rest. As the input power increases, there is no output and no idler excitation until the variation of elastance at frequency $4\omega_o$ is sufficient to generate an idler oscillation. Then this builds up, and if the input power is great enough, the elastance variation at $2\omega_o$ becomes so large that it excites a subharmonic oscillation at ω_o, the output. The idler excitation can occur only when

$$2\omega_o < m_4\omega_c, \tag{9.4}$$

and the oscillation at the output frequency can build up only when

$$\omega_o < m_2\omega_c. \tag{9.5}$$

Because of the limitation on the values of m_k, the divider cannot operate unless ω_o is less than $0.091\omega_c$.

The description of the build-up of the oscillations is quite difficult, and in Section 9.5 we treat only the large-signal steady-state condition. These curves given there are based on the results of Soheir T. Meleika,[4] and are given with her kind permission.

9.2. The Small-Signal Divide-by-n Circuit

Suppose that a varactor is pumped with current at frequency $n\omega_o$ and its harmonics, and we look for an output at frequency ω_o. As the oscillations build up (if they do), they pass from a small amplitude to a large amplitude. We now describe the behavior at small amplitudes by a linear small-signal theory.

The varactor elastance, as pumped, is

$$S(t) = \sum_{k=-\infty}^{+\infty} S_k e^{jk\omega_o t}, \tag{9.6}$$

with all S_k zero, except for S_0, S_n, S_{-n}, S_{2n}, S_{-2n}, S_{3n}, S_{-3n}, ... The small-signal current and voltage (that is, the total current and voltage less the pumping current and voltage) can also be written in similar Fourier series:

$$v(t) = \sum_{k=-\infty}^{+\infty} V_k e^{jk\omega_o t}, \tag{9.7}$$

$$i(t) = \sum_{k=-\infty}^{+\infty} I_k e^{jk\omega_o t} . \qquad (9.8)$$

The equation relating the small-signal voltage and current is Equation 4.40:

$$v(t) = S(t) \int i(t)\, dt + R_s i(t). \qquad (9.9)$$

Note that the small-signal equation is usually used where the frequencies of the small-signal variables are incommensurate with the pumping frequency, as for example in Chapters 5 and 6. Such a restriction is not necessary, however; here the small-signal frequencies are rationally related to the pumping frequency.

First we discuss the case with no idler, and then the case with one or more idler.

9.2.1. With No Idler

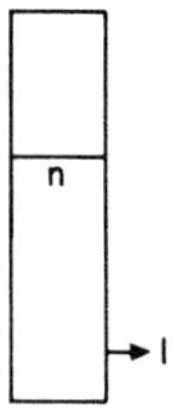

With no idler, the only small-signal current present is at frequency ω_o. Equation 9.9 predicts

$$V_1 = \left(R_s + \frac{S_0}{j\omega_o}\right) I_1 - \frac{S_2}{j\omega_o} I_1^* . \qquad (9.10)$$

Unless $S_2 \neq 0$ (that is, unless $n = 2$), the equivalent circuit at frequency ω_o of the varactor is merely a resistor R_s in series with an elastance S_0, and there is no way to extract power at that frequency. Therefore, unless $n = 2$, the divide-by-n circuit requires an idler to start.

9.2.2. With Idlers

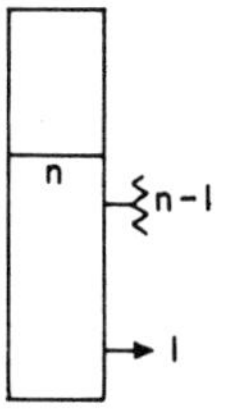

The divide-by-two circuit does not require an idler for small signals, but all other dividers do, as we have just shown. The idler or idlers can be put any of a variety of places.

If the varactor is excited at frequency $n\omega_o$, there can be an output at frequency ω_o if an idler at $(n - 1)\omega_o$ is present. This device behaves in a way similar to a nondegenerate parametric amplifier, with "signal" at ω_o, "idler" at $(n - 1)\omega_o$, and "pump" at $n\omega_o$. According to the analysis of Section 6.2, net power can be extracted from the parametric amplifier if and only if

$$\omega_o(n - 1)\omega_o < \frac{|S_n|^2}{R_s^2} = m_n^2\omega_c^2, \tag{9.11}$$

and by the same reasoning, this condition is both necessary and sufficient for small-amplitude subharmonic generation at frequency ω_o.

Unfortunately there is no physical effect (for small oscillations) that forces the phase of the output to be related to the phase of the pump, because in the expressions for the varactor impedance at the various frequencies, only the magnitude of S_n enters, and not its phase. Therefore, there is no physical effect to force the output frequency to be accurately $1/n$ times the input frequency. Such effects occur for large-signal operation but not for small signals. This subharmonic generator is, therefore, neither phase-sensitive nor frequency-stable. Slight mistuning of the circuits causes the oscillator to shift its output frequency slightly.

We have shown that if a divider requires an idler to start off, then placing the idler at frequency $(n - 1)\omega_o$ is not satisfactory, because the resulting oscillations are not frequency stable. But what about other positions for the idler frequencies?

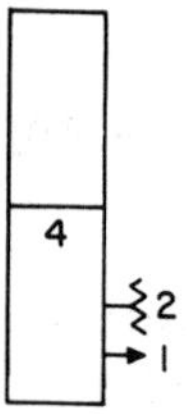

Consider the divide-by-four circuit. An idler is required; if this idler is at $3\omega_o$, the resulting oscillator is not frequency-stable. However, suppose the idler were at $2\omega_o$. The pumping causes growth of an output at $2\omega_o$, and in time this oscillation itself generates half its frequency, or the desired output frequency ω_o. An idler at $2\omega_o$ can therefore make the divide-by-four circuit frequency-stable. If an output at three-quarters of the pump frequency is desired, this can be obtained now, by mixing the currents at ω_o and $2\omega_o$; the resulting output frequency is stable.

This discussion indicates the following rule: It is possible, by proper placement of idlers, to generate phase-sensitive (and therefore frequency-stable) outputs at $1/2^k$ of the pump frequency (that is, 1/2, 1/4, 1/8, 1/16, 1/32, . . .). These divider circuits can therefore be frequency-stable. Similarly, among the rational-fraction multipliers, all circuits that multiply by $k/2^\ell$, for k and ℓ integers, can be made frequency-stable by proper choice of idler frequencies. For a further discussion of the idler requirement and a listing of possible idler configurations, see Section 8.3.

9.3. The Small-Signal Divide-by-Two Circuit

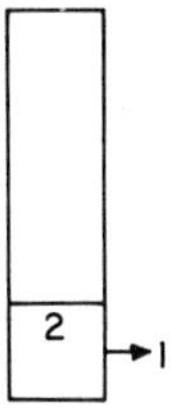

Of the frequency dividers, the divide-by-two circuit, shown in Figure 9.1, has special importance because it is phase-sensitive without any idler. The phase sensitivity is discussed in Section 9.3.1. The maximum rate of growth of the subharmonic oscillation is determined in Section 9.3.2, and finally, the necessary pump power is discussed in Section 9.3.3.

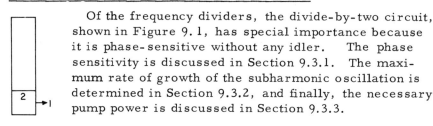

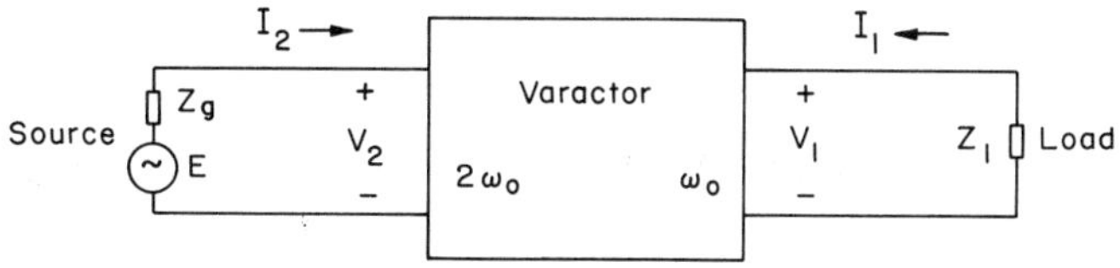

Figure 9.1. Divide-by-two circuit. Although the source and the load are shown separately, some physical elements may appear in both.

9.3.1. Phase Conditions for Oscillation

For the divide-by-two circuit, S_2 is nonzero; therefore, from Equation 9.10, the power output at frequency ω_o is

$$P_{out} = -2 \operatorname{Re} V_1 I_1^*$$

$$= 2|I_1|^2 \left(\frac{|S_2|}{\omega_o} \cos\theta - R_s \right)$$

$$= 2|I_1|^2 R_s \left(\frac{m_2 \omega_c}{\omega_o} \cos\theta - 1 \right), \qquad (9.12)$$

where the angle θ is the phase angle of S_2, minus 90°, minus twice the phase angle of I_1. Power can be drawn from the varactor at frequency ω_o only if the first term in the parentheses is larger than the second; in particular, only if $\omega_o < m_2\omega_c$. Also, note that the phases of I_1 for which power can be extracted are limited, because of the $\cos\theta$ term in Equation 9.12.

To interpret the angle θ more fully, consider the impedance at frequency ω_o seen by the varactor. If we let this impedance be Z_1, then

$$V_1 = -Z_1 I_1, \qquad (9.13)$$

and from Equation 9.10,

$$\left(Z_1 + R_s + \frac{S_0}{j\omega_o}\right) I_1 = \frac{S_2}{j\omega_o} I_1^* . \qquad (9.14)$$

Thus θ is also equal to the angle of the impedance

$$Z_1 + R_s + \frac{S_0}{j\omega_o} . \qquad (9.15)$$

Also, note from Equation 9.14 that unless $I_1 = 0$ (no oscillation),

$$\left| Z_1 + R_s + \frac{S_0}{j\omega_o} \right| = \frac{m_2\omega_c}{\omega_o} R_s . \qquad (9.16)$$

If the load impedance Z_1 is tuned, so that the quantity in Expression 9.15 is real, then $\cos\theta = 1$. Power is delivered (provided $I_1 \neq 0$) if $m_2\omega_c > \omega_o$. On the other hand, if the load is not tuned, not as much power is delivered (for the same current), and if the tuning is quite bad, no power at all is delivered. As the output frequency is raised toward $m_2\omega_c$, the tuning requirement becomes more and more critical.

It is interesting to examine Equation 9.16. We have assumed that the current is neither building up nor decaying, because it has the form of Equation 9.8 (with all $I_k = 0$ except I_1, I_2, I_{-1}, and I_{-2}). Equation 9.16 should be interpreted as specifying the value of load impedance that will just sustain oscillations of a constant value, neither letting them grow nor decay. If the load impedance is set so that Equation 9.16 is violated, then steady-state conditions do not prevail. If Z_1 is set too large, the oscillating current decays to zero, whereas if Z_1 is so small that the right-hand side of Equation 9.16 is larger than the left-hand side, then the oscillations grow in amplitude. The growth situation is discussed in more detail in Section 9.3.2.

Note from Equation 9.14 that the divide-by-two circuit has an output whose phase is not uniquely determined. Two possible phases that differ by 180° satisfy Equation 9.14. This fact leads to the use of this circuit as a bistable computer element, the parametron.[8,3,1,7]

9.3.2. Growing Oscillations

When the load impedance is smaller than the value given by Equation 9.16, the steady-state oscillations assumed in Section 9.3.1 do not exist. Rather, the oscillations grow exponentially with time. It is now proper to assume that the small-signal current is of the form

$$i(t) = I_1 e^{st} + I_1^* e^{s^* t}, \tag{9.17}$$

where s is a complex frequency

$$s = \sigma + j\omega_o. \tag{9.18}$$

If σ, the real part of s, is positive, the oscillations grow (and eventually get so large that the small-signal theory given here no longer applies). On the other hand, if σ is negative, then the oscillations decay and eventually vanish. Since σ determines the rate of growth, we shall try to determine how large σ can be.

The component of voltage V_1 at the complex frequency s is related to the component of current I_1 by

$$V_1 = \left(R_s + \frac{S_0}{s}\right) I_1 + \frac{S_2}{s^*} I_1^*, \tag{9.19}$$

which comes from Equation 9.9. Note that this equation is different from Equation 9.10, which holds for steady-state oscillations. Let us denote by Z the impedance seen by the varactor at the complex frequency s:

$$V_1 = -ZI_1 . \tag{9.20}$$

Then, from Equation 9.19,

$$\left(R_s + \frac{S_0}{s} + Z\right) I_1 = -\frac{S_2}{s^*} I_1^*. \tag{9.21}$$

If we use the definition of m_0,

$$m_0 = \frac{S_0}{S_{max} - S_{min}}, \tag{9.22}$$

Equation 9.21 becomes (by taking absolute values and multiplying each side by $|s|/R_s$)

$$\left| s + m_0\omega_c + \frac{sZ}{R_s} \right| = m_2\omega_c. \tag{9.23}$$

It is now pertinent to ask how large σ can be, for given values of m_0, m_2, ω_0, and ω_c, as Z is varied.

We assume that Z is the impedance of a passive network. It is well known that Z must therefore have a positive real part for $\sigma = 0$. It is not quite as well known that when σ is positive, the angle of Z must[2] be less than or equal to the angle of s. Thus, if we define α as the complement of the angle of s,

$$\tan \alpha = \frac{\sigma}{\omega_0}, \tag{9.24}$$

then Z(s) is restricted to lie in the region of the complex plane indicated in Figure 9.2. Therefore, the quantity

$$\frac{sZ}{R_s} \tag{9.25}$$

must lie in the complex plane in the shaded region of Figure 9.3. Only if the termination at frequency s is an inductor, does Expression 9.25 lie along the left-hand border of the shaded region in Figure 9.3.

The simplest way to solve Equation 9.23 for the highest value of σ corresponding to a passive load impedance Z is geometrically. The allowed region in the complex plane for the quantity

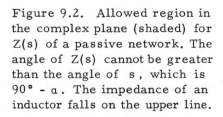

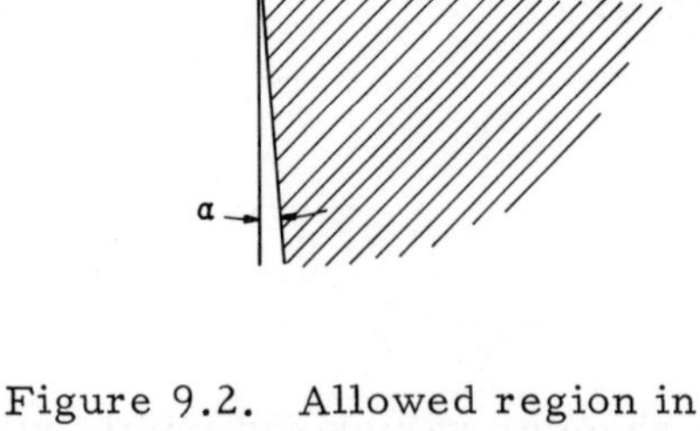

Figure 9.2. Allowed region in the complex plane (shaded) for Z(s) of a passive network. The angle of Z(s) cannot be greater than the angle of s, which is 90° - α. The impedance of an inductor falls on the upper line.

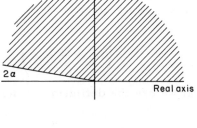

Figure 9.3. Allowed region in the complex plane (shaded) for sZ(s). For an inductance L, Z(s) = Ls, and sZ(s) lies on the left-hand boundary. Note that the angle indicated at the left is 2α, not α.

$$s + m_0\omega_c + \frac{sZ}{R_s} \qquad (9.26)$$

is shown shaded in Figure 9.4. If Equation 9.23 is to be obeyed, we must choose an impedance of such a value that the corre-

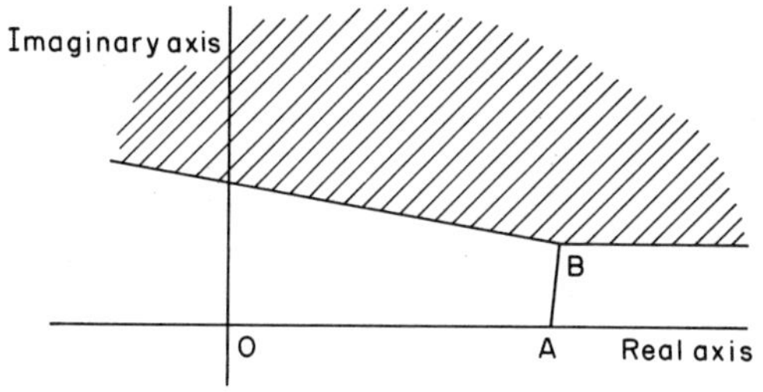

Figure 9.4. Allowed region in the complex plane for the quantity of Expression 9.26. The point A is $m_0\omega_c$; the point B is $m_0\omega_c + s$.

sponding point on Figure 9.4 is located a distance $m_2\omega_c$ from the origin. Since m_0 is greater than m_2, a circle of radius $m_2\omega_c$, centered at the origin, can intersect the boundary of the shaded region only to the left of the point B. To find the highest possible value for σ, we simply increase σ until the left-hand border is tangent to the circle, as indicated in Figure 9.5. Any higher value of σ would pull the line BC away from the circle.

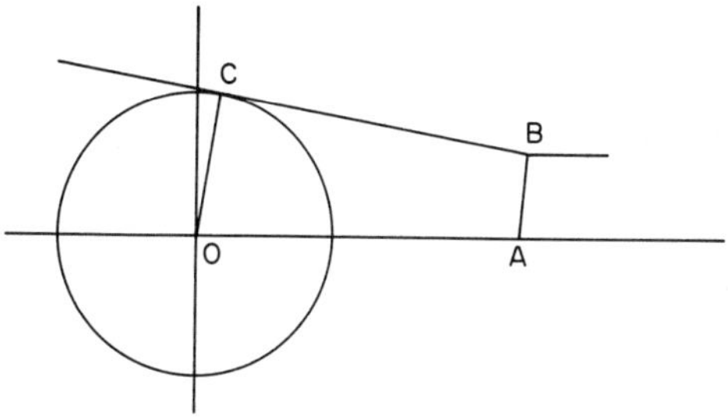

Figure 9.5. Use of an inductor to terminate the varactor at frequency s enables us to satisfy Equation 9.23 with the left-hand border of the allowed region. This termination yields the highest value of σ. The circle has radius $m_2\omega_c$.

In Figure 9.5, the distance OA is $m_0\omega_c$; the distance AB is $|s|$; the angle between AB and the vertical is α; the distance OC is $m_2\omega_c$; the angle between OC and the vertical is 2α.

Figure 9.6 shows an enlarged view of the line AB. We have extended line CB and dropped a perpendicular from A, which lands at the point D. Since the angle DAB is equal to α, the distance BD must be σ, and the distance AD must be ω_0.

If we draw a line through A parallel to the line BC, it intersects the line OC at right angles (say at the point E), at a distance $m_2\omega_c - \omega_0$ from the point O. Since the triangle OEA, shown in Figure 9.7, has acute angle 2α,

$$\sin 2\alpha = \frac{m_2\omega_c - \omega_0}{m_0\omega_c} \qquad (9.27)$$

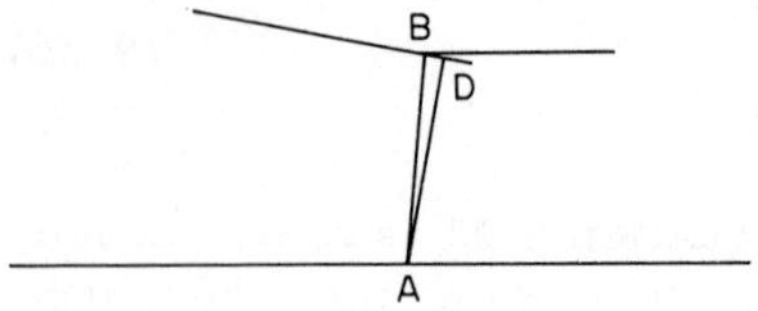

Figure 9.6. Demonstration that the point A is located a distance ω_0 from the line CB. The angle ADB is a right angle, and the angle DAB is α.

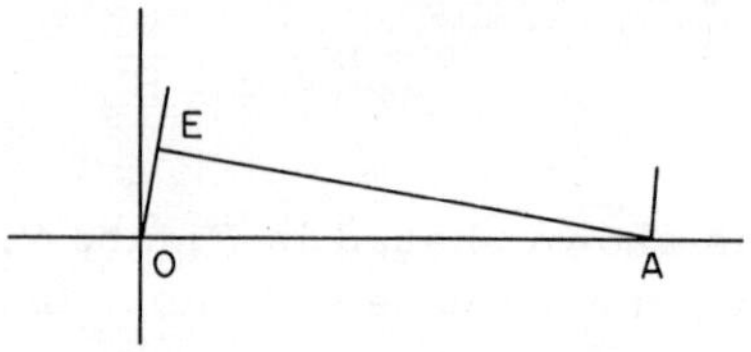

Figure 9.7. The triangle formed from the real axis, the line OC, and the line through A parallel to BC. The angle OAE is 2α, and the angle OEA is a right angle.

Furthermore, the length BC, which is the length AE minus σ, is equal to

$$\frac{|s|^2 L}{R_s}, \tag{9.28}$$

where L is the value of inductance to be used to achieve the maximum rate of growth σ.

From the trigonometric identity

$$\tan \alpha = \frac{1 - \sqrt{1 - \sin^2 2\alpha}}{\sin 2\alpha}, \tag{9.29}$$

and from Equation 9.24, we find an expression for σ

$$\sigma = \omega_0 \left[\frac{m_0 \omega_c}{m_2 \omega_c - \omega_0} - \sqrt{\left(\frac{m_0 \omega_c}{m_2 \omega_c - \omega_0} \right)^2 - 1} \right]; \tag{9.30}$$

and the value of inductance load that achieves this maximum σ is, from Expression 9.28 and Equation 9.30,

$$L = R_s \frac{m_2 \omega_c - \omega_0}{2\omega_0 m_0 \omega_c} \left[\frac{m_2 \omega_c - \omega_0}{\omega_0} \frac{\sqrt{(m_0 \omega_c)^2 - (m_2 \omega_c - \omega_0)^2}}{m_0 \omega_c - \sqrt{(m_0 \omega_c)^2 - (m_2 \omega_c - \omega_0)^2}} - 1 \right]$$

$$= \frac{R_s}{2\sigma} \left(\frac{m_2 \omega_c}{\omega_o} \frac{\omega_o^2 - \sigma^2}{\omega_o^2 + \sigma^2} - 1 \right). \tag{9.31}$$

It would be of interest to interpret L as the inductance that tunes out one of the characteristic capacitances of the varactor at one of the characteristic frequencies. However, L is smaller than the inductance necessary to tune out the elastance S_0 either at frequency ω_0 or frequency $\sqrt{\omega_0^2 + \sigma^2}$, and it is larger than the inductance that is necessary to tune out the elastance $S_0 - |S_2|$ at either of these two frequencies. For example, an alternate formula for L

$$L = \frac{S_0 - |S_2| \frac{\sigma}{\omega_0}}{\omega_0^2 + \sigma^2} \tag{9.32}$$

shows that the "effective elastance" that is tuned out by L at frequency $\sqrt{\omega_0^2 + \sigma^2}$ lies between S_0 and $S_0 - |S_2|$. If S_2 is small, the inductance tunes out an elastance approximately equal to S_0.

In plotting Equations 9.30 and 9.31, some assumption must be made concerning the ratio of m_0 to m_2. Pertinent results from Chapter 7 are summarized in Table 9.1, for current pumping, voltage pumping, and arbitrary waveform pumping (the values of m_2 and of m_0/m_2 shown for arbitrary waveform pumping cannot be achieved simultaneously). Table 9.1 indi-

Table 9.1. Brief Summary of Important Pumping Limits, Taken from Chapter 7

For sinusoidal voltage or current, the two limits shown can be achieved simultaneously if $S_{min} = 0$, and the limiting values of m_0 are achieved with self-bias. The arbitrary current limits cannot be achieved simultaneously.

	Graded-Junction Varactor	Abrupt-Junction Varactor
V Sinusoidal	$m_2 \leq 0.178$ $m_0 \geq 4m_2$	$m_2 \leq 0.212$ $m_0 \geq 3m_2$
I Sinusoidal	$m_2 \leq 0.212$ $m_0 \geq 3m_2$	$m_2 \leq 0.250$ $m_0 \geq 2m_2$
I Arbitrary	$m_2 < 0.318$ $m_0 > m_2$	$m_2 < 0.318$ $m_0 > m_2$

cates that current pumping is preferable to voltage pumping and, for the same cutoff frequency, an abrupt-junction varactor is better than a graded-junction.

For the limits of m_0/m_2 given in Table 9.1 (1, 2, 3, and 4), we show σ and σ/ω_o, as functions of ω_o, in Figures 9.8 and 9.9. The inductance is shown in Figure 9.10. Note from Figure

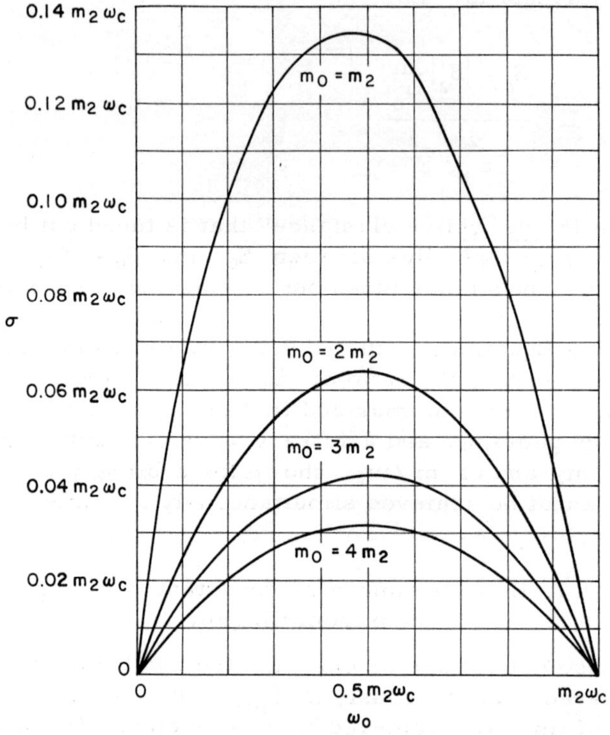

Figure 9.8. Maximum possible rate of growth σ, as a function of the output frequency ω_o, taken from Equation 9.30. Note that the largest values of σ are attained for a pump frequency $2\omega_o$ approximately equal to $m_2\omega_c$.

9.8 that there is an upper limit for σ, which is achieved with a pump frequency $2\omega_o$ approximately equal to $m_2\omega_c$. This frequency is approximately the optimum pump frequency encountered so often in Chapters 5 and 6.

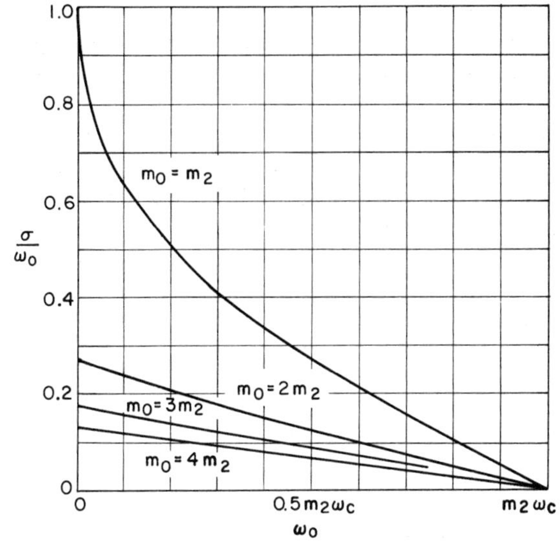

Figure 9.9. Maximum value of σ/ω_0, as a function of the output frequency ω_0, taken from Equation 9.30.

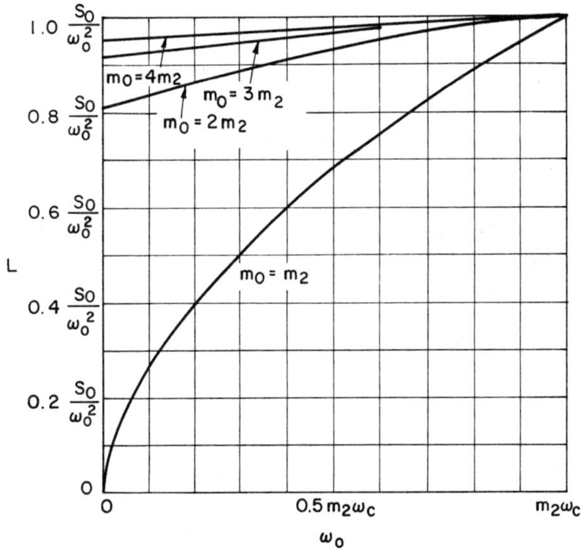

Figure 9.10. Inductance L necessary to achieve the fastest rise time, taken from Equation 9.31. Note that L is normalized with respect to the elastance S_0 and the frequency ω_0.

9.3.3. Pump Power

For typical varactors, the rise times predicted by Figures 9.8 and 9.9 are in the order of nanoseconds, small enough to be of interest for computer applications. However, the pump power required to maintain a high value of m_2 for such fast rise times may be large. The minimum pump power for a given value of m_2 is achieved with current pumping and self-bias; for graded-junction varactors this pump power is

$$P = 139 m_2^4 P_{norm} \frac{\omega_p^2}{\omega_c^2}, \tag{9.33}$$

and for abrupt-junction varactors,

$$P = 8 m_2^2 P_{norm} \frac{\omega_p^2}{\omega_c^2} \tag{9.34}$$

where $P_{norm} = (V_B + \phi)^2/R_s$ is the normalization power of the varactor, and $\omega_p = 2\omega_0$ is the pump frequency.‡

For a given value of σ/ω_0 (and therefore a given value of $\omega_0/m_2\omega_c$), Equations 9.33 and 9.34 show that the required pump power varies as the fourth power of m_2 for abrupt-junction varactors, and as the sixth power of m_2 for graded-junction varactors. Thus, the pump power varies as the inverse fourth (or sixth) power of the build-up time, so that even though extremely fast rise times require excessive power, moderate rise times may require orders of magnitude less power.

If the highest possible value of σ is desired, the varactor should be pumped to the maximum value of m_2 at the optimum pump frequency $m_2\omega_c$; the required pump power is $P_{norm}/32$ for abrupt-junction varactors, and $P_{norm}/79$ for graded-junction varactors.

9.3.4. Example

Suppose we have an abrupt-junction varactor with a cutoff frequency of 80 Gc and a normalization power of 16 watts. Suppose we require that the oscillations build up 400 times (that is, 52 db) in a time T. Then $e^{\sigma T}$ is 400, and therefore,

$$T = \frac{6}{\sigma}. \tag{9.35}$$

How small can T possibly be? Let us pump the varactor to

---

‡For simplicity, we assume $S_{min} << S_{max}$.

$m_2 = 0.25$ at the optimum pump frequency 20 Gc. Then the time T becomes 0.72×10^{-9} second, which is quite a small value.‡ During this interval the output has gone through only 7.2 periods at 10 Gc.

It is apparent that very fast operation is possible. Even if external resistance in the circuit and extra energy storage decreases the speed 10 times, the build-up will still occur in 7.2 nanoseconds, a speed that is short enough to be of considerable interest for computer applications.

Difficulties become apparent when we examine the pump power required to operate such a device. For our example, 500 milliwatts is required for each varactor. Aside from the fact that the varactor probably cannot safely dissipate this amount, the power supplies for a large-scale computer with several thousand parametrons would be large — much larger and (because alternating current is required) more complicated than would be required to operate the same number of tunnel diodes, for example.

To reduce the power required, we could pump to a value of m_2 one-fifth that used previously, or 0.05, at a frequency of 4 Gc. This would reduce the pump power by the fourth power of 5, or down to 0.8 mw. The build-up time would be 3.6 nanoseconds, and at the output frequency of 2 Gc, 7.2 cycles occur in this build-up. Even if the external losses and energy storage were such as to slow down this build-up by a factor of 10, it could still build up in 36 nanoseconds.

9.3.5. Conclusions

For three reasons we do not anticipate much application of varactors as parametrons in computers. First, a nominal amount of power is required to pump them. Suppose a computer requires 10^5 parametrons. If each one required 10 mw for pump power, we should need 1000 watts of rf power altogether. Although this amount of power could be obtained with fair efficiency, it is greater than what would be required by, say, bias circuits for tunnel diodes.

Second, and more important, the circuitry necessary to control and distribute this power would be complicated and costly. Some sort of high-power, fast-acting switches would be required to switch the power, and microwave techniques would have to be used to deliver the power to the varactors. Tunnel diodes can use dc bias, with simpler switching and distribution techniques.

A third difficulty is that of synthesizing a termination at the subharmonic frequency and of separating the pump and the subharmonics. The circuitry would be quite costly on a large scale.

---

‡For simplicity, we assume the initial rate of growth can be maintained even for large signals.

On the other hand, harmonic dividers have many applications in the field of frequency control and energy conversion. Both the small-signal theory, just described, and the large-signal steady-state theory of Section 9.4 are important.

9.4. The Large-Signal Divide-by-Two Circuit

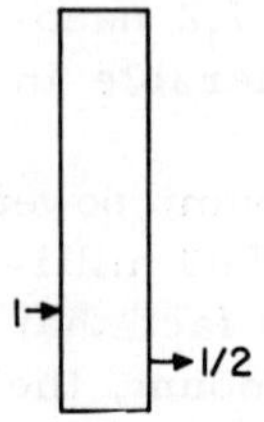

The large-signal nonlinear analysis is restricted to abrupt-junction varactors, for which solutions are obtainable in closed form. We assume that current flows at only the two frequencies $2\omega_o$ (input) and ω_o (output). With these restrictions, the nonlinear divide-by-two circuit can be solved exactly. After deriving the pertinent formulas, we discuss the limitation on the modulation ratios m_1 and m_2. The formulas are presented in graphical form in Section 9.4.3, for both low-power operation and high-power operation. In Section 9.4.4 we discuss synthesis of divide-by-two circuits.

9.4.1. Formulas

The steady-state operation of the varactor divide-by-two circuit is described by large-signal equations of the form of Equations 4.31. Because we assume the varactor has current at only frequencies ω_o and $2\omega_o$, all I_k vanish except for I_{-2}, I_{-1}, I_1, and I_2. Furthermore, because we restrict the analysis to abrupt-junction varactors, for which the elastance S is proportional to the charge q, all Fourier coefficients of elastance vanish from Equations 4.31 except for S_0, S_1, and S_2 (and, of course, S_{-1} and S_{-2}). The resulting equations for V_1 and V_2, the components of voltage at frequencies ω_o and $2\omega_o$, are

$$V_1 = \left(R_s + \frac{S_0}{j\omega_o}\right) I_1 + \frac{S_2}{j\omega_o} I_1^* + \frac{S_1^*}{j\omega_o} I_2 , \tag{9.36}$$

$$V_2 = \left(R_s + \frac{S_0}{j2\omega_o}\right) I_2 + \frac{S_1}{j2\omega_o} I_1 . \tag{9.37}$$

Because the elastance is proportional to the time integral of the current,

$$\frac{S_1^*}{S_2} = \frac{I_1^*}{-j\omega_o} \frac{j2\omega_o}{I_2} = -2 \frac{I_1^*}{I_2} , \tag{9.38}$$

and Equation 9.36 becomes

$$V_1 = \left(R_s + \frac{S_0}{j\omega_o}\right) I_1 - \frac{S_2}{j\omega_o} I_1^*$$

$$= \left(R_s + \frac{S_0}{j\omega_o}\right) I_1 + \frac{S_1^*}{j2\omega_o} I_2. \qquad (9.39)$$

Equations 9.37 and 9.39 also describe the varactor doubler; they are the same as Equations 8.21 and 8.24.

Let the load impedance seen by the varactor at frequency ω_o be Z_1; the load constraint is, therefore,

$$V_1 = -Z_1 I_1, \qquad (9.40)$$

so that Equation 9.39 becomes

$$\left(Z_1 + R_s + \frac{S_0}{j\omega_o}\right) I_1 = \frac{S_2}{j\omega_o} I_1^*. \qquad (9.41)$$

9.4.1(a). Load Impedance. By taking the absolute value of each side of Equation 9.41, we find

$$\frac{|S_2|}{\omega_o} = \left| Z_1 + R_s + \frac{S_0}{j\omega_o} \right| = \frac{m_2 \omega_c}{\omega_o} R_s, \qquad (9.42)$$

provided that I_1 is not zero.

Note that Equation 9.42 relates the load impedance to S_2, which depends on the drive level. Physically, these two quantities can be set independently. Thus Equation 9.42 should be interpreted as the necessary relationship between them if the operation is steady state. If m_2 is higher than the value of Equation 9.42, then the actual oscillations are growing, whereas if m_2 is smaller, the actual oscillations decay to some lower level. Consider what happens physically when the divider is turned on. First m_2 attains some initial value, and if the load impedance is small enough, the oscillations grow, in accordance with the theory of Section 9.3. One of two effects may limit these oscillations. First, the varactor may break down because of excessively high values of m_1 and m_2. Or second, the increased voltage V_2 may reduce m_2 (because of a finite source impedance) to a value consistent with Equation 9.42.

As in Section 9.3.1, let us denote by θ the phase angle of S_2, minus 90°, minus twice the phase angle of I_1. Then θ is also the phase angle of

$$Z_1 + R_s + \frac{S_0}{j\omega_0}, \tag{9.43}$$

and the load resistance $R_1 = \text{Re } Z_1$ is, from Equation 9.42,

$$R_1 = R_s \left(\frac{m_2 \omega_c}{\omega_0} \cos\theta - 1 \right). \tag{9.44}$$

9.4.1(b). Input Impedance. We can eliminate I_1 from Equations 9.37 and 9.39, to obtain

$$V_2 = Z_{in} I_2, \tag{9.45}$$

where

$$Z_{in} = R_s + \frac{S_0}{j2\omega_0} + \frac{m_1^2 \omega_c^2}{4\omega_0^2} \frac{R_s^2}{Z_1 + R_s + (S_0/j\omega_0)}$$

$$= R_s + \frac{S_0}{j2\omega_0} + \frac{m_1^2}{2m_2} \frac{\omega_c}{2\omega_0} R_s e^{-j\theta}, \tag{9.46}$$

where Equation 9.42 has been used. Note that Z_{in} is not the input impedance of the circuit, in the usual sense. The varactor is a nonlinear device and does not have an input impedance. The ratio of input voltage to input current is a well-defined concept (and we have called it Z_{in}), but its value depends on the drive level. We shall call Z_{in} the "input impedance," with the understanding that it is not really an impedance.

We shall call $R_{in} = \text{Re } Z_{in}$ the "input resistance," again with the understanding that since its value depends on the drive level, it is not a resistance in the usual sense:

$$R_{in} = R_s \left(1 + \frac{m_1^2}{2m_2} \frac{\omega_c}{2\omega_0} \cos\theta \right). \tag{9.47}$$

9.4.1(c). Input Power. The input power can be calculated from the input resistance:

$$P_{in} = 2 |I_2|^2 R_{in}. \tag{9.48}$$

We wish to relate this to the varactor normalization power. For the abrupt-junction varactor with normalization power P_{norm}

and maximum and minimum elastances S_{max} and S_{min}, Equation 8.40 holds, and can be used to find the alternate form of Equation 9.48:

$$P_{in} = 8P_{norm}\left(\frac{S_{max} - S_{min}}{S_{max} + S_{min}}\right)^2 \frac{\omega_o^2}{\omega_c^2} (2m_2)^2 \frac{R_{in}}{R_s}$$

$$= 8P_{norm}\left(\frac{S_{max} - S_{min}}{S_{max} + S_{min}}\right)^2 \frac{\omega_o^2}{\omega_c^2}\left(\frac{\omega_c}{\omega_o} m_1^2 m_2 \cos\theta + 4m_2^2\right), \tag{9.49}$$

where we have used Equation 9.47.

9.4.1(d). Output Power. The output power can be calculated in a similar way. It is

$$P_{out} = 2\,|I_1|^2 R_1, \tag{9.50}$$

and by use of Equations 8.39 and 9.44, we find

$$P_{out} = 8P_{norm}\left(\frac{S_{max} - S_{min}}{S_{max} + S_{min}}\right)^2 \frac{\omega_o^2}{\omega_c^2} m_1^2 \left(\frac{\omega_c}{\omega_o} m_2 \cos\theta - 1\right). \tag{9.51}$$

9.4.1(e). Dissipated Power. The power dissipated in the series resistance of the varactor can be calculated two ways. First, it is equal to

$$P_{diss} = 2R_s(|I_1|^2 + |I_2|^2); \tag{9.52}$$

second, it is simply $P_{in} - P_{out}$. Calculated either way,

$$P_{diss} = 8P_{norm}\left(\frac{S_{max} - S_{min}}{S_{max} + S_{min}}\right)^2 \frac{\omega_o^2}{\omega_c^2}(m_1^2 + 4m_2^2). \tag{9.53}$$

This formula is the same as Equation 8.44 for the abrupt-junction doubler.

9.4.1(f). Efficiency. The efficiency can be calculated two ways. First, it is the ratio of power output to power input,

$$\epsilon = \frac{P_{out}}{P_{in}}. \tag{9.54}$$

Alternatively, consider the input and load resistances. Of the power input, a certain fraction R_s/R_{in} is dissipated in the varactor series resistance. Of the remainder, only a fraction $R_1/(R_1 + R_s)$ actually reaches the load. Thus the efficiency is

$$\epsilon = \frac{R_{in} - R_s}{R_{in}} \frac{R_1}{R_1 + R_s}. \tag{9.55}$$

Calculated either way,

$$\epsilon = \frac{\dfrac{\omega_c}{2\omega_o} \cos\theta - \dfrac{1}{2m_2}}{\dfrac{\omega_c}{2\omega_o} \cos\theta + \dfrac{2m_2}{m_1^2}}. \tag{9.56}$$

9.4.1(g). Bias Voltage. It is usually necessary to bias the varactor with some average voltage V_0, which can easily be calculated in terms of the modulation ratios. For an abrupt-junction varactor, since the voltage v is proportional to the square of the elastance,

$$\frac{v + \phi}{V_B + \phi} = \left(\frac{S}{S_{max}}\right)^2, \tag{9.57}$$

where ϕ is the contact potential and not necessarily the minimum voltage V_{min}. The bias voltage is found by taking the average value of Equation 9.57:

$$\frac{V_0 + \phi}{V_B + \phi} = \left(\frac{S_{max} - S_{min}}{S_{max}}\right)^2 (m_0^2 + 2m_1^2 + 2m_2^2), \tag{9.58}$$

in agreement with Equation 8.49 for the abrupt-junction doubler.

9.4.1(h). Phase Condition. The formulas for input and load resistance, input and output power, and efficiency, all depend on the angle θ associated with the load impedance. Note that for fixed values of m_1 and m_2 and fixed frequency, the efficiency is highest when $\theta = 0$. We are therefore led to tune the load to make $\theta = 0$.

It is not clear that this procedure leads to the highest possible efficiency. The limit on the modulation ratios given in Section 9.4.2 depends on the angle θ, and for $\theta \neq 0$, somewhat larger values of m_1 and m_2 can be used. It is not known if these larger values, together with a nonzero value for θ, can give higher efficiencies. Nevertheless, for simplicity we henceforth assume that $\theta = 0$, recognizing that there may be a chance of a slight improvement over some of the limits to be derived.

9.4.2. Breakdown Limit for the Modulation Ratios

The modulation ratios m_1 and m_2 are restricted because for all time $S(t)$ must lie between S_{min} and S_{max}. The possible values of m_1 and m_2 depend on the phase angle between I_1 and I_2. By setting $\theta = 0$, in Section 9.4.1(h), we require that if I_1 is real and negative, so is I_2. Therefore, Q_1 and Q_2 can, without loss of generality, both be taken imaginary. Thus $S(t)$ is

$$\frac{S(t)}{S_{max} - S_{min}} = m_0 - 2m_1 \sin \omega_0 t - 2m_2 \sin 2\omega_0 t. \qquad (9.59)$$

To fit the largest possible values of m_1 and m_2, and still have $S_{min} < S(t) < S_{max}$, we must choose m_0 so that

$$S_0 = \frac{S_{max} + S_{min}}{2} . \qquad (9.60)$$

The restriction on m_1 and m_2 is then that

$$m_1 \sin \omega_0 t + m_2 \sin 2\omega_0 t \leq 0.25 \qquad (9.61)$$

for all time t. This condition is the same as that for the abrupt-junction doubler, Condition 8.52. The permissible values of m_1 and m_2 are exactly those of Figure 8.3, which shows the breakdown limit curve in the m_1, m_2-plane.

9.4.3. Large-Signal Solutions

Setting the angle θ to zero, as discussed in Section 9.4.1(h), we obtain the following formulas for load resistance, input resistance, input, output, and dissipated power, and efficiency:

$$R_1 = R_s \left(\frac{\omega_c}{\omega_0} m_2 - 1 \right), \qquad (9.62)$$

$$R_{in} = R_s \left(\frac{\omega_c}{2\omega_0} \frac{m_1^2}{2m_2} + 1 \right), \qquad (9.63)$$

$$P_{in} = 8P_{norm}\left(\frac{S_{max} - S_{min}}{S_{max} + S_{min}}\right)^2 \frac{\omega_o^2}{\omega_c^2}\left(\frac{\omega_c}{\omega_o} m_1^2 m_2 + 4m_2^2\right), \quad (9.64)$$

$$P_{out} = 8P_{norm}\left(\frac{S_{max} - S_{min}}{S_{max} + S_{min}}\right)^2 \frac{\omega_o^2}{\omega_c^2}\left(\frac{\omega_c}{\omega_o} m_1^2 m_2 - m_1^2\right), \quad (9.65)$$

$$P_{diss} = 8P_{norm}\left(\frac{S_{max} - S_{min}}{S_{max} + S_{min}}\right)^2 \frac{\omega_o^2}{\omega_c^2}(m_1^2 + 4m_2^2), \quad (9.66)$$

$$\epsilon = \frac{\dfrac{\omega_c}{2\omega_o} - \dfrac{1}{2m_2}}{\dfrac{\omega_c}{2\omega_o} + \dfrac{2m_2}{m_1^2}}. \quad (9.67)$$

For given load resistances and input drive circuits, the solution of these equations is a formidable task. Graphical solutions are the easiest to obtain. From the graphs to be given, optimized low-power and high-power dividers can be designed and their dynamic ranges computed.

The key to the solution is the observation that for each given frequency ω_o/ω_c, all the quantities in Equations 9.62 to 9.67 depend on only two parameters, m_1 and m_2. Although m_1 and m_2 are not ordinarily known quantities, in the m_1, m_2-plane we can still indicate values for each of the six quantities of interest. It is then a simple matter to determine what combinations of resistances, powers, and efficiency are possible, and to select the load impedance and drive circuit to achieve the optimum performance. This technique is similar to that used for the doubler equations, and in fact the graphs are similar in appearance.

As an example of this technique, we show in Figures 9.11 and 9.12 a typical plot of this sort, in the m_1-m_2 plane. In Figure 9.11 are the powers, and in Figure 9.12, the resistances and the efficiency. Both are for output frequency $\omega_o = 10^{-2}\omega_c$. In Appendix D a complete set of curves of this sort is given for a range of frequencies. For each frequency, the two sets of curves like Figures 9.11 and 9.12 are superimposed in different colors. For technical reasons, the axes correspond to m_2 and m_1^2, instead of m_2 and m_1, but since the lines of constant m_1 and m_2

are not shown anyway (their values are usually unimportant), the user of the graphs should not care.

Also shown in Figures 9.11 and 9.12 are two auxiliary curves. The one in Figure 9.12, the breakdown limit, denotes the region over which the diode can operate and not exceed the maximum or minimum elastances. Only the region to the left of this curve can be used. The other curve, in Figure 9.11, denotes a set of operating points that are optimum, in a sense to be discussed in Section 9.4.3(a).

Many preliminary design questions can be answered by referring to these graphs. For a given power input, what is the highest efficiency? For a specified power output, what is the minimum power input? What is the maximum possible efficiency of this divider? The maximum possible power output? For a fixed varactor dissipation rating, what is the maximum power output that can be achieved? Over what range of power input will the divider operate? What load impedances are appropriate for each of the optimum conditions?

These graphical solutions of Equations 9.62 to 9.67, given in detail in Appendix D, are very versatile. We now discuss three major applications of them: low-power optimization, high-power optimization, and dynamic range.

9.4.3(a). Low-Power Optimum. If the operating point is known to lie inside the line corresponding to the maximum and minimum elastance restriction, then it is often desirable to operate with maximum efficiency for a given power input or with minimum dissipation for a given power output, and so forth.

Consider a typical requirement of this sort. Suppose the varactor dissipation rating restricts the dissipated power to be less than a certain amount. What is the maximum power output that can be achieved with this limited dissipated power? From Figure 9.11, it is clear that such a maximum exists and that the operating point corresponding to it is the point on the dissipated-power curve that is tangent to a power-output curve. But if for a given dissipated power the power output is the greatest, then the efficiency must also be highest, and the power input the greatest. Therefore, at this single point on the dissipated-power curve, it is tangent not only to a power-output curve but also to an efficiency curve and to a power-input curve. Thus, if the varactor dissipation is limited, this one point simultaneously optimizes the efficiency, the power input, and the power output. It is clearly a very special point, and the collection of all such points, one for each value of dissipated power, is a curve that is very valuable. This curve, shown in Figure 9.11, is given by the formula

$$m_1^2 = 8m_2^2 \left(1 - \frac{\omega_o}{m_2\omega_c}\right). \qquad (9.68)$$

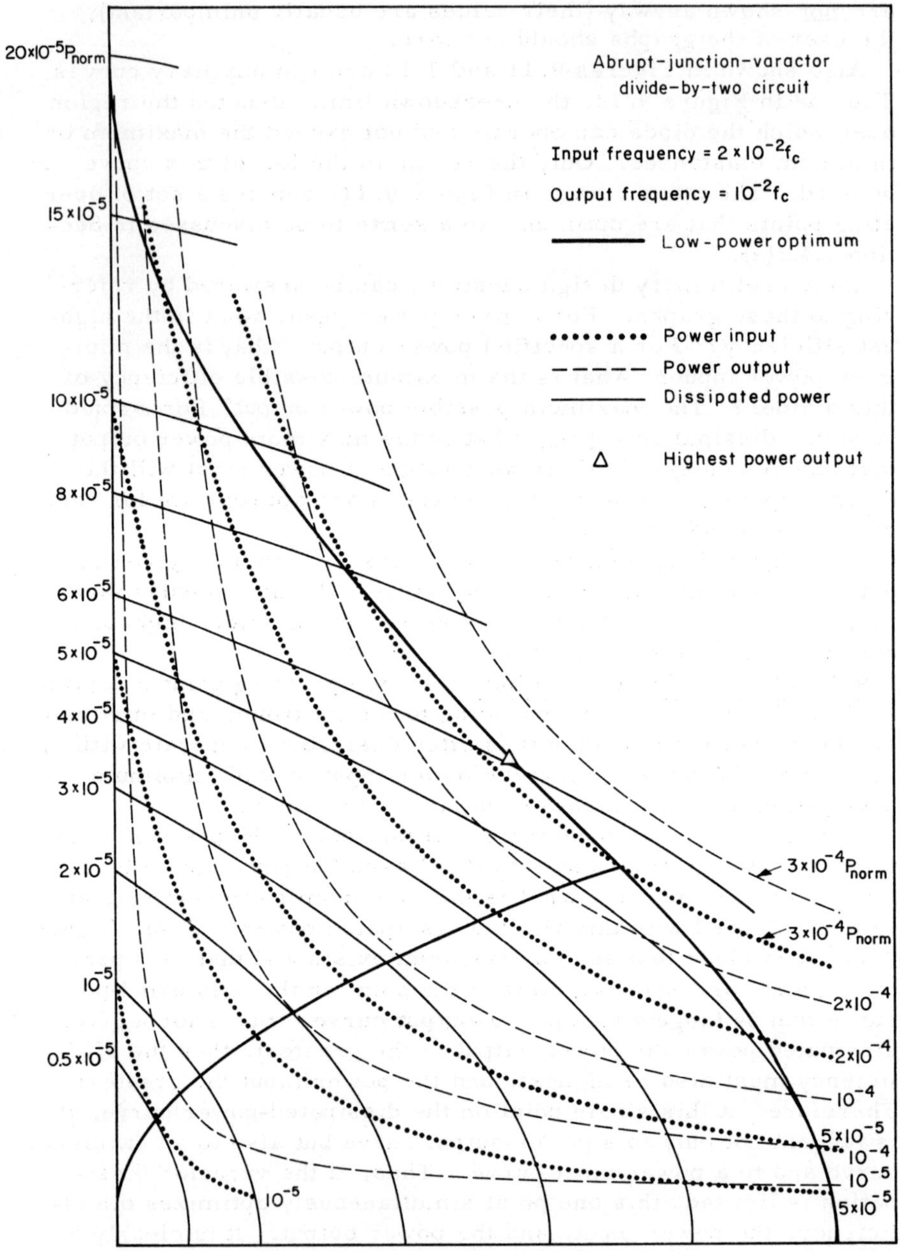

Figure 9.11. Plot of input, output, and dissipated power of a varactor divide-by-two circuit. It is assumed that $S_{min} = 0$. The heavy solid curve is the low-power-optimum curve discussed in Section 9.4.3(a), and the triangle is the point of highest power output, discussed in Section 9.4.3(c).

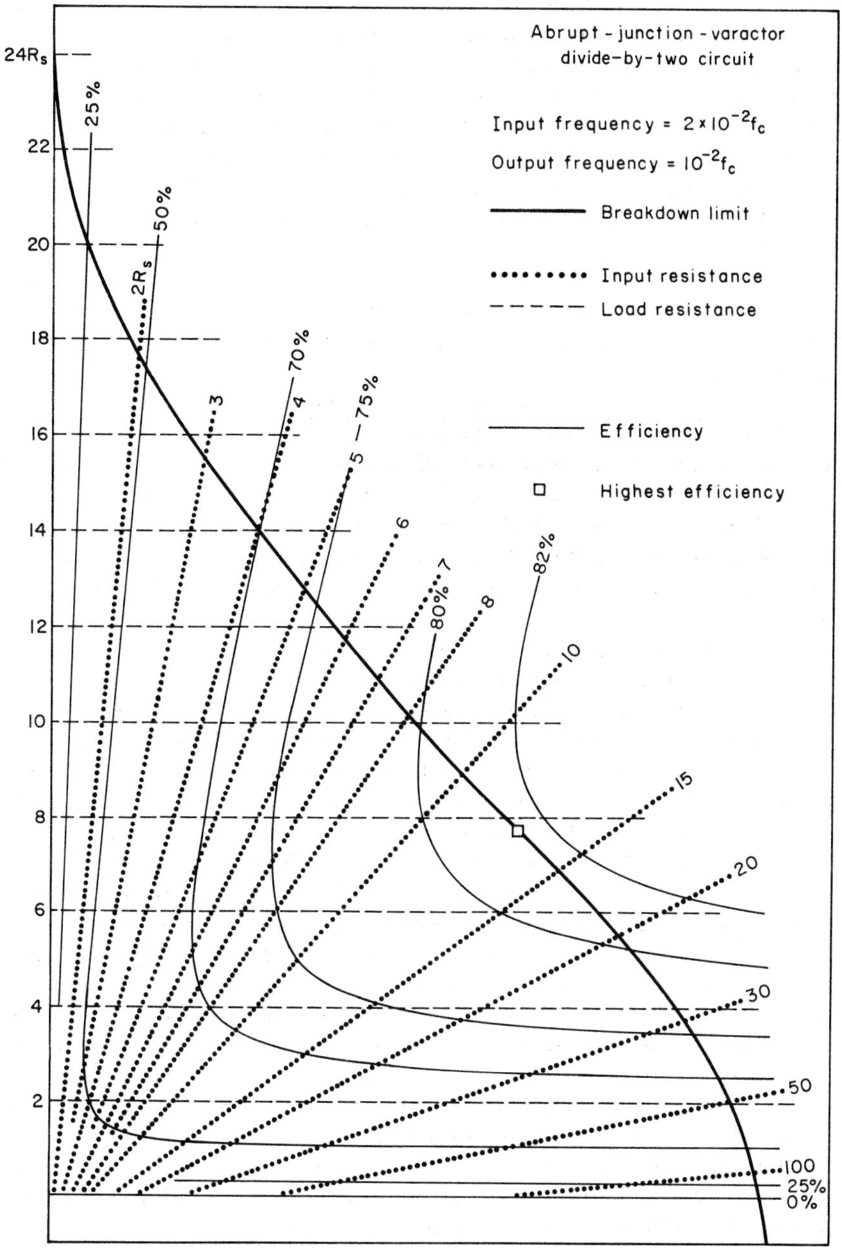

Figure 9.12. Plot of input and load resistances and efficiency of a varactor divide-by-two circuit. This plot should be superimposed upon Figure 9.11. The region of possible operation is to the left of the heavy solid breakdown-limit curve; the square is the point of maximum efficiency discussed in Section 9.4.3(b).

Now consider another problem. Suppose a certain power output is desired; what is the minimum power input? To answer this question, we look for that point on the power output curve that is tangent to an input power curve. But we have just seen that the four sets of curves (power input, dissipated power, power output, and efficiency) are all tangent to each other only along the curve of Equation 9.68. Therefore, to answer this question, merely look for the intersection of this curve with the desired output-power curve.

Note that the same curve, Equation 9.68, is used for all the low-power optimum conditions. The only restriction on the use of this curve is that the operating point must lie within the breakdown limit.

9.4.3(b). Maximum Possible Efficiency. What is the maximum possible efficiency for a given operating frequency? To answer this question, we merely search the space inside the breakdown limit and select that point with the maximum efficiency. This point lies, of course, on the breakdown-limit curve, and is denoted by a square in Figure 9.12 and in the graphs in Appendix D. From the graphs it is easy to determine the operating conditions at that point.

This operating point is of major importance, and many dividers should be made to operate there. We now show, as a function of frequency, the maximum possible efficiency and the associated operating conditions. In Figure 9.13 we give the efficiency and in Figure 9.14 the input, output, and dissipated power corresponding to maximum possible efficiency. In Figure 9.15 are the input and load resistances required, and in Figure 9.16 the bias voltage. Note that at low frequencies, where the effect of the series resistance is negligible, the efficiency is nearly 100 per cent, but at high frequencies it drops, and at frequency $0.25\omega_c$ it becomes zero. At low frequencies the power output and power input are approximately the same, whereas at high frequencies the power output goes through a maximum and then drops off to zero as the dissipated power and power input approach each other.

At low frequencies the input and load resistances are very large, whereas as ω_0 approaches $0.25\omega_c$, its maximum value, the load resistance approaches zero.

The bias voltage for maximum efficiency (and in fact the bias voltage for any point on the breakdown curve) is found by substituting Equation 9.60 into Equation 9.58:

$$\frac{V_0 + \phi}{V_B + \phi} = \left(\frac{S_{max} - S_{min}}{S_{max}}\right)^2 \left[\frac{1}{4} + 2(m_1^2 + m_2^2)\right] + \frac{S_{min}}{S_{max}}. \qquad (9.69)$$

In case $S_{min} << S_{max}$, only the quantity inside the square brack-

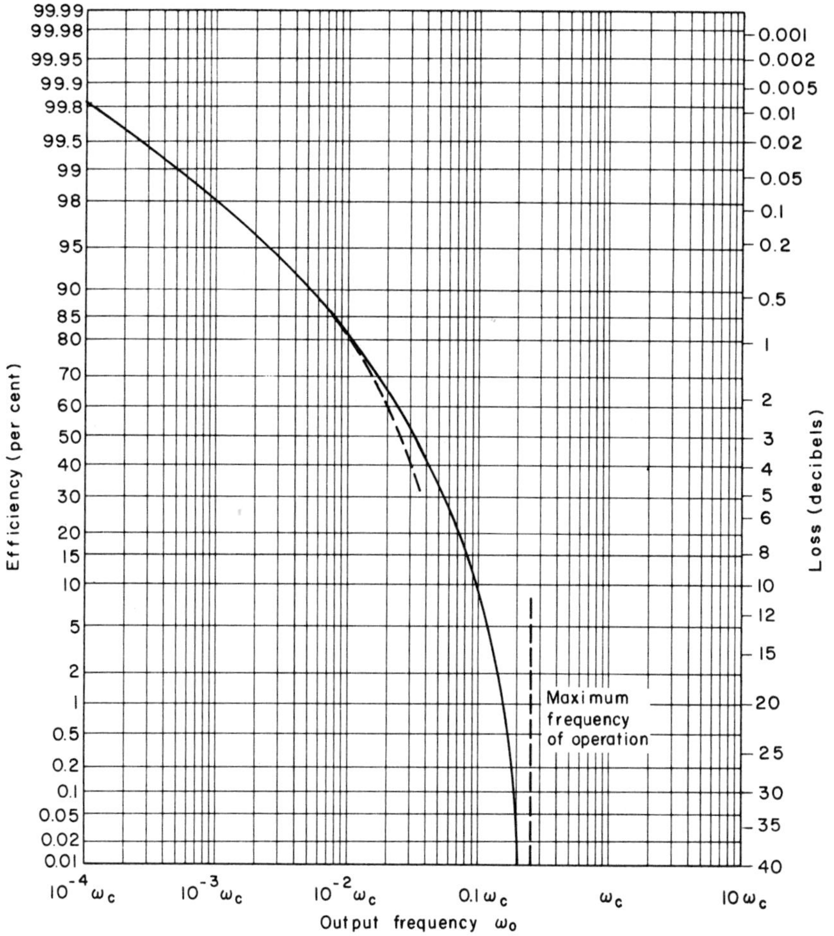

Figure 9.13. Maximum possible efficiency of an abrupt-junction divide-by-two circuit, as a function of output frequency ω_o. Since the varactor cannot divide above $0.25\omega_c$, the efficiency goes to zero there. The dashed curve is the asymptotic limit of Table 9.2.

ets remains on the right-hand side of Equation 9.69. It is this quantity that is plotted in Figure 9.16.

The low-frequency limiting behavior of the divide-by-two circuit is important. In Table 9.2 we show the asymptotic expressions and values for the cases of highest efficiency and also maximum power output.

Some nomographs, based on Figures 9.13 and 9.14, are given in Appendix D to facilitate the design of abrupt-junction divide-by-two circuits.

It is instructive to look at the elastance waveform under typical operation. This is shown in Figure 9.17 for the case of

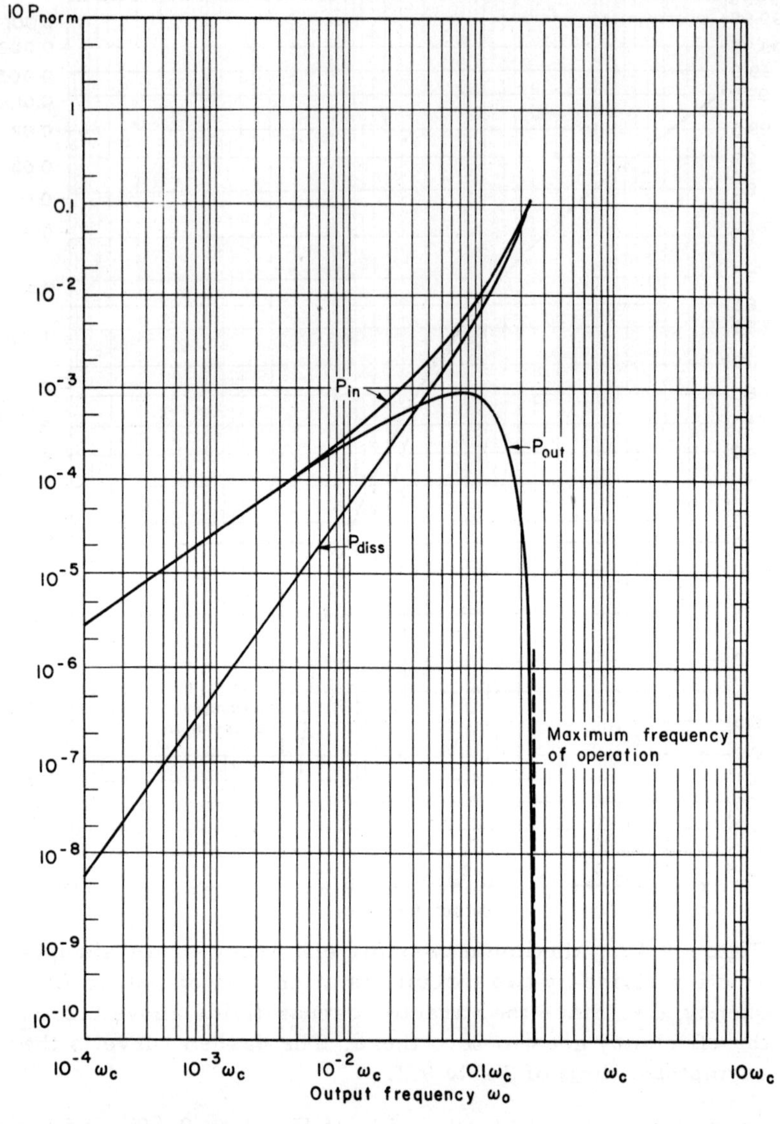

Figure 9.14. Input, output, and dissipated power of an abrupt-junction-varactor divide-by-two circuit when adjusted for optimum efficiency or optimum power output. Note that since the divider cannot operate for $\omega_o > 0.25\omega_c$, the power output goes to zero there.

maximum efficiency at low frequencies ($m_1 = 0.21$, $m_2 = 0.08$). Note that this is similar to Figure 8.10 for the elastance waveform of the doubler, except that it is upside down, or equivalently, it is produced from Figure 8.10 by a time reversal.

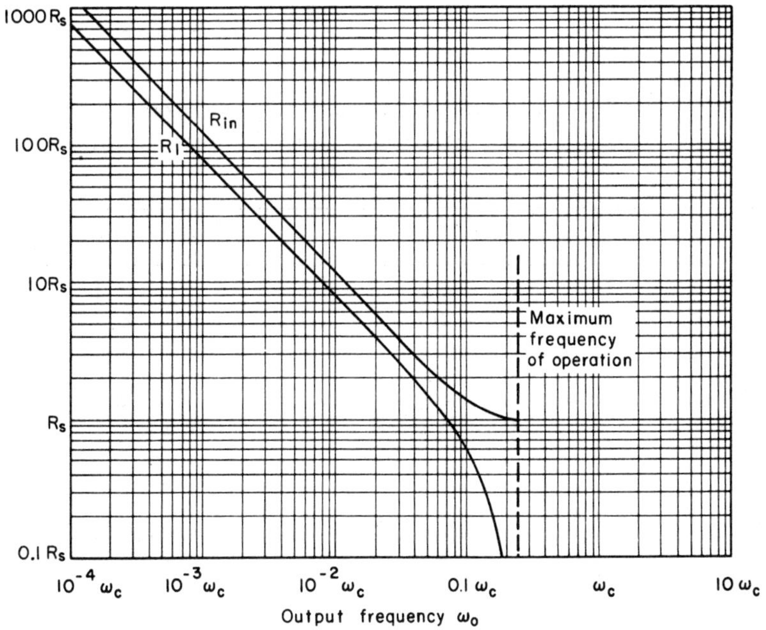

Figure 9.15. Input and load resistances to obtain the maximum efficiency. Since the input and load resistances for maximum power output lie between these two curves, for practical purposes at low frequencies both the power output and the efficiency can be maximized simultaneously.

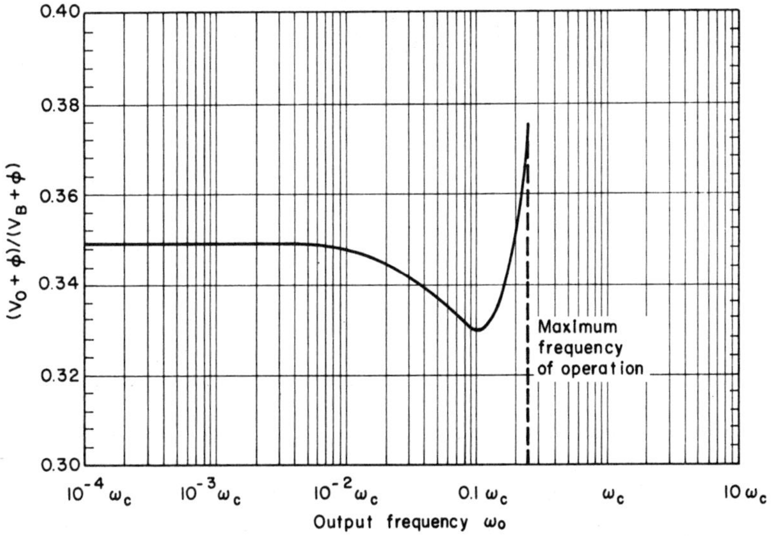

Figure 9.16. Bias voltage necessary to obtain maximum efficiency if $S_{min} = 0$. If $S_{min} \neq 0$, merely multiply the value shown by $(S_{max} - S_{min})^2/S_{max}^2$ and add S_{min}/S_{max}.

Table 9.2. Low-Frequency Behavior of the Divide-by-Two Circuit

Asymptotic limits for the abrupt-junction-varactor divide-by-two circuit at low frequencies, both for maximum efficiency and for maximum power output.

	Maximum ϵ	Maximum P_{out}
ϵ	$1 - 20\dfrac{\omega_o}{\omega_c}$	$1 - 21\dfrac{\omega_o}{\omega_c}$
R_{in}	$0.271\dfrac{S_{max}}{2\omega_o}$	$0.192\dfrac{S_{max}}{2\omega_o}$
R_1	$0.080\dfrac{S_{max}}{\omega_o}$	$0.096\dfrac{S_{max}}{\omega_o}$
P_{in}	$0.0277P_{norm}\dfrac{\omega_o}{\omega_c}$	$0.0285P_{norm}\dfrac{\omega_o}{\omega_c}$
P_{out}	$0.0277P_{norm}\dfrac{\omega_o}{\omega_c}$	$0.0285P_{norm}\dfrac{\omega_o}{\omega_c}$
P_{diss}	$0.551P_{norm}\left(\dfrac{\omega_o}{\omega_c}\right)^2$	$0.593P_{norm}\left(\dfrac{\omega_o}{\omega_c}\right)^2$
$\dfrac{V_0 + \phi}{V_B + \phi}$	0.349	0.343
m_1	0.208	0.192
m_2	0.080	0.096

It is interesting to note that there is an exact solution for the maximum efficiency. The efficiency, Equation 9.67, is restricted because m_1 and m_2 are limited by Condition 9.61. For maximum efficiency the equality is attained at one point in each cycle. If we differentiate Condition 9.61 and set the result to zero (to find the point at which the maximum occurs) and then resubstitute the resulting relation among m_1, m_2, and the

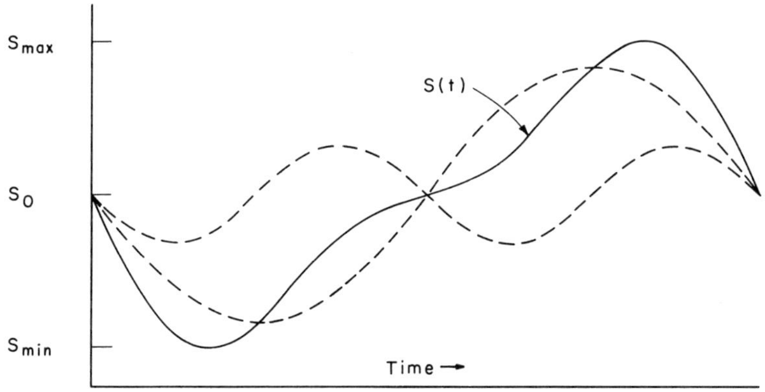

Figure 9.17. Elastance waveform of the low-frequency optimum-efficiency abrupt-junction-varactor divide-by-two circuit. The fundamental and second harmonic are shown individually, in dashed lines, and their sum S(t) is shown solid.

angle at which the maximum is achieved (in the first quadrant) into Condition 9.61, we can solve (in closed form) for a relationship between m_1 and m_2, Condition 8.56. Thus the efficiency can be expressed in terms of either m_1 or m_2 alone, and can then be differentiated with respect to m_1 or m_2, to find its maximum value.

9.4.3(c). Maximum Possible Power Output. It is clear from Figure 9.11 that the power output is limited at each frequency by the breakdown limit. The point of greatest possible power output, indicated by a triangle in Figure 9.11, is not far from the point of maximum efficiency. In fact, for practical purposes, the power output and efficiency can be optimized simultaneously by selecting the operating conditions given in Figures 9.13 to 9.16. The actual curves corresponding to maximum power output are so close to those in Figures 9.13 and 9.14 that they cannot be distinguished. The appropriate resistances are different, but not very much. As a practical matter, the load impedance may be varied over a wide range without affecting the efficiency or power output severely.

The maximum-power-output limit for low frequencies is indicated in Table 9.2.

9.4.3(d). Dynamic Range. A certain amount of input power must be used to overcome the series resistance loss if there is to be any subharmonic oscillation. On the other hand, there is a maximum amount of power that can be put into the varactor if the breakdown limit is not to be exceeded. The ratio of these two powers, roughly speaking, defines a dynamic range for the divider. Two such dynamic ranges are described now.

For a fixed load resistance R_1 and a given frequency, the value of m_2 and, therefore, the current I_2 are fixed by Equation 9.62. As a result, the power input is proportional to the input resistance R_{in}. If oscillations are barely maintained, the input resistance is simply R_s, whereas if the varactor is on the verge of clipping, the input resistance has a value somewhat greater than R_s; this value is easily read from the graphical solutions of Figure 9.12 and of Appendix D.

The dynamic range DR can be defined as the ratio of power input just before clipping to the minimum power input necessary to maintain oscillations. This ratio is therefore equal to

$$DR = \frac{(R_{in})_{max}}{R_s}, \tag{9.70}$$

where $(R_{in})_{max}$ is the maximum possible value of R_{in} for the given value of R_1. The portion of Figure 9.12 that has the resistance lines is repeated as Figure 9.18, and the location of the point to determine $(R_{in})_{max}$ is indicated in the graph, for $R_1 = 14R_s$. It is clear that for a given frequency and a given value of R_1, it is a simple procedure to determine graphically $(R_{in})_{max}$ and, therefore, DR. The dynamic range is clearly a function of R_1, and as R_1 increases, the dynamic range decreases. For example, for $\omega_0 = 10^{-2}\omega_c$, DR = 6 db for $R_1 = 14R_s$, and DR = 17 db for $R_1 = 2R_s$.

Another, similar quantity is the available dynamic range, which is defined differently. We suppose that the divider is driven from a source at frequency $2\omega_0$ with a fixed internal resistance R_g and an available power that can vary. Then, for a fixed load resistance R_1, the available dynamic range DR_{av} is defined as the ratio of (1) the maximum available power before clipping to (2) the minimum available power input to maintain oscillation. Since the available power varies as the square of the voltage, and the current is constant (because of the constant value of R_1), the available dynamic range is

$$DR_{av} = \left[\frac{R_g + (R_{in})_{max}}{R_g + R_s}\right]^2. \tag{9.71}$$

In many cases, the generator resistance is set to equal $(R_{in})_{max}$, so that the input is matched at the maximum power level. In that case, DR_{av} becomes

$$DR_{av} = 4\left[\frac{(R_{in})_{max}}{(R_{in})_{max} + R_s}\right]^2, \tag{9.72}$$

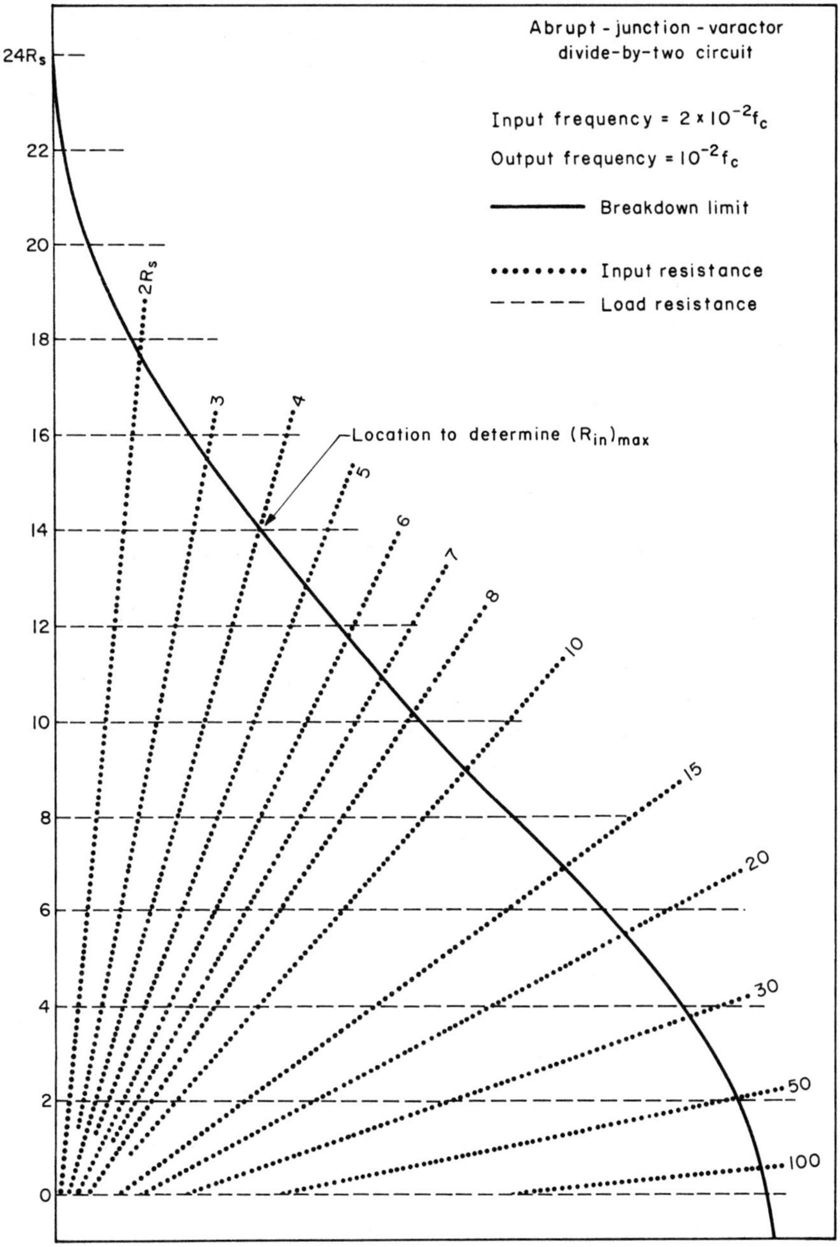

Figure 9.18. Resistance lines from Figure 9.12. For a given value of R_1, there is a maximum value of R_{in}, which is achieved at the breakdown limit. For example, corresponding to $R_1 = 14R_s$, $(R_{in})_{max}$ is seen to be approximately $4R_s$.

or about 6 db for low frequencies, where $(R_{in})_{max} >> R_s$.

The available dynamic range is a function of both the load and source resistances. In Figure 9.19 we show, for the particular case with DR = 4, how DR_{av} varies with generator resistance.

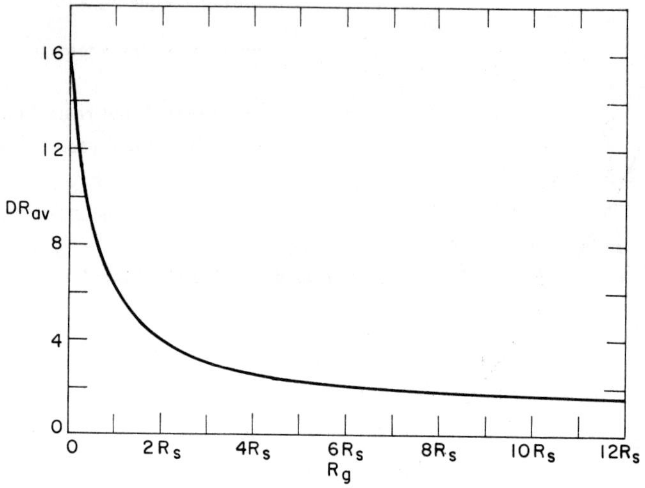

Figure 9.19. Available dynamic range as a function of generator resistance R_g, for the particular case with $(R_{in})_{max}/R_s = DR = 4$.

The available dynamic range is related to the dynamic range by the formula

$$DR_{av} = \left(\frac{\frac{R_g}{R_s} + DR}{\frac{R_g}{R_s} + 1} \right)^2 , \qquad (9.73)$$

or, when the generator resistance is adjusted to match $(R_{in})_{max}$,

$$DR_{av} = 4 \left(\frac{DR}{DR + 1} \right)^2 . \qquad (9.74)$$

9.4.4. Synthesis Techniques

Thus far we have not discussed the synthesis of dividers but have only derived fundamental limits on their operation. The derivation of these limits, however, indicates conditions that must be met to achieve the optimum performance. Now we discuss design of divide-by-two circuits specifically to fulfill these conditions. Because of their greater importance, we restrict this section to large-power dividers, operating in the steady state.

The design of dividers is similar to the design of doublers, in Section 8.4.4. There are two steps. First, in the preliminary design, the varactor parameters are chosen and the operating points determined. Second, a circuit is designed to separate the two frequencies ω_o and $2\omega_o$ while open-circuiting the varactor at certain other frequencies.

9.4.4(a). Preliminary Design. Since Section 9.4 is devoted to the preliminary design of varactor divide-by-two circuits, more discussion here would be superfluous. The operating characteristics such as input, output, and dissipated power, efficiency, and the like, are related to the varactor parameters. The graphs of Section 9.4 and Appendix D are designed to facilitate preliminary design; by their use this design can become a routine matter.

9.4.4(b). Separation of Frequencies. Once the operating conditions are known, a circuit must be designed to satisfy the assumptions of the analysis (those assumptions, at least, that lead to the fundamental limits) and to separate the frequencies. Specifically, for a divide-by-two circuit the conditions are the following:

1. The varactor should be open-circuited at frequency $3\omega_o$.
2. The varactor should be open-circuited at frequency $4\omega_o$.
3. The input and output frequencies $2\omega_o$ and ω_o should appear at separate terminals, and these terminals should have tuning adjustments.
4. The circuit should, for highest efficiency, be lossless.

These conditions are precisely the conditions for the synthesis of varactor doublers, discussed in Section 8.4.4. Therefore, all the circuits described there, when run backwards, can operate as dividers (provided the operating conditions are as specified by the divider preliminary design, not the doubler preliminary design). In particular, the circuits of Figures 8.12 to 8.14 are examples.

9.5. Divide-by-Four Circuit

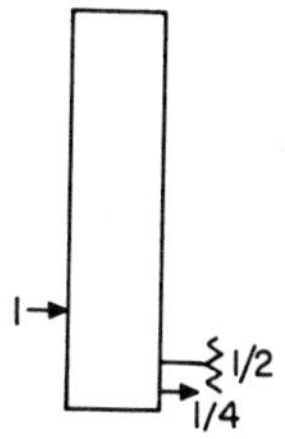

The divide-by-two circuit is the simplest divider. The next most simple is the divide-by-four circuit, which has one idler. We give here the large-signal theory of this divider, with some solutions for maximum drive.

The divide-by-four circuit can be thought of as a cascade of two divide-by-two circuits, using the same varactor. An input at frequency $4\omega_o$ generates a subharmonic at $2\omega_o$, and this grows in accordance with the theory given thus far. Then, this subharmonic itself generates an oscillation at half its frequency. This output, at ω_0, goes to a load.

9.5.1. Formulas

We assume that the idler termination at frequency $2\omega_0$ and the load termination at frequency ω_0 are tuned. This is analogous to setting θ to zero in Section 9.4.1(h). Then the load and idler impedances are expressed in terms of the elastance coefficients as in Equation 8.81:

$$\frac{R_1 + R_s}{R_s} = \frac{\omega_c}{\omega_0}\frac{M_1{*}(jM_2)}{M_1}, \tag{9.75}$$

and

$$\frac{R_2 + R_s}{R_s} = \frac{\omega_c}{4\omega_0}\left[\frac{2M_2{*}(jM_4)}{M_2} - \frac{M_1^2}{jM_2}\right]. \tag{9.76}$$

We see from Equation 9.75 that the phase angle of M_1 is either half the angle of jM_2 or 180° plus half this angle. When we use this information in Equation 9.76, we find that the angle of M_2 is either half the angle of jM_4 or 180° plus half the angle of jM_4. Therefore, if we choose our time origin to make jM_4 real and positive, it is clear that M_2 is real, either positive or negative, and that M_1 has one of these four angles: 45°, 135°, 225°, or 315°.

Recall that the divide-by-two circuit is characterized by two possible phases for the subharmonic oscillation. This is consistent with the fact that here M_2 has two possible phases. In the divide-by-four circuit, the output can have any of four possible phases, two corresponding to each of the two possible idler phases. Thus in principle the divide-by-four circuit could be used as a four-way-stable device.

When these phase relations are used, the formulas for load resistance R_1, idler resistance R_2, input resistance R_{in}, power input P_{in}, power output P_{out}, dissipated power P_{diss}, varactor dissipation $P_{diss,v}$, external idler dissipation $P_{diss,i}$, efficiency ϵ, and bias voltage V_0 given in Section 8.5 become

$$R_1 = R_s\left(\frac{\omega_c}{\omega_0}m_2 - 1\right), \tag{9.77}$$

$$R_2 = R_s\left[\frac{\omega_c}{4\omega_0}\left(2m_4 - \frac{m_1^2}{m_2}\right) - 1\right], \tag{9.78}$$

$$R_{in} = R_s \left(\frac{\omega_c}{8\omega_o} \frac{m_2^2}{m_4} + 1 \right), \tag{9.79}$$

$$P_{in} = 8P_{norm} \left(\frac{S_{max} - S_{min}}{S_{max} + S_{min}} \right)^2 \frac{\omega_o^2}{\omega_c^2} \left(\frac{2\omega_c}{\omega_o} m_2^2 m_4 + 16 m_4^2 \right), \tag{9.80}$$

$$P_{out} = 8P_{norm} \left(\frac{S_{max} - S_{min}}{S_{max} + S_{min}} \right)^2 \frac{\omega_o^2}{\omega_c^2} \left(\frac{\omega_c}{\omega_o} m_1^2 m_2 - m_1^2 \right), \tag{9.81}$$

$$P_{diss} = 8P_{norm} \left(\frac{S_{max} - S_{min}}{S_{max} + S_{min}} \right)^2 \frac{\omega_o^2}{\omega_c^2} \left[\frac{\omega_c}{\omega_o} (2m_2^2 m_4 - m_1^2 m_2) + m_1^2 + 16 m_4^2 \right], \tag{9.82}$$

$$P_{diss,v} = 8P_{norm} \left(\frac{S_{max} - S_{min}}{S_{max} + S_{min}} \right)^2 \frac{\omega_o^2}{\omega_c^2} (m_1^2 + 4m_2^2 + 16 m_4^2), \tag{9.83}$$

$$P_{diss,i} = 8P_{norm} \left(\frac{S_{max} - S_{min}}{S_{max} + S_{min}} \right)^2 \frac{\omega_o^2}{\omega_c^2} \left[\frac{\omega_c}{\omega_o} (2m_2^2 m_4 - m_1^2 m_2) - 4m_2^2 \right], \tag{9.84}$$

$$\epsilon = \frac{m_1}{4m_4} \frac{\dfrac{\omega_c}{\omega_o} m_1 m_2 - m_1}{\dfrac{\omega_c}{2\omega_o} m_2^2 + 4m_4}, \tag{9.85}$$

$$\frac{V_0 + \phi}{V_B + \phi} = \left(\frac{S_{max} - S_{min}}{S_{max}} \right)^2 (m_0^2 + 2m_1^2 + 2m_2^2 + 2m_4^2). \tag{9.86}$$

9.5.2. Technique of Solution

The technique of solution of these equations is similar to that for the multipliers treated in Chapter 8. If m_1, m_2, and m_4 are known, then all quantities of interest can be calculated from the formulas just given. On the other hand, if the problem is set up in such a way that different information is available, the values of m_k must be calculated. Often the simplest method of finding appropriate solutions is to try particular values of m_1, m_2, and m_4 until the desired solution is obtained.

When values of m_k are assumed, they must not be so large that the elastance waveform becomes lower than S_{min} or higher than S_{max} at any time. Thus, the values assumed must obey the inequality

$$m_1 \sin \omega_0 t + m_2 \sin 2\omega_0 t + m_4 \sin 4\omega_0 t \leq 0.25 \quad (9.87)$$

for all t. In addition, the values chosen must not yield negative values of R_1 or R_2 when used in Equations 9.77 and 9.78. Thus, in particular,

$$m_2 \omega_c > \omega_0 \quad (9.88)$$

and

$$m_4 \omega_c > 2\omega_0. \quad (9.89)$$

(Condition 9.88 is necessary and sufficient for R_1 to be positive; Condition 9.89 is necessary but not sufficient for R_2 to be positive.) Aside from these restrictions, any choice of m_1, m_2, and m_4 describes a possible operating point for the divide-by-four circuit.

Often the values of the m_k must be compatible with prescribed values of R_1 and R_2. In this case,

$$m_2 = \frac{\omega_0}{\omega_c} \frac{R_1 + R_s}{R_s}, \quad (9.90)$$

and m_1 and m_4 are related by the formula,

$$m_4 = \frac{2\omega_0}{\omega_c} \frac{R_2 + R_s}{R_s} + \frac{m_1^2 \omega_c}{2\omega_0} \frac{R_s}{R_1 + R_s}. \quad (9.91)$$

Equations 9.90 and 9.91 can be easily misinterpreted. They apply only when currents actually do flow at the three frequencies, and when this current is neither building up nor decaying. Thus, it is physically possible to build a divide-by-four circuit with certain values of R_1 and R_2, and then to drive it with a very low value of m_4, so that Equation 9.91 cannot hold. In this case the idler oscillations will not grow, or if already present they decay. If m_4 is now made larger than the first term on the right-hand side of Equation 9.91, idler oscillations can exist.

One of three effects then occurs. First, these oscillations may increase the input resistance so much that the source cannot supply such a large current, and an equilibrium is attained in which there are oscillations at $2\omega_0$ but none at ω_0. Second, the value of m_2 may become so large that the varactor breaks down before Equation 9.90 is met. Our theory does not tell us what happens in this case, but physically we expect this nonlinearity to prevent further growth of the oscillations. Third, it is possible that R_1 is low enough that Equation 9.90 is obeyed before the varactor breaks down. Then, oscillations at one-quarter the input frequency build up. In the steady state m_2 is held to the value of Equation 9.90, and m_1 is the proper value to make Equation 9.91 hold, provided these values of m_1, m_2, and m_4 obey Condition 9.87.

Perhaps the most important solutions of the divide-by-four circuit are for maximum drive, that is, when Condition 9.87 is barely met. A computer program to find these solutions (with the condition that $R_2 = 0$) has been written by Soheir T. Meleika.[4] We now give the most important of her results, with her kind permission.

9.5.3. Solutions for Maximum Drive

We give a few of Meleika's results for the abrupt-junction-varactor divide-by-four circuit with lossless idler termination and maximum efficiency. In Figures 9.20 to 9.23 we show, respectively, the maximum efficiency, the resulting input, output, and dissipated powers, the input and load resistances, and the bias voltage. In Figures 9.21 and 9.23 we have assumed that $S_{min} \ll S_{max}$; if not, the instructions accompanying Figures 8.40 and 8.42 can be followed. Since the solution depicted here is for maximum drive, the average elastance is given by Equation 9.60, and the bias voltage becomes, from Equation 9.86,

$$\frac{V_0 + \phi}{V_B + \phi} = \left(\frac{S_{max} - S_{min}}{S_{max}}\right)^2 \left[\frac{1}{4} + 2(m_1^2 + m_2^2 + m_4^2)\right] + \frac{S_{min}}{S_{max}}. \tag{9.92}$$

Figures 9.20 to 9.23 are similar to the divide-by-two circuit curves, Figures 9.13 to 9.16. At low frequencies the efficiency approaches 100 per cent, and the input and load resistances vary inversely with the frequency. Since the divider does not operate with output frequency ω_0 greater than $0.091\omega_c$, at this frequency the power output and efficiency drop to zero. At low frequencies the dissipated power varies as the square of the frequency, and the input and output powers are proportional to the frequency.

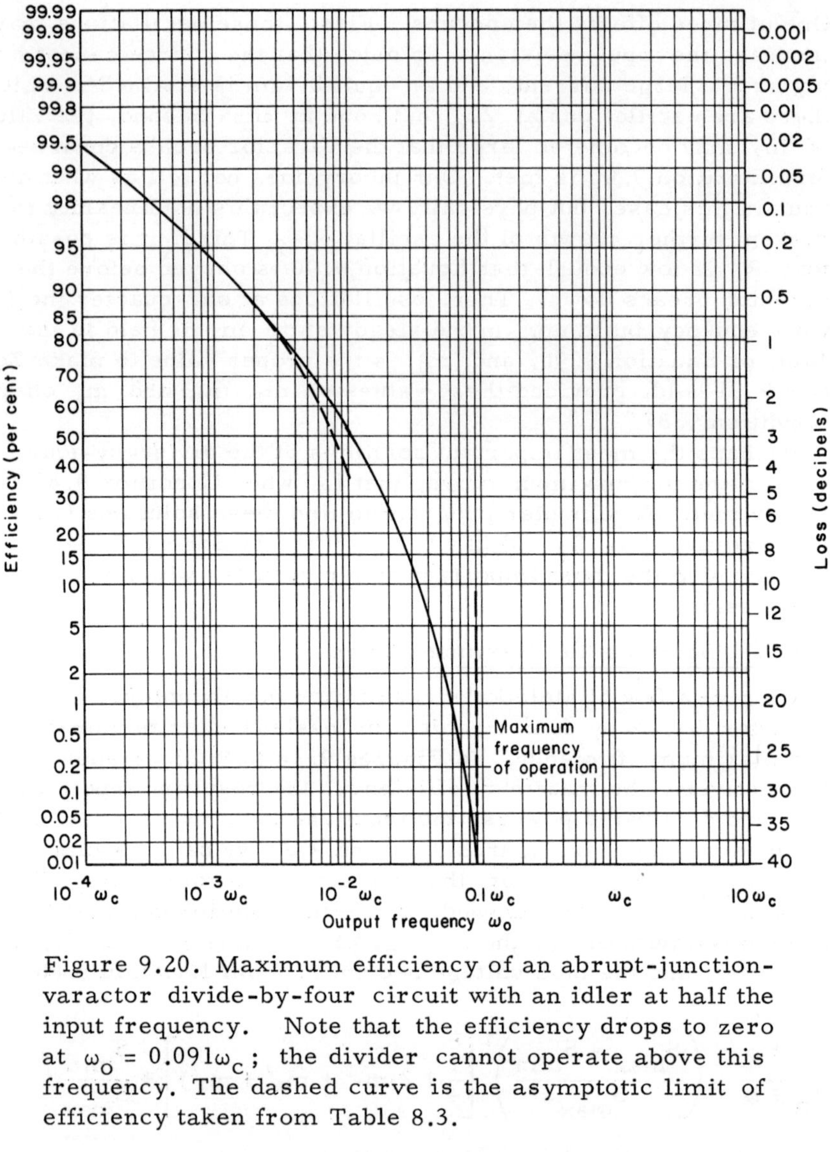

Figure 9.20. Maximum efficiency of an abrupt-junction-varactor divide-by-four circuit with an idler at half the input frequency. Note that the efficiency drops to zero at $\omega_o = 0.091\omega_c$; the divider cannot operate above this frequency. The dashed curve is the asymptotic limit of efficiency taken from Table 8.3.

Some nomographs are given in Appendix D, which can be used in the practical design of divide-by-four circuits with maximum efficiency. They are based on Figures 9.20 and 9.21.

If the power output, rather than the efficiency, is optimized, the resulting curves are very similar to Figures 9.20 to 9.23, and are not given here.

The low-frequency behavior of the divide-by-four circuit is indicated in Table 9.3, for the condition of maximum efficiency and the condition of maximum power output. For both cases we show the asymptotic limits of all parameters of interest. In the

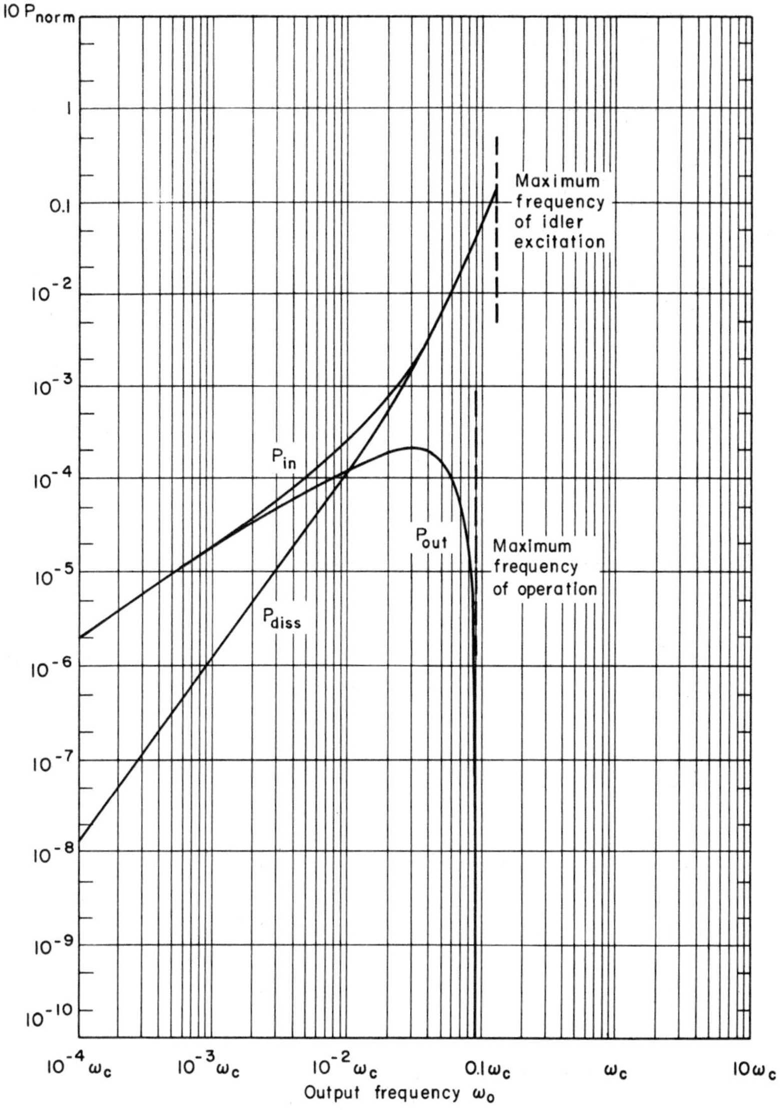

Figure 9.21. Input, output, and dissipated power of an abrupt-junction-varactor divide-by-four circuit adjusted for maximum efficiency.

formulas we assume that $S_{min} << S_{max}$ and that the idler termination is lossless and tuned. It is interesting to note that the low-frequency optimizations of the divide-by-two circuit occur for values of m_1, m_2, and m_4 that are the same as the values for the low-frequency limit of the quadrupler; this is true for both the maximum-efficiency case and the maximum-power-output case.

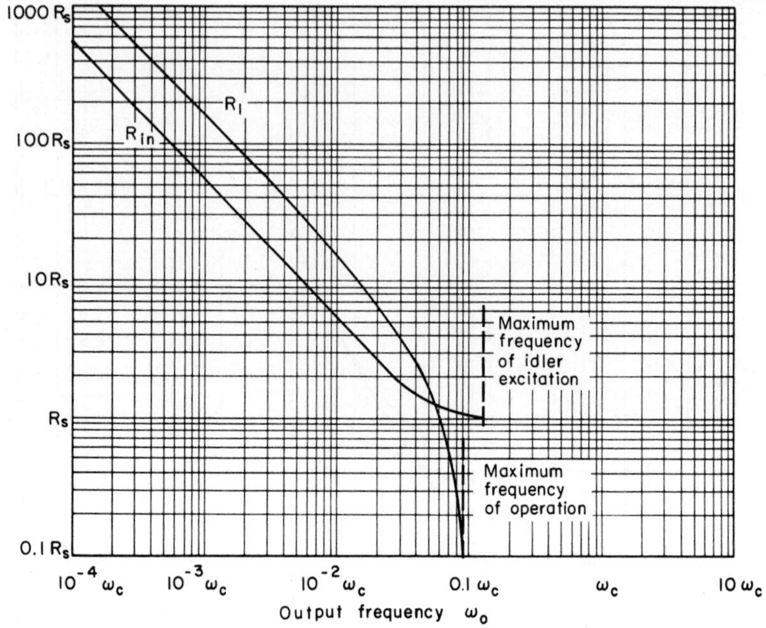

Figure 9.22. Input and load resistance to attain maximum efficiency of an abrupt-junction-varactor divide-by-four circuit.

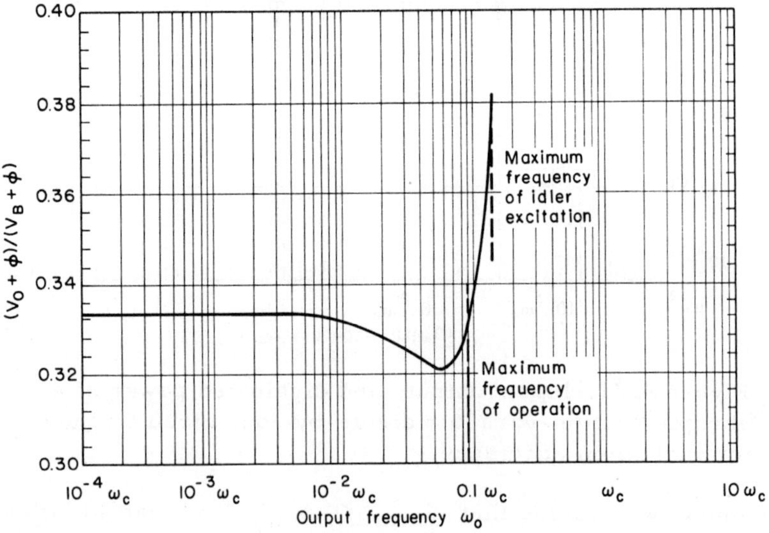

Figure 9.23. Bias voltage necessary to attain maximum efficiency of an abrupt-junction-varactor divide-by-four circuit, under the assumption that $S_{min} \ll S_{max}$.

Table 9.3. Low-Frequency Behavior of the Divide-by-Four Circuit

We show asymptotic formulas for cases of maximum efficiency and maximum power output, with $R_2 = 0$.

	Maximum ϵ	Maximum P_{out}
ϵ	$1 - 62.5\frac{\omega_o}{\omega_c}$	$1 - 66.2\frac{\omega_o}{\omega_c}$
R_{in}	$0.205\frac{S_{max}}{4\omega_o}$	$0.136\frac{S_{max}}{4\omega_o}$
R_1	$0.150\frac{S_{max}}{\omega_o}$	$0.136\frac{S_{max}}{\omega_o}$
P_{in}	$0.0196 P_{norm}\frac{\omega_o}{\omega_c}$	$0.0201 P_{norm}\frac{\omega_o}{\omega_c}$
P_{out}	$0.0196 P_{norm}\frac{\omega_o}{\omega_c}$	$0.0201 P_{norm}\frac{\omega_o}{\omega_c}$
P_{diss}	$1.23 P_{norm}\left(\frac{\omega_o}{\omega_c}\right)^2$	$1.33 P_{norm}\left(\frac{\omega_o}{\omega_c}\right)^2$
$\frac{V_0 + \phi}{V_B + \phi}$	0.334	0.333
m_1	0.128	0.136
m_2	0.150	0.136
m_4	0.055	0.068

9.6. Summary

Several important results were derived. Among the most significant are these:

1. Varactors can be used to generate oscillations at half of or a quarter of (or more generally at rational fractions of) an input frequency.
2. For small signals the divide-by-n circuit requires an idler unless $n = 2$.
3. For small oscillations divide-by-n circuits are not phase-sensitive unless n is of the form 2^{ℓ}.
4. The divide-by-two circuit has two stable phases of oscillation.
5. The divide-by-two circuit can operate only if $m_2\omega_c < \omega_o$, and even then only if the load resistance is small enough and if the load is well enough tuned (ω_o is the output frequency).
6. The varactor series resistance and average elastance limit the maximum rate of growth of half-frequency oscillations.
7. The abrupt-junction-varactor large-signal divide-by-two circuit has a maximum efficiency that approaches 100 per cent at low frequencies, but drops to zero at $\omega_o = 0.25\omega_c$, above which the divider will not operate.
8. This divider has a low-power optimization to obtain maximum efficiency, maximum power output, or minimum dissipation for a specified value of efficiency, power output, power input, or dissipation.
9. The abrupt-junction-varactor divide-by-four circuit has a maximum efficiency that approaches 100 per cent at low frequencies but drops to zero at $\omega_o = 0.091\omega_c$, above which the divider will not operate.

There are many aspects of dividers that we have omitted. We have neglected noise and bandwidth, as well as the effect of the breakdown limit on the subharmonic oscillations. We have not discussed the large-signal dividers with varactors other than abrupt-junction models. We have not investigated any of the interesting rational-fraction generators.

The rise-time capabilities of the divide-by-four circuit are quite interesting, but we have not investigated them. Finally, we have neglected almost completely the myriad of details connected with the synthesis of practical dividers.

REFERENCES

1. Goto, E., "The Parametron, a Digital Computing Element which Utilizes Parametric Oscillation," Proc. IRE, 47, 1304-1316 (August, 1959).
2. Guillemin, E. A., The Mathematics of Circuit Analysis, The Technology Press and John Wiley & Sons, Inc., New York (1949), p. 420.
3. Hilibrand, J., and W. R. Beam, "Semiconductor Diodes in Parametric Subharmonic Oscillators," RCA Rev., 20, 229-253 (June, 1959).
4. Meleika, Soheir T., "The Large-Signal Exact Solution for the Abrupt-Junction Divide-by-Four Circuit," S.M. Thesis, Department of Electrical Engineering, M.I.T., Cambridge (May, 1962).
5. Page, C. H., "Frequency Conversion With Positive Nonlinear Resistors," J. Res. Natl. Bur. Standards, 56, 179-182 (April, 1956).
6. Penfield, P., Jr., Frequency-Power Formulas, The Technology Press and John Wiley & Sons, Inc., New York (1960), Sec. 8.1.
7. Sterzer, F., "Microwave Parametric Subharmonic Oscillators for Digital Computing," Proc. IRE, 47, 1317-1324 (August, 1959).
8. Wigington, R. L., "A New Concept in Computing," Proc. IRE, 47, 516-523 (April, 1959).

Chapter 10

LARGE-SIGNAL FREQUENCY CONVERTERS AND AMPLIFIERS

There are many aspects of large-signal frequency converters and parametric amplifiers that deserve discussion. In Chapter 10 we cover only three, and these only very briefly. They are high-power upconverters, effect of variations of pump power on gain and phase of small-signal parametric amplifiers, and dynamic range of small-signal frequency converters and parametric amplifiers.

10.1. Introduction

Section 10.2 is devoted to large-power frequency converters, especially the upper-sideband upconverter. This device is of potential usefulness in multiplier chains. A common misbelief is that multiplier chains cannot be designed to yield prime-number multiples of the input frequency. However, power upconverters can be used as links in such multiplier chains to reach any harmonics, even primes. The parameters that describe abrupt-junction-varactor power upconverters are the input resistances and load resistance, the two input powers, the output power, the dissipated power, the bias voltage, and various efficiencies. In Section 10.2.1 these parameters are expressed in terms of the modulation ratios. The exact breakdown limit is not easily calculated; in Section 10.2.2 we discuss a very simple approximate condition on the modulation ratios. This condition leads naturally to a two-dimensional plot corresponding to maximum drive; this triangle chart is similar to the ternary phase diagrams employed in metallurgy.

The power upconverter can be adjusted to operate at maximum efficiency or maximum power output by suitably selecting the load impedance. In Section 10.2.3 we treat the output-power optimization. The various efficiency optimizations are not discussed because they depend on the relative frequencies.

In Section 10.3 we look into the stability of gain in small-signal parametric amplifiers as the pump power is varied. This problem is of practical importance because any modulation on the pump (such as noise or hum) may be transferred to the signal. Since the impedance of the varactor at signal frequency is independent of the phase of the pump, any phase noise carried by the pump does not, to first order, affect the signal. However, amplitude noise is transferred. In general, a change in pump amplitude (or power) affects the real part of the varactor impedance, whereas the associated change in average elastance of the varactor affects the reactive part. Thus, the phase of the output is altered by a change in the average elastance, and the magnitude of the output is altered by a change in the pump modulation ratio.

The most interesting result about these variations is that if the pump frequency is chosen correctly (close to the optimum pump frequencies derived in Chapters 5 and 6), the first-order variation in phase is zero. The change in phase caused by signal-frequency mistuning exactly cancels (to first order) the change in phase caused by idler-frequency mistuning. Thus, parametric amplifiers can be used in such phase-sensitive (but amplitude-insensitive) applications as FM receivers or tracking systems without (to first order) introducing any noise, hum, or other type of disturbance from the pump. The basic analysis of Section 10.3 applies to any small-signal parametric amplifiers, although in general the change in m_1, the pump modulation ratio, for arbitrary diodes is difficult to calculate. We calculate it explicitly only for abrupt-junction varactors.

In Section 10.4 we touch briefly on the question of dynamic range of small-signal amplifiers and converters. The small-signal theory of Chapters 5 and 6 does not indicate its own range of validity; for this, we need a nonlinear large-signal theory. The theory in Section 10.4 is limited to abrupt-junction varactors.

The dynamic range of such devices is usually defined as the ratio of some "minimum power" to some "maximum power." The exact meaning of these terms depends upon the application. In Section 10.4.1 we derive the low-power limit caused by noise. In Sections 10.4.2 and 10.4.3 we investigate two typical high-power limits. The first is a rather crude limit caused by breakdown. The second is a somewhat lower power limit for parametric amplifiers. This limit arises because of a change in gain of a parametric amplifier with increasing signal level. The origin of this saturation effect lies in the pump circuit. If the pump current is not regulated, it will tend to decrease as power is drawn from it for conversion to signal and idler frequencies. This decrease in turn decreases the gain. Formulas are given that relate the first-order changes in gain, varactor impedance at signal frequency, pump modulation ratio, and average elastance to changes in signal-frequency modulation ratio and, therefore, to signal power.

We have only scratched the surface as far as large-signal properties of varactor devices are concerned. We give only three analyses, and these not in complete detail. They serve only as examples, to aid the reader in making specific, detailed analyses of his own.

10.2. Large-Power Frequency Converters

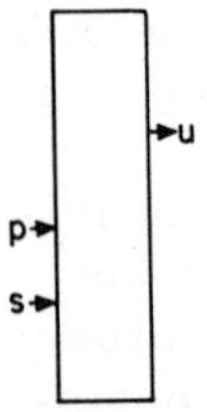

The large-signal nonlinear theory of frequency converters is important because of at least two applications. First, for signal processing, the actual operation of a frequency converter may involve signal amplitudes that are not small in comparison with the pump; therefore, the theory of Chapter 5 may not be valid. The nonlinear theory predicts the behavior at large signals, and indicates the range of validity of the small-signal theory. And second, frequency converters may be used to convert power for its own sake from one frequency to another. Power may be required at some frequency that is the sum of two frequencies at which power is already available. At what efficiency can it be converted?

As an example of this latter application, consider the problem of designing a multiply-by-five circuit. The 1-2-4-5 or 1-2-3-5 quintupler described in Chapter 8 could be used. However, these devices require two idlers; an alternate way of generating the fifth harmonic would be to employ a doubler and a tripler, each driven from the fundamental frequency, and then to use an upper-sideband power upconverter to sum the frequencies of the outputs from these two multipliers.

As an illustration of power frequency converters, we analyze the abrupt-junction-varactor upper-sideband upconverter. In Section 10.2.1 we derive its formulas, and in Section 10.2.2 we discuss the role of the breakdown limit. In Section 10.2.3 we investigate certain (but by no means all) optimizations.

10.2.1. Formulas

Large-signal upconverters are described by Equations 4.32, which holds whether or not ω_s and ω_p are commensurate. For the abrupt-junction-varactor upper-sideband upconverter, we allow current to flow only at frequencies ω_s, ω_p, and $\omega_u = \omega_s + \omega_p$, and since the elastance is proportional to the charge, the only elastance coefficients present are S_0, S_s, S_p, and S_u. For simplicity, we define M_u, M_s, and M_p as the normalized elastance coefficients and m_u, m_s, and m_p as their magnitudes:

$$M_k = \frac{S_k}{S_{max} - S_{min}}, \tag{10.1}$$

$$m_k = |M_k|, \tag{10.2}$$

for k = s, p, and u. If we neglect noise, Equations 4.32 reduce to

$$\begin{bmatrix} V_u^* \\ V_p^* \\ V_s^* \\ V_s \\ V_p \\ V_u \end{bmatrix} = \begin{bmatrix} R_s - \frac{S_0}{j\omega_u} & -\frac{S_s^*}{j\omega_u} & -\frac{S_p^*}{j\omega_u} & 0 & 0 & 0 \\ -\frac{S_s}{j\omega_p} & R_s - \frac{S_0}{j\omega_p} & 0 & -\frac{S_u^*}{j\omega_p} & 0 & 0 \\ -\frac{S_p}{j\omega_s} & 0 & R_s - \frac{S_0}{j\omega_s} & 0 & -\frac{S_u^*}{j\omega_s} & 0 \\ 0 & \frac{S_u}{j\omega_s} & 0 & R_s + \frac{S_0}{j\omega_s} & 0 & \frac{S_p^*}{j\omega_s} \\ 0 & 0 & \frac{S_u}{j\omega_p} & 0 & R_s + \frac{S_0}{j\omega_p} & \frac{S_s^*}{j\omega_p} \\ 0 & 0 & 0 & \frac{S_p}{j\omega_u} & \frac{S_s}{j\omega_u} & R_s + \frac{S_0}{j\omega_u} \end{bmatrix} + \begin{bmatrix} I_u^* \\ I_p^* \\ I_s^* \\ I_s \\ I_p \\ I_u \end{bmatrix}. \tag{10.3}$$

The varactor is excited at frequencies ω_s and ω_p, and is terminated in an impedance $Z_u = R_u + jX_u$ at the output frequency ω_u. Thus,

$$V_u = -Z_u I_u. \tag{10.4}$$

When we use the fact that the elastance coefficients are proportional to the charge coefficients, we find from substituting Equation 10.4 in Equations 10.3 that

$$Z_u + R_s + \frac{S_0}{j\omega_u} = R_s \frac{\omega_c}{\omega_u} \frac{(jM_s)(jM_p)}{jM_u}. \tag{10.5}$$

Let us denote by θ the phase angle of $Z_u + R_s + (S_0/j\omega_u)$. Thus,

$$R_u = R_s \left(\frac{\omega_c}{\omega_u} \frac{m_p m_s}{m_u} \cos\theta - 1 \right). \tag{10.6}$$

The ratios of voltage to current at the two frequencies ω_s and ω_p define "input impedances." Since these quantities depend upon the drive level, they are not really impedances in the usual sense:

$$Z_{in,s} = R_{in,s} + jX_{in,s}$$

$$= \frac{V_s}{I_s}$$

$$= R_s + \frac{S_0}{j\omega_s} + R_s \frac{\omega_c}{\omega_s} \frac{m_u m_p}{m_s} e^{-j\theta}, \tag{10.7}$$

$$Z_{in,p} = R_{in,p} + jX_{in,p}$$

$$= \frac{V_p}{I_p}$$

$$= R_s + \frac{S_0}{j\omega_p} + R_s \frac{\omega_c}{\omega_p} \frac{m_s m_u}{m_p} e^{-j\theta}. \tag{10.8}$$

If the output load is tuned so that $\theta = 0$, the two input resistances become

$$R_{in,s} = R_s \left(\frac{\omega_c}{\omega_s} \frac{m_p m_u}{m_s} + 1 \right), \tag{10.9}$$

$$R_{in,p} = R_s \left(\frac{\omega_c}{\omega_p} \frac{m_s m_u}{m_p} + 1 \right), \tag{10.10}$$

and the load resistance is

$$R_u = R_s \left(\frac{\omega_c}{\omega_u} \frac{m_s m_p}{m_u} - 1 \right). \tag{10.11}$$

These three equations can be combined in a variety of ways to eliminate m_u, m_p, or m_s in favor of the load resistance R_u. Thus, for example, the input resistance at signal frequency can be written in the form

$$R_{in,s} = R_s \left(\frac{m_p^2 \omega_c^2}{\omega_s \omega_u} \frac{R_s}{R_s + R_u} + 1 \right), \tag{10.12}$$

which is identical to the small-signal formula, Equation 5.37.

The input powers at the two frequencies ω_s and ω_p and the output power are

$$\begin{aligned} P_{in,s} &= 2R_{in,s}|I_s|^2 \\ &= 8P_{norm} \left(\frac{S_{max} - S_{min}}{S_{max} + S_{min}} \right)^2 \frac{\omega_s^2}{\omega_c^2} \frac{R_{in,s}}{R_s} m_s^2 \\ &= 8P_{norm} \left(\frac{S_{max} - S_{min}}{S_{max} + S_{min}} \right)^2 \frac{\omega_s^2}{\omega_c^2} \left(\frac{\omega_c}{\omega_s} m_s m_p m_u + m_s^2 \right), \end{aligned} \tag{10.13}$$

$$\begin{aligned} P_{in,p} &= 2R_{in,p}|I_p|^2 \\ &= 8P_{norm} \left(\frac{S_{max} - S_{min}}{S_{max} + S_{min}} \right)^2 \frac{\omega_p^2}{\omega_c^2} \frac{R_{in,p}}{R_s} m_p^2 \\ &= 8P_{norm} \left(\frac{S_{max} - S_{min}}{S_{max} + S_{min}} \right)^2 \frac{\omega_p^2}{\omega_c^2} \left(\frac{\omega_c}{\omega_p} m_s m_p m_u + m_p^2 \right), \end{aligned} \tag{10.14}$$

$$\begin{aligned} P_{out} &= 2R_u|I_u|^2 \\ &= 8P_{norm} \left(\frac{S_{max} - S_{min}}{S_{max} + S_{min}} \right)^2 \frac{\omega_u^2}{\omega_c^2} \frac{R_u}{R_s} m_u^2 \\ &= 8P_{norm} \left(\frac{S_{max} - S_{min}}{S_{max} + S_{min}} \right)^2 \frac{\omega_u^2}{\omega_c^2} \left(\frac{\omega_c}{\omega_u} m_s m_p m_u - m_u^2 \right), \end{aligned} \tag{10.15}$$

where the final line in each of these equations holds only if the load is tuned. Note that the output power approaches zero if m_s or m_u approaches zero (this is the small-signal case), and if m_p is small (this is the "no-pump" case). The power dissipated in the varactor is

$$P_{diss} = P_{in,s} + P_{in,p} - P_{out}$$

$$= 2R_s \left(|I_s|^2 + |I_p|^2 + |I_u|^2\right)$$

$$= 8P_{norm}\left(\frac{S_{max} - S_{min}}{S_{max} + S_{min}}\right)^2 \frac{\omega_s^2 m_s^2 + \omega_p^2 m_p^2 + \omega_u^2 m_u^2}{\omega_c^2}. \tag{10.16}$$

Many efficiencies can be defined for this upconverter. For example, the signal gain is the ratio of output power to signal input power,‡

$$\epsilon_{su} = \frac{P_{out}}{P_{in,s}}$$

$$= \frac{\omega_u}{\omega_s} \frac{m_s m_p m_u - \frac{\omega_u}{\omega_c} m_u^2}{m_s m_p m_u + \frac{\omega_s}{\omega_c} m_s^2}. \tag{10.17}$$

The pump efficiency can be defined as the ratio of power output to pump power input, ‡

$$\epsilon_{pu} = \frac{P_{out}}{P_{in,p}}$$

$$= \frac{\omega_u}{\omega_p} \frac{m_s m_p m_u - \frac{\omega_u}{\omega_c} m_u^2}{m_s m_p m_u + \frac{\omega_p}{\omega_c} m_p^2}. \tag{10.18}$$

The over-all efficiency ϵ_{spu} can be defined as the ratio of output to total input power, ‡

‡ The last lines in Equations 10.17 to 10.20 hold only for a tuned load.

$$\epsilon_{spu} = \frac{P_{out}}{P_{in,s} + P_{in,p}}$$

$$= \frac{\epsilon_{su}\epsilon_{pu}}{\epsilon_{su} + \epsilon_{pu}}$$

$$= \frac{m_s m_p m_u - \frac{\omega_u}{\omega_c} m_u^2}{m_s m_p m_u + \frac{\omega_s^2 m_s^2 + \omega_p^2 m_p^2}{\omega_u \omega_c}}. \qquad (10.19)$$

In addition, the ratio of the two input powers can be defined if thought to be useful:‡

$$\frac{P_{in,s}}{P_{in,p}} = \frac{\epsilon_{pu}}{\epsilon_{su}} = \frac{\omega_s}{\omega_p} \frac{m_s m_p m_u + \frac{\omega_s}{\omega_c} m_s^2}{m_s m_p m_u + \frac{\omega_p}{\omega_c} m_p^2}. \qquad (10.20)$$

The bias voltage is

$$\frac{V_0 + \phi}{V_B + \phi} = \left(\frac{S_{max} - S_{min}}{S_{max}}\right)^2 \left[m_0^2 + 2(m_s^2 + m_p^2 + m_u^2)\right]. \qquad (10.21)$$

10.2.2. Breakdown Limit

The fact that the elastance cannot be larger than S_{max} or smaller than S_{min} implies that the allowable values for m_0, m_s, m_p, and m_u are restricted. In particular,

$$\frac{S_{min}}{S_{max} - S_{min}} \leq \text{Re}\left(m_0 + 2M_s e^{j\omega_s t} + 2M_p e^{j\omega_p t} + 2M_u e^{j\omega_u t}\right)$$

$$\leq \frac{S_{max}}{S_{max} - S_{min}} \qquad (10.22)$$

‡The last lines in Equations 10.17 to 10.20 hold only for a tuned load.

for any value of t. Note that so far we have not specified whether or not ω_s and ω_p are commensurate or incommensurate, and we have not specified the phase relationships between M_s and M_p. In general, Condition 10.22 can be used to derive bounds on m_s, m_p, and m_u only when the relative frequencies and phases are known. We do not intend to investigate here these complicated restrictions.

If

$$m_s + m_p + m_u \leq 0.25, \tag{10.23}$$

then Condition 10.22 can be satisfied for all frequencies and all phases, for some value of m_0. Since some values of m_s, m_p, and m_u not satisfying Condition 10.23 nevertheless satisfy Condition 10.22, we must interpret Condition 10.23 as a sufficient but not necessary condition. Nevertheless, it is a convenient one, and we shall use it in Section 10.2.3.

A convenient way of representing Condition 10.23 is to plot, as a function of m_s and m_p, the maximum value of m_u. A triangle chart like Figure 10.1 is useful for this purpose. Each

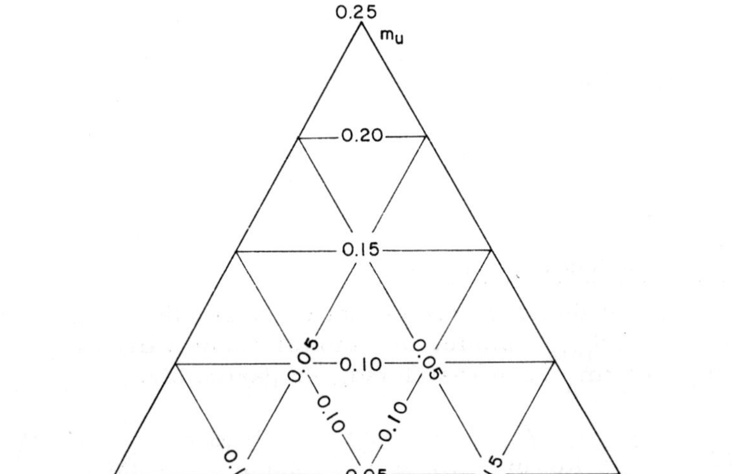

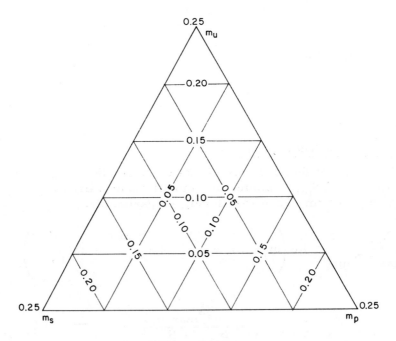

Figure 10.1. Chart showing maximum values of m_s, m_p, or m_u as a function of the values of the other two, subject to Condition 10.23.

point on this chart represents a set of values of m_s, m_p, and m_u for which Condition 10.23 is barely obeyed. It is convenient to represent operation of the upconverter by points on this chart.

10.2.3. Output-Power Optimization

The basic equations given thus far can be used to study many properties of upconverters, and they can be extended to cover upconverters with idlers, lower-sideband frequency converters, and so forth. As an illustration of their use, we consider the problem of designing a power upconverter with maximum power output. For simplicity, we constrain the values of m_p, m_s, and m_u by Condition 10.23 (for certain input phases or frequency values the output power might therefore be increased slightly).

The power output is given by Equation 10.15; to maximize this, for a specific varactor and a given output frequency ω_u, we must choose m_s, m_p, and m_u to maximize the quantity

$$m_s m_p m_u - \frac{\omega_u}{\omega_c} m_u^2. \qquad (10.24)$$

At low frequencies this is approximately $m_s m_p m_u$; when we maximize this subject to the constraint

$$m_s + m_p + m_u = 0.25, \qquad (10.25)$$

we find that

$$m_s = m_p = m_u = \frac{1}{12}. \qquad (10.26)$$

At higher frequencies Expression 10.24 is maximized by using equal values of m_p and m_s, and a smaller value of m_u:

$$m_s = m_p = \frac{1}{24} - \frac{2}{3}\frac{\omega_u}{\omega_c} + \sqrt{\frac{4}{9}\frac{\omega_u^2}{\omega_c^2} + \frac{1}{9}\frac{\omega_u}{\omega_c} + \frac{1}{576}}, \qquad (10.27)$$

$$m_u = \frac{1}{6} + \frac{4}{3}\frac{\omega_u}{\omega_c} - \sqrt{\frac{16}{9}\frac{\omega_u^2}{\omega_c^2} + \frac{4}{9}\frac{\omega_u}{\omega_c} + \frac{1}{144}}. \qquad (10.28)$$

These values can be used in the expressions given in Equations 10.9 to 10.21 to find the various efficiencies, powers, and resistances for maximum power output.

In the low-frequency case, Equation 10.26 implies that each resistance ($R_{in,s}$, $R_{in,p}$, and R_u) varies inversely with its respective frequency. The over-all efficiency approaches 1, and the input powers are in the ratio of the frequencies, as implied by the Manley-Rowe formulas. At high frequencies, on the other hand, $R_{in,s}$, $R_{in,p}$, and R_u all approach R_s. The output power approaches the limiting value $(1/2048)P_{norm}$ (if $S_{min} << S_{max}$). The values of both m_s and m_p approach $\frac{1}{8}$, and m_u approaches zero. Any interested reader can easily investigate this optimum-output-power condition more thoroughly.

The points representing maximum power output can be represented on the triangle chart; see Figure 10.2. At low frequen-

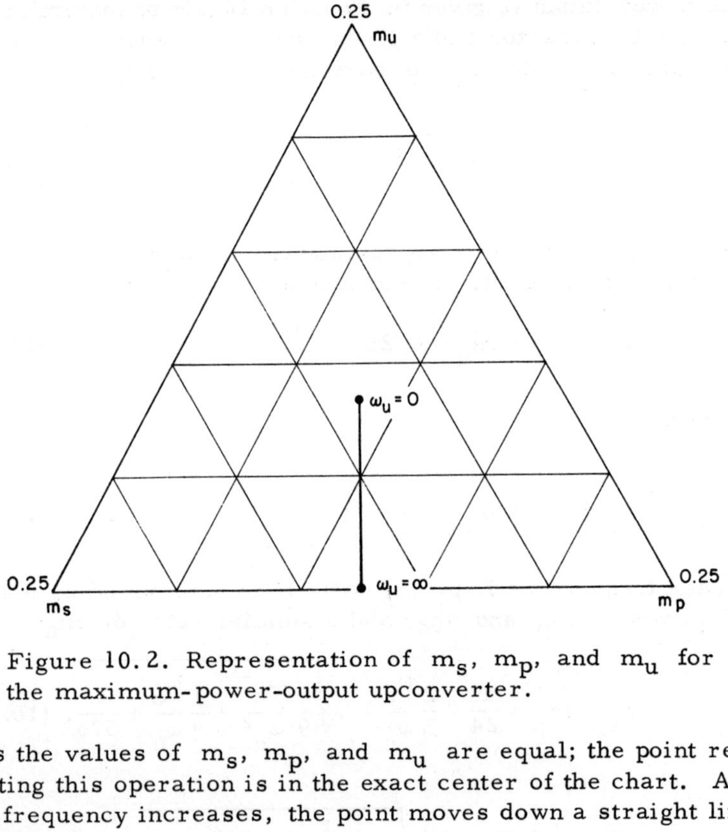

Figure 10.2. Representation of m_s, m_p, and m_u for the maximum-power-output upconverter.

cies the values of m_s, m_p, and m_u are equal; the point representing this operation is in the exact center of the chart. As the frequency increases, the point moves down a straight line toward the bottom edge.

10.3. Gain Stability in Parametric Amplifiers

Here we look briefly at the effect of variations in pump power on the gain and phase of a parametric amplifier. We assume the signal and idler excitations are small in amplitude, so that dynamic-range effects do not arise.

According to Section 4.6, a pumped varactor is characterized only by varactor parameters (such as S_{max}, P_{norm}, and ω_c) and the modulation ratios m_0, m_1, m_2, $\cdots$. Only the modulation ratios are affected by variations in pump power.

The impedance of the varactor at signal frequency Z is given by Equation 6.21 for the simplest nondegenerate parametric amplifier:

$$Z = R + jX$$

$$= R_s \left[1 - jm_0 \frac{\omega_c}{\omega_s} - \frac{m_1^2 \omega_c^2}{\omega_s \omega_i} \frac{R_s}{R_s + Z_i^* + jm_0 R_s(\omega_c/\omega_i)} \right]. \tag{10.29}$$

Note that the pumping affects this formula in three ways: first, through the second term; second, through m_1^2; and finally, through the idler tuning.

For the moment, let us consider a small variation of pump power δP about some desired power P_{des}, which produces small changes δm_1 and δm_0 about the desired values $m_{1,des}$ and $m_{0,des}$. Let us further assume that the idler termination is tuned at the desired operating point,

$$Z_i = R_i + jm_{0,des} R_s \frac{\omega_c}{\omega_i}. \tag{10.30}$$

Then the first-order changes in R and X from this desired operating point are

$$\delta Z = \delta R + j\,\delta X$$

$$= - \frac{m_1^2 \omega_c^2}{\omega_s \omega_i} \frac{R_s^2}{R_s + R_i} \frac{2\,\delta m_1}{m_1}$$

$$+ j \frac{\omega_c}{\omega_s} \left(\frac{m_1^2 \omega_c^2}{\omega_i^2} \frac{R_s^2}{(R_s + R_i)^2} - 1 \right) \delta m_0. \tag{10.31}$$

It is interesting to note that because the idler is tuned for the desired operating point, the change in average elastance affects only the imaginary part of the impedance, and the change in m_1 affects only the real part. Furthermore, the change in imaginary part (which usually is proportional to the change in phase of the amplified output) can be made zero by choosing the pump frequency properly. For the important case of a lossless idler termination ($R_i = 0$), δX is zero when

$$\omega_i = m_1 \omega_c . \tag{10.32}$$

Note that if $\omega_s << \omega_i$, this condition implies that the pump frequency should be at the optimum pump frequency $m_1 \omega_c$ discovered in Chapters 5 and 6. Clearly this pump frequency is "optimum" in more than one sense.

To apply Equation 10.31 to a specific case, suppose we have a circulator amplifier, such as the one described in Section C-3.1 of Appendix C. If the input is matched, the exchangeable gain is $|\Gamma|^2$, where the reflection coefficient Γ is

$$\Gamma = \frac{Z + jX_o - R_o}{Z + jX_o + R_o}, \tag{10.33}$$

where R_o is the circulator impedance (adjusted if necessary by a transformer), and we assume X_o is the reactance used to tune out the average elastance at signal frequency:

$$X_o = R_s m_{0,des} \frac{\omega_c}{\omega_s} . \tag{10.34}$$

The phase of the output depends upon the phase of Γ; thus if θ, the phase angle of Γ, changes by N degrees, the output phase will also change by N degrees. At the desired operating point, Γ is real. Because of a small change δZ in Z, Γ changes an amount $\delta\Gamma$:

$$\delta\Gamma = \frac{2R_o \, \delta Z}{(R_o + R)^2} . \tag{10.35}$$

The resulting first-order change in θ from its desired value of zero is

$$\delta\theta = \frac{G - 1}{2\sqrt{G}} \frac{\delta X}{R}, \tag{10.36}$$

where G is the gain $|\Gamma|^2$. The change in G is

$$\frac{\delta G}{G} = \frac{G - 1}{\sqrt{G}} \frac{\delta R}{R}. \qquad (10.37)$$

Thus the change in phase at the output can, to first order, be zero if $\delta X = 0$, that is, if the pump frequency is chosen properly. The change in gain, however, cannot be eliminated so easily, since it is related to δR and therefore, by Equation 10.31, to δm_1.

To extend this example, let us consider specifically an abrupt-junction varactor. The pump power is related to m_1 by Equation 7.95, so that

$$\frac{2\delta m_1}{m_1} = \frac{\delta P}{P}. \qquad (10.38)$$

Furthermore, if the bias voltage remains constant as the pump power changes, then Equation 7.77 shows that $m_0^2 + 2m_1^2$ must remain constant, so that δm_0 is

$$\delta m_0 = -2 \frac{m_1}{m_0} \delta m_1$$

$$= -\frac{m_1^2}{m_0} \frac{\delta P}{P}. \qquad (10.39)$$

These values can be used in Equations 10.31, 10.35, 10.36, and 10.37 to find the first-order changes in all quantities of interest in terms of δP. In particular, if G is high and if ω_s is much smaller than $m_1\omega_c$, the effect upon the gain is approximately

$$\frac{\delta G}{G} = \sqrt{G} \frac{\delta P}{P}. \qquad (10.40)$$

Therefore, for example, if the gain is 20 db, a variation of P amounting to 0.1 per cent would give a 1 per cent variation in gain. However, if the gain were adjusted to be 40 db, the same variation in P would give approximately 10 per cent change in gain. As in almost all negative-resistance amplifiers, the gain stability deteriorates with increasing gain, and very small amounts of modulation (for example, 60-cycle ripple) on the pump can heavily modulate any signal. If the optimum pump frequency is used, this is amplitude modulation alone, rather than phase modulation and amplitude modulation combined.

Equation 10.40 suggests that more stable amplifiers could be made by cascading two lower-gain stages. Thus two 20-db stages cascaded would have only about 2 per cent gain variation from a change in pump power of 0.1 per cent, rather than the 10 per cent change for a single 40-db stage.

So far we have examined only the effects of a small change in pump power. If the change is large, then higher-order terms are important. If the varactor is an abrupt-junction model, the effects of finite changes in P can be found in closed form, and for graded-junction varactors, the effects can be found from the graphs in Chapter 7. Thus, for example, for abrupt-junction varactors with, for simplicity, $S_{min} \ll S_{max}$, we find, from Equations 7.77 and 7.95,

$$m_1^2 = \frac{P}{8P_{norm}} \frac{\omega_c^2}{\omega_p^2} \tag{10.41}$$

and

$$m_0^2 = \frac{V_0 + \phi}{V_B + \phi} - \frac{P}{4P_{norm}} \frac{\omega_c^2}{\omega_p^2}. \tag{10.42}$$

We can substitute these into Equation 10.29 to obtain R and X explicitly as functions of P, V_0, Z_i, the frequencies, and the varactor parameters:

$$R = R(P, V_0, Z_i; \omega_p, \omega_s, \omega_i; R_s, V_B, \phi, \omega_c, P_{norm}), \tag{10.43}$$

$$X = X(P, V_0, Z_i; \omega_p, \omega_s, \omega_i; R_s, V_B, \phi, \omega_c, P_{norm}). \tag{10.44}$$

For finite changes in pump power and/or mistuned idler terminations, these formulas can be used to find the actual changes in gain and phase. For example, for the circulator amplifier, we have from Equation 10.33 that

$$\tan\theta = \frac{2R_o(X + X_o)}{(X + X_o)^2 + R^2 - R_o^2}, \tag{10.45}$$

$$G = \frac{(R_o - R)^2 + (X_o + X)^2}{(R_o + R)^2 + (X_o + X)^2}. \tag{10.46}$$

In this section we have examined only the stability of gain and phase with respect to variations in pump power. This theory can be extended to cover other types of variations, such as those caused by changes in temperature or aging.

10.4. Dynamic Range

Dynamic range is a useful concept partly because it is so vague. It means many different things to different people, and can be applied to many different systems in many different ways. Basically, it is the ratio (usually in decibels) of the maximum power to the minimum power. Calculations of dynamic ranges therefore must be based on specific assumptions about what constitutes the "maximum" and "minimum" powers.

For an amplifier or a supposedly linear device, the minimum power is usually set by the noise input, or the noise generated internally by the amplifier, or both. For other types of devices the criterion might be quite different. The maximum power is usually set by undesirable nonlinear effects of the device, such as saturation, reduction of gain, generation of harmonics of the signal, distortion, change in phase of the signal, or cross talk from other signals. It is clear, then, that the dynamic range appropriate for a communications receiver might be quite different from that for a missile-tracking receiver, although both devices might use the same sort of amplifiers. In all cases, the pertinent definition of maximum power rests on a clear description, from a system viewpoint, of the maximum tolerable nonlinear effects.

Abrupt-junction varactors are capable of operation with wide dynamic ranges, often greater than 120 or 140 decibels. As examples, we shall show now a few very simple calculations of dynamic range.

10.4.1. Minimum Power

Suppose the minimum power is dictated by the noise of the amplifier or frequency converter (with effective input noise temperature T) and the noise of an antenna (with temperature T_{ant}). It is usually convenient to assume that the minimum detectable signal has power equal to the noise power. Thus the minimum power would be

$$P_{min} = k(T + T_{ant})\, \Delta f. \tag{10.47}$$

If the device is a varactor parametric amplifier with a lossless, tuned idler termination and a pump at the optimum pump frequency, and if $\omega_s \ll \omega_p$, the noise temperature T is approximately

$$T = T_d \frac{2\omega_s}{m_p \omega_c}, \tag{10.48}$$

where m_p is the pump-frequency modulation ratio.

10.4.2. Crude Estimate of Maximum Power

One nonlinearity that limits the power in varactor amplifiers is breakdown. The varactor elastance must not exceed S_{max}, because intolerable breakdown noise and other undesirable effects appear. A crude estimate of the maximum power can be obtained for, say, an upconverter, by searching for the maximum transfer of power subject to the breakdown constraint. For example, at low frequencies, the analysis of Section 10.2 shows that this maximum occurs when

$$m_p = m_s = m_u = \frac{1}{12}. \tag{10.49}$$

The signal input power is‡

$$P_{max} = 4.6 \times 10^{-3} P_{norm} \frac{\omega_s}{\omega_c}. \tag{10.50}$$

Clearly, a single upconverter cannot be run from very small signals to this maximum-signal condition without special attention to keeping the pump drive and average elastance constant. Therefore, Equation 10.50 is probably an unachievable upper bound, but it yields a crude estimate for the maximum possible dynamic range:

$$DR = \frac{4.6 \times 10^{-3} P_{norm} \dfrac{\omega_s}{\omega_c}}{k\left(T_{ant} + 24 T_d \dfrac{\omega_s}{\omega_c}\right)\Delta f}. \tag{10.51}$$

Thus, if a room-temperature varactor with $P_{norm} = 200$ kw and $f_c = 6$ Gc is used to build an upconverter from 50 Mc with a bandwidth of 10 kc, to run from an antenna with noise temperature of 5000°K, the maximum dynamic range is 160 db. Even if our crude estimate is in error by as much as a factor of 100, the dynamic range would be 140 db, a value that is of interest for many communications problems.

10.4.3. Dynamic Range of Parametric Amplifiers

Suppose a small-signal parametric amplifier is operating with increasingly large signals. What nonlinear effects occur first?

‡We assume $S_{min} \ll S_{max}$.

Eventually, of course, the elastance components at the signal and idler frequencies become so large that the breakdown limit is exceeded. Long before then, however, other effects arise that reduce the gain, affect the phase of the output, and cause cross talk from other nearby signals. We wish to investigate now the first such effect to occur.

For simplicity, we analyze an abrupt-junction-varactor parametric amplifier. If current is restricted to flow only at frequencies ω_p, ω_s, and $\omega_i = \omega_p - \omega_s$, the large-signal Equations 4.32, without the noise terms, reduce to

$$\begin{bmatrix} V_p^* \\ V_i^* \\ V_s^* \\ V_s \\ V_i \\ V_p \end{bmatrix} = \begin{bmatrix} R_s - \frac{S_0}{j\omega_p} & -\frac{S_s^*}{j\omega_p} & -\frac{S_i^*}{j\omega_p} & 0 & 0 & 0 \\ -\frac{S_s}{j\omega_i} & R_s - \frac{S_0}{j\omega_i} & 0 & -\frac{S_p^*}{j\omega_i} & 0 & 0 \\ -\frac{S_i}{j\omega_s} & 0 & R_s - \frac{S_0}{j\omega_s} & 0 & -\frac{S_p^*}{j\omega_s} & 0 \\ 0 & \frac{S_p}{j\omega_s} & 0 & R_s + \frac{S_0}{j\omega_s} & 0 & \frac{S_i^*}{j\omega_s} \\ 0 & 0 & \frac{S_p}{j\omega_i} & 0 & R_s + \frac{S_0}{j\omega_i} & \frac{S_s^*}{j\omega_i} \\ 0 & 0 & 0 & \frac{S_i}{j\omega_p} & \frac{S_s}{j\omega_p} & R_s + \frac{S_0}{j\omega_p} \end{bmatrix} \begin{bmatrix} I_p^* \\ I_i^* \\ I_s^* \\ I_s \\ I_i \\ I_p \end{bmatrix} \tag{10.52}$$

We suppose the idler-frequency termination is an impedance Z_i; then the ratio of V_p to I_p defines a "pump input impedance" $Z_{in,p}$ (which depends on the signal level):

$$\begin{aligned} Z_{in,p} &= R_{in,p} + jX_{in,p} \\ &= R_s \left[1 - jm_0 \frac{\omega_c}{\omega_p} + \frac{m_s^2 \omega_c^2}{\omega_p \omega_i} \frac{R_s}{R_s + Z_i - jm_0 R_s(\omega_c/\omega_i)} \right], \end{aligned} \tag{10.53}$$

where m_s is the signal-frequency modulation ratio. The input impedance of the varactor at signal frequency is a function of the pump level:

$$Z = R + jX$$

$$= R_s \left[1 - jm_0 \frac{\omega_c}{\omega_s} - \frac{m_p^2 \omega_c^2}{\omega_s \omega_i} \frac{R_s}{R_s + Z_i^* + jm_0 R_s(\omega_c/\omega_i)} \right], \quad (10.54)$$

where m_p is the pump-frequency modulation ratio (called m_1 for the small-signal parametric amplifiers).

The idler-frequency modulation ratio m_i is related to the other two by the formula

$$m_i = m_s m_p \frac{\omega_c}{\omega_i} \frac{R_s}{|R_s + Z_i - jm_0 R_s(\omega_c/\omega_i)|}. \quad (10.55)$$

Note that Equation 10.54 for the impedance of the varactor at signal frequency has exactly the same form as Equation 10.29 for the small-signal amplifier (this is because of the linear charge-elastance relationship of the abrupt-junction varactor). The gain and phase of the amplifier can change only insofar as m_0 and m_p change. The first-order change in Z because of δm_0 and δm_p has been given as Equation 10.31, and we now have only to account for δm_0 and δm_p in terms of m_s or of the signal power.

The problem is now this: As the signal amplitude is increased from zero, how are m_p and m_0 affected? First, as m_s increases from zero, so does m_i, in accordance with Equation 10.55. Second, any such changes must be consistent with an assumed pump circuit. And third, these changes must be consistent with the bias voltage applied.

To calculate the first-order changes in m_p, we must account for the first-order changes in m_s^2 and m_i^2. Consider a typical pump circuit, shown in Figure 10.3. An inductance is used to

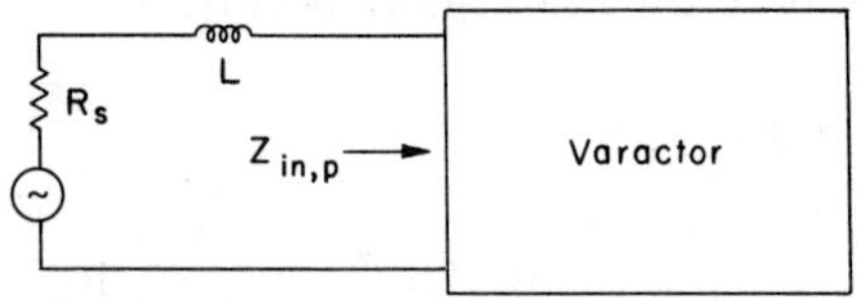

Figure 10.3. Typical pump circuit.

resonate the average elastance m_0, and for maximum power transfer the source resistance is equal to R_s. As the value of m_s^2 rises from zero, the pump-frequency varactor impedance $Z_{in,p}$ rises in accordance with Equation 10.53. The first-order change in the real part of $Z_{in,p}$ is, if the idler termination is tuned,

$$m_s^2 \frac{\omega_c^2}{\omega_p \omega_i} \frac{R_s^2}{R_s + R_i}, \tag{10.56}$$

and hence the first-order change δm_p is

$$\frac{\delta m_p}{m_p} = -\frac{1}{2} \frac{\omega_c^2}{\omega_p \omega_i} \frac{R_s}{R_s + R_i} m_s^2, \tag{10.57}$$

independent of δm_0. Other pump circuits would have different formulas instead of Equation 10.57. Furthermore, from the bias-voltage equation,

$$\frac{V_0 + \phi}{V_B + \phi} = m_0^2 + 2m_s^2 + 2m_i^2 + 2m_p^2, \tag{10.58}$$

we find that for the pump circuit of Figure 10.3 the first-order change in m_0 is‡

$$\delta m_0 = -\frac{m_s^2}{m_0} \left[1 + \frac{m_p^2 \omega_c^2}{\omega_i^2} \frac{R_s^2}{(R_s + R_i)^2} - \frac{m_p^2 \omega_c^2}{\omega_p \omega_i} \frac{R_s}{R_s + R_i} \right]. \tag{10.59}$$

These values of δm_p and δm_0 can be used in Equation 10.31 and in the other formulas of Section 10.3 to find the first-order changes in gain and phase of the parametric amplifier. In particular, if the pump frequency is chosen properly, so that

$$\omega_i = m_p \omega_c \frac{R_s^2}{R_s + R_i}, \tag{10.60}$$

the first-order change in phase is zero, and, incidentally, $m_i = m_s$ (if $\omega_s << \omega_p$, this is the optimum pump frequency again).

The first-order changes in Z, and therefore in gain of specific circuits, are now related to the value of m_s^2. They can also be related to the input power P_{in}. The net power generated at sig-

---

‡Note the partial cancellation within the brackets of Equation 10.59. If the varactor is pumped from a source with a very low internal resistance, instead of one that matches the varactor series resistance, then the last term within the brackets is twice as large, and δm_0 can be made very small by pumping at the optimum pump frequency.

nal frequency is

$$P_{in}(G - 1), \tag{10.61}$$

where G is the gain of the amplifier. But since this is also equal to $-2|I_s|^2R$,

$$m_s^2 = \frac{P_{in}}{P_{norm}} \frac{G-1}{8} \frac{R_s}{(-R)} \frac{\omega_c^2}{\omega_s^2}. \tag{10.62}$$

Therefore, all first-order changes can be expressed in terms of the signal input power P_{in}. Thus, for example, using Equation 10.37 to find the change in gain of a circulator amplifier in terms of δR, using Equation 10.31 to find δR in terms of δm_1 (or δm_p), using Equation 10.57 to find δm_p in terms of m_s^2, and finally using Equation 10.62 to find m_s^2 in terms of P_{in}, we find a relation between the first-order change in gain and the input signal power:

$$\frac{\delta G}{G} = -\frac{(G-1)^2}{8\sqrt{G}} \frac{m_p^2 \omega_c^6}{\omega_p \omega_i^2 \omega_s^3} \frac{R_s^2}{(R_s + R_i)^2} \frac{R_s^2}{R^2} \frac{P_{in}}{P_{norm}}. \tag{10.63}$$

The maximum input power P_{max} can now be found by specifying what is the maximum acceptable fractional change in gain $\delta G/G$. Thus, for example, at low frequencies and high gain we find

$$P_{max} = \frac{8P_{norm}}{G^{3/2}} \left(\frac{\delta G}{G}\right)_{max} \frac{m_p^2 \omega_p \omega_s}{\omega_c^2}. \tag{10.64}$$

This can also be related to the pump power, if desired, by Equation 7.95:

$$P_{max} = P_{pump} \left(\frac{\delta G}{G}\right)_{max} \frac{1}{G^{3/2}} \frac{\omega_s}{\omega_p}. \tag{10.65}$$

Now that P_{max} is known, the dynamic range can be calculated.

It is interesting to identify the physical cause of this first-order change in gain.‡ As the input power increases, the pump is

‡Blackwell and Kotzebue[1] also identify this cause as the dominant one for parametric amplifiers.

called upon to convert more and more power to the signal and idler frequencies. This causes a rise in the impedance of the varactor as seen by the pump source. The pump current and, therefore, m_p, decrease because of this. The change in m_p alters the resistance of the varactor at signal frequency. The accompanying change in m_0 affects the reactance of the signal-frequency impedance and, therefore, affects the phase of the output, but this change can be eliminated, to first order, by properly choosing the pump frequency. The first-order change in gain is caused entirely by a change in the pump current, this change in pump current being caused by the power the pump is called upon to supply.

Consider a slightly different case now. Suppose we wish to avoid cross talk between the desired signal and a strong signal on an adjacent frequency. If this strong signal is converted in the varactor, and if this conversion requires a significant amount of power, the value of m_p can be affected, and in particular can be modulated by any modulation on the strong signal. This modulation then appears on the desired output. Thus, Equation 10.63 can be interpreted as the change in gain caused by a signal with power P_{in} in the adjacent channel, provided this adjacent frequency is amplified by the amplifier.‡

We have given only the first-order theory of reduction of gain from changes in input power. Higher-order variations, and even the exact changes, can be calculated in a similar way.[2] Even among the first-order variations, we have investigated only a few of the interesting relationships.

REFERENCES

1. Blackwell, L. A., and K. L. Kotzebue, "Semiconductor-Diode Parametric Amplifiers," Prentice-Hall, Inc., Englewood Cliffs, N.J. (1961), Sec. 4.3.
2. Sohn, S. J., "The Dynamic Range of a Parametric Amplifier," S.M. Thesis, Department of Electrical Engineering, M.I.T., Cambridge (May, 1962).

‡The amount the adjacent channel is amplified depends upon the response of the amplifier at that frequency and, in particular, upon the behavior of any filters outside their passbands.

Appendix A

CALCULATION OF VARACTOR CAPACITANCE AND SERIES RESISTANCE

We now give formulas for calculating space-charge capacitance (or elastance) and series resistance for planar (one-dimensional) varactors with arbitrary doping distributions. After discussing the model for the p-n junction, we calculate the junction capacitance in terms of this model. Then, under the assumption that the depletion layer is well defined, we relate the capacitance to the charge. Next, the variations in series resistance are related to the variations in elastance. Finally, a few examples are given. We do not discuss diffusion capacitance, which is important for small biases.

The principal symbols are defined in Table A-1.

Table A-1. Symbols Used in Appendix A without Explicit Definition in the Text

k = Boltzmann's constant	D_h = hole diffusion constant
T_d = varactor temperature	ϵ = dielectric constant
e = magnitude of electronic charge	A = junction area
μ_e = electron mobility	$N(x)$ = doping density
μ_h = hole mobility	$n(x)$ = electron density
D_e = electron diffusion constant	$p(x)$ = hole density
	$\rho(x)$ = resistivity

A-1. Junction Model

For simplicity, we restrict the analysis to planar (one-dimensional) junctions. Although a few of the formulas are in vector notation, the only spatial variation of any variables is in the x-direction.

We suppose the doping distribution (the number of donors less the number of acceptors per unit volume) is $N(x)$, which is positive in the n-region, and negative in the p-region. Figure A-1 shows a typical doping distribution, together with a plot of electron and hole densities. The distribution of holes and electrons depends on the bias voltage.

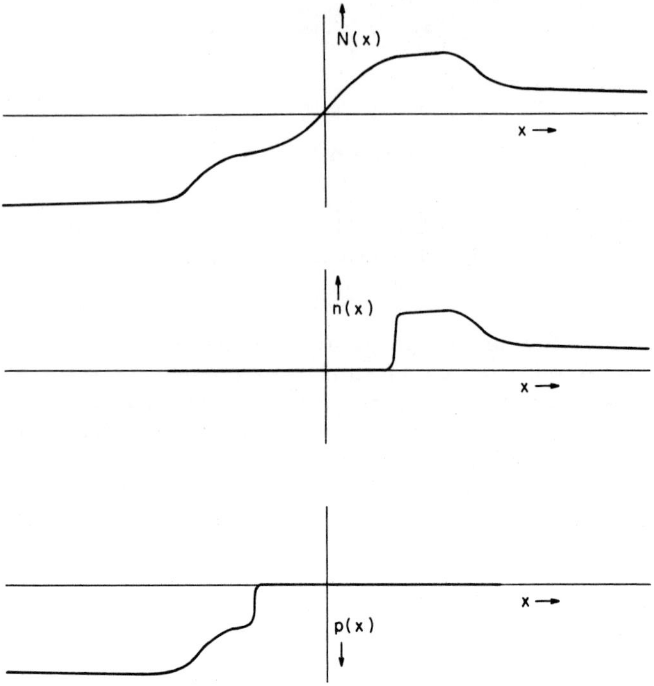

Figure A-1. Typical doping distribution, and distribution of holes and electrons. Note that the point $x = 0$ is taken at the junction, that is, where $N(x) = 0$.

To calculate the electron and hole distributions, we note that the electron and hole current densities‡ are‡‡

$$\bar{J}_e = e\mu_e n\bar{E} + eD_e \nabla n, \tag{A-1}$$

$$\bar{J}_h = e\mu_h p\bar{E} - eD_h \nabla p. \tag{A-2}$$

Each is made up of a drift component (the first term) and a diffusion component (the second term). It is often assumed‡‡‡ in junction analysis (and we assume so here) that except for minority carriers (that is, holes in regions where $p < n$, and electrons in regions where $n < p$), the net current densities $\bar{J}_e$ and $\bar{J}_h$ are much smaller than the individual drift and diffusion terms. This assumption is tantamount to saying that the carrier densities obey Boltzmann distributions, or that the "quasi-Fermi levels" of the carriers are uniform. For all purposes except calculation of currents, we equate the drift and diffusion components, to obtain formu-

‡Vectors are denoted by lines above the symbols.

‡‡For example, see Reference 7.

‡‡‡For example, see Reference 8.

las for the electric field $\overline{E}$:

$$\overline{E} = \frac{kT_d}{e} \frac{\nabla p}{p}, \tag{A-3}$$

$$\overline{E} = - \frac{kT_d}{e} \frac{\nabla n}{n}, \tag{A-4}$$

where we have used the Einstein relationships[15] ‡

$$\frac{D_e}{\mu_e} = \frac{D_h}{\mu_h}$$

$$= \frac{kT_d}{e}. \tag{A-5}$$

Equation A-3 is valid for $x < 0$, and Equation A-4 holds for $x > 0$.

The approximations of Equations A-3 and A-4 are exactly true in thermal equilibrium, because $\overline{J}_e$ and $\overline{J}_h$ both vanish. They hold under forward bias (the customary calculation of current in a junction is by means of the minority carriers on each side, since any other calculation would require subtracting nearly equal large numbers). The approximations are also valid under reverse bias, because aside from the small saturation current, $\overline{J}_e$ and $\overline{J}_h$ both vanish.

One of Maxwell's equations relates the electric field to the charge density $e(N + p - n)$:

$$\nabla \cdot \overline{E} = \frac{e}{\epsilon} (N + p - n). \tag{A-6}$$

In principle, $n(x)$ and $p(x)$ are obtained by simultaneously solving the nonlinear differential equations A-6, and A-3 and/or A-4. In practice, of course, this solution is seldom found exactly. The voltage v is then given (to within a constant) by the integral of the electric field, and the charge q is (to within a constant) the net charge removed from each side of the junction:

$$q = eA \int_0^{\infty} (N + p - n)\, dx$$

$$= -eA \int_{-\infty}^{0} (N + p - n)\, dx. \tag{A-7}$$

‡Superscript numerals denote references listed at the end of each appendix.

Note that Equation A-7 states that the charge removed from one side of the junction is the same as that removed from the other. The charge density and electric field are shown in Figure A-2 for a typical junction. In order to calculate capacitance, we wish to relate q and v.

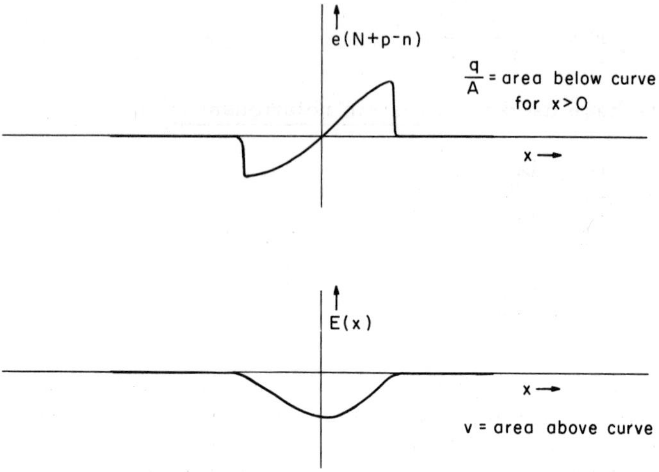

Figure A-2. Net charge density and electric field for the junction of Figure A-1.

We have shown, in Figures A-1 and A-2, a depletion region, in which n and p are small, and charge neutrality regions, in which N + p - n is small. Separating these regions are layers whose thicknesses are of the order of a shielding length

$$L_s = \sqrt{\frac{kT_d\epsilon}{e^2N}}, \tag{A-8}$$

where N is of the order of the doping density at the point in question. This shielding length has the physical significance of a length beyond which the electric field cannot penetrate a cloud of charges. If N were the intrinsic density n_i, then L_s would be the Debye length defined by Shockley.[13] Generally, however, L_s is smaller, because N is much greater than n_i.

A-2. Junction Capacitance

We now wish to show that for any doping distribution the incremental junction elastance of a planar varactor with a well-defined depletion layer is given by the familiar formula for the parallel-plate capacitor:

$$S = \frac{1}{C} = \frac{D}{\epsilon A}, \tag{A-9}$$

where D is the thickness of the depletion layer. The primary assumption to be made is that L_s is small, so that the depletion region is separated from the charge neutrality regions by distinct boundaries.‡

Suppose we remove from a back-biased varactor a small amount of negative charge δq from a point x_2 and the same amount of positive charge from a point x_1 (presumably, x_2 and x_1 are at the edges of the depletion layer). This produces a change in voltage δv, and the incremental elastance is, by definition, $S = \delta v / \delta q$.

To evaluate δv, we could evaluate $\delta(n - p)$ as a function of x, and calculate from Equation A-6 the change in electric field $\delta \overline{E}$. This, when integrated over x (both inside and outside the depletion layer), becomes the change in voltage δv. Each of these steps is linear; thus, once x_2 and x_1 are known, the steps to find δv are independent of the size of δq. The varactor elastance is nonlinear only because different values of x_2 and x_1 are appropriate for different biases.

Once x_2 and x_1 are known, the steps leading to δv are identical to those used in calculating the voltage across a parallel-plate capacitor with dielectric constant ϵ, area A, and separation $(x_2 - x_1)$. The ratio of δv to δq is therefore exactly equal to the elastance of such a fixed capacitor, which is well known to be (if fringing is neglected)

$$S = \frac{x_2 - x_1}{\epsilon A}. \tag{A-10}$$

The result of Equation A-10 is independent of the doping distribution $N(x)$, which may vary arbitrarily both inside and outside of the depletion layer.

Assuming that L_s is very small, so that the depletion layer is well defined, we can see clearly from Figures A-1 and A-2 that whenever charge is removed, it must come from the edges of the depletion layer. Thus, x_1 and x_2 must be the boundaries of the depletion layer, so that $x_2 - x_1 = D$, and Equation A-10 reduces to Equation A-9. On the other hand, if L_s is not small, then there is some uncertainty in the value to be assigned to x_1 and x_2.

For most varactors, the assumption of small L_s gives accurate results except when the depletion-layer thickness D is smaller than or of the same order of magnitude as L_s, or when there are sharp changes in doping density within a distance of approximately L_s from the edge of the depletion layer. Because of diffusion capacitance,[13, 17] our results are already known to be in error for very small depletion-layer thicknesses. When there are abrupt changes in doping density on one side of the junction, the effect of a finite L_s is merely to smooth out the otherwise sharp bends in the capacitance-voltage curve.

---

‡If L_s is not small, Equation A-9 still is valid if D is defined properly.

A-3. Variation of Capacitance

With no applied voltage, the depletion-layer thickness (which we shall call D) is of the same order of magnitude as L_s. Assuming that an appreciable back bias is applied (more than a few tenths of a volt), we can treat L_s as being much less than D. This assumption, which is made in virtually all similar analyses,[4,2,12,5,3,1] simplifies the problem so that a solution is possible. The solution we obtain is not valid for forward bias, but our analysis is already known to fail there because of diffusion capacitance.[13,17]

Under this approximation, the elastance is, from Section A-2,

$$S = \frac{D}{\epsilon A}. \tag{A-11}$$

The charge density is zero outside the depletion layer, and is eN(x) inside. Thus the charge q is given by two formulas, taken from Equation A-7,

$$\begin{aligned} q &= eA\int_0^{x_2} N(x)\,dx \\ &= -eA\int_{x_1}^{0} N(x)\,dx. \end{aligned} \tag{A-12}$$

This equation automatically relates x_1 to x_2. Since the depletion-layer thickness D is

$$D = x_2 - x_1, \tag{A-13}$$

whenever one of the three distances (x_1, x_2, or D) is known, the other two can be calculated with the aid of Equations A-12 and A-13. For convenience we define the reduced doping density N_r, a positive function of D,

$$\begin{aligned} N_r(D) &= \frac{N(x_1)N(x_2)}{N(x_1) - N(x_2)} \\ &= \frac{|N(x_1)|\,|N(x_2)|}{|N(x_1)| + |N(x_2)|}, \end{aligned} \tag{A-14}$$

where D, x_1, and x_2 are related as just described. Then

$$q = eA\int_0^{D} N_r(D)\,dD \tag{A-15}$$

in addition to the relations of Equation A-12.

Equation A-15 is important because it relates small changes in charge and depletion-layer thickness:

$$\delta q = eAN_r(D)\ \delta D, \tag{A-16}$$

and hence, by Equation A-11, small changes in q are related to small changes in S by

$$\delta S = \frac{\delta q}{e\epsilon A^2 N_r(D)}. \tag{A-17}$$

Using the definition of S, $\delta v = S\ \delta q$, we can relate changes in charge or voltage to changes in capacitance or elastance in a number of ways. We tabulate them by listing a variety of formulas for $N_r(D)$:

$$N_r(D) = \frac{1}{eA}\frac{\delta q}{\delta D} = \frac{1}{e\epsilon A^2}\frac{\delta q}{\delta S} = -\frac{C^2}{e\epsilon A^2}\frac{\delta q}{\delta C}$$

$$= \frac{\epsilon}{eD}\frac{\delta v}{\delta D} = \frac{1}{e\epsilon A^2 S}\frac{\delta v}{\delta S} = -\frac{C^3}{e\epsilon A^2}\frac{\delta v}{\delta C}$$

$$= \frac{2\epsilon}{e}\frac{\delta v}{\delta(D^2)} = \frac{2}{e\epsilon A^2}\frac{\delta v}{\delta(S^2)} = \frac{2}{e\epsilon A^2}\frac{\delta v}{\delta(1/C^2)}. \tag{A-18}$$

Back-bias capacitance measurements are often used to determine the doping distribution of varactors;[3] they are interpreted by the use of Equation A-18.

A-4. Series Resistance

When charge is supplied to the varactor junction, it travels only to the edge of the depletion layer. The series resistance it encounters can be found, to within an additive constant, by integrating the bulk resistivity of the semiconductor over the path of the current. Thus, for a planar diode with a well-defined depletion layer,

$$R_s = \text{constant} + \frac{1}{A}\int^{x_1} \rho(x)\ dx + \frac{1}{A}\int_{x_2} \rho(x)\ dx, \tag{A-19}$$

where the missing limits of integration are arbitrary because of the additive constant, and where $\rho(x)$ is the resistivity of the semiconductor, which, because of the doping distribution, depends on x. Observe that R_s depends on the depletion-layer thickness D through the variables x_1 and x_2. However, R_s is not a "nonlinear resistance" in the usual sense of the term, because it depends on the charge that has flowed through it, rather than on the instantaneous current. Thus R_s does not obey the frequency-power formulas that nonlinear resistances do.[9,11]

When the voltage across the junction (which is different from the voltage across the varactor, because of the series resistance) changes, the junction capacitance changes, as shown in Section A-3. The series resistance also changes; this change can be related to the change in elastance. From Equation A-19,

$$\delta R_s = -\frac{1}{A}\left[\rho(x_2)\,\delta x_2 - \rho(x_1)\,\delta x_1\right], \tag{A-20}$$

so that as D increases, R_s decreases, as expected. Changes in x_1, x_2, and D are related by Equations A-12 and A-13; therefore, eliminating δx_1 and δx_2, we find

$$\delta R_s = -\frac{\rho_{av}}{A}\,\delta D \tag{A-21}$$

or

$$\delta R_s = -\rho_{av}\epsilon\,\delta S, \tag{A-22}$$

where ρ_{av} lies between $\rho(x_1)$ and $\rho(x_2)$:

$$\rho_{av} = \frac{N(x_2)\rho(x_1) - N(x_1)\rho(x_2)}{N(x_2) - N(x_1)}$$

$$= \frac{|N(x_2)|\rho(x_1) + |N(x_1)|\rho(x_2)}{|N(x_2)| + |N(x_1)|}. \tag{A-23}$$

Equation A-22 relates changes in series resistance to changes in elastance; by use of Equation A-18, these may be related to changes in junction voltage or charge. The factor $\rho_{av}\epsilon$ has the dimensions of time, or inverse frequency; similar quantities are often called dielectric relaxation times.

A-5. Examples

We now wish to discuss some special cases of the formulas developed in Sections A-3 and A-4.

A-5.1. Doping Known on One Side

From Equation A-18, it is clear that capacitance-voltage measurements can help determine only $N_r(D)$, not the doping distribution on both sides of the junction. A given capacitance-voltage curve can be obtained from many doping distributions, in fact all that lead to the same $N_r(D)$.

However, if N(x) is known on one side of the junction (say for $x < 0$), then N(x) on the other side can be calculated once $N_r(D)$ is given. First, Equations A-12 and A-15 are used to express x_1 and

x_2 as functions of D, and then Equation A-14 is used to calculate $N(x_2)$.

With $N(x_1)$ and $N(x_2)$ known, the variations in R_s can be calculated from Equations A-21 and A-23, if desired.

This method is difficult to use in practice, not only because the integrals relating x_1, x_2, and D cannot always be expressed simply, but also because $N_r(D)$ must be known for very small D, whereas the analysis is known to fail for very small D.

A-5.2. One Side Heavily Doped

In many varactors (especially alloy type) one side of the junction is very heavily doped. For example, suppose that the p-side is heavily doped, so that $N(x_1)$ is very large. Consequently, x_1 is small, and therefore,

$$D = x_2, \tag{A-24}$$

$$N_r(D) = N(x_2), \tag{A-25}$$

and

$$\rho_{av} = \rho(x_2). \tag{A-26}$$

Thus the reduced doping density and average resistivity are those of the lightly doped side. Similarly, if the n-side is heavily doped, $D = -x_1 = |x_1|$, $N_r(D) = -N(x_1) = |N(x_1)|$, and $\rho_{av} = \rho(x_1)$.

A-5.3. Abrupt Junction

An abrupt junction is one for which $N(x)$ is equal to a constant N_D for $x > 0$, and a constant $-N_A$ for $x < 0$. Then

$$N_r = \frac{N_D N_A}{N_D + N_A} \tag{A-27}$$

is independent of D, and therefore, by Equation A-18, the elastance is proportional to the charge (plus a constant), as predicted in Section 4.2.1(a). Since $N(x)$ is evaluated only at the edges of the depletion layer, the elastance would be proportional to the charge (plus a constant) even for varactors that are not truly abrupt, provided only that the doping is uniform over all x_1 and x_2 encountered.[4]

If the resistivities of the two sides of the junction are ρ_D and ρ_A, then

$$\rho_{av} = \frac{N_D \rho_A + N_A \rho_D}{N_D + N_A}, \tag{A-28}$$

and the resistance is also seen, by Equation A-22, to be proportional to the charge (plus a constant). Integration of Equation A-22 yields the interesting formula,

$$R_s = R_{s,min} + \rho_{av}\epsilon(S_{max} - S), \tag{A-29}$$

where $R_{s,min}$ is the minimum series resistance, achieved at the breakdown voltage.

It is interesting to inquire what variations in series resistance are possible. If the variations are small, then the maximum fractional variation is

$$\frac{R_{s,max} - R_{s,min}}{R_s} = \rho_{av}\epsilon\omega_c, \tag{A-30}$$

where ω_c is the cutoff frequency, Equation 4.47. For a silicon varactor ($\epsilon \approx 10^{-10}$ farad/meter) with a base resistance of 0.1 ohm-cm, this becomes $f_c/1590$ Gc, or a rather small quantity, for presently available varactors (see Section 4.7.4).

A-5.4. Graded Junction

A graded junction is one for which $N(x)$ is proportional to x:

$$N(x) = ax. \tag{A-31}$$

Equations A-12 and A-13 indicate that $x_2 = -x_1 = D/2$, so that

$$N_r(D) = \frac{aD}{4}. \tag{A-32}$$

Thus, by Equation A-18, S^2 is proportional to the charge (plus a constant), as we predicted in Section 4.2.1(b).

Since $N(x_1) = -N(x_2)$, Equation A-23 predicts

$$\rho_{av} = \frac{\rho(x_1) + \rho(x_2)}{2}. \tag{A-33}$$

The value of ρ_{av} is not constant because of the doping distribution. If the resistivity of the material is inversely proportional to the doping density, then ρ_{av} is proportional to $1/D$:

$$\rho_{av} = \frac{1}{\alpha D}, \tag{A-34}$$

and integration of Equation A-22 yields

$$R_s = R_{s,min} + \frac{1}{\alpha A} \ln \frac{S_{max}}{S}. \tag{A-35}$$

The logarithmic dependence in Equation A-35 indicates that R_s can change significantly at small voltages.

A-5.5. Diffused Junction

Diffused-junction varactors were discussed briefly in Section 4.2.1(c); their capacitance-voltage characteristics have received special treatment.[12,5,1] At low bias voltages they appear to be graded-junction varactors, and at higher reverse biases they behave like abrupt-junction varactors, both in their capacitance and their series-resistance characteristics. High-quality diffused-junction varactors seldom operate in the abrupt-junction region.

Grown junctions, because of diffusion of impurities, also exhibit this dual behavior: At low voltages they act like graded junctions, and at high voltages, abrupt junctions.

REFERENCES

1. Cohen, J., "Transition Region Properties of Reverse-Biased Diffused p-n Junctions," IRE Trans. on Electron Devices, ED-9, 362-369 (September, 1961).
2. Giacoletto, L. J., "Junction Capacitance and Related Characteristics Using Graded Impurity Semiconductors," IRE Trans. on Electron Devices, ED-4, 207-215 (July, 1957).
3. Hilibrand, J., and R. D. Gold, "Determination of the Impurity Distribution in Junction Diodes From Capacitance-Voltage Measurements," RCA Rev., 21, 245-252 (June, 1960).
4. Kroemer, H., "The Apparent Contact Potential of a Pseudo-Abrupt P-N Junction," RCA Rev., 17, 515-521 (December, 1956).
5. Lawrence, H., and R. M. Warner, Jr., "Diffused Junction Depletion Layer Calculations," Bell System Tech. J., 39, 389-403 (March, 1960).
6. Middlebrook, R. D., An Introduction to Junction Transistor Theory, John Wiley & Sons, Inc., New York (1957).
7. Middlebrook, R. D., ibid., p. 88.
8. Middlebrook, R. D., ibid., p. 104.
9. Page, C. H., "Frequency Conversion With Positive Nonlinear Resistors," J. Research Natl. Bur. Standards, 56, 179-182 (April, 1956).
10. Penfield, P., Jr., Frequency-Power Formulas, The Technology Press and John Wiley & Sons, Inc., New York (1960).
11. Penfield, P., Jr., ibid., Chapter 5.
12. Scarlett, R. M., "Space-Charge Layer Width in Diffused Junctions," IRE Trans. on Electron Devices, ED-6, 405-408 (October, 1959).
13. Shockley, W., "The Theory of p-n Junctions in Semiconductors and p-n Junction Transistors," Bell System Tech. J., 28, 435-489 (July, 1949).
14. Shockley, W., Electrons and Holes in Semiconductors, D. Van Nostrand Co., Inc., New York (1950).
15. Shockley, W., ibid., Section 12.3.
16. Spenke, E., Electronic Semiconductors, McGraw-Hill Book Co., Inc., New York (1958).
17. Spenke, E., ibid., p. 108.

Appendix B

VARACTOR MANUFACTURERS

During the spring of 1962 each of the companies listed here indicated that they manufacture varactors commercially and were able to supply specification sheets describing their products. Besides these companies, many other firms have announced publicly (for example, through press releases, advertisements, directory listings, or convention displays) that they either are making varactors or have them under development, but such firms are not listed here unless they could substantiate their claim with printed specification sheets.

American Electronic Laboratories, Inc.
Richardson Road
Colmar, Pa.

Amperex Electronic Corporation
230 Duffy Avenue
Hicksville, L. I., N. Y.

The Bendix Corporation
Bendix Semiconductor Division
Holmdel, N. J.

Bomac Laboratories, Inc.
Salem Road
Beverly, Mass.

Hughes Aircraft Company
Semiconductor Division
Newport Beach, Calif.

Lignes Télégraphiques et Téléphoniques
89, Rue de la Faisanderie
Paris 16e, France

Micro State Electronics Corporation
152 Floral Avenue
Murray Hill, N. J.

Microwave Associates, Inc.
Burlington, Mass.

Northern Electric Co., Ltd.
Research and Development Laboratories
P. O. Box 3511, Station C
Ottawa, Ont., Canada

Pacific Semiconductors, Inc.
14520 Aviation Boulevard
Lawndale, Calif.

Philco Corporation
Lansdale Division
Church Road, Lansdale, Pa.

Raytheon Company
Semiconductor Division
900 Chelmsford Street
Lowell, Mass.

Semiconductor Devices, Inc.
875 West 15th Street
Newport Beach, Calif.

Sylvania Electric Products, Inc.
Semiconductor Division
100 Sylvan Road, Woburn, Mass.

Texas Instruments, Inc.
Components Division
P. O. Box 5012, Dallas 22, Tex.

Tokyo Shibaura Electric Co., Ltd.
1 Komukai, Kawasaki-shi, Japan

Tyco Semiconductor Corporation
Bear Hill, Waltham 54, Mass.

Appendix C

NEGATIVE-RESISTANCE AMPLIFIER CIRCUITS

Here we analyze in quite abstract terms the process of making a two-port amplifier from a one-port negative resistance. The discussion applies to any type of negative resistance, including tunnel diodes as well as nondegenerate varactor parametric amplifiers. In the latter case, we assume that we have a negative resistance to work with, or that the pump and idler circuits are already established, as for example by the techniques of Section 6.7.

The analytical technique applies to distributed as well as lumped circuitry, although for simplicity we discuss only lumped examples. Basically, the approach is to imbed the negative resistance in a three-port imbedding network and to use the remaining two ports as input and output. Obviously, the characteristics of the over-all amplifier depend both on the negative resistance and on the imbedding network. In Section C-1 we derive general expressions for input and output impedance, exchangeable gain, and noise temperature (and measure). Then in Section C-2 we assume this amplifier is followed by a noisy second stage. The formula for cascade noise temperature shows the relative contributions of the various noise mechanisms. Finally, in Section C-3 are a number of examples.

C-1. The Negative-Resistance Amplifier

A negative resistance is a one-port device; an amplifier has two ports, an input and an output. To make a negative-resistance amplifier, connect the negative resistance to one port of a three-port imbedding network, and use the remaining two ports as input and output. This imbedding is shown in Figure C-1.

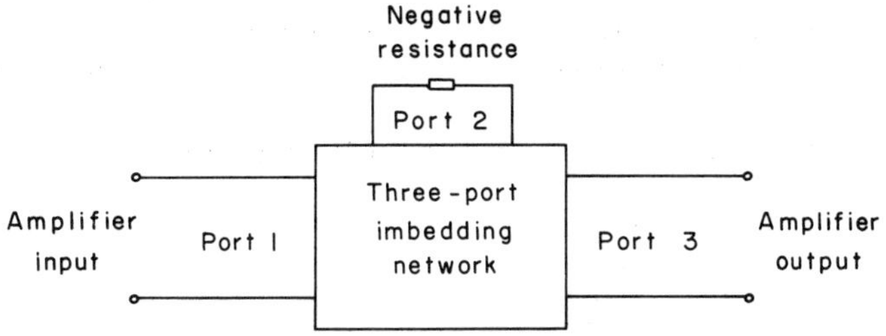

Figure C-1. Imbedding a negative resistance in a three-port network to obtain a two-port amplifier.

C-1.1. The Imbedding Network

Instead of describing the imbedding network by its impedance matrix or its admittance matrix, we prefer to use scattering variables,[2, 12‡] because they apply to all passive linear networks, including distributed systems such as microwave systems using transmission lines and waveguides. In terms of half-amplitude currents I_k (k = 1, 2, 3), voltages V_k, and normalization impedances Z_k (with positive real parts), we define incoming wave amplitudes

$$a_k = \frac{V_k + Z_k I_k}{\sqrt{Z_k + Z_k^*}} \tag{C-1}$$

and outgoing wave amplitudes

$$b_k = \frac{V_k - Z_k^* I_k}{\sqrt{Z_k + Z_k^*}}. \tag{C-2}$$

Then the linear threeport is described by a three-by-three scattering matrix $\underset{\sim}{S}$ and a set of equivalent wave generators b_{s1}, b_{s2} and b_{s3}:

$$\underset{\sim}{b} = \underset{\sim}{S}\underset{\sim}{a} + \underset{\sim}{b}_s, \tag{C-3}$$

where $\underset{\sim}{b}$, $\underset{\sim}{a}$, and $\underset{\sim}{b}_s$ are column matrices of, respectively, the b_k, a_k, and b_{sk} arranged in order. The b_{sk} sources account for signals or noise generated within the threeport; in our case we assume they represent thermal noise of temperature T_n generated inside the imbedding network.

The power into the network at each port is, from Equations C-1 and C-2,

$$2 \text{ Re } V_k I_k^* = |a_k|^2 - |b_k|^2, \tag{C-4}$$

where, if a_k and b_k represent random signals, a statistical averaging should be performed. Thus, the power into the entire threeport is (if noise generated within the network is neglected)

$$\begin{aligned} P &= |a_1|^2 + |a_2|^2 + |a_3|^2 - |b_1|^2 - |b_2|^2 - |b_3|^2 \\ &= \underset{\sim}{a}^\dagger \underset{\sim}{a} - \underset{\sim}{b}^\dagger \underset{\sim}{b} \\ &= \underset{\sim}{a}^\dagger (\underset{\sim}{1} - \underset{\sim}{S}^\dagger \underset{\sim}{S}) \underset{\sim}{a}, \end{aligned} \tag{C-5}$$

‡Superscript numerals denote references listed at the end of each appendix.

where the dagger represents the complex conjugate transpose. If the three-port imbedding network is lossless, then Equation C-5 must vanish for all choice of $\underset{\sim}{a}$. Therefore, $\underset{\sim}{S}$ is a unitary matrix, so that

$$0 = \underset{\sim}{1} - \underset{\sim}{S}^{\dagger}\underset{\sim}{S}$$

$$= \underset{\sim}{1} - \underset{\sim}{S}\underset{\sim}{S}^{\dagger} . \qquad \text{(C-6)}$$

On the other hand, if the network is passive, the quantity in Equation C-5 is never negative for any choice of $\underset{\sim}{a}$, so that the matrices $\underset{\sim}{1} - \underset{\sim}{S}^{\dagger}\underset{\sim}{S}$ and $\underset{\sim}{1} - \underset{\sim}{S}\underset{\sim}{S}^{\dagger}$ are both positive semidefinite matrices.[8]

We assume the threeport is in thermal equilibrium at temperature T_n, so that the $\underset{\sim}{b}_s$ matrix represents thermal noise. If the network is lossless, no thermal noise is generated; if the network is lossy, there is some noise generated, and in fact the loss is related to the equivalent wave generators by the formula[6]

$$\overline{\underset{\sim}{b}_s \underset{\sim}{b}_s^{\dagger}} = kT_n(\underset{\sim}{1} - \underset{\sim}{S}\underset{\sim}{S}^{\dagger})\,\Delta f , \qquad \text{(C-7)}$$

where the horizontal bar represents an average. Thus, if the network is lossless, $\underset{\sim}{b}_s = 0$; otherwise, not.

Note that we have not yet specified the normalization impedances Z_1, Z_2, and Z_3, except that they have positive real parts. The preceding formulas hold for any selection of Z_k, although in general when any Z_k is changed, all elements of $\underset{\sim}{S}$ and $\underset{\sim}{b}_s$ are changed.

C-1.2. The Terminations

We now turn to a description of the terminations of the three-port imbedding network.

At the input, we assume the termination includes an impedance and a signal or noise generator. In scattering variables, this termination has the equation

$$a_1 = \Gamma_1 b_1 + a_{s1} , \qquad \text{(C-8)}$$

where Γ_1, the input reflection coefficient, is less than 1 in magnitude, because the termination is assumed to be passive.‡ The available power from this termination is

$$\frac{|a_{s1}|^2}{1 - |\Gamma_1|^2} , \qquad \text{(C-9)}$$

where, for random signals or noise, an averaging must be performed.

‡If the input termination is not passive, the formulas still hold, with Expression C-9 interpreted as the exchangeable power.[3, 4, 5]

If the input is thermal noise at temperature T_0, then Expression C-9 is equal to $kT_0 \Delta f$.

The output of the amplifier is, we assume, connected to a load, or to a second stage, or to a meter of some sort; in any event there is some reflection coefficient Γ_3 and perhaps some source a_{s3}.

The termination at port 2 is the negative resistance, with reflection coefficient Γ_2 and noise source a_{s2}. Because this terminating impedance is negative,

$$|\Gamma_2| > 1. \tag{C-10}$$

The noise exchangeable power

$$P_e = \frac{\overline{|a_{s2}|^2}}{1 - |\Gamma_2|^2} \tag{C-11}$$

defines the temperature T (see Section 2.5.5)

$$\overline{|a_{s2}|^2} = (|\Gamma_2|^2 - 1)kT\,\Delta f. \tag{C-12}$$

This temperature T is what in Chapter 6 we called the noise temperature of the nondegenerate parametric amplifier. The only characteristics of the negative resistance that are needed now are Γ_2 and T.

In matrix notation, the three equations like Equation C-8 can be put into the form,

$$\underset{\sim}{a} = \underset{\sim}{\Gamma}\underset{\sim}{b} + \underset{\sim}{a}_s, \tag{C-13}$$

where

$$\underset{\sim}{\Gamma} = \begin{bmatrix} \Gamma_1 & 0 & 0 \\ 0 & \Gamma_2 & 0 \\ 0 & 0 & \Gamma_3 \end{bmatrix} \tag{C-14}$$

and $\underset{\sim}{a}_s$ is the column matrix consisting of the a_{sk} arranged in order. We find, from Equations C-3 and C-13,

$$\underset{\sim}{b} = (\underset{\sim}{1} - \underset{\sim}{S}\underset{\sim}{\Gamma})^{-1}\underset{\sim}{b}_s + \underset{\sim}{\Sigma}\underset{\sim}{a}_s, \tag{C-15}$$

where

$$\underset{\sim}{\Sigma} = (\underset{\sim}{1} - \underset{\sim}{S}\underset{\sim}{\Gamma})^{-1}\underset{\sim}{S} \tag{C-16}$$

is a convenient abbreviation.

We now are in a position to calculate the properties of the amplifier in terms of $\underset{\sim}{S}$, $\underset{\sim}{\Gamma}$, T, and T_n.

C-1.3. Input Impedance

To be consistent with our use of scattering variables, we calculate not the input impedance but instead the input reflection coefficient Γ_{in} of the threeport as terminated at ports 2 and 3. We neglect all noise and signals except a_{s1}. Then, from Equations C-15,

$$b_1 = \Sigma_{11} a_{s1} \tag{C-17}$$

and from Equations C-13,

$$a_1 = \Gamma_1 b_1 + a_{s1}, \tag{C-18}$$

so that

$$b_1 = \frac{\Sigma_{11}}{\Sigma_{11}\Gamma_1 + 1} a_1. \tag{C-19}$$

The reflection coefficient at the input is therefore

$$\Gamma_{in} = \frac{\Sigma_{11}}{\Sigma_{11}\Gamma_1 + 1}. \tag{C-20}$$

Contrary to its appearance, Γ_{in} is independent of Γ_1 but does depend on Γ_2 and Γ_3. An alternate form for Equation C-20 is derived from the 1,1 element of the matrix equation (which comes from Equations C-16)

$$\underset{\sim}{\Sigma}\underset{\sim}{\Gamma} + \underset{\sim}{1} = (\underset{\sim}{1} - \underset{\sim}{S}\underset{\sim}{\Gamma})^{-1}. \tag{C-21}$$

Thus,

$$\Gamma_{in} = \frac{\Sigma_{11}}{[(\underset{\sim}{1} - \underset{\sim}{S}\underset{\sim}{\Gamma})^{-1}]_{11}}$$

$$= \frac{\Sigma_{11}}{(\underset{\sim}{\Sigma}\underset{\sim}{S}^{-1})_{11}}. \tag{C-22}$$

C-1.4. Output Reflection Coefficient

In a similar way, the output reflection coefficient Γ_{out} can be calculated to be

$$\Gamma_{out} = \frac{\Sigma_{33}}{\Sigma_{33}\Gamma_3 + 1}$$

$$= \frac{\Sigma_{33}}{[(\underset{\sim}{1} - \underset{\sim}{S}\underset{\sim}{\Gamma})^{-1}]_{33}}$$

$$= \frac{\Sigma_{33}}{(\underset{\sim}{\Sigma}\underset{\sim}{S}^{-1})_{33}} . \qquad \text{(C-23)}$$

The output reflection coefficient is independent of Γ_3, but it does depend on Γ_1 and Γ_2. When a second stage is placed after the first, the source reflection coefficient it sees is simply Γ_{out}.

C-1.5. Exchangeable Gain

To calculate the exchangeable gain, we ignore all noise and signals except a_{s1}. The input exchangeable power $P_{e,in}$ is given by Expression C-9. The output exchangeable power is given in terms of b_3 as

$$P_{e,out} = \frac{|b_3|^2}{1 - |\Gamma_{out}|^2} |1 - \Gamma_3\Gamma_{out}|^2 . \qquad \text{(C-24)}$$

Thus, since $b_3 = \Sigma_{31}a_{s1}$, the exchangeable gain G_e is

$$G_e = \frac{P_{e,out}}{P_{e,in}}$$

$$= |\Sigma_{31}|^2 |1 - \Gamma_3\Gamma_{out}|^2 \frac{1 - |\Gamma_1|^2}{1 - |\Gamma_{out}|^2}$$

$$= \frac{|\Sigma_{31}|^2}{|\Sigma_{33}|^2} \frac{1 - |\Gamma_1|^2}{1 - |\Gamma_{out}|^2} |\Gamma_{out}|^2 . \qquad \text{(C-25)}$$

Note that G_e is independent of Γ_3, although it does depend on Γ_1 and Γ_2.

We now derive an alternate expression for G_e. From Equations C-16 and C-21, we can easily derive the formula

$$(\underset{\sim}{1} - \underset{\sim}{S}\underset{\sim}{\Gamma})^{-1}(\underset{\sim}{1} - \underset{\sim}{S}\underset{\sim}{S}^{\dagger})(\underset{\sim}{1} - \underset{\sim}{\Gamma}^{\dagger}\underset{\sim}{S}^{\dagger})^{-1} = \underset{\sim}{1} + \underset{\sim}{\Sigma}\underset{\sim}{\Gamma} + \underset{\sim}{\Gamma}^{\dagger}\underset{\sim}{\Sigma}^{\dagger} - \underset{\sim}{\Sigma}(\underset{\sim}{1} - \underset{\sim}{\Gamma}\,\underset{\sim}{\Gamma}^{\dagger})\underset{\sim}{\Sigma}^{\dagger}, \qquad \text{(C-26)}$$

of which the 3,3 element is

$$[(\underset{\sim}{1} - \underset{\approx}{S}\underset{\approx}{\Gamma})^{-1}(\underset{\sim}{1} - \underset{\approx}{S}\underset{\approx}{S}\dagger)(\underset{\sim}{1} - \underset{\approx}{\Gamma}\dagger\underset{\approx}{S}\dagger)^{-1}]_{33}$$

$$= 1 + \Sigma_{33}\Gamma_3 + \Sigma_{33}^*\Gamma_3^* - |\Sigma_{31}|^2(1 - |\Gamma_1|^2)$$

$$+ |\Sigma_{32}|^2(|\Gamma_2|^2 - 1) - |\Sigma_{33}|^2(1 - |\Gamma_3|^2). \qquad \text{(C-27)}$$

From Equation C-23, we find

$$\frac{1 - |\Gamma_{out}|^2}{|\Gamma_{out}|^2} = \frac{1 + \Sigma_{33}\Gamma_3 + \Sigma_{33}^*\Gamma_3^* - |\Sigma_{33}|^2(1 - |\Gamma_3|^2)}{|\Sigma_{33}|^2}. \qquad \text{(C-28)}$$

Use of Equations C-27 and C-28 in Equation C-25 yields

$$G_e = \frac{|\Sigma_{31}|^2(1 - |\Gamma_1|^2)}{1 + \Sigma_{33}\Gamma_3 + \Sigma_{33}^*\Gamma_3^* - |\Sigma_{33}|^2(1 - |\Gamma_3|^2)}$$

$$= \frac{|\Sigma_{31}|^2(1 - |\Gamma_1|^2)}{[(\underset{\sim}{1} - \underset{\approx}{S}\underset{\approx}{\Gamma})^{-1}(\underset{\sim}{1} - \underset{\approx}{S}\underset{\approx}{S}\dagger)(\underset{\sim}{1} - \underset{\approx}{\Gamma}\dagger\underset{\approx}{S}\dagger)^{-1}]_{33} + |\Sigma_{31}|^2(1 - |\Gamma_1|^2) - |\Sigma_{32}|^2(|\Gamma_2|^2 - 1)}.$$

(C-29)

This form, apparently more complicated, is given because if the imbedding network is lossless, the matrix element involving $\underset{\sim}{1} - \underset{\approx}{S}\underset{\approx}{S}\dagger$ vanishes.

C-1.6. Noise Figure

In calculating the noise figure of the amplifier, we consider $\underset{\sim}{b}_s$ and a_{s2} as internal noise sources and consider a_{s1} as input noise, and set a_{s3} to zero.‡ The over-all exchangeable power output is

$$\overline{|b_3|^2} \; \frac{|1 - \Gamma_3\Gamma_{out}|^2}{1 - |\Gamma_{out}|^2}, \qquad \text{(C-30)}$$

which is independent of Γ_3. From Equations C-15,

$$b_3 = [(\underset{\sim}{1} - \underset{\approx}{S}\underset{\approx}{\Gamma})^{-1}\underset{\sim}{b}_s]_3 + \Sigma_{31}a_{s1} + \Sigma_{32}a_{s2}, \qquad \text{(C-31)}$$

‡Noise from the output load does not, of course, contribute toward the single-stage noise temperature. See Section 2.5.7.

so that if the noise sources are uncorrelated,

$$\overline{|b_3|^2} = [(\underset{\sim}{1} - \underset{\sim}{S}\underset{\sim}{\Gamma})^{-1}\overline{\underset{\sim}{b}_s\underset{\sim}{b}_s{}^{\dagger}}(\underset{\sim}{1} - \underset{\sim}{\Gamma}^{\dagger}\underset{\sim}{S}^{\dagger})^{-1}]_{33}$$

$$+ |\Sigma_{32}|^2\overline{|a_{s2}|^2} + |\Sigma_{31}|^2\overline{|a_{s1}|^2}. \qquad \text{(C-32)}$$

The last term in Equation C-32 is caused by input noise at temperature T_0; the preceding two are internally generated. The excess noise figure is the ratio of noise power caused by internal sources to noise power caused by input thermal noise at temperature T_0, or simply

$$F_1 - 1 = \frac{T|\Sigma_{32}|^2(|\Gamma_2|^2 - 1) + T_n[(\underset{\sim}{1} - \underset{\sim}{S}\underset{\sim}{\Gamma})^{-1}(\underset{\sim}{1} - \underset{\sim}{S}\underset{\sim}{S}^{\dagger})(\underset{\sim}{1} - \underset{\sim}{\Gamma}^{\dagger}\underset{\sim}{S}^{\dagger})^{-1}]_{33}}{T_0|\Sigma_{31}|^2(1 - |\Gamma_1|^2)}, \qquad \text{(C-33)}$$

where we have used Equations C-7, Expression C-9, and Equation C-12. The noise temperature T_1 of this amplifier (the subscript 1 indicates the first stage) is simply T_0 times the expression in Equation C-33. Note that if the network is lossless, the T_n term disappears; then, for high exchangeable gain the noise temperature T_1 is approximately T.

C-1.7. Noise Measure

We may relate the noise figure to the gain in a meaningful way by using Haus and Adler's noise measure[4,5]

$$M_1 = \frac{F_1 - 1}{1 - \dfrac{1}{G_e}}. \qquad \text{(C-34)}$$

From Equations C-29 and C-33, we find

$$M_1 = \frac{T|\Sigma_{32}|^2(|\Gamma_2|^2 - 1) + T_n[(\underset{\sim}{1} - \underset{\sim}{S}\underset{\sim}{\Gamma})^{-1}(\underset{\sim}{1} - \underset{\sim}{S}\underset{\sim}{S}^{\dagger})(\underset{\sim}{1} - \underset{\sim}{\Gamma}^{\dagger}\underset{\sim}{S}^{\dagger})^{-1}]_{33}}{T_0\{|\Sigma_{32}|^2(|\Gamma_2|^2 - 1) - [(\underset{\sim}{1} - \underset{\sim}{S}\underset{\sim}{\Gamma})^{-1}(\underset{\sim}{1} - \underset{\sim}{S}\underset{\sim}{S}^{\dagger})(\underset{\sim}{1} - \underset{\sim}{\Gamma}^{\dagger}\underset{\sim}{S}^{\dagger})^{-1}]_{33}\}}. \qquad \text{(C-35)}$$

If the three-port imbedding network is lossless, then Equation C-35 shows[9] that $M_1 = T/T_0$; if the network is lossy, then M_1, if positive, is no less than T/T_0. On the other hand, if the output is decoupled from the negative resistance, the noise measure is $-T_n/T_0$, as for any passive twoport in thermal equilibrium at temperature T_n.

C-2. Coupling to a Second Stage

We have given in Section C-1 general formulas for the negative-resistance amplifier. In practice, most negative-resistance amplifiers are designed for use as preamplifiers, to be followed by a noisy second stage. It is important to know how to couple to the second stage for lowest noise, because the negative-resistance amplifier is usually used to improve the noise performance of the over-all amplifier.

In Section C-2.1 we discuss the noise properties of the second stage, and in Section C-2.2 we calculate the cascade noise temperature T_{12}.

C-2.1. Wave Representation of Second-Stage Noise

There are several schemes for representing noise in a linear twoport (or amplifier); all use two equivalent noise sources that are in general correlated. The most convenient for our purposes is the Bauer-Rothe[1, 10] wave model, which uses two uncorrelated generators.

This model is described in Section 2.5.1(e). We describe the second stage by scattering variables and a two-by-two scattering matrix $\underset{\sim}{S}_2$, using normalization impedances that are as yet unspecified. The two equivalent noise wave generators are placed at the input, as shown schematically in Figure C-2. The equations of the over-all amplifier are, in matrix form,

$$\begin{bmatrix} b_1 - b_n \\ \\ b_2 \end{bmatrix} = \underset{\sim}{S}_2 \begin{bmatrix} a_1 + a_n \\ \\ a_2 \end{bmatrix} . \qquad \text{(C-36)}$$

We have not yet specified the normalization impedances. As Z_1 is varied, all the elements of $\underset{\sim}{S}_2$ vary, and b_n and a_n also vary. One value of Z_1 can generally be chosen to make a_n and b_n uncorrelated, that is, to make

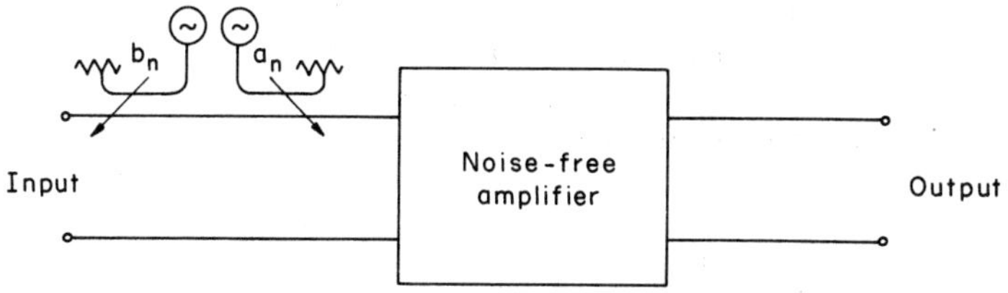

Figure C-2. Bauer-Rothe wave model of a noisy amplifier has a noise-free part and two uncorrelated wave generators at the input. This schematic diagram is accurate in the limit of small coupling coefficients between the "line" and the sources.

$$\overline{a_n b_n^*} = 0. \tag{C-37}$$

The simplicity of the expression for noise temperature, Equation C-40, arises from the fact that the two wave generators are uncorrelated.

The strength of a_n defines the temperature T_a:

$$T_a = \frac{\overline{|a_n|^2}}{k\,\Delta f}, \tag{C-38}$$

and the strength of b_n defines a temperature T_b:

$$T_b = \frac{\overline{|b_n|^2}}{k\,\Delta f}. \tag{C-39}$$

In terms of T_a and T_b, the noise temperature of the second stage is

$$T_2 = \frac{T_a + T_b|\Gamma_s|^2}{1 - |\Gamma_s|^2}, \tag{C-40}$$

where Γ_s is the source reflection coefficient calculated, of course, in terms of the normalization impedance that makes a_n and b_n uncorrelated. When the second stage described here is connected to the first stage described in Section C-1, Γ_s is simply the output reflection coefficient Γ_{out}, provided the same normalization impedance is used.

C-2.2. Cascade Noise Temperature

In practice the second stage often has high gain but poor noise performance by itself. We now calculate the cascade noise temperature of the combination, assuming that port 3 of the imbedding network in the first stage has a normalization impedance equal to the impedance used to make a_n and b_n uncorrelated.

The cascade noise temperature T_{12} is

$$T_{12} = T_1 + \frac{T_2}{G_e}, \tag{C-41}$$

where G_e is the exchangeable gain of the first stage. Thus, from Equations C-25, C-23, C-33, and C-40, we find

$$T_{12} = \frac{T|\Sigma_{32}|^2(|\Gamma_2|^2 - 1) + T_n[(\underset{\sim}{1} - \underset{\sim}{S}\underset{\sim}{\Gamma})^{-1}(\underset{\sim}{1} - \underset{\sim}{S}\underset{\sim}{S}^\dagger)(\underset{\sim}{1} - \underset{\sim}{\Gamma}^\dagger\underset{\sim}{S}^\dagger)^{-1}]_{33}}{|\Sigma_{31}|^2(1 - |\Gamma_1|^2)}$$

$$+ \frac{T_a|1 + \Sigma_{33}\Gamma_3|^2 + T_b|\Sigma_{33}|^2}{|\Sigma_{31}|^2(1 - |\Gamma_1|^2)} \qquad \text{(C-42)}$$

or, in view of Equation C-27,

$$T_{12} = \frac{a_1 T + a_2 T_n + a_3 T_a + a_4 T_b}{a_1 - a_2 + a_3 - a_4}, \qquad \text{(C-43)}$$

where a_1, a_2, a_3, and a_4 are, respectively, the factors multiplying T, T_n, T_a, and T_b in the numerator of Equation C-42. Equation C-43 shows that for best noise performance a_2 and a_4 should be set equal to zero (for example, by having the three-port imbedding network lossless, and by adjusting it so that $\Gamma_{out} = 0$). Then, since T is generally smaller than T_a, a_3 should be small for lowest T_{12}, as for example if G_e is high. Then T_{12} can be as close as desired to T. Therefore, even with a noisy second stage, the cascade noise temperature can be made to approach the noise temperature of the negative resistance.

C-3. Examples

The preceding theory is very general, and is only well understood through examples. We now discuss various types of imbedding networks and the resulting amplifiers, and we calculate the noise temperatures of the amplifiers followed by noisy second stages. The examples are the three-port circulator, the four-port circulator, the simple series connection, and the series connection with transformers.

C-3.1. Three-Port Circulator

A three-port circulator can be used as the three-port imbedding network, as shown in Figure C-3. The circulator has a scattering matrix

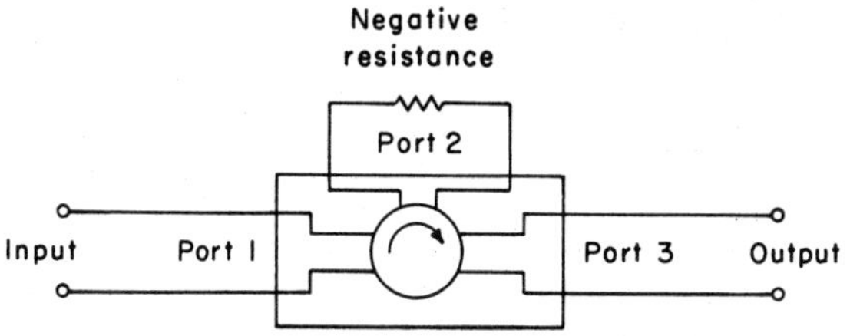

Figure C-3. Three-port circulator amplifier.

$$\underset{\sim}{S} = \begin{bmatrix} 0 & 0 & 1 \\ 1 & 0 & 0 \\ 0 & 1 & 0 \end{bmatrix} \qquad \text{(C-44)}$$

and is lossless, so that $\underset{\sim}{1} - \underset{\sim\sim}{SS\dagger} = 0$. The matrix $\underset{\sim}{\Sigma}$ is easily found to be

$$\underset{\sim}{\Sigma} = \frac{1}{1 - \Gamma_1\Gamma_2\Gamma_3} \begin{bmatrix} \Gamma_2\Gamma_3 & \Gamma_3 & 1 \\ 1 & \Gamma_1\Gamma_3 & \Gamma_1 \\ \Gamma_2 & 1 & \Gamma_1\Gamma_2 \end{bmatrix}. \qquad \text{(C-45)}$$

The input reflection coefficient Γ_{in} is clearly equal to

$$\Gamma_{in} = \Gamma_2\Gamma_3, \qquad \text{(C-46)}$$

and similarly, the output reflection coefficient is

$$\Gamma_{out} = \Gamma_1\Gamma_2. \qquad \text{(C-47)}$$

The exchangeable gain is

$$G_e = \frac{|\Gamma_2|^2(1 - |\Gamma_1|^2)}{1 - |\Gamma_1\Gamma_2|^2}, \qquad \text{(C-48)}$$

and the noise measure is

$$M_1 = \frac{T}{T_0} \qquad \text{(C-49)}$$

because the imbedding network is lossless.

Now we suppose that the second stage is connected to the output of the negative-resistance amplifier through a matching network to transform the characteristic impedance of the circulator into the impedance necessary to make a_n and b_n of the second stage uncorrelated. Then, from Equation C-42,

$$T_{12} = \frac{T(|\Gamma_2|^2 - 1) + T_a + T_b|\Gamma_1\Gamma_2|^2}{|\Gamma_2|^2(1 - |\Gamma_1|^2)}. \qquad \text{(C-50)}$$

This cascade noise temperature is obviously its smallest value when the source impedance is set so that $\Gamma_1 = 0$; then

$$(T_{12})_{min} = \frac{T(|\Gamma_2|^2 - 1) + T_a}{|\Gamma_2|^2}, \tag{C-51}$$

and as Γ_2 increases in magnitude, T_{12} approaches T. Thus the cascade noise temperature can be made as close to T as desired.‡

C-3.2. Four-Port Circulator

A four-port circulator can be used as the three-port imbedding network if the fourth port is terminated in some way, as shown in Figure C-4. Now we assume that the fourth port is terminated in

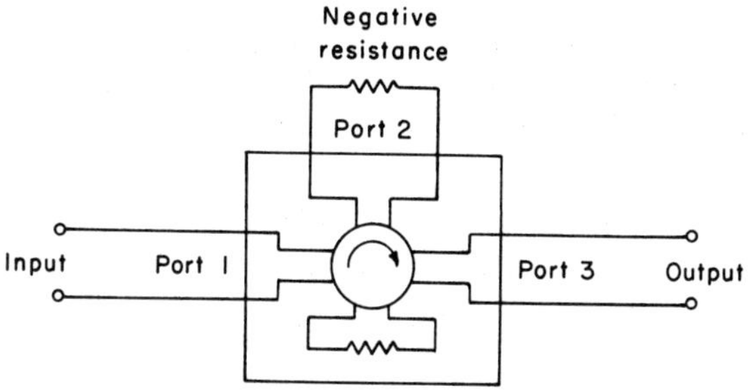

Figure C-4. Four-port circulator amplifier. In Section C-3.2 we assume the fourth circulator port is matched; in Section C-3.3 we do not.

a matched load, and in Section C-3.3 we assume the fourth port is not matched.

With the fourth port matched, the scattering matrix of the three-port is

$$\underset{\sim}{S} = \begin{bmatrix} 0 & 0 & 0 \\ 1 & 0 & 0 \\ 0 & 1 & 0 \end{bmatrix}, \tag{C-52}$$

and the matrix $\underset{\sim}{\Sigma}$ is easily calculated to be

$$\underset{\sim}{\Sigma} = \begin{bmatrix} 0 & 0 & 0 \\ 1 & 0 & 0 \\ \Gamma_2 & 1 & 0 \end{bmatrix}, \tag{C-53}$$

‡We have not mentioned any of the undesirable effects of such an adjustment. These include low bandwidth, incipient instability, and poor control of amplifier gain.

so that

$$\Gamma_{in} = \Gamma_{out} = 0, \tag{C-54}$$

$$G_e = |\Gamma_2|^2(1 - |\Gamma_1|^2), \tag{C-55}$$

and

$$T_1 = \frac{T(|\Gamma_2|^2 - 1) + T_n|\Gamma_1\Gamma_2|^2}{|\Gamma_2|^2(1 - |\Gamma_1|^2)}, \tag{C-56}$$

where T_n is the temperature of the matched circulator termination. The noise temperature is a minimum when the input is matched, that is, when $\Gamma_1 = 0$. The noise measure M_1 is

$$M_1 = \frac{T(|\Gamma_2|^2 - 1) + T_n|\Gamma_1\Gamma_2|^2}{T_0(|\Gamma_2|^2 - 1 - |\Gamma_1\Gamma_2|^2)}, \tag{C-57}$$

which attains its minimum value of T/T_0 when $\Gamma_1 = 0$.

If this amplifier is connected to a noisy second stage through a matching network to transform the characteristic impedance of the circulator to the normalization impedance for making the second stage a_n and b_n uncorrelated, then the cascade noise temperature is

$$T_{12} = \frac{T(|\Gamma_2|^2 - 1) + T_n|\Gamma_1\Gamma_2|^2 + T_a}{|\Gamma_2|^2(1 - |\Gamma_1|^2)}. \tag{C-58}$$

When the source is matched, that is, $\Gamma_1 = 0$, this noise temperature is the minimum

$$(T_{12})_{min} = \frac{T(|\Gamma_2|^2 - 1) + T_a}{|\Gamma_2|^2}, \tag{C-59}$$

which approaches T as $|\Gamma_2|$ becomes large.‡

‡We have not mentioned any of the undesirable effects of such an adjustment. These include low bandwidth, incipient instability, and poor control of amplifier gain.

C-3.3. Unmatched Four-Port Circulator

Suppose the three-port imbedding network consists of a four-port circulator with the fourth port terminated in a load with reflection coefficient Γ_4. The circuit is still that of Figure C-4, except that the extra port of the circulator is no longer matched. The three-by-three scattering matrix then has the form

$$\underset{\sim}{S} = \begin{bmatrix} 0 & 0 & \Gamma_4 \\ 1 & 0 & 0 \\ 0 & 1 & 0 \end{bmatrix}, \tag{C-60}$$

so that the matrix $\underset{\sim}{\Sigma}$ is

$$\underset{\sim}{\Sigma} = \frac{1}{1 - \Gamma_1\Gamma_2\Gamma_3\Gamma_4} \begin{bmatrix} \Gamma_2\Gamma_3\Gamma_4 & \Gamma_3\Gamma_4 & \Gamma_4 \\ 1 & \Gamma_1\Gamma_3\Gamma_4 & \Gamma_1\Gamma_4 \\ \Gamma_2 & 1 & \Gamma_1\Gamma_2\Gamma_4 \end{bmatrix}. \tag{C-61}$$

The amplifier parameters are

$$\Gamma_{in} = \Gamma_2\Gamma_3\Gamma_4 , \tag{C-62}$$

$$\Gamma_{out} = \Gamma_1\Gamma_2\Gamma_4 , \tag{C-63}$$

$$G_e = \frac{|\Gamma_2|^2(1 - |\Gamma_1|^2)}{1 - |\Gamma_1\Gamma_2\Gamma_4|^2} , \tag{C-64}$$

and

$$T_1 = \frac{T(|\Gamma_2|^2 - 1) + T_n|\Gamma_1\Gamma_2|^2(1 - |\Gamma_4|^2)}{|\Gamma_2|^2(1 - |\Gamma_1|^2)}, \tag{C-65}$$

where T_n is the temperature of the termination Γ_4. Note that the noise temperature is a minimum when the input is matched. The noise measure M_1 is

$$M_1 = \frac{T(|\Gamma_2|^2 - 1) + T_n|\Gamma_1\Gamma_2|^2(1 - |\Gamma_4|^2)}{T_0(|\Gamma_1\Gamma_2\Gamma_4|^2 - |\Gamma_1\Gamma_2|^2 + |\Gamma_2|^2 - 1)} . \tag{C-66}$$

The noise measure is minimized to the value T/T_0 by matching the source impedance.

If this amplifier is connected to a noisy second stage through an appropriate coupling network, the cascade noise temperature T_{12} is

$$T_{12} = \frac{T(|\Gamma_2|^2 - 1) + T_n|\Gamma_1\Gamma_2|^2(1 - |\Gamma_4|^2) + T_a + T_b|\Gamma_1\Gamma_2\Gamma_4|^2}{|\Gamma_2|^2(1 - |\Gamma_1|^2)} . \quad \text{(C-67)}$$

This cascade noise temperature reduces to its minimum value, equal to Equation C-59, when the source is matched.

Note that the formulas in this section reduce to those in Section C-3.1 for a three-port circulator when $|\Gamma_4| = 1$, and reduce to those in Section C-3.2 for a four-port circulator when $\Gamma_4 = 0$.

C-3.4. Series Connection

Consider now the series connection shown in Figure C-5. This simple circuit is often used in crude calculations, although there

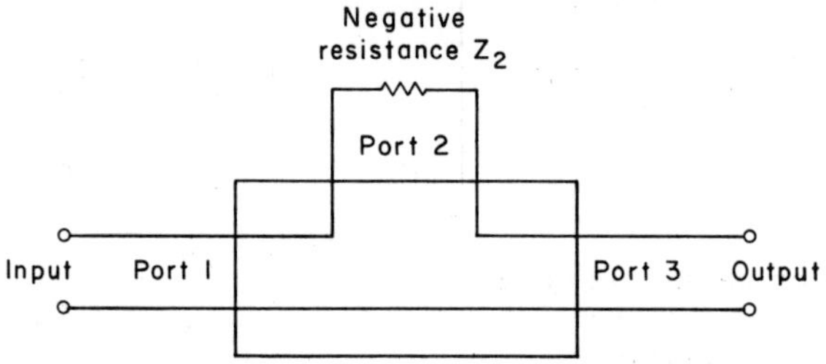

Figure C-5. Simple series connection. The impedance Z_2 has a real part R_2 that is negative.

is little evidence that any practical amplifiers are well described by this circuit. The simple parallel connection is a dual of this circuit, and will not be discussed separately.

This circuit is so simple that the elegant scattering-matrix technique is not necessary; instead we use impedance concepts. The input impedance Z_{in} is obviously

$$Z_{in} = Z_2 + Z_3, \quad \text{(C-68)}$$

where Z_2 is the impedance of the negative resistance, and therefore has a real part R_2 that is negative, and where Z_3 is the load impedance. Similarly, the output impedance is

$$Z_{out} = Z_1 + Z_2, \quad \text{(C-69)}$$

where Z_1 is the source impedance, with real part R_1. The exchangeable gain is

$$G_e = \frac{R_1}{R_1 + R_2}, \quad \text{(C-70)}$$

and the noise measure M_1 is

$$M_1 = \frac{T}{T_0}. \tag{C-71}$$

The noise temperature can be calculated from Equation C-34 to be

$$T_1 = T\frac{-R_2}{R_1}. \tag{C-72}$$

If this amplifier is connected to a noisy second stage, the cascade noise figure can be calculated easily. Let Z_ν (with real part R_ν) be the normalization impedance for which the noise wave representation in Section C-2.1 has uncorrelated generators. Then the reflection coefficient seen by the second stage is

$$\Gamma_{out} = \frac{Z_{out} - Z_\nu}{Z_{out} + Z_\nu^*}; \tag{C-73}$$

therefore, Equations C-40 and C-41 predict

$$T_{12} = T\frac{-R_2}{R_1} + \frac{T_a|Z_2 + Z_1 + Z_\nu^*|^2 + T_b|Z_2 + Z_1 - Z_\nu|^2}{4R_1R_\nu}. \tag{C-74}$$

This cascade noise temperature can be made to approach T if R_1 is set approximately equal to $-R_2$, and then both become large‡ compared to R_ν.

C-3.5. Series Connection with Transformers

The circuit of Figure C-6 is probably a better approximation to some amplifier configurations than the circuit of Figure C-5. It

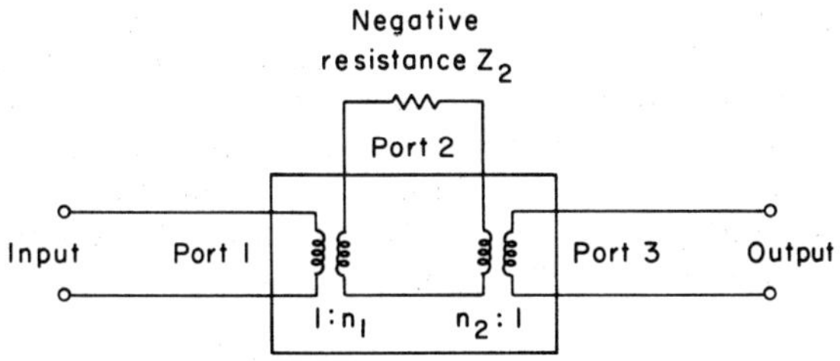

Figure C-6. Series connection with input and output transformers, assumed to be ideal.

‡We have not mentioned any of the undesirable effects of such an adjustment. These include low bandwidth, incipient instability, and poor control of amplifier gain.

is probably a reasonable equivalent circuit for a cavity-type amplifier with adjustable input and output probes.[11] In the analysis we assume that the transformers are ideal, with turns ratios n_1 and n_2. Then,

$$Z_{in} = \frac{Z_2 + n_2^2 Z_3}{n_1^2}, \tag{C-75}$$

$$Z_{out} = \frac{n_1^2 Z_1 + Z_2}{n_2^2}, \tag{C-76}$$

$$G_e = \frac{n_1^2 R_1}{n_1^2 R_1 + R_2}, \tag{C-77}$$

$$T_1 = T\frac{-R_2}{n_1^2 R_1}, \tag{C-78}$$

$$M_1 = \frac{T}{T_0}, \tag{C-79}$$

and if a noisy second stage is attached,

$$T_{12} = T\frac{-R_2}{n_1^2 R_1} + \frac{T_a |n_1^2 Z_1 + Z_2 + n_2^2 Z_\nu^*|^2 + T_b |n_1^2 Z_1 + Z_2 - n_2^2 Z_\nu|^2}{4 n_1^2 R_1 R_\nu}. \tag{C-80}$$

By adjusting n_1 and n_2, we can make T_{12} approach T as closely as desired.‡ For example, tune the input impedance Z_1 so that the quantities in the numerator of Equation C-79 are both real. Then simultaneously make G_e infinite and make n_2 approach zero. Alternatively, relate n_1 and n_2 by the condition that the factor multiplying T_b in Equation C-80 vanishes. Then, let n_2 approach zero and T_{12} approaches T.

‡We have not mentioned any of the undesirable effects of such an adjustment. These include low bandwidth, incipient instability, and poor control of amplifier gain.

REFERENCES

1. Bauer, H., and H. Rothe, "Der äquivalente Rauschvierpol als Wellenvierpol," Arch. der Elektrischen Übertragung, 10, 241-252 (1956).
2. Carlin, H. J., "The Scattering Matrix in Network Theory," IRE Trans. on Circuit Theory, CT-3, 88-97 (June, 1956).
3. Haus, H. A., and R. B. Adler, "An Extension of the Noise Figure Definition," Proc. IRE, 45, 690-691 (May, 1957).
4. Haus, H. A., and R. B. Adler, "Optimum Noise Performance of Linear Amplifiers," Proc. IRE, 46, 1517-1533 (August, 1958).
5. Haus, H. A., and R. B. Adler, Circuit Theory of Linear Noisy Networks, The Technology Press and John Wiley & Sons, Inc., New York (1959).
6. Haus, H. A., and R. B. Adler, ibid., Chapter 5.
7. Hildebrand, F. B., Methods of Applied Mathematics, Prentice-Hall, Inc., Englewood Cliffs, N.J. (1952).
8. Hildebrand, F. B., ibid., Chapter 1.
9. Penfield, P., Jr., "Noise in Negative-Resistance Amplifiers," IRE Trans. on Circuit Theory, CT-7, 166-170 (June, 1960).
10. Penfield, P., Jr., "Wave Representation of Amplifier Noise," IRE Trans. on Circuit Theory, CT-9, 84-86 (March, 1962).
11. Rafuse, R. P., "Measurement of Absolute Noise Performance of Parametric Amplifiers," Digest of Technical Papers, 1960 International Solid-State Circuits Conference, Philadelphia, Pa., (February 10-12, 1960).
12. Youla, D. C., "On Scattering Matrices Normalized to Complex Port Numbers," Proc. IRE, 49, 1221 (July, 1961).

Appendix D

DESIGN CHARTS AND NOMOGRAPHS

Many of the formulas and graphs derived in this book can be used directly in preliminary design of frequency converters, parametric amplifiers, multipliers, and dividers. To aid in their use, we give here some nomographs for the more important results. The first three deal with all applications. Then there are some for parametric amplifiers, and finally, some for multipliers and dividers.

Table of Contents
(Appendix D)

D-1. Diode Parameters

The nomograph on this page, Figure D-1, relates the following diode parameters: cutoff frequency f_c, series resistance R_s, voltage range $V_B + \phi$, minimum capacitance C_{min}, maximum elastance S_{max}, and normalization power P_{norm}. The nomograph is based on Equations 4.48 and 4.50:

$$f_c = \frac{S_{max} - S_{min}}{2\pi R_s}; \qquad P_{norm} = \frac{(V_B + \phi)^2}{R_s}.$$

It is assumed that $S_{min} \ll S_{max}$; if it is not, enter the value of $S_{max} - S_{min}$ in place of S_{max}, or $C_{max}C_{min}/(C_{max} - C_{min})$ in place of C_{min}.

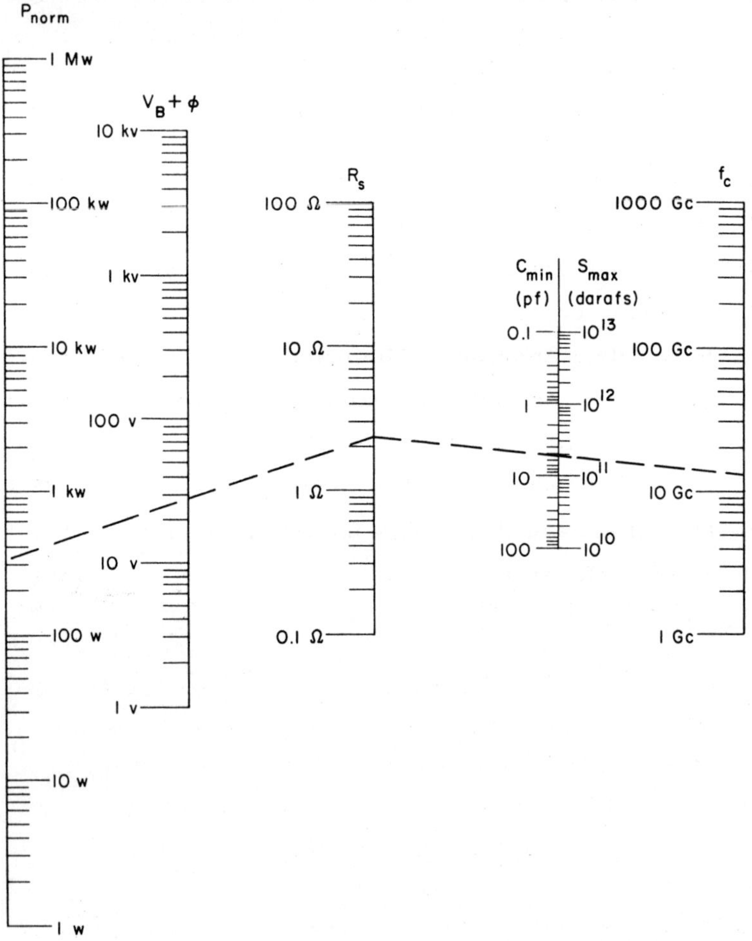

Figure D-1.

On this page, Figure D-1 is used to lead to diode space, a plot of f_c vs. P_{norm}. Several of the nomographs in Appendix D are built up around diode space.

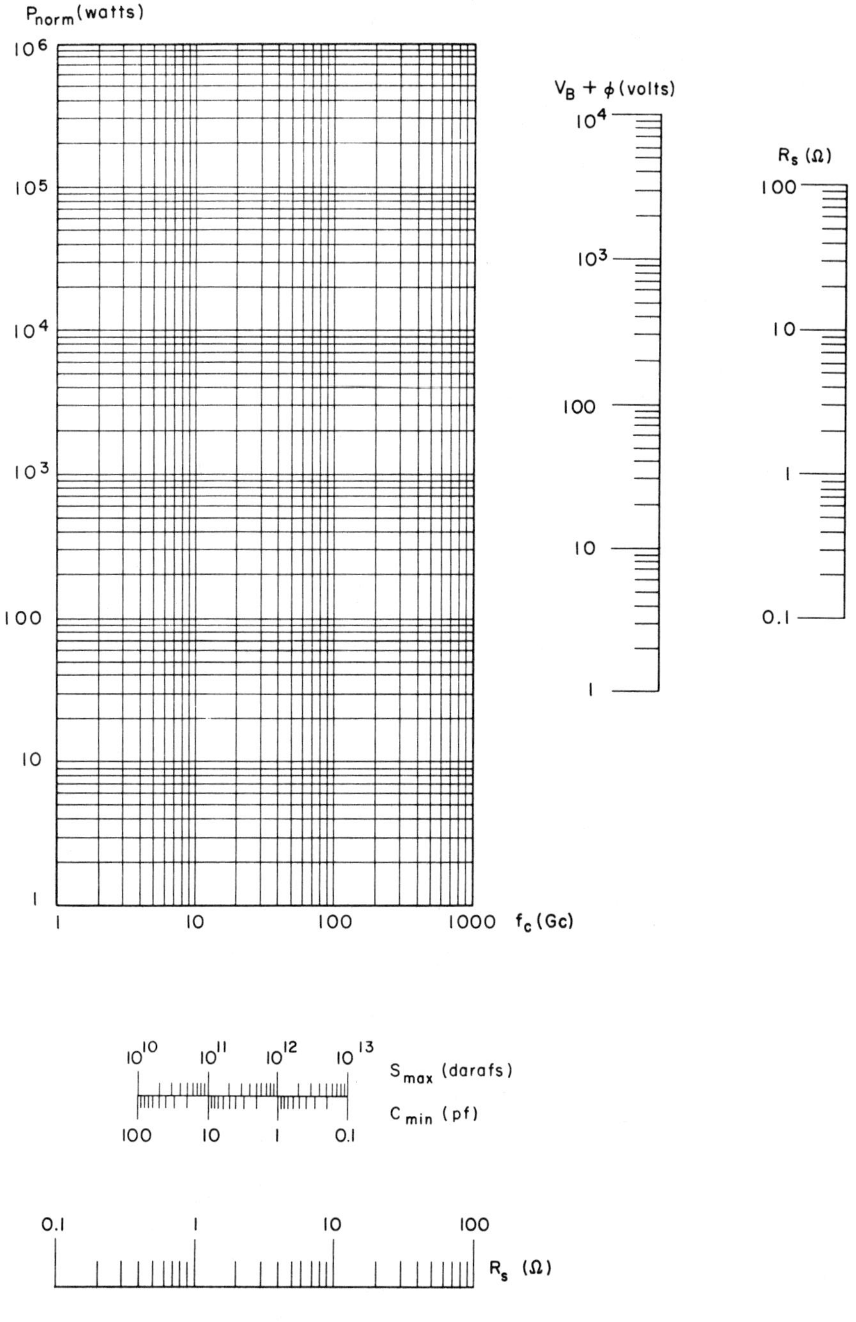

Figure D-2.

D-2. Bias Voltage

This nomograph, Figure D-3, shows the bias voltage $V_0 + \phi$ as a function of the voltage range $V_B + \phi$ and the normalized bias voltage $(V_0 + \phi)/(V_B + \phi)$, the parameter ordinarily calculated throughout this book.

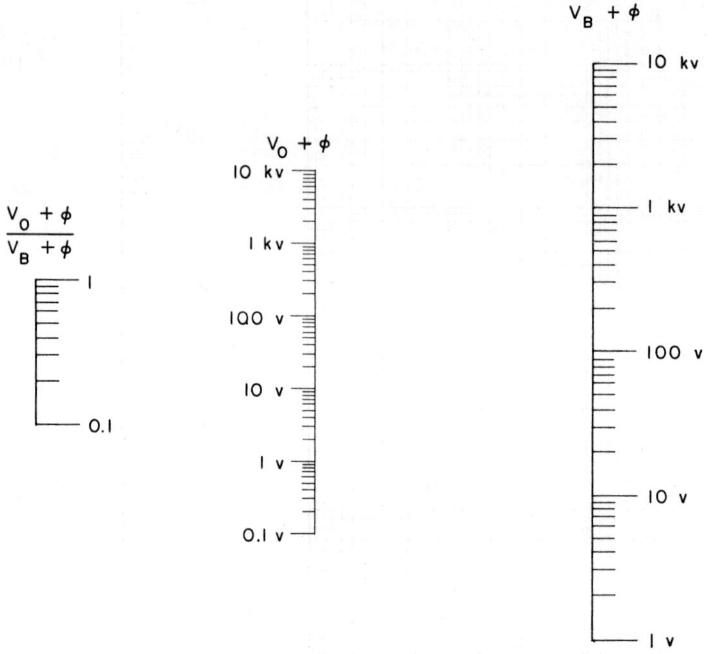

Figure D-3.

D-3. Reactance; Tuning

Figure D-4 can be used to find the average elastance S_0 from the value of m_0 and either C_{min} or S_{max}, and thence to find the reactance X of S_0 at any frequency f. The nomograph is based on the equations

$$S_0 = m_0(S_{max} - S_{min}); \qquad X = \frac{S_0}{2\pi f}.$$

Figure D-5 is a special case of Figure D-4 with $m_0 = 0.5$.

It is assumed that $S_{min} \ll S_{max}$; if not, enter the value of $S_{max} - S_{min}$ in place of S_{max}, or $C_{max}C_{min}/(C_{max} - C_{min})$ in place of C_{min}.

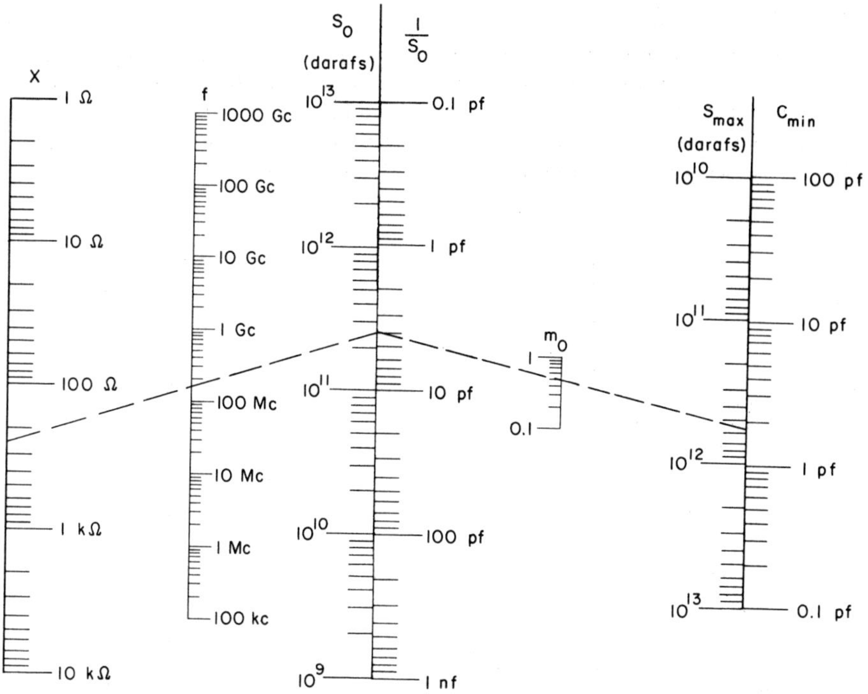

Figure D-4.

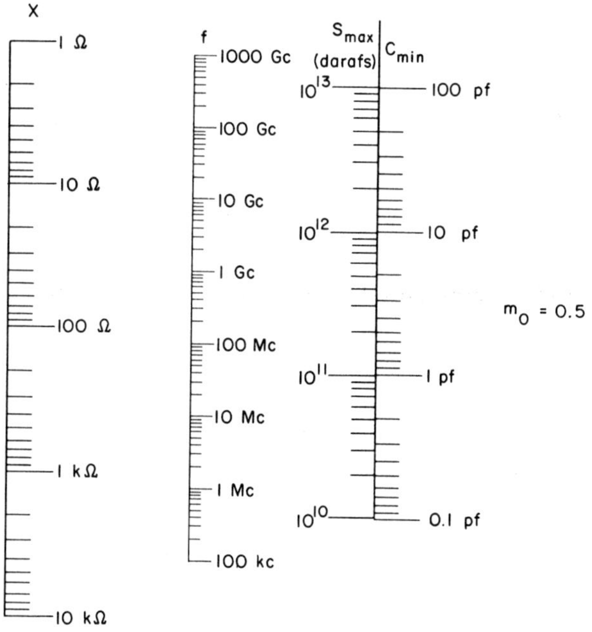

Figure D-5.

D-4. Pump Power

Figures D-6 and D-7 relate the pump power of current-pumped graded-junction or abrupt-junction varactors to the pump frequency, the modulation ratio m_1, and the varactor parameters. The nomographs are based on Equations 7.70 and 7.95. It is assumed that the graded-junction varactor is self-biased and that $S_{min} \ll S_{max}$.

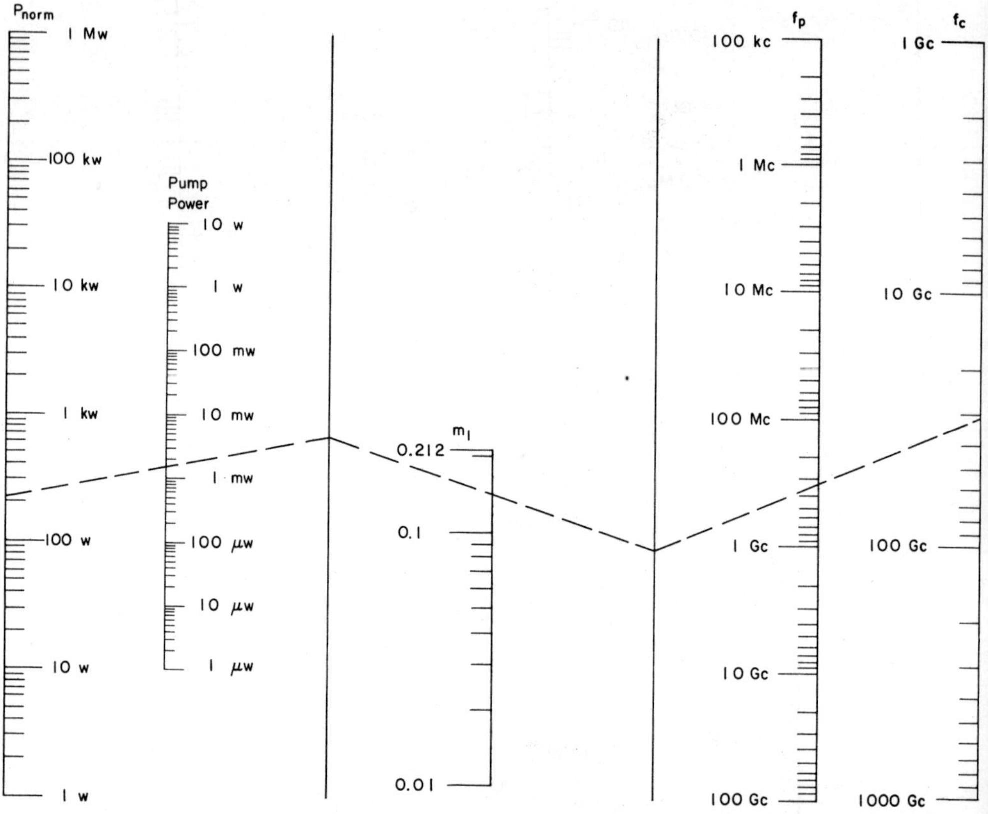

Figure D-6. Pump power for graded-junction varactors.

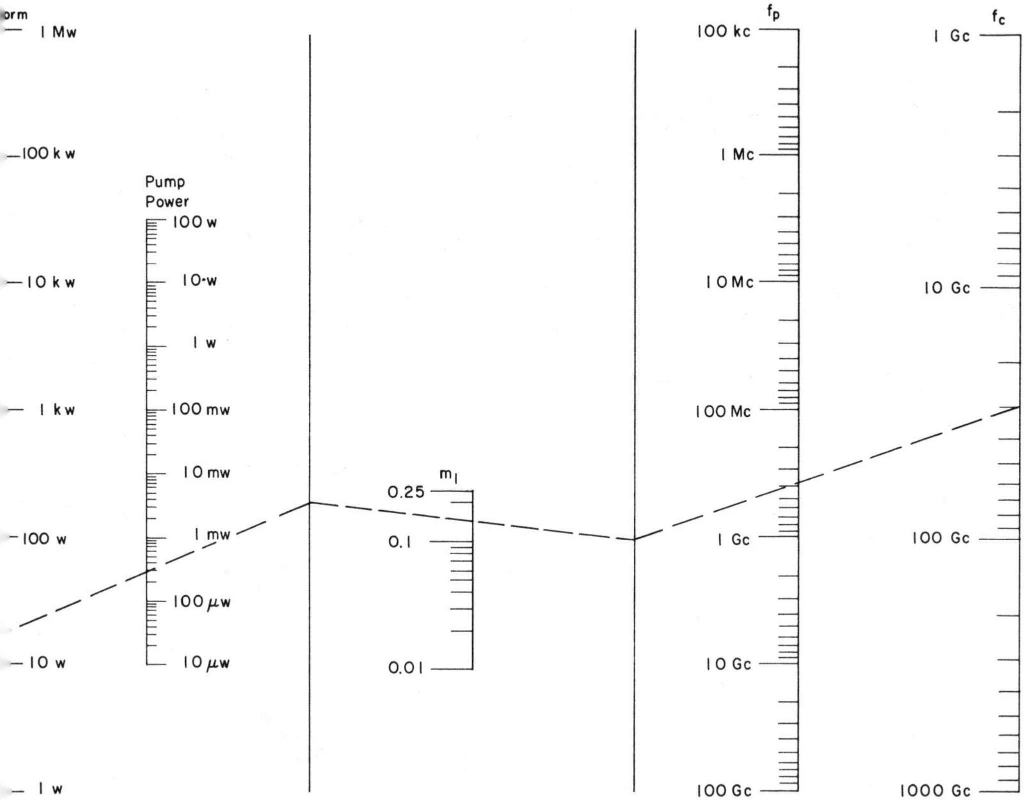

Figure D-7. Pump power for abrupt-junction varactors.

D-5. Parametric Amplifier — Minimum Noise

These two diode-space nomographs (pages 546 and 547) show the minimum noise temperature T_{min} and associated parameters for graded-junction (Figure D-8) or abrupt-junction (Figure D-9) varactors fully pumped with a sinusoidal current. First, use the bottom nomographs to find T_{min} from the cutoff frequency f_c and the signal frequency f_s. Then, the optimum pump frequency $f_{p,opt}$ can be found from the top nomographs, and the required pump power P from the side nomographs. It is assumed that the idler termination is lossless and tuned, that the varactor is at room temperature, and that $S_{min} << S_{max}$. Nomographs are based on Equations 6.105, 6.104, and 7.62 or 7.86. The bottom nomograph also applies to optimized frequency converters.

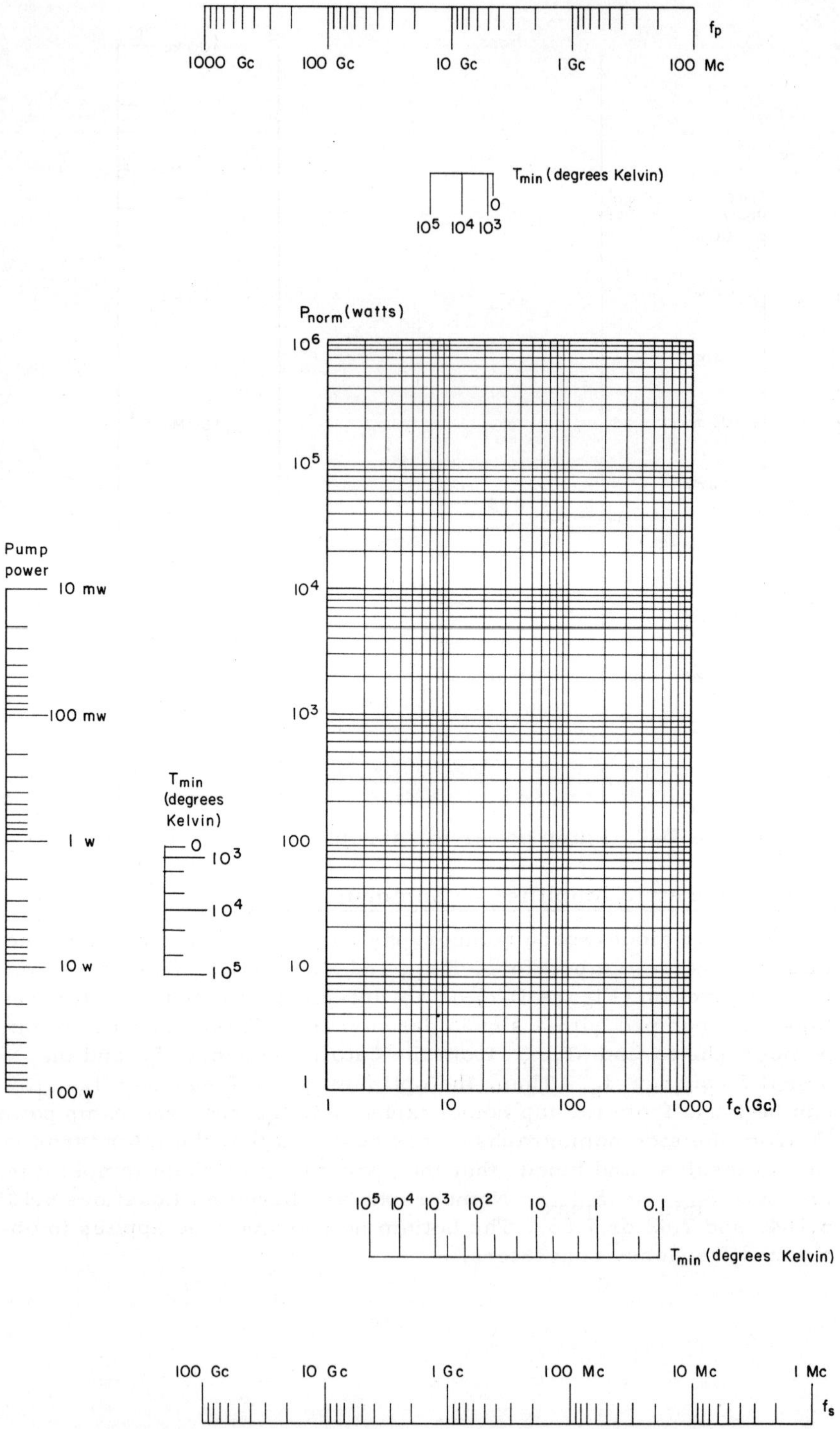

Figure D-8. Fully pumped graded-junction varactor.

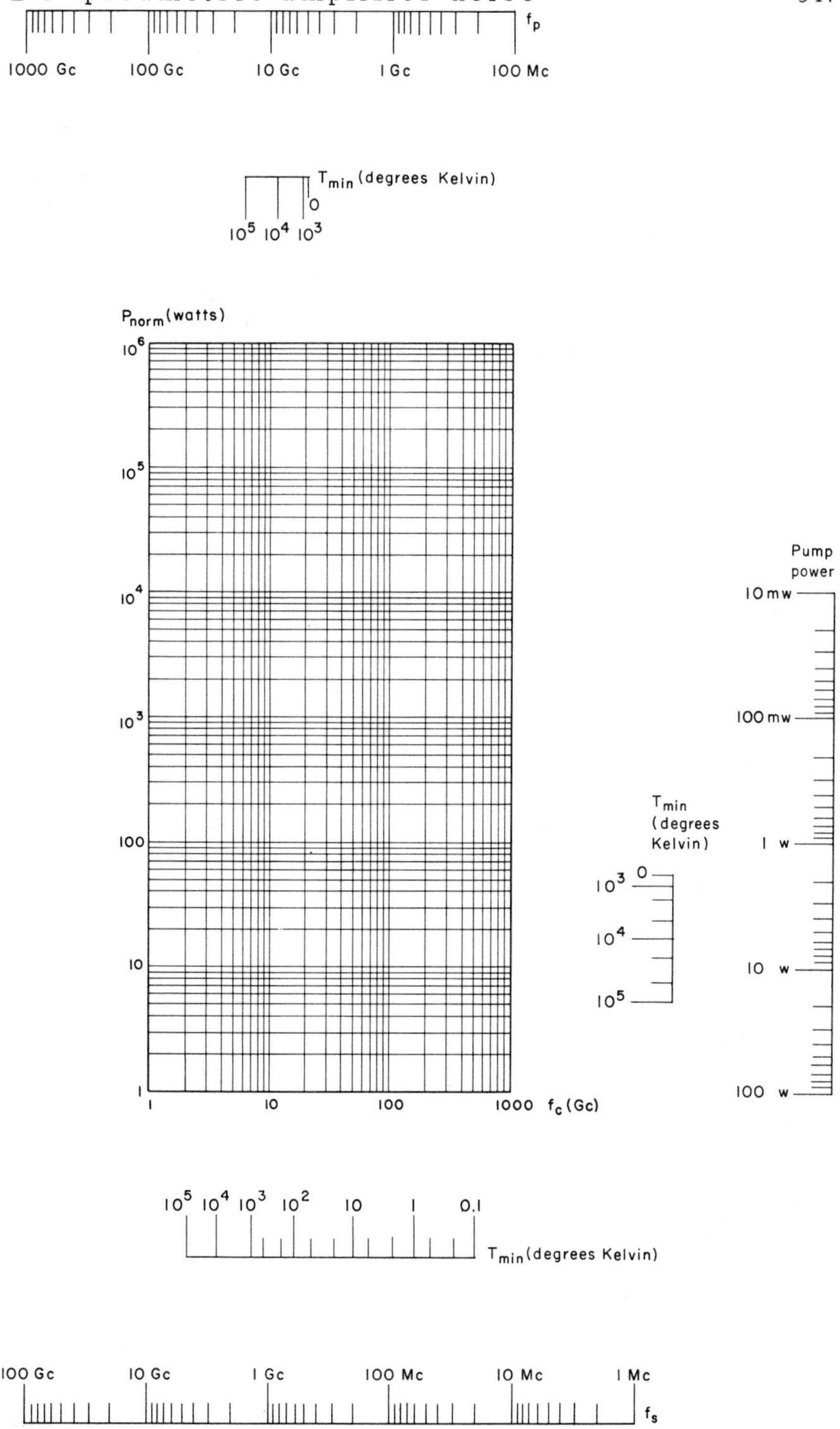

Figure D-9. Fully pumped abrupt-junction varactor.

D-6. Parametric Amplifier — Limited Pump Power

The graphs on pages 549-570 can be used to design optimum parametric amplifiers with limited pump power. They are based on the discussion of Section 7.7. It is assumed that the varactor, either graded-junction (pages 549-559) or abrupt-junction (pages 560-570), is pumped with a sinusoidal current, that it is at room temperature, and that the idler termination is lossless and tuned. We also assume that the graded-junction varactor is self-biased.

There is a separate graph for each signal frequency, and for the two types of varactors. Use the locator chart below to find the appropriate graph. For signal frequencies that are not plotted, interpolation can be used.

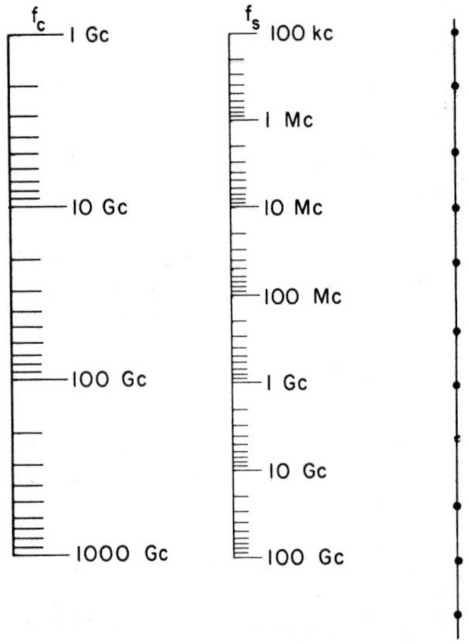

f_s/f_c	Graded-Junction Page	Abrupt-Junction Page
10^{-4}	549	560
2×10^{-4}	550	561
5×10^{-4}	551	562
10^{-3}	552	563
2×10^{-3}	553	564
5×10^{-3}	554	565
0.01	555	566
0.02	556	567
0.05	557	568
0.1	558	569
0.2	559	570

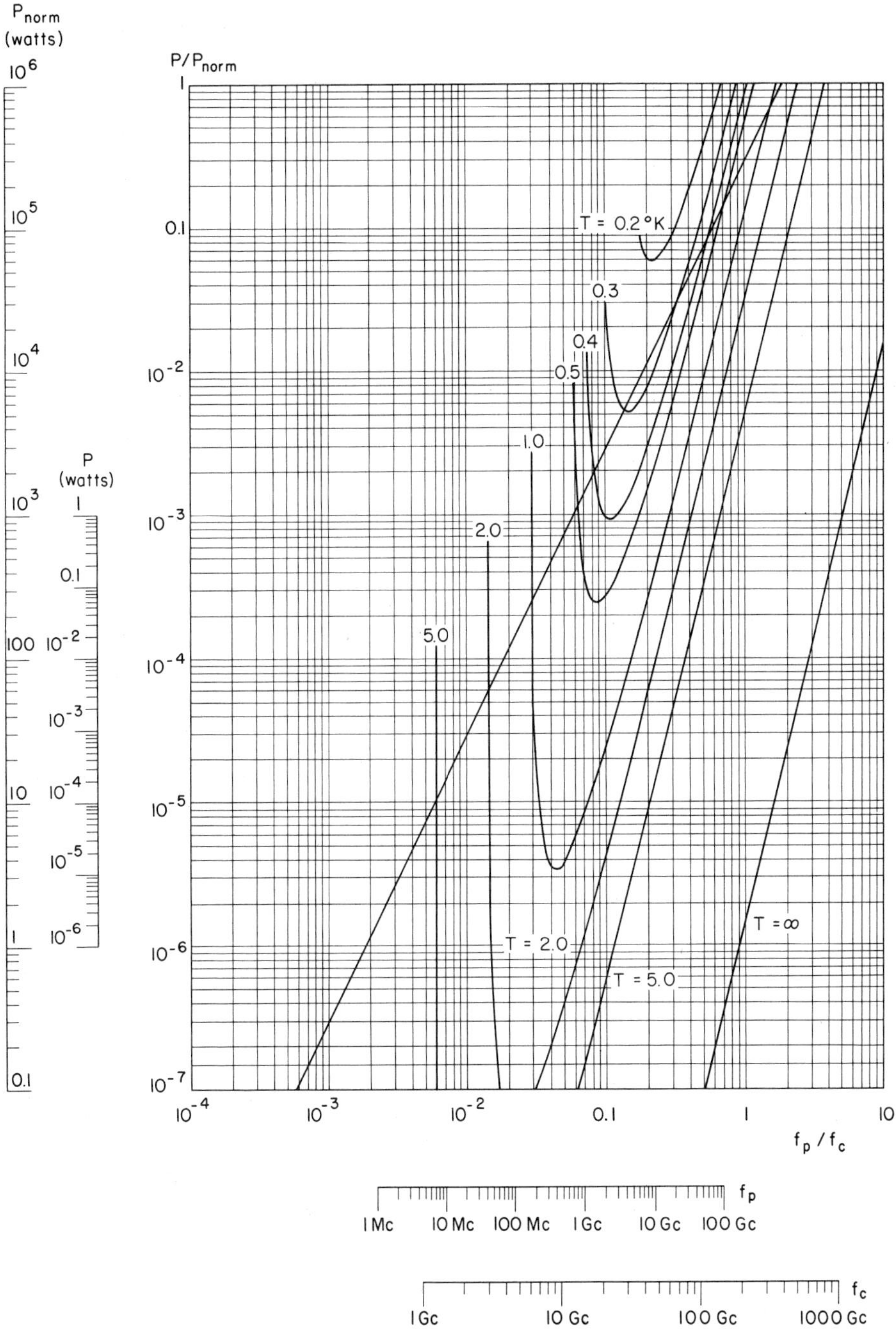

Figure D-10. Graded-junction-varactor parametric amplifier with limited pump power, for $f_s = 10^{-4} f_c$. It is assumed that $R_i = 0$ and $T_d = 290°K$.

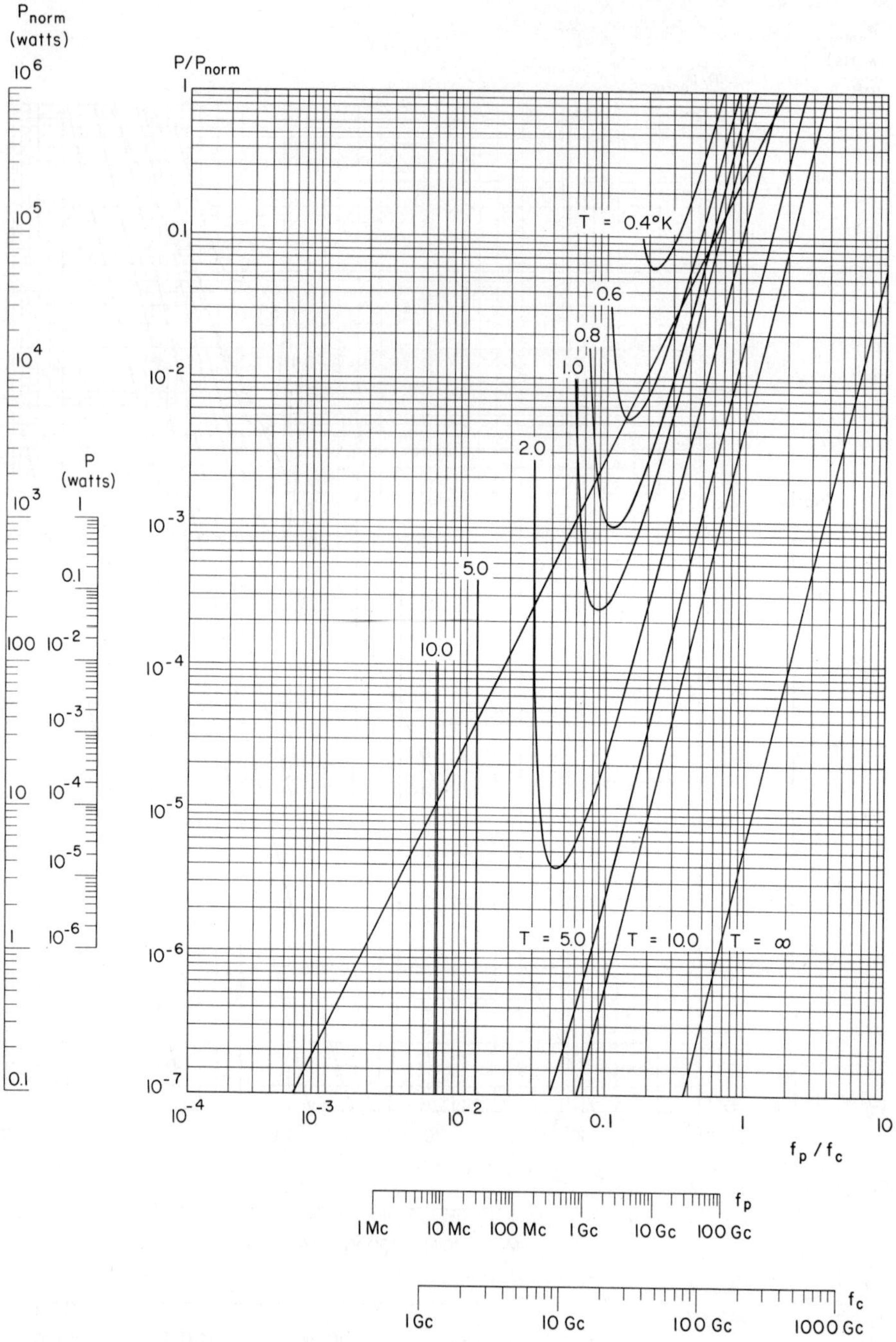

Figure D-11. Graded-junction-varactor parametric amplifier with limited pump power, for $f_s = 2 \times 10^{-4} f_c$. It is assumed that $R_i = 0$ and $T_d = 290^\circ$K.

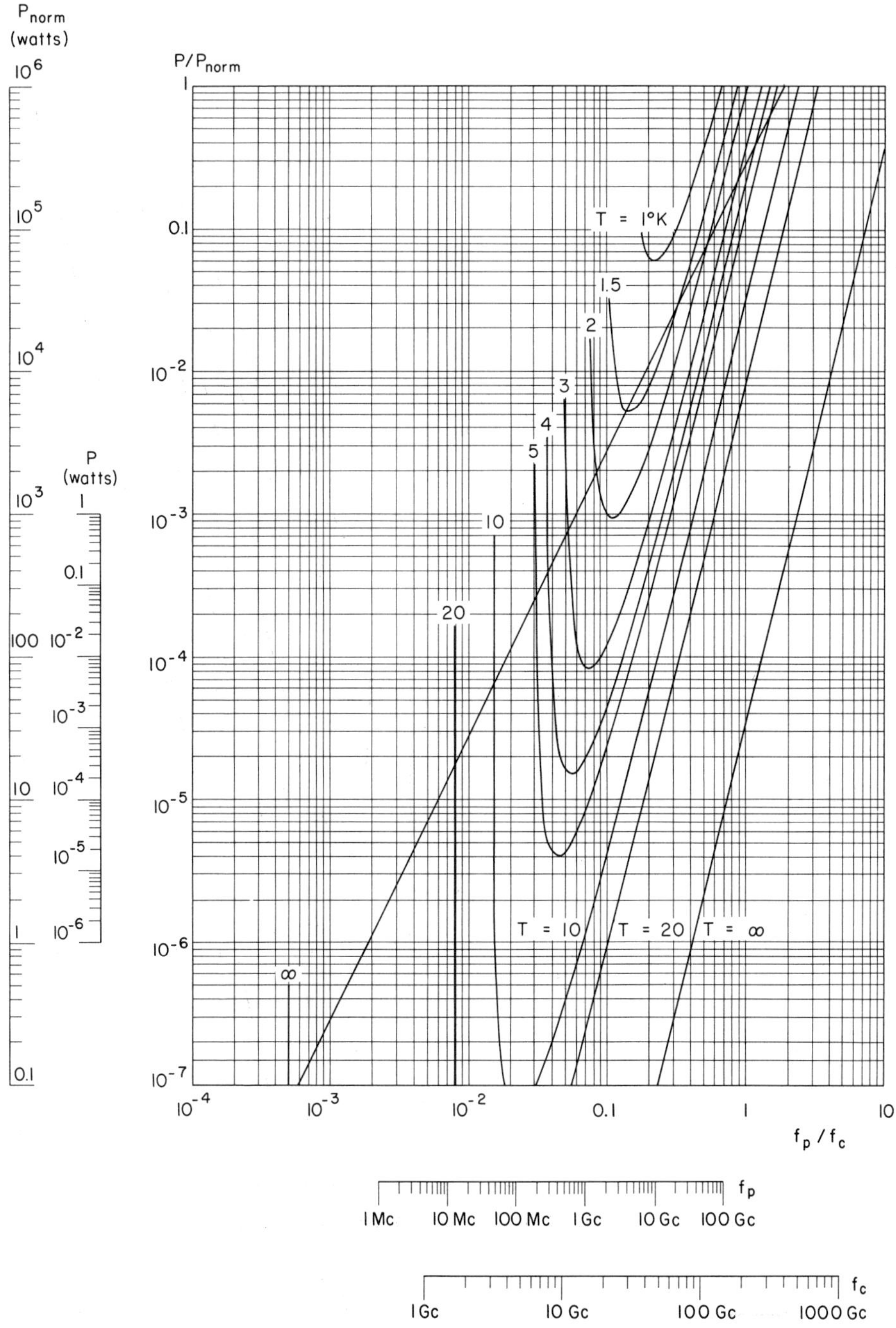

Figure D-12. Graded-junction-varactor parametric amplifier with limited pump power, for $f_s = 5 \times 10^{-4} f_c$. It is assumed that $R_i = 0$ and $T_d = 290°K$.

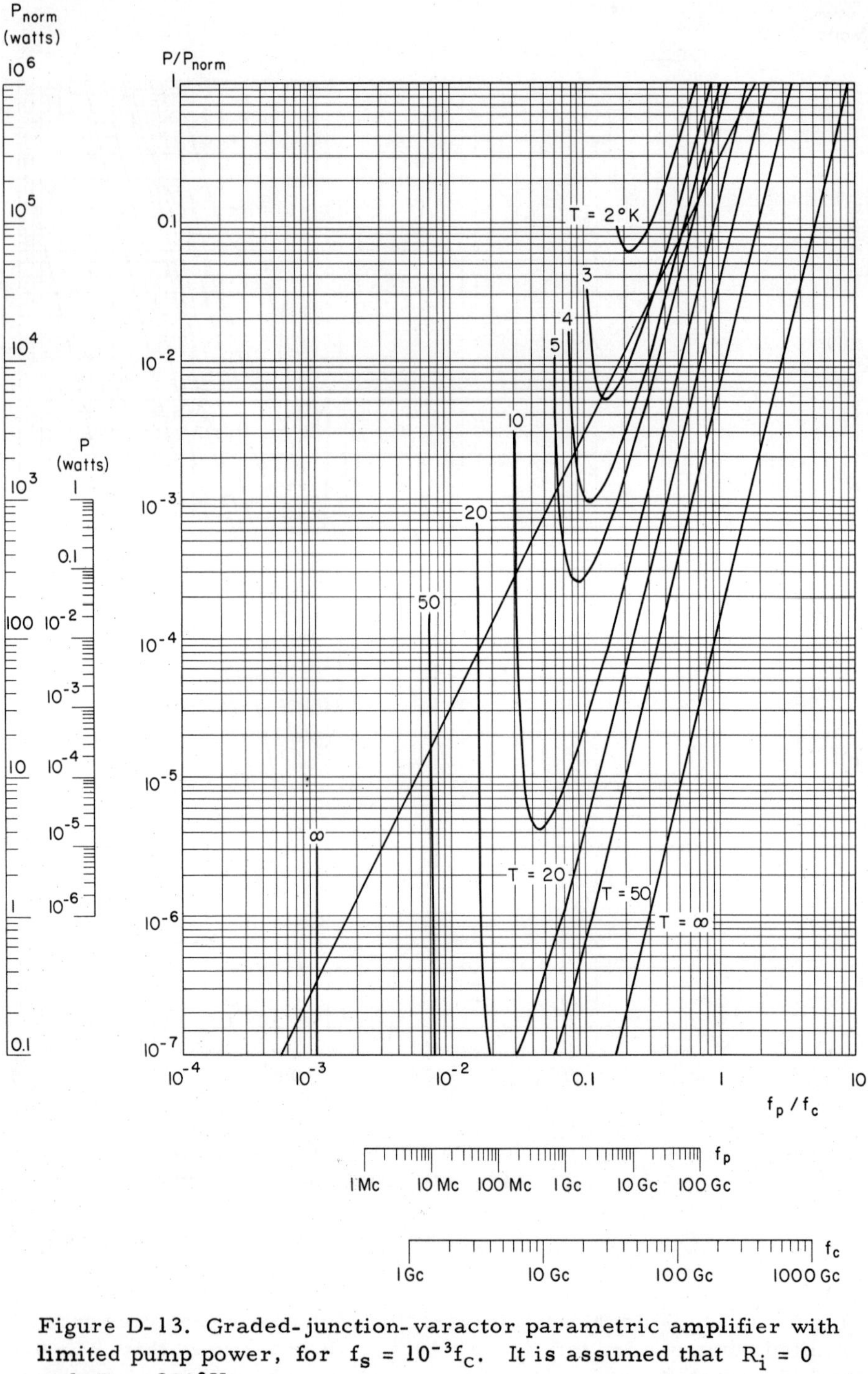

Figure D-13. Graded-junction-varactor parametric amplifier with limited pump power, for $f_s = 10^{-3} f_c$. It is assumed that $R_i = 0$ and $T_d = 290°K$.

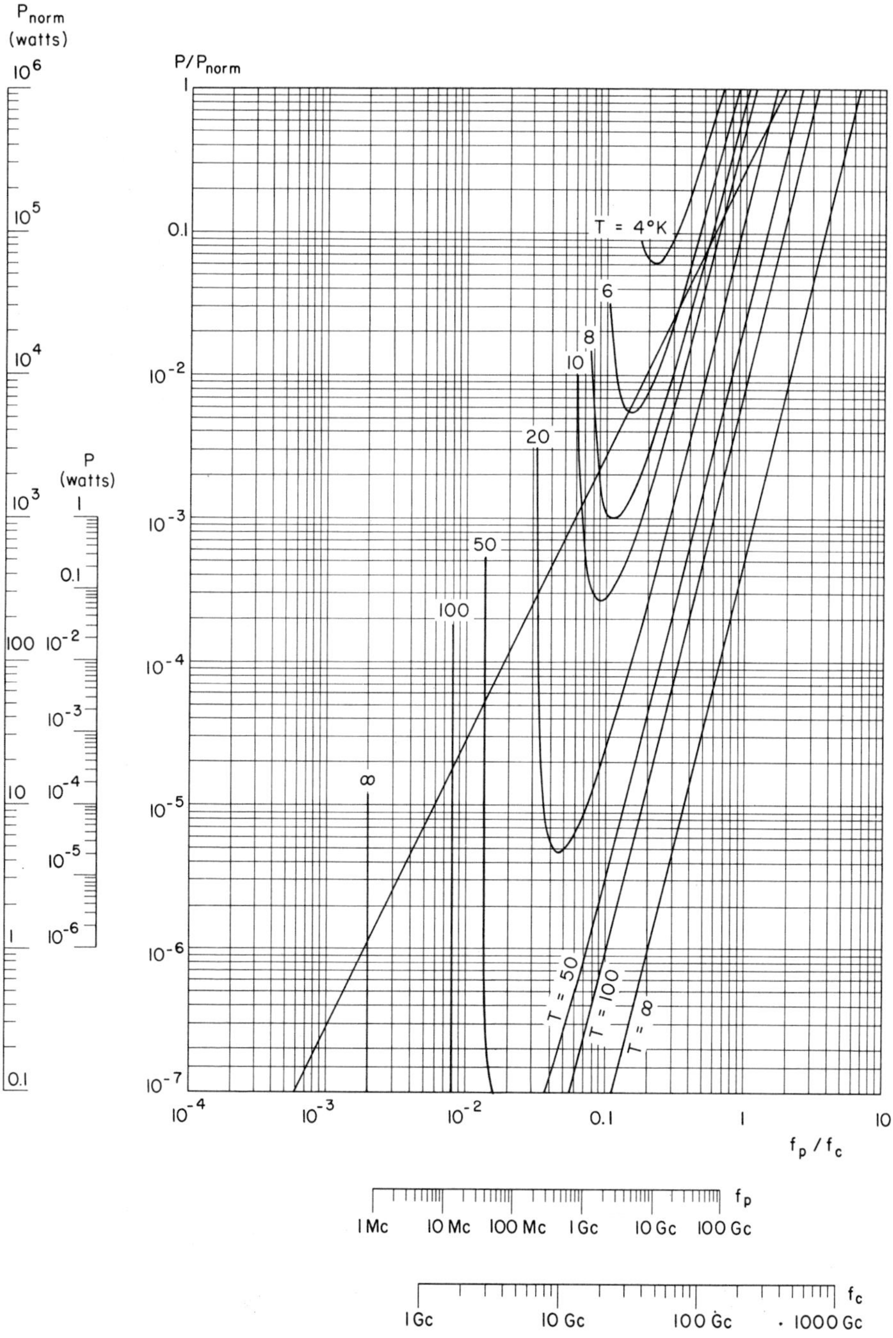

Figure D-14. Graded-junction-varactor parametric amplifier with limited pump power, for $f_s = 2 \times 10^{-3} f_c$. It is assumed that $R_i = 0$ and $T_d = 290°K$.

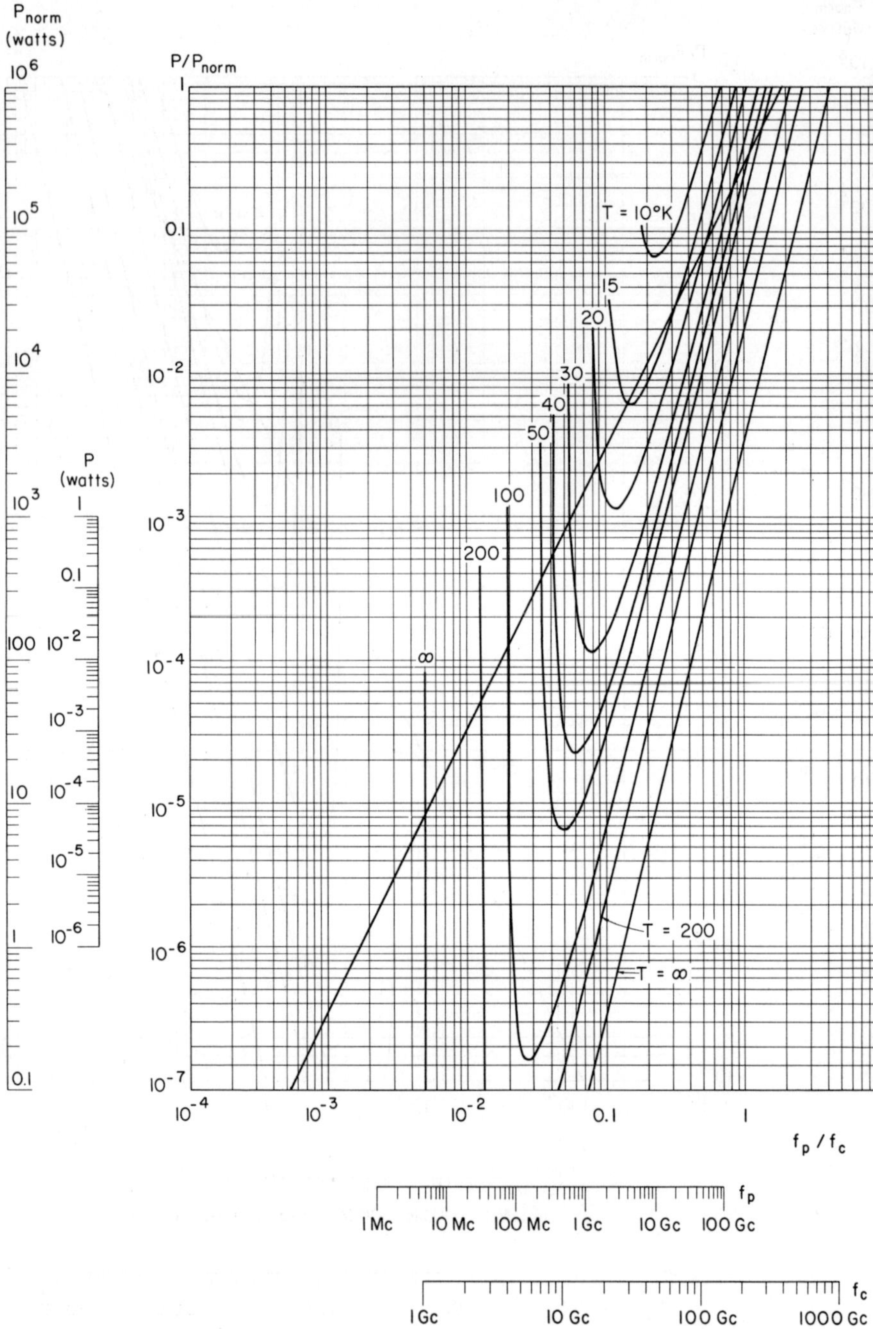

Figure D-15. Graded-junction-varactor parametric amplifier with limited pump power, for $f_s = 5 \times 10^{-3} f_c$. It is assumed that $R_i = 0$ and $T_d = 290°K$.

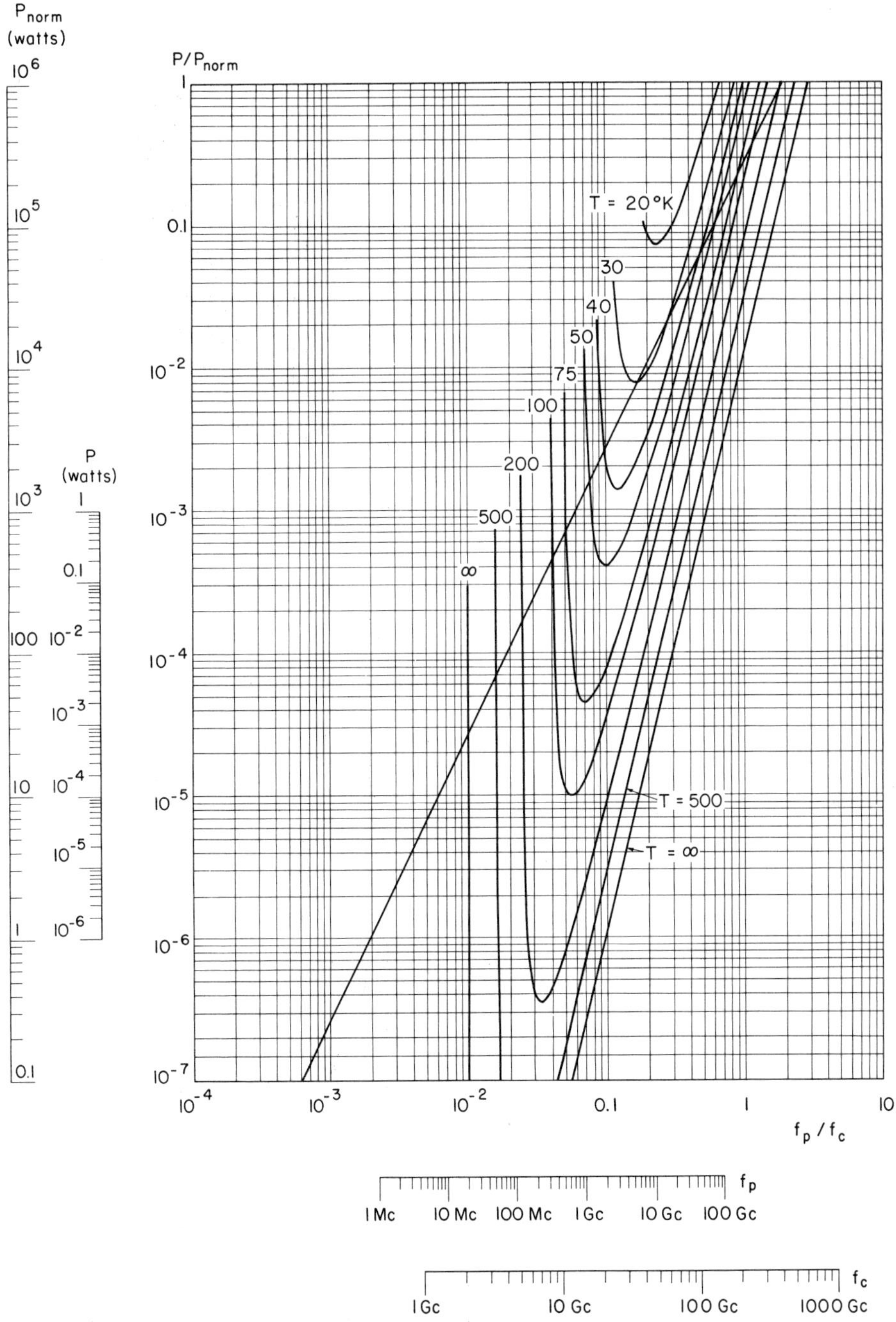

Figure D-16. Graded-junction-varactor parametric amplifier with limited pump power, for $f_s = 10^{-2} f_c$. It is assumed that $R_i = 0$ and $T_d = 290°K$.

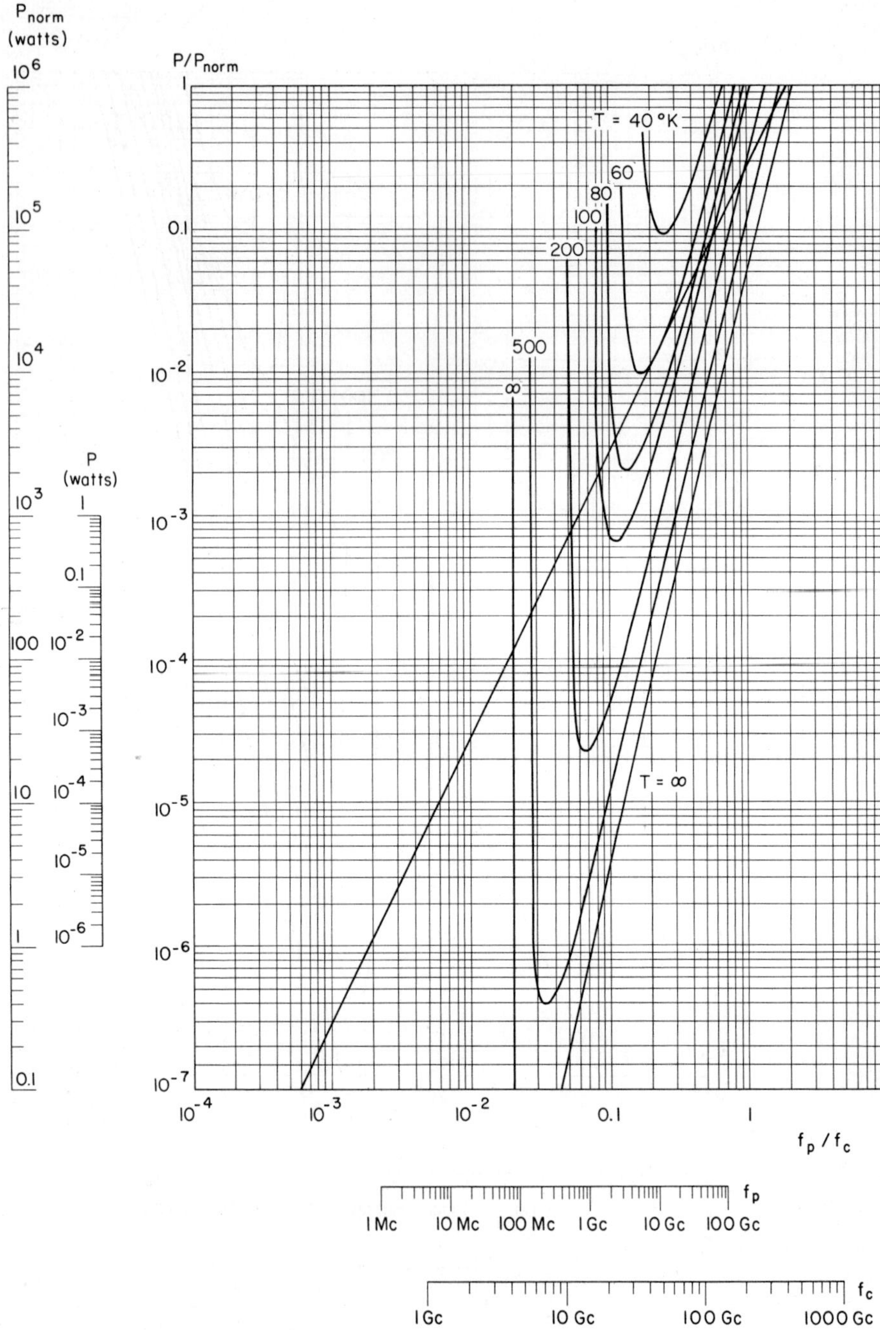

Figure D-17. Graded-junction-varactor parametric amplifier with limited pump power, for $f_s = 2 \times 10^{-2} f_c$. It is assumed that $R_i = 0$ and $T_d = 290°K$.

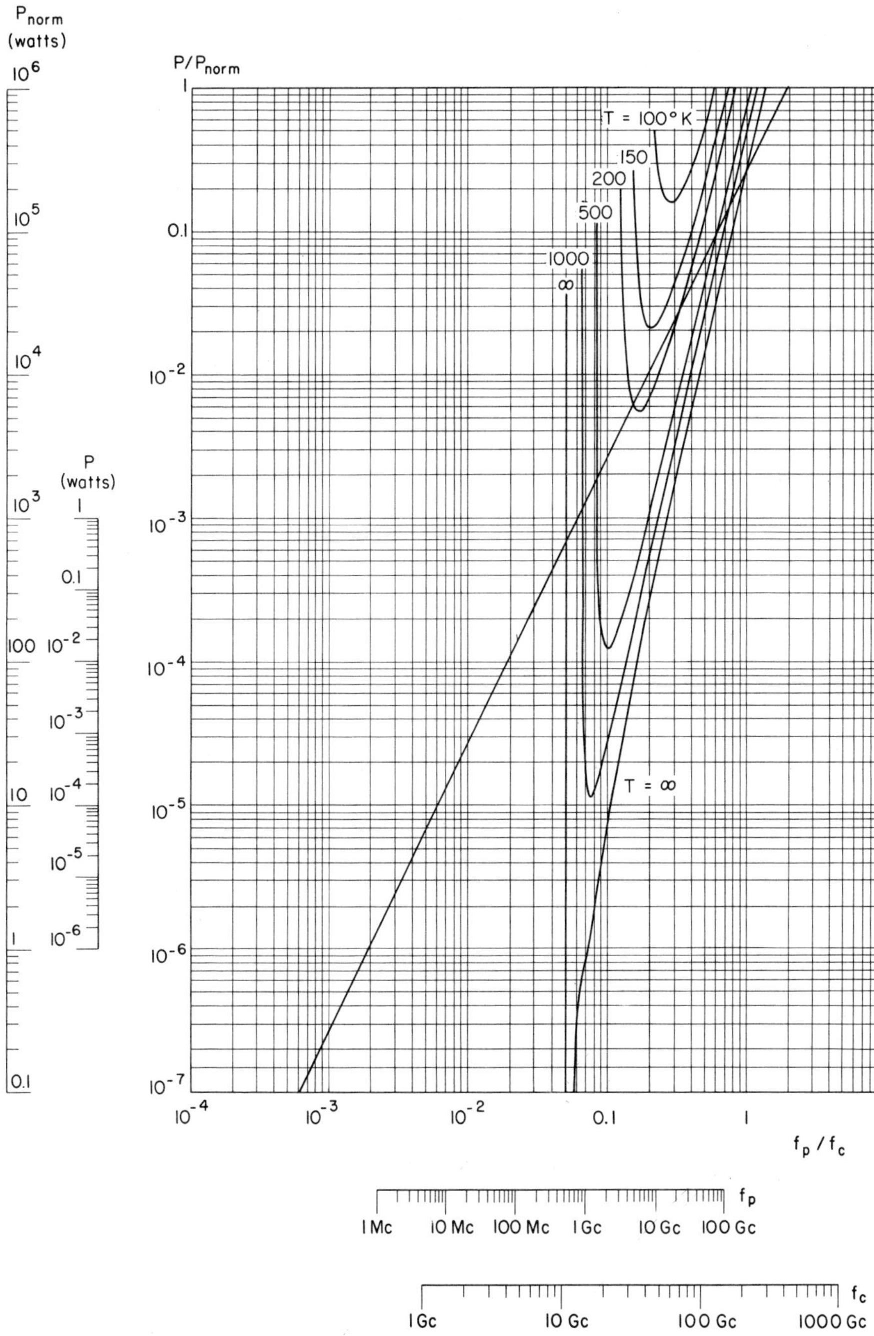

Figure D-18. Graded-junction-varactor parametric amplifier with limited pump power, for $f_s = 5 \times 10^{-2} f_c$. It is assumed that $R_i = 0$ and $T_d = 290°K$.

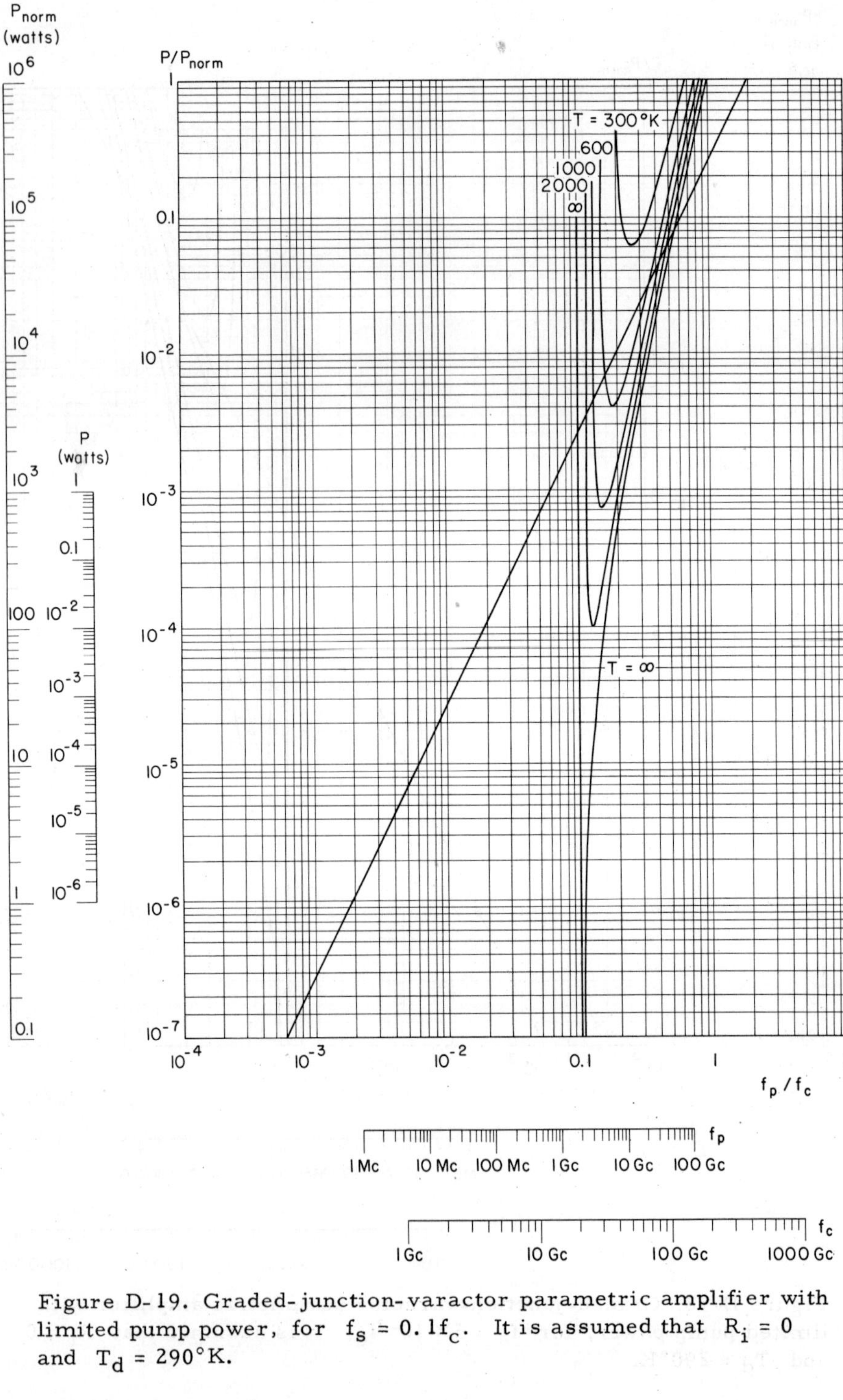

Figure D-19. Graded-junction-varactor parametric amplifier with limited pump power, for $f_s = 0.1f_c$. It is assumed that $R_i = 0$ and $T_d = 290°K$.

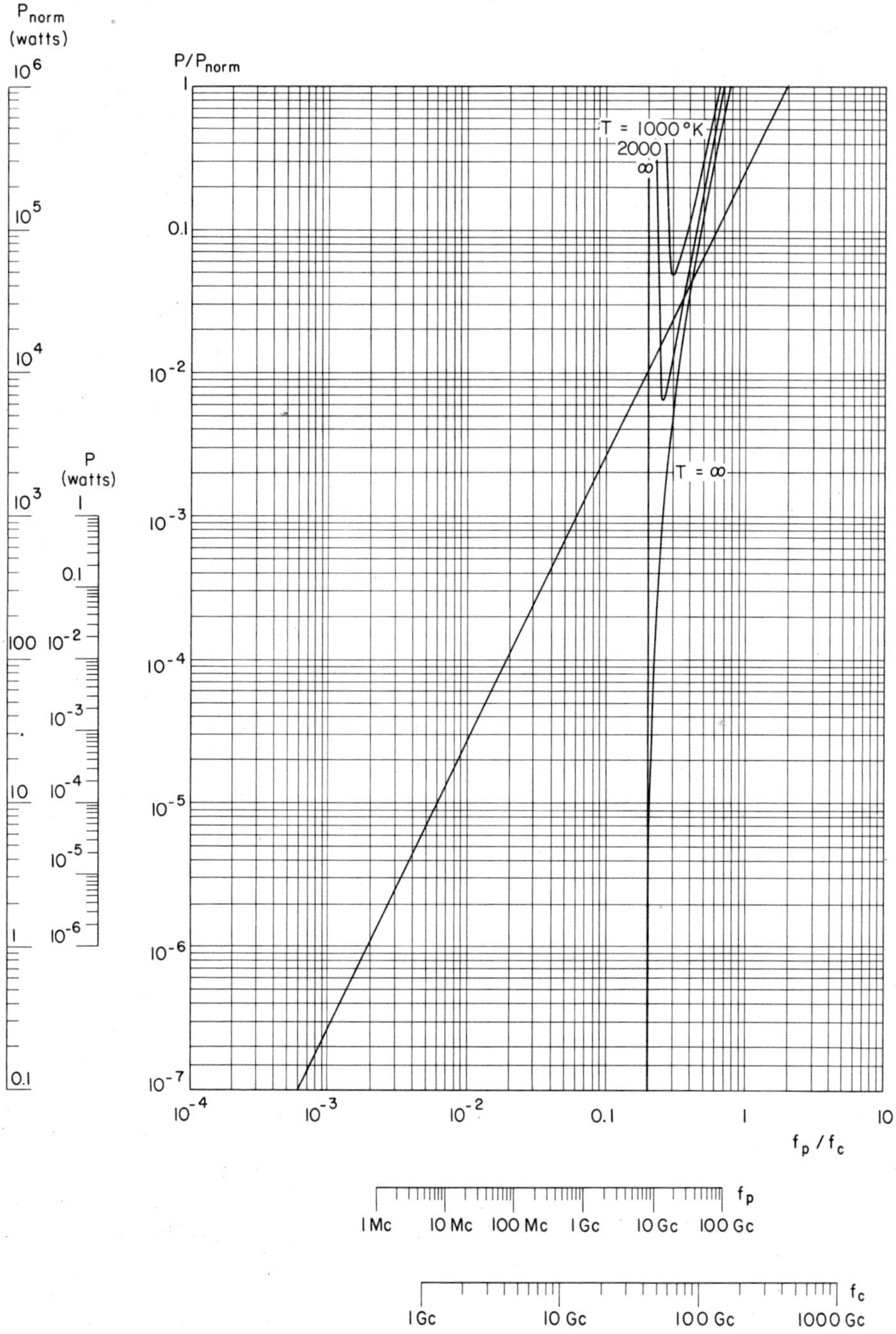

Figure D-20. Graded-junction-varactor parametric amplifier with limited pump power, for $f_s = 0.2f_c$. It is assumed that $R_i = 0$ and $T_d = 290°K$.

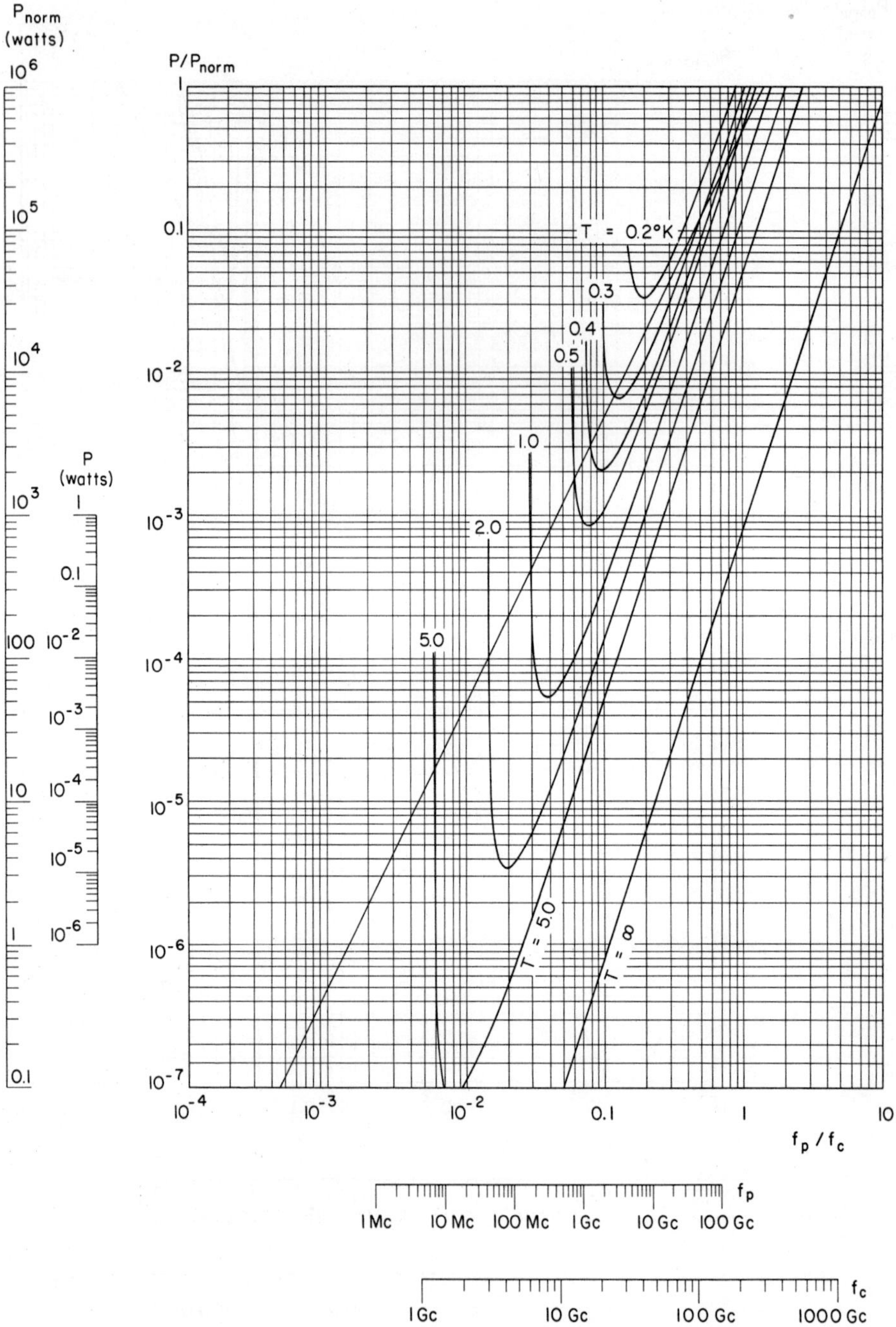

Figure D-21. Abrupt-junction-varactor parametric amplifier with limited pump power, for $f_s = 10^{-4} f_c$. It is assumed that $R_i = 0$ and $T_d = 290°K$.

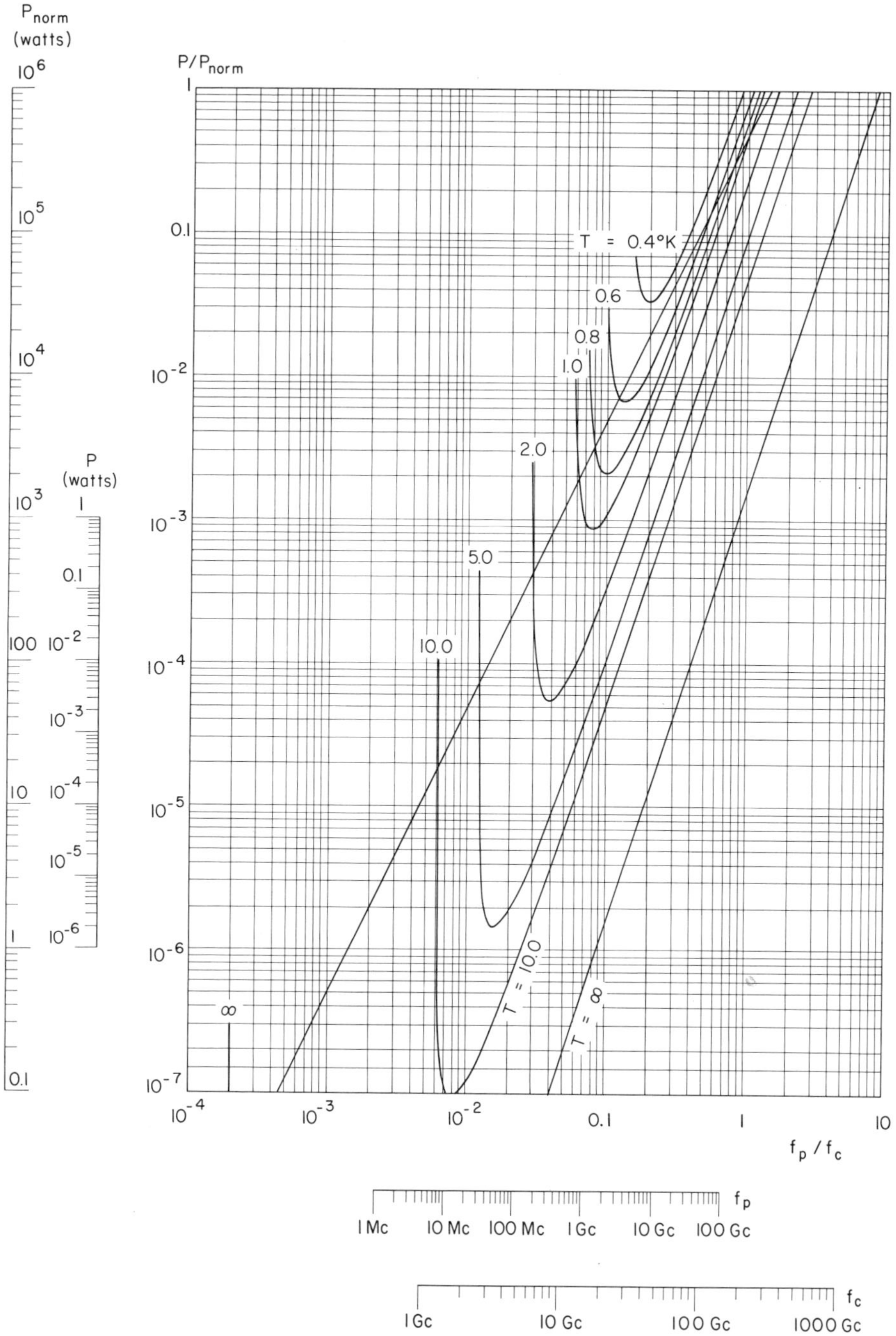

Figure D-22. Abrupt-junction-varactor parametric amplifier with limited pump power, for $f_s = 2 \times 10^{-4} f_c$. It is assumed that $R_i = 0$ and $T_d = 290°K$.

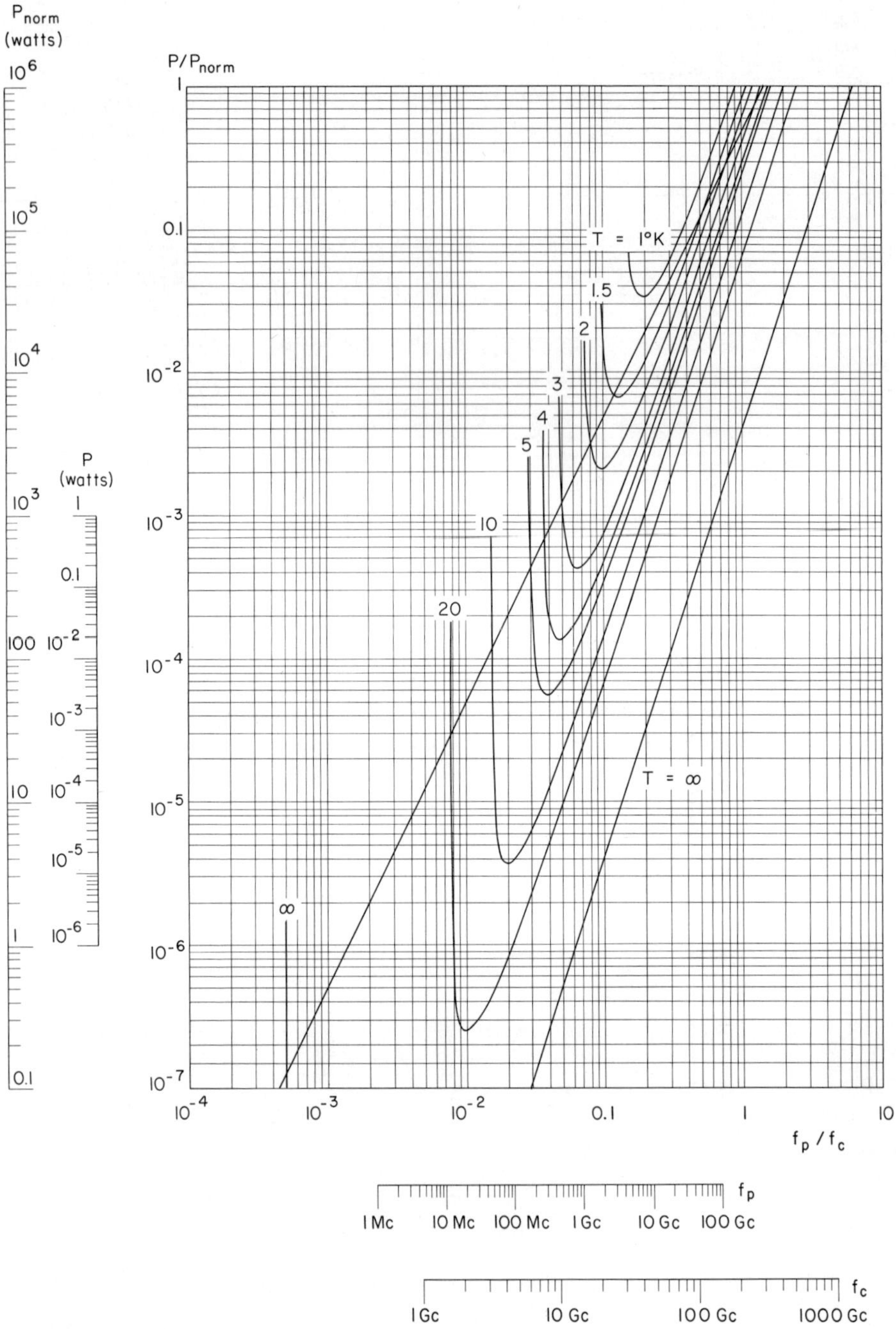

Figure D-23. Abrupt-junction-varactor parametric amplifier with limited pump power, for $f_s = 5 \times 10^{-4} f_c$. It is assumed that $R_i = 0$ and $T_d = 290°K$.

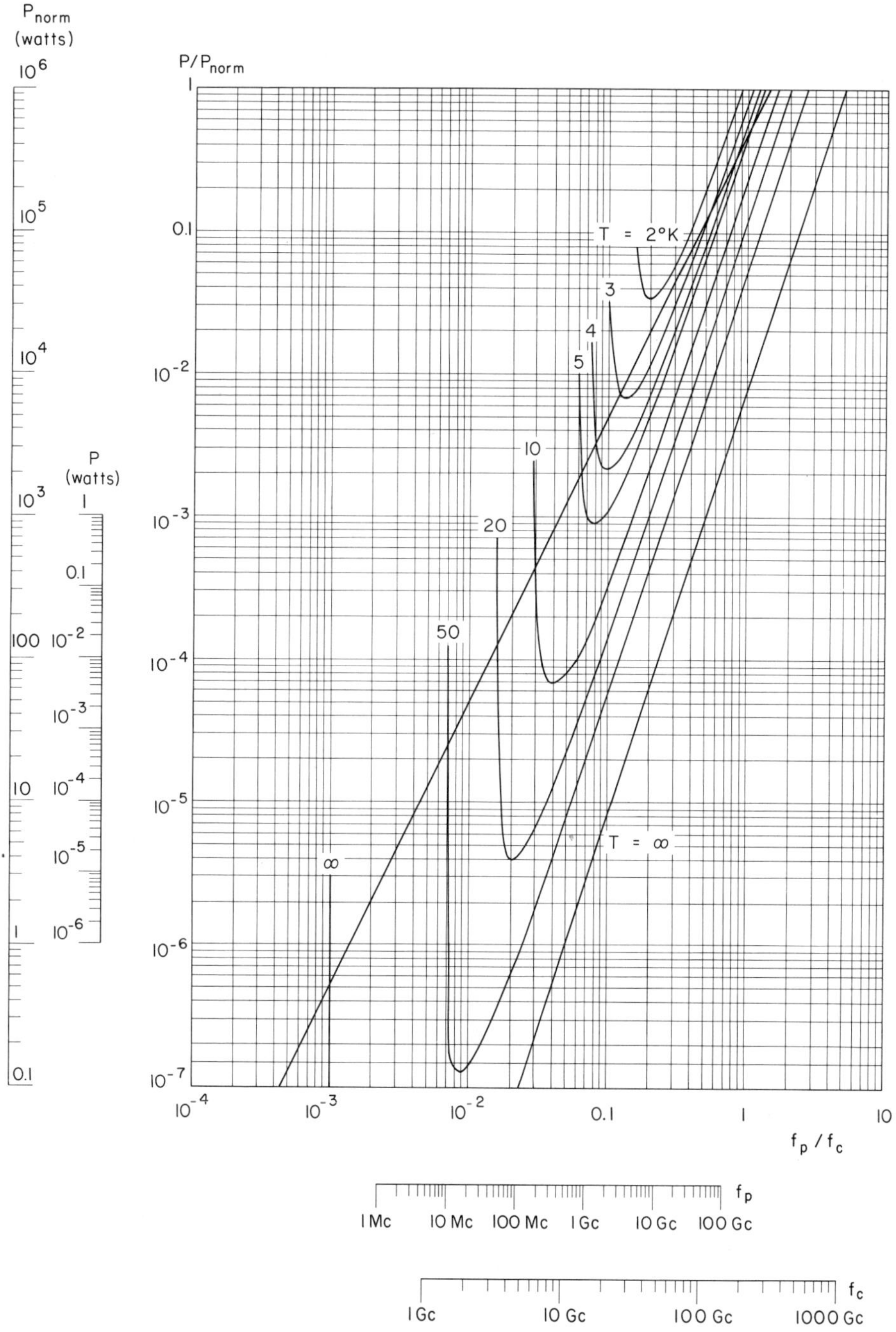

Figure D-24. Abrupt-junction-varactor parametric amplifier with limited pump power, for $f_s = 10^{-3} f_c$. It is assumed that $R_i = 0$ and $T_d = 290°K$.

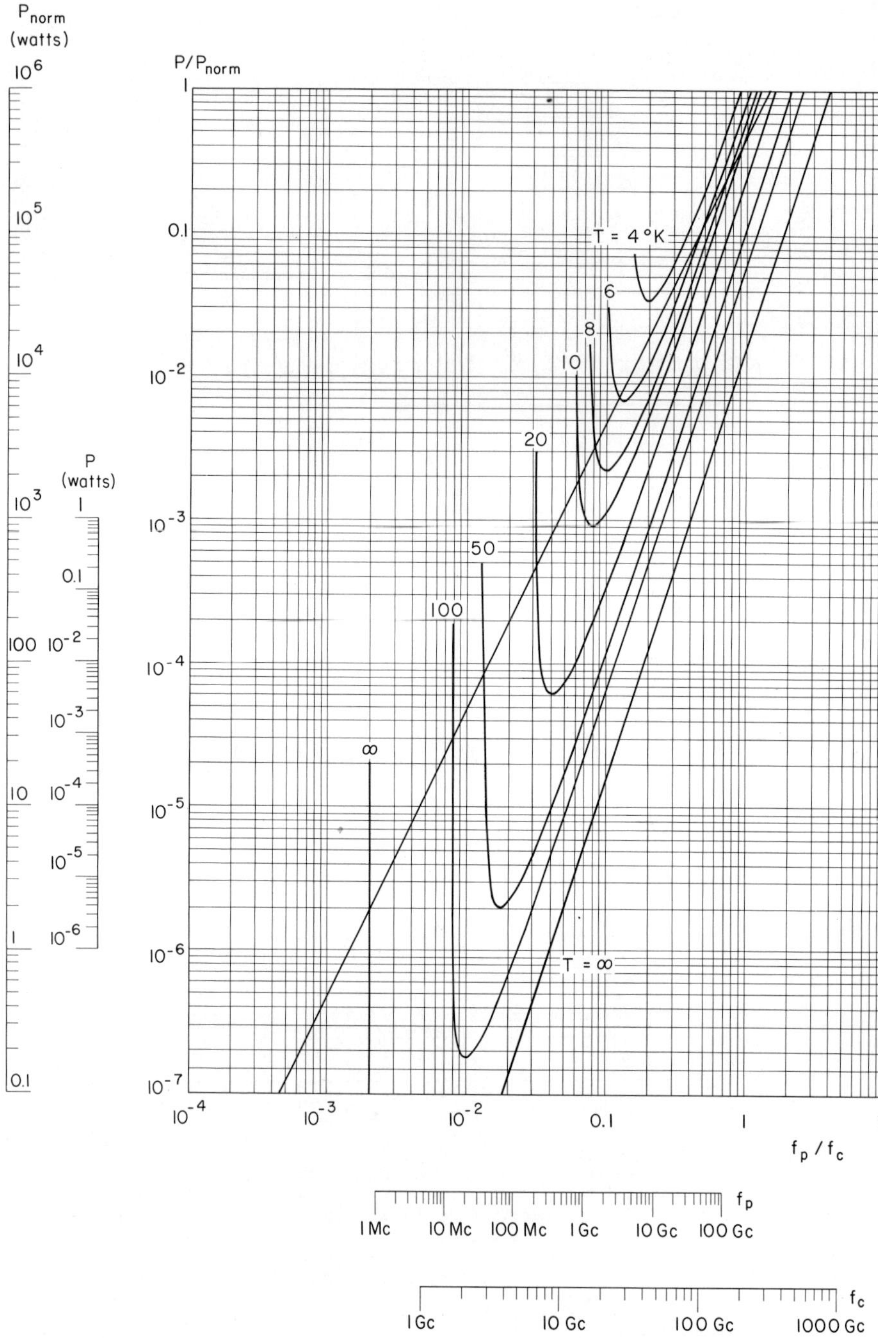

Figure D-25. Abrupt-junction-varactor parametric amplifier with limited pump power, for $f_s = 2 \times 10^{-3} f_c$. It is assumed that $R_i = 0$ and $T_d = 290°K$.

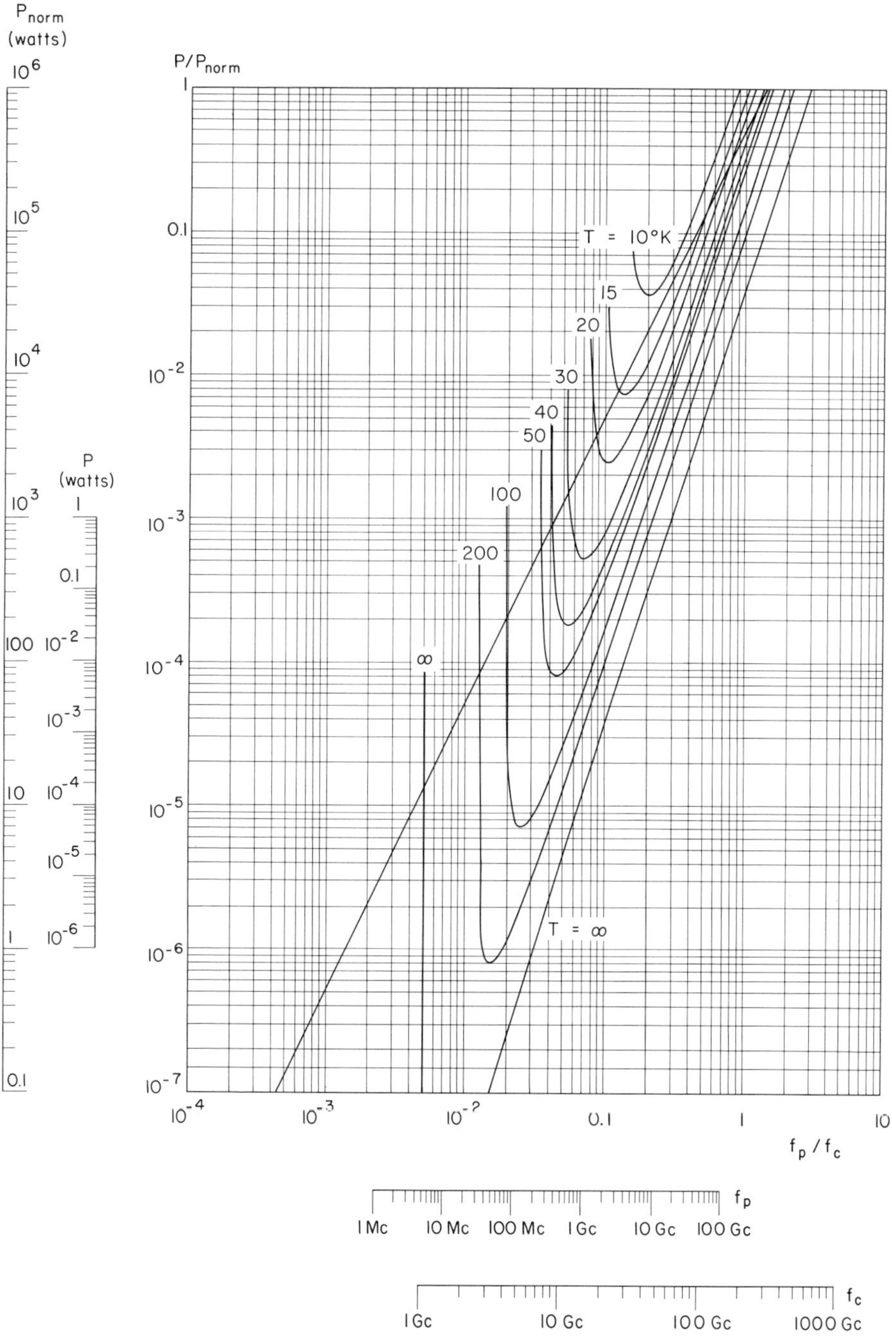

Figure D-26. Abrupt-junction-varactor parametric amplifier with limited pump power, for $f_s = 5 \times 10^{-3} f_c$. It is assumed that $R_i = 0$ and $T_d = 290°K$.

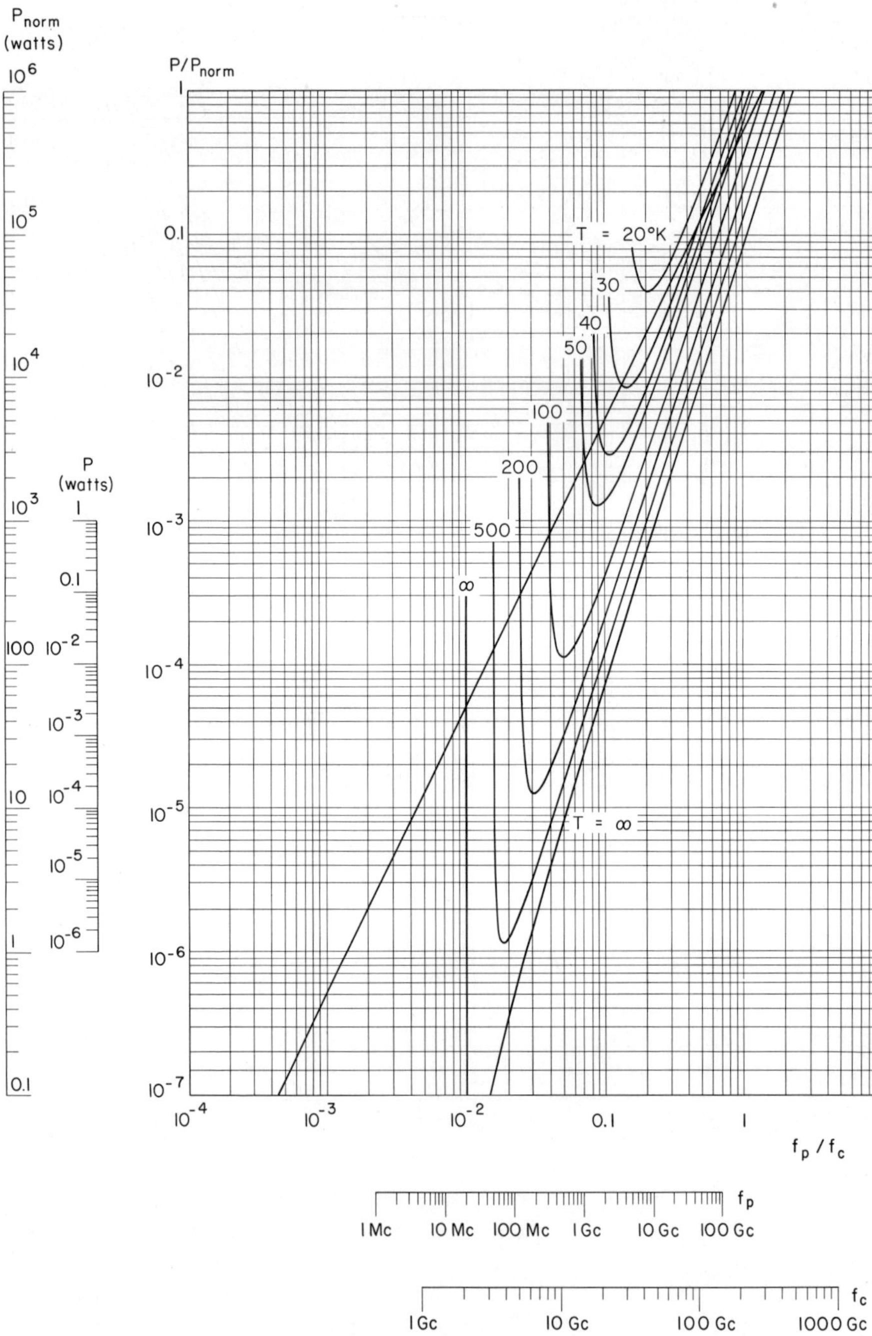

Figure D-27. Abrupt-junction-varactor parametric amplifier with limited pump power, for $f_s = 10^{-2} f_c$. It is assumed that $R_i = 0$ and $T_d = 290°K$.

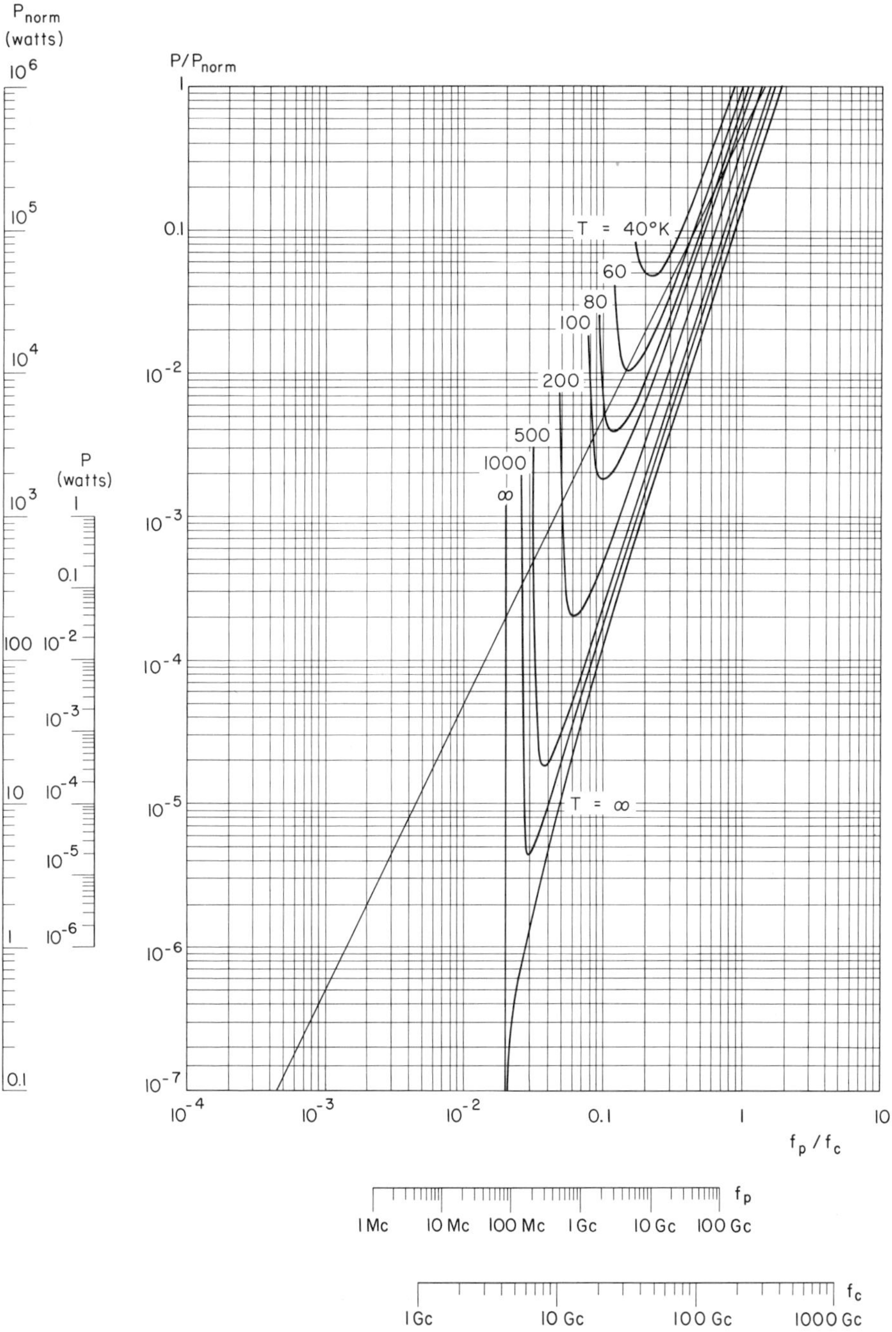

Figure D-28. Abrupt-junction-varactor parametric amplifier with limited pump power, for $f_s = 2 \times 10^{-2} f_c$. It is assumed that $R_i = 0$ and $T_d = 290°K$.

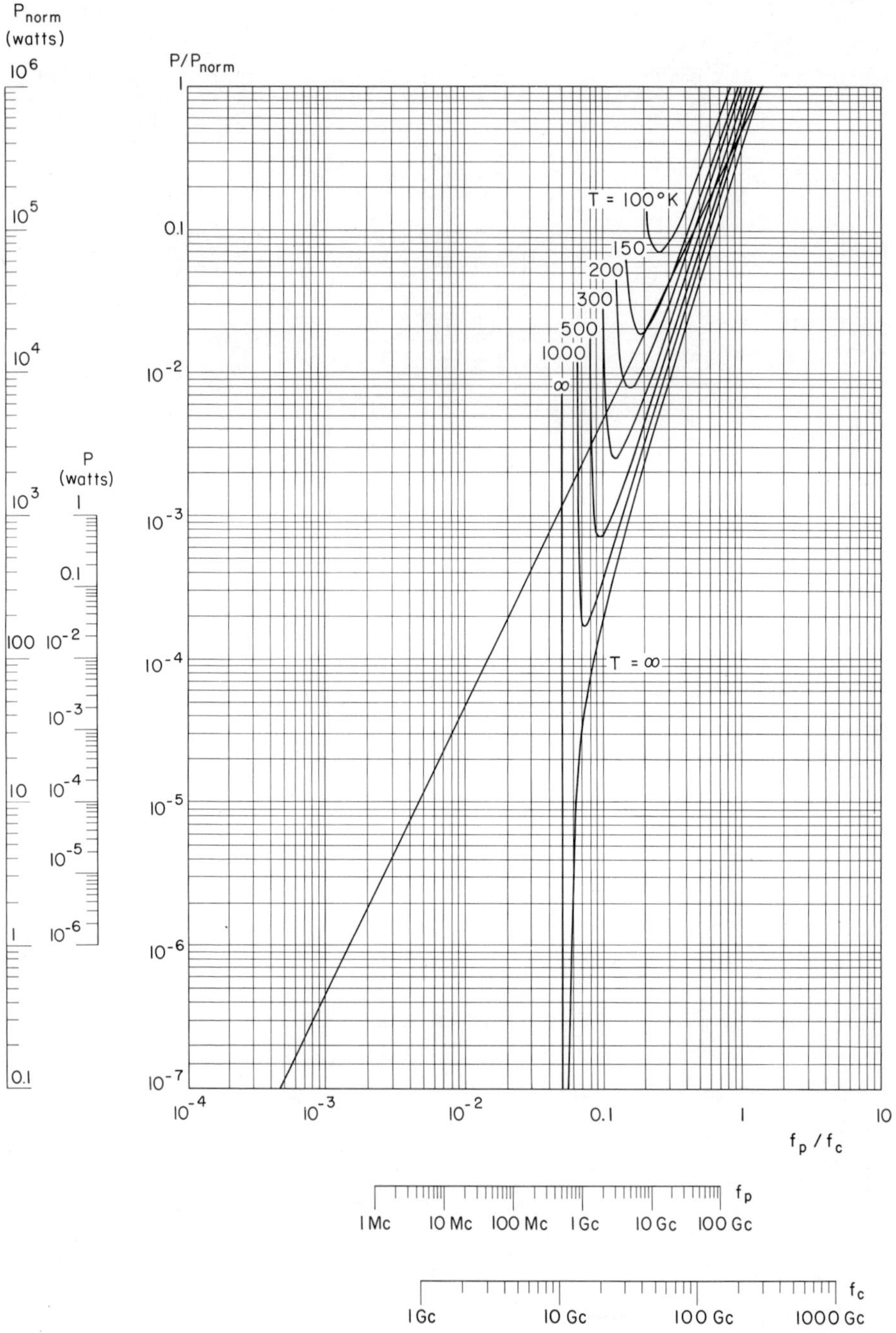

Figure D-29. Abrupt-junction-varactor parametric amplifier with limited pump power, for $f_s = 5 \times 10^{-2} f_c$. It is assumed that $R_i = 0$ and $T_d = 290°K$.

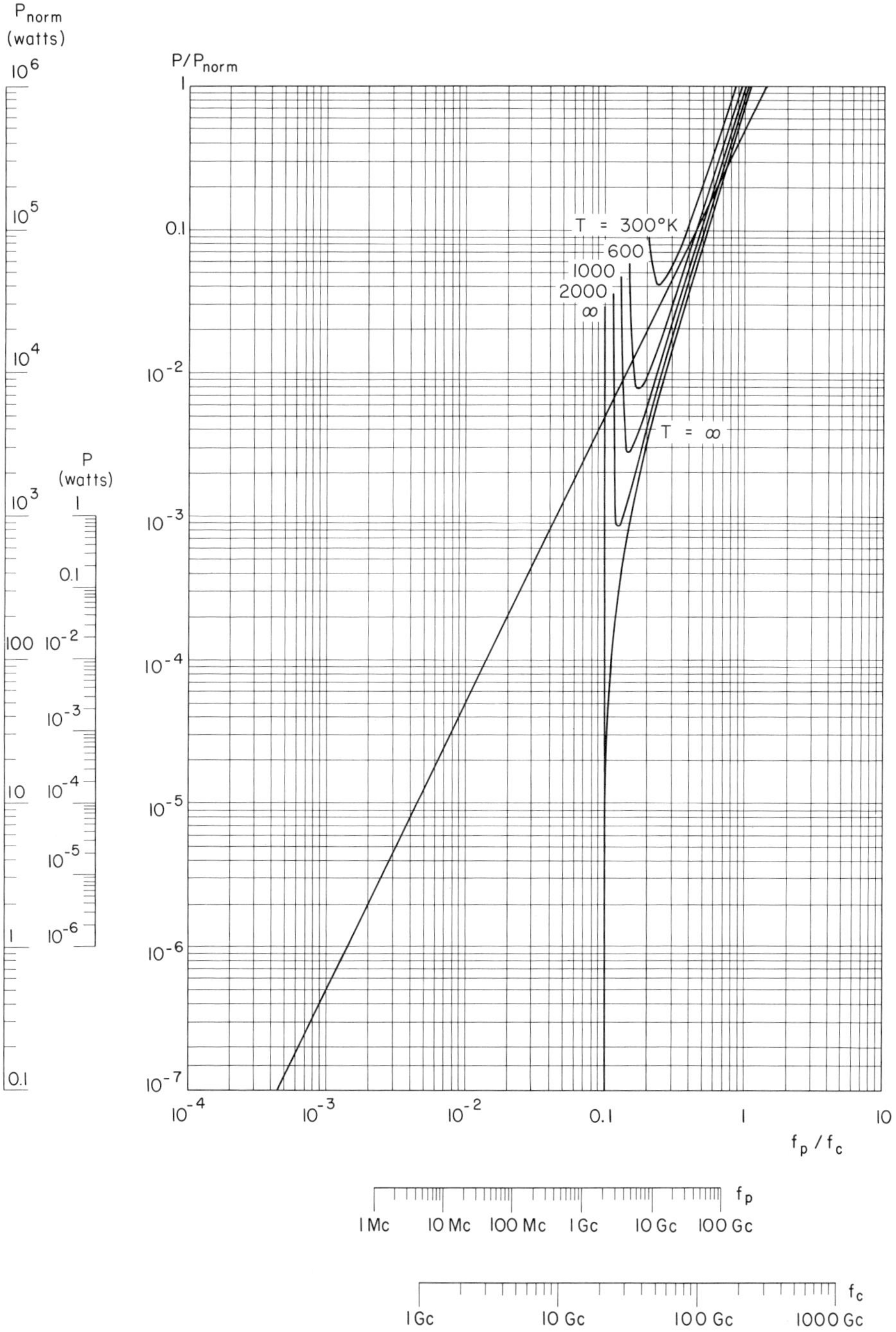

Figure D-30. Abrupt-junction-varactor parametric amplifier with limited pump power, for $f_s = 0.1f_c$. It is assumed that $R_i = 0$ and $T_d = 290°K$.

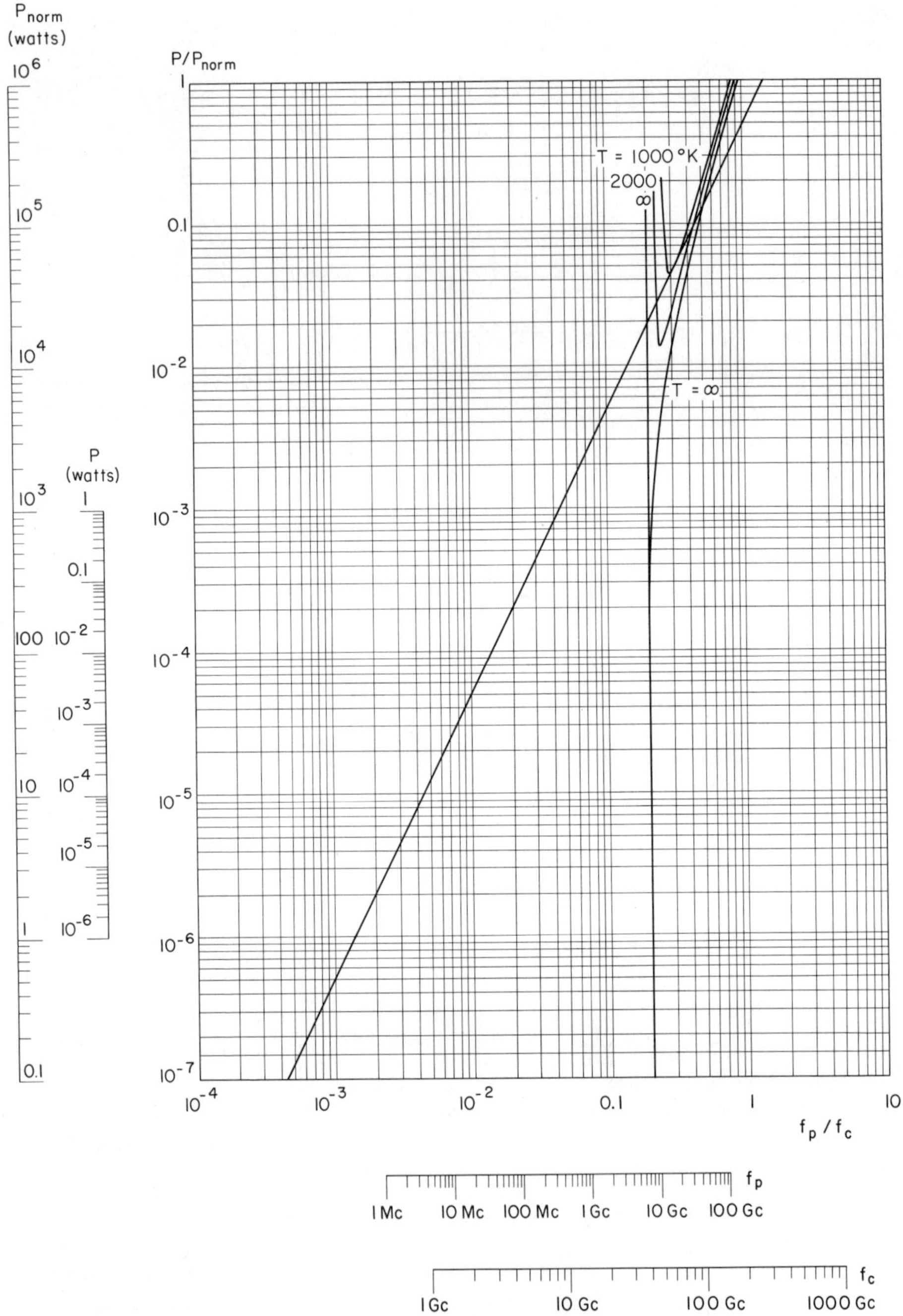

Figure D-31. Abrupt-junction-varactor parametric amplifier with limited pump power, for $f_s = 0.2f_c$. It is assumed that $R_i = 0$ and $T_d = 290°K$.

D-7. Doubler (Low-Power)

The graphs on pages 574-577 can be used to design abrupt-junction-varactor doublers with possibly less than maximum drive. They are based on the discussion in Section 8.4.

The powers are computed as if $S_{min} << S_{max}$; if it is not, merely multiply each of the powers by $(S_{max} - S_{min})^2/(S_{max} + S_{min})^2$.

Two service charts are given on pages 572 and 573: Figure D-32 is a plot of m_1 and m_2, and Figure D-33 is a plot of the maximum and minimum values of m_0. The magnitudes of input and output currents can be found from the formulas,

$$|I_1| = 2m_1 \frac{f_o}{f_c} \frac{V_B + \phi}{R_s} \frac{S_{max} - S_{min}}{S_{max} + S_{min}},$$

$$|I_2| = 4m_2 \frac{f_o}{f_c} \frac{V_B + \phi}{R_s} \frac{S_{max} - S_{min}}{S_{max} + S_{min}}.$$

There is a separate graph for each frequency. Use the locator chart below to find the appropriate graph. For other frequencies, interpolate between the graphs given.

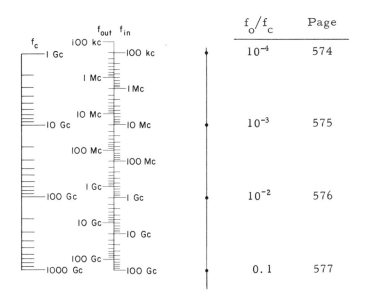

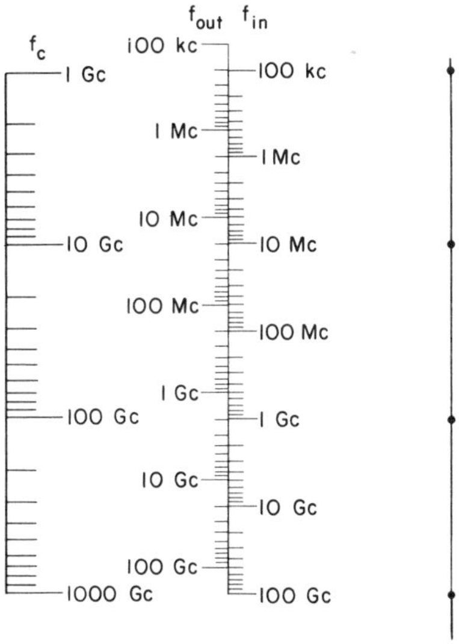

f_o/f_c	Page
10^{-4}	574
10^{-3}	575
10^{-2}	576
0.1	577

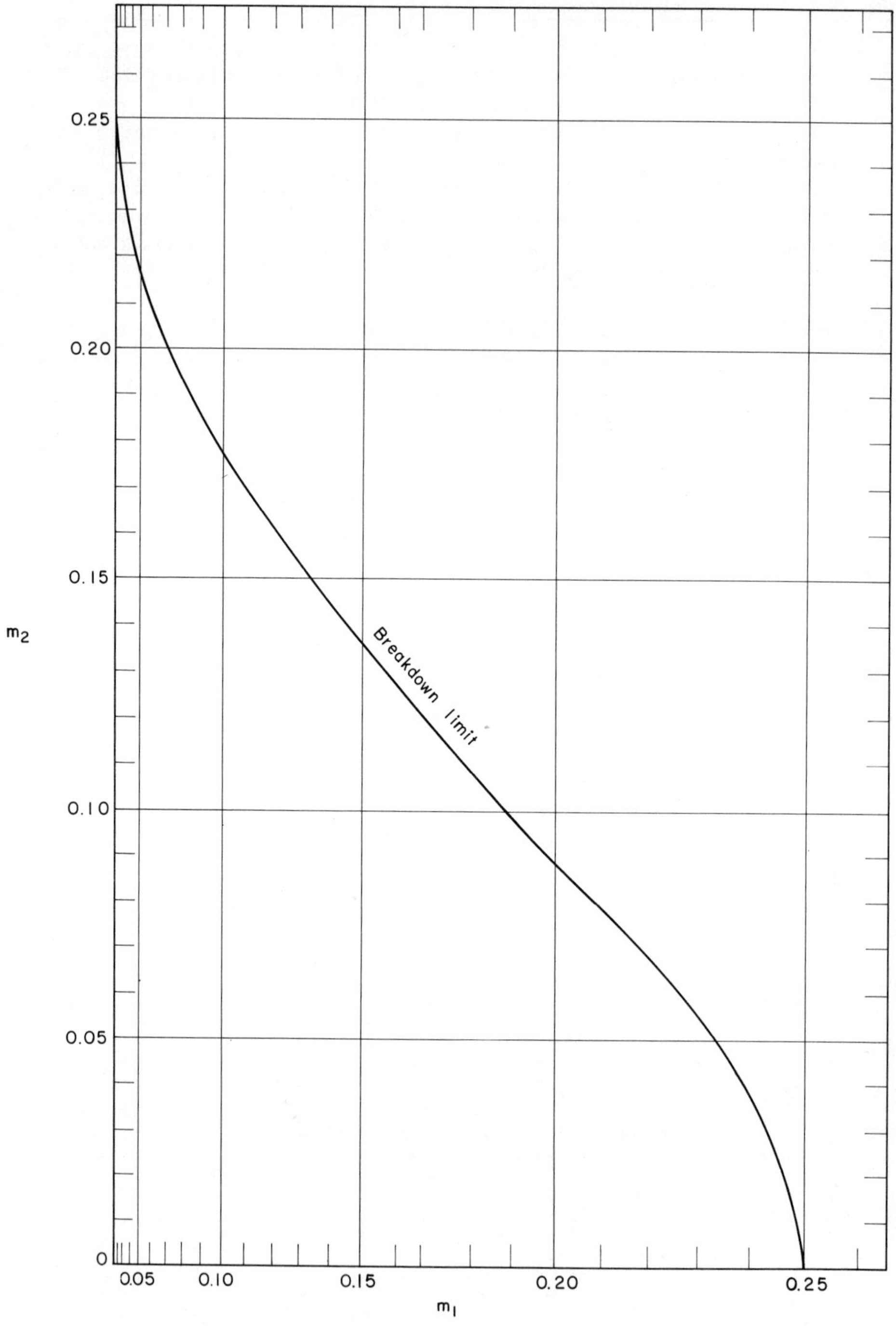

Figure D-32. Values of the modulation ratios m_1 and m_2.

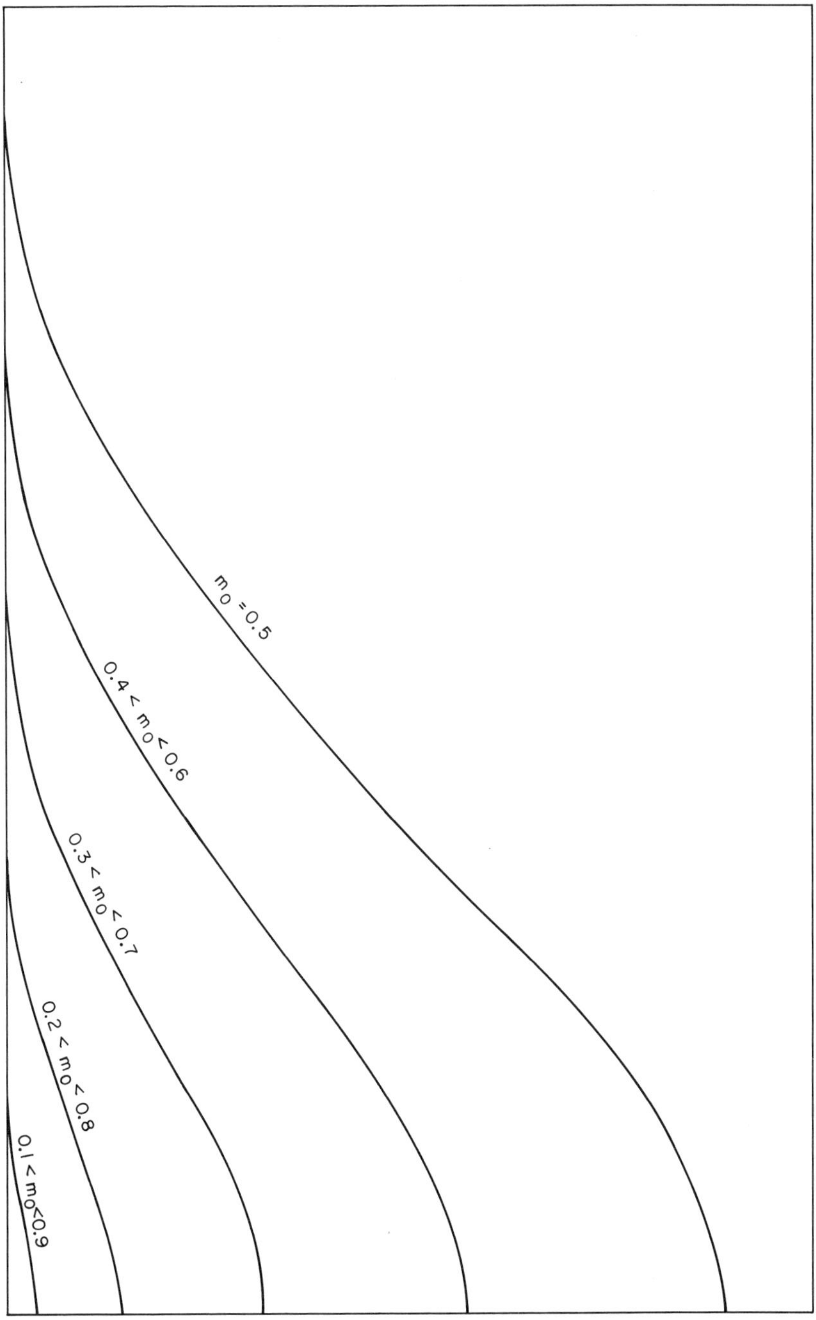

Figure D-33. To attain the values of m_1 and m_2 shown in Figure D-32 (opposite page), m_0 must lie within a certain range, as indicated by this figure. It is assumed that $S_{min} << S_{max}$; if it is not, then the quantity constrained by this figure is $m_0 - [S_{min}/(S_{max} - S_{min})]$, rather than merely m_0.

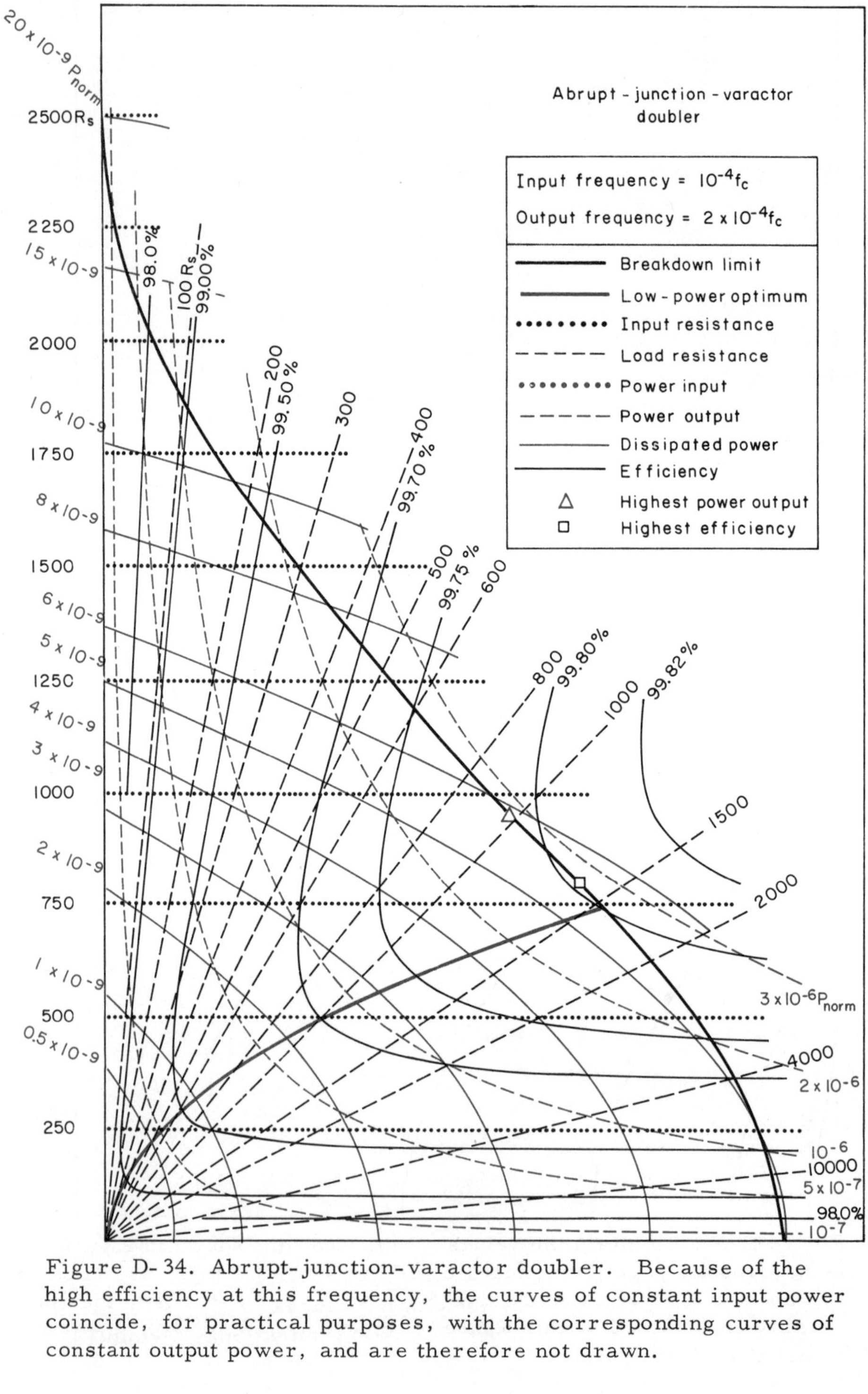

Figure D-34. Abrupt-junction-varactor doubler. Because of the high efficiency at this frequency, the curves of constant input power coincide, for practical purposes, with the corresponding curves of constant output power, and are therefore not drawn.

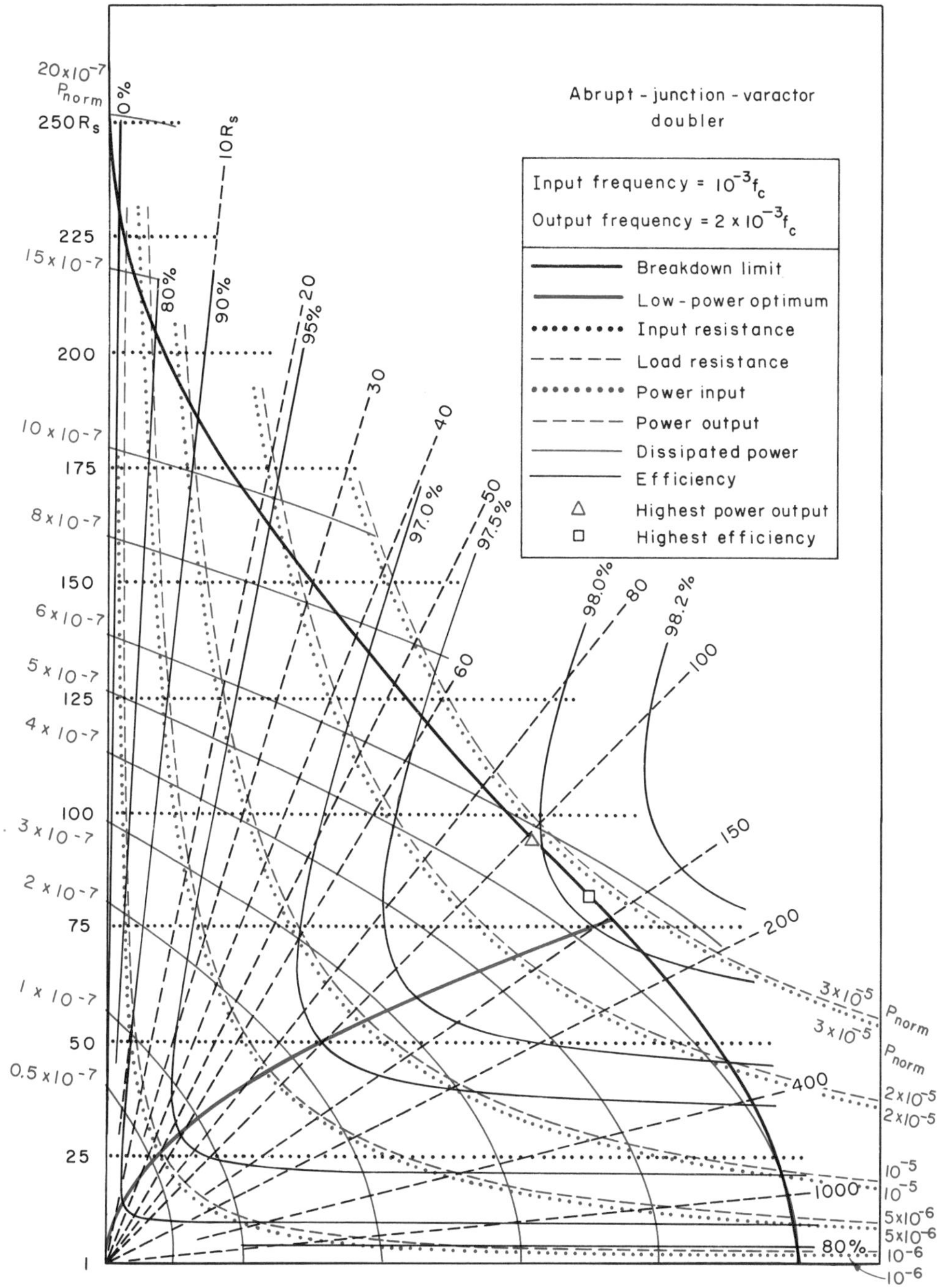

Figure D-35. Abrupt-junction-varactor doubler.

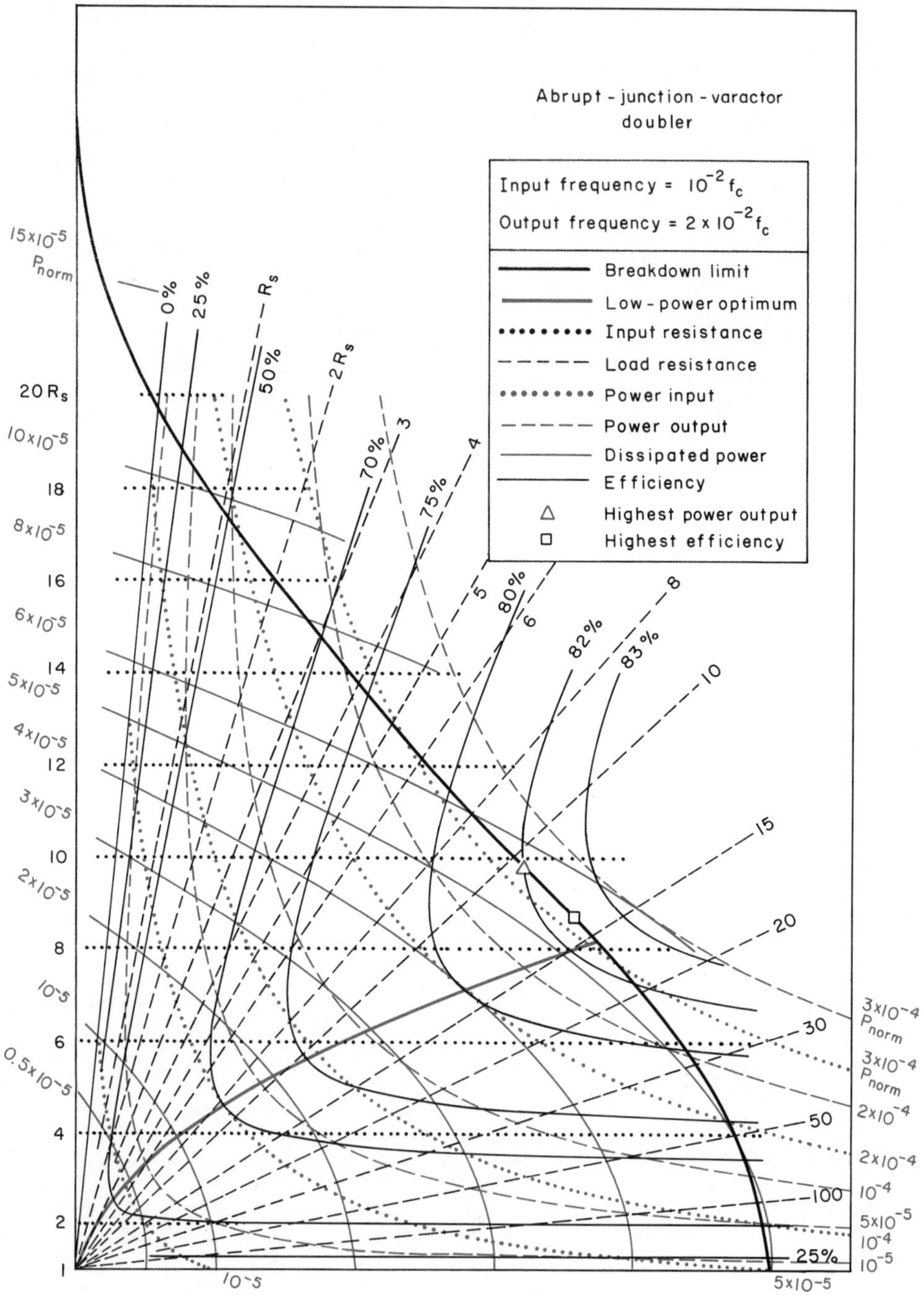

Figure D-36. Abrupt-junction-varactor doubler.

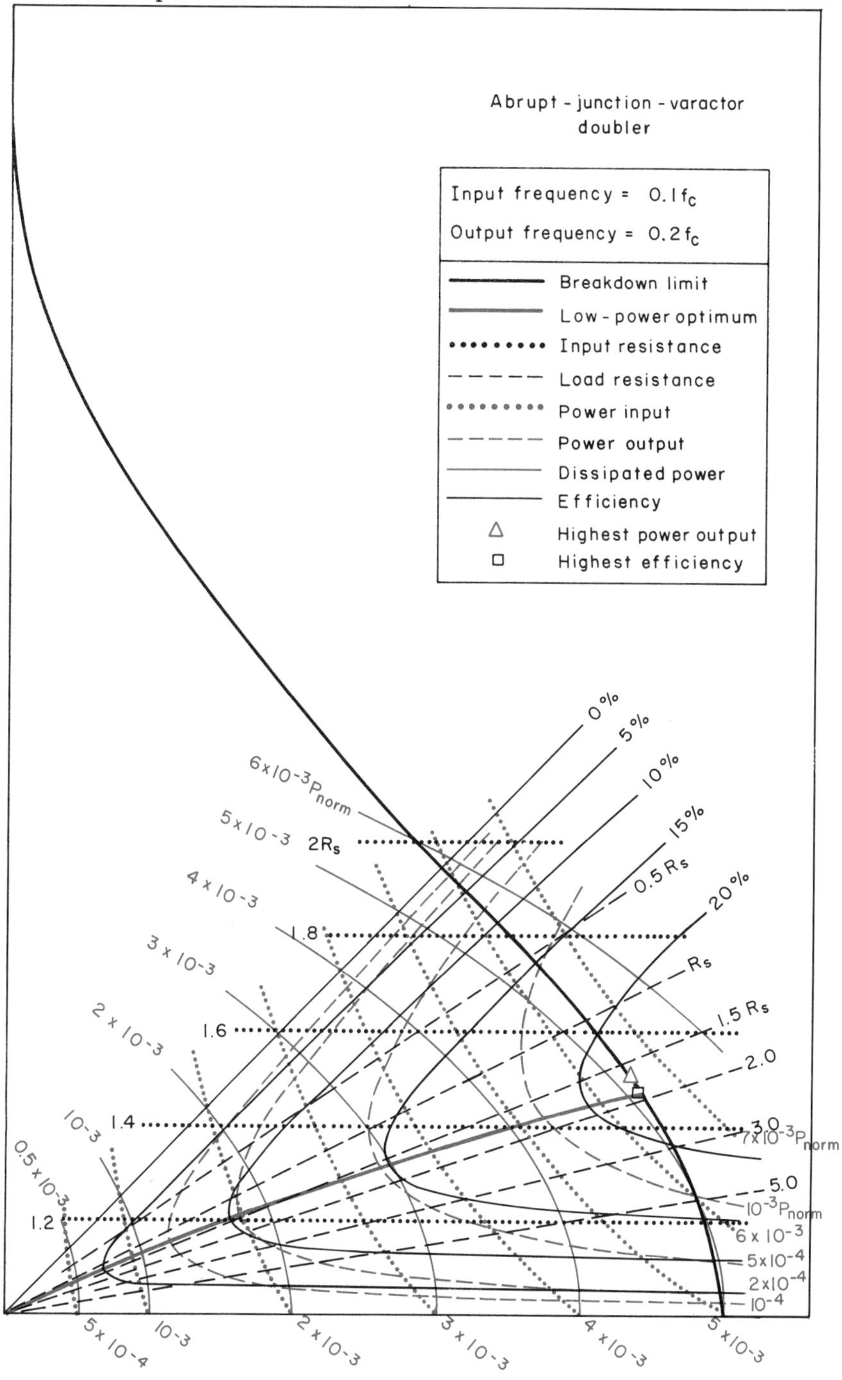

Figure D-37. Abrupt-junction-varactor doubler.

D-8. Divide-by-Two Circuit (Low-Power)

The graphs on pages 579-582 can be used to design abrupt-junction-varactor divide-by-two circuits with possibly less than maximum drive. They are based on the discussion of Section 9.4. The service charts, Figures D-32 and D-33, on pages 572 and 573, apply also to the divide-by-two circuit.

The powers are computed as if $S_{min} << S_{max}$; if it is not, merely multiply each of the powers by $(S_{max} - S_{min})^2/(S_{max} + S_{min})^2$.

There is a separate graph for each frequency. Use the locator chart below to find the appropriate graph. For other frequencies interpolation can be used.

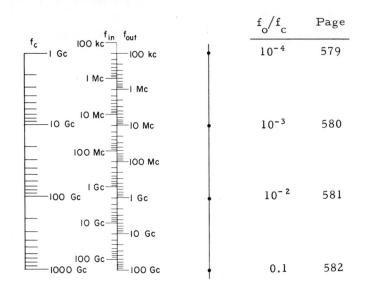

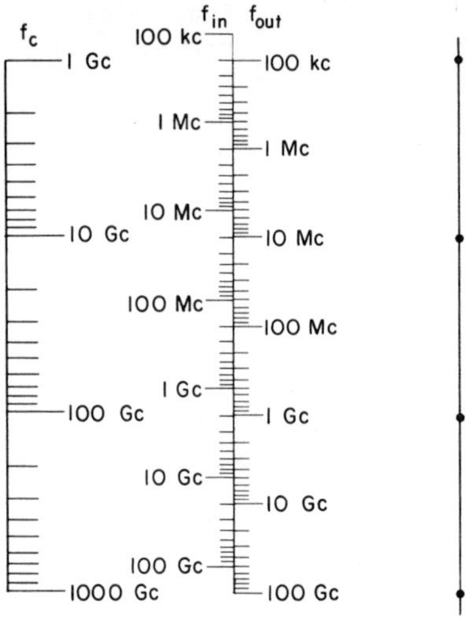

f_o/f_c	Page
10^{-4}	579
10^{-3}	580
10^{-2}	581
0.1	582

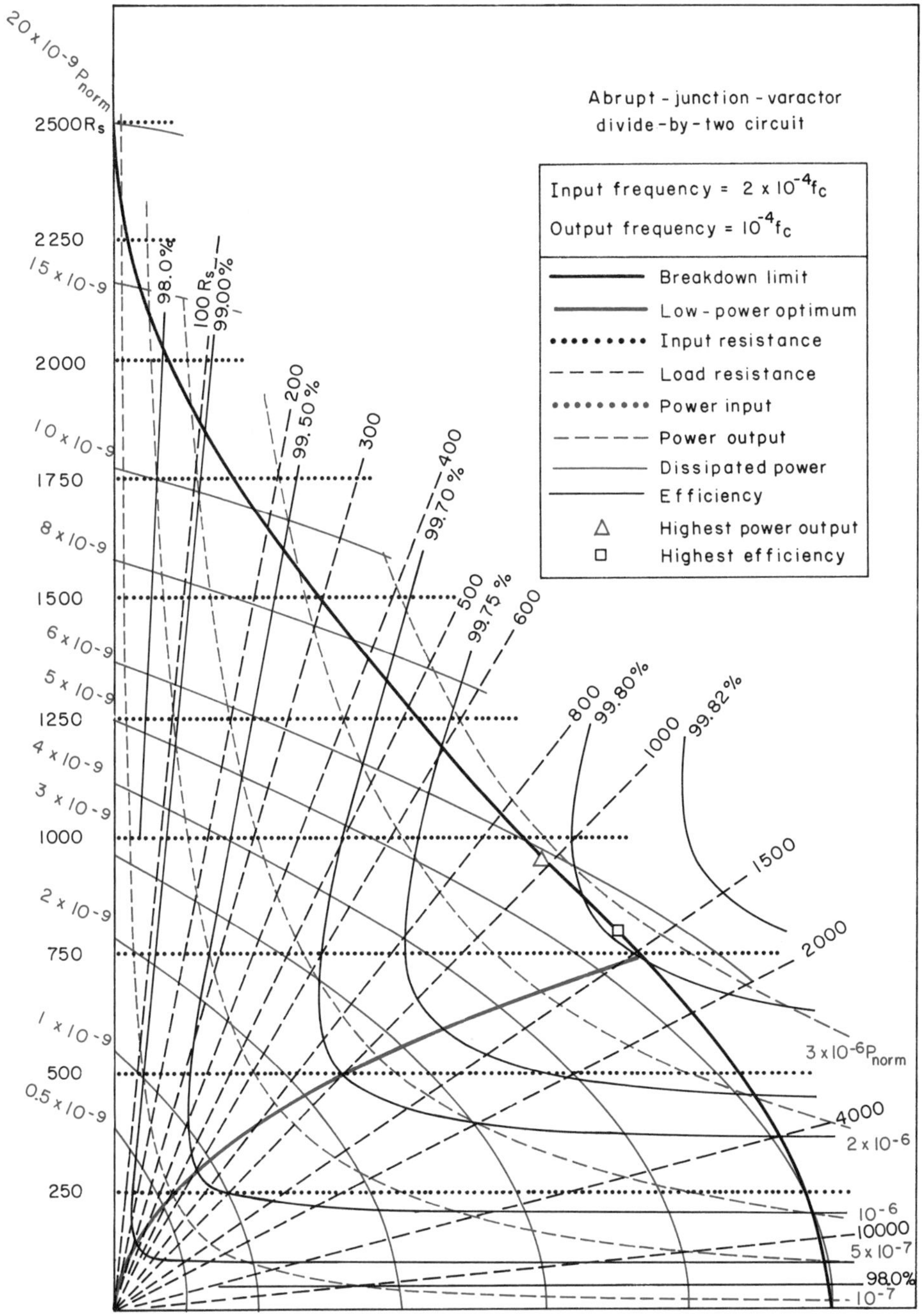

Figure D-38. Abrupt-junction-varactor divide-by-two circuit. Because of the high efficiency at this frequency, the curves of constant input power coincide, for practical purposes, with the corresponding curves of constant output power, and are therefore not drawn.

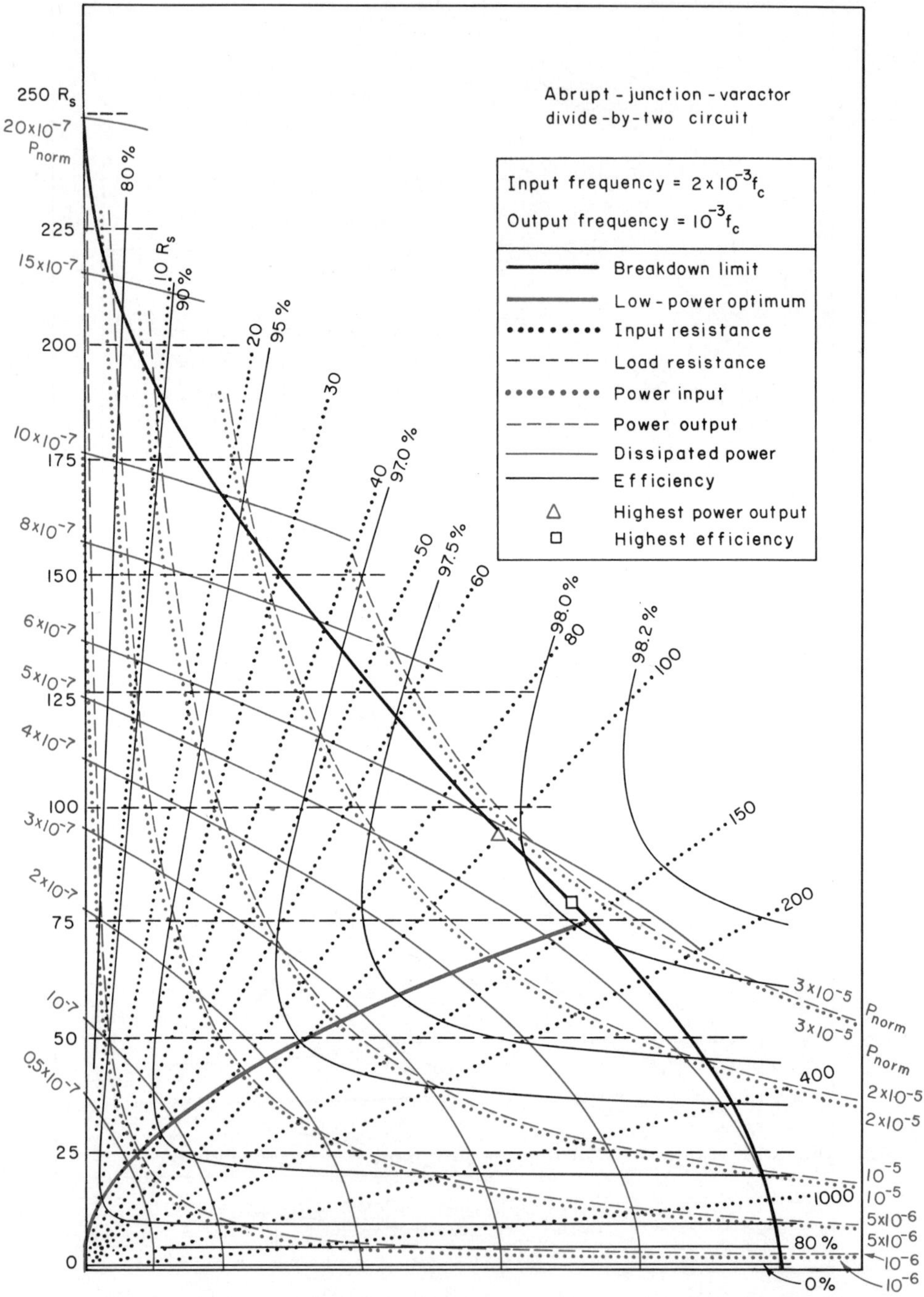

Figure D-39. Abrupt-junction-varactor divide-by-two circuit.

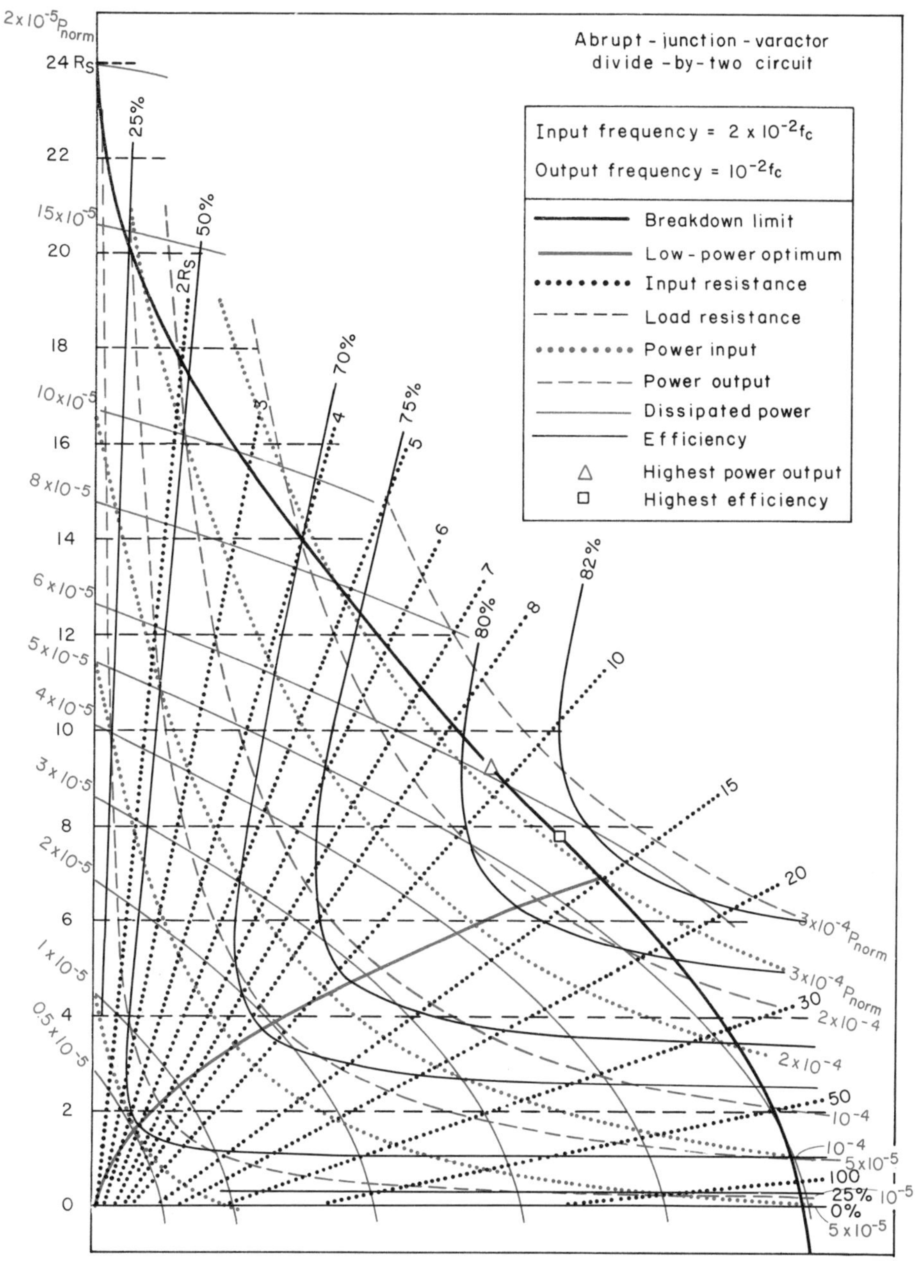

Figure D-40. Abrupt-junction-varactor divide-by-two circuit.

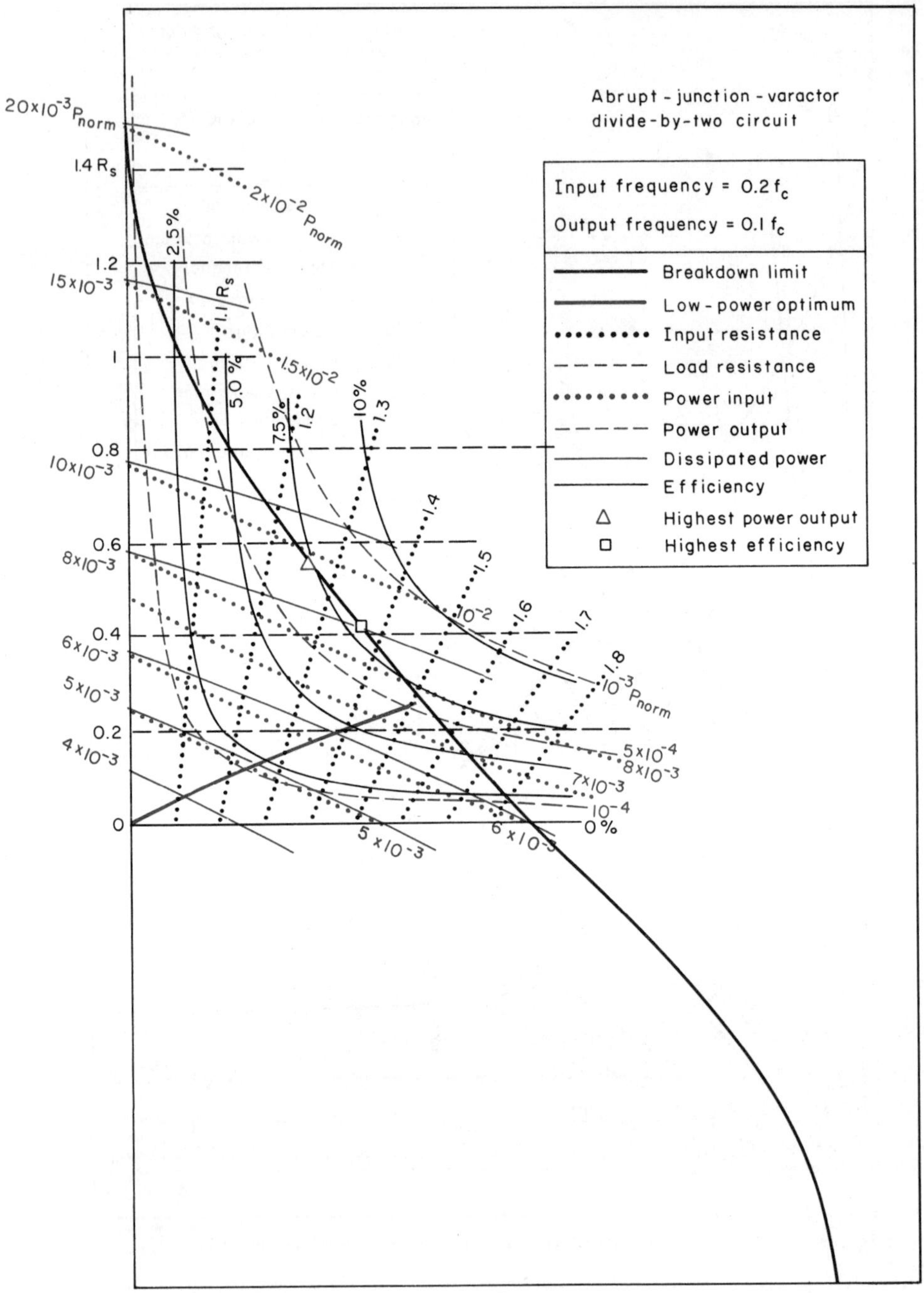

Figure D-41. Abrupt-junction-varactor divide-by-two circuit.

D-9. Multipliers and Dividers (Maximum Efficiency)

Figures D-42 to D-51 relate input power P_{in}, output power P_{out}, and efficiency ϵ to the varactor parameters f_c and P_{norm}, for all the multipliers and dividers analyzed in Chapters 8 and 9 and Appendix F. These charts were originated by S. L. Johnson, and are reproduced here with his kind permission.

It is assumed the varactor is fully driven and that the load is tuned and adjusted to maximize the efficiency, and that all idler terminations are lossless and tuned. We have assumed that $S_{min} << S_{max}$; if not, then all powers should be multiplied by $(S_{max} - S_{min})^2/(S_{max} + S_{min})^2$.

In each case, first use the nomograph at the bottom of diode space to find the efficiency, and then use this value of ϵ to find the input and output power. The dissipated power can be calculated using Figure D-52, on page 594.

All the multipliers and dividers (except the two doublers) have an output power that has a maximum value for a finite efficiency. Only the output power on the high-efficiency side of this maximum is given in Figures D-44 to D-51. At low efficiencies, the output power can be calculated most easily as the product of the input power (which is given at low efficiencies) times the efficiency.

The graded-junction-varactor doubler is on page 584. The abrupt-junction-varactor multipliers and dividers are located as follows:

Multiply by	Page
$\frac{1}{4}$	593
$\frac{1}{2}$	592
2	585
3 (1-2-3)	586
4 (1-2-4)	587

Multiply by	Page
5 (1-2-4-5)	588
6 (1-2-4-6)	589
6 (1-2-3-6)	590
8 (1-2-4-8)	591

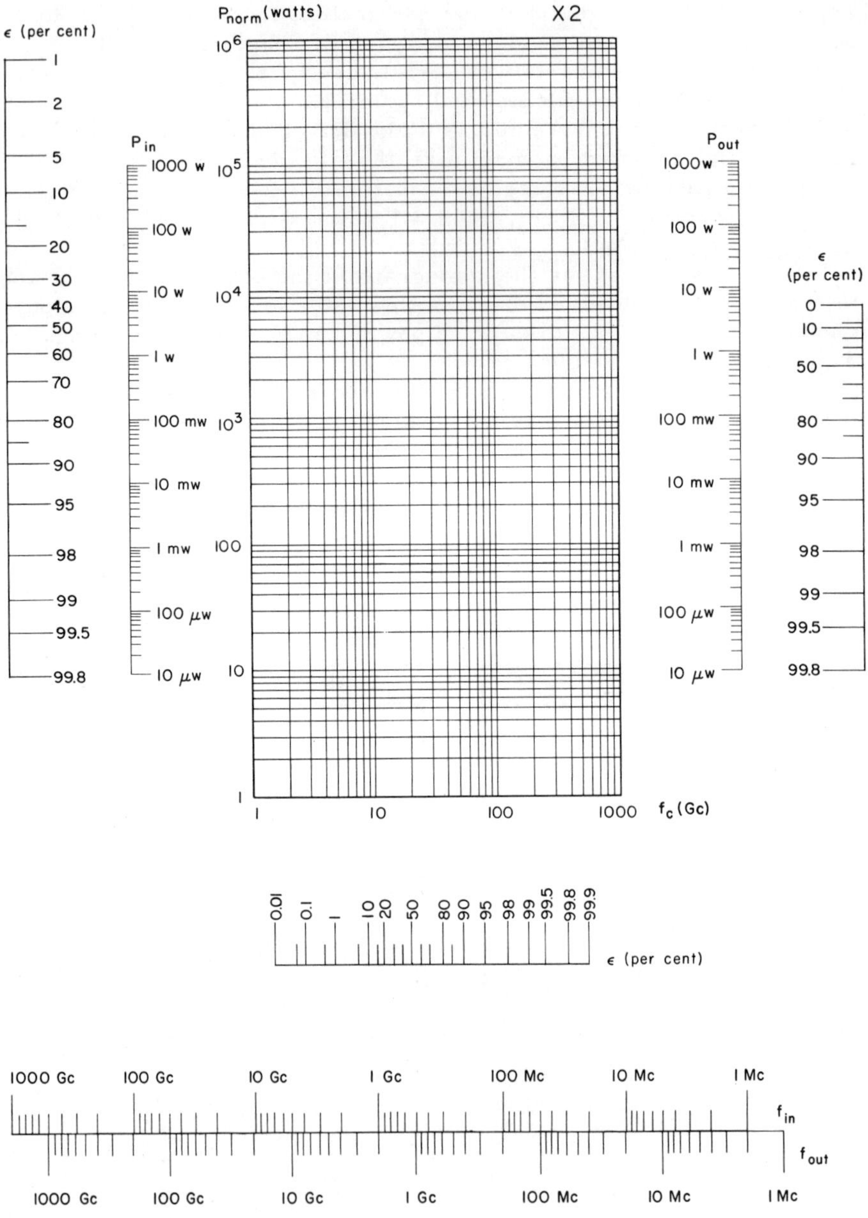

Figure D-42. Graded-junction-varactor doubler operated with maximum efficiency.

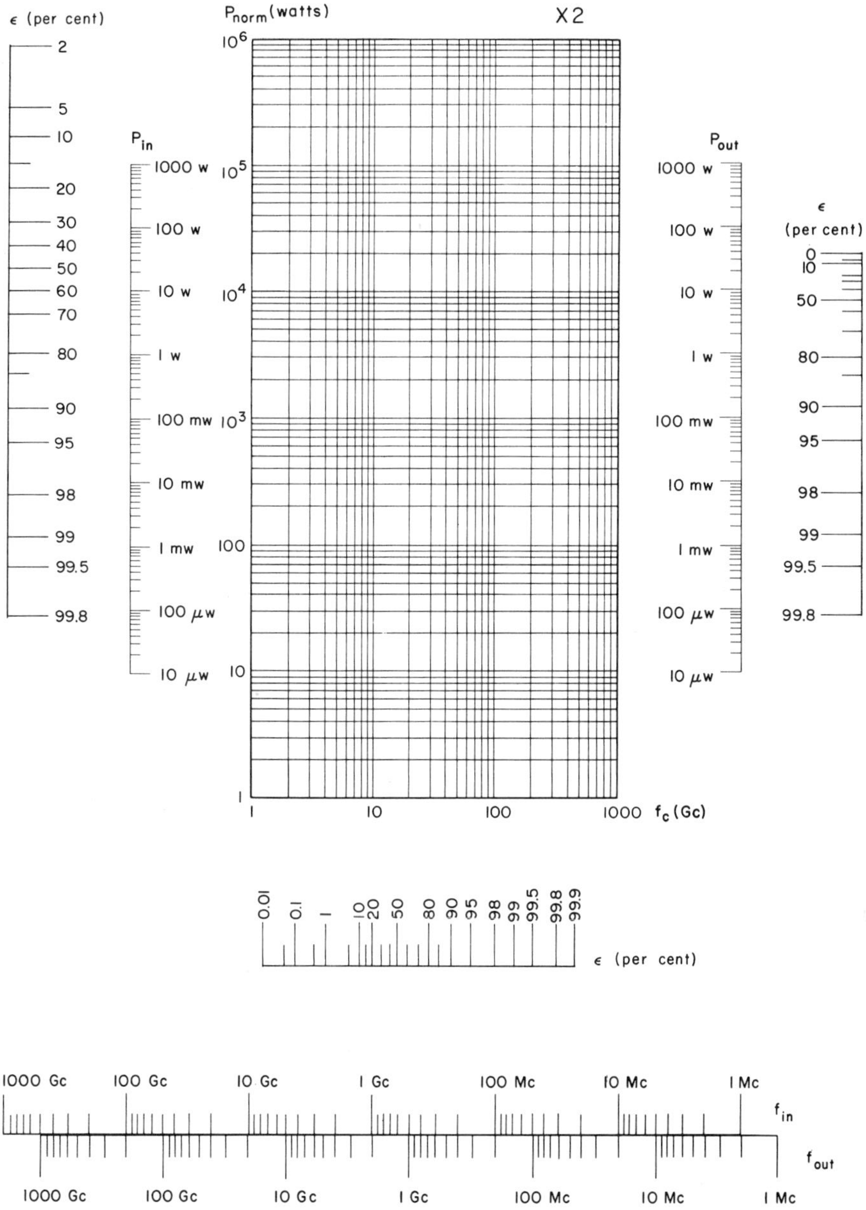

Figure D-43. Abrupt-junction-varactor doubler operated with maximum efficiency.

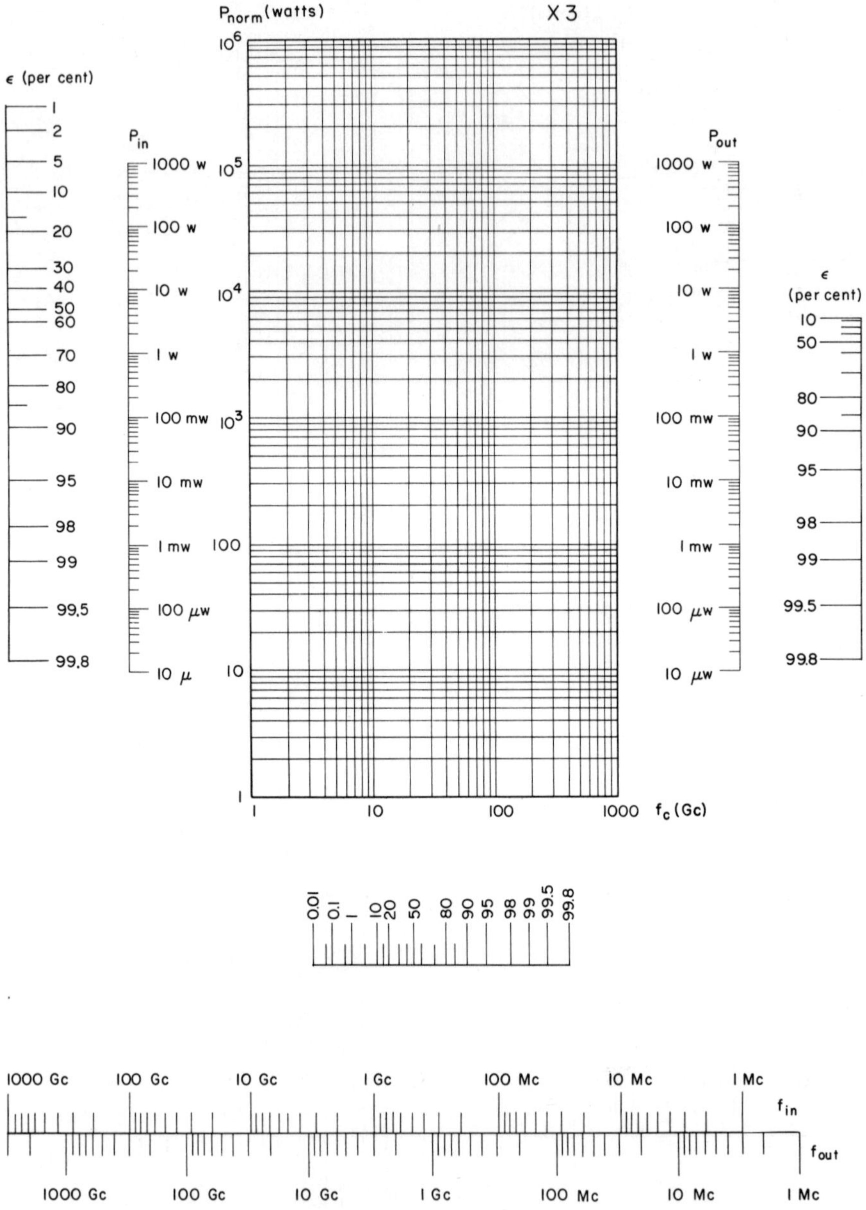

Figure D-44. Abrupt-junction-varactor 1-2-3 tripler operated with maximum efficiency. The idler termination at frequency $2f_o$ is tuned and lossless.

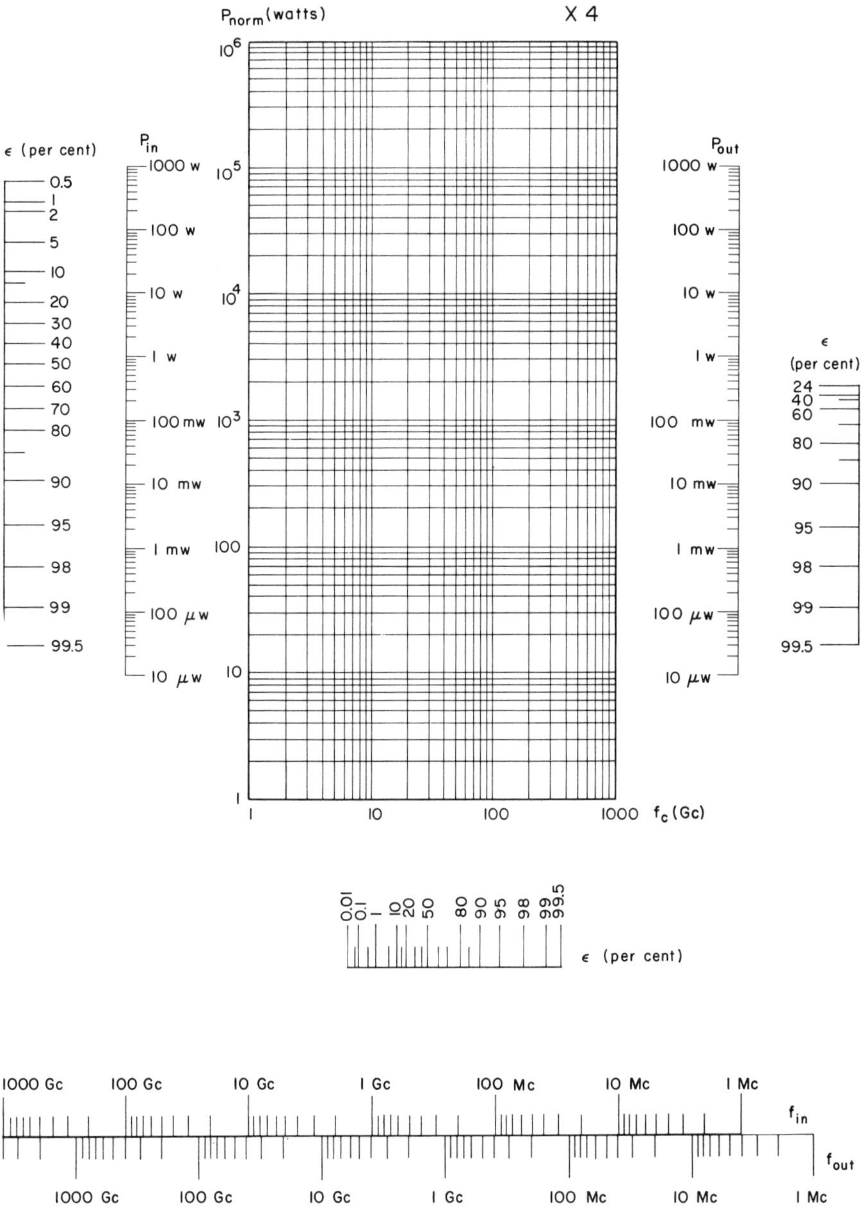

Figure D-45. Abrupt-junction-varactor 1-2-4 quadrupler operated with maximum efficiency. The idler termination at frequency $2f_o$ is tuned and lossless.

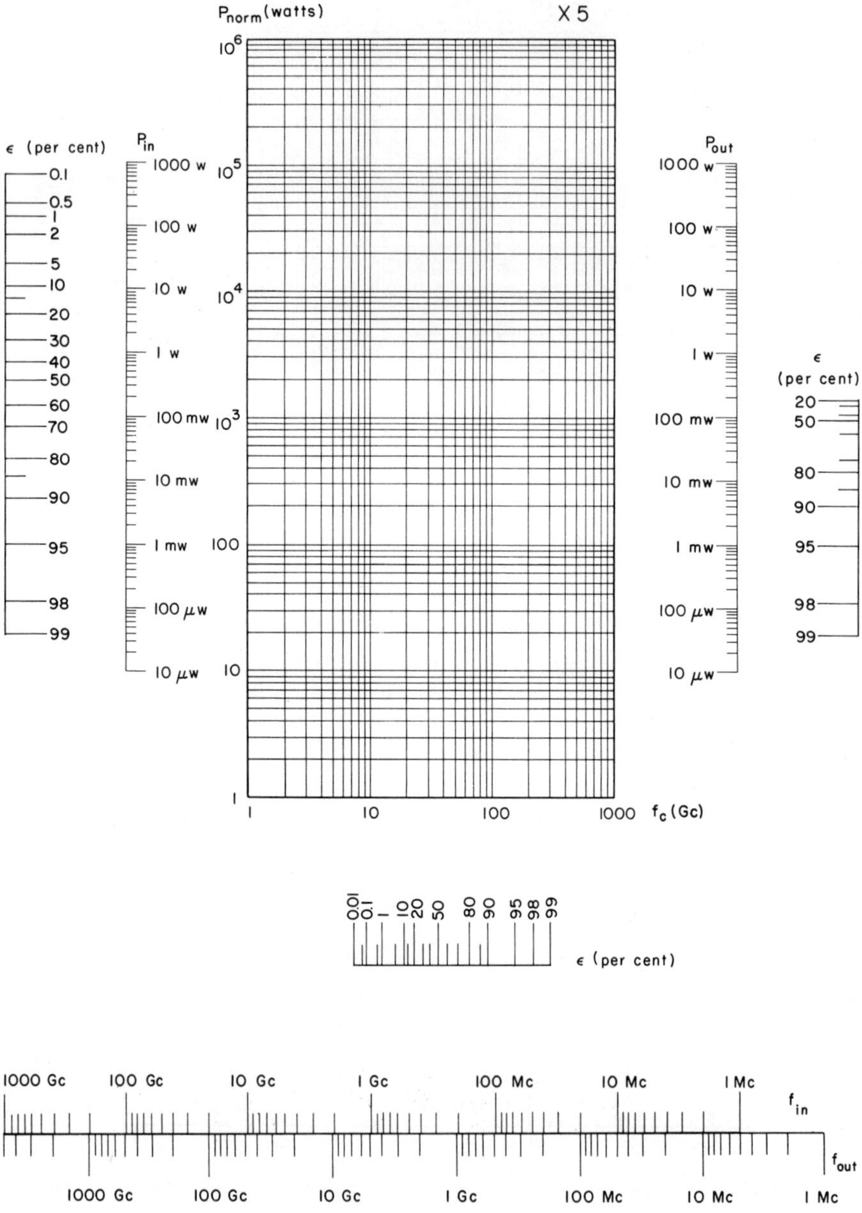

Figure D-46. Abrupt-junction-varactor 1-2-4-5 quintupler operated with maximum efficiency. The idler terminations at frequencies $2f_o$ and $4f_o$ are tuned and lossless.

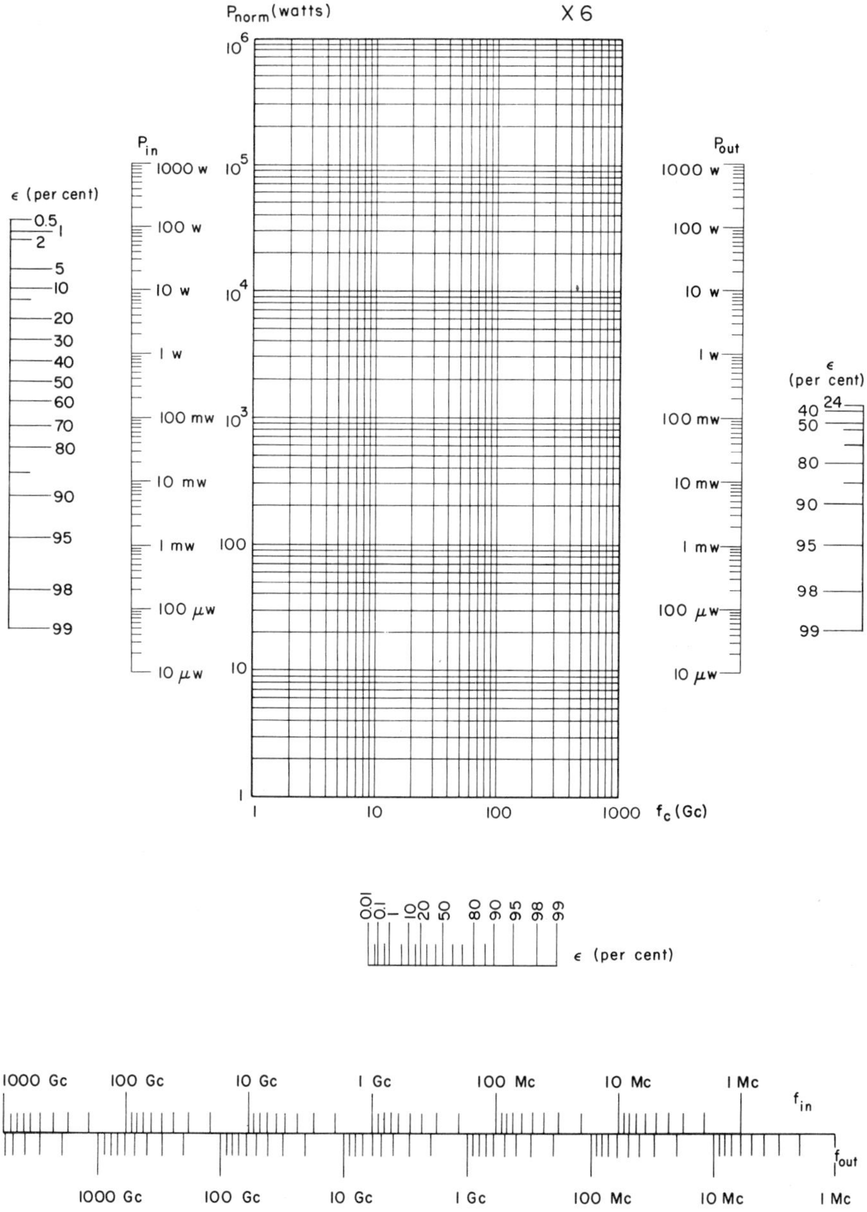

Figure D-47. Abrupt-junction-varactor 1-2-4-6 sextupler operated with maximum efficiency. The idler terminations at frequencies $2f_o$ and $4f_o$ are tuned and lossless.

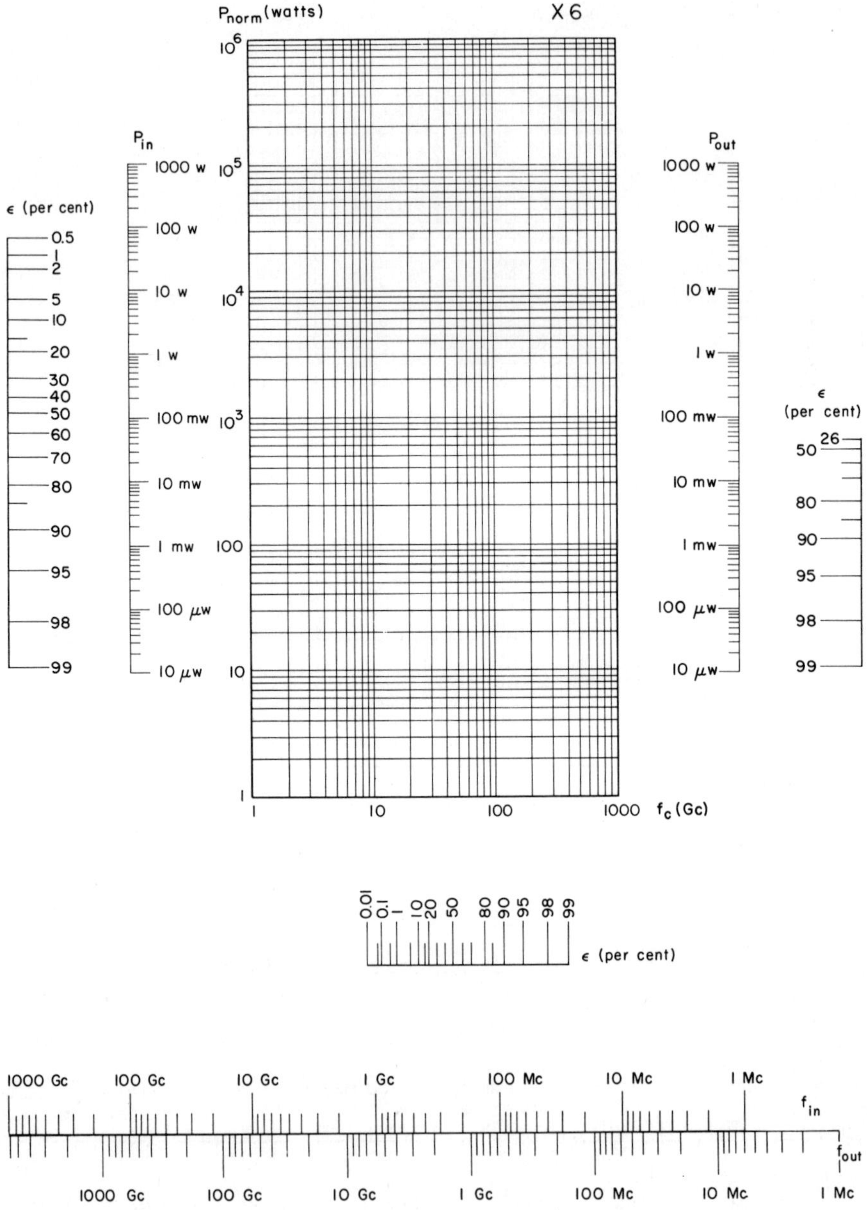

Figure D-48. Abrupt-junction-varactor 1-2-3-6 sextupler operated with maximum efficiency. The idler terminations at frequencies $2f_0$ and $3f_0$ are tuned and lossless.

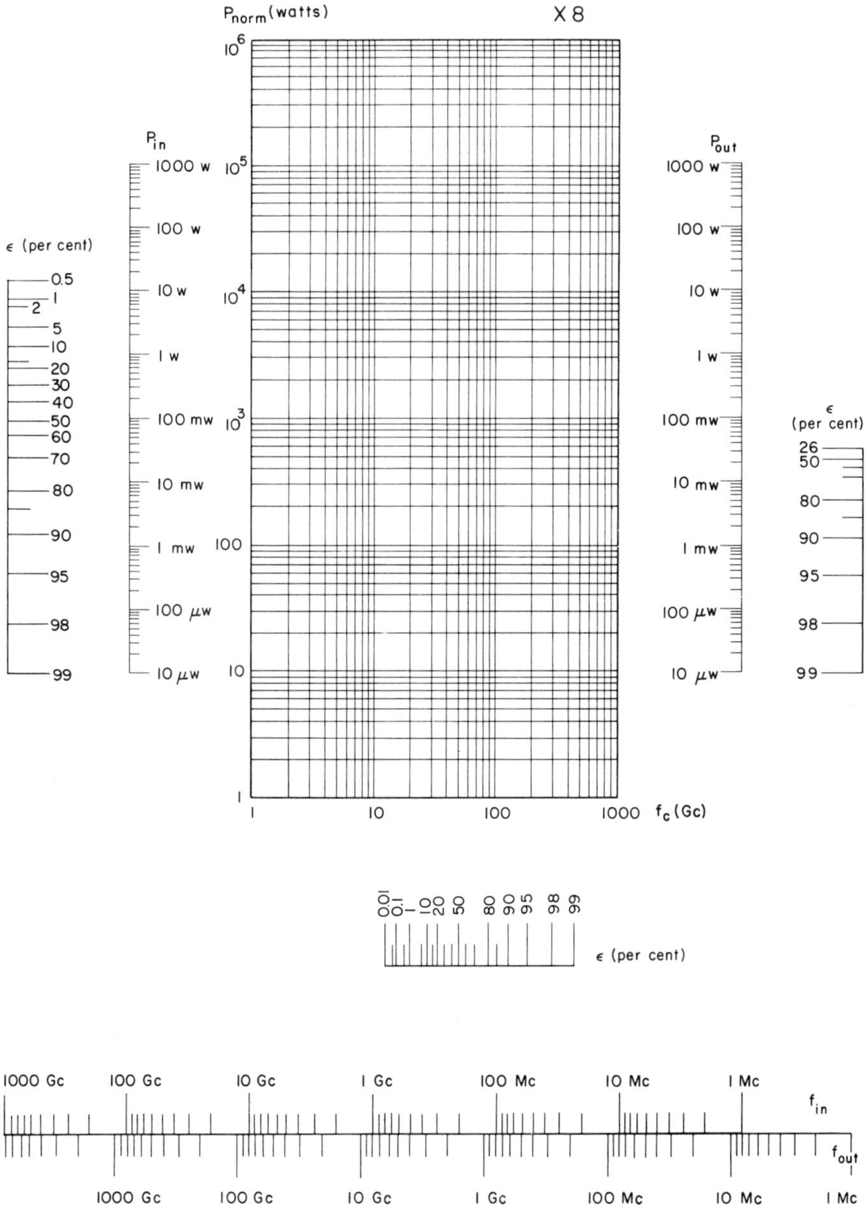

Figure D-49. Abrupt-junction-varactor 1-2-4-8 octupler operated with maximum efficiency. The idler terminations at frequencies $2f_0$ and $4f_0$ are tuned and lossless.

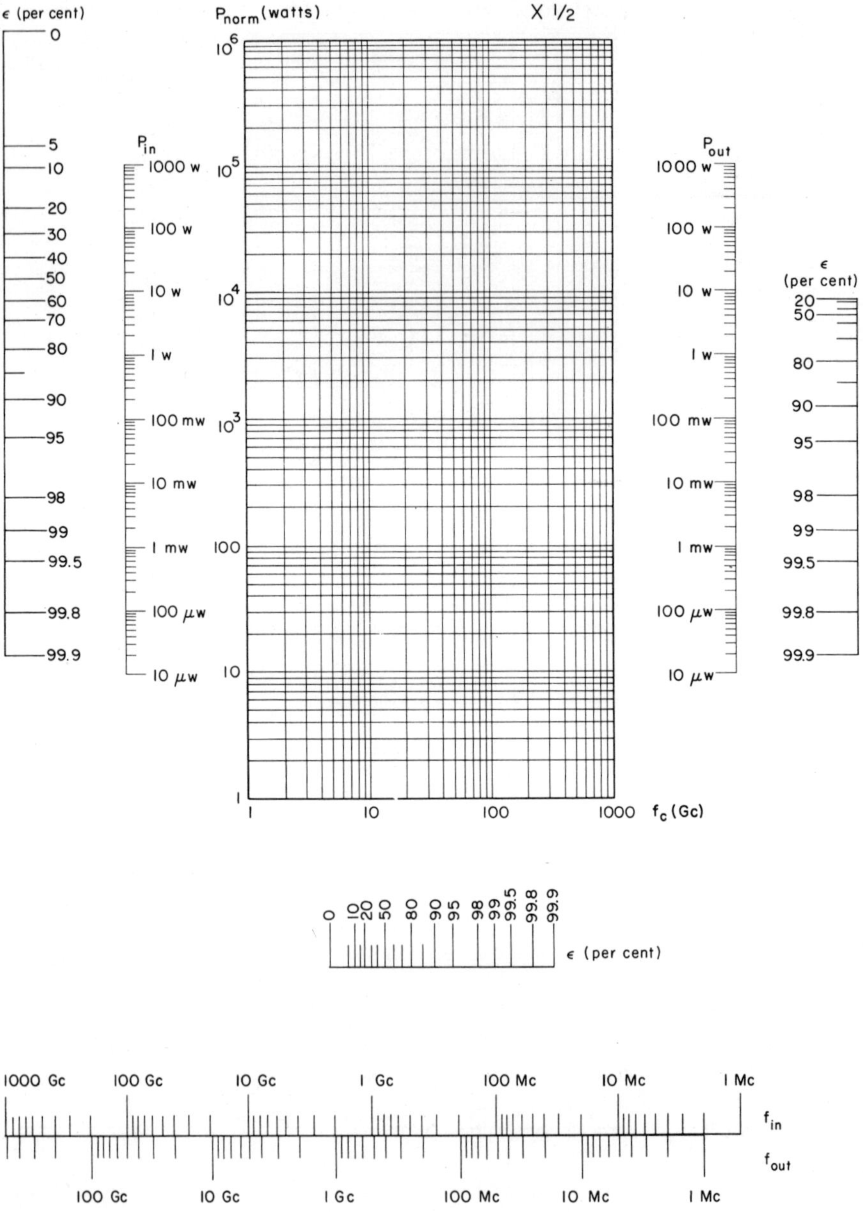

Figure D-50. Abrupt-junction-varactor divide-by-two circuit operated with maximum efficiency.

r P_{in}, output
re related by
ıe formulas

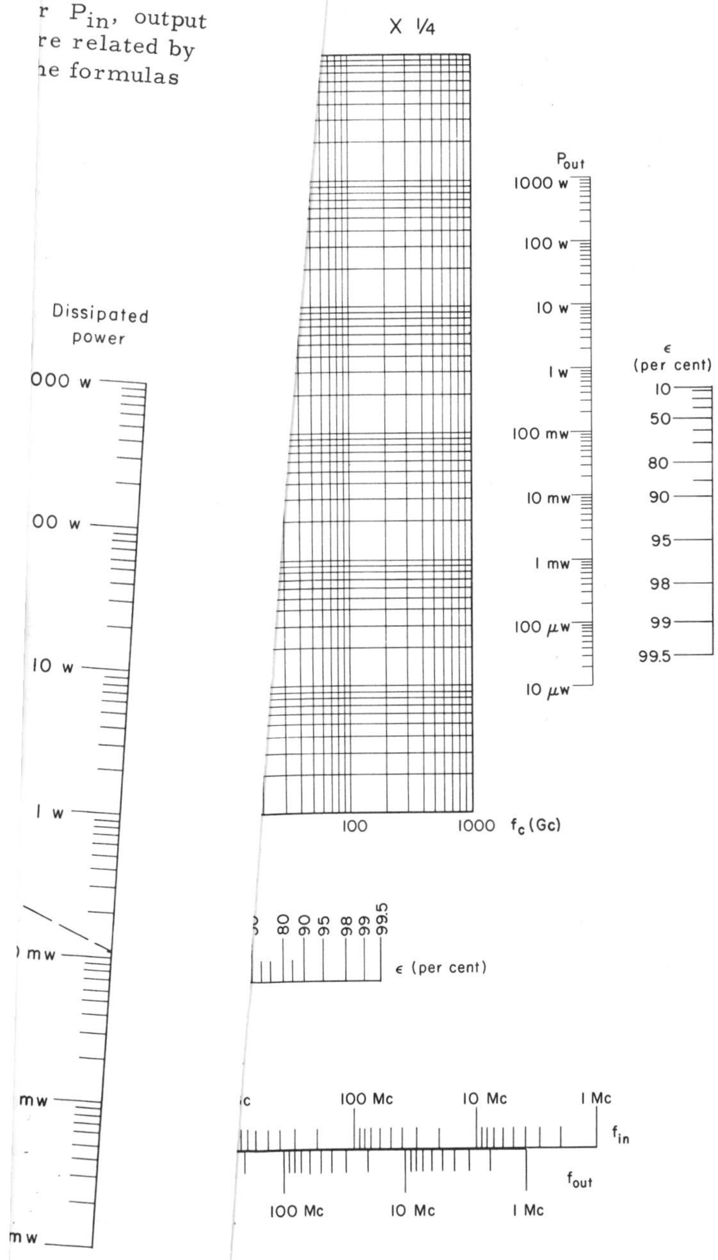

ıction-varactor divide-by-four cir-
ıum efficiency. The idler termina-
is tuned and lossless.

D-10. Power-Efficiency

For any of the multipliers or dividers, input powe
power P_{out}, dissipation P_{diss}, and efficiency ϵ a
the nomograph below, Figure D-52. It is based on t

$$P_{out} = \epsilon P_{in}; \qquad P_{diss} = (1 - \epsilon) P_{in}.$$

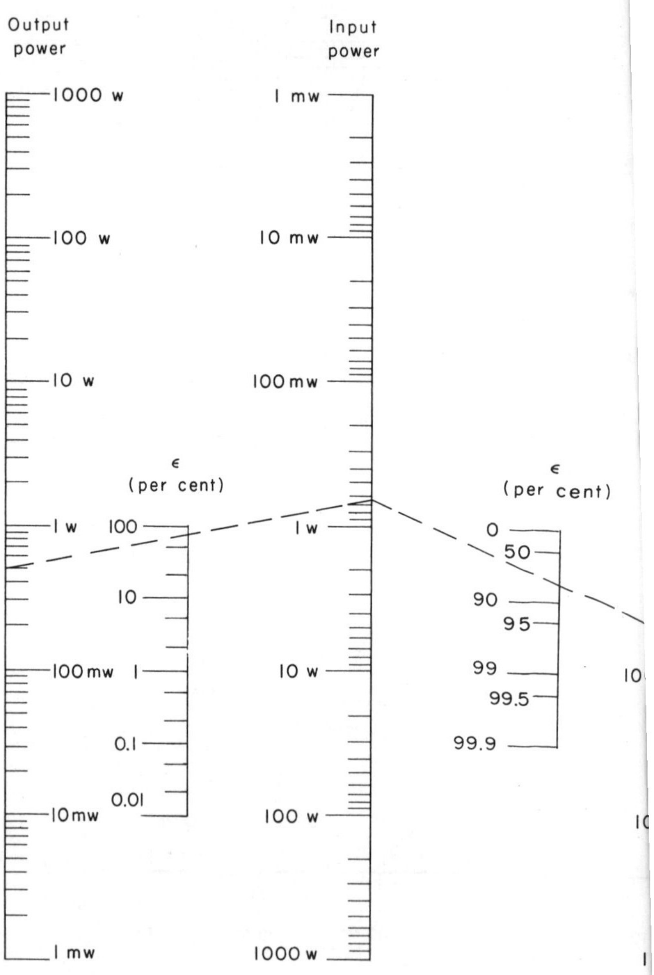

Figure D-52. Power-efficiency nomograph.

Appendix E

VOLTAGE TUNING

One important application of varactors that has not been discussed in this book is voltage tuning. The voltage-variable capacitance of the varactor is used to tune a resonant circuit from a slowly varying externally applied voltage V_0 (for use in automatic-frequency-control circuits, as voltage-variable reactances in frequency modulators, and in many other similar applications).

Most analyses of such circuits use lossless small-signal models for the varactor. Often, however, neither the loss nor the fact that large signals are present can be ignored without leading to serious engineering misjudgments. Even at small signals, the varactor loss limits the quality factor Q of the tuned circuit, and the inherent minimum elastance and stray capacitance further limit the frequency range possible. These effects are discussed in Section E-1. If the rf signal is large (in a sense to be indicated), further effects, discussed in Section E-2, must be considered.

E-1. Small-Signal Effects

There are two major limitations of practical varactors for voltage tuning when the rf power is small. These are (1) the degradation in quality factor Q caused by varactor loss and (2) the finite tuning range implied by the minimum elastance S_{min}.

Consider the varactor in a simple tuned circuit,‡ Figure E-1.

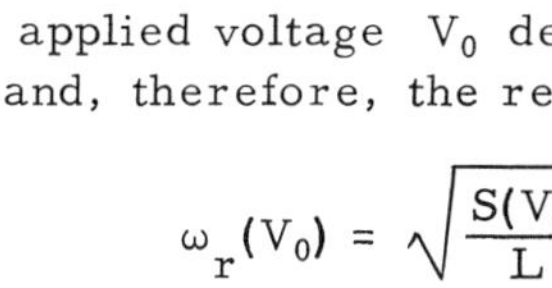

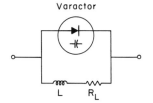

Figure E-1. Varactor tuning of a simple tuned circuit. The applied voltage V_0 is not shown.

The applied voltage V_0 determines the elastance $S(V_0)$ of the varactor and, therefore, the resonant frequency ω_r of the tuned circuit,

$$\omega_r(V_0) = \sqrt{\frac{S(V_0)}{L}}. \tag{E-1}$$

‡This circuit is chosen because the change in frequency is greatest for a given change in elastance. Other circuits with additional stored capacitive energy have resonant frequencies that do not vary as rapidly.

Since the varactor has a maximum elastance S_{max} at breakdown voltage V_B, the maximum resonant frequency is $\omega_{r,max}$:

$$\omega_{r,max} = \sqrt{\frac{S_{max}}{L}}. \tag{E-2}$$

The varactor also has a minimum elastance S_{min}, which is determined by the highest forward voltage that can be applied before forward conduction current and noise become intolerable. In many practical varactors, $S_{min} \ll S_{max}$, and can therefore be ignored, as far as parametric amplification, frequency conversion, and harmonic generation are concerned. However, for voltage tuning, we cannot neglect S_{min}, even though it is only 1 per cent or less of S_{max}, because it determines the minimum resonant frequency $\omega_{r,min}$:

$$\omega_{r,min} = \sqrt{\frac{S_{min}}{L}}. \tag{E-3}$$

As the bias voltage is increased, the resonant frequency ω_r increases from $\omega_{r,min}$ to $\omega_{r,max}$ as the elastance moves up from S_{min} to S_{max}.

One fundamental limitation on the frequency range is given by S_{min}. The ratio of maximum to minimum resonant frequency is a good indication of the ability of the varactor to tune over wide ranges. Thus,

$$\frac{\omega_{r,max}}{\omega_{r,min}} = \sqrt{\frac{S_{max}}{S_{min}}}. \tag{E-4}$$

For most applications, if S_{min} is, say, a hundredth of S_{max}, it can be neglected. However, for voltage tuning this S_{min} would restrict the tuning range to at most 10 to 1. If an abrupt-junction varactor is used, the ratio of S_{max} to S_{min} is itself the square root of the ratio of breakdown voltage to minimum voltage, so that the ratio of maximum to minimum resonant frequencies is

$$\frac{\omega_{r,max}}{\omega_{r,min}} = \sqrt[4]{\frac{V_B + \phi}{V_{min} + \phi}}. \tag{E-5}$$

Thus, a 260-volt breakdown varactor, if usable only to within 0.2 volt of ϕ, could yield only a 6-to-1 tuning ratio. Graded-junction varactors are even worse, because the elastance varies as the cube root of the voltage, so that

$$\frac{\omega_{r,max}}{\omega_{r,min}} = \sqrt[6]{\frac{V_B + \phi}{V_{min} + \phi}}. \tag{E-6}$$

Thus, if a 10-to-1 tuning ratio is required from a graded-junction varactor with a minimum voltage of 0.1 volt from the contact potential ϕ, the breakdown voltage would have to be 100 Kv, an unheard-of value.

This discussion indicates that for highest tuning ratio, a varactor with a very low value of S_{min} (relative to S_{max}) should be used. In addition, extra capacitive energy storage should be avoided. If the varactor is padded with a series capacitor, both S_{max} and S_{min} can be thought of as increased by the elastance of the padder, and the tuning range suffers. If the varactor is padded by a parallel capacitance, the maximum elastance from the combination is decreased, and the tuning ratio suffers.

Aside from the minimum elastance, the tuning range may be restricted by the loss of the varactor. The quality factor‡ Q of the tuned circuit is

$$Q = \frac{Q_L Q_v}{Q_L + Q_v}, \tag{E-7}$$

where Q_L is the Q of the inductor, at the resonant frequency. This is a function of the bias voltage V_0,

$$Q_L(V_0) = \frac{\omega_{res}(V_0)L}{R_L}, \tag{E-8}$$

where R_L is the resistance of the inductor, which may itself be a function of ω_{res}, for example, because of the skin effect. Similarly, Q_v is the Q of the varactor, and is therefore a function of the applied voltage,

$$Q_v(V_0) = \frac{S(V_0)}{\omega_{res}(V_0)R_s}, \tag{E-9}$$

where R_s is the series resistance of the varactor. Using Equation E-1, we find

$$Q_v(V_0) = \frac{\sqrt{S(V_0)L}}{R_s} \tag{E-10}$$

and

$$Q_v(V_0) = \frac{\omega_{res}(V_0)L}{R_s}. \tag{E-11}$$

---

‡Do not confuse the symbol Q with the symbol Q for charge, used elsewhere throughout this book.

Thus, the Q of the varactor varies linearly with resonant frequency. It has its maximum value Q_{max} at the maximum frequency $\omega_{r,max}$.

It may be that some minimum value for Q_v is tolerable. In that case, the minimum resonant frequency would be that frequency at which the specified minimum Q_v is reached. Thus,

$$\omega_{r,min} = \omega_{r,max} \frac{Q_{v,min}}{Q_{v,max}}, \tag{E-12}$$

where $Q_{v,min}$ is specified, and $\omega_{r,min}$ is the minimum resonant frequency for attaining that value of Q. Note that this $\omega_{r,min}$ may be different from the resonant frequency at S_{min}, Equation E-3. In some cases the minimum resonant frequency is dictated by the minimum elastance of the varactor, and in other cases by the minimum tolerable Q.

From Equation E-9, we can find the value of Q_v at the maximum frequency:

$$Q_{v,max} = \frac{\omega_c}{\omega_{r,max}}, \tag{E-13}$$

where we have assumed $S_{min} << S_{max}$, so that the varactor cutoff frequency is

$$\omega_c = \frac{S_{max}}{R_s}. \tag{E-14}$$

For example, suppose that a varactor with a 100-Gc cutoff frequency is used to obtain a maximum tuning frequency of 100 Mc. If the minimum tolerable Q of the circuit is 200, what is the tuning range? The value of $Q_{v,max}$, from Equation E-13, is 1000, so that from Equation E-12 the minimum frequency is 20 Mc. Thus, this varactor tunes from 20 to 100 Mc with a Q always greater than 200. To have this 5-to-1 variation in resonant frequency, the varactor must have a 25-to-1 variation in elastance available, so that S_{min} can be at most one twenty-fifth of S_{max}. On the other hand, if the maximum resonant frequency is changed to 10 Mc, the same value of $Q_{v,min}$ (200) is attained at 200 Kc, and the varactor can tune from 200 Kc to 10 Mc, a range of 50 to 1. To achieve this large frequency range, an extremely small value of S_{min} must be available; it cannot be more than $S_{max}/2500$. A varactor with a 2500-to-1 variation in elastance is most unusual. If, instead, the variation in elastance were only 100 to 1, the minimum resonant frequency would be determined by S_{min}, and would be only 1 Mc. The minimum Q would then be 1000, instead of 200, and we should not be utilizing the high cutoff frequency of the varactor. We could just as well have used a cheaper 20-Gc varactor.

In summary, the tuning range is determined by two conditions. First, the minimum elastance restricts it:

$$\frac{\omega_{r,max}}{\omega_{r,min}} \leq \sqrt{\frac{S_{max}}{S_{min}}}. \tag{E-15}$$

Second, it is restricted by the specified minimum tolerated Q of the circuit:

$$\frac{\omega_{r,max}}{\omega_{r,min}} \leq \frac{Q_{v,max}}{Q_{v,min}} = \frac{\omega_c}{\omega_{r,max}} \frac{1}{Q_{v,min}}. \tag{E-16}$$

If Equation E-15 determines the tuning range, the varactor being used has an unnecessarily high cutoff frequency. On the other hand, if Equation E-16 determines the tuning range, the varactor used has an unnecessarily low S_{min}. It is often wise to make both limits occur simultaneously; in that case, the cutoff frequency ω_c (assuming $S_{max} >> S_{min}$) and the ratio of maximum to minimum elastances are related by the formula

$$\omega_c = \sqrt{\frac{S_{max}}{S_{min}}}\, Q_{v,min}\,\omega_{r,max}. \tag{E-17}$$

E-2. Large-Signal Restrictions

If the rf power level in the varactor is large, then there are additional restrictions on the use of varactors, of which we mention three. First, the average elastance S_0, which determines the resonant frequency, depends upon drive level as well as bias voltage V_0. Second, for large rf power, the allowed variation of voltage V_0 is restricted; it can no longer range from $-\phi$ (or V_{min}) to V_B. Third, the rf power dissipated in the varactor may be too great, either because it exceeds the dissipation rating of the varactor, or because it lowers the efficiency of the tank circuit. Most of the pertinent calculations have already been completed in describing the pumping of varactors, and we shall rely heavily on the graphs in Chapter 7.

To be definite, we suppose the rf current through the varactor (either abrupt-junction or graded-junction, with $S_{min} << S_{max}$) is sinusoidal (for example, because of additional filters to block second and possibly higher-order harmonic currents). Let us define the amplitude parameter a as the ratio of peak-to-peak rf charge to the maximum stored charge of the varactor. Thus a is restricted to lie between zero (no rf excitation) and 1 (full rf excitation).

The average elastance S_0 (or, $m_0 = S_0/S_{max}$, the "normalized" average elastance) is, of course, a function of the tuning voltage

V_0. However, it is also a function of the rf power level, that is, a function of a. For example, consider a graded-junction varactor, biased so that‡ $V_0 + \phi = 0.424(V_B + \phi)$. Then m_0 varies as a function of a, as indicated on page 266 in Figure 7.8 (the "Fixed-bias" curve), going from a value of 0.752 at $a = 0$, to 0.637 at $a = 1$, for a drop of some 15 per cent.

In addition, if another value of V_0 had been chosen, the amplitude parameter a could not reach 1. The maximum value of a, as a function of V_0, is given on page 268 in Figure 7.9. For each combination of V_0 and a that falls within the allowable region in Figure 7.9, the corresponding value of m_0 can be calculated by the technique described in Section 7.4.1(d). Also, the rf dissipated power is related to a by Equation 7.64:

$$P = 0.281 a^2 P_{norm} \frac{\omega_{res}^2}{\omega_c^2}, \qquad \text{(E-18)}$$

where P_{norm} is the varactor normalization power $(V_B + \phi)^2/R_s$.

If a specific power level, and therefore a specific a, is necessary, the range of voltage V_0 is restricted, as indicated in Figure 7.9. Thus, the higher the power level, the smaller the range of tuning available.

On the other hand, suppose the varactor is an abrupt-junction varactor, with $S_{min} << S_{max}$. Then, the average elastance S_0 again depends upon both V_0 and a. If the voltage is fixed so that‡ $V_0 + \phi = 0.375(V_B + \phi)$, then m_0 as a function of a appears on page 274 in Figure 7.11 ("Fixed-bias" curve). From no rf drive to full excitation, the drop in m_0 is from 0.612 to 0.500, for a drop of about 18 per cent. If V_0 is different, then m_0 can be calculated from the closed-form expression, Equation 7.91:

$$m_0^2 = \frac{V_0 + \phi}{V_B + \phi} - \frac{a^2}{8}. \qquad \text{(E-19)}$$

For a given value of V_0, a is restricted according to Figure 7.12, page 277; similarly, for a given value of a, the possible range in V_0 is restricted according to Figure 7.12. The rf power dissipated is related to a by Equation 7.90:

$$P = 0.500 a^2 P_{norm} \frac{\omega_{res}^2}{\omega_c^2}, \qquad \text{(E-20)}$$

and Equations E-19 and E-20 can be combined to eliminate a:

---

‡This is the only bias voltage for which a can be as high as 1.

$$m_0^2 = \frac{V_0 + \phi}{V_B + \phi} - \frac{P}{4P_{norm}} \frac{\omega_c^2}{\omega_{res}^2}. \qquad (E\text{-}21)$$

In each case, the restriction on possible values of a and V_0 arises because of the requirement that the instantaneous voltage (the sum of the dc voltage and the instantaneous rf voltage) must lie between V_B and $-\phi$. Clearly, if the rf-voltage swing is large, then V_0 cannot move very far before either the positive or negative peaks exceed the possible voltage range. If the forward conduction mechanism of the varactor fails at the resonant frequency (see Section 4.1.4), then the restriction given by the left sides of the curves of Figures 7.9 and 7.12 can be slightly relaxed.

For abrupt-junction varactors, this same requirement can be easily expressed in terms of the instantaneous elastance. The average elastance must be high enough so that S(t) is never less than S_{min} (or zero, in our case). Thus, $a < 2m_0$. Similarly, to avoid exceeding the maximum elastance, $a < 2(1 - m_0)$. This restriction, shown in Figure E-2, is of particular importance for

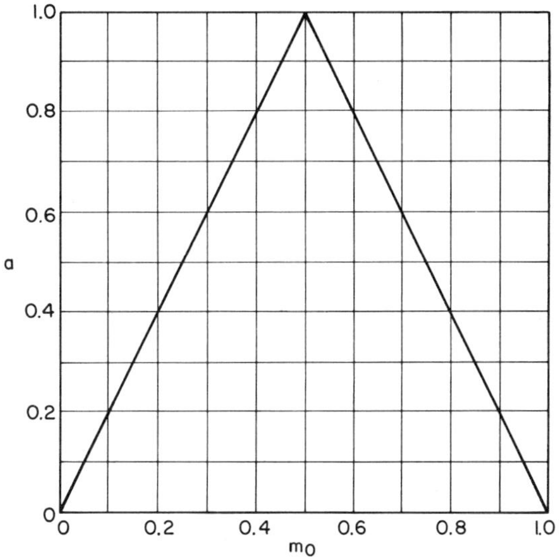

Figure E-2. Allowed combinations of a and m_0 must fall below this curve, for abrupt-junction varactors with sinusoidal rf current.

voltage tuning because it relates the rf excitation parameter a directly to m_0, which determines the resonant frequency, instead of through the voltage V_0. For a given excitation a, the maximum possible variation in m_0, and therefore in resonant frequency, is indicated directly in Figure E-2.

We have derived a few of the more important limits for high-power voltage tuning. First, the rf power dissipation, Equation E-18 or E-20, may be greater than the varactor dissipation rating. Second, the rf power dissipation may be so great that the circuit loss is intolerable. Third, for a given bias voltage, the value of m_0, and therefore the resonant frequency of the tuned circuit, depends upon the excitation level. Fourth, for moderately high power, the variation in m_0 (or in V_0) is restricted, in accordance with Figure 7.9, 7.12, or E-2. Fifth, for any given V_0, the maximum amplitude a is restricted by Figure 7.9 or 7.12.

Appendix F

GRADED-JUNCTION-VARACTOR MULTIPLIERS

In Chapter 8 we dealt primarily with abrupt-junction varactors. Here we give a much shorter, less complete discussion of multipliers made with graded-junction varactors. Some of these results are quoted in Section 8.12 and compared with the results for abrupt-junction varactors.

In Section F-1 we discuss the doubler and in Section F-2 the high-frequency limit of higher-order multipliers without idlers. In each case we assume that the varactor minimum elastance S_{min} is negligibly small.

F-1. Graded-Junction-Varactor Doubler

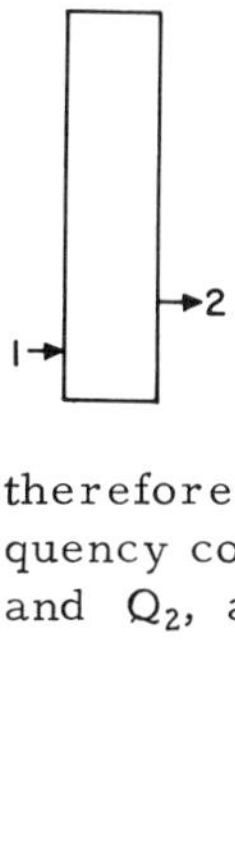

Here we describe a large-signal steady-state solution of the doubler. The calculations, much more difficult than those for the abrupt-junction-varactor doubler, were made on a digital computer by Marshall Greenspan.[1]‡ First we give the basic doubler formulas, and then give Greenspan's results, with his kind permission.

We assume that the current has components only at frequencies ω_o (input) and $2\omega_o$ (output). The charge therefore has only two components, but the elastance has other frequency components. Let us denote the charge coefficients as Q_1 and Q_2, and their normalized magnitudes as μ_1 and μ_2:

$$\mu_1 = \frac{|Q_1|}{Q_B + q_\phi}, \tag{F-1}$$

and

$$\mu_2 = \frac{|Q_2|}{Q_B + q_\phi}. \tag{F-2}$$

The abrupt-junction doubler was analyzed under the phase condition that the currents at the input and output frequencies were in phase. We make the same assumption here, although there is no real justification for it other than the fact that it simplifies the analysis. The condition that the charge lie between the charge at

‡Superscript numerals denote references listed at the end of each appendix.

breakdown Q_B and the charge at the contact potential q_ϕ implies that μ_1 and μ_2 are restricted in value:

$$\mu_1 \sin \omega_o t + \mu_2 \sin 2\omega_o t \leq 0.25. \qquad \text{(F-3)}$$

This restriction is identical in form with Condition 8.52 for the abrupt-junction-varactor doubler. Therefore, μ_1 and μ_2 are restricted by a plot exactly like that of Figure 8.3.

The Fourier coefficients of elastance and voltage can now be calculated from the formulas

$$S(t) = S_{max}(\mu_0 + 2\mu_1 \sin \omega_o t + 2\mu_2 \sin 2\omega_o t)^{\frac{1}{2}} \qquad \text{(F-4)}$$

and

$$v(t) + \phi = (V_B + \phi)(\mu_0 + 2\mu_1 \sin \omega_o t + 2\mu_2 \sin 2\omega_o t)^{\frac{3}{2}} + R_s i(t), \qquad \text{(F-5)}$$

where μ_0 is the average charge plus q_ϕ, divided by $(Q_B + q_\phi)$. These formulas are derived from Equations 4.11 and 4.12. The elastance coefficients are in general complex. The bias voltage V_0 and values of V_1 and V_2 can be found once μ_1, μ_2, and μ_0 are known.

From these voltage coefficients, we can calculate other parameters of the doubler. The "input impedance" is, as for all multipliers, a function of drive level:

$$Z_{in} = R_{in} + jX_{in}$$

$$= \frac{V_1}{I_1}. \qquad \text{(F-6)}$$

The load impedance necessary to maintain the assumed values of Q_1 and Q_2 is

$$Z_2 = R_2 + jX_2$$

$$= \frac{-V_2}{I_2}. \qquad \text{(F-7)}$$

The real parts of the input impedance and load impedance are called the input resistance and load resistance. For the abrupt-junction-varactor multipliers, the imaginary parts of the input and load resistance are merely the reactance of the average elastance S_0 at the pertinent frequency. For the graded-junction device, however, the imaginary parts X_{in} and X_2 are related neither to the average elastance nor to the average charge; X_{in} and X_2 must be solved for in the same way that R_{in} and R_2 are solved for. It is convenient, however, to think of the input reactance as that caused by some "effective" elastance $S_{eff,in}$, and similarly to think of the load reactance as that necessary to tune out another "effective elastance" $S_{eff,out}$:

$$X_{in} = -\frac{S_{eff,in}}{\omega_o}, \qquad (F\text{-}8)$$

$$X_2 = \frac{S_{eff,out}}{2\omega_o}. \qquad (F\text{-}9)$$

The power input is

$$P_{in} = 2 \text{ Re } V_1 I_1^*$$

$$= 2|I_1|^2 R_{in}, \qquad (F\text{-}10)$$

and the power output is

$$P_{out} = 2 \text{ Re } V_2 I_2^*$$

$$= 2|I_2|^2 R_2. \qquad (F\text{-}11)$$

The dissipated power can be calculated as either the difference between these two, or else as the power dissipated in the varactor series resistance:

$$P_{diss} = P_{in} - P_{out}$$

$$= 2R_s(|I_1|^2 + |I_2|^2). \qquad (F\text{-}12)$$

The efficiency is

$$\epsilon = \frac{P_{out}}{P_{in}}$$

$$= \frac{R_{in} - R_s}{R_{in}} \frac{R_2}{R_2 + R_s}. \qquad (F\text{-}13)$$

The highest efficiency is found when the varactor is driven over its full range, so that Condition F-3 is barely obeyed. Greenspan[1] has investigated the maximum-drive condition (as well as the underdriven case). The maximum efficiency (subject to the tuning assumption) is shown in Figure F-1. In Figure F-2 we show the input, output, and dissipated powers. These curves are qualitatively

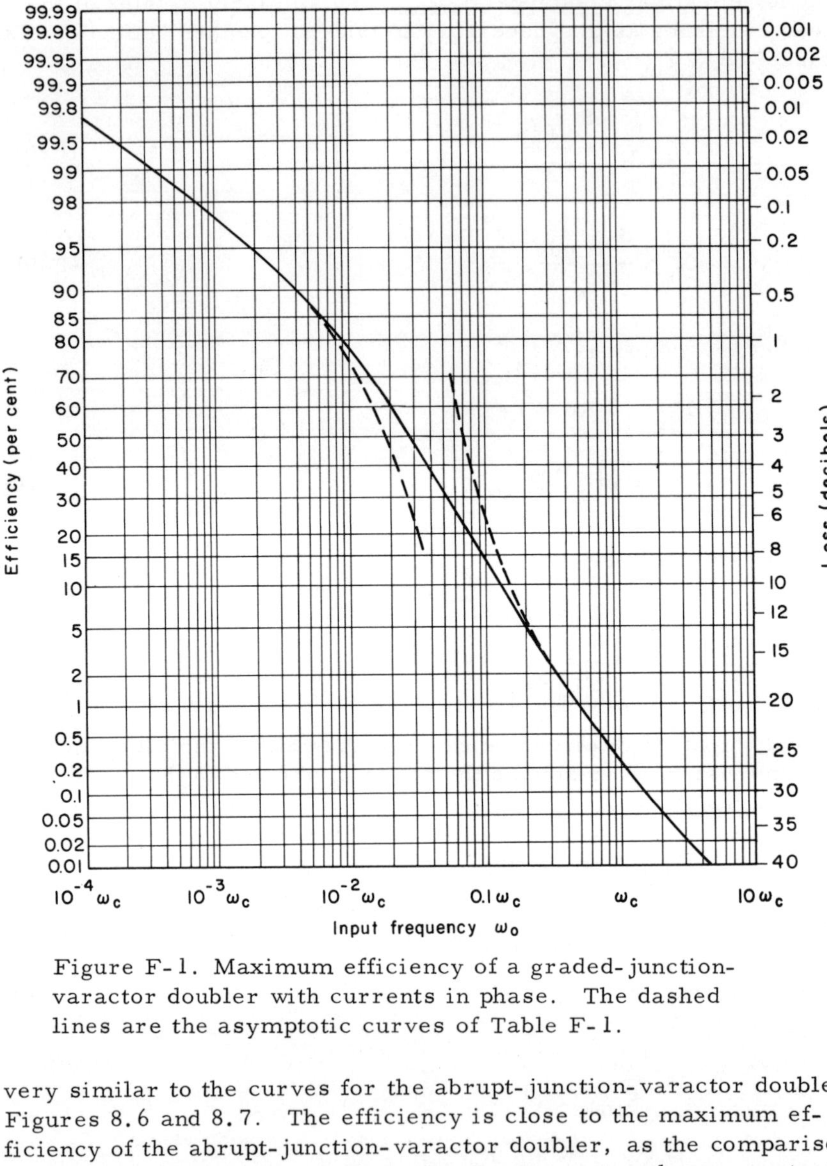

Figure F-1. Maximum efficiency of a graded-junction-varactor doubler with currents in phase. The dashed lines are the asymptotic curves of Table F-1.

very similar to the curves for the abrupt-junction-varactor doubler, Figures 8.6 and 8.7. The efficiency is close to the maximum efficiency of the abrupt-junction-varactor doubler, as the comparison of Figure 8.60 indicates. The output power approaches a constant value at high frequencies, and for low frequencies it varies linearly with the frequency.

Next, in Figure F-3 we show the input and load resistances necessary to attain highest efficiency, and then in Figure F-4 the input and output effective elastances $S_{eff,in}$ and $S_{eff,out}$. These effective elastances, which are different from each other and from the average elastance S_0, are plotted as fractions of S_{max}. To make computations using a capacitance, rather than an elastance, merely

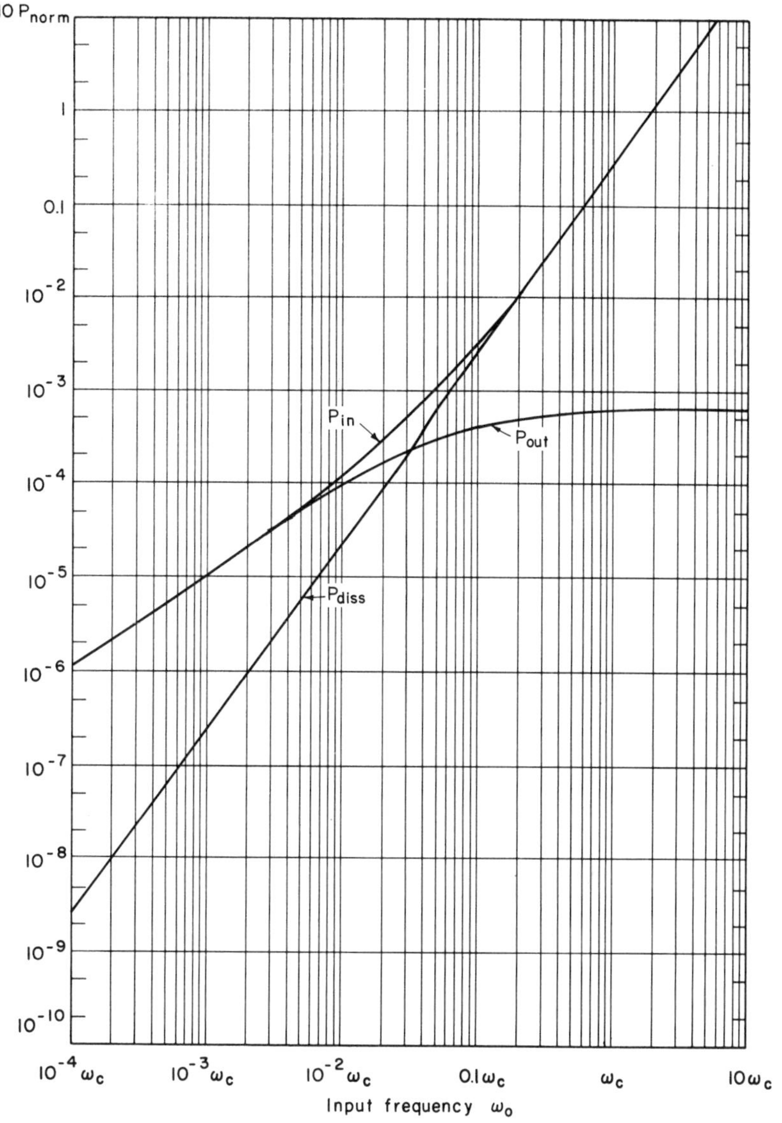

Figure F-2. Input, output, and dissipated powers of a graded-junction-varactor doubler with maximum efficiency.

use "effective capacitances" somewhat larger than C_{min}, as determined by the right-hand scale in Figure F-4.

In Figure F-5 is the bias voltage necessary for maximum efficiency.

Greenspan[1] has investigated the conditions of maximum power output, as well as those of maximum efficiency. The resulting curves of efficiency and powers are so nearly identical to Figures F-1 and F-2 that they cannot be distinguished. The input and load

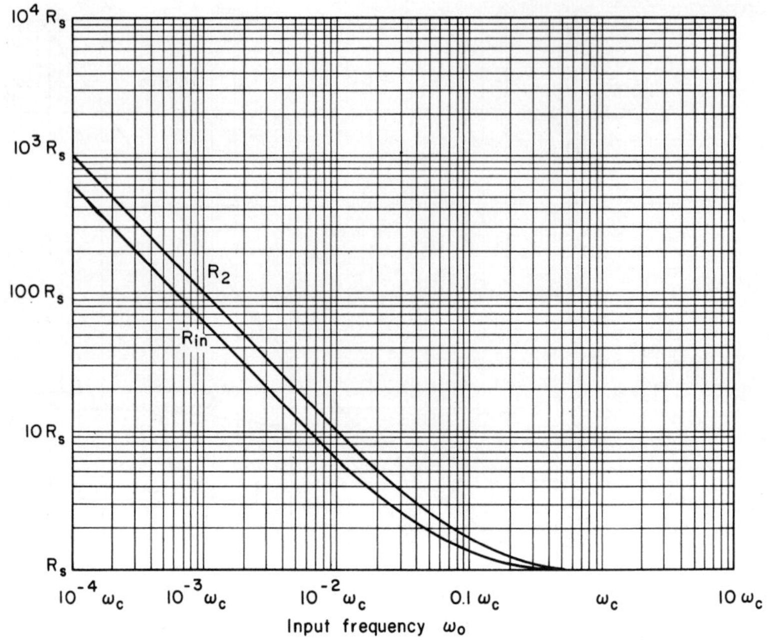

Figure F-3. Input and load resistances of a graded-junction-varactor doubler with maximum efficiency.

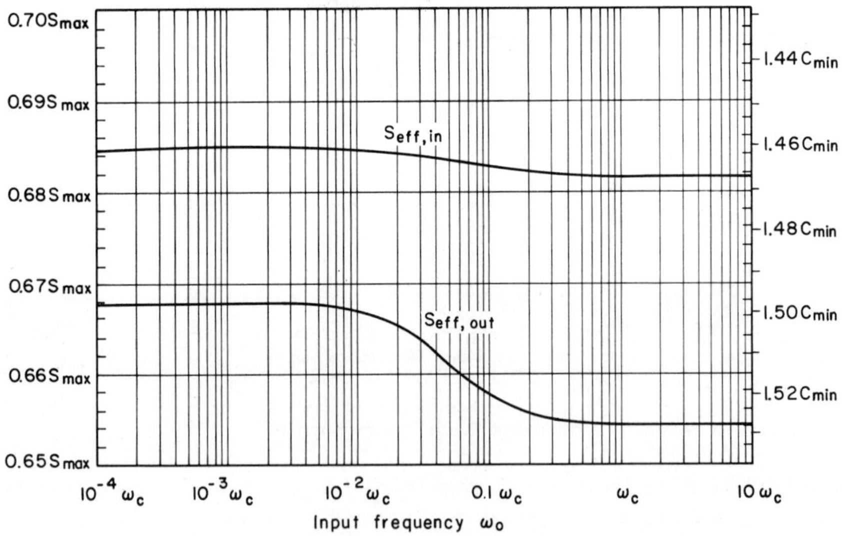

Figure F-4. Input and output effective elastances for a graded-junction-varactor doubler with maximum efficiency, as fractions of S_{max}. The effective capacitances corresponding to these effective elastances can be determined, as specific multiples of C_{min}, by using the right-hand scale.

Table F-1. Graded-Junction-Varactor Doubler

Asymptotic formulas for the graded-junction-varactor doubler at high frequencies and low frequencies. In each case the efficiency is maximized.

	Low-Frequency		High-Frequency
	Maximum ϵ	Maximum P_{out}	Maximum ϵ and P_{out}
ϵ	$1 - 25.6\frac{\omega_o}{\omega_c}$	$1 - 27.6\frac{\omega_o}{\omega_c}$	$0.00235\left(\frac{\omega_c}{\omega_o}\right)^2$
R_{in}	$0.0604\frac{S_{max}}{\omega_o}$	$0.0724\frac{S_{max}}{\omega_o}$	R_s
R_2	$0.204\frac{S_{max}}{2\omega_o}$	$0.145\frac{S_{max}}{2\omega_o}$	R_s
$S_{eff,in}$	$0.684\,S_{max}$	$0.684\,S_{max}$	$0.682\,S_{max}$
$S_{eff,out}$	$0.668\,S_{max}$	$0.673\,S_{max}$	$0.654\,S_{max}$
P_{in}	$0.0118P_{norm}\frac{\omega_o}{\omega_c}$	$0.0121P_{norm}\frac{\omega_o}{\omega_c}$	$0.281P_{norm}\left(\frac{\omega_o}{\omega_c}\right)^2$
P_{out}	$0.0118P_{norm}\frac{\omega_o}{\omega_c}$	$0.0121P_{norm}\frac{\omega_o}{\omega_c}$	$0.00066P_{norm}$
P_{diss}	$0.30P_{norm}\left(\frac{\omega_o}{\omega_c}\right)^2$	$0.33P_{norm}\left(\frac{\omega_o}{\omega_c}\right)^2$	$0.281P_{norm}\left(\frac{\omega_o}{\omega_c}\right)^2$
$\frac{V_0 + \phi}{V_B + \phi}$	0.409	0.405	0.424
μ_1	0.208	0.192	0.250
μ_2	0.080	0.096	$0.0061\frac{\omega_c}{\omega_o}$

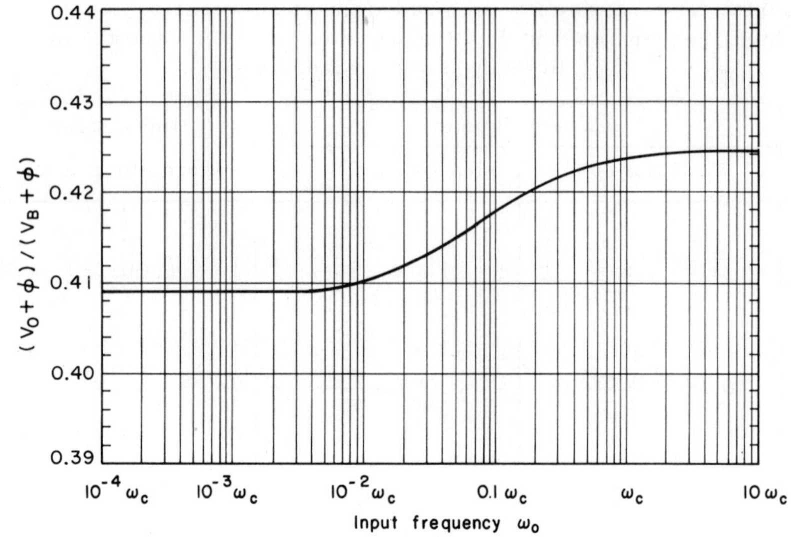

Figure F-5. Bias voltage of a graded-junction-varactor doubler with maximum efficiency.

resistances, the input and output effective elastances, and the bias voltage are slightly different.

The low-frequency and high-frequency behavior of the doubler is important. In Table F-1 we show this asymptotic behavior for conditions of both maximum efficiency and maximum power output. It is interesting to note from Table F-1 that for the same two conditions (maximum power and maximum efficiency) the low-frequency values for μ_1 and μ_2 are identical to the low-frequency values of m_1 and m_2 for the abrupt-junction-varactor doubler (see Tables 8.4 and 8.5). This fact suggests that perhaps these values (0.208, 0.080, 0.192, and 0.096) might apply to other varactor doublers. It also suggests that low-frequency, optimized, higher-order, graded-junction-varactor multipliers could be investigated by using the m-values pertinent to the corresponding abrupt-junction-varactor multipliers.

Greenspan[1] has also investigated the graded-junction-varactor doubler with less than full drive.

F-2. High-Frequency Limit

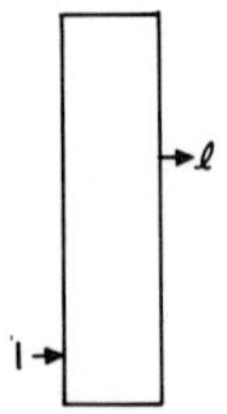

Here we calculate the high-frequency limit of any varactor multiplier without idlers, and then apply this result to graded-junction varactors.

In the high-frequency limit, power is put in the varactor at frequency ω_0, and the output at frequency $\ell\omega_0$ has a very small amplitude. If we assume that current flows only at frequencies ω_0 and $\ell\omega_0$, then the generated voltage at $\ell\omega_0$ is, from Equations 8.11,

$$E_\ell = \frac{1}{j\ell\omega_o}(S_{\ell-1}I_1 + S_{\ell+1}I_1^*), \tag{F-14}$$

where we have neglected the term $S_{2\ell}I_\ell^*$ because it is very much smaller than the others. Here $S_{\ell-1}$ and $S_{\ell+1}$ are the Fourier coefficients of elastance at frequencies $(\ell - 1)\omega_o$ and $(\ell + 1)\omega_o$, respectively. Since the output is much smaller in amplitude than the input, the elastance coefficients can be calculated by considering only the input current.

The magnitude of Equation F-14 is

$$|E_\ell| = \frac{\omega_c}{\ell\omega_o}|I_1||m_{\ell-1} \pm m_{\ell+1}|R_s. \tag{F-15}$$

The maximum power output and efficiency are obtained with a load resistance equal to R_s; this maximum power output is

$$P_{out} = \frac{|E_\ell|^2}{2R_s}. \tag{F-16}$$

The power input is the same as the power dissipated in the series resistance of the varactor:

$$P_{in} = 2R_s|I_1|^2. \tag{F-17}$$

The ratio of these two is the efficiency,

$$\epsilon = \frac{|E_\ell|^2}{4R_s^2|I_1|^2} = \frac{1}{4}\left(\frac{\omega_c}{\ell\omega_o}\right)^2 |m_{\ell-1} \pm m_{\ell+1}|^2. \tag{F-18}$$

This formula gives the high-frequency limit of efficiency. In general, without idlers this limit approaches zero inversely as the square of the frequency. For abrupt-junction multipliers (other than the doubler) the factor $|m_{\ell-1} \pm m_{\ell+1}|$ vanishes.

For graded-junction varactors the minus sign in Equation F-18 should be used. If the varactor is driven fully, then, according to Equation 7.60,

$$m_\ell = \frac{2}{\pi|4\ell^2 - 1|}, \tag{F-19}$$

and we find

$$\epsilon = \frac{\omega_c^2}{\omega_o^2}\left[\frac{16}{\pi(4\ell^2 - 1)(4\ell^2 - 9)}\right]^2 . \tag{F-20}$$

Since the input and dissipated power approach Equation 7.62,

$$P_{in} = 0.281 P_{norm}\frac{\omega_o^2}{\omega_c^2}, \tag{F-21}$$

the output power approaches the limiting value,

$$P_{out} = \frac{7.29 P_{norm}}{[(4\ell^2 - 1)(4\ell^2 - 9)]^2}. \tag{F-22}$$

The high-frequency limit for graded-junction varactors without idlers is described in Table F-2. All entries there are independent

Table F-2. High-Frequency Graded-Junction-Varactor Multipliers

High-frequency asymptotic expressions for graded-junction-varactor multipliers without idlers.

$$\epsilon \approx \left(\frac{\omega_c}{\omega_o}\right)^2 \left[\frac{16}{\pi(2\ell - 3)(2\ell - 1)(2\ell + 1)(2\ell + 3)}\right]^2$$

$$P_{in} \approx 0.281 P_{norm}\frac{\omega_o^2}{\omega_c^2}$$

$$P_{out} \approx 7.29 P_{norm}\left[\frac{1}{(2\ell - 3)(2\ell - 1)(2\ell + 1)(2\ell + 3)}\right]^2$$

$$P_{diss} \approx 0.281 P_{norm}\frac{\omega_o^2}{\omega_c^2}$$

$$R_{in} \approx R_s \qquad R_2 \approx R_s$$

$$\frac{V_0 + \phi}{V_B + \phi} \approx 0.424$$

of ℓ, the order of the harmonic being generated, except the efficiency and the power output. For specific values of ℓ, these can be found either from the formulas given, or else from Table F-3.

At high frequencies the efficiencies of all the graded-junction-varactor multipliers go to zero as the inverse square of the frequency. The high-order abrupt-junction varactor multipliers all have efficiencies that go to zero more rapidly. Therefore, at high

Table F-3. High-Frequency Graded-Junction-Varactor Multipliers

The power output and efficiency for specific multipliers can be found in this table or can be calculated from the formulas given in Table F-2.

	P_{out}	ϵ
$\times 2$	$6.62 \times 10^{-4} P_{norm}$	$2.35 \times 10^{-3} \left(\frac{\omega_c}{\omega_o}\right)^2$
$\times 3$	$8.17 \times 10^{-6} P_{norm}$	$2.91 \times 10^{-5} \left(\frac{\omega_c}{\omega_o}\right)^2$
$\times 4$	$6.08 \times 10^{-7} P_{norm}$	$2.16 \times 10^{-6} \left(\frac{\omega_c}{\omega_o}\right)^2$
$\times 5$	$8.99 \times 10^{-8} P_{norm}$	$3.20 \times 10^{-7} \left(\frac{\omega_c}{\omega_o}\right)^2$
$\times 6$	$1.96 \times 10^{-8} P_{norm}$	$6.96 \times 10^{-8} \left(\frac{\omega_c}{\omega_o}\right)^2$
$\times 7$	$5.49 \times 10^{-9} P_{norm}$	$1.95 \times 10^{-8} \left(\frac{\omega_c}{\omega_o}\right)^2$
$\times 8$	$1.84 \times 10^{-9} P_{norm}$	$6.54 \times 10^{-9} \left(\frac{\omega_c}{\omega_o}\right)^2$
$\times 9$	$7.05 \times 10^{-10} P_{norm}$	$2.51 \times 10^{-9} \left(\frac{\omega_c}{\omega_o}\right)^2$
$\times 10$	$3.00 \times 10^{-10} P_{norm}$	$1.07 \times 10^{-9} \left(\frac{\omega_c}{\omega_o}\right)^2$

enough frequencies, the graded-junction multipliers are preferable, although this is true only for frequencies so high that the efficiency is very low anyway.

For example, let us compare a graded-junction-varactor tripler without an idler to an abrupt-junction-varactor tripler with an idler at $2\omega_0$. Assume both varactors have the same cutoff frequency. The former is more efficient at output frequencies greater than approximately $1.5\omega_c$, although neither efficiency is very large (less than about 0.1 per cent). For higher-order multipliers the situation is approximately the same — for the same cutoff frequency, graded-junction varactors without idlers are better when the output frequency is larger than a certain frequency that lies between approximately ω_c and $2\omega_c$. The resulting efficiencies are very low.

REFERENCE

1. Greenspan, M., "The Graded-Junction Varactor Frequency Doubler," S.M. Thesis, Department of Electrical Engineering, M.I.T. (June, 1962).

NAME INDEX

SUBJECT INDEX